TEACHER'S WRAPAROUND EDITION

Teen Health
COURSE 2

Mary Bronson Merki, Ph.D.

Glencoe
McGraw-Hill

New York, New York   Columbus, Ohio   Woodland Hills, California   Peoria, Illinois

About Our Author

Introducing the Author

MARY BRONSON MERKI has taught health education in grades K-12, as well as health education methods classes at the undergraduate and graduate levels. As Health Education Specialist for the Dallas School District, Dr. Merki developed and implemented a districtwide health education program, *Skills for Living,* which was used as a model by the state education agency. She also helped develop and implement the district's Human Growth, Development and Sexuality program, which won the National PTA's Excellence in Education Award and earned her an honorary lifetime membership in the state PTA. Dr. Merki has assisted school districts throughout the country in developing local health education programs. In 1988, she was named the Texas Health Educator of the Year by the Texas Association for Health, Physical Education, Recreation, and Dance. Dr. Merki is also the author of *Glencoe Health,* a high school textbook adopted in school districts throughout the country. She currently teaches in a Texas public school where she was recently honored as Teacher of the Year. Dr. Merki completed her undergraduate work at Colorado State University in Fort Collins, Colorado. She earned her master's and doctoral degrees in health education at Texas Woman's University in Denton, Texas.

Glencoe/McGraw-Hill
A Division of The McGraw-Hill Companies

Copyright ©1999 by Glencoe/McGraw-Hill

All rights reserved. Except as permitted under the United States Copyright Act, no part of this publication may be reproduced or distributed in any form or by any means, or stored in a database or retrieval system, without prior written permission from the publisher.

Send all inquiries to:
Glencoe/McGraw-Hill
21600 Oxnard Street, Suite 500
Woodland Hills, California 91367

ISBN 0-02-653128-3   Teen Health Course 2 Student Text
ISBN 0-02-653129-1   Teen Health Course 2 Teacher's Wraparound Edition

Printed in the United States of America
3 4 5 6 7 8 9   004   06 05 04 03 02 01 00 99

Contributor and Reviewers

Editorial Consultant

PAMELA R. CONNOLLY
Subject Area Coordinator for Health Education,
   Pittsburgh Catholic Schools
Oakland Catholic High School,
   Diocese of Pittsburgh
Pittsburgh, Pennsylvania

Teacher Reviewers

BEVERLY J. BERKIN, CHES
Health Education Consultant
Bedford Corners, New York

DONNA BREITENSTEIN, ED.D.
Professor & Coordinator of Health Education,
Director of NC School Health Training Center
Appalachian State University
Boone, North Carolina

DEBRA PENLAND FORBES
Health Educator
Pleasanton Unified School District
Pleasanton, California

DEBRA C. HARRIS, PH.D.
HPE Dept. Chair, HPE Teacher
West Linn High School
West Linn, Oregon

LORI HART
Health Educator
West Shore Middle School
Milford, Connecticut

PEGGY V. JOHNS
Supervisor, Pre K-12 Health Education
Pinellas County Schools
Largo, Florida

DEBORAH L. TACKMANN, BS, M.E.P.D.
Health Education Instructor
North High School
Eau Claire, Wisconsin

ROBERT WANDBERG, PH.D.
Health Education
John F. Kennedy Senior High School
Bloomington, Minnesota

LYNN WESTBERG
Health Education/Department Head
Kearns High School
Kearns, Utah

BETTY ANNE WHITE
Health & Physical Education Teacher
Bryan Station Traditional Middle School
Lexington, Kentucky

Table of Contents

TEACHER'S
WRAPAROUND EDITION

About Our Author **T2**

Contributors and Reviewers **T3**

Table of Contents **T4**

NEW DIRECTIONS: Responding to Changes in Health Education **T5**

The Need for New Direction in Health Education
Teen Health Answers the Challenge
Teen Health Meets the National Health Education Standards
Teen Health Makes Connections
Teen Health Is Loaded with Activities
Teen Health Fits Your Classroom

PROGRAM RESOURCES: The *Teen Health* Integrated Learning System T7

Unifying Themes **T8**

Student Text **T9**

Organization of the Text
Chapter Features of the Text
Marginal Features of the Text

Student Modules **T11**

Conflict Resolution
Abstinence
Physical Activity

Teacher's Wraparound Edition . . . **T12**

Teacher's Classroom Resources **T14**

Professional Health Series **T16**

Dealing with Sensitive Issues
Promoting a Comprehensive School Health System
Cultural Diversity in the Health Classroom

Multimedia Resources **T17**

Teen Health Videodisc/VHS Series
Video Kit
Audiocassette Program (English/Spanish)
Testmaker
ABCNews InterActive™ Correlation Bar Code Guide

The Work Students Do **T18**

Decision-Making Activities
Critical Thinking and Student Investigations
Connections to Home and Community
Personal Inventories and Journals

Assessment **T20**

Performance Assessment Projects
Testing Programs

Meeting Student Diversity **T21**

Students with Language Diversity
Students with Varying Ability Levels and Learning Styles
Students with Special Needs

***Teen Health* and the National Health Standards** **T23**

Course Presentation **T26**

Scope and Sequence **T28**

Health Resources **T34**

New Directions

RESPONDING TO CHANGES IN HEALTH EDUCATION

The Need for New Direction in Health Education

Today, our lives are much less likely to be threatened by infectious diseases than they were in the past. However, our lives today are more likely to be threatened by "lifestyle diseases." More than 75 percent of the two million Americans who die each year are killed in accidents or by heart disease, stroke, cirrhosis of the liver, or emphysema. Individual behavior and health choices are contributing factors in many of these cases.

When we consider the well-being of our adolescent population, the situation becomes even more serious. They have an opportunity to live longer, healthier lives than earlier generations because of better nutrition, improved sanitation, the widespread use of vaccines and antibiotics, and highly sophisticated surgical techniques. Yet the rate of adolescent mortality is far too high. Accidents (frequently related to substance abuse), suicides, and homicides account for about 80 percent of all deaths among teenagers. In other words, the greatest cause of death among young people is social, mental, and emotional ill health.

Teen Health Answers the Challenge

Health education must do more than merely provide information. In order to be worthwhile, a health education program must provide information in such a way that it influences students to take positive action regarding their own health. *Teen Health* has been designed to meet this urgent objective.

In the program, students learn that good health habits can prevent illness. More importantly, as a motivating factor, they also learn that good health habits can improve the way they look, the way they perform in school and sports, the way they interact with others, and the way they feel about themselves.

Wellness, they discover, allows people to be at their very best, and wellness is simply a way of living each day that includes choices and decisions based on healthy attitudes.

Teen Health Meets The National Health Education Standards

The National Health Education Standards were created with the goal of improving educational achievement for students and improving health in the United States through the promotion of health literacy. Health literacy is defined as the capacity of individuals to obtain, interpret, and understand basic health information and services and the competence to use such information and services in ways which promote health. The National Health Standards apply the characteristics of a well-educated, literate person within the context of health. The health literate person is:

- a critical thinker and problem solver
- a responsible, productive citizen
- a self-directed learner
- an effective communicator

The Health Standards detail the knowledge and skills essential to the development of health literacy in the form of performance indicators.

New Directions

A lesson by lesson correlation of the *Teen Health* student text to the National Health Education Standards demonstrates how *Teen Health* meets these criteria.

Teen Health Makes Connections

It takes more than a few minutes each week in a health classroom for students to understand the benefits of making healthful choices. Our schools need to build healthy generations, and doing so requires the combined efforts of family, school, and community. This means that the health teacher must work collaboratively with parents, community, city-county government, education, health agencies, human service agencies, businesses, and volunteer health organizations to help provide meaningful experiences and appropriate services to benefit children and their quality of life. Throughout the *Teen Health* program students are given the opportunity to expand their learning beyond the classroom and into the community and to make connections that translate across curriculum areas.

Teen Health Is Loaded with Activities

Teen Health provides students with the opportunity to take an active and personal role in the learning process, a role that will allow them to become active partners in maintaining and improving their level of wellness.

Throughout the program students participate in simple experiments, personal inventories, and projects that help them learn how to make healthful choices in all areas of their daily lives—physical, mental and emotional, and social. As they acquire knowledge and decision-making skills, they also gain the confidence they need to take responsibility for their own health.

Teen Health Fits Your Classroom

Experiences that students bring to the classroom and the amount of instructional time vary from class to class and district to district. Therefore, *Teen Health* has been designed to help teachers address the needs of their students. Self-contained lessons give teachers the flexibility to pick and choose the topics that address their health classroom. The program is supplemented with modules covering key topics, such as sexuality and relationships; HIV/AIDS; violence prevention; and alcohol, drugs, and tobacco education—topics that present the highest risk for adolescent populations.

Program Resources

THE *TEEN HEALTH* INTEGRATED LEARNING SYSTEM

Teen Health offers an integrated learning system comprised of print and multimedia components. At the cornerstone of the program is the student text, with an array of print and electronic components supporting and complementing the teaching process. Four-color supplemental modules, blackline masters, videodiscs, and videos provide an integral and integrated learning system. As a whole, the system offers a multitude of learning experiences and options for today's health classroom.

The *Teen Health* Program Components

Student Edition
Teacher's Wraparound Edition
Teacher's Classroom Resources
Student Activities Workbook

Modules

Conflict Resolution
Abstinence
Physical Activity

Teacher's Classroom Resources

Concept Mapping Activities
Reteaching Activities
Enrichment Activities
Decision-Making Activities
Health Labs
Cross-Curriculum Activities
Reproducible Lesson Plans
Performance Assessment
Testing Program
Teaching Transparencies Binder

Multimedia Components

Teen Health Videodisc/VHS Series
Computerized Testing Program
Audiocassette Program (English)
Audiocassette Program (Spanish)
ABCNews InterActive™ Bar Code Guide
Video Kit

Student Diversity Strategies

Parent Letters and Activities
  (English/Spanish)
Spanish Summaries, Lesson Quizzes,
  and Activities

Professional Health Series

Promoting a Comprehensive School
  Health Program
Dealing with Sensitive Issues
Cultural Diversity in the Health Classroom

UNIFYING THEMES

Four unifying themes are woven throughout the *Teen Health* program. These include:

- *Acceptance of personal responsibility for lifelong health*
- *Respect for and promotion of the health of others*
- *An understanding of the process of growth and development*
- *Informed use of health-related information, products, and services*

The visibility of these themes within the student text helps underscore their importance. Every lesson review offers application activities in a section called "Applying Health Concepts," each one identified with the appropriate theme. The Teacher's Classroom Resources and Transparencies extend the thematic approach to health issues, again with identification of the theme. Furthermore, throughout the Teacher's Wraparound Edition you will find bottom columns that present background information and additional activities specifically tied to the four unifying themes—just look for bottom columns that display a shiny red apple.

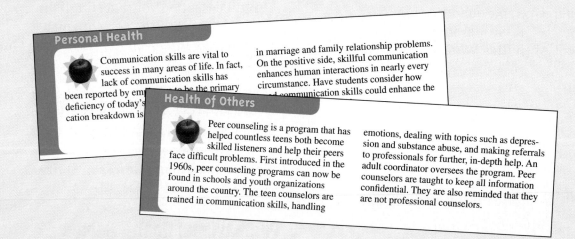

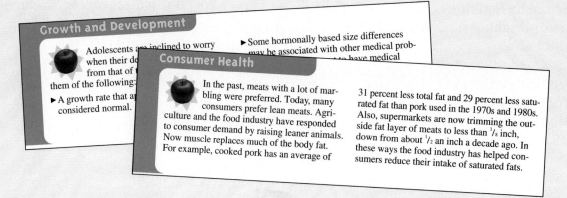

STUDENT TEXT

Teen Health is structured around 15 independent chapters, each containing several short, discrete lessons. This structure provides flexibility, allowing the teacher to tailor the course to his or her specific teaching methods and to the school's or district's health education framework.

Organization of the Text

Unit Opener. The text is divided into five units, which are divided into chapters and lessons. Each unit begins with a two-page spread that identifies the chapters contained within the unit.

Chapter Opener. Each chapter begins with a two-page spread that includes a full-page photograph and a dialogue from the perspective of young teens. In addition to the personal reflection, Student Expectations state the focus or objectives of the chapter. An activity, In Your Journal, directs students to gather information (by clipping articles, making notes on news stories or advertisements, or observing their own behavior over time) related to the chapter content. The information that students are told to gather here may become part of their portfolio activity or action plan activity in the chapter review.

Lessons. Lessons are four to six pages in length and organized for flexible use, allowing the teacher to pick and choose which lessons to teach. Each lesson opener lists the Words to Know—a list of important vocabulary words students will encounter in the lesson. Each lesson concludes with a three-part review that includes vocabulary questions, several recall and critical thinking questions, and activities in which students apply health concepts.

Chapter Review. Each chapter concludes with a two-page spread of chapter review material. The Chapter Summary provides the main points of each lesson's content in bulleted list form. Review Questions are organized according to lessons and assess students' abilities to recall basic factual information. Thinking Critically presents activities for each lesson that challenge students to use higher level thinking skills of analysis, synthesis, and evaluation.

The right-hand page of each Chapter Review contains the following elements:

- **Your Action Plan** involves students in setting short- and long-term goals, and following the action-plan model outlined in Chapter 1, formulating their own action plan to achieve those goals.

- **In Your Home and Community** provides students with a variety of activities they can pursue at home, in school, or in the

Student Text

community at large to promote the understanding of health issues and the practices of healthy behaviors.

- **Building Your Portfolio** offers a variety of activities that involve students in projects that can become part of their permanent health portfolio. Some of these draw on information gathered as part of the In Your Journal activity on the chapter opening page.

Chapter Features of the Text

The following features encourage discussion, promote critical thinking, and provide hands-on learning opportunities:

Making Healthy Decisions, second-person accounts, are friendly, reality-based, and focused on a situation that requires the student to make a decision. These features include consideration of the legal, family, moral, and safety aspects of the situation and the student's personal, family, and social values. Students are reminded of the decision-making model.

Life Skills present activities students can follow to develop an important health skill (examples are reading a nutrition label, listening effectively, and organizing a grass-roots public cleanup campaign).

Health Labs include science-based health activities and hands-on experiments that require little or no scientific equipment to illustrate various health concepts and to explore the wonder of the human body system.

Personal Inventories appear in some chapters and offer students the opportunity to assess their health attitudes and behaviors.

Teen Health Digest is a two-page feature appearing in each chapter and providing students with the opportunity to explore recent news and information. Each Digest has articles drawn from the following menu:

- **Consumer Focus** asks students to explore the truth and relevance of the kind of health information found in the media—reports of new studies or treatments, advertisements, product endorsements, and so on.
- **Teens Making a Difference** relates real-life stories of teens who are active in their families, school, or communities, promoting positive health behaviors.
- **People at Work** presents information about a real person working in a health-related career.
- **Sports and Recreation** profiles an athlete who provides a positive role model either through community involvement or by virtue of positive health behaviors; the article may also present valuable health information related to recreation.
- **Health Update** presents up-to-date information about health in the news.
- **Myths and Realities** explores common health-related misconceptions and presents students with the real facts.
- **Personal Trainer** presents information about nutrition and physical fitness.

Marginal Features of the Text

The marginal features contain a variety of supplementary snippets of information to increase students' interest. **Your Total Health** explores the interrelatedness of health by describing how the topic at hand in the text is connected to another health concern. Other features include:

- **Q & A** addresses questions of real concern to teens: Why does my voice crack? Is it really bad to pop my knuckles? Does coloring my hair really damage it?
- **Cultural Connections** explores how the customs and traditions of people in America's diverse ethnic groups can provide another perspective on how to live healthily.
- **Teen Issues** presents chapter concepts from the teen's perspective. It addresses teen habits, interests, and concerns. Its purpose is to make health "real" to young teens.

STUDENT MODULES

Teen Health Course 2 contains three supplemental modules to supply more in-depth information. Each module is four-color, and each is supplied as a Student Edition and a Teacher's Annotated Edition. The Teacher's Annotated Edition includes lesson plan material along with discussion questions and activities correlated to pages in the Student Module.

The modules are:

Abstinence

This module is designed to help students take a close look at their choices and decisions they make in their relationships and to recognize the value of practicing abstinence from sexual activity before marriage. The articles provide information about the consequences of high-risk behaviors, the skills of practicing decision making and effective communication skills, identifying personal, family, and community values, and controlling emotions. Students are empowered to take control of their lives and to make responsible decisions that will enable them to reach their lifetime goals.

Physical Activity

This module presents the benefits of physical activity and explores how individual and team sports can contribute to a person's physical, mental, and sometimes social health. Students assess their health-related and skills-related fitness levels. They learn how to set fitness goals that include endurance-building, strength-building, and flexibility-building exercises. The importance of nutrition for energy and health is presented, along with the helpful hints of how to avoid injury and use proper safety equipment when exercising.

Conflict Resolution

This module demonstrates how to resolve conflicts in peaceful ways. It provides skill-building activities that will give students practice in preventing, defusing, and avoiding conflicts. These include: learning to control anger; active listening; using "I messages" to communicate; how to work toward win-win solutions; and how to apply the techniques used by peer mediators.

Teacher's Wraparound Edition

TEACHER'S WRAPAROUND EDITION

The Teacher's Wraparound Edition provides teaching suggestions printed in the margin beside the actual student text page, which has been slightly reduced in size.

Chapter Planning Guide

To assist the teacher in course planning, a two-page Chapter Planning Guide has been inserted in the Teacher's Wraparound Edition immediately preceding the beginning of every chapter. The following types of information can be found on these pages.

- A list of **Glencoe Teaching Resources,** which are available for the specific chapter.
- A list of **Additional Resources,** including multimedia products and books and periodicals for the teacher and student.
- A correlation to the **National Health Education Standards** to the lesson material in each chapter.
- A correlation to the **ABCNews Interactive™** videodisc series *Understanding Ourselves.*

Chapter Opener Pages

On the two-page spread that begins each chapter, these elements are provided:

Chapter Overview. This brief description of the chapter and the specific lessons will be helpful in deciding which lessons to use.

Pathways Through the Program. This feature informs the teacher of how specific chapter content relates to other health content presented in the program. It enables the teacher to reinforce the interrelationship between a person's physical, emotional and mental, and social health.

Introducing the Chapter. Suggestions for using the chapter introduction are given.

Cooperative Learning Project. A cooperative learning project related to the Teen Health Digest is presented. A sampling of these types of projects includes implementing a health fair, creating a teen health newsletter, and developing a videotaped show. In all cases, technology options are provided.

Key to Ability Levels. This key provides a description of each of the three codes used throughout the lesson plan to identify activities designed for students of various abilities, as well as learning styles.

Lesson Pages

On the various lesson pages, these elements are provided:

Focus Activities and suggestions presented under this initial heading are designed to help you focus students' attention on the lesson material.

Lesson Objectives. Student objectives for the lesson are listed.

Motivator. This short activity is designed to assist students in "switching gears" from the previous class. It requires no teacher direction and can be completed while the teacher takes attendance.

Introducing the Lesson. This activity gives closure to the Motivator Activity and focuses students on the topic of the lesson.

Introducing Words to Know. An introductory vocabulary activity is provided to help students learn the lesson vocabulary and build vocabulary skills.

Teach The activities listed under this heading make up the suggested lesson plan. These are the activities and discussion topics recommended to develop the health content and academic skills. Strategies are grouped under content and skill headings for easy reference and are leveled to identify those that are most appropriate for students of varying ability levels and learning styles.

Assess Under this heading are suggestions for reviewing the lesson and evaluating student understanding.

Reteaching. Reteaching activities are provided for those students who have difficulty mastering the important lesson concepts.

Enrichment. Enrichment activities are provided for those students who are able and willing to explore the content in more depth.

Lesson Review. Answers to both categories of review questions are provided.

Close Under this heading are suggested ways to bring closure to the class and to recap the essence of the lesson.

In addition to the side column activities, each lesson displays, across the bottom of the page, a number of other teaching suggestions and activities. These include:

Technology Update
More About
In Your Journal
Meeting Student Diversity
Unifying themes of:
   Personal Health
   Health of Others
   Growth and Development
   Consumer Health
Home and Community Connection
Cooperative Learning

Teacher's Classroom Resources

TEACHER'S CLASSROOM RESOURCES

The Teacher's Classroom Resources contains teaching materials designed to reduce teacher preparation time and maximize students' learning time. The following booklets are provided to assist in meeting the individual needs and interests of your students.

Concept Mapping Activities. The *Concept Mapping Activities* book of the Teacher's Classroom Resources provides a developmental approach for students to practice concept mapping. They can be used to preview a lesson's content by visually relating the concepts to be learned and allowing students to read with purpose. As a review strategy, they reinforce main ideas and clarify relationships. The booklet provides one activity per lesson.

Reteaching Activities. The *Reteaching Activities* booklet is designed for students who need additional help in learning the concepts presented in the lesson. There is one activity for each lesson.

Enrichment Activities. The *Enrichment Activities* booklet in the Teacher's Classroom Resources allows students the opportunity to explore concepts further. There is one activity for each lesson.

Decision-Making Activities. Teaching the skills necessary to make wise decisions requires the opportunity for application. The *Decision-Making Activities* booklet provides two decision-making activities for each chapter. Role-plays and hypothetical situations enable students to practice and apply the decision-making process.

Cross-Curriculum Activities. The *Cross-Curriculum Activities* booklet provides students with activities that relate health information to the content of other subject areas, including literature, language arts, social studies, science, mathematics, art, and music. There are two activities for each chapter, each of which relates to a different subject area, with that content area clearly labeled on the sheet.

Health Labs. The *Health Lab* booklet in the Teacher's Classroom Resources gives students experience with making observations and hypotheses, collecting and recording data, and forming conclusions based on analysis and interpretation of experimental results. There is one Health Lab and a teacher page for each chapter in the text.

Parent Letters and Activities. Parental involvement is a critical component of a successful health program. The *Parent Letters and Activities* in the Teacher's Classroom Resources includes introductory teacher material about how to use these letters and how to inform parents or guardians of the instructional program and assessment techniques to be employed. This material appears in English and Spanish, and one letter is provided for each chapter.

Teacher's Classroom Resources

Performance Assessment. The activities in the *Performance Assessment* book enable teachers to assess learning in a way that requires students to manipulate information in flexible and creative ways. One activity is provided for each chapter. The booklet includes an introduction of how to use performance assessment in the classroom and rubrics and assessment lists to help the teacher evaluate students' work.

Testing Program. The *Testing Program* contains Lesson Quizzes and Chapter Tests. The Lesson Quizzes consist of a one-page quiz for each lesson in *Teen Health*. The Chapter Tests are two-page tests that cover the main concepts of the chapter.

Teaching Transparencies. Sixty-four overhead color transparencies provide motivating visuals that can be used to introduce or reinforce a lesson.

Student Activities Workbook. The *Student Activities Workbook* provides immediate reinforcement to the text. It contains the following types of activities: Chapter Study Guide, Applying Health Skills (one per lesson), and a Health Inventory (one per chapter). The Teacher's Annotated Edition of the *Student Activities Workbook* provides answers to all activities.

Professional Health Series

PROFESSIONAL HEALTH SERIES

Four booklets comprise the Glencoe Professional Health Series. These are intended to help teachers address the changing needs of their classroom and work cooperatively with their school, parental, and community organizations. The series includes:

Dealing with Sensitive Issues

This booklet prepares the educator to discuss issues with students that are generally considered sensitive by providing background information and teaching strategies. Included is information on how to communicate effectively, share knowledge, and promote skills. Classroom activities are given for each sensitive issue and cover such topics as depression, eating disorders, abuse, and sexuality.

Promoting a Comprehensive School Health System

This booklet has been developed to increase individual and community understanding of what comprehensive school health is and of the need for comprehensive school health programs throughout the United States. It provides methods to facilitate the assessment of local needs or concerns.

Cultural Diversity in the Health Classroom

This booklet focuses on how to teach a health education curriculum in a culturally diverse classroom. It answers questions such as: How can I teach more effectively in a culturally diverse classroom? How do cultural experiences affect students' interpretation of the subject matter? How can I prepare my students to live in a culturally diverse world?

Death and the Adolescent

This booklet describes how schools can function in responding quickly and effectively to deaths of students or staff members. It provides information on how death affects people, particularly teens, and the best ways to help them through the stages of mourning.

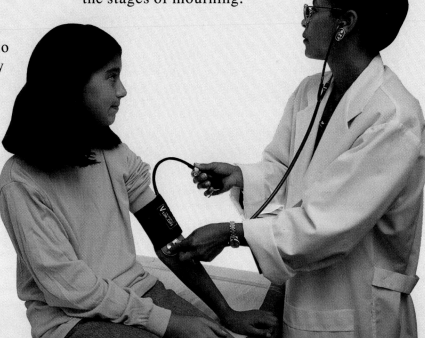

MULTIMEDIA RESOURCES

Teen Health Videodisc Series

A 45-minute CAV videodisc has been specifically produced to correlate with *Teen Health*. This videodisc series has English and Spanish narration and is also available in VHS format. It contains the following features:

Unit Launcher presents a survey of the core topics covered in the chapter in a lively, motivating, action-packed presentation. It provides an overview of the unit and stresses the dynamic relationship among topics.

Life Skills demonstrate techniques that develop lifelong health behaviors, such as communication techniques, dealing with peer pressure, refusal skills, and consumer skills.

Teen Issues feature a magazine-show style that documents how real teens addressed health-related issues and solved problems. These thought-provoking segments model positive health behavior.

Video Kit

This media kit includes compelling videos accompanied by an Instructor's Guide. The high-interest topics complement the student text and reinforce healthy life skills. Titles include:

Teenage Nutrition: Prevention of Obesity, Now and For a Lifetime
Cliques: Who's In? Who's Out?
Cool to be Me: Self-Esteem
All About Respect

Audiocassette Program (English/Spanish)

Audiocassettes provide summaries of chapter content for review, for reteaching, or for use when you do not have time to teach a particular chapter. Each summary is accompanied by a chapter activity and test based on the content of the audiocassettes. Spanish cassettes are also available.

Testmaker

This computer software test bank is available for Macintosh and IBM computers. It provides questions in various formats, with room to add questions of your own.

ABCNews InterActive™ Correlation Bar Code Guide

Teen Health has been correlated to the exceptional ABCNews InterActive™ health series, *Understanding Ourselves*. This series provides a truthful, nonjudgmental exploration of complex health issues. Each title includes one double-sided videodisc, closed captioning for hearing impaired students, and an English and Spanish narration of motion video segments. This videodisc series is also available in VHS format. The *Correlation Bar Code Guide* includes the following videodiscs: *AIDS, DRUGS AND SUBSTANCE ABUSE, TEENAGE SEXUALITY, ALCOHOL, TOBACCO, FOOD AND NUTRITION, VIOLENCE PREVENTION,* and *MAKING RESPONSIBLE DECISIONS*.

THE WORK STUDENTS DO

Decision-Making Activities

Decision making is a process in which a person selects from two or more possible choices. Decision making is a skill, and, like any other skill, it must be taught and then practiced over and over if it is to be mastered.

A number of skills are necessary to make decisions:

- the ability to identify options
- the ability to identify priorities
- the ability to identify and evaluate consequences
- the ability to make a choice and act on it

Adolescents are moving from concrete thinking to abstract thinking. This is a learned task. In other words, the ability to think abstractly doesn't just happen. Within any given class you will have students at various levels of thinking ability. Some will not be able to comprehend and apply the entire process described above. By practicing with a variety of examples, however, they can learn the process and develop the skill.

The skill of goal-setting is a critical part of helping students make responsible decisions, especially in the health-related areas of drugs and sexuality. Students who have set realistic short- and long-range goals and mapped out strategies to achieve these goals can weigh their alternatives to a given problem against these goals. Having goals also gives people some sense of personal control over their own lives. Students should realize they can make decisions that help or hinder them in achieving their goals. Work with your students on setting achievable short-term goals at first. These may be goals related to health behaviors. Achievement reinforces positive feelings of self-worth.

Critical Thinking and Student Investigations

Helping students think should be a primary goal of all education. Learning only easily testable facts is inadequate preparation for living in the twenty-first century. Certain assumptions underlie the teaching of thinking:

- All students are capable of higher level thinking.
- Thinking skills can be taught and learned.
- Thinking can be effectively taught within any content area.
- Appropriate expectations for logical thinking are based on physiological maturation, social experiences, and the knowledge level of the students.
- Learners can be taught to transfer thinking skills from a specific content area to a variety of new learnings.

How then do we teach thinking? Doesn't all learning require thinking? What is meant by higher level thinking? These questions and questions like them require teachers to reflect on the instructional strategies used in daily lessons and to include opportunities for students to reason and think about what is being learned.

All learning does require thinking. Benjamin Bloom's Taxonomy of the Cognitive Domain is probably the most widely recognized schema for levels of thinking. Questions, statements, and activities can be designed based on the taxonomy that will elicit the complexity of thinking congruent with the lesson's objective(s).

Each of Bloom's cognitive categories includes a list of a variety of thinking skills and indicates the kind of behavior students are to perform as the objectives or goals of specific learning tasks. Here are some examples:

- **Knowledge:** define, recognize, recall, identify, label, understand, examine, show, collect
- **Comprehension:** translate, interpret, explain, describe, summarize, extrapolate
- **Application:** apply, solve, experiment, show, predict
- **Analysis:** connect, relate, differentiate, classify, arrange, check, group, distinguish, organize, categorize, detect, compare, infer
- **Synthesis:** produce, propose, design, plan, combine, formulate, compose, hypothesize, construct
- **Evaluation:** appraise, judge, criticize, decide

Connections to Home and Community

The most successful health programs are those that result from cooperation between home and school. Parents should be informed of the health content and have an opportunity to review classroom materials. The *Teen Health* program has provided parent letters and activities in English and Spanish to encourage communication and cooperation between home and school.

A variety of health-related organizations and agencies can serve as resources for your health education classroom. Many organizations provide speakers on various health topics. Others offer resource materials such as pamphlets, posters, audiovisual materials, and computer software.

These organizations and agencies also offer opportunities for students to get involved in the community, as well as to practice their oral and written communication skills. Students may correspond with various organizations requesting information and literature. Or, they may participate in health-related activities. For example, the Cancer Society has numerous activities, including the Great American Smokeout. Other agencies sponsor walk-a-thons and health fairs. Encourage your students to participate in community activities and give extra credit for their participation.

Personal Inventories and Journals

Personal journals are effective tools that guide students to become critical thinkers and to evaluate their own health and well-being. This program incorporates journal writing activities in each chapter opener, and in various other places throughout the chapter. While this activity has great teaching value, issues of personal and family privacy should be of utmost concern to teachers. One challenge with such an activity is helping students to develop a system to ensure the confidentiality of their private journal materials. One method might be to assign numbers to each student's journal and collect them at the end of each class period. If such a system is not possible, suggest to students that they omit putting their names on journal materials and store their journal in a private place.

ASSESSMENT

Performance Assessment Projects

Activities in the *Performance Assessment* booklet in the Teacher's Classroom Resources provides activity sheets and projects that can be used for students to demonstrate their mastery of chapter content in a nontest situation. The first section offers teachers guidance on how to use performance assessments and suggests some additional assessment strategies. Rubrics are provided that enable teachers to evaluate students' work.

Testing Programs

The *Testing Program* booklet contains a Self-Inventory of Health Knowledge; Lesson Quizzes for each lesson in the Student Edition; Chapter Tests for each chapter in the Student Edition; and Answer Keys for the Self-Inventory, the Lesson Quizzes, and the Chapter Tests.

The Self-Inventory of Health Knowledge is an objective test that may be used as a pretest to determine students' level of health knowledge and understanding. It may also be used at the end of the course to determine how much the students' knowledge and understanding have increased.

The Self-Inventory contains a variety of questions designed to measure how well the students understand basic health concepts and the relationship their health decisions have to their level of wellness.

The Lesson Quizzes should be given at the completion of each lesson. They are each one page long. Each quiz is divided into two sections. Formats include matching items (with lettered words), fill-in completion items (with words in a box), multiple-choice items (with four possible answers), cluster true-false items, and modified true-false items.

Chapter Tests should be given at the completion of each chapter. They are each two pages long. Each test is divided into three sections. The first section is entitled "Reviewing Health Concepts," and the second is entitled "Applying Health Concepts." The third section of each Chapter Test is entitled "Thinking Critically About Health." This section always consists of two short-essay questions.

A computerized testing program, available for Macintosh and IBM, is also available and designed so teachers can include their own assessment items in the test bank.

MEETING STUDENT DIVERSITY

There are many effective teaching strategies teachers can use to create a positive learning environment—one that enhances instruction for every student.

Students with Language Diversity

Students with language diversity may not have trouble mastering the content. Their difficulty may be only with the language. The following guidelines can help you pace and reinforce instruction for these students.

Allow time for students to become familiar with the structure of English. Some students may know many English words but have difficulty with word order. For example, adjectives often precede the nouns they modify in English: "That is a yellow pencil." But in Spanish, the order would be "That is a pencil yellow." By accepting mistakes and praising students' efforts, you will provide an atmosphere in which students can experiment with English.

Remember that students can often understand more than they can express. Students may be able to use simple sentences but have trouble with figures of speech, idioms, and words with multiple meanings.

As students begin to develop writing skills, they may mix English with their native language. Accept these language mixtures. At this stage, confidence and enjoyment of writing are the most important goals.

Provide peer learning by grouping English-proficient students with students who have a limited proficiency. Encourage students to work in pairs or small heterogeneous groups to teach skills to one another.

Students with Varying Ability Levels and Learning Styles

This Teacher's Wraparound Edition provides many types of activities to support the concepts presented in the text. It is not intended that you use all of them. Instead, it is recommended that you select those activities that are most appropriate to the needs, abilities, interests, and maturity levels of your students and to your teaching methods. To assist you in this selection, each activity in the lesson plan is coded as follows:

> **L1** **Level 1** strategies should be within the ability range of all students. Often full class participation is required. These strategies usually require teacher direction.
>
> **L2** **Level 2** strategies are designed for average to above-average students or for small-group participation. Some teacher direction is necessary.
>
> **L3** **Level 3** strategies are designed for students able and willing to work independently. Minimal teacher direction is necessary.

Meeting Student Diversity

Students with Special Needs

Students with special needs are often mainstreamed in health classes. Because students with special needs are often slower to respond to questions, it is tempting to skip their responses when trying to meet class objectives. This deprives special needs students of equal opportunities.

In some cases you will need to alter your teaching methods when working with these students. When an activity requires students to write on the chalkboard, special needs students should be required to do so also, as long as they are physically and mentally capable. Students in wheelchairs may require a lowered chalkboard or the use of an overhead projector. For the visually impaired, an oral response is appropriate. When an activity requires class discussion or group work, special needs students should be called on to participate.

The following suggestions can help you structure your classroom to create a positive learning environment for all your students.

The Physically Challenged

- Adjust the room arrangement to fit students' needs.
- Encourage verbal activities and small group participation.
- Encourage these students to participate in physical activities within their capabilities.

The Learning Disabled

- Focus on the positive; point out students' strengths.
- Avoid assigning these students to seats around distracting students.
- Encourage small-group participation.
- Encourage students to express themselves verbally.
- Be very clear in the directions you provide.
- Provide these students with short, easy-to-read health handouts.
- Provide an organized environment in which expectations are clear—and realistic.
- Provide positive feedback when possible.

The Visually Impaired

- Seat students in good light near the front of the classroom.
- Whenever possible, provide hands-on experiences.
- Assign another student to assist the visually impaired student.
- Allow students to record assignments or complete them verbally.
- Tape-record each chapter of the textbook, or use the *Teen Health* audiocassette program.
- Encourage verbal participation.
- Administer tests orally.
- Allow students to move about freely so they can get close to charts or displays.

The Hearing Impaired

- Seat students near the front of the classroom.
- Use the chalkboards to highlight key terms and phrases.
- Guide students to the appropriate text page during discussions.
- Ask for a student volunteer to make copies of his or her notes for the hearing impaired students.
- Give written tests.
- Look directly at hearing impaired students when speaking. Talk normally.
- Learn sign language.
- Write directions for activities on the chalkboard.
- Encourage as much verbal interaction as possible.

TEEN HEALTH AND THE NATIONAL HEALTH EDUCATION STANDARDS

Correlation of National Health Education Standards

The National Health Education Standards were created with the goal of improving educational achievement for students and improving health in the United States through the promotion of health literacy. The Health Standards detail the knowledge and skills essential to the development of health literacy in the form of performance indicators. A lesson by lesson correlation of the *Teen Health* student text to the National Health Education Standards demonstrates how *Teen Health* meets these criteria. The following is a complete listing of the health standards.

National Health Education Standards

Health Education Standard 1:
Students will comprehend concepts related to health promotion and disease prevention.

Performance Indicators:
1.1 Explain the relationship between positive health behaviors and the prevention of injury, illness, disease, and premature death.
1.2 Describe the interrelationships of mental, emotional, social, and physical health during adolescence.
1.3 Explain how health is influenced by the interaction of body systems.
1.4 Describe how family and peers influence the health of adolescents.
1.5 Analyze how environment and personal health are interrelated.
1.6 Describe ways to reduce risks related to adolescent health problems.
1.7 Explain how appropriate health care can prevent premature death and disability.
1.8 Describe how lifestyle, pathogens, family history, and other risk factors are related to the cause or prevention of disease and other health problems.

Health Education Standard 2:
Students will demonstrate the ability to access valid health information and health-promoting products and services.

Performance Indicators:
2.1 Analyze the validity of health information, products, and services.
2.2 Demonstrate the ability to utilize resources from home, school, and community that provide valid health information.
2.3 Analyze how media influences the selection of health information and products.
2.4 Demonstrate the ability to locate health products and services.
2.5 Compare the costs and validity of health products.
2.6 Describe situations requiring professional health services.

Health Education Standard 3:
Students will demonstrate the ability to practice health-enhancing behaviors and reduce health risks.

Performance Indicators:
3.1 Explain the importance of assuming responsibility for personal health behaviors.
3.2 Analyze a personal health assessment to determine health strengths and risks.
3.3 Distinguish between safe and risky or harmful behaviors in relationships.

Health Standards Correlation

3.4 Demonstrate strategies to improve or maintain personal and family health.
3.5 Develop injury prevention and management strategies for personal and family health.
3.6 Demonstrate ways to avoid and reduce threatening situations.
3.7 Demonstrate strategies to manage stress.

Health Education Standard 4:
Students will analyze the influence of culture, media, technology, and other factors on health.

Performance Indicators:
4.1 Describe the influence of cultural beliefs on health behaviors and the use of health services.
4.2 Analyze how messages from media and other sources influence health behaviors.
4.3 Analyze the influence of technology on personal and family health.
4.4 Analyze how information from peers influences health.

Health Education Standard 5:
Students will demonstrate the ability to use interpersonal communication skills to enhance health.

Performance Indicators:
5.1 Demonstrate effective verbal and non-verbal communication skills to enhance health.
5.2 Describe how the behavior of family and peers affects interpersonal communication.
5.3 Demonstrate healthy ways to express needs, wants, and feelings.
5.4 Demonstrate ways to communicate care, consideration and respect of self and others.
5.5 Demonstrate communication skills to build and maintain healthy relationships.
5.6 Demonstrate refusal and negotiation skills to enhance health.
5.7 Analyze the possible causes of conflict among youth in schools and communities.
5.8 Demonstrate strategies to manage conflict in healthy ways.

Health Education Standard 6:
Students will demonstrate the ability to use goal-setting and decision-making skills to enhance health.

Performance Indicators:
6.1 Demonstrate the ability to apply a decision-making process to health issues and problems individually and collaboratively.
6.2 Analyze how health-related decisions are influenced by individual, family, and community values.
6.3 Predict how decisions regarding health behaviors have consequences for self and others.
6.4 Apply strategies and skills needed to attain personal health goals.
6.5 Describe how personal health goals are influenced by changing information, abilities, priorities, and responsibilities.
6.6 Develop a plan that addresses personal strengths, needs, and health risks.

Health Education Standard 7:
Students will demonstrate the ability to advocate for personal, family, and community health.

Performance Indicators:
7.1 Analyze various communication methods to accurately express health information and ideas.
7.2 Express information and opinions about health issues.
7.3 Identify barriers to effective communication of information, ideas, feelings, and opinions about health issues.
7.4 Demonstrate the ability to influence and support others in making positive health choices.
7.5 Demonstrate the ability to work cooperatively when advocating for healthy individuals, families, and schools.

Health Standards Correlation

How *Teen Health* Meets the Standards

At the beginning of each chapter, a correlation of the material in each chapter lesson to the National Health Education Standards has been provided. The correlation appears on the chapter planning guide.

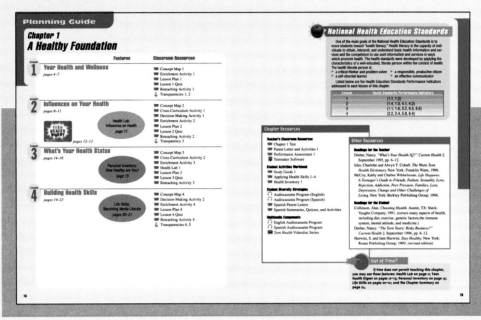

The health standards addressed in each lesson are indicated by their specific performance indicator number.

National Health Education Standards

One of the main goals of the National Health Education Standards is to move students toward "health literacy." Health literacy is the capacity of individuals to obtain, interpret, and understand basic health information and services and the competence to use such information and services in ways which promote health. The health standards were developed by applying the characteristics of a well-educated, literate person within the context of health. The health literate person is:

- a critical thinker and problem solver
- a responsible, productive citizen
- a self-directed learner
- an effective communicator

Listed below are the Health Education Standards Performance Indicators addressed in each lesson of this chapter.

Lesson	Health Standards Performance Indicators
1	(1.1, 1.2)
2	(1.4, 1.5, 4.1, 4.2)
3	(1.1, 1.6, 3.2, 6.5, 6.6)
4	(2.2, 2.4, 5.8, 6.4)

T25

Course Presentation

The structure of *Teen Health* allows for the material to be presented over the course of a full year, one semester, nine weeks, or six weeks. The chart below shows the recommended lessons to be taught and the amount of class time needed.

		Year Course	1 Semester Course	9 Week Course	6 Week Course
1 A Healthy Foundation	Lesson 1	2	1	1	1
	Lesson 2	2	1	1	1
	Lesson 3	2	2	X	X
	Lesson 4	3	1	X	X
2 Personal Responsibility and Decision Making	Lesson 1	2	1	1	1
	Lesson 2	2	1	X	X
	Lesson 3	2	1	1	1
	Lesson 4	3	1	1	½
	Lesson 5	3	1	1	½
3 Being a Health Care Consumer	Lesson 1	2	1	1	1
	Lesson 2	2	1	1	X
	Lesson 3	3	1	X	X
	Lesson 4	3	1	X	X
	Lesson 5	2	X	X	X
4 Food and Nutrition	Lesson 1	2	1	1	1
	Lesson 2	3	2	1	1
	Lesson 3	3	2	1	1
	Lesson 4	3	2	X	X
5 Physical Activity and Weight Management	Lesson 1	3	2	1	1
	Lesson 2	3	2	X	X
	Lesson 3	3	2	X	X
	Lesson 4	3	2	1	1
	Lesson 5	2	1	1	X
6 Sports and Conditioning	Lesson 1	3	2	X	X
	Lesson 2	3	2	1	1
	Lesson 3	3	2	1	1
	Lesson 4	2	X	X	X
7 Growth and Development	Lesson 1	3	1	1	1
	Lesson 2	3	1	X	X
	Lesson 3	2	1	1	1
	Lesson 4	2	1	1	1
	Lesson 5	2	1	1	X
	Lesson 6	2	1	X	X
8 Mental and Emotional Growth	Lesson 1	2	1	1	1
	Lesson 2	2	2	1	1
	Lesson 3	3	2	1	1
	Lesson 4	2	X	X	X
	Lesson 5	2	X	X	X

Scope and Sequence

		Year Course	1 Semester Course	9 Week Course	6 Week Course
9 Social Health: Families and Friends	Lesson 1 Lesson 2 Lesson 3 Lesson 4 Lesson 5	3 3 3 3 2	2 2 2 2 1	1 1 1 1 X	1 1 1 1 X
10 Tobacco	Lesson 1 Lesson 2 Lesson 3 Lesson 4	3 3 3 2	2 2 2 2	1 X 1 1	1 X 1 1
11 Alcohol and Drugs	Lesson 1 Lesson 2 Lesson 3 Lesson 4 Lesson 5 Lesson 6	3 3 3 3 2 2	2 2 X 2 2 X	1 1 X X 1 X	1 1 X X 1 X
12 Understanding Communicable Diseases	Lesson 1 Lesson 2 Lesson 3 Lesson 4 Lesson 5	2 3 3 2 3	1 1 1 1 1	1 X 1 X 1	X X X X X
13 Understanding Noncommunicable Dieseases	Lesson 1 Lesson 2 Lesson 3 Lesson 4	3 3 3 2	1 1 1 1	1 1 1 X	X X X X
14 Preventing Violence and Abuse	Lesson 1 Lesson 2 Lesson 3	3 3 3	1 1 1	1 1 1	X X X
15 Safety and Recreation	Lesson 1 Lesson 2 Lesson 3 Lesson 4 Lesson 5 Lesson 6	2 2 2 2 2 2	1 1 1 1 1 1	1 1 X X 1 X	X X X X 1 X
	Total days	**180**	**90**	**45**	**30**

Scope and Sequence

Content Strands	Chapter 1 A Healthy Foundation	Chapter 2 Personal Responsibility and Decision Making
Personal Health	▪ Health Triangle of physical, social, and mental/emotional health (1) ▪ Recognizing health and wellness (1) ▪ Evaluating your health status (3) ▪ Prevention and risk behaviors (3) ▪ Making choices to increase level of wellness (3) ▪ Creating a Personal Wellness Contract (3) ▪ Practicing daily personal hygiene (4)	▪ Making positive health choices (1) ▪ Establishing good health habits (2) ▪ Developing a healthful habit (2) ▪ Breaking harmful habits (2) ▪ Recognizing health risks (3) ▪ Avoiding risks through your actions and attitudes (3) ▪ Decision-making process as a health skill (4) ▪ Steps of the decision-making process (4) ▪ Developing an action plan to reach goals (5)
Consumer and Community Health	▪ How your friends affect your health (2) ▪ Evaluating health messages in the media (2) ▪ Effectiveness of antibacterial soaps (TD) ▪ Becoming media literate (4) ▪ Finding reliable information sources (4)	▪ Volunteer work and responsibility (1) ▪ Affecting the health of others through your actions (1) ▪ Establishing positive social habits (2) ▪ Social consequences of risky behavior (3) ▪ Influence of society on values (4) ▪ Helping others in your community (TD)
Injury Prevention and Safety	▪ Using an exercise video safely (TD) ▪ Examining risks to your physical health (4)	▪ Air bags and car safety rules (TD) ▪ Being prepared to handle emergencies (1) ▪ Life-threatening habits (2) ▪ Using safety belts and bicycle helmets (3) ▪ Using objective information to evaluate risks (3) ▪ Cumulative risks and chance of injury (3)
Alcohol, Tobacco, and Other Drugs	▪ Avoiding tobacco, alcohol, and drugs to maintain physical health (1)	▪ Identifying smoking, drinking, and drug use as life-threatening habits (2) ▪ Smoking as a risk behavior (3) ▪ Tobacco advertising that targets teens (3)
Nutrition	▪ Importance of a well-balanced diet (1) ▪ Common food allergies (1) ▪ Nutrition and cancer prevention (TD) ▪ Exercise and appetite (TD)	▪ Substituting healthful foods as snacks (2) ▪ Change bad eating habits early in life (TD)
Environmental Health	▪ How environment affects your health (2) ▪ Examining the pollution in your air and water (2)	
Family Living	▪ Identifying the major traits passed on from parents to children (2) ▪ Hereditary traits as a health guidepost (2)	▪ The family's role in health management (1) ▪ Changing harmful habits with family help (2) ▪ Value formation and the family (4)
Individual Growth and Development	▪ Ways to maintain mental/emotional and social health (1) ▪ Creating a Personal Wellness Contract (3) ▪ Developing time-management skills (3) ▪ Developing good communication skills (4) ▪ Building self-management skills (4) ▪ Refusal skills and upholding values (4) ▪ Balancing work and social life (TD)	▪ The meaning of responsibility (1) ▪ Leadership skills (1) ▪ Emotional control for effective communication (1) ▪ Mental/emotional consequences of risk behavior (3) ▪ Major vs. minor decisions (4) ▪ Role of values in decision making (4) ▪ Benefits of setting and pursuing goals (5) ▪ Short term vs. long term goals (5)
Communicable and Chronic Disease		▪ Healthful habits to prevent diabetes (TD)

Key: (1), (2), (3) … indicates lesson number. (TD) refers to Teen Health Digest.

Scope and Sequence

Chapter 3 Being a Health Care Consumer	Chapter 4 Food and Nutrition	Chapter 5 Physical Activity and Weight Management
■ Being a wise consumer of health information (1) ■ Choosing personal products wisely (2) ■ The parts of a tooth and proper dental hygiene (3) ■ Tooth decay, plaque, and tartar (3) ■ The parts of the skin and its functions (3) ■ Acne and proper skin hygiene (3) ■ How to care for hair (3) ■ The parts of the eye and eyecare (4) ■ The parts of the ear and ear care (4)	■ Making healthful daily food choices (1) ■ Difference in recommended daily servings for boys and girls (1) ■ Fiber and a healthy diet (2) ■ Managing weight through exercise (2) ■ Avoiding excess sugar, sodium, fats, and cholesterol (2) ■ Caring for the digestive and excretory systems (4) ■ Role of digestive and excretory organs in maintaining health (4)	■ Benefits of fitness to physical health (1) ■ Strength and flexibility of muscles and joints (1) ■ Aerobic and anaerobic exercise (1) ■ Parts and functions of the circulatory system (2) ■ Functions of blood and plasma (2) ■ Functions of skeletal and muscular systems (3) ■ Determining ideal weight by the Body Mass Index (5) ■ Healthy weight loss and gain (5)
■ Peer influence on buying decisions (1) ■ Understanding and evaluating the role of advertising in purchasing decisions (1) ■ Assessing product claims (1) ■ Health care specialists and facilities (5) ■ Health insurance options (5) ■ Working as a health care volunteer (TD)	■ Making healthful decisions in fast-food restaurants (2) ■ Encouraging a friend to eat a healthful breakfast (3) ■ Youth in action fighting hunger (TD) ■ Tips for buying perishables (TD)	■ How personal fitness affects others (1) ■ Benefits of fitness to social health (1) ■ Choosing exercise equipment (4) ■ Working out with friends (4) ■ Credibility in diet advertisements (TD)
■ Effectiveness of sunscreens (3) ■ Preventing tooth decay with dental sealants (TD) ■ Body piercing as a health risk (TD) ■ Indoor tanning and skin cancer (TD)	■ Understanding the harmful effects of bulimia (4) ■ Cranberry juice and the prevention of kidney and bladder problems (4) ■ Rinsing hands and food for protection against harmful germs (TD)	■ Sensible precautions for a safe workout (4) ■ Stretching safely to prevent injury (4) ■ RICE — four ways to treat an injury (4) ■ Health dangers of anorexia and bulimia (5) ■ Health risks of obesity (5)
■ Role of rehab centers in overcoming drug abuse (5)	■ The link between alcohol consumption and cirrhosis of the liver (4)	■ How smoking harms the circulatory system (2)
■ The link between sweets and cavities (3) ■ Calcium and healthy teeth (3) ■ Nutrition and healthy skin (3)	■ Using the food guide pyramid to achieve healthy nutrition (1) ■ Six types of nutrients (2) ■ Nutrient density in snack foods (3) ■ How the body absorbs nutrients (4)	■ Nutrition and care of the skeletal and muscular systems (3) ■ Calories burned in exercise (5) ■ When and what to eat before exercise (TD)
	■ Influence of geographical region upon food choices and availability (1)	■ How weather affects your workout (4)
■ Influence of family on purchasing decisions (1)	■ Family background and food choices (1)	
■ Developing skills of comparison shopping (2) ■ Reading product labels (2) ■ Visiting an optometrist or ophthalmologist (4) ■ Dentistry as a career choice (TD)	■ Nutrient intake and healthy development (2) ■ Reading a nutrition label (2) ■ Eating a good breakfast as a lifelong habit (3) ■ Link between sugar and hyperactive behavior (TD) ■ Calcium and bone development (TD)	■ Everyday fitness — a life goal (1) ■ Effect of stress on heart and blood vessels (2) ■ Limiting tensions through exercise (2) ■ Planning frequency, intensity, and duration of workouts (4) ■ Assessing your fitness program and progress (4) ■ Helping a friend with an eating disorder (5)
	■ Fiber and disease prevention (2)	

Scope and Sequence

Content Strands	Chapter 6 Sports and Conditioning	Chapter 7 Growth and Development
Personal Health	■ Achieving fitness through individual and team sports (1) ■ Value of competition (1) ■ Importance of water consumption (2) ■ Balanced fitness through cross-training (3) ■ Setting goals for conditioning (3) ■ Simple exercise for strengthening the hips and lower back (TD)	■ The endocrine system's role in regulating body functions (2) ■ How the body controls hormones (2) ■ Parts and functions of the male and female reproductive systems (3), (4) ■ Maintaining healthy male and female reproductive systems (3), (4) ■ Understanding the menstrual cycle (4) ■ Understanding fertilization (4) ■ Good prenatal care (5)
Consumer and Community Health	■ Social interaction and team sports (1) ■ Choosing athletic shoes wisely (TD)	■ Helping others cope with loss (6) ■ Career profile of a hospice worker (TD) ■ Teens helping people in need (TD) ■ Increase in mortality rates (TD) ■ Products that promise to retard aging (TD)
Injury Prevention and Safety	■ Avoiding sports injuries (2)	■ Performing testicular and breast self-exams (3) ■ Pap tests and early detection of cancer (4)
Alcohol, Tobacco, and Other Drugs	■ Health dangers of anabolic steroids (2)	■ Adverse effects of alcohol, tobacco, and drugs on fetal development (5)
Nutrition	■ Carbohydrates as fuel food (2) ■ Energy supplied by a balanced diet (2)	■ Good nutrition to ensure fetal health (5)
Environmental Health		■ Healthy prenatal environment (5) ■ Impact of environment on baby brain development (TD)
Family Living	■ Balancing school, sports, and family life (4) ■ Donating an organ to a family member (TD)	■ Adolescence and independence from family (1) ■ Genetics and heredity (5) ■ Heredity and genetic disorders (5)
Individual Growth and Development	■ Sports and lifetime fitness (1) ■ Self-discipline and individual sports (1) ■ Controlling emotions during sports activities (2) ■ Helping a friend avoid overtraining (3) ■ Improved mental/emotional health through sports (3) ■ Warning signs of overcommitment and burnout (4) ■ Podiatry as a career choice (TD)	■ Physical changes during adolescence (1) ■ Mental, emotional, and social development (1) ■ Dealing with conflicting feelings (1) ■ Controlling the symptoms of PMS (4) ■ Fertilization and fetal growth (5) ■ Stages of childhood and adulthood (6) ■ Five stages of grief (6)
Communicable and Chronic Disease		■ Disorders of the endocrine system (2) ■ Disorders of the reproductive system (3)

Key: (1), (2), (3) … indicates lesson number. (TD) refers to Teen Health Digest.

Scope and Sequence

Chapter 8 Mental and Emotional Health	Chapter 9 Social Health: Families and Friends	Chapter 10 Tobacco
■ Good mental health habits (1) ■ Self-esteem and mental/physical health (2) ■ Stress and fatigue (3) ■ Understanding defense mechanisms (3) ■ Understanding mental disorders (4) ■ Warning signs of suicide (4) ■ Breathing exercises to reduce stress (TD)	■ Understanding the types of communication (1) ■ Symptoms of anger (5)	■ Smokeless tobacco and its dangers (1) ■ Damaging effects of tobacco to the body (1) ■ How smoking affects breathing (1) ■ Parts and functions of the respiratory system (2) ■ Symptoms of tobacco withdrawal (3) ■ Reasons to be tobacco free (4) ■ Tips for quitting smoking (4)
■ How your mental health affects those around you (1) ■ Teachers, school counselors, and other sources of help (5) ■ History of the Special Olympics (TD) ■ Self-image, self-doubt, and the tactics of advertising (TD) ■ Turning teens away from suicide (TD)	■ Importance of friends (3) ■ Role of peer counselors (5) ■ Cellular phone for the deaf (TD) ■ Career profile: Therapist for children (TD) ■ Making friends in your community (TD)	■ Estimating annual cost of cigarettes (1) ■ Costs to society (1) ■ Reasons teen use tobacco (3) ■ Nonsmokers' rights (4) ■ The beauty myths of smoking ads (TD) ■ Tobacco advertising and sports events (TD)
■ Personal safety and self-esteem (2)	■ Abstinence from sex, drugs, and alcohol (4)	
■ Relationship between low self-esteem and alcohol and drug abuse (2)	■ Peer pressure to use tobacco, drugs, and alcohol (3)	■ Harmful substances in tobacco (1) ■ Physiological and psychological dependence on tobacco (3) ■ Working out for recovery (TD)
■ Good nutrition and stress management (3)		■ Choosing healthy snacks when quitting smoking (4)
■ Environment as a factor in shaping personality (1) ■ Seasonal affective disorder (the Winter blues) (4)		■ How tobacco affects nonsmokers (4)
■ Heredity as a factor in shaping personality (1) ■ How family shapes values through behavior and actions (1) ■ Influence of family on self-esteem (2) ■ Parents and family as a support system (5)	■ How families nurture the individual members (2) ■ Facing family changes and challenges (2) ■ Tips on strengthening family bonds (2) ■ Family responsibilities and communication (TD)	■ How parents' smoking habits can affect their children (3)
■ Benefits of a positive outlook (1) ■ Behavior's role in shaping personality (1) ■ Responsibility for personal behavior (1) ■ Behavior that signifies self-esteem (2) ■ Improving your self-esteem (2) ■ Time-management and stress reduction (3) ■ Long-term stress management (3) ■ Helping a friend who talks about suicide (4) ■ Recognizing when to seek outside help (5) ■ Music Therapist as a career choice (TD)	■ Communicating effectively (1) ■ Tips on giving and taking criticism (1) ■ Compromise and conflict resolution (1) ■ Personal Inventory for being a good friend (3) ■ Recognizing positive and negative peer pressure (3) ■ Effective refusal skills (4) ■ Responses to pressure statements (4) ■ Arguments and their causes (5) ■ Conflict prevention and recognition (5) ■ Conflict resolution skills (non-violent confrontation) (5)	■ Analyzing a smoking advertisement (3) ■ Overcoming peer pressure to smoke (4) ■ Negative impact of smoking on fetal development (4) ■ Cancer causing chemicals in smoke (1) ■ Cancer risk of pipe and cigar smoke (1) ■ Diseases of the respiratory system (2) ■ Secondhand smoke and cancer (4)

Scope and Sequence

Contents Strands	Chapter 11 Alcohol and Drugs	Chapter 12 Understanding Communicable Diseases
Personal Health	■ Long-term and short-term effects of alcohol on the body (1) ■ Understanding blood alcohol levels (1) ■ Signs of alcoholism (1) ■ Medicine and the proper use of drugs (2) ■ The functions and parts of the nervous system (3)	■ How germs are spread (1) ■ The body's lines of defense against germs (1) ■ Personal hygiene and protection against the common cold (3) ■ Vaccination schedules (3) ■ Protection against the spread of disease (4)
Consumer and Community Health	■ Support groups for alcohol abusers (1) ■ Understanding a prescription label (2) ■ Reading over-the-counter drug labels (2) ■ Students Against Destructive Decisions (4) ■ Effects of drugs and alcohol on others (4) ■ Alcohol and drug treatment centers (6) ■ Addiction recovery (6) ■ Alcohol advertising on the Internet (TD)	■ Quarantine and chicken pox, measles, and mumps (3) ■ Ways to protect others from germs (4) ■ Value of AIDS education and research (5) ■ Examining over-the-counter cold remedies (TD) ■ Twelve year-old AIDS activist (TD) ■ Career profile: Diseases detective (TD)
Injury Prevention and Safety	■ Safety precautions to prevent nervous system damage (3) ■ Fetal Alcohol Syndrome (3) ■ Physical risks of drug and alcohol abuse (4) ■ Drinking and driving (4)	
Alcohol, Tobacco, and Other Drugs	■ Chemicals in alcohol (1) ■ Misuse and abuse of drugs (2) ■ Street drugs: marijuana, stimulants (amphetamines, cocaine, etc.), depressants, inhalants, and narcotics (2)	■ Avoidance of alcohol, tobacco, and drugs to resist disease (4)
Nutrition		■ Good nutrition and prevention of communicable disease (4) ■ Feeding a cold (TD)
Environmental Health		
Family Living	■ Help for alcoholics and their family (1) ■ Risks to the family (4) ■ Drinking and behavioral changes (TD)	■ Benefits of sexual abstinence before marriage (5)
Individual Growth and Development	■ Personal Inventory of attitudes toward alcohol (1) ■ Helping someone with a drinking problem (1) or a drug problem (6) ■ Effects of drug use on the body (2) ■ Mental/emotional consequences of alcohol and drug abuse (4) ■ Saying no to substance abuse (5)	■ Personal Inventory of good health habits (4) ■ Abstinence from sexual activity to protect against STDs (5) ■ Facts and fictions about how HIV is spread (5)
Communicable and Chronic Disease	■ Exposure to deadly infections through drug use (2) ■ Disorders and diseases of the nervous system (3) ■ Addiction and withdrawal (6)	■ Types of germs (1) ■ Difference between communicable and noncommunicable diseases (1) ■ The immune system's reactions to germs (2) ■ Symptoms and treatment of mononucleosis, hepatitis, and tuberculosis (3) ■ Understanding STDs (5) ■ Understanding HIV/AIDS (5)

Key: (1), (2), (3) … indicates lesson number. (TD) refers to Teen Health Digest.

Scope and Sequence

Chapter 13 Understanding Noncommunicable Diseases	Chapter 14 Preventing Violence and Abuse	Chapter 15 Safety and Recreation
■ Regular physical check-ups and disease prevention (2) ■ Treating allergies (3) ■ Taking medication for chronic conditions (3)	■ Signs of abuse (2)	■ Becoming safety conscious (1) ■ Personal fire safety (2) ■ Understanding traffic signs (3) ■ Tips for crime protection (3) ■ Bicycle, skating, and pedestrian safety (3) ■ Weather emergencies and natural disasters (4)
■ Medical treatments for heart disease (1) ■ Medical treatments for cancer (2) ■ New drug for diabetics (TD) ■ Career profile: Physical therapist (TD)	■ Violence in society (1) ■ Community measures for violence prevention (1) ■ Gangs and gang-related violence (1) ■ Violence in school (1) ■ Community resources for abusers and their victims (3) ■ Career profile: Social worker (TD)	■ Safety at school (2) ■ Community courses in CPR (5) ■ State and federal environmental agencies (6) ■ Career profile: Emergency medical technician (TD) ■ Myths and realities about safety belts (TD) ■ Buying safe toys for children (TD)
	■ Safety tips to protect against violent crime (1) ■ Gun safety and causes of gun-related deaths among young people (1) ■ Aikido as self-defense (TD)	■ Breaking the accident chain (1) ■ Falls, poisonings, electrical shocks, and gun accidents (2) ■ Rescue breathing (5) ■ Controlling bleeding (5) ■ First aid for choking victims (5) ■ First aid for burns, poisonings, and broken bones (5)
■ Alcohol, tobacco, and drug use as causes of noncommunicable disease (1)	■ Link between alcohol, drugs, and violence (1) ■ Relationship between alcoholism/drug addiction and abuse (2)	
■ Good nutrition as a "heart smart" habit (1) ■ Dietary guidelines to lower cancer risk (2) ■ High-starch, low-fiber diet and diabetes (4)		
■ Environmental causes of noncommunicable diseases (1)		■ Causes of air and water pollution (6) ■ Preventing pollution (6) ■ Land pollution and its causes (6)
■ Inherited diseases (1) ■ Hereditary risks of diabetes (4)	■ Importance of good communication skills in the family (2) ■ Rating TV programs (TD)	■ Planning a fire escape route (2) ■ Safety measures for the home (2) ■ Safety guidelines for guns in the home (2)
■ "Heart smart" habits (1) ■ Healthy lifestyle decisions (2) ■ Coping with chronic conditions (3) ■ Coping with a relative who has Alzheimer's disease (4) ■ Ice hockey champion and his battle with Hodgkin's disease (TD)	■ Resisting peer pressure to fight (1) ■ Types and causes of abuse (2) ■ Deciding to report abuse (2) ■ Mental/emotional effects of abuse (2) ■ Breaking the cycle of abuse (3) ■ Danger of remaining silent about abuse (3) ■ Myths and realities about abuse (TD)	■ Preventing accidents when baby-sitting (1) ■ Safety rules for outdoor activities (3) ■ Safety during floods, earthquakes, tornadoes, hurricanes, and blizzards (4) ■ ABCs of First Aid (5) ■ Teens starting a non-profit organization (TD)
■ Definitions of noncommunicable and chronic disease (1) ■ Understanding heart disease (1) ■ Understanding cancer and recognizing the warning signs (2) ■ Understanding allergies and asthma (3) ■ Understanding arthritis and diabetes (4) ■ Understanding Alzheimer's disease (4) ■ Chronic fatigue syndrome (TD)		

HEALTH RESOURCES

Action on Smoking and Health
2013 H Street NW
Washington, DC 20006

Administration on Aging
330 Independence Avenue SW
Washington, DC 20201

Alcoholics Anonymous
Central Office
15 E. 26th St.
New York, NY 10010–1501

Al-Anon/Alateen Family Group
Headquarters
1600 Corporate Landing Parkway
Virginia Beach, VA 23454–5617

American Academy of Pediatrics
141 Northwest Point Road
Elk Grove Village, IL 60007–1098

American Automobile Association,
Foundation for Traffic Safety
1440 New York Avenue NW
Suite 201
Washington, DC 20005

American Cancer Society
1599 Clifton Road
Atlanta, GA 30329

American Counseling Association
5999 Stevenson Avenue
Alexandria, VA 22304

American Dental Association
211 East Chicago Avenue
Chicago, IL 60611

American Heart Association
7272 Greenville Avenue
Dallas, TX 75231–4596

American Institute of Nutrition
9650 Rockville Pike
Bethesda, MD 20814

American Insurance Association,
Engineering and Safety Service
85 John Street
New York, NY 10038

American Lung Association
1740 Broadway
New York, NY 10019

American Medical Association
515 North State Street
Chicago, IL 60610

American Optometric Association
243 North Lindbergh
St. Louis, MO 63141

American Society of Safety Engineers
1800 E. Oakton Street
Des Plaines, IL 60018–2187

Asthma and Allergy Foundation of
America
1125 15th Street, Suite 502
Washington, DC 20005

Centers for Disease Control and
Prevention (CDC)
1600 Clifton Rd., NE
Atlanta, GA 30333

Council on Environmental Quality
Old Executive Office Building
Room 360
Washington, DC 20502

ERIC—Clearinghouse on Teacher
Education—Health Education
One Dupont Circle NW, Suite 610
Washington, DC 20036–1186

Food and Drug Administration
Office of Consumer Affairs
5600 Fishers Lane
Rockville, MD 20857

Juvenile Diabetes Foundation
The Diabetes Research Foundation
120 Wall Street
New York, NY 10005–4001

March of Dimes Birth Defects
Foundation
1275 Mamaroneck Avenue
White Plains, NY 10605

Metropolitan Society for Crippled
Children and Adults
287 North Avenue
Mt. Clemens, MI 48043

Mothers Against Drunk Driving
511 East John Carpenter Freeway
Irving, TX 75062

National Arthritis and Musculoskeletal
and Skin Diseases Information
Clearinghouse
1 AMS Circle
Bethesda, MD 20892–3675

National Cancer Institute,
Office of Cancer Communications
Building 31, Room 10A24
9000 Rockville Pike
Bethesda, MD 20892

National Center for Health Statistics
Centers for Disease Control and
Prevention
6525 Belcrest Road
Hyattsville, MD 20782

National Clearinghouse for Alcohol and
Drug Information
P.O. Box 2345
Rockville, MD 20847–2345

National Congress of Parents-Teachers
Association Alcohol Education
Program
330 North Wabash Ave., Suite 2100
Chicago, IL 60611–3690

National Council on Alcoholism and
Drug Dependence
12 West 21st Street
New York, NY 10010

National Dairy Council
10255 West Higgins Rd.
Rosemont, IL 60018–5606

National Fire Protection Association
1 Batterymarch Park
Quincy, MA 02269–9101

National Health Information Center
P.O. Box 1133
Washington, DC 20013–1133

National Institute of Allergy and
Infectious Diseases
Building 31, Room 7A-50
31 Center Drive, MSC 2520
Bethesda, MD 20892–2520

National Institute of Mental Health
5600 Fishers Lane, Room 7C-02
Rockville, MD 20857–8030

National Safety Council
1121 Spring Lake Drive
Itasca, IL 60143–3201

National Wildlife Federation
8925 Leesburg Pike
Vienna, VA 22184

Office on Smoking and Health
Centers for Disease Control
and Prevention
Mailstop K 50
4770 Buford Highway NE
Atlanta, GA 30341–3724

Parents Magazine
685 Third Avenue
New York, NY 10017

Sierra Club
730 Polk Street
San Francisco, CA 94109–7813

Students Against Destructive Decisions
255 Main Street
Marlborough, MA 01752–5505

United Cerebral Palsy, Inc.
10 Waterside Plaza
New York City, NY 10010–2602

USDA Food and Nutrition Information
and Education Resources Center
National Agricultural Library,
U.S. Department of Agriculture
10301 Baltimore Boulevard, Room 304
Beltsville, MD 20705

Wheelchair Sports U.S.A.
3595 East Fountain Blvd., Suite L-1
Colorado Springs, CO 80910–1740

TEEN Health
COURSE 2

TEEN *Health*
COURSE 2

Mary Bronson Merki, Ph.D.

Glencoe McGraw-Hill

New York, New York
Columbus, Ohio
Woodland Hills, California
Peoria, Illinois

Meet the Author

Mary Bronson Merki has taught health education in grades K–12, as well as health education methods classes at the undergraduate and graduate levels. As Health Education Specialist for the Dallas School District, Dr. Merki developed and implemented a district-wide health education program, *Skills for Living,* which was used as a model by the state education agency. She also helped develop and implement the district's Human Growth, Development, and Sexuality program, which won the National PTA's Excellence in Education Award and earned her an honorary lifetime membership in the state PTA. Dr. Merki has assisted school districts throughout the country in developing local health education programs. In 1988 she was named the Texas Health Educator of the Year by the Texas Association for Health, Physical Education, Recreation, and Dance. Dr. Merki is also the author of *Glencoe Health,* a high school textbook adopted in school districts throughout the country. She currently teaches in a Texas public school, where she was recently honored as Teacher of the Year. Dr. Merki completed her undergraduate work at Colorado State University in Fort Collins, Colorado. She earned her master's and doctoral degrees in health education at Texas Woman's University in Denton, Texas.

Editorial and production services
provided by Visual Education Corporation, Princeton, NJ.

Design by Bill Smith Studio, New York, NY.

Glencoe/McGraw-Hill
A Division of The McGraw-Hill Companies

Copyright © 1999 by Glencoe/McGraw-Hill.
All rights reserved. Except as permitted under the United States Copyright Act, no part of this publication may be reproduced or distributed in any form or by any means, or stored in a database or retrieval system, without prior written permission from the publisher.

Send all inquiries to:
Glencoe/McGraw-Hill
21600 Oxnard Street, Suite 500
Woodland Hills, California 91367

ISBN 0-02-653128-3 (Course 2 Student Text)
ISBN 0-02-653129-1 (Course 2 Teacher's Wraparound Edition)

Printed in the United States of America.

1 2 3 4 5 6 7 8 9   003   06 05 04 03 02 01 00 99 98

Health Consultants

Unit 1
Choosing a Healthy Life

Becky J. Smith, Ph.D.; C.H.E.S.
Reston, Virginia

Robert Frye, M.P.H.
Health Education Consultant
Apex, North Carolina

Alice B. Pappas, Ph.D.; R.N.
Associate Professor
Associate Dean for Academic Affairs
Baylor University School of Nursing
Dallas, Texas

Unit 2
Nutrition and Fitness

Cathy Strain, M.S.; R.D.; C.D.
Associate Professor
Marian College
Indianapolis, Indiana

Kim Shibinski, M.S.
Department of Physical Education and Health
    Promotion
Columbia College
Columbia, South Carolina

Ronald G. Knowlton, Ph.D.
Professor and Chair, Physical Education Department
Southern Illinois University
Carbondale, Illinois

Unit 3
Understanding Yourself and Others

Janice Livingston, R.N.; M.Ed.; M.S.; A.R.N.P.
Professor
Central Florida Community College
Ocala, Florida

Howard Steven Shapiro, M.D.
Assistant Professor of Psychiatry
University of Southern California School of Medicine
    and Senior Attending Physician
Cedars Sinai Medical Center
Los Angeles, California

E. Laurette Taylor, Ph.D.
Associate Professor
Department of Health and Sport Sciences
The University of Oklahoma
Norman, Oklahoma

Unit 4
Protecting Your Health

Robert D. Russell, Ed.D.
Professor of Health Education Emeritus
Southern Illinois University
Carbondale, Illinois

Richard L. Papenfuss, Ph.D.
Associate Professor of Health Education
Arizona Health Sciences Center
University of Arizona
Tucson, Arizona

David M. Allen, M.D.
Infectious Disease Associates
Dallas, Texas

Mark Dignan, Ph.D.; M.P.H.
Chair, Center for Community Studies
AMC Cancer Research Center
Denver, Colorado

Unit 5
Personal Safety

Richard J. Shuntich, Ph.D.
College of Social and Behavioral Sciences
Department of Psychology
Eastern Kentucky University
Richmond, Kentucky

David Sleet, Ph.D.
Associate Director for Science
Unintentional Injury Prevention
Centers for Disease Control and Prevention (CDC)
Atlanta, Georgia

Teacher Reviewers

Reviewer for Entire Book

Pamela R. Connolly
Subject Area Coordinator for Health Education,
    Pittsburgh Catholic Schools
Oakland Catholic High School, Diocese of Pittsburgh
Pittsburgh, Pennsylvania

Unit 1
Choosing a Healthy Life

Beverly J. Berkin, C.H.E.S.
Health Education Consultant
Bedford Corners, New York

Lori Hart
Health Educator
West Shore Middle School
Milford, Connecticut

Unit 2
Nutrition and Fitness

Robert Wandberg, Ph.D.
Health Education
John F. Kennedy Senior High School
Bloomington, Minnesota

Debra C. Harris, Ph.D.
HPE Department Chair, HPE Teacher
West Linn High School
West Linn, Oregon

Unit 3
Understanding Yourself and Others

Betty Anne White
Health & Physical Education Teacher
Bryan Station Traditional Middle School
Lexington, Kentucky

Donna Breitenstein, Ed.D.
Professor & Coordinator of Health Education,
Director of NC School Health Training Center
Appalachian State University
Boone, North Carolina

Unit 4
Protecting Your Health

Deborah L. Tackmann, B.S.; M.E.P.D.
Health Education Instructor
North High School
Eau Claire, Wisconsin

Peggy V. Johns
Supervisor, Pre-K–12 Health Education
Pinellas County Schools
Largo, Florida

Beverely J. Berkin, C.H.E.S.
Health Education Consultant
Bedford Corners, New York

Lynn Westberg
Health Education Department Head
Kearns High School
Kearns, Utah

Unit 5
Personal Safety

Debra Penland Forbes
Health Educator
Pleasanton Unified School District
Pleasanton, California

Contents

Unit 1
Choosing a Healthy Life

Chapter 1 A Healthy Foundation		**2**
Lesson 1 Your Health and Wellness		**4**
Lesson 2 Influences on Your Health		**8**
▶ Health Lab: Influences on Health		11
✹ **Teen Health Digest**		**12**
Lesson 3 What's Your Health Status?		**14**
▶ Personal Inventory: How Healthy Are You?		15
Lesson 4 Building Health Skills		**19**
▶ Life Skills: Becoming Media Literate		20

Chapter 2 Personal Responsibility and Decision Making		**26**
Lesson 1 Taking Responsibility for Your Actions		**28**
▶ Life Skills: Skills in Leadership		30
Lesson 2 Recognizing and Managing Habits		**34**
▶ Health Lab: The Effects of Repetition		36
Lesson 3 Health Risks and Your Behavior		**39**
✹ **Teen Health Digest**		**44**
Lesson 4 Making Responsible Decisions		**46**
▶ Making Healthy Decisions: Making a Difficult Choice		50
Lesson 5 Setting Goals and Making Action Plans		**52**

Chapter 3 Being a Health Care Consumer		**58**
Lesson 1 Making Consumer Choices		**60**
▶ Life Skills: Being a Wise Consumer of Health Information		62
Lesson 2 Buying Personal Products		**65**
▶ Making Healthy Decisions: Reaching a Purchasing Decision		67
Lesson 3 Caring for Your Teeth, Skin, and Hair		**69**
▶ Health Lab: How Well Do Sunscreens Protect?		72
Lesson 4 Caring for Your Eyes and Ears		**76**
✹ **Teen Health Digest**		**80**
Lesson 5 Health Care Providers		**82**

Contents   vii

Unit 2
Nutrition and Fitness

Chapter 4 Food and Nutrition — 90
Lesson 1 The Food Guide Pyramid — 92
Lesson 2 Nutrients for Health and Wellness — 96
▶ Life Skills: Making Healthful Choices at Fast-Food Restaurants — 100
Lesson 3 Healthful Meal Planning — 102
▶ Making Healthy Decisions: Making Breakfast Important — 104
✹ **Teen Health Digest** — 106
Lesson 4 The Digestive and Excretory Systems — 108
▶ Health Lab: The Digestion of Protein — 110

Chapter 5 Physical Activity and Weight Management — 116
Lesson 1 Physical Fitness and You — 118
▶ Life Skills: Everyday Fitness — 120
Lesson 2 The Circulatory System — 123
Lesson 3 The Skeletal and Muscular Systems — 128
▶ Health Lab: Pulling Pairs — 132
✹ **Teen Health Digest** — 134
Lesson 4 Planning a Fitness Program — 136
Lesson 5 Weight Management — 141
▶ Making Healthy Decisions: Helping a Friend with an Eating Disorder — 144

Chapter 6 Sports and Conditioning — 148
Lesson 1 Individual and Team Sports — 150
▶ Life Skills: Teamwork — 153
Lesson 2 Sports and Physical Wellness — 154
Lesson 3 Conditioning Goals and Techniques — 159
▶ Health Lab: Getting into Condition — 162
✹ **Teen Health Digest** — 164
Lesson 4 Balancing School, Sports, and Home Life — 166
▶ Making Healthy Decisions: Balancing a Life — 169

Unit 3
Understanding Yourself and Others

Chapter 7 Growth and Development — 174

- **Lesson 1 Adolescence** — 176
  - ▶ Life Skills: Dealing with Conflicting Feelings — 179
- **Lesson 2 The Endocrine System** — 182
- **Lesson 3 The Male Reproductive System** — 186
- **Lesson 4 The Female Reproductive System** — 190
  - ✹ Teen Health Digest — 194
- **Lesson 5 Human Development** — 196
  - ▶ Health Lab: Your Unique Fingerprints — 198
- **Lesson 6 Life Stages** — 201
  - ▶ Making Healthy Decisions: Helping a Friend Deal with Loss — 204

Chapter 8 Mental and Emotional Health — 208

- **Lesson 1 What Is Mental Health?** — 210
- **Lesson 2 Building Positive Self-Esteem** — 214
  - ▶ Health Lab: The Qualities You Admire — 215
- **Lesson 3 Managing Stress** — 219
  - ▶ Life Skills: Time Management — 221
  - ▶ Personal Inventory: Life Changes and Stress — 223
  - ✹ Teen Health Digest — 226
- **Lesson 4 Mental Disorders** — 228
- **Lesson 5 Sources of Help** — 232
  - ▶ Making Healthy Decisions: Deciding Whether to Tell — 234

Chapter 9 Social Health: Families and Friends — 238

- **Lesson 1 Communication Skills** — 240
  - ▶ Life Skills: Giving and Taking Criticism — 242
- **Lesson 2 Understanding Family Relationships** — 245
- **Lesson 3 Friendships and Peer Pressure** — 249
  - ▶ Personal Inventory: Are You a Good Friend? — 250
  - ▶ Making Healthy Decisions: Facing Peer Pressure — 252
- **Lesson 4 Abstinence and Refusal Skills** — 254
  - ✹ Teen Health Digest — 258
- **Lesson 5 Resolving Conflicts at Home and at School** — 260
  - ▶ Health Lab: Mediating a Conflict — 262

Unit 4
Protecting Your Health

Chapter 10 Tobacco — 270

Lesson 1 **What Tobacco Does to Your Body** — 272
▶ Health Lab: Smoking and Breathing — 275

Lesson 2 **The Respiratory System** — 276
✹ Teen Health Digest — 280

Lesson 3 **Tobacco Addiction** — 282
▶ Life Skills: Analyzing a Media Message — 285

Lesson 4 **Choosing to Be Tobacco Free** — 287
▶ Making Healthy Decisions: Overcoming the Pressure to Smoke — 288

Chapter 11 Alcohol and Drugs — 294

Lesson 1 **Use and Abuse of Alcohol** — 296
▶ Personal Inventory: Attitudes About Alcohol — 298
▶ Life Skills: Helping Someone with a Drinking Problem — 299

Lesson 2 **Use and Abuse of Drugs** — 301

Lesson 3 **The Nervous System** — 307
▶ Health Lab: Nervous System Tricks — 310
✹ Teen Health Digest — 312

Lesson 4 **Risks Involved with Alcohol and Drug Use** — 314

Lesson 5 **Avoiding Substance Abuse** — 319
▶ Making Healthy Decisions: Offering Alternatives — 322

Lesson 6 **Addiction and Recovery** — 324

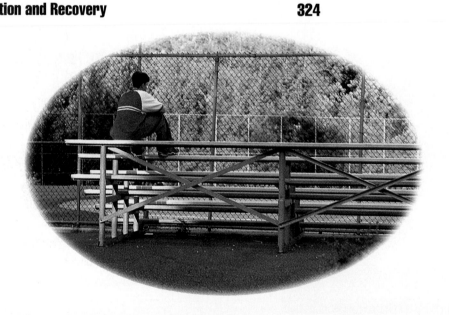

Unit 4
Protecting Your Health

Chapter 12 Understanding Communicable Diseases — 330

Lesson 1 Causes of Communicable Diseases — 332
- Health Lab: Observing Bacteria — 335

Lesson 2 The Immune System — 336

Lesson 3 Common Communicable Diseases — 340

Lesson 4 Avoiding Common Communicable Diseases — 344
- Life Skills: Being a Reliable Information Source — 346
- Personal Inventory: Good Health Habits — 347
- ✸ Teen Health Digest — 348

Lesson 5 Sexually Transmitted Diseases and HIV/AIDS — 350
- Making Healthy Decisions: Deciding Whether to Speak Up — 357

Chapter 13 Understanding Noncommunicable Diseases — 360

Lesson 1 Understanding Heart Disease — 362

Lesson 2 Understanding Cancer — 367
- Life Skills: Dietary Guidelines to Lower Cancer Risk — 370

Lesson 3 Allergies and Asthma — 373
- Health Lab: Determining Lung Capacity — 376
- ✸ Teen Health Digest — 378

Lesson 4 Other Noncommunicable Diseases — 380
- Making Healthy Decisions: Coping with a Relative Who Has Alzheimer's Disease — 384

Contents xi

Unit 5
Personal Safety

Chapter 14 Preventing Violence and Abuse 390
- **Lesson 1 Dealing with Violence** 392
  - ▶ Life Skills: Protecting Yourself 394
  - ✹ **Teen Health Digest** 398
- **Lesson 2 Understanding Abuse** 400
  - ▶ Making Healthy Decisions: Deciding to Report Abuse 402
- **Lesson 3 Finding Help** 404
  - ▶ Health Lab: Abuse and the Media 406

Chapter 15 Safety and Recreation 410
- **Lesson 1 Building Safe Habits** 412
  - ▶ Making Healthy Decisions: Preventing Accidents While Baby-sitting 414
- **Lesson 2 Safety at Home and in School** 416
  - ▶ Life Skills: How to Extinguish Fires Correctly 418
- **Lesson 3 Safety on the Road and Outdoors** 422
  - ✹ **Teen Health Digest** 428
- **Lesson 4 Safety in Weather Emergencies** 430
- **Lesson 5 First Aid** 435
- **Lesson 6 Protecting Our Planet** 440
  - ▶ Health Lab: The Dangers of Water Pollution 443

Glossary 446
Glosario 457
Index 468

xii Contents

Features

Health Lab

Influences on Health	11	Mediating a Conflict	262
The Effects of Repetition	36	Smoking and Breathing	275
How Well Do Sunscreens Protect?	72	Nervous System Tricks	310
The Digestion of Protein	110	Observing Bacteria	335
Pulling Pairs	132	Determining Lung Capacity	376
Getting into Condition	162	Abuse and the Media	406
Your Unique Fingerprints	198	The Dangers of Water Pollution	443
The Qualities You Admire	215		

Life Skills

Becoming Media Literate	20	Giving and Taking Criticism	242
Skills in Leadership	30	Analyzing a Media Message	285
Being a Wise Consumer of Health Information	62	Helping Someone with a Drinking Problem	299
Making Healthful Choices at Fast-Food Restaurants	100	Being a Reliable Information Source	346
Everyday Fitness	120	Dietary Guidelines to Lower Cancer Risk	370
Teamwork	153	Protecting Yourself	394
Dealing with Conflicting Feelings	179	How to Extinguish Fires Correctly	418
Time Management	221		

Making Healthy Decisions

Making a Difficult Choice	50	Overcoming the Pressure to Smoke	288
Reaching a Purchasing Decision	67	Offering Alternatives	322
Making Breakfast Important	104	Deciding Whether to Speak Up	357
Helping a Friend with an Eating Disorder	144	Coping with a Relative Who Has Alzheimer's Disease	384
Balancing a Life	169	Deciding to Report Abuse	402
Helping a Friend Deal with Loss	204	Preventing Accidents While Baby-sitting	414
Deciding Whether to Tell	234		
Facing Peer Pressure	252		

Personal Inventory

How Healthy Are You?	15	Attitudes About Alcohol	298
Life Changes and Stress	223	Good Health Habits	347
Are You a Good Friend?	250		

People at Work

Open Wide . . .	80
On the Road to Podiatry	165
Providing Care and Comfort	194
Music Therapist	226
A Therapist for Children	259
Country Counselor	312
Disease Detective	349
Physical Therapist	379
Helping People	399
A Lifesaving Job	428

HEALTH UPDATE

Young and Healthy	13
Is Your Heart at Risk?	45
Decay? No Way!	80
Safe Fruits and Vegetables	107
Time to Eat	135
Senior Power	195
New Phones for the Hearing Impaired	258
Smoke Less, See More	281
Danger to the Unborn	313
Fleeing the Flu	349
No More Shots?	379

CON$UMER FOCU$

Handwashing for Health?	12
Are Air Bags Safe?	45
Fashion Statement or Health Risk?	81
Good Bargains, Good Nutrition	107
Don't Fall for These Ads	134
Choose Your Shoes	164
Eating to Stay Young	195
Self-Image Bait	227
Smoking Out the Beauty Myths	281
On-Line Targets?	312
Health Hype	348
Rating TV Programs	398
Safe Toys for Tots	429

Teens Making a Difference

Y.O.U.T.H. Matters	45
A Teen for Others	81
Strength in Numbers	106
A Lifesaving Decision	165
Bringing People Together	194
Choosing Life	227
A Friend in Need	259
Tricks and Talks	313
Hoping to Save a Life	348
Cancer Crusader	378
A Dramatic 47 Seconds	399
Tree Musketeers	429

Myths and Realities

The More I Exercise, the Hungrier I Get.	13
Taking Control of Your Health	44
Seeing the Light	81
The Truth About Sugar	106
Realistic Expectations	135
Determined at Birth?	195
The Family Balancing Act	259
It's Smart to Be a (Cold) Turkey	280
Tough Guy?	313
Feed a Cold? Not Too Much	349
A Mysterious Illness	379
The Truth About Abuse	398
The Facts About Safety Belts	428

Sports and Recreation

A Champ with a Sense of Balance	12
A Camp for Kids	44
Maine's Main Woman	135
A "Rhodes" Runner	164
Very Special Athletes	227
Big-Time Players, Lifetime Friends	258
Getting Tobacco Off the Field	281
Super Mario	378
A Very Un-Martial Art	399
Cycling Tips from a Coach	429

Personal Trainer

Safe Video Routines	13
Make a Commitment with Calcium!	107
Water Workout	134
Back to Basics	165
Take It Easy	226
The Recovery Workout	280

Teen Health Digest   XV

Unit 1
Choosing a Healthy Life

UNIT OBJECTIVES

Students will learn about the three essential aspects of health and explore ways of protecting and promoting their own health and wellness. Students will focus on the process of making responsible decisions and setting healthful goals, and they will examine their own roles as responsible health care consumers.

UNIT OVERVIEW

Chapter 1 A Healthy Foundation

This chapter introduces students to a more complete understanding of health and wellness. It guides students in recognizing the importance of guarding and developing their own physical health, mental/emotional health, and social health.

Chapter 2 Personal Responsibility and Decision Making

In this chapter, students consider their own growing responsibility, both for themselves and for other people. They learn approaches to decision making and goal setting, and learn to identify and develop healthy habits.

Chapter 3 Being a Health Care Consumer

Chapter 3 helps students develop their skills as consumers of health products and health services. It helps them understand labels on personal products, evaluate advertising, and practice proper care of their teeth, skin, hair, eyes, and ears. The chapter also discusses decisions about health care providers and health insurance.

Bulletin Board Suggestion

It's Your Choice Cut a large equilateral triangle from construction paper, and label the sides *Physical Health, Mental and Emotional Health,* and *Social Health.* Post this triangle in the middle of a bulletin board, and refer to it as students examine the relationships between the three aspects of health. As students consider the decision-making process and the development of healthy habits, ask them to add to the bulletin board. Have them bring in pictures from magazines, cartoons, their own drawings or photographs, and/or words and phrases—all depicting or suggesting choices that teens have to make. Let students post their contributions on the bulletin board, next to the side of the health triangle they consider most relevant.

Unit 1
Choosing a Healthy Life

Chapter 1 — A Healthy Foundation

Chapter 2 — Personal Responsibility and Decision Making

Chapter 3 — Being a Health Care Consumer

Unit 1

INTRODUCING THE UNIT

Write the following question on the board: *How are you today?* Let several volunteers respond orally, or ask all of the students to write brief responses. Then ask: What do your responses tell about your health? Do the following responses tell something about a person's health? Why or why not?

- I feel terrific.
- I'm so excited! I'm going to a great party this weekend.
- I think I'm coming down with a cold.
- I guess I feel OK, but I've been awfully lonely since my best friend moved.
- School is getting me down. I find I just can't seem to concentrate.

After students have discussed their ideas, explain that *all* these responses relate to people's health. Tell students that they will learn to understand various aspects of health as they begin studying the chapters in this text.

VIDEODISC/VHS

Teen Health Course 2

You may wish to use video segment 2, "Personal Responsibility," to show students some habits that will benefit their health throughout their lives.

Videodisc Side 1, Chapter 2
Personal Responsibility

Search Chapter 2, Play to 3

DEALING WITH SENSITIVE ISSUES

The Bigger Picture When you deal with sensitive issues in the classroom, don't forget that parents, the school board, and the community may be concerned about what you teach and how you teach it. You can minimize their concerns by following these guidelines: Familiarize yourself with the regulations and guidelines in your school district about course content, parental permissions, and related issues. In your classes, emphasize abstinence from sex as well as from alcohol and illegal drugs. Avoid *preaching* to students about sensitive issues. Avoid classroom activities and behaviors that may be misinterpreted. Always thank parents for their concern.

Planning Guide

Chapter 1
A Healthy Foundation

	Features	Classroom Resources
Lesson 1 **Your Health and Wellness** *pages 4–7*		Concept Map 1 Enrichment Activity 1 Lesson Plan 1 Lesson 1 Quiz Reteaching Activity 1 Transparencies 1, 2
Lesson 2 **Influences on Your Health** *pages 8–11*	Health Lab: Influences on Health *page 11* TEEN HEALTH DIGEST *pages 12–13*	Concept Map 2 Cross-Curriculum Activity 1 Decision-Making Activity 1 Enrichment Activity 2 Lesson Plan 2 Lesson 2 Quiz Reteaching Activity 2 Transparency 3
Lesson 3 **What's Your Health Status?** *pages 14–18*	Personal Inventory: How Healthy Are You? *page 15*	Concept Map 3 Cross-Curriculum Activity 2 Enrichment Activity 3 Health Lab 1 Lesson Plan 3 Lesson 3 Quiz Reteaching Activity 3
Lesson 4 **Building Health Skills** *pages 19–23*	Life Skills: Becoming Media Literate *pages 20–21*	Concept Map 4 Decision-Making Activity 2 Enrichment Activity 4 Lesson Plan 4 Lesson 4 Quiz Reteaching Activity 4 Transparencies 4, 5

National Health Education Standards

One of the main goals of the National Health Education Standards is to move students toward "health literacy." Health literacy is the capacity of individuals to obtain, interpret, and understand basic health information and services and the competence to use such information and services in ways which promote health. The health standards were developed by applying the characteristics of a well-educated, literate person within the context of health. The health literate person is:

- a critical thinker and problem solver
- a self-directed learner
- a responsible, productive citizen
- an effective communicator

Listed below are the Health Education Standards Performance Indicators addressed in each lesson of this chapter.

Lesson	Health Standards Performance Indicators
1	(1.1, 1.2)
2	(1.4, 1.5, 4.1, 4.2)
3	(1.1, 1.6, 3.2, 6.5, 6.6)
4	(2.2, 2.4, 5.8, 6.4)

Chapter Resources

Teacher's Classroom Resources
- Chapter 1 Test
- Parent Letter and Activities 1
- Performance Assessment 1
- Testmaker Software

Student Activities Workbook
- Study Guide 1
- Applying Health Skills 1–4
- Health Inventory 1

Student Diversity Strategies
- Audiocassette Program (English)
- Audiocassette Program (Spanish)
- Spanish Parent Letters
- Spanish Summaries, Quizzes, and Activities

Multimedia Components
- English Audiocassette Program
- Spanish Audiocassette Program
- *Teen Health* Videodisc/VHS Series

Other Resources

Readings for the Teacher
Dreher, Nancy. "What's Your Health IQ?" Current Health 2, September 1995, pp. 6–12.

Isler, Charlotte and Alwyn T. Cohall. *The Watts Teen Health Dictionary,* New York: Franklin Watts, 1996.

McCoy, Kathy, and Charles Wibbelsman. *Life Happens: A Teenager's Guide to Friends, Failure, Sexuality, Love, Rejection, Addiction, Peer Pressure, Families, Loss, Depression, Change and Other Challenges of Living.* New York: Berkley Publishing Group, 1996.

Readings for the Student
Collinson, Alan. *Choosing Health.* Austin, TX: Steck-Vaughn Company, 1991.

Dreher, Nancy. "The Teen Years: Risky Business?" Current Health 2, September 1996, pp. 6–12.

Hurwitz, S., and Jane Hurwitz. *Stay Healthy.* New York: Rosen Publishing Group, 1993 (revised edition).

Out of Time?

If time does not permit teaching this chapter, you may use these features: Health Lab on page 11; Teen Health Digest on pages 12–13; Personal Inventory on page 15; Life Skills on pages 20–21; and the Chapter Summary on page 24.

Chapter 1
A Healthy Foundation

CHAPTER OVERVIEW

Chapter 1 introduces students to an expanded understanding of health and helps them begin to consider the status of their own health.

Lesson 1 defines both health and wellness and explains the three basic aspects of health.

Lesson 2 discusses the influences of heredity, environment, and other external factors on an individual's health.

Lesson 3 guides students in assessing their own health status and in developing their own personal wellness contract.

Lesson 4 helps students examine steps they can take to promote and protect their own health.

PATHWAYS THROUGH THE PROGRAM

Chapter 1 introduces students both to the *Teen Health* course and to an understanding of the health triangle—the interrelationship of their physical health, mental/emotional health, and social health. Students will apply the knowledge developed in this chapter throughout the following lessons of the course; they will also learn how to apply this knowledge directly to their lives, developing and protecting all aspects of their health.

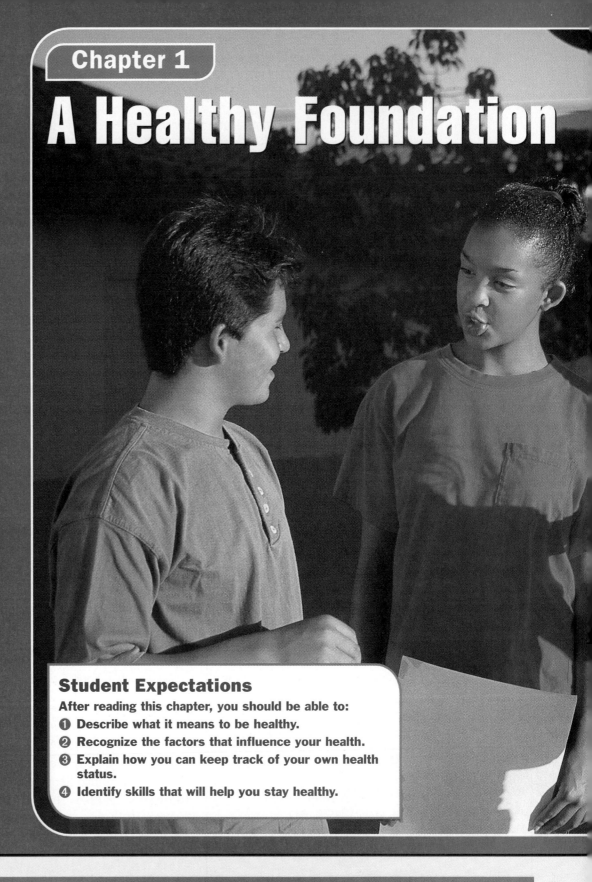

Chapter 1
A Healthy Foundation

Student Expectations
After reading this chapter, you should be able to:
1. Describe what it means to be healthy.
2. Recognize the factors that influence your health.
3. Explain how you can keep track of your own health status.
4. Identify skills that will help you stay healthy.

Key to Ability Levels

Teaching strategies that appear throughout the chapter have been identified by one of three codes to indicate their suitability for students of varying learning styles and abilities.

L1 **Level 1** strategies should be within the ability range of all students. Full class participation is often required. Teacher direction is usually needed.

L2 **Level 2** strategies are for average to above-average students or for small groups. Some teacher direction is necessary.

L3 **Level 3** strategies are designed for students who are able and willing to work independently. Minimal teacher direction is required.

Teen Chat Group

Megan: Where's Brittany? I thought she was going to help put up posters for the jazz band competition.

Dee: She's home sick.

Rafael: Again? She was out a couple of days last week. What's going on?

Dee: I think she's way too busy. She's on the basketball team, she babysits twice a week, and she volunteers for almost every school project. It's starting to get to her, I think.

Megan: She never even has time to sit down with us and eat lunch anymore. She just grabs a drink and runs off somewhere. No wonder she's sick all the time.

Rafael: I think I'll stop by and see her on my way home. Maybe I can help her figure out how to make some changes in her schedule.

in your journal

In each chapter, you will be asked to write entries in your journal. These are your private reflections and thoughts and are for your use only. Read the dialogue on this page. Have you ever wondered if you should take better care of your health? Start your private journal entries on your overall health and wellness by answering these questions:

- What advice would you give to Brittany?
- What does being healthy mean to you?
- What steps do you take to protect your health?
- What kinds of health habits would you like to improve?

When you reach the end of the chapter, you will use your journal entries to make an action plan.

Chapter 1: A Healthy Foundation 3

Chapter 1

INTRODUCING THE CHAPTER

▶ Prepare a bulletin board using illustrated magazine articles and/or magazine ads that relate to the three basic aspects of health. For example, articles from teen magazines might cover such topics as avoiding smoking, dealing with anger, and getting along with others. Give students an opportunity to examine the pictures and scan the articles on the bulletin board. Then lead the class in a short discussion: What main ideas are presented on the board? How do these ideas relate to health? What would be a good title for this bulletin board? You may want to conclude the discussion by having a few volunteers add an appropriate title to the display.

▶ Ask students to write brief responses to these questions: What is health? What do you think being healthy involves? Have students save their answers; after they have studied the entire chapter, guide students in reviewing their responses and in revising their ideas.

▶ Give students an opportunity to leaf through the lessons of Chapter 1, noting especially the titles, headings, charts, and photographs.

Cooperative Learning Project

Three-Ring Health Circus

The Teen Health Digest on pages 12 and 13 provides students with high-interest articles related to the content of this chapter.

The material in the Teacher's Wraparound Edition provides suggestions for a class project in which students will plan and present a series of "ìactsî" designed to heighten audience awareness of the three aspects of health.

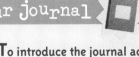

in your journal

To introduce the journal activities in this course, help students distinguish between a diary (a record of events and experiences) and a journal (a record of thoughts, feelings, and reactions to events and experiences). Emphasize that students' entries in their Health Journals are completely private; they are intended to help students think about their own experiences and develop their own ideas.

In preparation for this first journal activity, let students work in groups to improvise their own conversations with Brittany. Ideas expressed in these conversations will spark students' thinking for their journal writing.

3

Lesson 1
Your Health and Wellness

Focus

LESSON OBJECTIVES

After studying this lesson, students should be able to
- define the terms *health* and *wellness*.
- identify the three aspects of health.
- discuss the importance of keeping the three aspects of health in balance.

MOTIVATOR

Write the following question on the board: *How healthy are you?* Have students list at least five conditions or behaviors that indicate the state of their own health.

INTRODUCING THE LESSON

Ask volunteers to share the lists they made in response to the Motivator question. Encourage students to discuss what each listed condition or behavior indicates about a person's health. If most of the items on the volunteers' lists relate to physical health, suggest specific behaviors—laughs a lot, feels confident, keeps trying, talks about problems, for example—and help students discuss how these behaviors relate to health.

INTRODUCING WORDS TO KNOW

Have students look through the lesson to find the definitions (in **bold type** and *italics*) of the words *health* and *wellness*. Let students work independently or in small groups to use each letter of *health* as the first letter in a word or short phrase that relates to good health. For example: **h**appy; **e**nergetic; **a**ccepting responsibility; **l**earning every day; etc. Do the same with *wellness*.

Lesson 1: Your Health and Wellness

This lesson will help you find answers to questions that teens often ask about personal health and wellness. For example:
- What does it mean to be healthy?
- How can I keep my physical, mental/emotional, and social health in balance?
- What is the relationship between health and wellness?

Words to Know
health
emotions
wellness

Being Healthy

What does it mean to be healthy? Think about that question as you read about Roy and Karen. Roy is on the track team. He is in excellent physical shape. Yet he finds he has trouble controlling his temper. As a result, his friendships rarely last. Karen is a good student. She also writes for the school newspaper. She often stays up late studying, however, and sometimes skips breakfast. As a result, Karen feels tired much of the time. How would you rate Roy's and Karen's health?

Good health involves every part of your life. **Health** is *a combination of physical, mental/emotional, and social well-being.* Your physical health has to do with the condition of your body. Your mental/emotional health involves your thoughts and **emotions,** which are *feelings such as love, joy, or fear.* Your social health involves the ways in which you relate to other people. Each of these aspects of health deserves a closer look.

Your Physical Health

Maintaining physical health involves taking care of your body. There are many ways to do this.

- Eat a well-balanced diet.
    - Stay fit through regular exercise.
        - Get at least eight hours of sleep each night.
            - Keep your body, teeth, and hair clean.
                - Have regular medical and dental check-ups.
                    - Avoid tobacco, alcohol, and drugs.

PHYSICAL HEALTH

Chapter 1: A Healthy Foundation

Lesson 1 Resources

Teacher's Classroom Resources
- 📁 Concept Map 1
- 📁 Enrichment Activity 1
- 📁 Lesson Plan 1
- 📁 Lesson 1 Quiz
- 📁 Reteaching Activity 1
- Transparencies 1, 2

Student Activities Workbook
- 📁 Study Guide 1
- 📁 Applying Health Skills 1

Your Mental/Emotional Health

Maintaining mental and emotional health means taking care of your mind. Here are some ways to keep yourself mentally and emotionally healthy.

- Face problems with a positive and realistic attitude.
- Express your feelings clearly and calmly to others.
- Set priorities so that you will not be overwhelmed by school, responsibilities at home, or athletic activities.
- See your mistakes as opportunities to learn, grow, and change.
- Recognize your weak areas and try to improve them.
- Develop effective decision-making skills.
- Use your talents effectively.
- Accept responsibility for your actions.
- Stand up for your beliefs and values.
- Remain open to learning new information.

interNET CONNECTION
Tap into the Web for health resources geared to your interests and concerns.
http://www.glencoe.com/sec/health

Your Social Health

Maintaining social health means taking care of the ways in which you get along with other people. You can do this in the following ways.

- Have a friendly, open attitude toward other people.
- Learn to communicate effectively.
- Respect and care for family members.
- Be a loyal, truthful, and dependable friend.
- Honor other people's feelings.
- Respect other people's property.
- Learn to disagree without arguing.
- Learn to resolve conflicts effectively.
- Give support and help when it is needed.

in your journal

Reread the guidelines for mental and emotional health. In your journal, make a list of the guidelines that you follow regularly. Make a second list of the guidelines that you do not follow regularly. Write a paragraph about the positive effects you would receive if you began following the guidelines that you do not follow now.

Lesson 1: Your Health and Wellness 5

Chapter 1 Lesson 1

Teach

L1 Discussing
Encourage students to discuss why Roy and Karen (on page 4) are not truly healthy. Why is being in excellent physical shape not enough to make Roy healthy? Why is being a good student not enough to make Karen healthy? What advice would you give Roy and Karen?

L2 Brainstorming
Have students work in small groups to brainstorm lists of school-based resources that can help students develop each aspect of their own good health. For example, students might list intramural sports for developing physical health, school counselors for developing mental/emotional health, and computer or drama clubs for developing social health.

L2 Language Arts
Let students work with partners to generate a list of synonyms or near-synonyms for the word *healthy*. Then have them note any differences in meaning among the words.

Would You Believe?

Laughter may actually be "good medicine." Individual experiences and scientific experiments document the close relationship between good emotional health, as expressed in humor and laughter, and good physical health. In Anatomy of an Illness, an account of his struggle with a severe and painful connective tissue disease, Norman Cousins wrote, "I made the joyous discovery that ten minutes of genuine belly laughter had an anesthetic effect and would give me at least two hours of pain-free sleep."

Personal Health

The three aspects of health are closely interrelated. In teens, the interrelationship may be most evident in emotional and social health. Teens with high self-esteem, self-confidence, and a positive outlook (indications of good emotional health) are most likely to form and sustain friendships (indications of good social health). Satisfying friendships, in turn, foster teens' confidence and self-esteem.

Students who appear to be having difficulty with emotional and social health can be encouraged to focus both on themselves (What do you like about yourself? What do you do well?) and on others (How can you help other people? What interests can you share with others?).

Chapter 1 Lesson 1

Visual Learning
Use the diagram in Figure 1.1 to help students understand the relationship between the three aspects of health. Ask questions such as:

- What are some examples of health that each side of the triangle represents?
- How are the three sides of the triangle related to each other? Why do you think they are all the same length?
- How would the triangle be affected if one of the sides were very large? If one were very small? If one were taken out? What does this tell you about health?

L3 Creating
Ask students to create small flip books with a series of drawings that illustrate the concept of wellness in one person's life. (Each flip book requires at least 15 small pages of lightweight paper; the illustrations are closely related so that viewers see an animated cartoon when they flip quickly through the book.) Let students share their flip books with the rest of the class. Guide the class in discussing the relationship of each drawing to the term *health* and the relationship of the animated series to the term *wellness*.

VIDEODISC/VHS

Teen Health Course 2
You may wish to use video segment 1, "A Healthy Foundation," to show how overall wellness requires balancing the three aspects of health.

Videodisc Side 1, Chapter 1
A Healthy Foundation

Search Chapter 1, Play to 2

Staying in Balance

It's easy to concentrate on one aspect of health and neglect the others. That's what happened to Roy and Karen, the teens you read about at the beginning of the lesson. A totally healthy person, however, keeps physical, mental/emotional, and social health in proper balance. To help you achieve a balance of the three aspects of health, think of your health as a triangle with equal sides, as shown in **Figure 1.1.** To achieve and maintain *total* personal health, make sure that you give attention to all three sides of your health triangle.

Figure 1.1
The Health Triangle
You can keep your health in balance by thinking of it as a triangle with three sides.

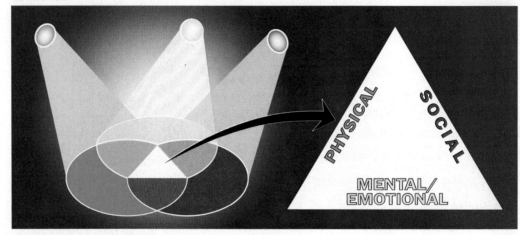

What Is Wellness?

Keeping the three parts of your health triangle in balance is the best way to achieve **wellness,** *an overall state of well-being involving regular behaviors that have a positive result over time.* How do health and wellness relate to each other? Think of health as a snapshot—a picture of how you are doing physically, mentally, emotionally, and socially at a particular moment. Your level of wellness, on the other hand, is more like a film that shows how these aspects of health interact over a period of time. The choices about health that you make every moment become part of your long-term level of wellness.

What does it mean to have a high level of wellness? **Figure 1.2** shows one way to understand levels of wellness. It shows various conditions and circumstances and where they might be placed on a wellness meter. Most people would prefer to enjoy a high level of wellness. The decisions you make now can help you maintain or achieve that goal throughout your life.

Did You Know?

I Can't Eat That!

People who have food allergies are extremely sensitive to substances in certain foods. These foods may cause tingling or swelling of the mouth, stomach upset, or diarrhea. The following foods often cause food allergies:

- milk products
- eggs
- seafood
- nuts
- wheat products

6 Chapter 1: A Healthy Foundation

Cooperative Learning

Constructing Models Instruct students to form small groups to plan and construct three-dimensional models of the health triangle. Have group members begin by brainstorming various ideas for materials, construction methods, and meaningful pictures or graphics that might decorate the sides of the triangle. Emphasize the importance of working together as you guide students in considering all ideas, making final decisions, and carrying out their plans for building the models.

Provide time for each group to share its completed model with the rest of the class. Then lead a discussion of their responses to this question: What effect did this cooperative effort have on your health?

Figure 1.2
Measuring Your Level of Wellness
Where would you place yourself on this wellness meter?

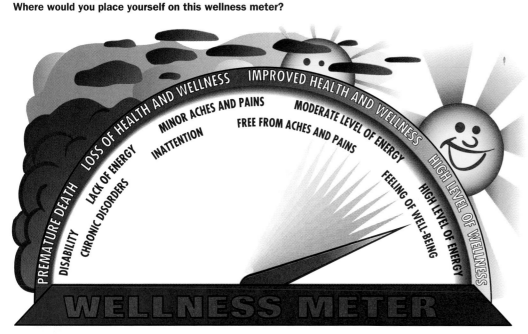

Review — Lesson 1

Using complete sentences, answer the following questions on a separate sheet of paper.

Reviewing Terms and Facts

1. **Vocabulary** Define the term *emotions*.
2. **Vocabulary** Explain the meaning of *wellness*. Then use the term in an original sentence.
3. **List** Identify the three aspects of health that must be balanced in order to achieve wellness.
4. **Restate** Explain the relationship between health and wellness.

Thinking Critically

5. **Analyze** Reread the descriptions of Roy and Karen on page 4. Explain why their health triangles seem to be out of balance. Draw a diagram to illustrate each situation.
6. **Explain** Many of the health guidelines in this lesson (pages 4 and 5) are related to each other. For example, developing effective decision-making skills will help you avoid using tobacco, alcohol, and drugs. Select guidelines from two different aspects of health. Explain how the guidelines are related.

Applying Health Concepts

7. **Health of Others** Based on what you have learned, make a poster to persuade younger students to make the right choices to achieve lifelong wellness. Use specific facts from the lesson to make your poster as informative and helpful as possible.

Lesson 1: Your Health and Wellness

Promoting Comprehensive School Health

Working Together As a health teacher and a member of the school community, you are concerned about all aspects of your students' health. You recognize, however, that even in working with other teachers and school personnel, you cannot meet all the health needs of every student. Still, schools have an essential role to play. They can work collaboratively with other groups—families, city and county governments, local businesses, social agencies, and religious groups—to develop a comprehensive health program that will foster teen health and development. For more information about participating in the development of such a program, consult *Promoting a Comprehensive School Health Program* in the TCR.

Chapter 1 Lesson 1

Assess

EVALUATING THE LESSON
Assign Reviewing Terms and Facts and Thinking Critically to review the lesson; then assign the Lesson 1 Quiz in the TCR.

LESSON 1 REVIEW

Answers to Reviewing Terms and Facts

1. Emotions are feelings such as love, joy, or fear.
2. Wellness is an overall state of well-being involving regular behaviors that have a positive result over time.
3. The three aspects of health to balance are physical, mental/emotional, and social.
4. Balanced health choices over time lead to a high overall level of wellness. Choices a young person makes regarding the protection of health will affect his or her wellness levels throughout life.

Answers to Thinking Critically

5. Roy concentrates on his physical health, but his mental/emotional health and social health suffer due to his temper. Although Karen is mentally healthy, she is risking her physical health by not getting enough rest and not eating breakfast.
6. Responses will vary.

RETEACHING
▶ Assign Concept Map 1 in the TCR.
▶ Have students complete Reteaching Activity 1 in the TCR.

ENRICHMENT
Assign Enrichment Activity 1 in the TCR.

Close

In small groups, have students identify a behavior or attitude they want to develop.

Lesson 2

Influences on Your Health

Focus

LESSON OBJECTIVES

After studying this lesson, students should be able to
- identify the external factors that affect their health.
- define *heredity* and discuss the influences of heredity on health.
- define *environment* and discuss the influences of environment on health.
- explain how their health choices may be influenced by other external factors.

MOTIVATOR

Display a poster or a magazine photograph of a family that includes at least one teen, two parents, and two or more grandparents. Point out a teen member of the family and ask: What traits do you think this teen has inherited from other members of his or her family? Give students five minutes to list as many traits as possible.

INTRODUCING THE LESSON

Have students share and compare the lists they made in response to the Motivator activity. Encourage a brief discussion, posing questions such as these: From which family member do you think the teen inherited each trait? How can you be sure a specific trait is the result of heredity rather than of environment?

INTRODUCING WORDS TO KNOW

Help students find and briefly discuss the definitions of the terms *heredity* and *environment*. Let students share what they already know about the effects of heredity and environment on the development of individuals.

Lesson 2

Influences on Your Health

This lesson will help you find answers to questions that teens often ask about what affects their health. For example:
▶ How do heredity and my environment affect my health?
▶ How do the people around me affect my health?

Words to Know

heredity
environment

in your journal

In your journal, list three things that influence your health. Are the influences positive or negative?

What Affects Your Health?

As you have learned, the choices you make and the ways you think and act have a strong effect on your total health. These are parts of your life over which you have quite a bit of control. There are, however, outside factors that influence your health. To make the best choices, you will need to understand what these factors are and what you can do about them.

Putting the Puzzle Together

Every person is an individual. Each life unfolds in its own way, guided and molded by both personal choices and outside factors. Think of a person as a one-of-a-kind jigsaw puzzle. The factors that influence that person's health are the pieces of the puzzle. Each person's puzzle has different pieces that, when put together, form a unique picture, as shown in **Figure 1.3**.

Figure 1.3
Pieces of a Puzzle
Many elements come together to create the person you are.

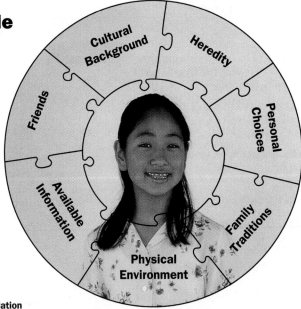

8 Chapter 1: A Healthy Foundation

Lesson 2 Resources

Teacher's Classroom Resources
- Concept Map 2
- Cross-Curriculum Activity 1
- Decision-Making Activity 1
- Enrichment Activity 2
- Lesson Plan 2
- Lesson 2 Quiz
- Reteaching Activity 2
- Transparency 3

Student Activities Workbook
- Study Guide 1
- Applying Health Skills 2

Heredity and Environment

Heredity (huh·RED·i·tee) is *the passing on of traits from biological parents to children.* Some traits passed on through heredity are the following:

- Skin, eye, and hair color
- Body build
- Growth patterns
- The tendency to get certain diseases

Certain traits are passed on from parents to children.

Hereditary traits are "built in"—you can't choose them or avoid them. However, recognizing the traits that you have already inherited from your parents can help you make crucial decisions about your health. If, for example, members of your family have had a high number of dental cavities, you should pay extra attention to your dental care. Similarly, people who know that heart disease or cancer has affected several family members should learn all they can about reducing their own risks of developing these particular diseases.

Your **environment** (en·VY·ruhn·ment) is *all the living and nonliving things that surround you.* Part of your environment is your physical environment, including:

- **Your home, neighborhood, and school.** Do you live in a large city or a small town? Are you surrounded by farms, by other homes, or by shops and restaurants?
- **The air you breathe and the water you drink.** Is the air clean, or do you live an area with heavy pollution? Does your water come from a well or is there a public water supply?

Like your hereditary traits, choices about where you live and the school you attend may be beyond your control. You still have choices to make, however. You can decide to be as healthy as possible within your physical environment. This involves recognizing characteristics of your environment that can be harmful to your overall health and taking steps to protect yourself. For example, if you live in a hot, sunny climate, you will need to take special precautions to protect your skin from the sun. Think also about how you affect the environment that you share with others. Are there ways that you and your family, for example, can avoid contributing to air or water pollution?

Social Studies Connection

Environmental Features

Describe the geography of the area where you live. Is there a body of water nearby? Are you in a city? Is the area mountainous or flat?

Based on your description, write ten features of your physical environment that might influence the health of people living there. For at least two of the features, tell how a person in your area can behave to prevent illness or injury.

Lesson 2: Influences on Your Health 9

More About...

Heredity Scientists around the world have undertaken an extensive exploration of the genetic factors that contribute to human development. The U.S. Department of Energy and the National Institutes of Health are coordinating the U.S. Human Genome Project, a 15-year effort to identify and map all the genes on every chromosome in the human body. (The term *genome* refers to all the genetic material in the chromosomes of a particular organism.) The U.S. project is part of a worldwide investigation into genetics. Over 1,000 scientists representing more than 50 countries are members of the Human Genome Organization (HUGO). Thus far, results of these studies indicate that there may be a specific gene for nearly every human ability.

Chapter 1 Lesson 2

Teach

L1 Discussing
Encourage students to discuss the specific elements of their local environment: their homes, their neighborhoods, and their school. In what sense do they all share the same environment? How is each student's environment different from that of other students?

Visual Learning
Focus students' attention on the picture of the puzzle in Figure 1.3. Start a brief discussion by asking: What does each piece of the puzzle represent? How do the pieces fit together?

L1 Critical Thinking
Explain to students that genetic tests are now being used to determine whether individuals are likely to develop specific forms of cancer that have affected other family members. (Examples include cancer of the breast, pancreas, and colon, but new tests are being introduced at a rapid rate.) Encourage students to discuss the benefits and the possible drawbacks of such tests: What difficulties might arise from *knowing* that you are likely to develop a specific kind of cancer by a certain age? If you were a candidate for such a test, would you choose to be tested? Why or why not?

L3 Interviewing
Ask a group of volunteers to interview one or more health care providers about the kind of family health history they usually take from their patients. What specific questions do the health care providers ask? What diseases or other health problems are of particular interest? Which family relationships do they consider relevant to their patients' health? Have students report their findings.

9

Chapter 1 Lesson 2

L3 Applying Knowledge

Encourage interested students to create their own family-health trees. Students should employ the standard branching format found in family trees; however, rather than writing the names of their relatives, students should write brief notes about the health of each known relative. Ask students to develop and use a color code to identify health problems that are or may be inherited.

L1 Analyzing

Guide students in discussing violence as an environmental health factor. To what extent is violence part of the environment? How does violence affect people's physical, mental/emotional, and social health? What, if anything, can students do to control violence within their own community? In other communities?

L2 Demonstrating

Have students bring to class magazine or newspaper ads that might influence teens' health-related decisions. Let students meet in small groups to analyze these ads.

Visual Learning

Guide students in reading and discussing Figure 1.4. Encourage them to identify specific ways in which culture and traditions, friends, and media influence their own health choices.

Other Influences

Along with your heredity and physical environment, many other factors in your life influence your health and your health choices. **Figure 1.4** illustrates some of the most important factors.

Figure 1.4
Influences on Your Health Choices
How will knowing about these influences affect the decisions you make about your health?

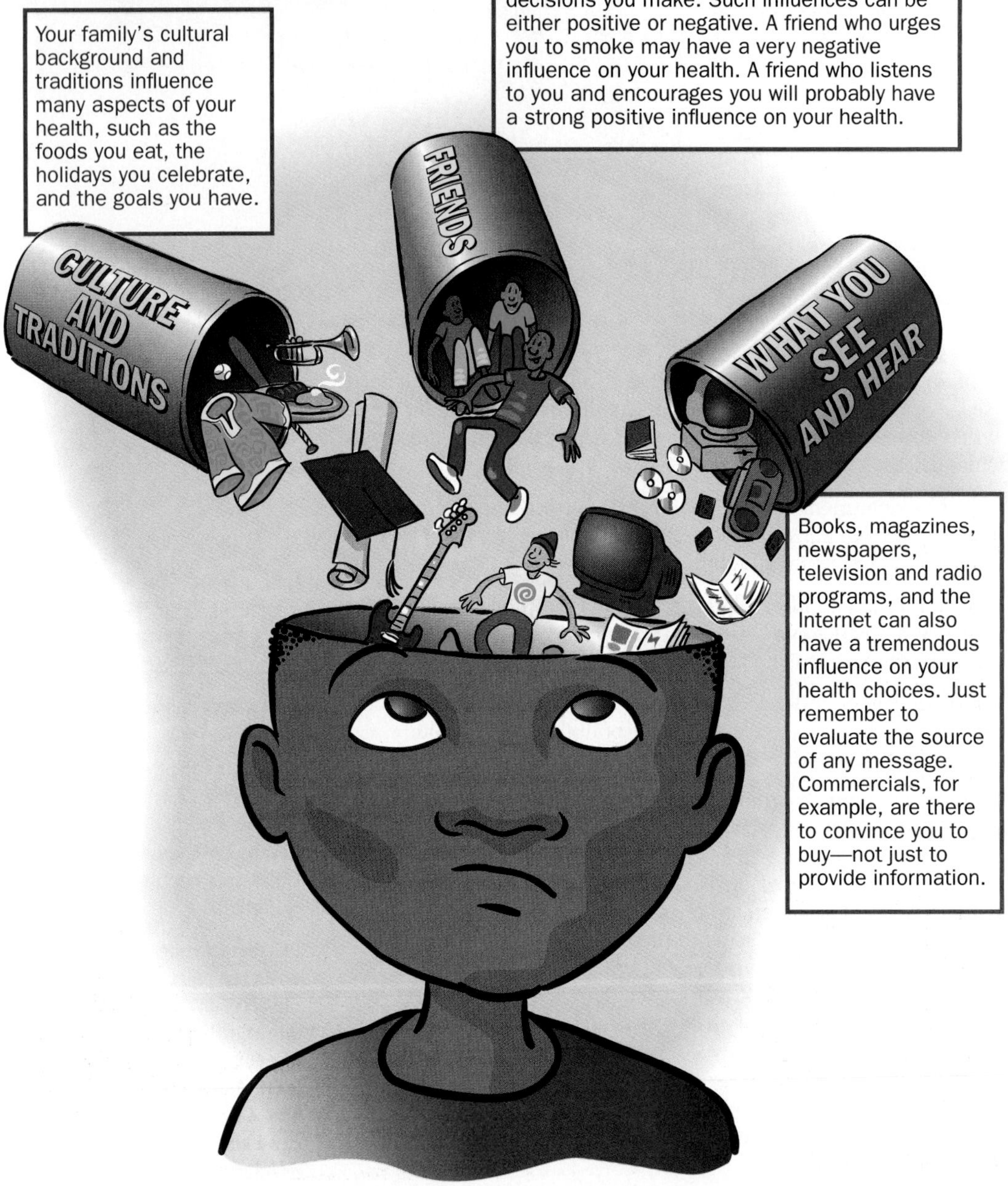

Your family's cultural background and traditions influence many aspects of your health, such as the foods you eat, the holidays you celebrate, and the goals you have.

Your friends can have a major influence on the decisions you make. Such influences can be either positive or negative. A friend who urges you to smoke may have a very negative influence on your health. A friend who listens to you and encourages you will probably have a strong positive influence on your health.

Books, magazines, newspapers, television and radio programs, and the Internet can also have a tremendous influence on your health choices. Just remember to evaluate the source of any message. Commercials, for example, are there to convince you to buy—not just to provide information.

10   Chapter 1: A Healthy Foundation

HEALTH LAB

Influences on Health

Time Needed

1 week, if time allows

Supplies

none

Focus on the Health Lab

Ask students to share their own experiences with surveys. Who usually conducts surveys? What is the purpose of most surveys?

Explain that students will develop their own surveys for the special purpose of gathering information about people around them. They will use the survey questions to examine the factors that may most strongly influence people's health.

Have students work in small groups to develop their survey questions. Encourage them to refer to the text and to their own experiences for guidance in thinking of relevant questions. You may find it helpful to let all the groups share their survey questions with the entire class; this may provide students with additional questions to add to their list.

Review

Using complete sentences, answer the following questions on a separate sheet of paper.

Reviewing Terms and Facts

1. **Vocabulary** Define the terms *heredity* and *environment*.
2. **Restate** Describe how you can use information about hereditary traits to make health decisions.
3. **List** Identify three factors, other than heredity and physical environment, that can have an influence on a person's health.

Thinking Critically

4. **Explain** Why are there laws about allowable levels of environmental pollution?
5. **Evaluate** Choose one of the categories of influences in **Figure 1.4**. Describe a possible positive and a possible negative health influence in that category.

Applying Health Concepts

6. **Personal Health** With a group of four students, create a skit about the factors that influence personal health. Create a scene or a series of scenes in which a teen is influenced by one or more of the factors described in **Figure 1.4**. Show how the teen responds and how it affects his or her health. Perform your skit for the class.

HEALTH LAB

Influences on Health

Introduction: Recognizing the factors that influence your health can provide valuable background information as you make choices and take actions that affect your overall health and wellness.

Objective: With a group of three to five classmates, design, conduct, and analyze the results of a survey about how people view influences on their health.

Materials and Method: As a class, come up with a list of 15 to 20 factors that might have an influence on people's health. (Possible factors to include are diet, religious beliefs, inherited tendencies toward disease, the opinions of friends, air quality, or neighborhood crime.) Work with your own group to create a short survey that asks people to choose from this list the five most important influences. Suggest that people then rank the factors they have chosen from 1 to 5 in order of importance. Conduct the survey on students and adults outside your class.

Observations and Analysis: After conducting your survey, analyze the results. You may want to display the results in graph form. Which factors do the people you surveyed consider most important? Which do they consider least important? How do your results compare with the results of your classmates? In what areas do you think people should be more aware of the influences on their health? How could health professionals help people become more aware of how to make wise choices about health?

Lesson 2: Influences on Your Health

Chapter 1 Lesson 2

Assess

EVALUATING THE LESSON
Assign Reviewing Terms and Facts and Thinking Critically to review the lesson; then assign the Lesson 2 Quiz.

LESSON 2 REVIEW

Answers to Reviewing Terms and Facts
1. Heredity is the passing on of traits from biological parents to their children. Your environment is made up of all the living and nonliving things that surround you.
2. Responses will vary.
3. Any three: cultural background, family traditions, friends, available information and personal choices.

Answers to Thinking Critically
4. The laws are designed to protect people from unknowingly being exposed to high levels of pollution.
5. Responses will vary.

RETEACHING

▶ Assign Concept Map 2.
▶ Have students complete Reteaching Activity 2 in the TCR.

ENRICHMENT

Assign Enrichment Activity 2 in the TCR.

Close

Give each student an opportunity to identify the most important influence on his or her health.

Understanding Objectives
Students should become more aware of the importance of identifying outside factors that influence people's health.

Observation and Analysis
Once students have obtained responses to their survey questions, have them meet with their groups to discuss the survey results and respond to the Health Lab questions. Have students record the results of their surveys in their journals.

Further Investigation
Encourage students to share their survey questions and results with family members. In some families, discussions of hereditary and environmental risk factors may grow out of this experience.

Also, ask students for suggestions on how other people might benefit from responding to this kind of survey. Volunteers may be interested in incorporating the survey questions into an article for the school newspaper.

Assessing the Work Students Do
Have students write paragraphs explaining what they learned from the survey and how they hope to use that information.

Focus

The Teen Health Digest articles can be used in two ways: as an individual activity for reflection and enrichment or as a cooperative learning activity as described below.

Motivator

Allow a volunteer to draw a large health triangle on the board, labeling the three sides of the figure. Ask students to keep the health triangle in mind as they read the five Teen Health Digest articles. Then have students refer to the triangle as they discuss the aspect or aspects of health on which each article focuses.

Cooperative Learning Project

Setting Goals

Have students work as a class to plan and present a "three-ring circus" on health. The three rings will correspond to the three basic aspects of health. Within each ring, students will put on short skits, give demonstrations, sing original songs, dance, or create other kinds of performances to promote awareness of the behaviors and attitudes that promote good health.

Delegating Tasks

Let students divide themselves into three groups, one for each aspect of health: physical, mental/emotional, and social. Remind students of the importance of working cooperatively with other group members to share and discuss ideas, make decisions, and create a fun and educational performance out of those ideas.

TEEN HEALTH DIGEST

Sports and Recreation
A Champ with a Sense of Balance

In 1996, at the age of 15, California teen Michelle Kwan won the World Figure Skating Championship. Such early success might tempt an athlete to focus solely on his or her sport. Yet Kwan realizes that there is more to life than her skating career.

Kwan, an excellent student, gives careful attention to her tutored courses in biology, algebra, English, French, and history. She also understands the importance of her social health. She remains physically and emotionally close to her family. Kwan depends on her parents for comfort and advice.

Michelle Kwan looks forward to putting her skates away someday and going to college. "A lot of skaters look at skating and nothing else," she says. "Hello! There's a life in between!"

Chapter 1: A Healthy Foundation

CON$UMER FOCU$

Handwashing for Health?

Q: Do those antibacterial soaps I see advertised really work? Can I protect myself against germs by using them?

A: Yes and no. The soaps really *do* contain ingredients that kill bacteria. They do an especially good job on the germs that produce odor, and they can also kill some of the germs that spread illness. To get the full benefits, however, you have to wash in the right way. That means rubbing your hands briskly under warm water for at least 20 seconds. Even after rinsing, your hands may retain an antibacterial coating that protects against future germs.

Try This:
Use a timer to practice washing your hands in the way described. Try to use the method each time you wash.

TECHNOLOGY OPTION

Computer Some students may want to use computers to compose and play original music for acts in the three-ring health circus. Others may be interested in using computers to generate banners or other artwork to use as backdrops. Still others may want to use the Internet to search for information to be included in their songs or skits, and many students may prefer to use word processing programs to compose, edit, and revise scripts or other written materials. Provide encouragement and assistance; if appropriate, ask another teacher or a parent volunteer to work with these students to help facilitate their activities and to increase their understanding of various computer capabilities.

HEALTH UPDATE
Young and Healthy

Young Americans are eating in ways that reduce their cancer risk. A 1996 American Cancer Society report showed that the students surveyed were doing better than adults in some areas of nutrition. "The proportion of students who ate at least five servings of fruits and vegetables daily doubled between 1993 and 1995," the report noted.

The American Cancer Society recommends having at least five servings of fruits and vegetables daily as well as limiting fat intake and getting regular physical activity. These guidelines are also recommended for preventing other diseases, such as heart disease and diabetes.

Try This: For the next five days, write down how many times you eat fruits and vegetables, as well as how much you eat. If you aren't getting enough of these foods, make a plan to get what you need.

Myths and Realities
The More I Exercise, the Hungrier I Get

Q. I thought that exercise would dull my appetite, but I'm actually hungrier after I exercise. Is that unusual?

A. The answer seems to depend on whether you are male or female. Recent studies in England and elsewhere found that although exercising seemed to make men lose their appetites, it had the opposite effect on women. Exercising remains a crucial element of any weight management program. If exercising gives you an appetite, however, be sure to snack on healthful, low-calorie foods.

Personal Trainer
Safe Video Routines

You say your friend bought an exercise video and the two of you want to try it out? Keep these tips in mind.

1. **Protect your feet.** If the video includes jumping exercises, be sure you wear athletic shoes. Don't exercise in socks or bare feet.
2. **Clear an area.** Make sure you have cleared enough space in front of the TV. Try to exercise away from tables, lamps, and other furniture.
3. **Don't overdo it.** Don't worry about keeping up with the instructor on the tape. You haven't been doing this as long as he or she has.
4. **Don't forget to pause.** The beauty of a video is that you can stop anytime. Stop if you feel any pain or if you feel you are exercising too hard.

Teen Health Digest 13

Chapter 1

The Work Students Do

Following are suggestions for implementing the project:
1. Have students build on the information in the Teen Health Digest articles, adding other ideas that they have developed while studying Chapter 1.
2. Encourage students to make use of their own special abilities and interests in preparing their acts. Some group members may wish to write and direct a skit, but not perform in it. Others may want to sing or dance, but not be responsible for any artwork. Students who are interested in playing musical instruments or using computers or video cameras should be encouraged to incorporate their abilities into the group's productions.
3. Remind students that their groups should work together in ways that protect and promote all three aspects of their own health. How might this consideration affect the ways they work? The deadlines they set? The cooperation they foster among group members?
4. Have the entire class work together to develop and implement a plan for sharing their health circus with an outside audience—other students at school, parents and other family members, students at an elementary school, residents of a retirement center, or other community members.

Assessing the Work Students Do

Ask the members of each group to write a brief evaluation of their own work on the project. The evaluation should identify at least two strengths of the group and at least two areas in which the group needs improvement.

If possible, also talk briefly with each student about the project.

Meeting Student Diversity

Learning Styles The variety of activities involved in the health circus provide a good opportunity for students to select the kind of work that is most appropriate for them. You may want to have short conferences with some of the students to help them select the activities that are best suited to their individual learning styles.

Language Diversity Suggest that students who are fluent in a language other than English might present skits or songs in that language. This would give them an opportunity to express themselves comfortably and to reach others who speak that language, bringing health awareness to an even wider audience.

13

Lesson 3
What's Your Health Status?

Focus

LESSON OBJECTIVES

After studying this lesson, students should be able to

- discuss why they themselves are the most important influence on their own health.
- identify examples of risk behaviors.
- explain how self-assessment can help them take an active role in making health- and safety-related choices.
- discuss what a personal wellness contract is and how they can use these contracts to improve their own health.

MOTIVATOR

Ask students to list at least ten health problems they think are preventable. Remind them to consider all three aspects of their health: physical, mental/emotional, and social.

INTRODUCING THE LESSON

Have volunteers share some of the preventable health problems they listed during the Motivator activity. Guide students in a discussion: What might cause each specific problem? How can you be responsible in preventing each problem? What kinds of assistance do you think you might need in preventing each problem? Encourage students to consider the importance of prevention: Which is usually easier, dealing with a problem or preventing a problem? Why?

INTRODUCING WORDS TO KNOW

Ask students to identify the verbs that form the bases of the three Words to Know: *prevent, behave, assess.* What does each verb mean? How do these verbs help you better understand the meaning of the Words to Know?

Lesson 3
What's Your Health Status?

This lesson will help you find answers to questions that teens often ask about the status of their health. For example:

▶ Why should I check my health status?
▶ What do I need to do to be healthy?
▶ How can I plan for a high level of wellness?

Words to Know

prevention
risk behavior
self-assessment

Q & A

Listen to Grandma!

Q: When I was little, my grandmother used to make chicken soup to help me get better from a cold. Aren't cold medicines more reliable than chicken soup?

A: Your grandmother is in good company. Chicken soup is a folk remedy all over the world, from Italy to Greece to China. Scientists comparing various liquids have found that chicken soup actually seems to have a beneficial effect on colds.

Where You Are and Where You're Going

As you have learned, your heredity, environment, and the people with whom you interact are all major influences on your overall health and wellness. However, such factors are not the *only* influences. In fact, the most important influence on your health is *you.*

One of the keys to overall wellness is the prevention of health problems. **Prevention** involves *taking steps to make sure that something unhealthy does not happen.* For example, you wash your hands to prevent illnesses caused by germs. Most importantly, prevention involves avoiding risk behaviors. A **risk behavior** is *an action or choice that may cause injury or harm to you or others.* Smoking cigarettes is a risk behavior because it can lead to lung disease and many types of cancer. Riding a bike without a helmet is a risk behavior because it can lead to head injury.

When you were a young child, you depended on trusted adults such as parents, teachers, and counselors to "set the rules." Now that you are a teen, you are taking a more active role in the choices that you make for your own safety and wellness. One of the most effective ways to begin taking that active role is **self-assessment,** or *careful examination and evaluation of your own patterns of behavior.* Honest self-assessment can give you a clear picture of your health status.

As a teen, you can have a positive influence on the health choices of young children.

14

Lesson 3 Resources

Teacher's Classroom Resources
- Concept Map 3
- Cross-Curriculum Activity 2
- Enrichment Activity 3
- Health Lab 1
- Lesson Plan 3

- Lesson 3 Quiz
- Reteaching Activity 3

Student Activities Workbook
- Study Guide 1
- Applying Health Skills 3

Personal Inventory
HOW HEALTHY ARE YOU?

Use this self-assessment form to examine your health status. It should help you evaluate how your choices and actions influence your health and wellness.

On a separate sheet of paper, write yes or no to tell whether each statement describes you.

Physical Health
1. I get at least eight hours of sleep each night.
2. I eat a well-balanced diet, including a healthful breakfast every day.
3. I wear a seat belt in cars and protective gear when bicycling or playing sports.
4. I keep my body, teeth, and hair clean.
5. I do not use tobacco, alcohol, or drugs.
6. I exercise regularly.
7. I do not skip meals or use harsh diet plans to try to lose weight.
8. I have regular check-ups with my doctor and dentist.
9. I am aware of hereditary illnesses within my family and take steps to protect my health.
10. I do not plan to engage in sexual activity before marriage.

Mental/Emotional Health
1. I generally like and accept who I am.
2. I can accept helpful criticism.
3. I can express my feelings clearly and calmly, even when I'm angry or sad.
4. I do not blame others for my mistakes.
5. I accept that I will make mistakes, and I try to learn from them.
6. I can stand up for my own values.
7. I can face problems calmly.
8. I have at least one hobby that I enjoy.
9. I enjoy learning new information and acquiring new skills.
10. I feel that people like me.

Social Health
1. I have at least one or two close friends.
2. I respect and care for my family.
3. I have a friendly, open attitude when I meet new people.
4. I work well in a group.
5. I feel that my friends know that I am truthful and dependable.
6. I can disagree without arguing.
7. I am willing to give and get support from others when needed.
8. I am a good listener.
9. I can confidently say no when people ask me to do something harmful or wrong.
10. I respect the right of others to have opinions that may differ from mine.

Give yourself one point for every yes. Total the number of yes responses in each of the three areas. Then check your score with the following ranking.

In each section, a score of 9–10 is very good, 7–8 is good, and 5–6 is fair. If you scored 0–4 in any area, you need to take a more active role in avoiding risk behaviors that can harm your health.

Lesson 3: What's Your Health Status? **15**

Chapter 1 Lesson 3

Teach

L1 Discussing
Have students respond to this statement from the text: *The most important influence on your health is you.* Ask questions such as these: Do you think there are any exceptions to this statement? What do you and other teens do to influence each aspect of your own health?

L1 Critical Thinking
Guide students in discussing risk behaviors: What specific risk behaviors are teens likely to engage in? Since risk behaviors can cause harm, why don't all teens avoid them? What outside factors might encourage teens to engage in risk behaviors? How can teens resist the influence of these factors? In what ways, if any, do you think the risk itself attracts teens to engage in such behaviors? Encourage students to explain and support their ideas.

Teacher Talk
Dealing with Sensitive Issues

Young teens may feel uncomfortable discussing issues that relate to their personal health. Emphasize the importance of responding to the Personal Inventory: How Healthy Are You? on page 15, honestly and privately. Remind students that this is a personal inventory designed for their own use; they do not need to share their responses with anyone. If appropriate, suggest that students who are concerned about their personal inventories speak privately with you, a school counselor, a parent, or another trusted adult.

Growth and Development

Adolescence often involves emotional ups and downs—the teen mood swings familiar to many teachers and parents. These changes in emotional outlook are part of growing up, but they can be difficult for young people to handle.

Talking openly with a trusted peer or adult often helps teens who are feeling down. So does getting enough rest, exercising, and eating a healthy diet.

However, teens who feel sad or hopeless for as long as two weeks, without any positive mood changes, may be suffering from clinical depression. Withdrawal, inactivity, and loss of appetite can be further signs of depression. Clinically depressed teens need professional help.

Chapter 1 Lesson 3

L1 Discussing
Help students discuss the importance of taking an active role in making their own choices about safety and wellness: How do the rules that parents, teachers, and counselors set down for you in the past affect the choices you make in the present? Are there some choices that still should be left up to your parents or guardians? Why or why not? What are the advantages of making your own health and safety decisions for yourself as you mature? What are the disadvantages, if any?

L2 Journal Writing
Ask students to think about the trusted adults with whom they might discuss their health status. In their private journals, have them identify the adult with whom they would most like to talk and the topics they would most like to discuss. Then encourage students to use this journal entry as a motivator to initiate such a discussion.

L3 Writing Letters
Have interested students select a few health-related groups (the Juvenile Diabetes Foundation or the Asthma and Allergy Association of America, for example) and write short business letters to those organizations, asking them for brochures and other printed information. Allow these students to share the organizations' responses with the class.

L2 Creating Posters
Let students work in small groups to identify a single statement from the Personal Inventory on page 15 that they feel is worth sharing with others (such as *always wear a seat belt in cars, face problems calmly,* or *be a good listener*). Have group members work cooperatively to make a poster encouraging other students to make that positive health choice. Display the completed posters in a hallway or another prominent area of the school.

in your journal

Take a look at the results of your personal inventory. List five steps you could take to improve your overall health. Read over your list, and then rank the items from one to five in order of importance.

Improving Your Health

The Personal Inventory on the previous page is just one way you might assess your own health status. Here are some other ways.

- Talk with a trusted adult, such as a parent.
- Discuss your health status with a health care professional.
- Ask your school nurse for information about health care topics.
- Read recommendations in health books and publications from groups such as the American Heart Association or the American Cancer Society, paying attention to how closely your health behaviors follow their recommendations.

All of these methods have the same goal—to help you understand the choices you are making right now and how those choices will affect your level of wellness. Your ultimate goal is to make choices that will increase your overall level of wellness.

What Have I Learned?

Through self-assessment you can learn about your own behaviors and possible risks you are taking. Part of this process involves making judgments. Look again at the Personal Inventory in this lesson. You will notice that the risks represented in the statements are not all equal. For example, skipping breakfast one morning is a poor health choice, but it is not a choice that will put your life in danger. Not wearing a seat belt, however, *is* life threatening. Every time you ride in a car without wearing a seat belt you are in danger of being seriously injured or killed.

What's next? You've assessed the status of your health. You've found some areas in which you could improve. You've thought about which areas are the most essential for you to work on right away. Now you're ready for the next and most important step. You're ready to use the information in a positive way by making a personal wellness contract.

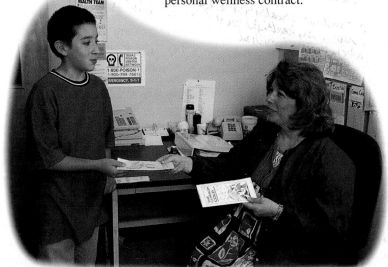

Doctors, school nurses, and other health professionals can point you to good sources of accurate and reliable health information that will help you make wise choices.

16 Chapter 1: A Healthy Foundation

Consumer Health

Misinformation is one of the factors that work against teens in their effort to follow a healthy eating plan. Many students say, "I don't have to watch what I eat—I take vitamin supplements."

Unfortunately, supplements cannot make up for healthful foods that may be missing from teens' diets, or for the unhealthy foods that may be included in those diets. Nutrients in supplements are not absorbed by the body as completely as those in food. Bring to class several empty bottles of vitamin supplements. Encourage students to read the labels carefully, noting the nutrients provided. How might they get these nutrients as part of their daily diets?

Creating a Personal Wellness Contract

To help you create a wellness contract, ask yourself the following questions and jot down all the ideas you have for possible answers.

- According to my self-assessment, what positive behaviors am I engaged in regarding my health?
- What risk behaviors are putting my health in jeopardy?
- What additional health issues do I want to work on?
- Which areas should I work on first?
- How might changing some of my behaviors make me happier and healthier?
- What specific steps might I take to change each behavior?

Look back at the ideas you have written. Use those ideas to make a personal wellness contract. Think of this as a real contract—an agreement you will abide by—with yourself. In the contract, specify what you will do in order to improve your health and wellness. **Figure 1.5** shows one form that this contract might take.

Q & A

It's Never Too Late

Q: Most mornings, I'm too rushed for breakfast! What can I do?

A: Try a piece of whole wheat bread and this instant, delicious shake, mixed in a blender or shaken in a lidded jar:
- 8 ounces plain nonfat yogurt
- 1 banana, chopped or mashed
- 1 cup orange juice

Figure 1.5
A Personal Wellness Contract
What is the value of writing down your promises and dating the contract?

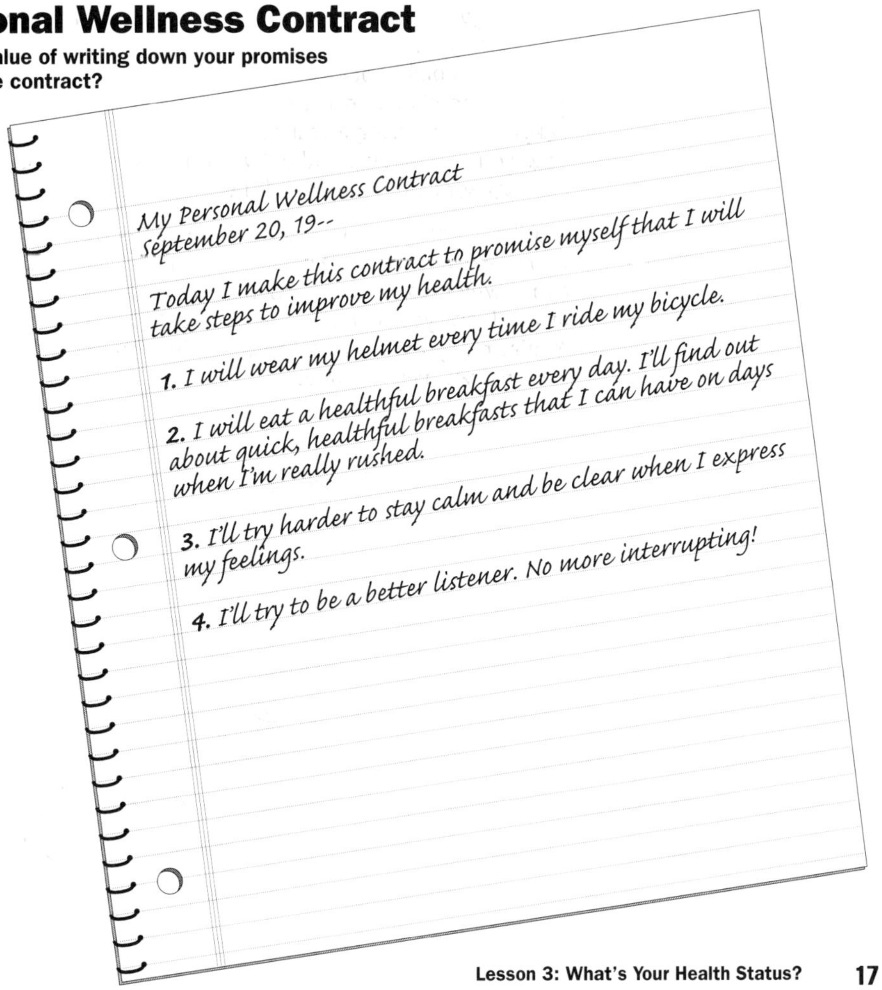

My Personal Wellness Contract
September 20, 19--

Today I make this contract to promise myself that I will take steps to improve my health.

1. *I will wear my helmet every time I ride my bicycle.*
2. *I will eat a healthful breakfast every day. I'll find out about quick, healthful breakfasts that I can have on days when I'm really rushed.*
3. *I'll try harder to stay calm and be clear when I express my feelings.*
4. *I'll try to be a better listener. No more interrupting!*

Lesson 3: What's Your Health Status?

Chapter 1 Lesson 3

L1 Discussing
Discuss why using information in a positive way is the most important step in the process of monitoring and improving health. Ask students: In what situations have you gathered useful information and then failed to put that information to use? Why didn't you use the information? What was the result? How will a self-assessment affect your health if you don't put your ideas into action?

L1 Language Arts
Ask the class: What are some of the positive health choices teens can make independently? Of these choices, which do you consider most important? Why? Have students write a short essay on the choice they consider most important.

L2 Social Studies
Let several volunteers work together to gather information about the health status of teens in other countries around the world. Suggest that they compile statistics about diseases, life expectancy, and average height and weight to indicate the general health of various populations. Have the volunteers choose a suitable method for sharing this information with the rest of the class.

L3 Creating Documents
Encourage interested students to create unique versions of their wellness contracts. They might wish to use calligraphy, design patterns or other decorations, "antiquing" techniques, or computer programs to fashion their special documents with a personal flare.

More About...

Commitment Making—and sticking to—a personal wellness contract is a very mature undertaking. Teens will probably agree that a promise to oneself is of the utmost importance. However, keeping such a promise can be very difficult. For most people, regardless of age, an effective support system is essential. Weight-loss advisers, for example, recommend dieting with a buddy and enlisting the encouragement of friends and family members. Similarly, students who share their wellness goals with family and close friends are likely to experience more success in fulfilling their personal wellness contracts.

Chapter 1 Lesson 3

Assess

EVALUATING THE LESSON
Assign Reviewing Terms and Facts and Thinking Critically to review the lesson; then assign the Lesson 3 Quiz in the TCR to evaluate students' understanding.

LESSON 3 REVIEW

Answers to Reviewing Terms and Facts
1. A risk behavior is an action or choice that may cause injury or harm. Examples will vary.
2. Responses will vary. Possible answers include talking with a parent, discussing health status with a health care professional, and reading health publications.
3. The purpose of a personal wellness contract is to remind you of your health goals and how you can achieve them.

Answers to Thinking Critically
4. Responses will vary.
5. Responses will vary. Possible answers include taking a deep breath or counting to 10 before speaking, trying to understand another person's point of view, and keeping your voice even and low.

RETEACHING
- Assign Concept Map 3 in the TCR.
- Have students complete Reteaching Activity 3 in the TCR.

ENRICHMENT
Assign Enrichment Activity 3 in the TCR.

Close
Ask students to identify the most important change they want to make to improve and/or protect their health.

Teen Issues
Control the Clock! ACTIVITY

Do you ever feel that you don't have time to manage all your activities and responsibilities? Try taking control of the clock! Make a daily schedule. Give yourself a realistic amount of time for each task that you must do, and don't forget to schedule in a reward—time for fun and relaxation after those tasks are completed. Experiment with your daily schedule until you find the right balance of task time and reward time.

Using the Contract
The rest is up to you! Stick to your contract and take an active role in improving your health. Try sharing your contract with a friend, and ask your friend to share his or hers with you. How can the two of you help each other to fulfill the contracts? By following through, you can become positive influences on each other's health.

A friend's encouragement can have a powerful effect.

Lesson 3 Review

Using complete sentences, answer the following questions on a separate sheet of paper.

Reviewing Terms and Facts
1. **Vocabulary** Define the term *risk behavior*. Give two examples.
2. **List** Identify two ways, other than a personal inventory, to perform self-assessment.
3. **Recall** What is the purpose of a personal wellness contract?

Thinking Critically
4. **Describe** Choose a sport or other physical activity that you enjoy. Describe specific preventive steps you might take to avoid injuries or accidents in that activity.
5. **Recommend** Reread the contract shown on page 17. What specific steps might the teen take to try to stay calm and be clear when expressing his or her feelings?

Applying Health Concepts
6. **Personal Health** For the next week, keep a log of your progress as you work to fulfill your contract with yourself. Describe specific actions or choices you take, and tell whether or not you feel that they are helpful.
7. **Growth and Development** Make a poster that would help early elementary children (ages six to eight) become aware of the factors that influence their health. The poster could for example, talk about healthful eating habits or safety on bicycles. If you choose to discuss negative influences, avoid making the poster frightening for this age group.

18 Chapter 1: A Healthy Foundation

Health of Others

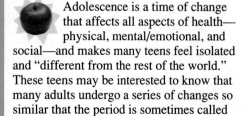

Adolescence is a time of change that affects all aspects of health—physical, mental/emotional, and social—and makes many teens feel isolated and "different from the rest of the world." These teens may be interested to know that many adults undergo a series of changes so similar that the period is sometimes called "middlescence." Middle age brings physical changes that some adults find disturbing and uncomfortable. It can also be a time of questioning values and commitments, of changing jobs or even careers, of reviewing and revising social ties. You might encourage interested students to read about middlescence or to interview willing middle-age adults.

Building Health Skills

Lesson 4

This lesson will help you find answers to questions that teens often ask about taking charge of their health. For example:

▶ What steps can I take to be in charge of my own health?
▶ What skills can I develop in the three areas of health?
▶ How can I avoid health risks?

Taking Charge of Your Own Health

Making a personal wellness contract is the first step in taking charge of your own health. The next step is to *follow* the plan. To do that, you will need to understand and practice specific skills. One way to think about the skills is to remember the health triangle. Some skills are most important for your physical health. Others will help improve your mental/emotional health. Still others will enhance your social health. This lesson provides an overview of these skills. The rest of the textbook explains how they can be used in your daily life.

Skills for Building Physical Health

You can develop many skills that will help you build and maintain physical health. Some of these skills will protect you from immediate illness or injury. Others will increase your level of physical wellness far into the future.

Staying Safe

Take safety precautions to avoid risks that can lead to injuries. For example, obey traffic rules when bicycling, riding in a car, or walking along the side of a road. Don't go out alone after dark.

Staying safe also involves practicing personal **hygiene** (HY·jeen), or *cleanliness*. Daily hygiene includes care of the skin, teeth, nails, and hair. A clean body is the best line of defense against germs that cause illness.

Words to Know

hygiene
self-management
values
refusal skills

Cultural Connections

Yoga for Fitness and Relaxation

Yoga is a system of mental and physical exercise. It was developed centuries ago by Hindus in India. In Sanskrit, the classical language of India, yoga means "discipline." Most yoga exercises are simple to do and combine coordinated movements and postures with breathing techniques and meditation. Many people throughout the world practice yoga to improve fitness, reduce tension, and improve overall health.

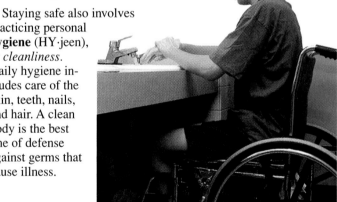

Washing your hands is one of the simplest and most effective ways to prevent the spread of illness.

Lesson 4: Building Health Skills **19**

Lesson 4 Resources

Teacher's Classroom Resources
- Concept Map 4
- Decision-Making Activity 2
- Enrichment Activity 4
- Lesson Plan 4
- Lesson 4 Quiz
- Reteaching Activity 4
- Transparencies 4, 5

Student Activities Workbook
- Study Guide 1
- Applying Health Skills 4
- Health Inventory 1

Lesson 4
Building Health Skills

FOCUS

LESSON OBJECTIVES

After studying this lesson, students should be able to

▶ discuss the importance of taking charge of their own health.
▶ identify skills they can develop to build and maintain their own physical, mental/emotional, and social health.
▶ identify ways to avoid health risks.

MOTIVATOR

Direct students' attention to the title of the lesson. Give them three minutes to list as many health skills as they can think of.

INTRODUCING THE LESSON

Encourage students to share and discuss the health skills they listed in response to the Motivator activity. Ask: How does each skill relate to your health? Why is it important for you to be responsible for developing each skill? Point out that, according to the American Cancer Society, health skills developed in childhood persist into adult life; this is true for both positive and negative health habits. Guide students in discussing implications of this tendancy.

INTRODUCING WORDS TO KNOW

Let students work in pairs to discuss their understanding of the Words to Know. Then have the members of each pair alternate scanning Lesson 4 to find and read the specific definition for each term. How close was their understanding to the meaning of the words?

19

Chapter 1 Lesson 4

Teach

L1 Recalling
Ask students to recall the health triangle: What are its three basic aspects? What does each involve? How are the three aspects interrelated?

L2 Writing Lists
Give the class five minutes to list as many safety precautions as possible. Students may work independently or with partners.

> **Visual Learning**
> Discuss with students the photo at the bottom of page 19. What is the teen doing? When is it important to wash your hands or face? Why?

L2 Consumer Skills
Ask students to bring to class magazine ads for various hygiene products designed for the special care of skin, teeth, nails, and hair. Then, in groups or as a class, let students discuss the advertised products: What is the special purpose of this product? Would a generic product serve that purpose equally well? How much does the product cost? Are similar products available at a lower price?

L2 Evaluating
Let students work in small groups to evaluate their school lunch options. What healthful food choices are available, either in the school lunch program or as lunch-box items from home?

Literature Connection

Knowing Yourself ACTIVITY

In William Shakespeare's classic drama "Hamlet," a character gives the following advice to a young man: "To thine own self be true." In more modern and informal language, this sentence might translate to "Be honest with yourself."

Write a letter to a student who is younger than you and share this advice. Point out the positive effects it can have on emotional and mental health. Include examples to make your meaning clear.

Eating Well and Staying Fit

Other important physical skills involve choosing healthful foods and staying active. Learn as much as you can about food and good health. Eat a balanced diet that is rich in whole grains, fruits, vegetables, lean meats, and dairy products. Avoid foods that are high in fat. In addition, put your body to work every day. Whenever possible, walk! Choose the stairs instead of the escalator. Join a sports team or start a regular exercise group with your friends.

Skills for Building Mental and Emotional Health

The following skills are ones that will help you mentally and emotionally. They will allow you to understand yourself and others. They will also give you access to information available from experts in health. Finally, the skills in this section will help you feel good about yourself and about what you can do with your life.

Knowing Yourself

Get into the habit of honest self-assessment. Examine and evaluate your choices and actions. Understand that everyone—including you—makes mistakes. Don't let mistakes get you down, however. Use them instead in a positive way—to learn, grow, and change.

LIFE SKILLS
Becoming Media Literate

*A*n important point to remember whenever you are gathering facts is that not all information is equally valid. In other words, not all sources of information—books, articles, and data on the Internet—can be trusted. Here are some tips to follow when gathering information.

▶ DO learn to use such research tools as the library computer or card catalog, *The Reader's Guide to Periodical Literature*, and Internet search engines. These tools will help you find information from a variety of useful sources. A teacher or librarian can show you how to begin.

▶ DO check your sources. Check the credentials of the author and any people whom the author quotes. Are these people experts on the topic presented? Then check the accuracy of the author's sources and findings. Is the information based on reliable scientific studies or reports?

▶ DON'T randomly surf the Internet, or gather health-related information from a public message board or chat room.

20 Chapter 1: A Healthy Foundation

LIFE SKILLS
Becoming Media Literate

Focus on Life Skills
Use the Life Skills title, Becoming Media Literate, to help students preview the activity. Ask: What are media? (Some students may be surprised to discover that media include more than television and radio.) What does it mean to be literate? What do you think media literacy is?

Let volunteers read aloud the listed DOs and DON'Ts, and guide the class in discussing each. Why is this discussion important? What experiences have you had with the advice offered? If you need specific kinds of help, whom might you ask?

Encourage students to work with partners in completing the Follow-Up Activity. Post the completed charts in the classroom where other students can view and discuss them.

Making Health Connections
Finding and evaluating health-related information is an important part of assuming responsibility for your own health. As teens mature, they become

20

Locating Helpful Information

Today's world provides access to all kinds of information. Develop good fact-finding skills so that you can gather the information you need to make healthful decisions. Are you aware of the resources available at home, at school, and in your community for valid health information? **Figure 1.6** summarizes some of them.

Figure 1.6
How to Find the Facts
Many information sources are available to you.

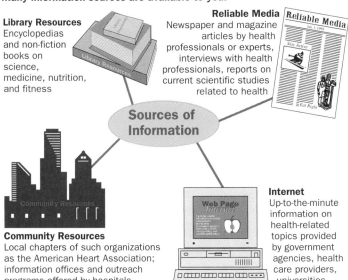

Library Resources
Encyclopedias and non-fiction books on science, medicine, nutrition, and fitness

Reliable Media
Newspaper and magazine articles by health professionals or experts, interviews with health professionals, reports on current scientific studies related to health

Community Resources
Local chapters of such organizations as the American Heart Association; information offices and outreach programs offered by hospitals, clinics, and universities; government offices, such as the local department of health

Internet
Up-to-the-minute information on health-related topics provided by government agencies, health care providers, universities, and scientific publications

Q & A

Who Do I Believe?

Q: It seems like every time one report comes out telling me that something is bad for me, a couple of months later another researcher finds something different. Who do I trust?

A: It's not unusual for two scientists performing the same types of experiments to come up with conflicting results. In today's world, news travels so quickly that sometimes preliminary results are reported as final. Also, in a quick "sound bite," full explanations of a study are impossible. What does this mean? Before making any changes in your lifestyle, check out the information with a health professional. He or she can probably put the information in the proper perspective.

Make sure that your information is coming from experts whom you can trust.

▶ DO seek information from the Web sites of such reliable sources as universities, libraries, and government agencies such as the Centers for Disease Control and Prevention (CDC), the National Institutes of Health (NIH), and the Environmental Protection Agency (EPA).

▶ DO seek information from reliable media sources such as *Consumer Reports* and the news and science sections of responsible newspapers.

▶ DO seek information from such television sources as news, science, or health-related programs.

▶ DO seek information from the Web sites and printed materials of recognized national organizations such as the American Cancer Society, the American Medical Association, and the American Heart Association.

▶ DON'T seek factual information from advertisements in newspapers and magazines or on television, or from any Internet Web site that offers a product for sale.

▶ DO run a "back-up" check on the reliability of any source of information by getting the advice of a trusted adult, such as a parent, a teacher, or a librarian.

Follow-up Activity

Select a health-related topic, such as immunization, bicycle safety, air pollution, or dietary fiber. Then use the tips listed above to find reliable sources of information on your topic. Make a chart to report on and evaluate at least five sources that you find.

Lesson 4: Building Health Skills

Chapter 1 Lesson 4

L1 Discussing

Ask students to share and discuss their own experiences in finding, evaluating, and using information on the Internet. Caution students that information on the Internet requires careful screening for reliability, because anyone can open a Web site.

L3 Gathering Information

Have a group of volunteers collect brochures, fliers, and other informational publications from local health organizations. Then let them select a way to share these resources with other students, such as creating a class bulletin board, a notebook, or a set of labeled files. Be sure all students have time to browse through the collection.

L3 Interviewing

Tell students to interview adult family members or friends about the problems, stresses, and conflicts they may have faced as teens. What did the adult find difficult as a teen? What perspective does the adult now have on those problems? Have students write short reports about their interviews, preserving the interviewees' anonymity if desired.

L1 Guest Speaker

Ask a school counselor or a local practicing psychologist to speak with the class about the importance of teens taking responsibility for themselves.

L1 Journal Writing

Pose these questions to students: What are your own most important values? Where do they come from? How do they affect the way you live? Encourage students to record their responses as journal entries.

L2 Role Playing

Let students work with partners to role-play situations in which teens can protect their health by saying no to their peers.

Teen Issues

Disagreeing Without Conflict

Sometimes the hardest thing to do when someone disagrees with you is to stay cool. Everyone's opinions are influenced by facts as well as emotions. During a disagreement, strong emotions like fear or pride can ignite a discussion just as a match ignites paper. Remain aware of your emotions and the roles they play in your approach to disagreements. Preventing a fire is often much easier than putting one out!

Having a Healthy Attitude

Develop the skill of **self-management,** *the ability to take care of your overall health and to take control of your behavior and actions.* Start building self-management skills by learning to like and accept yourself. Take pride in your strengths and accomplishments, and be aware of the things about yourself that you'd like to improve upon. After all, everybody has both strengths and weaknesses. Instead of dwelling on the weaknesses, focus on developing the strengths.

Remember that no one leads a totally trouble-free life. Everyone must face problems, stress, and occasional conflicts. However, you can prevent some problems by identifying and avoiding unnecessary risks and risk behaviors. You can manage other problems by learning how to stay calm as you recognize, discuss, and solve them.

Taking Responsibility

Now that you are a teen, you are taking more and more responsibility for yourself. Being responsible means taking charge. For example, it means completing necessary tasks such as schoolwork and household chores without being reminded to do so. It also means making decisions that uphold your **values,** *the beliefs that guide the way a person lives, such as beliefs about what is right and wrong and what is most important.*

Skills for Building Social Health

Do you stop and think about how important your family and friends are? They are the people you spend time with and who share your interests and ideas. They can be valuable resources because they can help and guide you in making good decisions. Your relationships with family and friends depend on good social skills. People with good social skills communicate effectively, know how to say no, and have a positive attitude.

Being able to talk openly and honestly with friends becomes more important in the teen years.

Listening and Talking

Communication skills involve much more than being able to speak clearly. People also get their messages across by facial expressions, tone of voice, and choice of words. People even communicate by the way they listen. Effective communication skills are one way to prevent misunderstandings. Skills in communication will allow you to give support to others when they need it. In addition, effective communication skills will allow you to express your own feelings clearly.

Home and Community Connection

Networking: About the Net If your students have access to the Internet at home, their families may wish to share ideas about encouraging teens' use of certain Web sites and restricting their use of others. Suggest that students discuss the use of the Internet with adult family members. Open communication may alleviate many parental concerns.

Still, some parents and guardians might be interested to know about blocking devices that can limit the range of options open to teen Web surfers. Encourage these adults to communicate with each other about Internet blocking devices through informal conversations, via E-mail, at a scheduled meeting, or through a newsletter.

Refusal Skills

A very important group of health-building skills includes **refusal skills,** which are *ways to say no effectively.* Understanding and respecting your values—what's right and wrong and what's important to you—will help you to say no when you need to. If a friend urges you to do something that you feel is wrong, your values will help you to refuse. True friends will respect your decision.

Making a Difference

Learning how to be a positive, reliable influence on someone else's life is extremely valuable. Each time you help a friend develop a new skill, make a healthy decision, or prevent or manage a problem, you make a positive difference in that friend's life.

in your journal

Reread the Personal Wellness Contract you created in Lesson 3. In your journal, write how you might change or add to your contract in order to improve it. Explain what specific health-building skills you might work on in order to strengthen your overall health status.

Helping someone else makes you feel good about yourself.

Review Lesson 4

Using complete sentences, answer the following questions on a separate sheet of paper.

Reviewing Terms and Facts

1. **Vocabulary** Define the term *hygiene.* Then give two examples of important steps to take each day for personal hygiene.
2. **List** Identify three reliable sources for factual information related to health.
3. **Explain** Explain the roles of values and self-management in helping a person to be emotionally healthy.

Thinking Critically

4. **Describe** Select one of the health-building skills summarized in this lesson. Write a short paragraph that describes how developing the skill can help a person gain a high level of physical, mental/emotional, or social health.
5. **Integrate** Think of a situation in which you would use skills from each of the three aspects of health (physical, mental/emotional, social). Describe the situation.

Applying Health Concepts

6. **Consumer Health** Make a list of five community resources for reliable health-related information. Combine your list with those of your classmates to create a class list of community information sources. Ask a school administrator how your class list of resources might be used by others in your school.

Lesson 4: Building Health Skills   **23**

Cultural Awareness In discussing communication skills with the class, it is important to keep in mind the various cultural influences on your students' lives. Students from traditional Asian cultures or from certain Native American cultures, for example, may be uncomfortable looking other people in the eye, speaking forcefully, or breaking into a conversation; they may have learned that these behaviors show disrespect of others. Many teens whose families uphold these traditional norms and attitudes learn to live in "two different worlds," that of the family and that of their peers. Some, however, may still want or need to adhere to their families' standards.

Chapter 1 Review

CHECKING COMPREHENSION

Use the Chapter Summary and the Chapter 1 Review to help students reexamine the most important ideas presented in Chapter 1. Encourage students to ask questions and add details as appropriate.

CHAPTER 1 REVIEW ANSWERS

Reviewing Key Terms and Concepts

1. Health is a combination of physical, mental/emotional, and social well-being.
2. Descriptions will vary. Students may mention a feeling of well-being, high energy, and freedom from disease.
3. Examples will vary. Heredity can influence a person's health through inherited tendencies toward specific diseases; environment can influence a person's health through exposure to air or water pollution.
4. Responses will vary. For example, students might identify their ability to exert control over the influence of friends by saying no to risk behaviors.
5. Prevention involves taking steps to make sure that something unhealthy does not happen.
6. Self-assessment helps a person identify the areas of health that need improvement; these are the areas that the individual will target in a personal wellness contract.
7. Self-management is the ability to take care of your overall health and to take control of your behavior and actions.
8. Refusal skills are ways to say no effectively.

Chapter 1 Review

Chapter Summary

- **Lesson 1** Health is a combination of physical, mental/emotional, and social well-being. A person can achieve a high level of wellness by taking care of these three aspects of health over time.
- **Lesson 2** Each person's overall health is influenced by factors such as heredity and environment as well as culture and family traditions, friends, and available information.
- **Lesson 3** Individuals should practice self-assessment to check the status of their own health and avoid risk behaviors.
- **Lesson 4** Developing specific health-building skills is important in taking effective control over personal health and wellness.

Reviewing Key Terms and Concepts

Using complete sentences, answer the following questions on a separate sheet of paper.

Lesson 1
1. What is the meaning of the term *health*?
2. Describe a high level of wellness.

Lesson 2
3. Provide examples of the ways both heredity and environment can influence a person's overall health.
4. Describe how you can exert control over one outside influence on your health.

Lesson 3
5. What is *prevention*?
6. How does self-assessment help you to create a personal wellness contract?

Lesson 4
7. What does *self-management* mean?
8. Describe what is meant by the term *refusal skills*.

Thinking Critically

Using complete sentences, answer the following questions on a separate sheet of paper.

9. **Analyze** How might high personal levels of wellness among individuals benefit society as a whole?
10. **Synthesize** Imagine that you are a teen who moves from a small town to a large city. What environmental changes might you experience, and how might those changes affect your overall health?
11. **Infer** What effect do a person's values have on her or his physical, mental/emotional, and social health?
12. **Explain** Discuss the role that taking responsibility for your actions plays in achieving a high level of wellness.

Chapter 1: A Healthy Foundation

Meeting Student Diversity

Language Diversity
Use the following suggestions to help students who have difficulty with English:

- Pair these learners with native speakers of English who can restate the Chapter Summary in language that helps students comprehend important concepts.
- Direct auditory learners or those students with language diversity to the Teen Health Audiocassette Program. Available in English and Spanish, this component provides an audio and written summary of the chapter.

Chapter 1 Review

Your Action Plan

To achieve a high level of wellness, it is important to assess your own health habits. You have begun that work in creating a personal wellness contract. Now it is time to extend your work.

Step 1 Review the journal entries that you created throughout Chapter 1. Summarize your entries, highlighting points that will help you strengthen your personal wellness contract.

Step 2 Based on your summary, write a thoughtful letter to yourself, describing what you need to work on.

Use the letter to make a weekly schedule. Provide time, each day of the week, to achieve the steps in your personal wellness contract. When you feel you have achieved a goal, check it off and reward yourself for what you have accomplished. Enjoy a movie or spend some time just doing what you want.

In Your Home and Community

1. **Personal Health** Select one family member or friend as a wellness partner. Using your own personal wellness contract as a model, encourage your partner to create a similar overall plan and daily schedule. Set aside time to work on your plans together.

2. **Community Resources** Conduct research on one health-related group in your community. Through a telephone interview, find out how that group promotes wellness in your community. Report on your research in your class.

Building Your Portfolio

1. **List of Health Tips** Create a list of Top Ten Tips for Achieving Wellness. Be sure that the tips cover the three aspects of health and that they represent what you feel are the most important highlights. Add the list to your portfolio.

2. **Profile** Write a profile for an imaginary person—The Healthiest Person Alive. Describe his or her qualities in ways that show how the individual has achieved a superior level of wellness. Add the description to your portfolio.

3. **Personal Assessment** Look through all the activities and projects you did for this chapter. Choose one or two that you would like to include in your portfolio.

Performance Assessment

▶ **Self-evaluation** Direct students to review the activities that are provided throughout the chapter. Encourage each student to select one finished product or activity that demonstrates his or her best work for the chapter. Have students explain what they learned and how the examples they selected show their progress.

▶ **Teacher's Classroom Resources** Assign Performance Assessment 1, "Total Health and Wellness," in the TCR.

Chapter 1 Review

Thinking Critically

9. Possible response: People with high levels of wellness are productive members of society who contribute positively to the lives of others.
10. Possible response: Negative effects of the move may include higher levels of both stress and pollution; positive effects may include more to do and better access to medical care.
11. Possible response: A person who values fitness will exercise regularly; a person who values self-expression will be able to share emotions openly without hurting himself or herself or others; a person who values friendship will enjoy the benefits of social contact.
12. Possible response: People who take responsibility for their own actions will make plans to improve all aspects of their health over time. For example, they will plan to have a healthful diet because they know their present dietary habits will affect their physical health in the future.

RETEACHING

Assign Study Guide 1 in the Student Activities Workbook.

EVALUATE

▶ Use the reproducible Chapter 1 Test in the TCR, or construct your own test using the Testmaker Software.
▶ Use Performance Assessment 1 in the TCR.

EXTENSION

In small groups, let students select articles from current newspapers or magazines. How do the events reported in each article relate to health? Which aspects of health are directly involved in these events? Why? Then tell each group to select one article to be included on a bulletin board; have the members prepare a written or pictorial explanation of the relationship between the article and the health triangle.

Planning Guide

Chapter 2
Personal Responsibility and Decision Making

	Features	**Classroom Resources**

Lesson 1

Taking Responsibility for Your Actions
pages 28–33

Life Skills:
Skills in Leadership
pages 30–31

- Concept Map 5
- Enrichment Activity 5
- Lesson Plan 1
- Lesson 1 Quiz
- Reteaching Activity 5
- Transparency 6

Lesson 2

Recognizing and Managing Habits
pages 34–38

Health Lab:
The Effects of Repetition
pages 36–37

- Concept Map 6
- Enrichment Activity 6
- Health Lab 2
- Lesson Plan 2
- Lesson 2 Quiz
- Reteaching Activity 6
- Transparency 7

Lesson 3

Health Risks and Your Behavior
pages 39–43

TEEN HEALTH DIGEST
pages 44–45

- Concept Map 7
- Decision-Making Activity 3
- Enrichment Activity 7
- Lesson Plan 3
- Lesson 3 Quiz
- Reteaching Activity 7
- Transparency 8

Lesson 4

Making Responsible Decisions
pages 46–51

Making
Healthy Decisions:
Making a Difficult Choice
pages 50–51

- Concept Map 8
- Cross-Curriculum Activity 3
- Decision-Making Activity 4
- Enrichment Activity 8
- Lesson Plan 4
- Lesson 4 Quiz
- Reteaching Activity 8
- Transparency 9

Lesson 5

Setting Goals and Making Action Plans
pages 52–55

- Concept Map 9
- Cross-Curriculum Activity 4
- Enrichment Activity 9
- Lesson Plan 5
- Lesson 5 Quiz
- Reteaching Activity 9

National Health Education Standards

One of the main goals of the National Health Education Standards is to move students toward "health literacy." Health literacy is the capacity of individuals to obtain, interpret, and understand basic health information and services and the competence to use such information and services in ways which promote health. The health standards were developed by applying the characteristics of a well-educated, literate person within the context of health. The health literate person is:

- a critical thinker and problem solver
- a self-directed learner
- a responsible, productive citizen
- an effective communicator

Listed below are the Health Education Standards Performance Indicators addressed in each lesson of this chapter.

Lesson	Health Standards Performance Indicators
1	(3.1, 5.3, 6.3)
2	(1.6, 3.1, 3.4, 6.1, 6.4)
3	(1.1, 1.6, 3.1, 6.3)
4	(3.1, 6.1, 6.2, 6.3)
5	(6.4, 6.5, 6.6)

ABCNEWS InterActive Videodisc Series

You may wish to use the following videodiscs with this chapter: *Making Responsible Decisions* and *Violence Prevention.* Use the *ABCNews InterActive™ Correlation Bar Code Guide* for title reference. Also available in VHS format.

Chapter Resources

Teacher's Classroom Resources
- Chapter 2 Test
- Parent Letter and Activities 2
- Performance Assessment 2
- Testmaker Software

Student Activities Workbook
- Study Guide 2
- Applying Health Skills 5–9
- Health Inventory 2

Student Diversity Strategies
- Audiocassette Program (English)
- Audiocassette Program (Spanish)
- Spanish Parent Letters
- Spanish Summaries, Quizzes, and Activities

Multimedia Components
- English Audiocassette Program
- Spanish Audiocassette Program
- *Teen Health* Videodisc/VHS Series
- *Teen Health* Video Kit: *Natural Born Leaders*

Other Resources

Readings for the Teacher

Caroselli, Marlene, and David Harris. *Risk-Taking: 50 Ways to Turn Risks into Rewards.* Mission, KS: Skillpath Publications, 1993.

Dreher, Nancy. "The Teen Years: Risky Business?" Current Health 2, September 1996, pp. 6–12.

Greene, Lawrence J. *Lifesmart Kid: Teaching Your Child to Use Good Judgment in Every Situation.* Rocklin, CA: Prima Publishing, 1995.

Karnes, Frances A., and Suzanne M. Bean. *Girls and Young Women Leading the Way: 20 True Stories about Leadership.* Minneapolis, MN: Free Spirit Publishing, Inc., 1993.

Readings for the Student

Hoose, Phillip. *It's Our World, Too! Stories of Young People Who Are Making a Difference.* Boston, MA: Little, Brown and Company, 1993.

Meltzer, Milton. *Who Cares? Millions Do...A Book About Altruism.* New York: Walker and Company, 1994.

Milos, Rita. *Discovering How to Make Good Choices.* New York: Rosen Publishing Group, 1992.

Steins, Richard. *Morality.* New York: Rosen Publishing Group, 1992.

Out of Time?

If time does not permit teaching this chapter, you may use these features: Life Skills on pages 30–31; Health Lab on pages 36–37; Teen Health Digest on pages 44–45; Making Healthy Decisions on pages 50–51; and the Chapter Summary on page 56.

Chapter 2
Personal Responsibility and Decision Making

CHAPTER OVERVIEW
Chapter 2 helps students understand their own developing responsibility for themselves and others. It guides them in establishing and practicing health skills, including avoiding risk behaviors, making responsible decisions, and setting useful goals.

Lesson 1 explains what it means to be responsible and helps students identify some of the responsibilities they are assuming as teens.

Lesson 2 guides students in recognizing and modifying their own habits, with special attention to good health habits.

Lesson 3 discusses the risks that teens face and helps students understand what they can do to avoid unsafe behavior.

Lesson 4 helps students recognize decisions that affect their health and introduces the six-step decision-making process.

Lesson 5 explains the importance of setting and pursuing goals and discusses the use of an action plan for reaching goals.

PATHWAYS THROUGH THE PROGRAM
Learning how and when to make responsible decisions is an essential part of maturing and developing healthful habits. Teens who are capable of taking responsibility for their own health, choosing to say no to unsafe behavior, and setting both short- and long-term goals will protect and enhance their total health and the health of others.

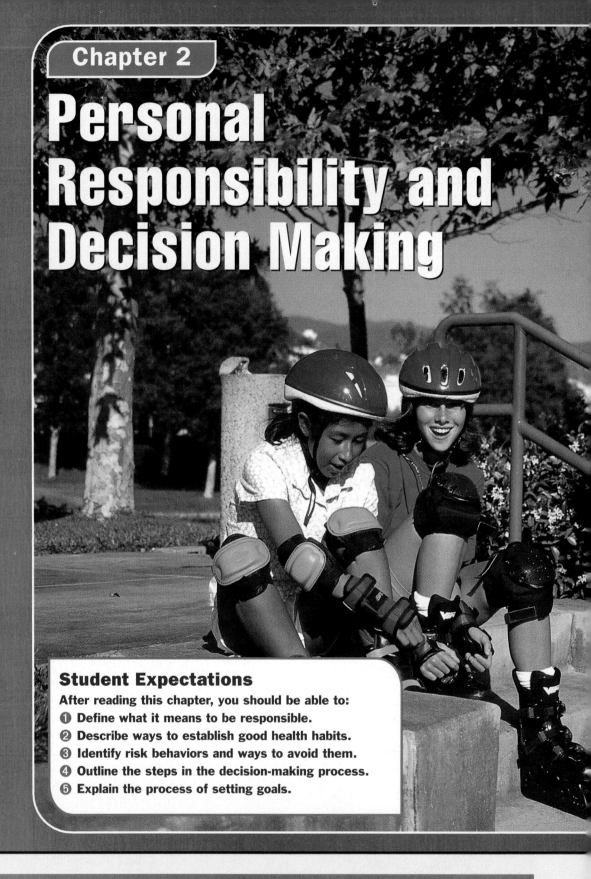

Chapter 2
Personal Responsibility and Decision Making

Student Expectations
After reading this chapter, you should be able to:
1. Define what it means to be responsible.
2. Describe ways to establish good health habits.
3. Identify risk behaviors and ways to avoid them.
4. Outline the steps in the decision-making process.
5. Explain the process of setting goals.

Key to Ability Levels

Teaching strategies that appear throughout the chapter have been identified by one of three codes to give you an idea of their suitability for students of varying learning styles and abilities.

L1 Level 1 strategies should be within the ability range of all students. Full class participation is often required. Teacher direction is usually needed.

L2 Level 2 strategies are for average to above-average students or for small groups. Some teacher direction is necessary.

L3 Level 3 strategies are designed for students able and willing to work independently. Minimal teacher direction is necessary.

Teen Chat Group

Melissa: Come on, David. Put on your helmet and let's get going. I have to be home for dinner at six, and I want to have lots of time for skating.

David: How can you two wear these helmets? I think I'll just leave mine off. I feel like such a dork in it.

Kendra: You'll look like a bigger dork if you fall and smash your head on the pavement!

Melissa: Yeah, David. My cousin wasn't wearing a helmet when she fell off her bike. She hit her head really hard and had to stay in the hospital for a whole week. The doctor said she was lucky to be alive.

David: All right, I'll wear it, but it's such a pain to put on all of this safety gear.

Kendra: Just think of it as a habit like brushing your teeth or wearing a seat belt. Pretty soon you'll just do it without even thinking!

in your journal

Read the dialogue on this page. Have you ever taken risks with your health or safety? Start your private journal entries on taking responsibility for your health by responding to these questions.

▶ How would you have convinced David to wear his helmet?
▶ What choices have you made that show that you take responsibility for your own health?
▶ How do your health decisions affect your friends and family members?

When you reach the end of the chapter, you will use your journal entries to make an action plan.

Chapter 2: Personal Responsibility and Decision Making 27

Chapter 2

INTRODUCING THE CHAPTER

▶ Display several photographs of groups of teens engaged in different activities; for example, include teens talking casually, playing volleyball, in-line skating, or reading together. Let students describe each picture. Then ask: Which group would you be interested in joining? Why? When you make new friends, what choices do you sometimes have to make about other, more established friendships? About spending time with family members and with friends? Tell students they will learn more about decisions and responsibility in Chapter 2.

▶ Read aloud a short selection from a novel the students are likely to find engaging. (*Hatchet* by Gary Paulsen and *Number the Stars* by Lois Lowry are two possibilities.) Choose a passage in which a character is faced with a difficult decision, and stop reading before that character makes a final choice. What do you think the character will do? What effects do you think the character's decision will have? Conclude by telling students that decision making is one of the skills explained in Chapter 2.

Cooperative Learning Project

"It's Your Choice": The Game Show

The Teen Health Digest on pages 44 and 45 provides students with high-interest articles related to the content of this chapter.

The material in the Teacher's Wraparound Edition presents suggestions for a class project in which students will plan and present a game show with teen "contestants" who have to make healthful choices in difficult situations.

---

in your journal

Before they read the Teen Chat Group conversation, show students a safety helmet and ask: How many of you wear this kind of helmet? If you don't wear one, what risks are you taking? After students have shared their own ideas, let volunteers assume the roles of the three pictured teens and dramatize their conversation. Help students recognize that the teens are discussing their own positions on assuming responsibility for themselves, on developing and maintaining good habits, and on making wise decisions.

Have students form small groups and discuss their responses to the In Your Journal questions. After these discussions, have students write their own private journal entries.

27

Lesson 1
Taking Responsibility for Your Actions

Focus

LESSON OBJECTIVES

After studying this lesson, students should be able to
- discuss what is involved in being a responsible person.
- explain how they can take responsibility for their own health.
- describe how their own actions may affect the health of other people.
- discuss the importance of understanding and accepting the consequences of their own actions.

MOTIVATOR

Write the following sentence on the board: *It's your responsibility.* Let students read the sentence silently and then write at least five phrases or sentences that it brings to mind.

INTRODUCING THE LESSON

Ask several volunteers to read aloud the Motivator sentence. Help students recognize that different emphasis and intonation can give the sentence various meanings. Then ask other volunteers to share the phrases or sentences they wrote. Ask: How do you feel about taking on responsibilities? What are some of the benefits and some of the challenges of being responsible?

INTRODUCING WORDS TO KNOW

Have students read the two Words to Know and use the Glossary to find the formal definition of each. Ask students to list other related concepts, referring to a dictionary if necessary.

Lesson 1
Taking Responsibility for Your Actions

This lesson will help you find answers to questions that teens often ask about showing responsibility. For example:

▶ What does it mean to be responsible?
▶ What responsibilities do I have for my own health?
▶ What responsibilities do I have for the health of others?

Words to Know

responsibility
consequences

Being Responsible

Do you think you are a responsible person? What does it really mean to be responsible? **Responsibility** is *the ability to make choices and to accept the results of those choices.* One aspect of being responsible involves living up to what is expected of you. For example, at home you may be expected to keep your room clean, wash the dishes, or help care for a younger brother or sister. At school you are expected to participate in class and turn in your homework on time. Showing responsibility also means helping others and being dependable and reliable.

As you grow up, you take on greater responsibility. You also gain more freedom. You have probably noticed that people expect more of you now than they did a few years ago. With this added responsibility usually comes greater freedom and the opportunity to make your own decisions. For example, your parents probably expect you to keep track of your own school assignments, but they may also allow you to make more of your own decisions about clothing purchases. They may expect you to be responsible for the way your friends behave when they are at your home, but they may also allow you to stay at home without supervision for longer periods than before.

It is easier to act responsibly when you believe that your efforts can make a difference. For example, suppose that you are upset with a decision your parents have made. You might choose to talk calmly with them instead of getting angry and slamming the door. You will be more likely to make this decision if you believe that talking will help them to understand your point of view. In many parts of your life, recognizing the benefits of responsible action will help you choose positive behavior.

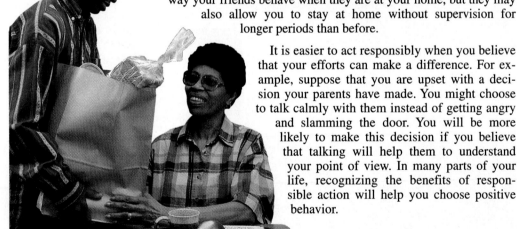

Sometimes you can see the positive results of your actions right away.

Chapter 2: Personal Responsibility and Decision Making

Lesson 1 Resources

Teacher's Classroom Resources
- 📁 Concept Map 5
- 📁 Enrichment Activity 5
- 📁 Lesson Plan 1
- 📁 Lesson 1 Quiz
- 📁 Reteaching Activity 5
- 🖨 Transparency 6

Student Activities Workbook
- 📁 Study Guide 2
- 📁 Applying Health Skills 5

Responsibility for Your Own Health

As a teen, you have an important responsibility: your own health. Accepting this responsibility means that you will make decisions that will promote your good health. These decisions involve

- your physical health.
- your habits.
- the activities in which you participate.
- your mental and emotional health.
- your social health.

interNET CONNECTION
Check cyberspace for ideas on how to manage habits, risks, and goals responsibly.
http://www.glencoe.com/sec/health

Healthful decisions will promote your overall wellness. Making these decisions shows that you take your own health seriously. With a high level of wellness, you will be able to make positive contributions to your family, friends, and community.

Making Positive Choices

To be responsible, you need to make positive choices. Every day you are presented with many options. Some decisions, such as what to eat for lunch, may seem routine. You know that other decisions, such as choosing not to smoke cigarettes, will have lifelong benefits. Whether they are large or small, however, all choices that affect your health are important.

As a teen, you can make wise choices that will have a positive effect on many areas of your health. Five of these areas are shown in **Figure 2.1**. In each of these areas, you will make better choices if you seek reliable information. For example, consider choices about diet. You could read a book on nutrition, search the Internet for nutritional information, or take an after-school class on preparing healthful snack foods. Applying what you learn to your everyday eating habits will help you stay healthier.

Figure 2.1
Making Healthful Choices
Can you think of other areas in which you could make positive health choices?

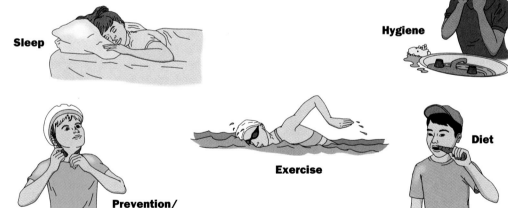

Sleep • Hygiene • Prevention/Risk Protection • Exercise • Diet

Lesson 1: Taking Responsibility for Your Actions   29

Consumer Health

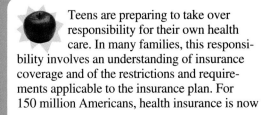

Teens are preparing to take over responsibility for their own health care. In many families, this responsibility involves an understanding of insurance coverage and of the restrictions and requirements applicable to the insurance plan. For 150 million Americans, health insurance is now provided by health maintenance organizations (HMOs), and this number is expected to grow. Encourage students to work in small groups to gather information about HMOs: When and how do they provide care? What restrictions may be involved? What are the costs? What are the advantages and disadvantages of HMOs?

Chapter 2   Lesson 1

Teach

L1 Discussing
The text poses this question: *Do you think you are a responsible person?* Ask students to ponder questions such as these: How can you recognize a responsible person? What actions and attitudes mark a person as responsible? Can anyone be a responsible person all the time? Why or why not? After a brief discussion, remind students that they should consider the text question privately.

Visual Learning
Tell students to describe and discuss the photograph on page 28. What action has the teen taken? What are the results of his action? How do you think his actions affect his relationship with the person he is helping? Guide students in discussing some of the mental and emotional benefits of responsible actions.

L1 Critical Thinking
Help students consider the relationship between increased responsibilities and increased freedom: Why do teens who accept more responsibilities also gain more freedom? What is likely to happen to a person with many responsibilities and very little freedom? To a person with few responsibilities and a great deal of freedom?

L1 Discussing
Have students describe and discuss each teen in Figure 2.1: What is the teen doing? What choice has the teen made? How is that choice going to affect his or her health now and in the future?

Chapter 2 Lesson 1

L1 Discussing
Help students consider the areas in which they make decisions to promote their own health. Ask: What specific choices do you make now about your physical health? (*when to bathe, brush teeth, wash hands*) About your mental and emotional health? (*what problems to discuss, when and how to get help*) About your social health? (*what friends to have, what activities to participate in*)

L1 Comprehending
Guide students in discussing their school responsibilities: What responsibilities do you have at school? In what sense are those responsibilities choices?

L1 Creating Posters
Ask students to identify short-term and long-term benefits of deciding not to use tobacco, alcohol, and other drugs. Record their ideas on the board. Have students work with partners or in small groups to create posters that incorporate these benefits in an anti-use message.

L2 Math
Have interested volunteers survey classmates about their responsibilities or their freedoms. Volunteers might ask what new freedom (such as staying home without supervision) each student has gained in recent months, or whether or not students have a particular responsibility (such as washing dishes every night). Once the volunteers have gathered information from 20 or more students, have them select the most appropriate graph (bar, line, or pie) and present their findings in this form.

in your journal

Make a list of your abilities. Put a star next to each ability that you think you could use to improve your health. For each ability you have starred, provide an explanation.

Did You Know?

Responsibility at Work

On April 27–29 of 1997, President Bill Clinton held the Presidents' Summit in Philadelphia. Appearing with former presidents George Bush and Jimmy Carter, Clinton promoted the idea of serving one's community as a volunteer. Volunteering is one way to use your growing freedom and responsibility to benefit everyone.

Knowing and Using Your Abilities

Another way in which you can take responsibility for your health is to use your abilities to their best advantage. Let's say, for example, that you often help prepare meals for your family. You can use your cooking ability to make sure that the meals taste good and are nutritious. If you are good at communicating, you can use this ability to join a peer counseling program. If you are naturally athletic, you can use this ability to teach younger neighbors specific sports skills.

How do you know what your abilities are? Think about the activities that you enjoy. Which ones do you think you are best at? Which ones do you receive compliments on? Ask family members and friends for their opinions. In what areas do they think you are skillful?

In addition to using your own abilities, you can also make use of the abilities of others. Everyone needs help and encouragement from other people. You have access to many people who can assist you. Look to the following people for advice, answers to questions, and help in managing aspects of your health:

■ Family members—parents, grandparents, aunts and uncles, older brothers and sisters

■ Educational leaders—teachers, guidance counselors

■ Health care professionals—pediatrician, your family's physician, school nurse, psychologist

■ Members of the clergy—priest, minister, rabbi, mullah

■ Other trusted adults—neighbors, scout leaders, coaches, friends of the family

LIFE SKILLS
Skills in Leadership

*A*s you get older, you will probably take on more leadership roles. For example, you might head a peer mediation team or help organize a fund-raiser for the American Heart Association. Skills in leadership will help you do these jobs effectively. What are the qualities of a good leader? Here are five key leadership skills:

▶ **Ability to communicate clearly.** Effective leaders make clear what is needed and how it might be done. Remember, however, that good communication involves listening as well as speaking.

▶ **Ability to identify the strengths of others.** Effective leaders understand that every person is different. They recognize what each person can contribute and work to encourage all.

▶ **Ability to be assertive.** Effective leaders are assertive—they stand up for themselves in a respectful way. Assertive people express their feelings politely, but without apologizing.

▶ **Ability to solve problems.** Good leaders look for solutions to problems. They search for solutions that will satisfy as many people as possible.

30 Chapter 2: Personal Responsibility and Decision Making

LIFE SKILLS
Skills in Leadership

Focus on Life Skills

Ask: What are some of the roles that student leaders assume in our school? In our community? Think about the teens and young adults whom you consider good leaders. What special qualities do they have? What kinds of contributions do they make?

Encourage students to recall experiences in which they have themselves been leaders. Remind them that everyone has opportunities for leadership and everyone can develop the skills of an effective leader.

After this introductory conversation, have students read and discuss the Life Skills feature.

Making Health Connections

Everyday life requires each individual to function well within various groups, both as a member and as a leader. Developing leadership skills can help students gain new self-confidence and will enhance both their mental/emotional health and their social health.

Responsibility for the Health of Others

Besides being responsible for your own health, you also have a responsibility for the health and well-being of others. Many of your actions and the choices you make affect other people. You might think of any action you take as a stone skipping across a pond. This concept is illustrated in **Figure 2.2**.

Figure 2.2
Actions Have Effects
Your actions may affect the health of others. Those effects may in turn create additional effects. Think about the following example.

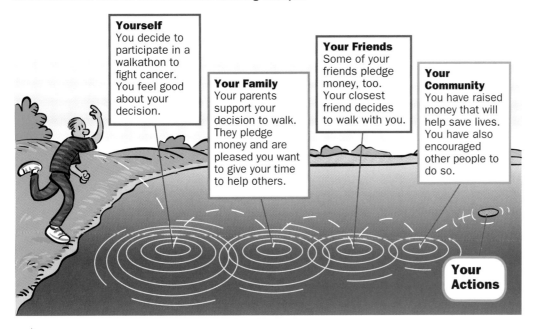

Yourself — You decide to participate in a walkathon to fight cancer. You feel good about your decision.

Your Family — Your parents support your decision to walk. They pledge money and are pleased you want to give your time to help others.

Your Friends — Some of your friends pledge money, too. Your closest friend decides to walk with you.

Your Community — You have raised money that will help save lives. You have also encouraged other people to do so.

Your Actions

▶ **Ability to control emotions.** Good leaders can communicate their feelings calmly. Everyone feels strong emotions such as fear and anger, but effective leaders are able to control the way they express these emotions.

Follow-up Activity
Identify a health-related activity that might be accomplished in your school or community. Write a short story about a person who takes a leadership role in that type of activity. Show how the main character displays at least three of the skills described in this feature.

Chapter 2 Lesson 1

L2 Presenting Skits
Prepare a set of index cards, each listing a positive health choice, such as *wear a seat belt, eat breakfast, choose smoke-free environments,* and *snack on fresh fruit.* Have students choose partners, and give each pair of students an index card. Ask each pair to plan and present a short skit in which one or two teens make the health choice identified on the card.

L1 Discussing
Discuss with students the importance of identifying and using their own particular abilities: Who is the best judge of your abilities? (*You are*) How do you think this choice will affect your mental/emotional health? Your social health?

Visual Learning
Guide students in using Figure 2.2 to understand the effects their actions can have on others: What is the initial action? How does it affect the person who takes it, as well as that person's family, friends, and community? Let students suggest a negative choice a teen might make. Have them refer to the picture as they explain the effects that choice would have. Repeat the process with another positive choice.

L1 Journal Writing
Have students write private journal entries, naming at least six adults they would feel comfortable turning to for help and encouragement.

Meeting Student Diversity
Students who feel different—or who are considered different by their peers—are likely to have difficulty assuming leadership roles. Emphasize to the class that differences in physical ability, for example, do not affect individuals' leadership potential. During various class and group projects, assign students to share leadership responsibilities equally; avoid letting a small group of capable students function as academic and/or social leaders. When students write their Follow-up Activity stories, encourage them to think of themselves as the leading character.

Assessing the Work Students Do
Encourage students to share their completed stories in groups or with the entire class. Meet with each student for a brief review of the information and a personal assessment of his or her learning. Assign credit based on students' participation, effort, and progress.

Chapter 2   Lesson 1

L1 Comprehending

To be sure students understand the term *consequences,* refer them again to Figure 2.2. What were the consequences of the initial action? Who experienced those consequences?

L1 Discussing

Ask students: How do you think Ryan felt when he realized that he had forgotten to mow the lawn? (*He may have felt sorry that he hadn't fulfilled his responsibility and worried about what the consequences might be.*) What do you think Ryan learned from the consequences of his forgetfulness? (*He learned to take care of his chores first, before going out with his friends.*) How do consequences help shape your behavior? (*Thinking about possible negative consequences can help you avoid irresponsible behaviors.*) Have the students explain their responses.

L1 Critical Thinking

Most consequences occur naturally; some are imposed by other people. Have students give examples of each. Ask: What are some benefits and drawbacks of each of these types of consequence?

L1 Brainstorming

Divide the class into two groups. Have the members of each group work together to brainstorm a list of things teens can do to protect themselves. One group can focus on protection from illness, and the other can focus on protection from injuries and accidents.

L2 Reporting

Ask a group of volunteers to compile a list of local classes where teens can learn about handling emergencies. These might include first-aid classes, babysitting classes, CPR classes, and others. Encourage the volunteers to include as much relevant information as possible, such as dates, times, and fees. Then have the volunteers duplicate the list for distribution to other students.

Your Total Health

What's Luck Got to Do with It?

Can you make your own luck? Psychologists believe that there is some truth in this idea. People who believe that bad things "just happen" to them, for example, are less likely to make an effort to improve their lives. One way to make your own luck is to learn from experience. If, for example, you seem to be involved in many arguments, stop and think about why this might be happening. Is it bad luck, or are you taking unnecessary chances—for example, by stopping to watch or take sides in a fight between classmates? By taking action to protect yourself, you may change your luck.

Living with Consequences

As you have just seen, your actions have effects on both yourself and others. Even when choices are made without thinking, they have **consequences,** which are *the effects or results of actions.* Consider Ryan's story. Ryan was expected to mow the lawn on Saturday. He was playing basketball at a friend's house, however, and forgot. Ryan's brother, Scott, got stuck mowing the lawn instead. Ryan's parents decided that he would have to take over one of Scott's responsibilities, which was to baby-sit their younger sister on Sunday. As a result, Ryan wasn't able to play in his softball game. His teammates had to find a last-minute replacement for him. Ryan's action (forgetting to mow the lawn) had consequences that affected several people. These people included Ryan, his brother, his parents, his sister, and his softball teammates.

Consequences may be positive or negative. In some cases, the same action can have both positive *and* negative consequences. Suppose that you and your friend both tried out for the lead role in the school play. Your friend gets the part—a positive consequence for her. You are happy for her (a positive consequence), but you are also personally disappointed (a negative consequence).

Accepting the consequences of your actions is an important part of becoming an adult. Being responsible involves both enjoying the positive consequences of your actions and accepting the negative ones. For example:

- Taking responsibility allows you to take credit for a positive outcome, but it also means that you may have to take the blame for a negative one.
- Taking responsibility means that you can be proud when you live up to your ideals, but that you will feel disappointment when you do not.
- Taking responsibility means that you can feel good when you make a friend happy, but that you will feel sad when you disappoint someone.

One of the benefits of taking responsibility is that it allows you to enjoy the positive feelings that come with success.

Chapter 2: Personal Responsibility and Decision Making

Health of Others

Teens should understand that they can help protect their own health, as well as the health of those around them, by limiting the volume of the music they play or listen to. Any exposure to sounds over 90 decibels is likely to present a risk of hearing damage. It is common for teens to play CDs or radios near this volume on a regular basis.

Even louder is the typical rock concert, where music may be played at or above 100 decibels. Experts recommend that concertgoers who will be near the stage take along earplugs. Experts also warn that the volume should be turned down anytime people outside a closed room or car can hear the music being played inside.

Factors You Can't Control

No matter how responsible you are, some things are beyond your control. Your heredity and your environment may be such factors. You may have a tendency toward certain diseases because of your heredity. Your environment may contain pollution or be prone to tornadoes. In addition, other people's decisions, actions, and errors are generally beyond your control.

Becoming a mature adult involves recognizing all that you can accomplish. It also includes realizing that you cannot control everything. Luck and chance, however, are not the deciding factors in your life. You can manage your life by following these rules:

- **Be careful.** First, do all you can to protect yourself against illness and accidents. Don't take unnecessary chances. For example, wear a bike helmet, practice good hygiene, and never drive with someone who has been drinking.

- **Be prepared.** If you know how to handle emergencies, you will be better able to deal with them if they occur. Know what to do if you or another person should become ill or experience an injury.

Taking a class in first aid is one way to show responsibility for yourself and others.

Review — Lesson 1

Using complete sentences, answer the following questions on a separate sheet of paper.

Reviewing Terms and Facts

1. **Vocabulary** Define the term *responsibility*. Give three examples of responsibilities that a teen might have at home.
2. **List** Name five areas of health in which you can make positive choices.
3. **Vocabulary** Explain the meaning of the term *consequences*. Then use the term in an original sentence.
4. **Identify** List three factors over which you do not have control.

Thinking Critically

5. **Explain** Describe a situation in which you saw positive consequences of an action you took.
6. **Evaluate** Suppose that you decide to join a soccer team. You will have practice several times a week and games on the weekends. Outline the possible positive and negative consequences of your decision.

Applying Health Concepts

7. **Health of Others** Make a poster that shows ways in which teens can make wise choices about their own health. Include at least one example from each of the following areas: physical health, habits, activities, mental and emotional health, and social health.

Lesson 1: Taking Responsibility for Your Actions 33

Personal Health

Teens cannot grow into emotionally healthy adults unless they are permitted and even encouraged to accept the consequences of their choices. Families and communities that try to protect children and adolescents from these consequences actually put all aspects of their health—physical, mental/emotional, and social—at risk. Students who have been excused from accepting responsibility for themselves are easy to identify; they're quick to assert that "It's not fair" or "It's not my fault" when a small problem develops. These students need to be encouraged to accept responsibility in appropriate steps, until they are in a position to acknowledge both the credit and the blame for their age-appropriate choices.

Chapter 2 Lesson 1

Assess

EVALUATING THE LESSON

Assign Reviewing Terms and Facts and Thinking Critically on page 33 to review the lesson; then assign the Lesson 1 Quiz in the TCR to evaluate students' understanding.

LESSON 1 REVIEW

Answers to Reviewing Terms and Facts

1. Responsibility is the ability to make choices and to accept the results of those choices. Examples will vary.
2. Five areas of health in which you can make positive choices are diet, exercise, sleep, hygiene, and prevention/risk protection.
3. Consequences are the effects or results of actions. Sentences will vary.
4. Factors over which you do not have control include (any three) heredity, environment, decisions of others, actions of others, and errors made by others.

Answers to Thinking Critically

5. Responses will vary.
6. Responses will vary.

RETEACHING

▶ Assign Concept Map 5 in the TCR.
▶ Have students complete Reteaching Activity 5 in the TCR.

ENRICHMENT

Assign Enrichment Activity 5 in the TCR.

Close

Have students write and complete the following sentence: *For me, the best part about becoming a more responsible person is...*

Lesson 2
Recognizing and Managing Habits

Focus

LESSON OBJECTIVES

After studying this lesson, students should be able to
- explain what a habit is.
- discuss how they can form healthful habits.
- discuss how they can change harmful habits.

MOTIVATOR

Ask students to list at least three positive health habits and at least three negative health habits that affect their short-term and long-term health.

INTRODUCING THE LESSON

Let volunteers read aloud their lists of health habits. Ask: What makes you think that behavior is a habit? What makes it a positive (or a negative) health habit?

Help students generalize about their lists: Which positive habits have been mentioned by several members of the class? Which negative habits have been included on many lists? Based on the lists you have written, what do you think a habit is? Conclude the discussion by telling students that Lesson 2 presents more information about health habits and how they can be developed or changed.

INTRODUCING WORDS TO KNOW

Have volunteers read aloud the Words to Know, and encourage students to use their own base of knowledge to suggest definitions for the terms. Then have students record each Word to Know, along with its Glossary definition, in their notebooks. Suggest that students add small mnemonic pictures beside the definitions.

34

Lesson 2
Recognizing and Managing Habits

This lesson will help you find answers to questions that teens often ask about personal habits. For example:

▶ What is a habit?
▶ How can I establish good health habits?
▶ How can I change harmful habits?

Words to Know

habit
life-threatening
life-altering

What Is a Habit?

A **habit** is *a pattern of behavior repeated frequently enough to be performed almost without thinking.* Everyone has habits. Some habits, such as hair twirling or finger tapping, are minor and do not affect your health. Others, such as daily bathing and regular exercise, form an important part of the responsible lifestyle that you create for yourself.

Because habits are such a normal part of your day-to-day life, you may not even be aware of them. Habits, however, grow in strength with frequent repetition. Good habits that are formed early in life can have lifelong benefits. Making a habit of eating plenty of fruits and vegetables, for example, can help you stay well and even help prevent certain diseases. On the other hand, harmful habits formed early in life can be difficult to change. That's why it is a good idea to start thinking about your habits now.

Although you may not realize it, your habits affect other people as well. Your harmless finger tapping may annoy a classmate. Suppose, on the other hand, that you make it a habit to be supportive of your friends or to greet people in a friendly manner. This behavior helps you socially because it enables you to be a good friend. It also helps other people by making them feel comfortable and relaxed.

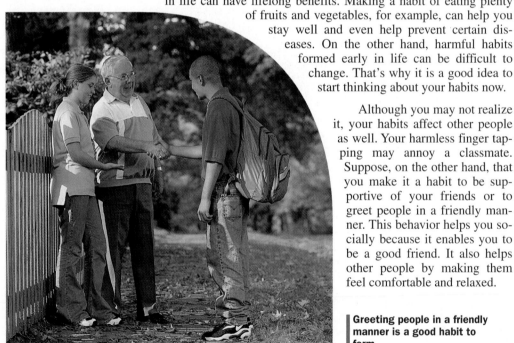

Greeting people in a friendly manner is a good habit to form.

34   Chapter 2: Personal Responsibility and Decision Making

Lesson 2 Resources

Teacher's Classroom Resources
- Concept Map 6
- Enrichment Activity 6
- Health Lab 2
- Lesson Plan 2
- Lesson 2 Quiz
- Reteaching Activity 6
- Transparency 7

Student Activities Workbook
- Study Guide 2
- Applying Health Skills 6

Forming Healthful Habits

The key to establishing good health habits is repetition. The more often you perform an action, the more natural it will seem. Soon the action will become a part of your everyday life—something you don't even think about anymore. **Figure 2.3** illustrates how to form a healthful habit, such as exercising for 20 minutes each day.

Figure 2.3
Developing a Habit
Here are the steps you can take to form a healthful habit.

1. Choosing
The first step is to decide what you want to achieve. Having a clear idea of the end result you want will help you reach it. Wanting to be in better shape or to become skilled at a sport are examples of good goals.

2. Remembering
At first, you may have to make a real effort just to remember to perform the action. Reminders from parents, teachers, and friends can be helpful. Making your own reminder list or schedule can also keep you on target.

3. Performing
The next step is to make an effort to do a good job of performing the action. This can take time and energy. You need to make a commitment to yourself to stick with it.

4. Repeating
As time passes, you get used to remembering and performing the action. It becomes more and more natural. You may begin to plan around your exercise sessions. You know that you are doing something good for your health.

5. Growing
Your overall feeling of well-being increases. In addition, your confidence grows because you know that you are in control of this area of your health.

Lesson 2: Recognizing and Managing Habits   **35**

Personal Health

A common teen habit may represent a double threat to health. Watching too much TV prevents teens from involving themselves in activities or social interactions; it also exposes teens to levels of violence that may influence their values and their concepts of appropriate behavior. Repeated studies have shown that watching violent acts on TV results in an immunity to the horrors of violence. As a result, children and teens become unable to comprehend the drastic consequences of real-life violent behavior. TV violence also appears to increase general levels of aggressiveness among children and teens, even when violence is not a part of their home and school environment.

Chapter 2 Lesson 2

Teach

L1 Discussing
Guide students in discussing a wide range of habits: Which habits, that appear to be minor, could have a devastating effect on one's health or on the health of others? (*leaving skates in a walkway or on stairs where someone could trip*) What are some of the health habits you learned in childhood? How do those habits affect your life now?

L1 Identifying Examples
Let students work together to generate a list of habits that can help teens achieve physical and mental/emotional health. (*Exercise each day; use time management skills; learn to control anger.*) Ask: Which of these habits do you already have? Which do you think you should try to develop?

L1 Discussing
Encourage students to discuss the people in the photograph on page 34: How does the teen's friendly greeting make his friend and her grandfather feel? How do you think the teen feels?

Visual Learning
Use Figure 2.3 to help students discuss and comprehend the steps in forming a healthful habit: What is the teen in the picture doing? What choice has she made? What effort has she made to remember? How is she performing the action she wants to make a habit? How will repeating that action help make exercising a habit? How do you think this habit will help the teen grow?

L2 Applying Information
Have each student select one healthful habit he or she would like to develop. Then, using Figure 2.3 as a guide, have students write an action plan for developing that habit.

Chapter 2 Lesson 2

L1 Identifying Examples

Ask students to suggest harmful habits. Record their suggestions on the board. Ask: Which of these are life-threatening habits? Why? Which are harmful without being life threatening?

L3 Researching

Have a group of volunteers research addictions: To what kinds of substances can people become addicted? What are the effects of these addictions? What percentage of addicted people are able to recover? Ask these volunteers to prepare a chart or poster presenting their findings.

L1 Identifying Examples

Have students work in four groups to generate lists of harmful—but not life-threatening—habits that teens might want to change. Ask each group to focus on one of these areas: diet, exercise, hygiene, or accident prevention.

L2 Reporting

What local organizations provide help in breaking life-threatening habits, such as smoking, drinking alcohol, or abusing drugs? Let several volunteers compile a list and share it with the rest of the class.

L2 Math

Have each student select a positive health habit, and survey 20 people about that habit. How many have practiced the habit for two or more years? How many have developed the habit over the past two years? How many do not practice the habit? Ask students to draw an appropriate graph presenting the results of the survey.

Cultural Connections

Getting Accustomed ACTIVITY!

A custom is similar to a habit, except that it is usually associated with a cultural group. In the United States, for example, it is common to shake hands when you are introduced to a person. In other societies, such as Japan, the custom is to bow.

Look in a reference source to find out about the customs of other cultures that affect health. Share your findings with the class.

Changing Harmful Habits

Everyone has habits that they would like to break. Perhaps you interrupt people when they are talking, or you put off doing your homework until the last minute. Although these behaviors may be bothersome to you or to others, they are not life-threatening. A **life-threatening** habit is one that *may cause death*. Riding in a car without fastening your seat belt is a life-threatening habit. Using tobacco, drinking alcohol, and taking other drugs are also life-threatening habits. They can also become addictions. With an addiction, a person develops a physical or psychological need for a substance.

It is important to learn how to break habits that may threaten your health or life. If you have developed habits that could become a serious problem, such as smoking, drinking, or using drugs, you should talk to a parent or another trusted adult. It is very difficult to break this type of habit on your own. You can, however, work toward changing other, less serious behaviors. Here are some tips to follow.

- **Identify the habit.** If you don't know what your harmful habits might be, ask a family member or friend. If you have more than one harmful habit, choose the one that you would most like to change. Trying to change all of your habits at once may cause you to become frustrated and give up.

- **Analyze the habit.** Think about your habit. Why do you do it? When do you do it? Try to change the cause of the behavior. Maybe you don't wear your bicycle helmet because you are always in a rush. If you start allowing yourself a few extra minutes, you will have time to put on your helmet.

HEALTH LAB
The Effects of Repetition

Introduction: The more often you perform a task, the easier it is to do. The principles demonstrated in this simple experiment can be applied to forming and practicing healthful habits.

Objective: Work with a partner to repeat a difficult task until it becomes easier. In this way, you will see how habits can be formed.

Materials and Method: Each set of partners will need a stopwatch or a watch with a second hand, sheets of unlined paper, pencils, and a mirror. On separate sheets of paper, each of you should draw a simple maze using parallel lines.

Photocopy or trace each maze on four more sheets of paper so that each partner has five copies of the same maze.

Put one copy of your maze on a desk. Hold a small mirror in front of the maze while your partner sits opposite you. Experiment with tilting the mirror until your partner can see the maze in the mirror. He or she should then solve the maze while looking at the mirror image only. (Make sure that the person is not looking down at the maze itself.) Use the stopwatch or watch to see how long it takes your partner to trace the maze correctly with a pencil.

36 Chapter 2: Personal Responsibility and Decision Making

HEALTH LAB

Effects of Repetition

Time Needed

1 class period

Supplies

▶ stopwatch or watch with a second hand
▶ unlined paper
▶ pencils
▶ medium-size mirror

Focus on the Health Lab

Encourage students to share their own experiences in establishing positive habits: What was the habit? What efforts did you make to develop that habit? How long did it take for the particular action or activity to become a habit? Also, encourage them to discuss their efforts to break negative habits: Which is harder to change, a relatively new habit or a long-established one? Why? After this introductory discussion, let students choose partners and read the Health Lab feature together.

Understanding Objectives

Students will recognize that repeating an action

- **Consider the future.** Ask yourself this question: Could this habit eventually hurt me or someone else? Thinking about the possible consequences may help you realize the importance of kicking the habit now.
- **Set goals.** Set a short-term goal. You might decide, for example, that you will eat a healthful breakfast every morning. At the end of several weeks, reward yourself if you have reached your goal. As each week goes by, it will be easier to follow your new, healthful habit.
- **Ask for help.** Perhaps you can ask a friend to give you a secret signal when she sees you biting your nails. Maybe your parents can avoid keeping candy in the house, so that you will not be tempted to eat it. Allow others to offer their encouragement.
- **Find a substitute behavior.** If you usually eat chips or ice cream when you are doing homework, try snacking on apple slices or carrot sticks instead. Over time, you will associate these more healthful foods with homework time.
- **Be patient.** Keep in mind that you have probably been practicing your harmful habit for awhile. It will take time to break the habit. If you slip back into an old behavior, don't be too hard on yourself. Just get back on track and try not to slip again!

in your journal

Are there any habits that you would like to develop to improve your health? Perhaps you want to establish a regular exercise program or eat more nutritious foods. In your journal, describe the healthful habit you would like to have, why you would like to have it, and how you plan to develop it.

Finding a substitute activity can help you form a new, more healthful habit.

Repeat the test with the other four copies of the maze, keeping track of the time it takes to complete each test. Then switch places so that you are completing your partner's maze while he or she is holding the mirror and timing you.

Observations and Analysis:
After you and your partner have completed your tests, discuss your findings with the class. Was the maze easier to finish after you had done it several times? Did you complete it faster each time? What does this experiment suggest to you about how repetition can help you establish good habits?

Lesson 2: Recognizing and Managing Habits 37

Chapter 2 Lesson 2

L1 Discussing
Help students discuss the importance of focusing on short-term goals: What problems are you likely to face if you set only long-term goals?

L1 Discussing
Guide students in discussing the role a positive attitude plays in changing habits: Why is it important to be patient with yourself when you're trying to change a harmful habit? How do you think you could best support a friend who was trying to change a habit?

L3 Researching
Have volunteers use library resources to find out about compulsive behaviors: What causes these behaviors? How is this disorder treated? Let the students summarize their findings in written reports.

VIDEODISC/VHS

Teen Health Course 2
You may wish to use video segment 2, "Personal Responsibility," to show students some habits that will benefit their health throughout their lives.

Videodisc Side 1, Chapter 2
Personal Responsibility

Search Chapter 2, Play to 3

Would You Believe?
What habits bother the parents of teens? In a recent survey, the top-rated negative habit among teens was failing to keep their rooms clean.

---

makes doing that action easier and more natural. This activity will help them appreciate the importance of consciously repeating behaviors and attitudes that they want to make into personal habits.

Observation and Analysis
Go over the specific activity instructions with the class to be sure everyone understands them. Then have the pairs work independently, offering specific assistance or correction as needed. When the class discusses the results of the lab, encourage students to be specific in their responses, and allow several volunteers to describe their experiences.

Further Investigation
Suggest that each student talk with a parent or other adult family member about a habit that the adult has changed recently: What habit did the adult want to change? Why? What approach did the adult use?

Assessing the Work Students Do
Let students write short summaries of how this Health Lab activity helps them understand the importance of repetition in building positive habits.

37

Chapter 2 Lesson 2

Assess

EVALUATING THE LESSON
Assign Reviewing Terms and Facts and Thinking Critically on page 38 to review the lesson; then assign the Lesson 2 Quiz in the TCR to evaluate students' understanding.

LESSON 2 REVIEW

Answers to Reviewing Terms and Facts
1. Responses will vary.
2. The five steps to developing a healthful habit are choosing, remembering, performing, repeating, and growing.
3. Tips for changing a harmful habit include (any five) identifying the habit, analyzing the habit, considering the future, setting goals, asking for help, finding a substitute behavior, and being patient.
4. Life-altering habits are capable of changing a person's day-to-day existence; life-threatening habits may cause death.

Answers to Thinking Critically
5. Responses will vary.
6. Responses will vary.

RETEACHING
▶ Assign Concept Map 6 in the TCR.
▶ Have students complete Reteaching Activity 6 in the TCR.

ENRICHMENT
Assign Enrichment Activity 6 in the TCR.

Close

Let each student answer this question, either orally or in writing: What is the most important fact about habits that you learned from reading this lesson?

Going Too Far
Even good habits can become harmful. People sometimes become too strongly focused on a certain behavior. For example, a person may be overly concerned with cleanliness and germs. That person might wash his hands over and over again or take many showers in one day. Another person may become so worried about getting cavities that she brushes her teeth until her gums bleed. A focus that is normal for one teen may become harmful for another. For instance, teens are typically concerned about their looks. Sometimes, however, a teen becomes so anxious about weight loss and exercise that little else seems important. Such an intense focus can lead to serious health problems.

Don't be afraid to ask for help in breaking a habit that you think may be taking over your life. A counselor will listen to your concerns and be ready to help.

Actions that begin as habits sometimes take control of a person. The individual may not feel able to change the habit. In these cases, the habit can become **life-altering,** or *capable of changing a person's day-to-day existence.* If you think that you or a friend might have a life-altering habit, it is best to talk to a parent or another trusted adult. In many cases, professional help is needed to break the habit.

Lesson 2 Review

Using complete sentences, answer the following questions on a separate sheet of paper.

Reviewing Terms and Facts
1. **Give Examples** List two examples of healthful habits and two examples of harmful habits.
2. **Identify** List the five steps to developing a healthful habit.
3. **List** Identify five tips for changing a harmful habit.
4. **Vocabulary** Explain the difference between *life-altering* and *life-threatening* habits.

Thinking Critically
5. **Hypothesize** Why might it be a good idea to examine your own habits?
6. **Analyze** What kinds of harmful habits do you think are most common among people your age? Choose two of these habits and explain how they could negatively influence a person's future health.

Applying Health Concepts
7. **Consumer Health** Examine products (in a store or in magazine advertisements) that claim to help people stop harmful habits such as nail biting or overeating. Are the claims believable? How do the products say they will work? Do you think that the products could be harmful? Write a short report about what you have found.

38 Chapter 2: Personal Responsibility and Decision Making

Growth and Development

Young teens are in the midst of establishing and developing many different habits. Among these should be positive spending and saving habits. American teens have a large amount of disposable income: more than $100 billion, which comes from jobs, allowances, and other sources. This is in addition to the billions families spend on or for their teen members. Teens should be encouraged to learn basic economic strategies, including comparison shopping, goal setting, contributing to savings plans, and making basic investments.

Health Risks and Your Behavior

Lesson 3

This lesson will help you find answers to questions that teens often ask about taking risks. For example:

- What is a risk behavior?
- How do I know if something is really a risk?
- What are cumulative risks?
- How can I avoid unnecessary risk behaviors?

Recognizing Risks

Risks—the chances that something harmful may happen—come in many forms. You are probably aware of the risks associated with using alcohol and other drugs. Did you know, however, that you also create health risks by eating foods that are high in fats and by not getting enough exercise? An important part of taking responsibility for your health is avoiding risk behaviors. As you learned in Chapter 1, a risk behavior is an action or choice that may cause injury or harm to you or to others. Harmful habits usually include several risk behaviors.

Health risks may have a variety of consequences. Some consequences affect only you and may not be especially dangerous. If you skip breakfast one morning, for example, you will probably feel hungry and have little energy. However, your actions will most likely not have major consequences for you or anyone else. Some risk behaviors, on the other hand, carry very undesirable consequences. These consequences may include serious injury and **mortality** (mor·TA·luh·tee), or *death*.

Consider Ben's story. One of his friends invited him to go in-line skating. Ben was showing off, and he lost his balance and shattered several bones in his wrist. Because of his risk behavior, Ben suffered serious consequences:

- **Physical consequences.** Ben now has a badly broken wrist. It was painful when it occurred and for some time afterward. He will have to wear a cast for many weeks.

Words to Know

mortality
subjective
objective
hazard
precaution
cumulative risks

You should pay attention to warnings that tell you about how to avoid risk behaviors.

Lesson 3: Health Risks and Your Behavior

Lesson 3 Resources

Teacher's Classroom Resources
- Concept Map 7
- Decision-Making Activity 3
- Enrichment Activity 7
- Lesson Plan 3
- Lesson 3 Quiz
- Reteaching Activity 7
- Transparency 8

Student Activities Workbook
- Study Guide 2
- Applying Health Skills 7

Lesson 3
Health Risks and Your Behavior

Focus

LESSON OBJECTIVES

After studying this lesson, students should be able to

- explain what risks are and how they can be identified.
- discuss the difference between subjective information about risks and objective information about risks.
- identify attitudes and actions that will help them avoid taking risks.
- discuss the importance of acting responsibly to reduce the risks in their lives.

MOTIVATOR

Have students sketch pictures of one or more teens in an unsafe, high-risk situation.

INTRODUCING THE LESSON

Let students share and discuss their sketches. Ask: What unsafe situation are the teens in each picture facing? How do you think they got into this high-risk situation? What do you think is the safest way for them to get out of the situation? After all the students have had a chance to contribute, explain that Lesson 3 presents more information on risks and includes facts about identifying and avoiding risk behavior.

INTRODUCING WORDS TO KNOW

Guide students in reading aloud the Words to Know, and ask volunteers to find and read aloud the definitions provided in the lesson. Have students work with partners to write original sentences using four of the Words to Know.

39

Chapter 2 Lesson 3

Teach

L1 Discussing

Help students consider how to evaluate risks: What risks did you face in walking or riding to school this morning? What unreasonable risks did you avoid taking on the way to school?

L1 Identifying Examples

Let students suggest specific risk behaviors with consequences that include mortality. Then ask: Why do you think people engage in these risk behaviors?

L1 Brainstorming

After discussing the photograph on page 39, divide the class into four or five groups. Have the groups list as many other warning signs as they can. After the groups have shared and compared their lists, lead a discussion of ap-propriate responses to those signs.

L1 Discussing

Discuss with students the risks that Ben took: What could Ben have done to reduce his risk of injury? Do you think he still would have enjoyed in-line skating with his friend? What do you imagine Ben has learned from the consequences of this risk-taking?

Visual Learning

Ask students to study the graphs in Figure 2.4: Which risk behavior did the highest percentage of ninth-graders engage in? Why do you think so many took that risk? What are the possible consequences of each of these risk behaviors? How could peer pressure help reduce the number of teens who take these risks?

L2 Critical Thinking

Have students write paragraphs explaining their responses to this question: Why is committing any kind of crime a high-risk behavior?

in your journal

Have you ever chosen to engage in a risky recreational activity? Describe the risk in your journal. Did the risk have physical, academic, social, or financial consequences? Explain how you could avoid taking that risk in the future.

- **Social consequences.** Ben will have to sit out the rest of the basketball season. His team, on which he plays forward, is disappointed.
- **Mental and emotional consequences.** Ben feels depressed about not playing basketball and sorry about disappointing his team. He feels foolish about showing off.
- **Academic consequences.** Ben had to take several days off from school after he broke his wrist. He did poorly on several tests because of the work he missed.
- **Financial consequences.** Ben's parents had to pay for expensive medical care. Their insurance covered most costs, but some were not covered.

In addition, some risks are illegal. Suppose, for example, that a person climbs over a fence to get into a construction site. That person is taking a health risk by being in a dangerous area. He or she is also committing a crime by breaking into private property. Legal consequences may result.

Who's Taking Chances?

Are teens taking risks? The answer seems to be yes. A survey sponsored by the Centers for Disease Control (CDC) asked more than 10,000 high school students about risk behaviors they had engaged in, either at one time or within a specified period preceding the survey. **Figure 2.4** shows some of the results for ninth graders—the youngest group of students surveyed.

Figure 2.4
Youth Risk Behavior, 1995

These graphs show the percentage of ninth graders surveyed who had engaged in these risk behaviors. How could taking each of these risks harm a person's health?

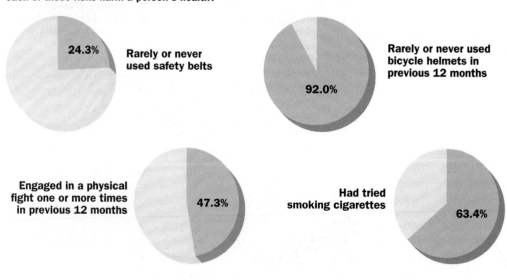

Teens and Risk We all face various risks as part of our everyday lives. One common risk is the possibility of being involved in an automobile accident. For teens, however, the risk rises dramatically, as the following figures show:

▶ Among children 5 to 15 years of age, 5.9 percent of deaths are caused by motor vehicle accidents.

▶ Among teens and young adults 16 to 24 years of age, the figure rises to 30.8 percent.

▶ Among adults age 25 and up, the rate drops to 16 percent.

Young teens should recognize these risks. They can begin acting responsibly by learning about safe behaviors and habits for drivers and passengers and by avoiding risky situations.

Is It Really a Risk?

Why do teens take risks? Some young people take risks because they believe that nothing bad can happen to them. Others may question whether certain actions are really risks. They may believe, for example, that people who are involved in accidents are just unlucky. This idea is **subjective,** which means that it *comes from a person's own views and beliefs, not necessarily from facts.*

When considering risks, it is much better to use **objective** information, which is *based on facts.* The following is an example of subjective versus objective thinking.

- Subjective (involving a person's own views): "Lots of people smoke cigarettes, so how can it be harmful?"
- Objective (based on facts): "Smokers are ten times more likely to get lung cancer than nonsmokers."

Objective information can help you act responsibly so that you can prevent many injuries and illnesses. Looking out for **hazards,** or *potential sources of danger,* is a good habit to develop. Hazards occur in many different forms, including high levels of dietary fat, the violent behavior of others, and alcohol use. If you are aware of hazards, you can take precautions against them. A **precaution** is *an action taken to avoid danger.*

Some hazards or risk factors by themselves may not seem very dangerous. If, for example, your blood pressure is a little high, you may not be especially worried about your health. However, if you have high blood pressure and you are overweight, your risk for heart disease increases. If you don't get much exercise, your risk increases even more. The greater the number of risk factors you have, the more likely you will be to develop a particular disease.

Groups of risks like the ones just mentioned are called **cumulative** (KYOO·myuh·luh·tiv) **risks.** These are *related risks that increase in effect with each added risk.* Cumulative risks are associated with injuries as well as diseases. Riding a bicycle without a helmet, for example, is *one* risk factor. If you are also riding on a busy street (second risk factor), and it is raining (third risk factor), your chance of serious injury increases greatly.

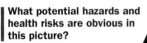
What potential hazards and health risks are obvious in this picture?

Teen Issues

Targeting Teens ACTIVITY

The Food and Drug Administration (FDA) has been concerned for years about the effects of tobacco product advertising aimed at children and teens. Their research suggested that such advertisements encouraged young people to develop the habit of smoking, which caused them to face long-term health risks. A regulation that took effect in August 1996 prohibited the sale or giveaway of such items as caps, jackets, or gym bags that carry tobacco brand names or logos. The regulation also prohibited tobacco brand name sponsorship of sports teams and sporting events. Do you think giveaways encourage young people to risk their health by smoking? Defend your opinion in a paragraph.

Growth and Development

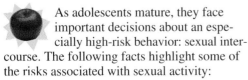

As adolescents mature, they face important decisions about an especially high-risk behavior: sexual intercourse. The following facts highlight some of the risks associated with sexual activity:

- Each year, more than 10 percent of female teens between the ages of 15 and 19 become pregnant.
- Of the teens who become pregnant a first time, nearly 20 percent will become pregnant again within a year.
- Nearly all teen pregnancies are unplanned.
- Every year, 3 million teens contract an STD.
- It is believed that 1.5 million Americans are infected with HIV; many of these individuals are teens.

Chapter 2 Lesson 3

L1 Journal Writing
Ask students to write private journal entries in response to these questions: Which of the risk behaviors in Figure 2.4 do you consider most dangerous? What do you plan to do to avoid that behavior? Which of the risk behaviors do you engage in or consider engaging in? What do you think you should do to alter your risk-taking behavior?

L2 Applying Knowledge
Let students work with partners or in small groups to record two subjective reasons teens might give for engaging in each risk behavior shown in Figure 2.4. Then have them list at least two objective reasons for avoiding each risk behavior.

L1 Discussing
Encourage students to explain, in their own words, the meaning and the use of the words *subjective* and *objective.* What is the difference between a subjective view and an objective view? Which is more reliable? Why? In what situations is it reasonable to rely on your own subjective view? Why?

L2 Language Arts
Encourage students to discuss nonfiction books or articles they have read about high-risk adventures, such as sailing alone around the world or climbing Mount Everest. Why did these adventurers take such risks? What consequences did they accept? What comments did they make on the risks they took?

L2 Guest Speaker
Ask a family nurse practitioner or other health care provider to speak with the class about cumulative risks.

Visual Learning
Challenge students to identify as many hazards and health risks as possible in the photograph on page 41. What cumulative risks can they identify?

Chapter 2 Lesson 3

L1 Discussing
Have volunteers describe the teen in Figure 2.5: How is she protected? Guide students in reading and discussing each of the attitudes and actions that can protect them against risk-taking.

L1 Brainstorming
Let the class brainstorm a list of hazards that teens face. Record students' ideas on the board. Then ask students to identify and discuss precautions they can take to protect themselves from those hazards.

L2 Language Arts
Assign two or three students to find the meaning of the word *invulnerability* and to read about the connection between adolescence and a sense of invulnerability. Let these students explain their findings by preparing and presenting a short skit demonstrating the connection between teens' sense of invulnerability and risk behaviors.

L1 Applying Life Skills
Let volunteers demonstrate how they might resist pressure from others to engage in risk behaviors. Ask several students to persist in urging each volunteer to join in a specific behavior, and encourage the volunteer to demonstrate effective refusals.

L2 Creating Ads
Have students work in small groups to plan, write, and illustrate magazine-style ads urging teens to avoid risk behaviors.

L1 Critical Thinking
Ask students to consider the kinds of support teens need to resist risk behaviors: How can friends help one another make responsible choices about avoiding risks? What role should parents play in supporting their teens' responsible choices? What do you think teachers and other school staff members can do to encourage safe, responsible behavior?

Avoiding Risks

The best way to prevent the problems associated with risks is to act safely and practice healthful habits. For example, although riding in a car can be risky, choosing to ride with a responsible driver and to wear a seat belt reduces many of those risks. Although walking home from school could put you at risk of physical attack, choosing a safe route and walking with a friend makes such problems much less likely. **Figure 2.5** presents other ways to avoid risks. Getting in the habit of asking yourself these questions will help you make decisions that are good for your health.

Figure 2.5
How to Avoid Risks
Think of each of these attitudes and actions as a piece of armor to help protect you against risks that could threaten your health and your life.

A Resist pressure from others. Am I doing this because I really want to or because someone else wants me to? If I think that this is dangerous, have I expressed that thought clearly?

B Stay away from risk takers. Who is involved? Are these people who regularly take part in dangerous behavior?

C Pay attention to what you are doing. Is my mind on something else? Am I feeling tired, upset, or rushed?

D Know your limits. Can I really do this without hurting myself? Is this within my abilities? Am I in control of the situation?

E Consider the consequences. What is likely to happen to me or to someone else if I do this? Am I prepared for the possible consequences?

F Consider other options. What else can I do instead? Would a different choice be safer?

Chapter 2: Personal Responsibility and Decision Making

Home and Community Connection

Firearms in the Home Family-owned guns represent a serious risk to the health and safety of adolescents and children. Each day in this country, ten teens and children are killed by handguns; these deaths result from suicides, homicides, and accidents. Many more teens and children are wounded in such incidents.

The best way to avoid the risks posed by firearms is to keep guns of all kinds out of the home. Students should be reminded that if they encounter a gun—at a friend's home or anywhere else—they should not touch it. If anyone else handles it, they should leave the situation.

Managing Your Life

Part of being responsible means taking an active role in managing your own health. You can still have fun and enjoy life while considering your own safety and the safety of others. After all, most accidents and injuries don't "just happen," and they are not caused by luck or fate. If you act responsibly and avoid risks, you can prevent most injuries. By practicing good health habits, you can prevent many illnesses.

in your journal

Think about the ways to avoid risks that you learned in this lesson. In your journal, make a list of the health and safety rules that you already follow. Then make a list of the risk behaviors that you would like to change. Describe how you will change just one of these behaviors.

Learning how to play a sport correctly is one way to decrease risks.

Review — Lesson 3

Using complete sentences, answer the following questions on a separate sheet of paper.

Reviewing Terms and Facts

1. **Identify** List five types of consequences that may result from risk behavior.
2. **Compare** What is the difference between *subjective* and *objective* information about risks?
3. **Vocabulary** What is the difference between a *hazard* and a *precaution?* Give one example of each.
4. **Vocabulary** Define the term *cumulative risks*. Then give an example of cumulative risks.
5. **Recall** Identify four ways to avoid risks.

Thinking Critically

6. **Predict** What might the consequences be if a person went for a boat ride and did not wear a life jacket?
7. **Describe** How do risk factors become cumulative risks?

Applying Health Concepts

8. **Growth and Development** With one of your classmates, plan a skit for younger children. Show them how they can reduce risks while doing the following: crossing the street, bicycling, and riding in a car.

Lesson 3: Health Risks and Your Behavior 43

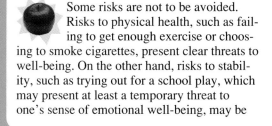

Some risks are not to be avoided. Risks to physical health, such as failing to get enough exercise or choosing to smoke cigarettes, present clear threats to well-being. On the other hand, risks to stability, such as trying out for a school play, which may present at least a temporary threat to one's sense of emotional well-being, may be an essential component of growth. All change involves some risk: Am I brave enough to try to make a new friend? Do I have the courage to set a high goal for myself and work hard to reach that goal? Guide students to recognize the positive risk involved in change and growth.

Chapter 2 Lesson 3

Assess

EVALUATING THE LESSON

Assign Reviewing Terms and Facts and Thinking Critically on page 43 to review the lesson; then assign the Lesson 3 Quiz in the TCR.

LESSON 3 REVIEW

Answers to Reviewing Terms and Facts

1. Consequences are physical, social, mental/emotional, academic, and financial.
2. Objective information is based on facts; subjective information is based on a person's own views and beliefs.
3. A hazard is a potential source of danger; a precaution is an action taken to avoid danger. Examples will vary.
4. Cumulative risks are related risks that increase in effect with each added risk. Examples will vary.
5. Ways to avoid risks include any four listed in Figure 2.5.

Answers to Thinking Critically

6. The consequences might be serious injury or death from drowning.
7. Risk factors become cumulative risks when a person has more than one risk factor that can lead to a particular disease or type of injury.

RETEACHING

▶ Assign Concept Map 7 in the TCR.
▶ Have students complete Reteaching Activity 7 in the TCR.

ENRICHMENT

Assign Enrichment Activity 7 in the TCR.

Close

Ask students to list five hazards they face in their everyday lives and five risks they intend never to take.

Focus

The Teen Health Digest articles can be used in two ways: as an individual activity for reflection and enrichment or as a cooperative learning activity as described below.

Motivator

Guide the class in a brief discussion of the choices teens face and some of the healthful decisions they have made. Have students read the Teen Health Digest articles, either independently or in small groups. Ask: What choices are presented in these articles? What healthful decisions are described? How are these decisions relevant to your own life?

Cooperative Learning Project

"It's Your Choice": The Game Show

Setting Goals

Students will work together to create, develop, and put on their own game show called "It's Your Choice." They will plan a format for the show, prepare questions, create a game show set, and make prizes to be awarded to the contestants.

Delegating Tasks

All the students will discuss and evaluate possible formats for the game show. They will work together to select the format they want to use. Then the class will divide into four groups to work on these aspects of the show: writing questions; making a simple set for the show; collecting or creating prizes; preparing to serve as hosts or be contestants.

TEEN HEALTH DIGEST

Sports and Recreation
A Camp for Kids

Andrea Jaeger became a professional tennis player when she was only 14 years old. In the early 1980s, she was one of the world's top ten female players. After 1984, however, a shoulder injury caused Jaeger's tennis game to decline.

In 1989, Jaeger was hit by a drunk driver. While recovering from the accident, she had time to think about her life. She realized that visiting sick children had always been rewarding to her. That was when Jaeger decided to start the Kids' Stuff Foundation. Through it, she opened a camp for children who have life-threatening illnesses.

Each year, 100 children spend a week at the camp in Aspen, Colorado. They take part in such activities as horseback riding, white-water rafting, and crafts. At camp, the kids can forget about their problems and just have fun. Making the children happy makes Andrea Jaeger happy, too.

Myths and Realities

Taking Control of Your Health

Q. Two of my grandparents have diabetes. That means I'll get it too, right?

A. Because diabetes runs in your family, you do have a higher risk than many other people. However, healthful habits can improve your chances of avoiding the disease.

- **Watch your weight.** Many people who have diabetes are overweight. By maintaining a healthy weight, you can reduce your risk.
- **Eat healthful foods.** A low-fat diet rich in fruits, vegetables, and grains will help keep you healthy.
- **Exercise regularly.** Studies show that regular exercise helps prevent diabetes. Besides helping to control weight, exercise lowers blood sugar levels.
- **Don't smoke.** Smoking cigarettes increases your risk of diabetes.

TECHNOLOGY OPTIONS

Computer Students with computer skills and access to a word processing program can compose and revise game show questions on the computer. Working on the computer allows students to review one another's work and ideas. It also enables students to edit, change, and reorganize their questions easily.

Video Recorder If the class has access to a video camera, students may want to tape the final production of their game show. The tape will make it possible to share the project with a wider audience, including family members, other students, and faculty and staff members unable to attend the "live show."

Teens Making a Difference

Y.O.U.T.H. Matters

Helping others is important to Lenti Smith. At an early age, this teen realized that she could do more with an organization than she could by herself. Smith and some friends started Youth Organization Unites to Help (Y.O.U.T.H.).

Y.O.U.T.H has sponsored many projects to help others. The group has visited elderly people in nursing homes and recorded books on tape for people who are visually impaired. Group members have also collected canned goods for hurricane victims and picked up litter.

Try This: List some projects in your community that could be done by a group. Share your ideas with your class, and consider making one a class project.

HEALTH UPDATE

Is Your Heart at Risk?

Do you enjoy foods such as fast-food cheeseburgers and milkshakes? If you think that you don't have to worry about your diet until you get older, think again. Recent studies show that young people with diets high in fatty foods increase their risk of heart disease in adulthood. Changing bad eating habits now can help you lead a longer and healthier life.

Try This: Divide a sheet of paper into two columns. In the first, list fatty foods that you like, such as ice cream. In the second, list more healthful choices, such as nonfat frozen yogurt.

CON$UMER FOCU$

Are Air Bags Safe?

Q. I've been hearing news reports about the dangers of car air bags. Aren't air bags supposed to save lives?

A. Between 1986 and 1996, air bags saved more than 1,600 lives. However, air bags have also caused some injuries and deaths. Most of the problems occurred when people were not wearing seat belts and when small children were in the front seat. To stay safe, follow these rules when riding in a car.

- *Always* wear your seat belt.
- Whenever possible, sit in the back.
- If you must sit in the front, push the seat back as far as you possibly can.
- Make sure that children aged 12 and under sit in the back. Small children should be in safety seats.

Teen Health Digest 45

Chapter 2

The Work Students Do

Following are suggestions for implementing the project:
1. Remind students to plan the format of their show in detail before they prepare other aspects of the show. How will questions be posed? How will responses be evaluated? How will points or prizes be awarded? How many contestants will be involved? How many hosts? What will be the host's responsibilities?
2. Encourage the members of each group to work together cooperatively and productively. If necessary, remind students that the show is a production of the entire class, not of any individual or small group.
3. Suggest that students who are developing game show questions use the Teen Health Digest articles as a starting point. Then have them draw upon their own ideas and experiences, as well as information in the text, to come up with other questions.
4. Provide time for students to conduct a "dress rehearsal."
5. Let students select an audience for the final production of their game show. They may choose to invite another class, parents and other family members, or people from the community. Remind students to publicize the show and/or to issue invitations with enough advance notice.

Assessing the Work Students Do

After students have put their game show on for an audience, lead them in a short evaluative discussion. Then ask students to write personal evaluations, noting particularly their own strengths and weaknesses in working with other group members and explaining what they learned from participating in this project.

Meeting Student Diversity

Language Diversity
Students who are still developing English skills should be encouraged to participate in every aspect of this cooperative project. You may want to have them work with capable and helpful native speakers. If appropriate for your students and for their audience, the class may want to plan and produce a bilingual game show.

Hearing Impaired If family members who will be attending the game show production are hearing impaired, one or more students may know American Sign Language. In this case, provide time for students to prepare to sign the questions and responses, along with announcements and jokes, presented as part of the game show.

Lesson 4

Making Responsible Decisions

Focus

LESSON OBJECTIVES

After studying this lesson, students should be able to
- discuss the kinds of decisions, both major and minor, they are required to make.
- explain the importance of considering their values when they make decisions.
- outline the six steps in the decision-making process.
- explain why some decisions are connected to others.

MOTIVATOR

Allow five minutes for students to list all the decisions—whether important or relatively unimportant—they have made so far today. Have them list as many decisions as possible.

INTRODUCING THE LESSON

Ask students how many decisions they were able to list. Then let a few volunteers read their lists aloud. Pose questions such as these: What are some of the easiest decisions you have to make? What truly difficult decisions have you already made? What important decisions lie in your future? Tell students that, as they read Lesson 4, they will learn approaches to decision making that will help them deal successfully with even the most challenging decisions.

INTRODUCING WORDS TO KNOW

Guide students in discussing their current understanding of Words to Know. Then have students record the two terms, along with their formal definitions, in their notebooks.

46

Lesson 4 Making Responsible Decisions

This lesson will help you find answers to questions that teens often ask about decision making. For example:
- What kinds of decisions affect my health?
- How do my values affect my decisions?
- What are the steps in the decision-making process?

Words to Know

decision making
criteria

Teen Issues

Silent Decisions, Poor Decisions ACTIVITY

In a recent antidrug campaign aimed at parents of teens, this message was used: Silence is acceptance. What do you think is meant by this? Can silence be a type of decision—one you have not made very well? For instance, suppose that several of your friends are making fun of another student. You don't join in, but you don't stop them, either. How does this example illustrate the meaning of "Silence is acceptance"? Write a paragraph explaining your answer, and tell how the decision-making process could have helped you make a better decision.

Facing Decisions

Although you may not even realize it, you use decision-making skills every day. **Decision making** is *the process of making a choice or finding a solution.* You make decisions in every area of your life—your health, family, friends, school, activities, and so on. The decisions you make can expose you or others to risks, or they can decrease the risk of accident or injury. Learning how to make wise decisions is part of becoming a responsible adult.

Minor and Major Decisions

Many of the decisions you make are simple, and you make them almost automatically. These are minor decisions. They often involve only you, and they have few possible negative consequences. Minor decisions might include what to wear or whether to walk or ride your bike to school.

From time to time, you will be faced with major decisions. Major decisions in your future might include choices about your education or whether you will enter military service. These kinds of major decisions require serious thought and should be made in an orderly way.

When making any decision, you should always consider the possible consequences, or results of your actions. On the surface, for example, choosing whether or not to wear a bicycle helmet may seem like a minor decision. Yet it involves major health risks and consequences. Because you risk serious injury or even death by not wearing a helmet, this choice is an important one.

What healthful decision is this teen making?

46

Lesson 4 Resources

Teacher's Classroom Resources
- Concept Map 8
- Cross-Curriculum Activity 3
- Decision-Making Activity 4
- Enrichment Activity 8
- Lesson Plan 4
- Lesson 4 Quiz
- Reteaching Activity 8
- Transparency 9

Student Activities Workbook
- Study Guide 2
- Applying Health Skills 8

Thinking About Values

When making decisions, you will want to consider your values, or beliefs. Values such as honesty and self-respect are an important part of your personal identity and self-image. If you make a choice that goes against your values, your actions will have a negative effect on your self-image. On the other hand, making decisions based on your values will help you feel good about yourself.

For example, keeping promises and being honest are important values. Suppose that you have promised your parents that you will attend all of your scheduled track practices. One day, however, a couple of your friends tell you about their plans to meet at one person's house after school. They ask you to come along. In order to go, you would have to miss practice. When you point this out, one friend suggests that you could just tell your parents that you went to the friend's home *after* track practice. By following this advice, you would go against your own values. You would probably feel guilty. Following your values—telling your friends that you'll come along another day—will make you feel good about keeping your promises.

Where do values come from? People develop values from many sources:

- **Family.** You learn values from your parents, grandparents, older brothers and sisters, and other relatives. If your family values physical fitness, for example, you will probably also value it.
- **Religious beliefs.** Some of your values, especially about what is right and wrong, may be based on your religious beliefs.
- **Cultural heritage and the society in which you live.** You may share values with other members of the cultural group to which you belong. In addition, you may share the values of people in your society. Many Americans share a belief in democracy and personal freedom.
- **Personal experiences.** All the places where you have lived, the people you have known, and the things you have accomplished influence your values.

Cultural Connections

Your Values and You ACTIVITY

An important part of the decision-making process is considering your values. Some of your values come from your cultural heritage. Talk to your parents or other relatives about your family's cultural heritage. Find out more about customs and values from your cultural background by looking in reference books.

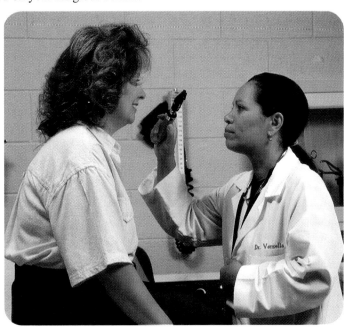

Your choice of profession may be based on what is important to you. Many physicians chose medicine because they value helping others.

Lesson 4: Making Responsible Decisions   **47**

Meeting Student Diversity

Cultural Awareness
For many individuals, personal values develop from religious teachings. Although the religions practiced around the world differ in many ways, they do have certain values in common. The "Golden Rule" (Treat other people as you would like to be treated.) is probably familiar to most students in one form or another, as it is a central teaching of many religions. Ask students to write their own version of the "Golden Rule." Have volunteers share what they have written with the class.

Chapter 2 Lesson 4

Teach

L1 Comprehending
Help students recognize that, as adolescents, they are making more and more decisions—and increasingly important decisions. Ask: What decisions do you make for yourself now that you didn't make a few years ago? What decisions do you expect to make as an adult that you don't make now as a teen?

L1 Discussing
Tell students to imagine they will hold an important political office in their adult life—they might be president, secretary of state, or a state governor. Ask: What kinds of major decisions would you have to make in order to reach that office? What kinds of major decisions would you expect to face while in office?

Visual Learning
Ask students to identify the choice being made by the teen in the picture near the bottom of page 46. What other foods has she decided not to eat? What effect will these decisions have on her health?

L1 Identifying Examples
Ask students to identify values. Record students' suggestions on the board. Then ask of each listed item: How is this value likely to affect your decision making?

L2 Writing Narratives
Have the class write short stories (or skits) about teens who face a decision similar to the example regarding track practice. Suggest that students note the most responsible decision and the values that come into play. Students might perform their skits for the class or post their work on a class bulletin board.

47

Chapter 2 Lesson 4

L1 Discussing

Ask: How do families pass along their values? (*modeling behaviors, storytelling, instruction*) At what age do children begin reflecting the values of their parents and other older family members? (*very early childhood*) At what stage in life are children likely to question the values of their parents? (*adolescence*) What kinds of conflicts sometimes result from this questioning? (*disagreements*) How can these conflicts be successfully resolved? (*discussion*)

L1 Analyzing

Encourage students to discuss their responses to these questions: Does everyone have values? Do all people have the same values? What kinds of conflicts sometimes result from differences in values?

L3 Researching

Some students may be interested in learning more about the discipline of ethics. Encourage them to do limited research and to write short reports of their findings.

L1 Critical Thinking

Point out that in the swimming example, none of the HELP criteria is met. How might a decision become more complicated if most of the HELP criteria are met—for example, in a situation that is healthful, ethical, and legal, but of which your parents would not approve? Is any of these criteria more important than the others? If so, which one? Why?

L2 Presenting Skits

Have students form groups of three or four. Instruct group members to work together in planning and practicing a skit in which one teen applies the HELP criteria. Provide time for all the groups to perform their skits for the rest of the class.

48

 Language Arts Connection

Making Choices ACTIVITY

Choose a story you have read in language arts class or a television drama or movie you have seen recently. Think about one character in that story, drama, or film who was faced with an important decision. How did that person make a choice? Did he or she carefully consider the options, or was the decision made on impulse? What were the consequences of the choice? Write several paragraphs describing the choice you would have made if faced with the same decision.

Warnings are put up in order to protect your health and safety. What risks might these teens be taking by ignoring this warning?

in your journal

List a decision you might make in the next month—to attend a particular party or to join a sports team, for example. Then list at least three values that might be used as criteria in your decision-making process. Tell which value or values will be most important to you in making this decision.

48   Chapter 2: Personal Responsibility and Decision Making

Applying Criteria

Your values provide you with **criteria** (kry·TIR·ee·uh), or *standards on which to base your decisions*. When making a decision, you can use these criteria to evaluate the situation. After you make a choice, you can also use your criteria to evaluate the outcome of your decision.

Consider this situation. You place a high value on your health and safety. You also try to do what is right. Some of your friends want you to go swimming with them in a lake at night. Swimming is prohibited in this lake at night because no lifeguards are on duty. On the basis of your values, you could evaluate the situation by applying the following HELP criteria:

- **H (Healthful).** Might swimming here lead to an injury?
- **E (Ethical).** Something that is ethical is what you consider right, according to your values and morals. Is it right to go swimming here?
- **L (Legal).** Since swimming is prohibited at night, am I breaking the law by swimming here?
- **P (Parent approved).** Would my parents approve of my swimming here?

In this case, using the above criteria to evaluate the situation would help you make the decision not to join your friends—and, in fact, to try to convince them to swim in a safer place or choose another activity.

The Decision-Making Process

The decision-making process doesn't have to be overwhelming. It can be easier if you break it down into the six steps illustrated in **Figure 2.6.** The next time you are faced with a decision, try asking yourself the questions shown in this illustration. Although you will normally use this skill only on major decisions, start practicing with small ones. Then, when you need to make an important choice, you will be familiar with the steps.

Promoting Comprehensive School Health

Are students encouraged to make responsible decisions at school? Do they face realistic consequences, both positive and negative, of their actions? Are they encouraged to evaluate possible outcomes, to weigh options, and to consider the impact their actions will have on themselves and others? These are questions that relate to everyday classroom conduct, to administrative policies, and to the input parents and community groups may have on the school and the students. They are also questions that are essential as groups work together to develop their shared vision of a comprehensive school health program. For more information, consult *Promoting a Comprehensive School Health Program* in the TCR.

Figure 2.6
The Steps of the Decision-Making Process
There are six steps in the decision-making process.

1 State the Situation
What is the problem? How did the problem occur? Who is involved? How much time do I have to make my decision?

2 List the Options
How could I solve the problem? What are my choices? With whom can I talk about the problem?

3 Weigh the Possible Outcomes
What are the consequences of each option? How will my choice affect me both now and in the future? Whom else will it affect, and how?

4 Consider Your Values
How does each of my options fit in with my value system? How will my values influence my decision?

5 Make a Decision and Act
What choice shall I make? What do I need to do to follow through on my decision?

6 Evaluate the Decision
What were the consequences of my decision? Did the results turn out as I planned? If I had to do it again, would I make the same choice?

Lesson 4: Making Responsible Decisions

Chapter 2 Lesson 4

L1 Discussing
Encourage students to answer the question in the caption beside the photograph on page 48. Ask: What other warning signs are intended to protect your health and safety? What risks would you be taking if you ignored those warnings?

L1 Comprehending
Guide students in reading aloud and discussing the six steps of the decision-making process in Figure 2.6. Let volunteers restate the steps in their own words and describe and explain the pictures that accompany them. About each step, ask: Why is this step important? How do you think skipping this step would affect your decision and your feelings about your decision?

L1 Applying Knowledge
Ask students to suggest various kinds of decisions teens have to make. Record their ideas on the board. Help the class choose one relatively simple decision (choosing between two options, for example) and one more complicated decision. Then guide the students through the decision-making process for each example. Help students recognize that they will not all reach the same decision. Ask them why this is so. Also, help students realize that, in most instances, they will have to wait for a period of time before they can evaluate the consequences of their decision (Step 6).

VIDEODISC/VHS

Teen Health Course 2
You may wish to use video segment 11, "Alcohol and Decision Making," in which a teen uses decision-making skills to avoid being pressured to use alcohol.

Videodisc Side 2, Chapter 11
Alcohol and Decision Making

Search Chapter 11, Play to 12

Meeting Student Diversity

Learning Differences
Young teens are especially energetic and often impulsive; not all inattentive behavior at school should be considered a sign of attention disorders. Students who *consistently* display the following characteristics, however, may need help:

▶ Difficulty in organizing books and papers.
▶ Tendency to call out during class.
▶ Difficulty waiting for their turn.
▶ Likely to make careless mistakes.
▶ Distracted more easily and more frequently than other members of the class.

These students may have attention deficit disorder or attention deficit hyperactivity disorder. A consultation with the school's nurse or psychologist and/or with the students' parents should be considered.

Chapter 2 Lesson 4

L1 Discussing

As students read about chains of decisions, help them consider several specific examples: If you had decided not to do your homework last night, what other decisions would you have to make? What might happen if you decided to get up half an hour earlier each morning so you could exercise? What chain of decisions might result? Let students suggest other simple decisions that might lead to a chain of related decisions.

L2 Creating Models

Let students work with partners or in small groups to make chains from looped strips of construction paper. On each strip of paper, have students write a decision that might be part of a chain of decisions. Display the completed chains in the classroom.

Visual Learning

Ask students to describe the scene in the photograph on page 50: What decision has this teen made? How and why do you think he made that decision? What other decisions do you think he had to make as a result?

Assess

EVALUATING THE LESSON

Assign Reviewing Terms and Facts and Thinking Critically on page 51 to review the lesson; then use the Lesson 4 Quiz in the TCR to evaluate students' understanding.

Chains of Decisions

In many cases, you will make decisions that lead to other decisions. These chains of decisions are a natural part of life. If you decide to join the basketball team, for example, you may also have to make decisions about other activities. Basketball practice may take place at the same time as your gymnastics lessons. Should you drop gymnastics, or can you take lessons at a different time? Weekend basketball games may conflict with your responsibilities at home. How will you get your chores done? You may have to decide on ways to rearrange your weekend schedule.

With practice, decision making will become easier. The six-step process will help you make wise choices. Every time you successfully make an important decision, you show that you are a mature and responsible individual.

Part of the decision-making process involves balancing family responsibilities with the other things you want to do.

MAKING HEALTHY DECISIONS
Making a Difficult Choice

Justina has been offered a fantastic opportunity. Her aunt is going to Paris, France, on vacation. She would like Justina to come along. Justina has begun taking French this year in middle school and is very excited about the idea of going to Paris.

Unfortunately, the two-week period when her aunt is going to Paris is the same two-week period when she usually goes to summer camp. For the past three years, Justina and her friend Alison have gone to camp together. Alison and Justina were best friends in Indiana, where Justina used to live before her family moved here to Florida.

Justina's parents want to leave the decision about what to do up to Justina. However, she doesn't know what she wants to do. On one hand, she looks forward to camp every year. Most years, the weeks at camp are the best part of her summer. It is the only time that she gets to see Alison anymore. To make it more difficult, Justina knows that Alison is expecting her to go.

On the other hand, a trip to Paris is a once-in-a-lifetime opportunity. Justina will disappoint her aunt and miss out on going if she goes to camp. Justina decides to use the six-step decision-making process to solve her problem:

❶ **State the situation**
❷ **List the options**
❸ **Weigh the possible outcomes**

50   Chapter 2: Personal Responsibility and Decision Making

MAKING HEALTHY DECISIONS
Making a Difficult Choice

Focus on Healthy Decisions

Ask students: How easy is it to decide whether you want to take advantage of a terrific opportunity? What happens to your decision making when one such opportunity conflicts with another? Then have students read and discuss the Making Healthy Decisions feature.

Activity

Use the following suggestions to discuss Follow-up Activity 1:
Step 1   Justina must decide whether to go to France with her aunt or go to camp with her friend.
Step 2   She can go to France and miss camp, or she can go to camp and miss France.
Step 3   If Justina goes to France, she will have an exciting travel opportunity, but she will disappoint her friend and perhaps put their friendship at risk.

50

Review — Lesson 4

Using complete sentences, answer the following questions on a separate sheet of paper.

Reviewing Terms and Facts

1. **Give Examples** What are two examples of minor decisions and two examples of major ones?
2. **Identify** List four sources from which people get their values.
3. **Vocabulary** Define the term *criteria*. Then use it in an original sentence.
4. **List** What are the six steps of the decision-making process?

Thinking Critically

5. **Explain** Why should you think about your values before making a major decision?
6. **Predict** What might the consequences be of deciding not to study for a test?
7. **Analyze** Which of the six steps in the decision-making process do you think is most important? Explain your answer.

Applying Health Concepts

8. **Health of Others** Think about how you might adapt the decision-making model for younger children, ages nine to ten. Make a colorful poster that could help them work through a similar process, but on a simpler level. Share your poster with your class.

❹ **Consider your values**
❺ **Make a decision and act**
❻ **Evaluate the decision**

Follow-up Activities

1. Apply the six steps of the decision-making process to Justina's story.
2. What types of values are involved in Justina's decision?
3. With a partner, role-play a scene in which Justina discusses her decision either with Alison or with her aunt. (The discussion should be with whichever person she will have to disappoint.)

Lesson 4: Making Responsible Decisions   51

Lesson 5

Setting Goals and Making Action Plans

This lesson will help you find answers to questions that teens often ask about goals and action plans. For example:
- What are the benefits of setting goals?
- Why do I need both short-term and long-term goals?
- What is an action plan?

Words to Know

self-esteem
long-term goal
short-term goal
action plan

Q & A

Career Goals

Q: My best friend has always liked animals and knows that he wants to be a veterinarian. My career goals seem to change from one week to the next. Is that unusual?

A: Not at all. During the teen years, you are developing your interests, skills, and abilities. As these change, your goals and ambitions change, too. Many people do not make a career choice until after they finish high school or college.

Why Have Goals?

You may wonder why having goals is important. Goals help give direction to your behavior and a pattern to your decisions. A goal is also one way to measure your success. You can look at goals as milestones on a journey. They allow you to evaluate how far you have traveled and how far you have left to go.

Some goals may be easy to achieve, while others are very challenging. Suppose, for example, that your goal is to improve your grade in science class. Getting an A on one science project may not be a problem for you. Getting an A in science for the whole year, however, may be very difficult. In either case, achieving the goal that you have set for yourself is a rewarding experience.

What short-term goal do you think this teen has set for herself?

52   Chapter 2: Personal Responsibility and Decision Making

Lesson 5

Setting Goals and Making Action Plans

Focus

LESSON OBJECTIVES

After studying this lesson, students should be able to
- discuss the importance of having goals.
- explain the positive effect that meeting goals can have on self-esteem.
- distinguish between long-term goals and short-term goals, and explain the relationship between them.
- explain how action plans can be used as an aid to reaching goals, and identify the six steps in an action plan.

MOTIVATOR

Ask students to imagine their lives ten years from now. Ask: What will you be doing? Where will you be living? Have students jot down notes or write sentences in response.

INTRODUCING THE LESSON

Let several students describe the lives they imagine for themselves ten years in the future. Then ask: Based on your descriptions, what are your goals for the future? How do you think you will be able to reach those goals? Explain that Lesson 5 will help them learn more about how to set and reach goals.

INTRODUCING WORDS TO KNOW

In their notebooks, have students write each listed Word to Know and a phrase what they think the term means. Then have them write the formal definition of each term as presented in the lesson.

Lesson 5 Resources

Teacher's Classroom Resources
- 📁 Concept Map 9
- 📁 Cross-Curriculum Activity 4
- 📁 Enrichment Activity 9
- 📁 Lesson Plan 5
- 📁 Lesson 5 Quiz
- 📁 Reteaching Activity 9

Student Activities Workbook
- 📁 Study Guide 2
- 📁 Applying Health Skills 9
- 📁 Health Inventory 2

Goals and Self-Esteem

Merely having goals won't make you reach them automatically. You will have to put in both thought and effort to achieve what you want. When you set a goal, you make a promise to yourself that you will work to reach it. Part of that work involves planning. The rest involves carrying out your plans and overcoming any obstacles that may come along.

Although it may sometimes seem difficult to achieve your goals, your efforts will be worthwhile. Meeting your goals will have a positive effect on your **self-esteem,** or *confidence in yourself.* Knowing that you have reached some goals in the past will help you improve your self-esteem. It will also give you the confidence to try to reach new goals in the future. Suppose, for instance, that you have achieved a goal of learning to control your temper. Next, you might feel ready to work with a peer mediation group to help others work out their disagreements in healthy ways.

Types of Goals

Time is an important element in the process of setting goals. Some goals will take much longer to achieve than others. A **long-term goal** is *a goal that you plan to reach over an extended length of time.* Examples of long-term goals include getting into college or becoming a professional golfer. These goals would take months or even years to achieve. Long-term goals, however, usually mean much more to you than other goals.

Some goals do not take much time. A **short-term goal** is *a goal that you can reach right away.* Examples of short-term goals include writing a book report or cleaning your room. Setting a series of short-term goals is a good way to achieve a long-term goal. If your long-term goal was to save money to go on the eighth-grade class trip, you could follow the steps shown in **Figure 2.7** on the next page.

in your journal

Think about an important goal that you achieved recently. How did reaching your goal make you feel? Did it give you the confidence to set new goals for the future? Describe your experience in your journal.

Teen Issues

Conflicting Goals

At any given time, you probably have several goals that you would like to achieve. Sometimes you may find that two or more of these goals are in conflict. For example, you may not be able to take both karate and guitar lessons. When this happens, it is helpful to make a list of your goals and number them in order of their importance to you.

You may not reach all of your goals, but every success makes you feel good about yourself.

Lesson 5: Setting Goals and Making Action Plans   53

More About...

Setting Goals For nearly all teens, setting goals and striving toward them are healthy behaviors. However, in a few cases, teens become obsessed with specific goals. Working obsessively toward achieving goals may be harmful to their physical health or may be a symptom of poor emotional health. One obvious example is a teen whose goal of becoming thin and fit leads to anorexia. Less obvious may be a student who concentrates solely on academic success. On the one hand, this kind of focus may reflect the student's most important interests. On the other hand, it may mask a lack of friends and even casual social contacts, or it may hide a fear of revealing aspects of the self or the family.

Chapter 2   Lesson 5

Teach

L1 Creating Illustrations
Let students work in groups to illustrate goals as milestones on a journey.

L1 Critical Thinking
Encourage students to share their responses to these questions: How do you think setting a goal that you don't meet affects your self-esteem? What positive effects could this experience have? Why?

Visual Learning
Have a volunteer read aloud the question in the photo caption on page 52. Encourage students to respond by identifying several short-term goals the teen has set for herself. Ask about the teen's possible long-term goals (related to the short-term goals), and again encourage a variety of responses.

L1 Listing Examples
Have students work in groups to brainstorm a list of behaviors and attitudes that are signs of high self-esteem.

L1 Applying Life Skills
Ask students to recall the images of their future that they discussed in the Motivator and Introducing the Lesson activities. Have each student select one of his or her long-term goals and list at least five short-term goals that might help in achieving that long-term goal.

L1 Listing Examples
Urge students to list at least five short-term goals that relate to their schoolwork during the next week. What long-term goals will each lead to?

L2 Guest Speaker
Invite an adult from the community (perhaps the parent of a student) to speak about short- and long-term goals in his or her business.

53

Chapter 2 Lesson 5

L1 Discussing
Help students consider and discuss the short-term goals shown in Figure 2.7: How does each short-term goal relate to the teen's long-term goal?

L2 Interviewing
Have students interview adults about the long-term goals they have set for themselves: What long-term goals did they set while they were still in school? How did they try to reach those goals? With what success? What long-term goals did they set as younger adults? How have those long-term goals changed (if at all)? What motivated those changes? Let students write essays explaining what they learned from the adults' experiences.

L3 Career Education
Ask students to investigate specific careers that they may want to pursue. Encourage them to research the education and training required for a career that interests them. Then have them write a set of short-term goals that will help them meet this long-term goal.

L1 Analyzing
Guide students in reading and discussing the action plan steps shown in Figure 2.8: Which step do you consider most important? Why? What might happen to the plan if you skip one of the steps?

L2 Applying Knowledge
Have students work as a class or in two or three large groups to plan and make a video about goal setting for teens. (If video equipment is not available, let students undertake another relatively complex project together.) Students should work together to develop an action plan for creating their video. Point out that they might benefit from revising their action plan as they proceed with the project. Once the video has been completed, arrange for students to share it with others. Then help students discuss the benefits of making and following (and sometimes revising) an action plan.

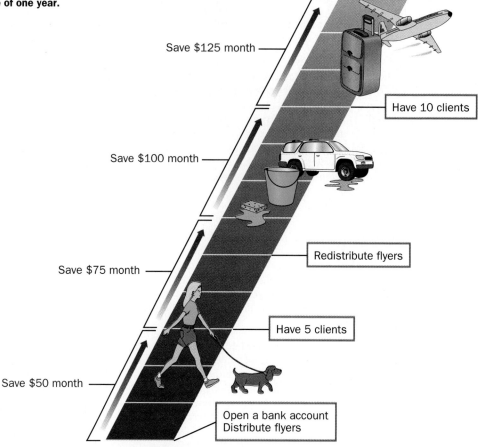

Figure 2.7
Achieving a Long-Term Goal
Reaching a long-term goal involves setting and meeting many short-term goals. This teen plans to work on her goal over the course of one year.

Long-Term Goal: To save enough money to go on the eighth-grade class trip.

- Save $125 month
- Have 10 clients
- Save $100 month
- Redistribute flyers
- Save $75 month
- Have 5 clients
- Save $50 month
- Open a bank account / Distribute flyers

Start a car-washing and dog-walking service.

Goal, Plans, Action!

Imagine that you are a film director shooting a new movie. The actors are ready, the stage is set, and the cameras are rolling. There's only one problem—there's no script! Everything comes to a standstill because no one knows what to do next. In much the same way that a director needs a script to shoot a movie, you need an action plan to achieve your goals. An **action plan** is *a series of steps for reaching your goal.*

Creating an action plan is the first step you should take after you decide on a goal. Your action plan will help you organize your efforts, manage your time, and achieve the results you want. **Figure 2.8** describes this process.

54 Chapter 2: Personal Responsibility and Decision Making

Meeting Student Diversity

Hearing Impaired To be sure students with hearing impairment are included in the discussion of goal setting and action plans, check the classroom arrangements. Any student with hearing difficulties should be able to see the faces of all speakers. Meet with the student in private and ask how well he or she has been able to follow class discussions.

Vision Impaired To help visually impaired students comprehend the steps in developing an action plan as shown in Figure 2.8 (as well as other graphic presentations in the text), be sure volunteers read the information and caption aloud and describe all photographs and drawings. In some cases, you might ask volunteers to prepare a large-print poster of a chart or list, to be displayed for the benefit of all students.

Figure 2.8
Creating an Action Plan

An action plan is like a script. Just as a script tells film actors what to do next, your action plan will provide step-by-step instructions for achieving your goal.

Step 1 Decide on your goal and write it down.
Step 2 List the steps you will take to reach your goal.
Step 3 Identify sources of help and support.
Step 4 Set a specific time period to reach your goal.
Step 5 Set up checkpoints to see how well you are doing.
Step 6 Reward yourself after you have achieved your goal.

My Action Plan

1 Goal: To be able to run three miles without stopping to walk or rest.

2 (a) I will begin by running/walking 1 mile, working up to 3 miles of running and walking.
(b) First, I will increase the mileage I cover, and then I will concentrate on walking less and running more.

3 Sources of help: my brother and my friend (who will exercise with me).

4 I have eight weeks to work toward my goal.

5 In two weeks: run/walk 2 miles. In four weeks: run/walk 3 miles. In six weeks: run 2 miles; run/walk 1 mile. In eight weeks: run 3 miles without walking or stopping.

6 If I achieve my goal, I will buy myself a video game.

Review Lesson 5

Using complete sentences, answer the following questions on a separate sheet of paper.

Reviewing Terms and Facts

1. **Restate** What is the purpose of setting goals?
2. **Vocabulary** Define the term *self-esteem*. Use it in an original sentence.
3. **Identify** Give examples of two long-term goals and two short-term goals.
4. **List** Identify the six steps involved in creating an action plan.

Thinking Critically

5. **Predict** What might happen if you never set specific goals for yourself?
6. **Analyze** Why is it a good idea to break down a long-term goal into a series of short-term goals?

Applying Health Concepts

7. **Personal Health** Think of a short-term goal that you would like to achieve. Following the steps to create an action plan, describe how you would go about reaching your goal.

Lesson 5: Setting Goals and Making Action Plans 55

Cooperative Learning

Discussing Biographies Let small groups of students work together to select an accomplished individual to study. Suggest that group members consider inventors, explorers, politicians, scientists, and athletes before deciding on a single person. Have each group member find and read a biography or autobiography of that person (or a chapter or article, if appropriate). Then let students compare and discuss what they read: What goals did this individual set? What action plan did he or she apparently decide on? How successful was the person in meeting goals? How did success affect his or her later efforts? How did he or she respond when unable to meet goals?

Chapter 2 Lesson 5

Assess

EVALUATING THE LESSON
Assign Reviewing Terms and Facts and Thinking Critically on page 55 to review the lesson; then assign the Lesson 5 Quiz in the TCR.

LESSON 5 REVIEW
Answers to Reviewing Terms and Facts
1. Responses will vary.
2. Self-esteem is confidence in yourself. Sentences will vary.
3. Responses will vary.
4. The six steps in creating an action plan are: decide on your goal and write it down; list the steps you will take to reach your goal; identify sources of help and support; set a specific time period to reach your goal; set up checkpoints to see how well you are doing; reward yourself after you have achieved your goal.

Answers to Thinking Critically
5. Responses will vary.
6. Responses will vary. Possible response: A series of short-term goals provides a sense of repeated success as the individual works toward his or her long-term goals.

RETEACHING
▶ Assign Concept Map 9 in the TCR.
▶ Have students complete Reteaching Activity 9 in the TCR.

ENRICHMENT
Assign Enrichment Activity 9 in the TCR.

Close

Ask students: What goal can you set for yourselves as a class? Help students determine a goal they can reach by the end of the year, and have them work together to write an action plan for meeting that goal.

55

Chapter 2 Review

CHECKING COMPREHENSION

Use the Chapter Summary and the Chapter 2 Review to help students go over the most important ideas presented in Chapter 2. Encourage students to ask questions and add details as appropriate.

CHAPTER 2 REVIEW ANSWERS

Reviewing Key Terms and Concepts

1. You are responsible for your own well-being in the areas of physical health, habits, activities, mental and emotional health, and social health.
2. Responses will vary. Possible response: An individual who smokes harms the health of the people around him or her.
3. Examples will vary. Possible responses: At home, keeping your belongings neatly put away makes life easier for other family members; tapping your fingers may annoy classmates.
4. The more often you perform an action, the more likely you are to make that action a habit.
5. Mortality means death.
6. Examples will vary. Possible response: subjective—Lots of people smoke, so it can't be a harmful habit; objective—Smokers are ten times more likely than non-smokers to get lung cancer.
7. Decision making is the process of making a choice or finding a solution.
8. When you consider possible outcomes, you think about what might result from various decisions, and you weigh the effects of those decisions on yourself and others.
9. A short-term goal is a goal that you can reach soon. A long-term goal is a goal that you plan to reach over an extended period of time.

Chapter 2 Review

Chapter Summary

▶ **Lesson 1** Becoming a teen means taking more responsibility for your behavior. The choices you make may have consequences that will affect your own health and the health of other people.

▶ **Lesson 2** Establishing good health habits during the teen years will have benefits that last a lifetime. Harmful habits, on the other hand, can be life-altering or life-threatening.

▶ **Lesson 3** You can protect your health by recognizing risks and avoiding them whenever possible.

▶ **Lesson 4** You are faced with a variety of decisions every day. By using the six steps of the decision-making process, you will be able to make responsible choices.

▶ **Lesson 5** Setting short-term and long-term goals helps give direction to your life. Creating an action plan will help you achieve your goals.

Reviewing Key Terms and Concepts

Using complete sentences, answer the following questions on a separate sheet of paper.

Lesson 1

1. In what areas are you responsible for your own health?
2. Explain ways in which your actions can affect the health of others.

Lesson 2

3. Provide two examples of ways your habits affect other people.
4. Explain the importance of repetition in establishing good health habits.

Lesson 3

5. What is the meaning of the term *mortality*?

6. Give an example of a subjective statement about a risk and an objective statement about the same risk.

Lesson 4

7. What does *decision making* mean?
8. What does it mean to consider possible outcomes when you are making a decision?

Lesson 5

9. What is the difference between a short-term goal and a long-term goal?
10. What is an action plan?

Thinking Critically

Using complete sentences, answer the following questions on a separate sheet of paper.

11. **Analyze** Explain how avoiding risk behaviors is an essential part of taking responsibility for your own health.
12. **Infer** Name some lifelong benefits of developing the habit of exercising on a regular basis.
13. **Synthesize** Describe how you could use the six-step decision-making process to choose a long-term goal.

Chapter 2: Personal Responsibility and Decision Making

Meeting Student Diversity

Language Diversity
Use the following suggestions to help students who have difficulty with English:

▶ Pair those learners with native speakers of English who can restate the Chapter Summary in language that helps students comprehend important concepts.

▶ Direct auditory learners or those students with language diversity to the Teen Health Audiocassette Program. Available in English and Spanish, this component provides an audio and written summary of the chapter.

National Health Education Standards

One of the main goals of the National Health Education Standards is to move students toward "health literacy." Health literacy is the capacity of individuals to obtain, interpret, and understand basic health information and services and the competence to use such information and services in ways which promote health. The health standards were developed by applying the characteristics of a well-educated, literate person within the context of health. The health literate person is:

- a critical thinker and problem solver
- a self-directed learner
- a responsible, productive citizen
- an effective communicator

Listed below are the Health Education Standards Performance Indicators addressed in each lesson of this chapter.

Lesson	Health Standards Performance Indicators
1	(2.1, 2.3, 4.2, 6.2)
2	(2.1, 2.5, 6.1)
3	(1.1, 1.6, 3.1)
4	(1.1, 1.6, 1.7)
5	(2.2, 2.4, 2.6)

ABCNEWS InterActive Videodisc Series

You may wish to use the following videodiscs with this chapter: *Teenage Sexuality,* side one, video segments 10, 11; *Food and Nutrition,* side two, video segments 5, 6; *Alcohol,* side one, video segment 14. Use the *ABCNews InterActive™ Correlation Bar Code Guide* for title reference. Also available in VHS format.

Chapter Resources

Teacher's Classroom Resources
- Chapter 3 Test
- Parent Letter and Activities 3
- Performance Assessment 3
- Testmaker Software

Student Activities Workbook
- Study Guide 3
- Applying Health Skills 10–14
- Health Inventory 3

Student Diversity Strategies
- Audiocassette Program (English)
- Audiocassette Program (Spanish)
- Spanish Parent Letters
- Spanish Summaries, Quizzes, and Activities

Multimedia Components
- English Audiocassette Program
- Spanish Audiocassette Program
- *Teen Health* Videodisc/VHS Series

Other Resources

Readings for the Teacher
MacKie, Rona M. *Healthy Skin: The Facts.* Oxford and New York: Oxford University Press, 1992.

Siegel, Dorothy S. *Dental Health.* New York: Chelsea House, 1994 (revised edition).

Slap, Gail B., M.D., and Martha M. Jablow. *Teenage Health Care: The First Comprehensive Family Guide for the Preteen to Young Adult Years.* New York: Pocket Books, 1994.

Readings for the Student
Hammerslough, Jane. *Everything You Need to Know About Skin Care.* New York: Rosen Publishing Group, 1994.

McCoy, Kathleen, and Charles Wibbelsman. *The Teenage Body Book.* New York: Berkley Publishers, 1992 (revised edition).

Monroe, Judy. "The Net: A Treasure Trove of Health Info?" Current Health 2, February 1997, pp. 30–31.

Out of Time?

If time does not permit teaching this chapter, you may use these features: Life Skills on pages 62–63; Making Healthy Decisions on page 67; Health Lab on page 72; Teen Health Digest on pages 80–81; and the Chapter Summary on page 86.

Chapter 3
Being a Health Care Consumer

CHAPTER OVERVIEW

Chapter 3 helps students explore their own role as consumers of health products and health care.

Lesson 1 explains what it means to be a consumer and helps students understand how advertising and other factors influence the decisions they will make as consumers.

Lesson 2 explains the information listed on personal product labels and discusses how students can make wise decisions about purchasing personal products.

Lesson 3 presents information that will help students protect and maintain the health of their teeth, skin, hair, and eyes.

Lesson 4 helps students understand the practices and procedures they can follow to protect the health of their eyes and ears.

Lesson 5 discusses consumers' access to various kinds of health care professionals and different forms of health insurance.

PATHWAYS THROUGH THE PROGRAM

Young teens are already active consumers of health products and of health care. The amount of money they have available to spend and their responsibilities for choosing health-related products and services will grow rapidly in the coming years. This chapter provides a solid foundation of information and techniques that students can use to develop their skills as wise, careful health consumers.

Chapter 3
Being a Health Care Consumer

Student Expectations
After reading this chapter, you should be able to:
1. Identify factors that influence your consumer decisions.
2. Explain how to compare and choose personal products.
3. Describe how to keep your teeth, skin, and hair healthy.
4. Explain how to take care of your eyes and ears.
5. Compare different sources of health care.

Key to Ability Levels

Teaching strategies that appear throughout the chapter have been identified by one of three codes to give you an idea of their suitability for students of varying learning styles and abilities.

L1 **Level 1** strategies should be within the ability range of all students. Full class participation is often required. Teacher direction is usually needed.

L2 **Level 2** strategies are for average to above-average students or for small groups. Some teacher direction is necessary.

L3 **Level 3** strategies are designed for students able and willing to work independently. Minimal teacher direction is necessary.

Teen Chat Group

Becky: I can't believe it!

Arianna: What's the problem?

Becky: I bought that new No-Frizz shampoo that was advertised in my magazine and used it last night. It smells great, but it doesn't work. My hair still looks frizzy.

Arianna: Are you sure you used it right?

Becky: I followed the directions exactly. What a waste of time. And it was expensive! The commercials make it look so good.

Arianna: It might not make a difference in one night. Maybe you should try it for a week and see if it works any better.

Becky: You might be right. At least then I wouldn't waste *all* of it. I guess the models in those commercials didn't get hair like that overnight.

Arianna: In fact, I bet their hair looked like that to begin with!

in your journal

Read the dialogue on this page. Have you ever had an experience like Becky's? Start your private journal entries by answering these questions:

- What suggestions would you make to Becky if you were taking part in this conversation?
- What factors influence your choices about what personal products to buy?
- How can people make wise buying decisions?

When you reach the end of the chapter, you will use your journal entries to make an action plan.

Chapter 3: Being a Health Care Consumer

Chapter 3

INTRODUCING THE CHAPTER

▶ Write the following questions on the board: What is a consumer? What role do you play as a consumer in our economy? Let students jot down their responses and then share their ideas in a class discussion. As the discussion develops, you may want to ask: Are any of you consumers? What kind of consumers are you? How will your roles as consumers change as you grow older? Conclude the discussion by telling students that they will learn more about being consumers, especially consumers of health products and health care, as they read the lessons of Chapter 3.

▶ Ask students how they spend their money: What products and services do you buy every week or every month? What special products or services do you buy occasionally? List their responses on the board. Then ask: Which of these products and services are related to your health? How do they help keep you healthy? Explain that students will learn more about selecting and buying health-related products as they study Chapter 3.

Cooperative Learning Project

Health Careers Day

The Teen Health Digest on pages 80 and 81 provides students with high-interest articles related to the content of this chapter.

The material in the Teacher's Wraparound Edition presents suggestions for a class project in which students plan and put on a Health Careers Day, offering the entire student body an opportunity to learn about different careers in health and health-related fields.

in your journal

As they examine their own interests, emotions, and purchasing practices in their journals, students can lay a foundation on which to build a personal understanding of themselves as consumers. They will come to understand and deal with advertising, appealing packaging, and peer pressure.

To prepare for writing these entries, let students share their responses to the Teen Chat Group and to the journal questions in informal discussion groups. Remind students that there are no "right answers" to these journal questions; rather, the questions should be considered prompts to help students explore their own ideas and understanding.

Lesson 1
Making Consumer Choices

Focus

LESSON OBJECTIVES

After studying this lesson, students should be able to
- discuss their own roles as consumers.
- explain how consumers can understand and use advertising.

MOTIVATOR

Ask students to make two lists: one of things they have bought for themselves in the past two weeks, the other of things that parents or other adults have bought for them at their request.

INTRODUCING THE LESSON

Let several volunteers read their lists aloud. Guide students in discussing selected items on the lists: What made you want to buy that particular item? What factors affected your choice? How many brands—or even other items—did you consider? Conclude by pointing out to students that, as people who make purchases, they are consumers. Now they have the responsibility of learning more about the consumer choices they can make and the external factors that influence those choices.

INTRODUCING WORDS TO KNOW

Guide students in reading aloud the Words to Know and in skimming the lesson to find the definition of each. Have students form six groups, and assign each group one of the terms. Group members will plan and present a short skit that helps explain the meaning of their assigned Word to Know.

Lesson 1
Making Consumer Choices

This lesson will help you find answers to questions that teens often ask about making wise purchasing decisions. For example:
▶ What does it mean to be a consumer?
▶ What factors influence my decisions as a consumer?
▶ How can I evaluate advertising messages?

Words to Know

consumer
advertisement
media
mass media
infomercial
endorsement

Did You Know?

Teen Purchasing Power

In 1996, 12- to 19-year-olds spent an estimated $103 billion, according to a research firm called Teenage Research Unlimited. The majority of this spending money came from parents, from gifts, and from part-time jobs.

You, the Consumer

How do you spend your money? On clothing? On CDs? On food? You probably buy these items and many more. Like other teens, you're a consumer. A **consumer** (kuhn·SOO·mer) is *anyone who buys goods and services.*

Besides being a consumer of goods and services, you're also a consumer of information. Every day you read and hear statements about all sorts of products and services. Some of these statements are true. Others are false or misleading. Being a wise consumer means evaluating information carefully and making purchasing decisions based on facts. Such careful evaluation is especially important when you purchase health products and services.

Using Your Purchasing Power Wisely

As you get older, you have more money to spend. Along with your increased purchasing power comes increased responsibility. Your purchasing decisions will affect you, your family, your friends, and even the world around you.

To choose products and services wisely, you have to be an informed consumer. You need to gather and evaluate the facts, make comparisons, and weigh your options. You also need to be aware of the various factors that influence your decisions. **Figure 3.1** on the next page shows some of these factors.

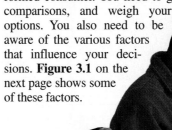

Consumer Reports and other magazines can help you compare products before you make purchases.

60 Chapter 3: Being a Health Care Consumer

Lesson 1 Resources

Teacher's Classroom Resources
- Concept Map 10
- Cross-Curriculum Activity 5
- Decision-Making Activity 5
- Enrichment Activity 10
- Lesson Plan 1
- Lesson 1 Quiz
- Reteaching Activity 10
- Transparency 10

Student Activities Workbook
- Study Guide 3
- Applying Health Skills 10

Media Messages

The various methods for communicating information are called **media. Mass media** are *media that can reach large groups of people.* Here are some examples of mass media:

- Newspapers and magazines
- Television and radio
- Movies
- Books
- Recordings
- The Internet

interNET CONNECTION
Surf the net for consumer tips on products and services to keep you healthy and safe.
http://www.glencoe.com/sec/health

All of these media are very powerful. They can convey facts and opinions to you and to millions of other people around the world. They can entertain and amuse you. They often try to persuade you to think or act in a particular way. Sometimes they can do all of these things at once.

Figure 3.1
Why Do You Buy?
Many different factors affect your decision to buy a product or use a service.

- Hey! Aren't those the shoes Gail Devers wore when she ran in the Olympics?
- But those aren't the ones I saw advertised in this month's *Sports Today*.
- I heard they're pretty expensive.
- Yes. My older sister suggested I buy them.
- Nearly everybody on my team wears the ones you're talking about. But I chose these instead.
- They are, but the salesperson showed me how well made they are. These shoes are going to have to last a couple of seasons.

① **Your Values.** Your personal beliefs about what is important play a big part in your purchasing decisions.

② **Your Family Background.** Your culture and family help shape your buying decisions.

③ **The Power of Advertising.** Commercials and **advertisements**, *messages designed to get consumers to buy a product or service,* have a strong influence on your buying choices. Remember that ads are sales tools, meant to persuade you to buy a product or service.

④ **The Influence of Your Peers.** The opinions of friends may be the first factor you consider when making a buying decision. However, what's right for someone else isn't necessarily right for you.

⑤ **The Cost of a Product or Service.** Price can be a determining factor. However, more expensive does not necessarily mean better. Wise consumers shop for both quality and value.

⑥ **The Advice of Salespeople.** Store employees can help you make a purchasing decision, but the final choice should be based on your own needs and wants—and on facts. Remember that salespeople will tell you anything just to make the sale.

Lesson 1: Making Consumer Choices   61

Chapter 3 Lesson 1

Teach

L1 Applying Knowledge
What health products and services do you buy now? What health products and services do adults select and buy for you? How can you prepare to make responsible choices about being a healthcare consumer?

L1 Discussing
Guide students in describing and discussing the photograph at the bottom of page 60: What sources is the girl in the photo consulting? Have you or members of your family ever used those sources? For what purpose? With what result?

Teacher Talk
Dealing with Sensitive Issues

Inevitably, some students will have more money to spend than others. These variations often reflect not only different family incomes but also different family values. Teens, who are unusually aware of brand names and "must have" products, may find these differences difficult to confront. It may be best to keep discussions of spending, budgeting, and making consumer choices as general as possible.

L2 Journal Writing
Remind students that understanding and maintaining one's own values is an important part of becoming mature and responsible. Ask students to list at least eight of their personal values (such as honesty) in their journals. Have them pick at least three items on the list, and note how each value influences their purchasing decisions.

Cooperative Learning

Math Activity How do teens spend their money? Let students work in groups to agree on an answer to this question by discussing their own spending habits and those of other teens they know. (Remind students that they will be making educated guesses and considering all suggestions, not conducting a scientific survey.)

Have the group members plan and draw a large pie graph to show the percentage of money an "average teen" spends on each kind of product or service. Let groups display and discuss their completed graphs: How similar are the graphs? What accounts for the differences?

Chapter 3 Lesson 1

L1 Discussing
In small discussion groups, challenge students to identify and evaluate the influences that their culture and family exert on their attitudes toward making purchases. Help students recognize that many families have more than one cultural background; family adults may have various—and perhaps even opposing—buying attitudes.

L3 Researching
Ask a group of interested volunteers to learn more about specific advertising techniques (such as bandwagon, plain folks, and testimonial). Let these volunteers share their findings with the rest of the class, presenting magazine ads or other examples to support their explanations.

L2 Family and Consumer Science
Ask students to collect ads—from local newspapers and mailings, for example—that offer the same products at different prices. How great are the differences in price? What accounts for those differences?

L1 Discussing
Guide students in discussing the influence salespeople can have on their purchasing decisions: How are teens often treated in stores? What do you think might account for that treatment? How do you usually respond when a salesperson helps you or offers you advice?

Your Total Health

A Better Alternative ACTIVITY

Before buying or using a health product, ask yourself whether a change in your lifestyle or diet could produce the same results you're hoping to get from that product. For example, including enough milk and other dairy products in your diet will eliminate the need for calcium supplements. List some changes you could make in your own lifestyle or diet that would benefit your health.

Advertising is a central part of most media. Magazines overflow with advertisements. Television programs are interrupted by commercials. More and more advertising appears on the Internet.

Advertisements can be very useful. They can make you aware of various products and provide information for your purchasing decisions. However, advertisements can also be misleading. They may exaggerate positive aspects of a product and omit negative ones. They may blend fact and opinion in such a way that you can't distinguish between the two. To make wise purchasing decisions, you need to know the facts. You must also recognize the difference between your needs (the goods or services you must have to live) and your wants (the goods and services you would like to have but that aren't essential). Sometimes advertisements try to convince you that your wants and your needs are the same thing.

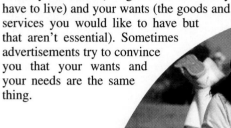

What kinds of information do you think are being communicated to this teen through this magazine?

LIFE SKILLS
Being a Wise Consumer of Health Information

Every day you're bombarded with information about health care products. This information comes from many sources: newspapers and magazines, television and radio, friends and family. To protect yourself, be skeptical. Don't believe everything you read or hear, no matter how much you may *want* to believe it. Here are some guidelines to help you become a wise information consumer.

- Be wary of any product that promises "miracle" results in "no time at all." Body building, weight loss, and other physical changes do not occur overnight. Also, be wary of vague claims, such as "increased energy."
- Don't rely on information printed on packages and in pamphlets and mail-order catalogs. Such data may not be reliable. Similarly, don't believe testimonials attributed to "thousands of satisfied customers." Rarely, if ever, can you talk to any of these individuals to find out how satisfied they really were.
- Beware of products claiming "scientific breakthroughs." True breakthroughs are rare and would be reported in the news.
- Remember that medical studies differ. Some involve a small number of patients who were observed for a short time. They are less reliable than long-term studies that involve more patients.
- When you read a research report, ask yourself if a clear cause-effect relationship has been established. Some studies "suggest" a link, but don't actually prove it.
- Don't start—or stop—using a product or medicine just because of research findings reported in the media. Discuss the findings with your doctor, or consult other reliable, factual reference

62 Chapter 3: Being a Health Care Consumer

LIFE SKILLS
Being a Wise Consumer of Health Information

Focus on Life Skills
Ask students to identify and discuss some of the health care products their families have at home—pain relievers, cold medications, vitamin and mineral supplements, and so on: Which brands do you usually buy? Which family members choose the products and the brands? What factors influence those choices? Ask whether some students take prescription health care products, such as allergy medications: Who makes the decisions about which kind of medication and what dosage you take? After this introduction, guide students in reading and discussing the Life Skills feature.

Making Health Connections
Selecting and using appropriate health care products can be central to protecting and maintaining good physical health. In addition, assuming

Understanding Advertising Methods

Whether in print or on television or radio, advertisements usually fall into one of two groups: informational ads and image ads. Both types of advertisements have the same basic purpose: to convince you to buy a product or service.

Informational ads rely mainly on facts. They may use statistics or charts to back their claims, or they may include the words of experts in the field. Ads that use phrases such as "nine out of ten doctors recommend" or "90 percent more effective" are usually informational ads.

A special kind of informational ad is the infomercial. An **infomercial** (IN·foh·mer·shuhl) is *a longer TV commercial whose main purpose appears to be to present information rather than to sell a product.* Infomercials report on everything from political candidates to the latest in home workout equipment. Many are misleading because they look like other types of television programs. An infomercial *is* an advertisement, however, even if it resembles a factual account or a scientific explanation.

Image ads link a product or service to a desirable image. They may feature a glamorous model or famous athlete giving an **endorsement** (en·DOR·smuhnt), *a statement of approval.* Image ads may show carefree skiers or blissfully happy couples. Their message—stated or implied—is usually the same: *This could be you!*

Language Arts Connection

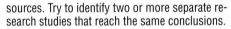

Article or Advertisement? ACTIVITY

Sometimes the line between magazine articles and advertisements becomes unclear. Perhaps you have read magazines that include "special sections" that provide information and sell products at the same time. A section on fitness, for example, may give helpful exercise hints while suggesting the use of exercise equipment made by a certain manufacturer. Critics say that these types of advertisements are misleading. What do you think? Look for examples of this kind of special section in magazines. How are they like advertisements? How are they like articles?

sources. Try to identify two or more separate research studies that reach the same conclusions.

▶ Watch for studies that add to—or contradict—previous information. Some side effects may not become known for years. Substances that seem safe at first may later be taken off the market.

▶ Scan the fine print on bottles and packages to find out if a product's health claims have been reviewed or approved by the Food and Drug Administration. Remember, however, that not all products are regulated in this way.

Follow-up Activity

Compare the packaging of nonprescription cold remedies. What claims do the various products make? Which statements are factual? Which are nothing more than advertising? Which words—such as new or improved—do you see repeatedly? Summarize your findings in a brief report.

Lesson 1: Making Consumer Choices

Chapter 3 Lesson 1

L1 Critical Thinking
Help students consider the power exerted by various media: What makes the media so powerful? Who or what has the power, and over whom? What can you, as a consumer, do to avoid being unduly influenced by the media?

L3 Social Studies
Let volunteers research the history of advertising: What were the earliest forms of advertising? When and where were they used? What have been the most significant developments in advertising since then? Ask the volunteers to share their findings with the other students.

L2 Demonstrating
Have the class work together to create a two-part bulletin board: one part with informational ads and the other with image ads. Encourage all the students to contribute magazine and newspaper ads to the display, posting the ads in the appropriate section of the board.

VIDEODISC/VHS
Teen Health Course 2
You may wish to use video segment 3, "Making Consumer Choices," to show how teens can analyze advertising and use comparison shopping techniques to get good value on the consumer choices they make.

Videodisc Side 1, Chapter 3
Making Consumer Choices

Search Chapter 3, Play to 4

---

responsibility for this aspect of self-care can enhance students' mental/emotional health. Help students recognize that they are gaining important skills they will use throughout their lives.

Meeting Student Diversity
Aural learners or students with reading-related learning difficulties may have trouble comprehending the listed guidelines. Let capable volunteers read the guidelines aloud, one at a time, for the consideration of the entire class, or pair students who have difficulties with other students who are able readers, and let the partners work together on both reading and writing the report.

Assessing the Work Students Do
Have each student bring to class an ad (from a magazine or newspaper) for a health care product. Let students work in small groups to analyze the ads and decide what choice they, as wise consumers, would make about purchasing that product. Assign credit based on students' participation and attitude.

Chapter 3 Lesson 1

Assess

EVALUATING THE LESSON

Assign Reviewing Terms and Facts and Thinking Critically on page 64 to review the lesson; then assign the Lesson 1 Quiz in the TCR to evaluate students' understanding.

LESSON 1 REVIEW

Answers to Reviewing Terms and Facts

1. Values, family background, advertising, the influence of peers, cost, and the advice of salespeople influence buying decisions.
2. Mass media are media (the various methods for communicating information) that can reach large numbers of people. Examples (any two): newspapers, magazines, television, radio, movies, books, recordings, and the Internet.
3. An endorsement is a statement of approval of an advertised product (such as that given by a model or an athlete).

Answers to Thinking Critically

4. Responses will vary.
5. Responses will vary.

RETEACHING

▶ Assign Concept Map 10 in the TCR.
▶ Have students complete Reteaching Activity 10 in the TCR.

ENRICHMENT

Assign Enrichment Activity 10 in the TCR.

Close

Let each student respond to this question, either orally or in writing: How has your understanding of yourself as a consumer changed after studying this lesson?

Being an Aware Consumer

Being a consumer is a little like being a detective. You examine the claims that manufacturers make about the products that are available to you and ask questions. Avoid accepting advertising at face value. **Figure 3.2** suggests some questions to ask yourself when evaluating advertisements.

Figure 3.2
Analyzing Advertising Claims
Some advertisements don't make their claims openly—they just imply that something is true. Learn to ask the right questions.

Ask yourself: Can this claim be true, or is it an exaggeration? Even if many people do use this product, does that mean it's what I want?

Ask yourself: Is this claim based on fact or emotional appeal? What does it really mean? Does my social life really depend on what brand of sugarless gum I use?

New **Fresh** is the best deodorant there is!

Ask yourself: What does "best" really mean? What factual proof does the manufacturer provide that this product is better than others?

Ask yourself: Can any article of clothing help me play "like a pro"? What does an endorsement from a paid celebrity or actor mean? Does this perhaps only raise the price of the product?

Lesson 1 Review

Using complete sentences, answer the following questions on a separate sheet of paper.

Reviewing Terms and Facts

1. **List** Identify six factors that influence your buying decisions.
2. **Vocabulary** Explain the term *mass media*. Give two examples.
3. **Vocabulary** What is an *endorsement*?

Thinking Critically

4. **Analyze** Where might you look to find factual information about a product or service? List two possible places to look.
5. **Create** Examine several advertisements and think of two questions you might ask yourself before buying the advertised products.

Applying Health Concepts

6. **Consumer Health** Choose an advertisement for a personal care or health product from a magazine aimed at teens. Write a paragraph explaining how the advertisement tries to persuade readers to buy the product.

Chapter 3: Being a Health Care Consumer

Cooperative Learning

Creating Ads Have students work in small groups to "create" a new brand of a product they will market to teens—Shimmer Shampoo or Sun-Up Soda, for example. Then encourage group members to explore various advertising campaigns that might appeal to young consumers. Students should agree on a specific technique or combination of techniques and incorporate these in a full-page magazine ad. Encourage students to create as polished an ad as possible, using computer-generated art and/or display type if available. Let the members of each group share their ad with the class. Encourage other students to identify the advertising techniques used and to evaluate the effectiveness of each ad, perhaps using a prepared rubric.

Buying Personal Products

Lesson 2

This lesson will help you find answers to questions that teens often ask about buying personal products. For example:

▶ What factors should I consider when choosing personal products?
▶ What information can I find by reading product labels?
▶ How can I choose personal products wisely?

Decisions, Decisions

Consumers spend millions of dollars each year on personal products, which include toothpaste, shampoo, and skin creams. Companies spend millions of dollars, too—trying to convince you to buy *their* products. How can you choose wisely? First, remember that you must take responsibility for collecting information and evaluating products. **Figure 3.3** shows important questions you should consider when making choices.

Words to Know

quackery
comparison shopping
discount store
generic products
warranty

Figure 3.3
Choosing Personal Products Wisely

Products lined up on store shelves may *look* good, but wise consumers are not fooled by appearances. They ask smart questions and think carefully about the answers.

☑ Does the product offer the results I am looking for?
☑ Is the product safe? Can it harm me—or others?
☑ What benefits can I realistically expect from the product?
☑ Is the product worth the price? Is there an equivalent product that costs less?
☑ What sets the product apart from other, similar products?
☑ Have I used other items made by the same company? Was I satisfied?

Speech bubbles: "I cost less!" "I'll make you more popular!" "Don't leave the store without checking me out!" "Here's the finest money can buy!" "I'm on sale today!"

Lesson 2: Buying Personal Products

Lesson 2 Resources

Teacher's Classroom Resources
- Concept Map 11
- Cross-Curriculum Activity 6
- Enrichment Activity 11
- Lesson Plan 2
- Lesson 2 Quiz
- Reteaching Activity 11
- Transparencies 11, 12

Student Activities Workbook
- Study Guide 3
- Applying Health Skills 11

Lesson 2
Buying Personal Products

Focus

LESSON OBJECTIVES
After studying this lesson, students should be able to

▶ explain how to make wise decisions about buying personal products.
▶ read and understand the information on personal product labels.
▶ discuss the importance of comparison shopping.

MOTIVATOR
Display four or five new, same-sized containers of toothpaste (or another personal care product). For this survey explain that all the products cost the same. Have students write their responses to these questions: Which product will you buy? Why?

INTRODUCING THE LESSON
Hold up each container and ask students to raise their hands if they chose to buy that product. Ask several volunteers to explain the reasons for their choices: How are their choices similar? How are they different? Are any reasons for choosing one brand of toothpaste "better" than other reasons? Why or why not? Explain to students that they will learn more about selecting and buying personal care products (such as toothpaste) when they study Lesson 2.

INTRODUCING WORDS TO KNOW
Help students read aloud the five Words to Know and their formal definitions, as given in the Glossary. Then let students work with partners to write original narrative paragraphs including all the listed Words to Know.

Chapter 3   Lesson 2

Teach

Visual Learning
Guide the class in describing and discussing Figure 3.3. Which personal products are most likely to "call out" to you? Why? How does advertising help a product look attractive? How does packaging affect a product's appeal? If you were the consumer in this picture, what would you do?

L1 Environmentalism
Help students consider environment-related questions they might ask when choosing products, including questions related to excess packaging, recycled and/or recyclable packaging materials, and biodegradable products.

L2 Applying Knowledge
Ask students to collect ads or brochures for the kinds of worthless products and treatments described here. Let them collect the ads either on a bulletin board or in a notebook. Provide time for students to discuss and analyze the ads: How can you recognize that each is an example of quackery?

L1 Discussing
Discuss with students the kinds of warnings that appear on product labels. In addition to themselves, who must students consider when reading product warnings? Why is it especially important to consider young children in reading these warnings?

Teen Issues

Unrealistic Images

Magazine advertisements and television commercials frequently show very thin models with perfect-looking hair, skin, and teeth. Even these models, who are real people, may not look this way in real life. Photos may be altered or touched up in a variety of ways. Discuss with your classmates why such ads are not just unrealistic but potentially harmful to a teen's self-image.

Be a cautious consumer. "Amazing" products may not work, and some may even be harmful. Investigate before you buy.

Math Connection

Using Unit Pricing

Because product packages contain varying amounts, making price comparisons can be tricky. Always check the unit price. For example, a 6-ounce package selling for $3 has a unit price of $0.50 per ounce: $3.00 divided by 6. An 8-ounce package selling for $3.60 has a unit price of $0.45 per ounce: $3.60 divided by 8. Larger packages often have a lower unit price than smaller ones.

66   Chapter 3: Being a Health Care Consumer

Wise Buying

Quackery (KWAK·uh·ree) is *the sale of worthless products and treatments through false claims.* Such products and treatments may be advertised in the mass media, sold through the mail, or offered over the phone. Their makers may claim that they cure or prevent diseases or other health problems. While some such products and treatments may not be completely worthless, their value is far less than the seller claims. Here are some tips to help you avoid quackery and be a wise buyer of health products.

■ Remember that product claims that sound too good to be true usually are. Be especially careful of claims concerning beauty aids, diets, and "miracle" products.

■ Don't be taken in by impressive-sounding words, "secret ingredients," or glowing testimonials by unknown people. Such advertising techniques may reflect nothing more than the creativity of advertisement writers.

■ Beware of "free" samples for which you're asked to send money "to cover shipping and handling costs."

■ Check with your doctor, a pharmacist, or another reliable source of information before buying or using products. This is especially important for products that you take into your body, such as dietary supplements.

Reading Product Labels

The labels that appear on personal products include valuable information. If you pay careful attention to this information, it will help you make smart buying decisions. Labels can also help you make safe use of the products you buy.

MAKING HEALTHY DECISIONS

Reaching a Purchasing Decision

Focus on Healthy Decisions

Display two or three informational ads for different kinds of bicycles. Then present this situation: You've decided to make bike riding part of your fitness program, and you've saved enough money to buy yourself a new bicycle. Your choices are shown in these ads. Which bike will you buy?

After volunteers have shared their responses, ask: How did you reach your decision?

Activity

Use the following suggestions to discuss Follow-up Activity 1:

Step 1 Kim has to decide which kind of in-line skates to buy.

Step 2 She can choose between Brand A and Brand Z. She might also choose to put off her decision and not buy skates right now.

Step 3 If she buys Brand A, Kim knows she will have all the features she wants in her new skates.

Figure 3.4 shows the kinds of information that labels typically provide and suggests how you can use that information wisely. If any information on a product label seems confusing or incomplete, speak with your doctor or pharmacist or check with the manufacturer.

Figure 3.4
Label Information
Product labels can tell you a great deal, so you should take the time to read them.

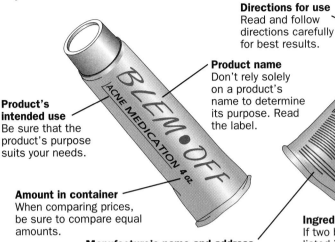

Product's intended use
Be sure that the product's purpose suits your needs.

Amount in container
When comparing prices, be sure to compare equal amounts.

Manufacture's name and address
Contact the manufacturer if you have a problem or question. Some product labels also include phone numbers.

Directions for use
Read and follow directions carefully for best results.

Product name
Don't rely solely on a product's name to determine its purpose. Read the label.

Control number
If a problem develops, the manufacturer may ask you to provide this identifying number.

Warnings
To protect yourself and others, pay close attention to warnings. Even safe products can be hazardous if they are not used properly.

Ingredients
If two brands have the same ingredients listed in the same order, they will probably have the same effects. If you are allergic to a particular substance, you will want to see if the product contains it.

MAKING HEALTHY DECISIONS

Reaching a Purchasing Decision

Kim takes physical fitness seriously. She is a member of the track team at school, and she has decided to buy in-line skates to help her stay in shape at home.

After shopping around, Kim has narrowed down her choice to two models. One model has every feature she wants. It's manufactured by a company that Kim knows has an excellent reputation. The XL-62s, however, have one big drawback—their price. The skates are very expensive.

By contrast, the MX-25s are much more reasonably priced. However, they lack the extra padding that Kim wants. In addition, she's never heard of the company that makes these skates. A salesperson has assured her, though, that the company does make a high-quality product.

To make her decision, Kim uses the decision-making process:

1. **State the situation**
2. **List the options**
3. **Weigh the possible outcomes**
4. **Consider your values**
5. **Make a decision and act**
6. **Evaluate the decision**

Follow-up Activities

1. Apply the six steps of the decision-making process to Kim's dilemma.
2. List at least three other actions Kim could take before making a final decision. Explain how each action would help Kim make a good decision.

Lesson 2: Buying Personal Products

Chapter 3 Lesson 2

Assess

EVALUATING THE LESSON
Assign Reviewing Terms and Facts and Thinking Critically on page 68 to review the lesson; then assign the Lesson 2 Quiz in the TCR to evaluate students' understanding.

LESSON 2 REVIEW

Answers to Reviewing Terms and Facts
1. Quackery is the sale of worthless products and treatments through false claims.
2. Responses will vary.
3. A warranty is a company's or store's written agreement to repair a product or refund your money if the product is defective.

Answers to Thinking Critically
4. Responses will vary.
5. Not reading the label could cause you to use the product improperly and perhaps harm yourself and others; it could also result in your using a product with an ingredient to which you are allergic.

RETEACHING
▶ Assign Concept Map 11 in the TCR.
▶ Have students complete Reteaching Activity 11 in the TCR.

ENRICHMENT
Assign Enrichment Activity 11 in the TCR.

Close
Ask students to write one- or two-sentence responses to this question: After studying this lesson, what is the most important change you plan to make in your practice of buying personal products?

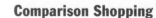

Wise consumers gather information before making a purchasing decision. Think of a hair styling or personal care product you would like to buy. In your journal, list some sources of information about the product. Put a check next to sources you consider particularly reliable.

Comparison Shopping

Make a habit of **comparison** (kuhm·PEHR·i·suhn) **shopping,** or *comparing products, evaluating their benefits, and choosing the products that offer the best value.* Consider these factors:

- **Cost.** Compare prices of the same brand in different stores. Check newspapers for sales. Look for **discount stores,** *stores that offer special reduced prices.* Also look for **generic** (juh·NEHR·ik) **products**—*goods sold in plain packages at lower prices than brand name goods.* Generic products can save you money when they are equal in quality to brand-name items.
- **Features.** Avoid paying for features that you don't need. However, *do* pay for features that you will find especially useful.
- **Quality.** Well-made products generally outlast those that are poorly made. A cheap product is no bargain if it falls apart.
- **Warranty.** Before you buy a costly product, ask about the **warranty.** A warranty is *a company's or store's written agreement to repair a product or refund your money if the product does not work properly.*

Find out about store policies *before* you buy. For example, will the store return your money if you're dissatisfied with a product, or only give credit?

Lesson 2 Review

Using complete sentences, answer the following questions on a separate sheet of paper.

Reviewing Terms and Facts
1. **Vocabulary** Define the term *quackery.*
2. **Give Examples** List two ways in which the information on product labels can help you.
3. **Vocabulary** What is a *warranty?*

Thinking Critically
4. **Apply** Choose a shampoo or sunblock that you use and explain what factors you considered in selecting that brand.
5. **Analyze** How might *not* reading a product's label be dangerous?

Applying Health Concepts
6. **Consumer Health** Choose two leading brands of the same product, such as toothpaste or deodorant. Make a chart comparing the two based on what you've read in this lesson. Use your chart to explain why you would choose one over the other.

68 Chapter 3: Being a Health Care Consumer

TECHNOLOGY UPDATE

Part of making wise purchasing decisions is gathering relevant information. The Internet can be a useful tool for gathering such information. Have students select a personal product and research the product on-line. The process is simple: Select a search engine, enter the name of the product, and skim or read the articles that the search engine lists. Encourage students to do this kind of research with partners if possible, and review with them the "safety procedures" for Internet use. Together, students can discuss the source and reliability of each article they read. Emphasize the importance of asking questions such as these: What is the purpose of this article? Is it informative or persuasive? What biases can I expect to find in this article? Why?

Caring for Your Teeth, Skin, and Hair

Lesson 3

This lesson will help you find answers to questions that teens often ask about teeth, skin, and hair. For example:

▶ How can I keep my teeth healthy?
▶ How should I clean and protect my skin?
▶ How do I care for my hair?

Healthy Teeth

If you're like most teens, you pay attention to your appearance. You probably spend time—and money—making sure that your teeth, skin, and hair look their best. Such personal care is important, not just because it makes you feel good about how you look, but because it is sensible personal hygiene.

Think about your teeth for a moment. Clean, healthy teeth not only add to your smile but also enable you to chew food and speak clearly. **Figure 3.5** shows how a tooth looks inside and out.

Words to Know

fluoride
plaque
tartar
cavity
epidermis
melanin
dermis
pores
dandruff
head lice

Figure 3.5
The Parts of a Tooth
Understanding what's inside your teeth will help you take care of them more effectively.

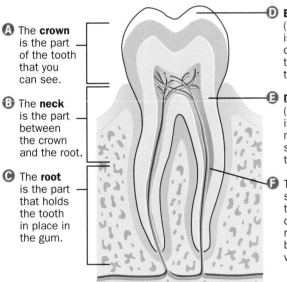

A The **crown** is the part of the tooth that you can see.

B The **neck** is the part between the crown and the root.

C The **root** is the part that holds the tooth in place in the gum.

D **Enamel** (ee·NA·muhl) is the hard outer layer that covers the crown.

E **Dentin** (DEN·tin) is bonelike material surrounding the pulp.

F The **pulp** is soft, sensitive tissue containing nerves and blood vessels.

Did You Know?

Hard as... Teeth?

Tooth enamel is the hardest substance in the human body. It's so hard that a dentist's drill must turn at a rate of more than 8,000 times per second in order to drill into a tooth.

Lesson 3: Caring for Your Teeth, Skin, and Hair

Lesson 3
Caring for Your Teeth, Skin, and Hair

Focus

LESSON OBJECTIVES

After studying this lesson, students should be able to

▶ identify the parts of a tooth and explain how to keep teeth healthy.
▶ describe the functions of the skin and explain how to clean and protect their skin.
▶ explain the structure of hair and discuss how to keep the hair and scalp healthy.

MOTIVATOR

Ask students: What do you do every day to keep your teeth healthy? your skin? your hair? Let students sketch or write their responses.

INTRODUCING THE LESSON

Encourage volunteers to share their sketches or to read aloud their written responses to the Motivator questions. Then ask students why they think it is important to keep their teeth, skin, and hair clean and healthy: How can these practices affect your physical, mental/emotional, and social health?

INTRODUCING WORDS TO KNOW

With partners, have students read and discuss the Words to Know. Then have the pairs prepare 20 word cards, writing one Word to Know on each of ten cards and the formal definition (as presented in the lesson) on each of the remaining ten cards. Let students use these word cards to play Vocabulary Concentration, matching each Word to Know with its definition.

Lesson 3 Resources

Teacher's Classroom Resources
- Concept Map 12
- Decision-Making Activity 6
- Enrichment Activity 12
- Health Lab 3
- Lesson Plan 3
- Lesson 3 Quiz
- Reteaching Activity 12
- Transparency 13

Student Activities Workbook
- Study Guide 3
- Applying Health Skills 12

Chapter 3 Lesson 3

Teach

L1 Discussing

Guide students in a discussion of the impact that clean, healthy teeth can have on a person's appearance: How do you think good dental care can affect your physical health now and in the future? (*false teeth*) your mental/emotional health? (*self-esteem*)

Visual Learning

Help students use Figure 3.5 to study the parts of a tooth. Let a volunteer read each label aloud as students point to the corresponding portion of the drawing. Then ask: Which of these labeled parts can you see on your own teeth? (*crown, enamel*) Which part is most likely to be involved in a toothache? (*pulp*)

L2 Science

Ask interested volunteers to learn more about fluoride: What is it? Where does it occur naturally? How does it help teeth resist decay? In addition to toothpaste, what other sources of fluoride are available? What problems are associated with too much fluoride? Let these volunteers prepare and display a chart or poster summarizing their findings.

L1 Comprehending

Ask students to explain how rinsing the mouth with water can help promote dental health. (*Rinsing with water can remove bits of food that might otherwise remain trapped on or between teeth and contribute to plaque.*)

Health Minute

Sodas—as well as coffee and tea—can cause yellowing of the teeth. Try drinking water instead.

Your Total Health
Sweet Talk ACTIVITY

Limiting your intake of foods that are high in sugar will reduce your likelihood of getting cavities and also help you control your weight. If you chew gum, chew sugarless gum. Make a list of healthful snacks you might eat in place of sugary treats.

Caring for Your Teeth

Caring for your teeth means keeping them clean and healthy. If you neglect your teeth, you allow the process of tooth decay to begin. This process is shown in **Figure 3.6**.

To fight tooth decay and keep your teeth and gums in good shape, follow these guidelines:

- **Brush your teeth regularly.** Brush at least twice a day. If possible, brush as soon as possible after each meal. If you can't brush after every meal, at least rinse your mouth with water. Regular brushing is especially important if you wear braces, which can trap bits of food. Choose a toothpaste that contains **fluoride** (FLAWR·eyed), *a substance that helps teeth resist decay*. Use a soft-bristled toothbrush, and replace it with a new one every few months. You should also replace your toothbrush after you have had a bad cold or a sore throat.

Figure 3.6
What Causes Tooth Decay

Decay is a process that can damage or even destroy teeth. You can prevent this process by taking care of your teeth.

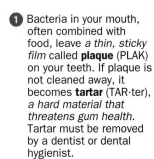

① Bacteria in your mouth, often combined with food, leave *a thin, sticky film* called **plaque** (PLAK) on your teeth. If plaque is not cleaned away, it becomes **tartar** (TAR·ter), *a hard material that threatens gum health*. Tartar must be removed by a dentist or dental hygienist.

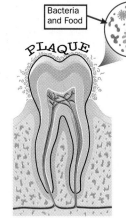

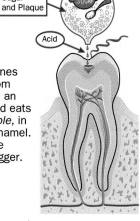

② Plaque combines with sugar from foods to form an acid. The acid eats a **cavity** or *hole*, in the tooth's enamel. Over time, the cavity gets bigger.

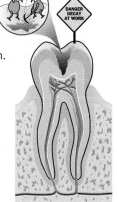

③ The decay spreads, invading the tooth's dentin.

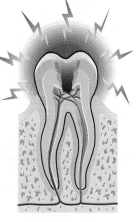

④ The decay then spreads to the pulp. If the cavity exposes a nerve, you are likely to have a toothache.

70 Chapter 3: Being a Health Care Consumer

Health of Others

Good dental care is important for everyone—even for babies whose teeth have not yet erupted. Many pediatric dentists recommend "brushing" a baby's gums with a soft, wet washcloth. This practice can become part of the baby's daily washing and can be continued even after the teeth have come through the gums.

Once a young child is old enough to say "I want to do it myself," the job of the parent or caregiver does not end. After the child has brushed, an adult should brush the child's teeth again, to make sure the mouth has been thoroughly cleaned.

- **Use dental floss.** Floss at least once a day to clean between your teeth. **Figure 3.7** shows you how. Floss can remove trapped food that brushing alone cannot reach. Floss comes in several different types. Your dentist can recommend the best type for you.
- **Cut down on sugar.** Limit your intake of foods that are mostly sugar, such as candy and cookies.
- **Eat a balanced diet.** Include plenty of healthful foods, such as fruits and vegetables, and foods high in calcium, such as milk and cheese. You can have regular snacks, but don't snack all day. Constant eating will bathe your teeth in cavity-producing acids.
- **Have regular dental checkups.** Visit your dentist regularly. Most dentists see their patients twice a year. Your dentist and dental hygienist can help you keep your teeth clean. They can also spot tooth and gum problems before they become serious. Dental clinics offer full dental services for those who cannot afford a private dentist.

in your journal

How well do you take care of your teeth? In your journal, list the actions you always take now to keep your teeth healthy. Then make a second list of actions you can take in the future to improve your approach to tooth care.

Figure 3.7
Brushing and Flossing Your Teeth

Proper brushing and flossing will keep your teeth healthy. The first photo shows a person's teeth after he has just chewed a special tablet that reveals areas of the teeth that need cleaning. After proper brushing and flossing, his teeth are bright and clean.

How to Brush
Ⓐ Brush the outer surfaces of your upper and lower teeth. Use a combination of up-and-down strokes and small circular or side-to-side strokes.

Ⓑ Thoroughly brush all chewing surfaces.

Ⓒ Brush the inside surfaces of your upper and lower teeth.

Ⓓ Brush your tongue.

Ⓔ Rinse your mouth.

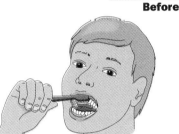

Before

After

How to Use Dental Floss
Ⓐ Wrap about 18 inches of floss around the middle finger of each hand.

Ⓑ Grip the floss tightly between thumb and forefinger.

Ⓒ Slide the floss back and forth between teeth until it touches your gumline.

Ⓓ Forming a C with the floss around each tooth, keep sliding the floss back and forth gently as you move it up and down along the side of the tooth.

Ⓔ Do the same for all of your teeth, using a clean section of floss for each one.

Ⓕ When you've finished, rinse your mouth.

Lesson 3: Caring for Your Teeth, Skin, and Hair

Chapter 3 Lesson 3

L1 Discussing
Ask students to identify high-sugar foods that are not good for dental health. Then have them suggest specific snacks and desserts that they can substitute for sugary foods.

L1 Comprehending
Ask students: Is your skin the same all over your body? What are some of the differences? (*differences in color, texture, and sensitivity*) What is the most obvious way in which any individual's skin differs from other people's skin? (*color*) What effect does that difference have on the functions of the skin? (*none*)

L2 Reporting
Ask an interested volunteer to learn more about melanin: Where and how is it produced? How does it function in the body? In addition to the skin, where else is melanin found? (*in hair*) What other organisms produce melanin? Let this volunteer prepare and share a poster or present a brief oral report.

Visual Learning
Guide students in reading and discussing the labels for the parts of the skin, Figure 3.9. Ask students to name and point to the skin features in the epidermis (pores), in the dermis (blood vessels, nerve ending, hair follicle, sweat gland, oil gland), and below the dermis (fat tissue).

Healthy Skin

Your skin is an organ of your body, like your lungs or heart. In fact, the skin is the largest body organ. Because it performs several key functions, it is also one of the most important. **Figure 3.8** describes some of these functions.

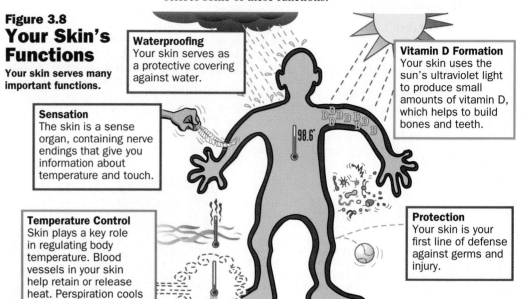

Figure 3.8 Your Skin's Functions
Your skin serves many important functions.

Waterproofing Your skin serves as a protective covering against water.

Vitamin D Formation Your skin uses the sun's ultraviolet light to produce small amounts of vitamin D, which helps to build bones and teeth.

Sensation The skin is a sense organ, containing nerve endings that give you information about temperature and touch.

Protection Your skin is your first line of defense against germs and injury.

Temperature Control Skin plays a key role in regulating body temperature. Blood vessels in your skin help retain or release heat. Perspiration cools your body.

HEALTH LAB
How Well Do Sunscreens Protect?

Introduction: Using sunscreen or sunblock lotion helps protect against the effects of the sun. Not all lotions protect equally.

Objective: Compare the protection offered by three sunscreen lotions with different SPFs, or sun protection factors.

Materials and Method: You'll need three different lotions: one with an SPF of 2–4, one with an SPF of 15, and one with an SPF of 30 or more. You'll also need masking tape; a clear plastic folder; photographic developing paper; 3 tablespoons of concentrated dechlorination solution (about 18 percent sodium thiosulfate); 1 cup of distilled water; a shallow plastic pan; and tweezers.

Divide the clear top sheet of the folder into three equal parts with strips of masking tape. Coat each part with a different lotion. Keep track of which lotion you apply to which part.

In a darkened room, slide a sheet of developing paper into the folder. The glossy side should face up, under the top sheet. Place the closed folder, lotion side up, outside in sunshine for 5 to 10 minutes.

Combine the dechlorination solution and distilled water in the pan. Next, use tweezers to put the exposed developing paper into the solution for 3 seconds. Take the paper out, rinse it with cold water, and let it dry.

Observations and Analysis: Observe the differences in color among the three sections of the paper. Which lotion provided the most protection? Which provided the least? What happened to the paper under the masking tape?

72 Chapter 3: Being a Health Care Consumer

HEALTH LAB
How Well Do Sunscreens Protect?

Time Needed
2 half-hour sessions on consecutive days

Supplies
- 3 bottles of sunscreen or sunblock lotion with these SPF ratings: 2; 4; 15; 30 or above
- masking tape
- clear plastic folder
- photographic developing paper
- 3 tablespoons of concentrated dechlorination solution (about 18 percent sodium thiosulfate)
- 1 cup of distilled water
- shallow plastic pan
- tweezers

Focus on the Health Lab
Display several containers of sunscreen or sunblock. Encourage several volunteers to share their own

The Inside Story

Your skin is composed of two main layers. The **epidermis** (e·puh·DER·mis) is *the outermost layer of skin*. New cells made in the epidermis continuously replace old cells, which are lost from the surface of the skin. Cells in the epidermis make **melanin** (MEL·uh·nin), *the substance that gives your skin its color.*

Beneath the epidermis is the **dermis** (DER·mis). The dermis is *the skin's inner layer that contains blood vessels, nerve endings, and hair follicles.* The dermis also contains two kinds of glands. Sweat glands let perspiration escape through **pores,** or *tiny openings in the skin.* Oil glands produce oils to keep skin soft. Under the dermis is a layer of fat tissue. **Figure 3.9** shows the parts of the skin.

Did You Know?

The Skin's Weight

The skin covering the human body weighs between 6 and 10 pounds. If it were laid out on a flat surface, it would cover an area of about 21 square feet.

Figure 3.9
Parts of the Skin
The skin is constantly replacing and repairing itself.

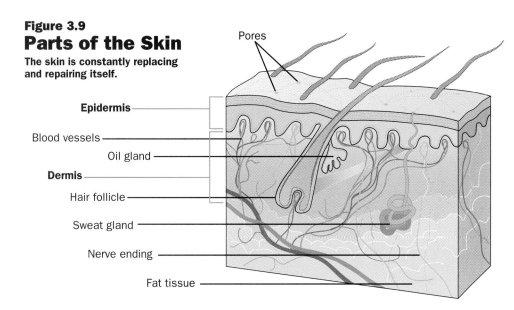

Caring for Your Skin

Taking care of your skin helps you stay healthy and keeps you looking your best. Here are some skin care tips:

- **Keep clean.** Take a bath or shower daily. Keeping your skin clean is particularly important during your early teen years, when sweat glands become more active. Bacteria mixing with perspiration can cause an unpleasant odor, especially under your arms, where there are many sweat glands. Washing with soap removes bacteria and excess oils from your skin. You can also combat body odor by using a deodorant or antiperspirant.

- **Take care of yourself.** Eat a healthful diet. Foods such as milk, green and yellow vegetables, and eggs are high in vitamin A, a vitamin that helps your skin stay healthy. In addition, get regular exercise and plenty of rest.

Lesson 3: Caring for Your Teeth, Skin, and Hair

Chapter 3 Lesson 3

L2 Family and Consumer Sciences

Ask students to collect advertisements (from magazines and newspapers, for example) for deodorants and antiperspirants. Have them meet in groups to compare and discuss the ads: How does each ad try to appeal to consumers? What benefits are highlighted for each product? Which of the products, if any, would you choose to buy? Why?

L1 Comprehending

Ask students to consider the specific ways to *Take care of yourself* (on this page): How do these tips for promoting the health of your skin relate to the dietary guidelines for promoting good dental health? How do they relate to suggestions for establishing good overall physical health through proper diet and exercise?

L1 Discussing

Encourage students to share their own experiences with choosing and using sunscreens and with getting sunburned. Guide students in recognizing that having dark skin is not in itself a protection against sunburn. Skin sensitivity to ultraviolet rays varies with each individual.

L2 Researching

Tell students that half of all new cancers in the United States are skin cancers. Ask students to work with partners or in small groups to research various forms of skin cancers. How are they diagnosed? What treatments are recommended? With what outcomes? Let students summarize their findings in brief written reports.

---

experiences in avoiding sunburn by using such products. Ask: What are the short-term and long-term consequences of failing to use sunscreen and sunblock?

Understanding Objectives

Students will observe the protection that sunscreens and/or sunblocks offer against the effects of the sun's rays. They will compare the levels of protection provided by products with different SPF ratings.

Observation and Analysis

Divide the class into groups of three or four to perform the experiment. If supplies are limited, let the groups take turns preparing their folders. Once their paper is dry, have group members work together to observe the results and to discuss responses to the questions.

Further Investigation

Suggest that students read more about the meaning of SPF ratings and about dermatologists' recommendations for using sunscreens and sunblocks.

Assessing the Work Students Do

Ask each student to write a paragraph explaining what he or she learned from this Health Lab activity.

Chapter 3 Lesson 3

L2 Creating Posters

Working in groups, have students plan and make posters urging teens to protect their skin against exposure to the sun. Display the completed posters in a hallway or another public area of the school.

L1 Guest Speaker

Invite the school nurse or another health care professional to talk with the class about acne. What are the best skin care practices? What effect does diet have on acne? How can teens deal successfully with the concerns or even anxiety that acne often causes? Provide time for students to ask the speaker questions.

L2 Math

Have students use the numbers presented in Figure 3.10 to answer these questions: What fraction of your total hair follicles are in your scalp? *(100,000/5,000,000 = 1/50)* What percentage of your hair follicles are in your scalp? *(1 ÷ 50 = .02 = 2%)*

L1 Family and Consumer Sciences

Bring to class several empty containers of different brands of shampoo, or ask students to bring in their own containers. In groups, have students read the labels and compare the ingredients of various brands. Then ask: What do you think accounts for the wide variety of prices charged by shampoo manufacturers?

L2 Reporting

Let several volunteers work together to learn more about head lice: How do head lice spread from one person to another? Why can head lice create special problems in schools and other areas where young children spend time together? Why is it important to get rid of head lice completely? What products are recommended for eliminating head lice? Ask these volunteers to share their findings in a brief oral report.

Language Arts Connection

Have You Heard the Expression . . . ? ACTIVITY

Many common expressions refer to teeth, skin, or hair. For example, people escape "by the skin of their teeth" or "get into each other's hair." Make a list of expressions you've heard that use these words. See if you can figure out how these expressions may have come to have these meanings.

■ **Guard against the sun.** Ultraviolet light from the sun can damage your skin and increase your risk of getting skin cancer. Experts warn that one severe sunburn during the first 15 years of life can double the risk for skin cancer. Try to keep out of the sun between 10:00 a.m. and 3:00 p.m., the period when ultraviolet rays are most intense. If you do spend time in the sun, use sunscreen or sunblock lotion to protect your skin. Choose a lotion with an SPF (sun protection factor) of 15 or higher. Follow the instructions on the label carefully.

Keep in mind that the sun does not affect everyone equally. For example, fair-skinned people, whose skins have less melanin, will get sunburned more easily than darker skinned people. Also, remember that ultraviolet rays are invisible. They are present—and dangerous—even on overcast days.

Acne

Many teens have to cope with a skin condition called acne. This condition occurs when active oil glands cause pores to become clogged. Pimples, whiteheads, or blackheads may result. Acne often appears on the face, but may also affect the neck, back, and shoulders.

To fight acne, gently wash the affected area at least twice daily with mild soap and warm water. Avoid touching, picking at, or rubbing the area, and don't apply heavy makeup or creams. If you are concerned about your skin, talk with your doctor or other primary health care provider, who may refer you to a dermatologist (duhr·muh·TAH·luh·jist), a physician who specializes in treating the skin and its diseases.

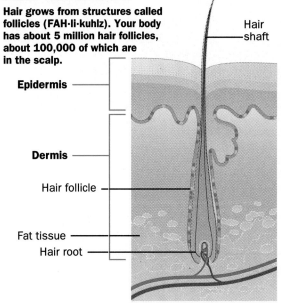

Figure 3.10
Hair Structure

Hair grows from structures called follicles (FAH·li·kuhlz). Your body has about 5 million hair follicles, about 100,000 of which are in the scalp.

- Epidermis
- Dermis
- Hair follicle
- Fat tissue
- Hair root
- Hair shaft

Healthy Hair

Your hair is one of the first features people notice. Keeping it clean and healthy is well worth the time and effort you invest. Hair care is an important part of good personal grooming.

Hair color comes from melanin, just as skin color does. Whether your hair is straight, wavy, or curly depends on the shape of the hair shaft, the part of the hair that you can see (see **Figure 3.10**). This visible portion is made of dead cells. New hair cells grow beneath the skin surface.

74 Chapter 3: Being a Health Care Consumer

Personal Health

 Nearly all teens are likely to be affected by acne to some degree. Remind students that 90% of American teenagers have acne and that acne problems can be controlled. Careful washing—not scrubbing—and over-the-counter topical medications are best for most cases of teenage acne. Teens with serious acne may need to use topical antibiotics or even oral medications.

For some teens, the psychological effects of acne can be devastating. A teen's perception of his or her acne condition (even if an objective observer sees "just a mild case") can lead to withdrawal from social contacts, anger, loneliness, confusion, and depression.

Caring for Your Hair

Daily brushing keeps your hair healthy by removing dirt. Brushing also spreads oils down hair shafts, making hair shiny. You should also wash your hair regularly, using a gentle shampoo. How often you should wash depends on whether your hair tends to be dry or oily. Most people need to wash their hair at least twice a week. If possible, allow your hair to air dry. Blow dryers can make your hair brittle, particularly if you use a high-heat setting.

Several problems can occur with the hair and with the scalp, the skin beneath the hair. One is **dandruff,** *a flaking of the outer layer of dead skin cells on the scalp.* This condition is usually caused by dry skin. You may be able to control dandruff through regular shampooing. Special dandruff shampoos are also available. In certain cases, your doctor may prescribe a medicated shampoo.

Another hair and scalp problem involves **head lice,** *parasitic insects that live in the hair and cause itching.* Head lice are very easy to catch from other people. To avoid the problem, don't share combs, brushes, or hats with others. If you do get head lice, you'll need to use a special shampoo to kill them. You'll also have to wash all bedding, towels, and clothes that you have used. Other members of your family, even pets, may need to be treated, too.

Teens often try out different hairstyles as a means of self-expression.

Review Lesson 3

Using complete sentences, answer the following questions on a separate sheet of paper.

Reviewing Terms and Facts

1. **Vocabulary** Define the terms *plaque* and *tartar.*
2. **Recall** Describe five of the functions of the skin.
3. **Vocabulary** What are the *epidermis* and *dermis?*

Thinking Critically

4. **Hypothesize** A friend tells you, "I don't have to go to the dentist because I've never had a cavity." Do you agree or disagree with the conclusion your friend has made. Why?
5. **Apply** You're going to vacation at the beach. What actions can you take to protect your skin?

Applying Health Concepts

6. **Growth and Development** Head lice spread easily among groups of younger children, especially in schools. Work with a partner to create a poster for children, suggesting ways to avoid getting head lice.
7. **Consumer Health** Compare three brands of shampoo sold in stores. How are they alike? How do they differ? Prepare an oral or written report explaining how a consumer might go about deciding which of those three brands to buy.

Lesson 3: Caring for Your Teeth, Skin, and Hair

More About...

Head Lice It is important to communicate to students and family members some basic facts about head lice:

▶ Head lice are not an indication of dirty hair or of bad hygiene. Anyone can have head lice.
▶ Head lice move from one scalp to another— that is, they're contagious. For this reason, it is important to treat lice as soon as they are identified. It is also important to notify friends, other family members, and the school.
▶ Head lice may cause itching, but some individuals have no indications that they have been infected with lice. The only way to determine whether a person has head lice is to check his or her hair and scalp.

Chapter 3 Lesson 3

Assess

EVALUATING THE LESSON

Assign Reviewing Terms and Facts and Thinking Critically on page 75 to review the lesson; then assign the Lesson 3 Quiz in the TCR to evaluate students' understanding.

LESSON 3 REVIEW

Answers to Reviewing Terms and Facts

1. Plaque is a thin, sticky film left on the teeth by bacteria and food. If plaque is not cleaned away, it can harden into tartar.
2. Functions of the skin include waterproofing, sensation, temperature control, protection, and the formation of vitamin D.
3. The epidermis is the outermost layer of skin. The dermis is the skin's inner layer that contains blood vessels, nerve endings, and hair follicles.

Answers to Thinking Critically

4. Responses will vary.
5. Use sunscreen or sunblock lotion to protect skin; minimize sun exposure between 10:00 a.m. and 3:00 p.m.

RETEACHING

▶ Assign Concept Map 12 in the TCR.
▶ Have students complete Reteaching Activity 12 in the TCR.

ENRICHMENT

Assign Enrichment Activity 12 in the TCR.

Close

Draw a health triangle on the board, labeling the three sides physical health, mental/emotional health, and social health. Ask: How can taking care of your teeth, skin, and hair affect all three elements of your health?

Lesson 4
Caring for Your Eyes and Ears

Focus

LESSON OBJECTIVES

After studying this lesson, students should be able to
- explain how the eyes work.
- list eye care tips and discuss how to deal with vision problems.
- explain how the ears work.
- discuss practices to protect the health of the ears.

MOTIVATOR

Display several pairs of sunglasses, or show several photographs of teens wearing sunglasses. Ask: Why do people wear sunglasses? Have students list as many reasons as they can think of.

INTRODUCING THE LESSON

Allow several volunteers to share their responses to the Motivator question. Encourage students to recognize that people may have a variety of reasons for wearing sunglasses, and emphasize that the most important reason is to protect the health of the eyes. Ask: What else can you do to protect your eyes? After several suggestions, tell students that they will learn more about their eyes—and their ears—and about protecting the health of these important sense organs.

INTRODUCING WORDS TO KNOW

Have students form small groups to read and discuss the Words to Know, using the Glossary to check and correct their understanding of each term. Then have group members work together to draw one or more word webs to illustrate the relationships among these Words to Know.

Lesson 4 Caring for Your Eyes and Ears

This lesson will help you find answers to questions that teens often ask about caring for their eyes and ears. For example:

▶ How can I keep my eyes healthy?
▶ What should I know about wearing glasses or contacts?
▶ How can I protect my ears from loud noises?

Words to Know

cornea
iris
pupil
lens
retina
optic nerve
optometrist
ophthalmologist
astigmatism
decibel

How Your Eyes Work

Your eyes tell you about the world—about light, darkness, shapes, colors, and movement. The data gathered by your eyes is interpreted by your brain, allowing you to recognize your friend coming toward you in the hallway or the words on this page. To see how your eyes work, look at **Figure 3.11**.

Figure 3.11
The Eye

The eye is like a camera. It takes in light and focuses it to create an image. This image is sent to the brain, which "develops" the picture.

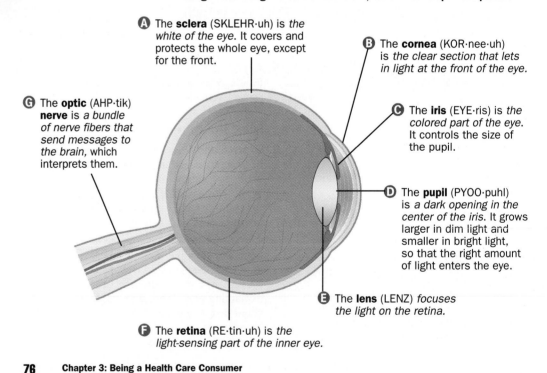

ⓐ The **sclera** (SKLEHR·uh) is *the white of the eye*. It covers and protects the whole eye, except for the front.

ⓑ The **cornea** (KOR·nee·uh) is *the clear section that lets in light at the front of the eye*.

ⓒ The **iris** (EYE·ris) is *the colored part of the eye*. It controls the size of the pupil.

ⓓ The **pupil** (PYOO·puhl) is *a dark opening in the center of the iris*. It grows larger in dim light and smaller in bright light, so that the right amount of light enters the eye.

ⓔ The **lens** (LENZ) focuses the light on the retina.

ⓕ The **retina** (RE·tin·uh) is *the light-sensing part of the inner eye*.

ⓖ The **optic** (AHP·tik) **nerve** is *a bundle of nerve fibers that send messages to the brain*, which interprets them.

Chapter 3: Being a Health Care Consumer

Lesson 4 Resources

Teacher's Classroom Resources
- Concept Map 13
- Enrichment Activity 13
- Lesson Plan 4
- Lesson 4 Quiz
- Reteaching Activity 13

Student Activities Workbook
- Study Guide 3
- Applying Health Skills 13

Caring for Your Eyes

Because your eyes tell you so much about your world, you will want to protect them. To care for your eyes, follow these tips.

- Make sure that you have enough light. Read and watch television in a well-lighted room. If necessary, use a reading lamp.

- Avoid having too much light. Sit at least 6 feet away from the television set. Do not look directly at the sun or at any other bright light. When you are outside, wear sunglasses that protect your eyes from ultraviolet (UV) rays.

- Don't rub your eyes. If dirt gets into your eyes, rubbing them may scratch the cornea. Instead, rinse your eyes with cool water.

- Protect your eyes from injury. Wear protective equipment when playing sports, such as baseball or hockey, that could result in eye injury. Also, wear protective glasses or goggles if you work with power tools or chemicals.

- Have your eyes examined regularly by an eye care professional. An **optometrist** (ahp·TAH·muh·trist) *is trained to examine the eyes for vision problems and to prescribe corrective lenses.* An **ophthalmologist** (ahf·thahl·MAH·luh·jist) is *a physician who specializes in the structure, functions, and diseases of the eye.* If you wear either glasses or contact lenses, you should have your eyes checked once a year. Otherwise, have them checked every two years.

In Your Journal

Do you think that you take good care of your eyes? Name one habit you have now that promotes good eye care. Then list two things you might start doing immediately that would help you take better care of your eyes.

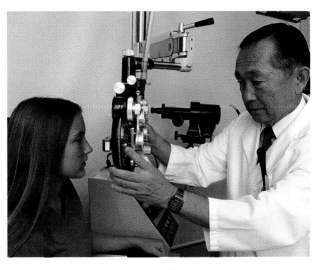

Regular eye checkups can detect vision problems before they become serious.

Problems with Vision

When you have your eyes examined, the doctor will check for several possible problems. Your vision will be tested with an eye chart, and the parts of your eye will be examined to see if they are working properly. The doctor will also check for glaucoma (glaw·KOH·muh), a disease in which fluid builds up in the eyes, causing pressure that can destroy the optic nerve.

Three common problems that may be detected are nearsightedness, farsightedness, and astigmatism. With nearsightedness, only objects close to the eye can be seen clearly. Farsightedness is just the opposite. Faraway objects are seen clearly, but nearby objects appear blurry. **Astigmatism** (uh·STIG·muh·tiz·uhm) is *an eye condition in which images are distorted.* Both faraway and close-up objects appear wavy or blurry.

Lesson 4: Caring for Your Eyes and Ears   77

More About...

Natural Protection of the Eyes
The structure of the skull and face offers natural protection to the eyes:

▶ The eyelashes and eyebrows help prevent too much light from entering the eye.

▶ The eyelid closes to prevent excessive light from meeting the eye and to protect the eye from approaching objects.

▶ The eyelid blinks—it closes automatically and regularly. This frequent, brief closing of the eyelid wipes the eye clean.

▶ Tear fluid washes over the eye in small amounts with every blink.

Chapter 3 Lesson 4

L1 Guest Speaker

If your school screens students for vision problems, ask the school nurse or other professional who manages this screening to speak with the class. What specific tests are included in the screening? What kinds of problems will the screening detect? How does the school respond when a possible vision problem is detected in a student? After the presentation, allow time for the class to ask questions.

L2 Demonstrating

Let volunteers demonstrate and/or describe the correct daily care for glasses and for specific kinds of contact lenses (hard lenses, soft lenses, long-term lenses, for example).

Visual Learning

Guide students in identifying and discussing the structures of the ear, as shown in Figure 3.12. Which structure is part of the outer ear? (*external auditory canal*) Which structures are part of the middle ear? (*eardrum, hammer, anvil, stirrup*) Which are part of the inner ear? (*oval window, Eustachian tube, semicircular canals, auditory nerve, cochlea*)

L2 Science

Pose these questions for students' consideration: How does sound travel? How do differences in volume and pitch affect the size and shape of sound waves? How fast do sound waves travel? Is the speed of sound waves affected by volume or pitch? What other questions do you have about sound and how it travels? Let students research the answers to some or all of the questions, and ask them to share their findings in groups.

Q & A

Contact Solutions

Q: I'd like to get contact lenses. What kinds of lenses are there, and which is best?

A: There are three basic types of contact lenses. Rigid gas-permeable (or "hard") lenses are easy to take care of, but many people find them uncomfortable. Soft lenses are easier to get used to, but they are harder to clean. Extended-wear lenses are worn for 30 days and then thrown away. However, it is possible to forget about them and wear them too long. If you are not sure which type is best for you, ask your doctor.

Figure 3.12
The Ear
Some of the structures of the ear are not directly concerned with hearing. The Eustachian (yoo-STAY-shuhn) tube leads from the back of the eardrum to the throat. This tube keeps the air pressure equal on both sides of the eardrum. The semicircular canals are filled with fluid and tiny hairs. These hairs send messages through nerves to the brain, helping your body keep its balance.

Correcting Vision Problems

Many vision problems can be corrected with eyeglasses or contact lenses. Both types of lenses correct the focusing problems of the eye. Eyeglasses, however, are mounted on frames, while contact lenses rest on the cornea. If you wear contact lenses, it is important to care for them properly. Here are some points to remember.

- Clean your lenses every day. Different types of lenses require different methods of cleaning. Follow your doctor's instructions for proper cleaning and storage of your lenses.
- Always wash your hands before you handle your lenses.
- Always remove your contact lenses before going to sleep, unless you have been told by your eye doctor that it is safe to wear them to bed.
- Replace your lenses if they are torn, cracked, or warped. Damaged lenses can scratch the cornea of your eye. These scratches can lead to eye infections or even to blindness.

How Your Ears Work

Your ears allow you to hear and help you keep your balance. Their structure is quite complex. **Figure 3.12** shows the structures of the ear, which are usually organized into three parts. The way these parts work together is explained below.

- The outer ear is shaped like a cup to pick up sound waves, which are vibrations in the air. The sound waves travel through the external auditory canal.
- In the middle ear, these waves make the eardrum vibrate. The vibrations of the eardrum move three tiny bones called the hammer, the anvil, and the stirrup. These bones carry the vibrations to the oval window.
- In the inner ear, the oval window causes the fluid in the cochlea (KOK·lee·uh) to move. Tiny hairs lining the cochlea vibrate in response, sending electrical messages to the auditory nerve. These messages travel to the brain, which identifies the sound.

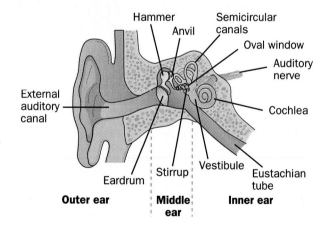

Cooperative Learning

Math Activity Present the following fact: The average person blinks 12 times every minute. Have students meet in small groups to plan short skits. Each skit should present a word problem that involves the number of times people blink. Simple skits might involve one person waking up at a certain time and falling asleep at another time—how many times did the person blink during that waking period? More complicated skits might involve a group of people spending time together—how many times in all did they blink?

Let each group perform its skit for the rest of the class. After each skit, have students write out their solution to the math problem presented in the skit.

Caring for Your Ears

One way to care for your ears is to protect them from loud sounds. *The unit for measuring the loudness of sound waves* is the **decibel.** Normal conversation is about 60 decibels. Sounds over 125 decibels are loud enough to be painful. Lower levels of sound can also harm the ears if the sounds continue for a long time. Some tips for protecting your ears from noise and preventing other problems include the following.

- Keep the volume fairly low on your radio and television. This is especially important if you are using headphones.

- Wear ear plugs or other hearing protection if you are going to be exposed to loud, prolonged noise, such as the noise from a power lawnmower.

- Clean the outside of your ear with a wet washcloth. Do not put a cotton swab or any other object into your ear canal. If ear wax builds up and becomes a problem, see a doctor to have it removed.

- On cold days, wear earmuffs or a hat that covers your ears. Cold air can irritate your middle ear. It can also cause frostbite on your outer ear.

- See a doctor promptly if you have pain in the ear, an ear infection, or a hearing problem.

Science Connection

Over There

How do you tell which direction a sound comes from? The sound reaches the closer ear about 1/1500 of a second sooner than it reaches the ear that is farther away. Also, the sound is slightly louder in the closer ear. These small differences are analyzed by the brain instantaneously.

Adjust the volume on headphones or earphones to avoid ear damage.

Review Lesson 4

Using complete sentences, answer the following questions on a separate sheet of paper.

Reviewing Terms and Facts

1. **Vocabulary** Identify the *cornea, iris, pupil, lens,* and *retina.*
2. **Define** Explain what an *optometrist* and an *ophthalmologist* are.
3. **Vocabulary** What is a *decibel?*

Thinking Critically

4. **Evaluate** What kinds of factors might you consider when you choose an eye care professional?

5. **Relate** What types of activities or situations in your life might be harmful to your hearing? What could you do to protect your hearing?

Applying Health Concepts

6. **Consumer Health** Write an advertisement for sunglasses, ear plugs, or any other type of protective eyewear or earwear. Explain what the product does and why it is important to use it. Include an illustration if you wish. Display the advertisement in your classroom.

Lesson 4: Caring for Your Eyes and Ears   **79**

More About...

Contact Lenses Teens—and other eyewear consumers—should be aware of the advantages and disadvantages of contact lenses.

▶ Advantage: In many cases, contact lenses can correct vision even more effectively than glasses can.

▶ Advantage: Contact lenses permit good peripheral vision, which is usually distorted for those who wear glasses.

▶ Disadvantage: Individuals with dry or sensitive eyes may be unable to adjust to wearing contact lenses.

▶ Disadvantage: Wearing contact lenses requires a commitment of time and effort.

Chapter 3 Lesson 4

Assess

EVALUATING THE LESSON

Assign Reviewing Terms and Facts and Thinking Critically on page 79 to review the lesson; then assign the Lesson 4 Quiz in the TCR to evaluate students' understanding.

LESSON 4 REVIEW

Answers to Reviewing Terms and Facts

1. The cornea is the clear section that lets in light at the front of the eye. The iris is the colored part of the eye. The lens is the part of the eye that focuses light on the retina. The retina is the light-sensing part of the inner eye.
2. An optometrist examines the eyes for vision problems and prescribes corrective lenses. An ophthalmologist specializes in the structure, functions, and diseases of the eye.
3. A decibel is a unit for measuring the loudness of sound waves.

Answers to Thinking Critically

4. Responses will vary.
5. Responses will vary.

RETEACHING

▶ Assign Concept Map 13 in the TCR.
▶ Have students complete Reteaching Activity 13 in the TCR.

ENRICHMENT

Assign Enrichment Activity 13 in the TCR.

Close

Ask students to spend one minute using their eyes to observe details about the classroom. Then have them spend one minute listening for sounds in the classroom. Ask them to identify what they saw and heard.

79

Teen Health Digest

Focus
The Teen Health Digest articles can be used in two ways: as an individual activity for reflection and enrichment, or as a cooperative learning activity as described below.

Motivator
Ask students to review their own roles as health care consumers. What questions can they ask themselves when they are choosing health care services and products? Suggest that students keep these questions in mind as they read the Teen Health Digest articles. After each article, let volunteers explain how the information in the article relates to their own activities as consumers.

Cooperative Learning Project

Health Careers Day

Setting Goals

Explain that all the class members will work together to plan and put on a Health Careers Day for their entire school. Adults working in many different health-related fields will be available to talk about their work, to discuss their education and training, and to describe some of the satisfactions and advantages of their work.

Delegating Tasks

Let the class work as a large group or in several smaller groups to generate a list of speakers they will invite to participate in the Health Careers Day. Then have each student volunteer to be on one of the following committees: Invitations, Publicity, or Arrangements and Set-Up.

Teen Health Digest

People at Work

Open Wide . . .

Job: Dentist
Responsibilities: To care for patients' teeth, gums, and mouths; teach patients how to keep their teeth healthy; fill cavities; repair or remove teeth; replace missing teeth.
Education and Training: Six to ten years beyond high school, including four years at dental college.
Workplace: Dentists may practice individually or as a group. Many have the help of a dental assistant or dental hygienist.
Positive: High prestige; good earning potential; can set own work schedule.
Negative: Long hours standing up; must deal with patients who are nervous or upset; contact with patients' germs.

Try This:
Gather some information on the history of dentistry. Compare the role of dentists in the past with the way they work today.

Health Update

Decay? No Way!

Keeping your teeth clean and limiting sugary snacks will help you fight the battle against tooth decay. Another way to protect your teeth is with dental sealants.

Dental sealants are plastic coatings that your dentist can apply to the chewing surfaces of your teeth. They form a protective covering, shielding tooth enamel from the effects of decay-causing bacteria.

Sealants are not put on all tooth surfaces, so you'll still need to practice good dental hygiene. However, sealants are one possible way that you can keep cavities at bay.

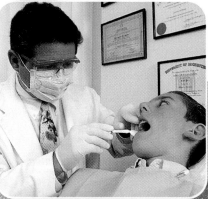

80 Chapter 3: Being a Health Care Consumer

TECHNOLOGY OPTIONS

Photography Students who are interested in photography and have access to cameras may want to make a photographic record of the Health Careers Day. Suggest that the photographers take pictures of each invited speaker, as well as candid shots of some of the guests. Then let interested students work as a group to arrange the photos in an album and add appropriate captions.

Computers Students may be able to use school or home computers as part of this cooperative project. They might use word-processing programs to create formal invitations that can be sent to speakers. In addition, they may be able to use banner programs to make advertising posters and signs to be displayed during the Health Careers Day.

CON$UMER FOCU$

Fashion Statement or Health Risk?

In recent years, body piercing has become increasingly popular. Many people are not aware, however, of the health risks associated with body piercing.

Piercing equipment must be sterile to avoid causing infection. A minor infection that develops after a piercing can be controlled by cleansing the pierced area properly with an antibacterial soap or other cleanser. Equipment that is not sterile, however, can create much more serious problems. These include the passing on of viruses such as HIV, which causes AIDS, and the hepatitis virus.

Because of the potential dangers, some states are working to set health standards for piercing. These standards would regulate techniques and require piercing specialists to be licensed.

Myths and Realities

Seeing the Light

Q. I've heard that indoor tanning, using a sunlamp or tanning bed, is a safe way to get a tan. Is that true?

A. No. The light rays from sunlamps and tanning beds actually penetrate your skin more deeply than the burning rays of the sun. Studies have shown that indoor tanning devices may cause early aging and skin cancer. In fact, cancer from indoor tanning devices can be more severe—and grow faster—than cancer from sunlight. Your eyes can also be damaged by this type of light.

Teens Making a Difference

A Teen for Others

Most voluntary health groups could not work successfully without thousands of dedicated volunteers. One such volunteer is Stephen Massey, a student from Arizona. He has been a volunteer for the March of Dimes Birth Defects Foundation since he was in eighth grade. Massey has done everything from collecting items for mothers and babies who live in crisis centers to helping organize a walkathon fund raiser. His work paid off: in that walkathon, his high school raised more than $21,000, the most collected by any high school.

Try This:
Interview someone who works as a volunteer for a health organization. Ask the person what kind of tasks are involved, and why he or she decided to volunteer.

Teen Health Digest 81

Chapter 3

The Work Students Do

Following are ideas for implementing the activity.

1. Guide students in generating as complete a list as possible of potential speakers. Help students consider local adults other than physicians, dentists, and nurses, such as pharmacists, health insurance agents, accupressurists, and so on. Encourage students to ask their parents or other adults for suggestions as well.
2. Members of the Arrangements and Set-Up Committee should begin by establishing a definite date, time, and place for the Health Careers Day. Members of the Invitations and Publicity committees cannot proceed without this information.
3. Members of the Invitations Committee should send written invitations to each adult who might speak at the Career Day. Also, whenever possible, ask a student who is acquainted with (or related to) the adult to speak with him or her in person or by phone.
4. In addition to advertising the Health Careers Day at school, members of the Publicity Committee may want to put up posters or other notices on community bulletin boards.
5. Be sure students write thank-you letters to all the adults who participate in the Health Careers Day.

Assessing the Work Students Do

Meet briefly but regularly with each committee to be sure that group members are working cooperatively and are making appropriate progress in their planning. After the Health Careers Day, let the members of each committee work together to write a summary and evaluation of their contribution to the project.

Meeting Student Diversity

Language Diversity If the school community includes groups of students and family members who speak languages other than English, students proficient in those languages can serve as interpreters during the Health Careers Day. In addition, you might encourage students to invite speakers whose first language is not English; they may be willing to speak with some of the students or other visitors in their native languages.

Hearing Impaired Encourage students to invite as speakers several adults who work with hearing-impaired clients and patients. If possible, arrange for an American Sign Language interpreter to provide interpretation at the Health Careers Day.

81

Lesson 5

Health Care Providers

This lesson will help you find answers to questions that teens often ask about health care. For example:

- What kinds of health care workers can help me stay well?
- Where can I find help if I become ill or injured?
- Why is health insurance important?

Words to Know

primary care provider
specialist
health insurance
managed care
health maintenance organization (HMO)
Medicare
Medicaid

Prevention and Treatment

When was the last time you visited a doctor? Perhaps you had a bad cough and a fever, or maybe you sprained your ankle playing sports. Doctors treat these and many other kinds of illnesses and injuries. Doctors and other health professionals also try to *prevent* illness and injury. Health care workers, including dental hygienists, nurses, counselors, health teachers, and dietitians work to educate people and help them stay healthy. In addition, there are many voluntary organizations—such as the American Heart Association and the American Cancer Society—that offer health-related information and services. People donate time and money to these groups. The donated funds help to pay for medical research.

When you become ill or have an injury, you probably seek health care at a doctor's office, clinic, or hospital. Any of a wide range of health care workers may help you. Some provide general care, while others have special training to handle particular medical problems.

 in your journal

Washing your hands is one way to prevent the spread of germs. What other kinds of "preventive medicine" do you practice? In your journal, make a list of everyday actions you can take that will help keep yourself and others healthy.

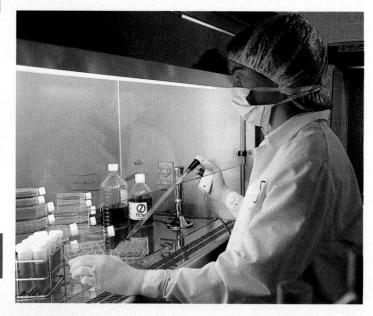

Medical research goes on in many different settings to help prevent and cure disease.

82 Chapter 3: Being a Health Care Consumer

General and Specialized Care

Primary care providers are *the doctors and other health professionals who provide checkups and general care.* For some problems, a doctor may suggest a **specialist** (SPE·she·list). A specialist is *a doctor trained to treat particular types of patients or health matters.* **Figure 3.13** shows some common specialists.

Figure 3.13
Medical Specialists
Which of these specialists are already familiar to you?

Specialist	Specialty
allergist	asthma, hay fever, and other allergies
cardiologist	the heart, its functions, and its diseases
dermatologist	the skin and its diseases
gynecologist	women's health care and diseases
ophthalmologist	the eye, its functions, and its diseases
orthodontist	irregularities of the teeth and jaw
orthopedist	bones, joints, and muscles and their injuries and diseases
pediatrician	infant and child health care and diseases
plastic surgeon	cosmetic surgery to repair damage
psychiatrist	mental and emotional problems
urologist	the urinary system and its diseases

Health Care Facilities

Health care is available in many different places. In addition to doctors' offices, clinics, and hospitals, health care facilities include the following.

- Birthing centers are an alternative to hospitals for giving birth.
- Nursing homes take care of sick, disabled, and elderly people who require special care. Assisted living (or assisted care) facilities are residences for people who need extra help with bathing or other tasks of daily living.
- Drug treatment centers offer help to people with drug-related problems.
- Rehabilitation centers aid people who need special help to recover from serious illness or injury.
- Hospices provide care and support for terminally ill patients.

Cultural Connections
Alternative Medicine

A traditional practice in Chinese medicine is acupuncture (A·kyuh·pung·cher). This treatment involves the inserting of needles into parts of the body. As used in Chinese medicine, acupuncture is a holistic treatment designed to aid the body's own sources of energy. Western practitioners, however, more often use acupuncture to relieve pain. Some therapists have also had success in using acupuncture to help people quit smoking and withdraw from other addictive substances.

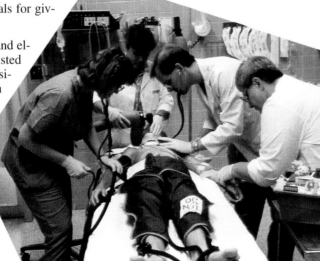

Hospital emergency rooms provide immediate medical care, especially for victims of accidents or sudden illnesses.

Home and Community Connection

Local Specialists Ask students to contact and interview medical specialists who practice in their community. Remind students to prepare for these interviews by researching the particular specialty and by writing several open-ended questions to ask during the interview. Encourage students to find out about the doctor's medical training and current practice, as well as the interests and goals that led to the choice of his or her particular field of medicine. How is this specialty different from general medical practice? from other specialties? What does the doctor consider to be the particular advantages and disadvantages of his or her specialty? After the interviews, have students write short reports and/or discuss their findings with the rest of the class.

Chapter 3 Lesson 5

Teach

L1 Discussing
Lead a class discussion on the differences between treating and preventing illnesses and injuries: Which is easier and more comfortable—prevention or treatment? Which is more cost-effective? What steps can you take to prevent illnesses and injuries? (*diet, exercise, sleep, caution*) How can health professionals help you in your prevention efforts?

L3 Investigating
Let interested students learn more about the relatively new professions of nurse practitioner and physician's assistant: What training do these professionals have? What health care responsibilities can they assume? Which forms of treatment are they not qualified to practice? What special contribution can they make to consumers' health care? Ask the volunteers to discuss their findings with the rest of the class.

L1 Discussing
Encourage students to discuss their own knowledge of, and experiences with, the specialists listed in Figure 3.11.

L2 Language Arts
Let students work in groups to identify the components in each specialist name: What meaning does the root carry? What meaning does the suffix add? How are terms combined? Students might compare, for example, the names and roles of a cardiologist and a cardiovascular surgeon.

L2 Career Education
What careers are available to people who are interested in working in birthing centers? in nursing homes? in drug treatment centers? in rehabilitation centers? in hospices? Let each student select one kind of facility and investigate at least two careers relevant to that setting.

83

Chapter 3 Lesson 5

L1 Life Skills
Help students discuss the kinds of volunteer opportunities available at different health care facilities: What could you offer to people in a nursing home, for example? How would you benefit from volunteering in a health care facility?

L2 Critical Thinking
Encourage students to explore the benefits and problems of organ transplants: How do organ transplants save lives? Do all patients enjoy long or healthy lives after a transplant? Why not? Is it possible to be truly fair in choosing organ recipients? What methods of selecting recipients would you suggest? Why?

L1 Comprehending
To help students understand employer-supported health insurance, ask: What kinds of jobs usually offer health insurance as a benefit? If, a few years from now, you get a part-time job at a fast-food restaurant, will the job likely include health insurance? Why not?

L3 Reporting
Let students work in groups to investigate both the benefits and the problems associated with managed care plans. Ask them to write short, balanced reports on the pros and cons of managed care.

L1 Discussing
Help students consider what happens to people who have no health insurance: What kinds of preventive health care are they likely to receive? How is this likely to affect the costs of the treatment they do receive?

L2 Current Events
Let students work together to investigate the government's current efforts to make health care more affordable.

---

in your journal
You may be surprised at how many different health care providers you and your family have called upon. With other family members, brainstorm a list of health care workers who have served your family over the past five years. Which of these workers are specialists?

Paying for Health Care

Health services are available in many forms. Doctors may work on their own in a private practice or join with others to create a group practice. They may also work for clinics. Medical care is often very expensive. Surgery and hospital stays, for example, typically cost thousands of dollars.

A number of factors account for the rising cost of health care. High-technology equipment and advanced medical procedures, such as organ transplants, are very expensive. Providing care for people with AIDS has made enormous demands on the health care system. In addition, health care providers and medical facilities pay high insurance premiums to protect themselves against possible lawsuits. Drug companies invest huge sums of money to develop and test new medicines and treatments.

Health Insurance

To pay for health care, many people buy **health insurance** (in·SHUR·uhns). This is *a plan in which people pay a set fee to an insurance company in return for the company's agreement to pay some or most medical costs.* People may buy health insurance on their own or as part of a group. Group insurance is offered mainly through people's employers. Some companies pay all or part of the cost of health insurance for their workers. An employee's health insurance may also cover some of the medical expenses for the employee's family.

Many people receive health services through **managed care** plans. Such plans *emphasize preventive medicine and work to manage the cost and quality of health care.* A common form of managed care is the **health maintenance** (MAYN·tuh·nuhns) **organization,** or **HMO.** An HMO *offers its members the services of many different types of health care providers.* Members pay a monthly or yearly fee instead of paying for individual services.

Millions of Americans, however, are not covered by health insurance. Their employers may not offer it—or may pay only a small part of the cost—and they cannot afford insurance on an individual basis. The government and the health care industry are looking for ways to make health care more affordable. None of the types of health insurance currently available is perfect. A lower cost usually means fewer services, while the higher cost of private insurance is out of reach for millions of American families.

Modern medical technology helps doctors diagnose and treat patients. However, this new equipment is costly.

84   Chapter 3: Being a Health Care Consumer

Personal Health

Most teens see pediatricians or family primary care providers. Some, however, may have health providers who specialize in adolescent medicine. These health professionals focus on issues that relate to all aspects of the health of adolescents, including the physical changes during puberty, and risk behaviors. These physicians are represented by the Society for Adolescent Medicine (SAM), a multidisciplinary organization of professionals committed to improving the physical and psychosocial health and well-being of all adolescents. The American Medical Association also supports health professionals who work with adolescents; its program is called GAPS, Guidelines for Adolescent Preventive Services.

Government Programs

Federal, state, and local governments each play a role in health care. For example, all states and most cities have health departments. These departments work to maintain community health standards. They also provide health-related information and services to the public.

Local health department workers promote community health by talking with people and distributing information.

The federal government has two insurance programs to help pay the cost of health care. **Medicare** (MED·i·kehr) *provides health insurance to people who are 65 years old or over.* Medicare covers hospital care, but patients must buy other insurance to pay doctors' bills and other costs.

Medicaid (MED·i·kayd) is *a public health insurance program for low-income families and individuals.* In general, Medicaid offers coverage for young children, people whose personal or family income is below a certain level, and people who are disabled. The federal and state governments both support this program. Because Medicaid programs are run by state governments, however, rules vary from state to state.

Review — Lesson 5

Using complete sentences, answer the following questions on a separate sheet of paper.

Reviewing Terms and Facts

1. **Vocabulary** What is the difference between a *primary care provider* and a *specialist?*
2. **List** Identify three factors that help explain the rising cost of health care.
3. **Vocabulary** What is a *health maintenance organization (HMO)?*
4. **Recall** Describe two health insurance programs provided by the federal or state governments.

Thinking Critically

5. **Explain** How do primary care providers and specialists work together to treat patients?
6. **Analyze** Why is it important for a family to be covered by health insurance?

Applying Health Concepts

7. **Personal Health** Draw a map of the route from your home to the nearest hospital emergency room. How could you get there in a hurry if you had to?
8. **Health of Others** Contact the American Heart Association, the American Diabetes Association, or another voluntary health organization. Ask about the organization's purpose and activities. Share the information with your class.

Lesson 5: Health Care Providers   85

Comsumer Health

Although most students at this age are unlikely to be making their own choice of health care providers, they will soon have that responsibility for themselves and perhaps for other family members. These guidelines can help anyone looking for a new health care provider:

▶ Begin by checking your health insurance. What limits does your policy place on your selection?
▶ Ask health care professionals whom you know and trust for their recommendations.
▶ Meet and talk with the health care provider you are considering. Ask questions about health issues that concern you.

Chapter 3 Review

CHECKING COMPREHENSION

Use the Chapter Summary and the Chapter 3 Review to go over the most important ideas presented in Chapter 3. Encourage students to ask questions and add details as appropriate.

CHAPTER 3 REVIEW ANSWERS

Reviewing Key Terms and Concepts

1. Responses will vary. Students should point out that they can evaluate information carefully, understand the kinds of claims advertisements are likely to present, and make purchasing decisions based on facts.
2. An infomercial is a TV commercial that is usually a half hour long. It appears to present information, but it is actually selling a product.
3. Comparison shopping involves comparing products, evaluating their benefits, and choosing those that offer the best value.
4. Discount stores offer reduced prices. Generic products are sold in plain packages at lower prices than brand name goods.
5. Responses will vary.
6. Melanin is the substance that gives skin its color.
7. The optic nerve sends messages from the eye to the brain.
8. Objects are seen clearly near at hand for the nearsighted, far away for the farsighted. Astigmatism distorts images near and far.
9. Responses will vary. Possible response: You could seek medical care at a doctor's office, a clinic, or a hospital.
10. Managed care refers to insurance plans that emphasize preventive medicine and try to manage the cost and quality of health care.

Chapter 3 Review

Chapter Summary

▶ **Lesson 1** To make wise consumer choices, you need to gather and evaluate information and recognize how advertising and other factors influence your decision making.

▶ **Lesson 2** Always compare personal products carefully before you buy. Product labels provide helpful information.

▶ **Lesson 3** Caring for your teeth, skin, and hair involves brushing and flossing, regular dentist visits (teeth); good hygiene, protection from the sun (skin); and regular shampooing (hair).

▶ **Lesson 4** Caring for your eyes and ears includes regular eye checkups and protecting your ears from loud sounds.

▶ **Lesson 5** Many different health care professionals treat illnesses and injuries and work to prevent them. Health insurance helps people pay for health care.

Reviewing Key Terms and Concepts

Using complete sentences, answer the following questions on a separate sheet of paper.

Lesson 1
1. What can you do to be a wise consumer?
2. What is an *infomercial*?

Lesson 2
3. What is *comparison shopping*?
4. How can knowing about discount stores and generic products help you get good value for your money?

Lesson 3
5. What actions can you take to prevent cavities?

6. What is *melanin*?

Lesson 4
7. What is the *optic nerve*?
8. Describe nearsightedness, farsightedness, and astigmatism.

Lesson 5
9. If you were ill or injured, where could you go for medical care? Give several examples.
10. What is *managed care*?

Thinking Critically

Using complete sentences, answer the following questions on a separate sheet of paper.

11. **Synthesize** How can analyzing advertising claims help you make wise purchasing decisions?
12. **Apply** Describe two potentially serious health consequences that might result from having poor consumer skills.
13. **Explain** How are healthy teeth the result of both good hygiene and good diet?
14. **Predict** How might allowing eye or ear problems to go untreated affect other areas of your health?
15. **Assess** Do you think that companies should be required to provide health insurance for employees and their families? Why or why not?

86 Chapter 3: Being a Health Care Consumer

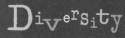 **Meeting Student Diversity**

Language Diversity Use the following suggestions to help students who have difficulty with English:

▶ Pair those learners with native speakers of English who can restate the Chapter Summary in language that helps students comprehend important concepts.

▶ Direct auditory learners or those students with language diversity to the Teen Health Audiocassette Program. Available in English and Spanish, this component provides an audio and written summary of the chapter. 🎧

Chapter 3 Review

Your Action Plan

Worthwhile goals always require effort. For example, it takes effort to stay healthy and to look your best. It also takes effort to be a wise consumer.

Step 1 Review your journal entries for this chapter. What do they suggest about personal goals you might set for yourself?

Step 2 Set some short-term and long-term personal goals for yourself. A possible short-term goal might be to compare at least three different brands before making your next product purchase. A long-term goal might be limiting your skin's exposure to the sun.

Step 3 Alongside each of your goals, list specific steps you can take to accomplish the goal.

Periodically, check your progress. Feel good about what you've accomplished, and make note of what you still need to do.

In Your Home and Community

1. **Health of Others** Find out from your teacher, school nurse, or school counselor what opportunities you might create to help younger students learn how to take proper care of their teeth, skin, hair, eyes, and ears. For example, students from your school might conduct a "Care for Yourself" clinic for elementary school children. The clinic could combine live demonstrations with informative posters and handouts.

2. **Consumer Health** Prepare a "Smart Shopper" newsletter for your local community. Include tips for comparison shopping, analyzing advertisements, and reading product labels. Find out how you can post or distribute your newsletter.

Building Your Portfolio

1. **Analysis of Advertisements** Clip several magazine advertisements for personal products. Attach each ad to a sheet of paper. Below the ad, write your analysis of how the advertiser is trying to persuade you to buy the product. Add your analyses to your portfolio.

2. **Careers in Health Care** A variety of careers exist in the health care field. Check the employment section of the newspaper, or look at the *Occupational Outlook Handbook* at the library. Choose two health care jobs that sound interesting. Research their qualifications and responsibilities, and write a summary of your findings. Put the summary in your portfolio.

3. **Personal Assessment** Look through all the activities and projects you did for this chapter. Choose one or two that you would like to include in your portfolio.

Chapter 3: Chapter Review **87**

Performance Assessment

▶ **Self-evaluation** Direct students to review the activities that are provided throughout the chapter. Encourage each student to select one finished product or activity that demonstrates his or her best work for the chapter. Have students explain what they learned and how the examples they selected show their progress.

▶ **Teacher's Classroom Resources** Assign Performance Assessment 3, "Advertising Campaign for Health Care Consumers," in the TCR.

Chapter 3 Review

Thinking Critically
Responses will vary. Possible responses are given:

11. Analyzing claims can help you distinguish between what an ad implies a product can do and what the product actually does.

12. A person who is taken in by quackery could fail to get adequate treatment for a serious condition or might use a product that is harmful. A person who fails to read a label might use a product incorrectly, resulting in illness or injury.

13. To keep teeth healthy, you need to brush and floss them regularly (good hygiene). In addition, you must consume a good diet, eat foods high in nutrients that build strong teeth, and avoid sugary foods that are especially likely to lead to cavities.

14. If you don't take care of your eyes and ears, you may not be as aware of your surroundings as you should be; you could have an accident as a result.

15. Responses will vary.

RETEACHING
Assign Study Guide 3 in the Student Activities Workbook.

EVALUATE
▶ Use the reproducible Chapter 3 Test in the TCR, or construct your own test using the Testmaker Software.

▶ Use Performance Assessment 3 in the TCR.

EXTENSION
Encourage students to work together to learn about consumerism: What is this movement? When and how did it begin? What government agencies and private organizations seek to protect the rights and promote the interests of health care consumers?

87

Unit 2
Nutrition and Fitness

UNIT OBJECTIVES

Students will learn about the important roles nutrition and exercise play in the development and maintenance of total health. They will learn how to make healthy food choices, how to integrate regular physical activity into their lives, and how to make responsible decisions about sports and conditioning. They will also learn about several of the main body systems.

UNIT OVERVIEW

Chapter 4 Food and Nutrition

Chapter 4 introduces the Food Guide Pyramid and helps students understand how they can use the Pyramid as the basis for planning a healthful diet. The chapter also presents information about the human digestive and excretory systems.

Chapter 5 Physical Activity and Weight Management

In Chapter 5, the students explore the many advantages of developing and maintaining physical fitness. They also learn about the structure and function of the human circulatory, skeletal, and muscular systems.

Chapter 6 Sports and Conditioning

Chapter 6 gives students an opportunity to explore various sports and to develop their own plans for conditioning. It also helps students understand how they can balance the various responsibilities of their busy daily lives.

Bulletin Board Suggestion

Keep Moving! Healthy physical activity takes many forms. Collect photographs from magazines or drawings that show teens engaged in a wide variety of sports and other physical activities. If at all possible, include active teens with physical disabilities. Use the bulletin board to help students explore the wide variety of ways in which they can "keep moving" and thus maintain or improve their physical fitness. Encourage students to suggest other activities that might be shown on the board, and ask volunteers to find and add appropriate pictures.

Unit 2
Nutrition and Fitness

Chapter 4 Food and Nutrition

Chapter 5 Physical Activity and Weight Management

Chapter 6 Sports and Conditioning

Unit 2

INTRODUCING THE UNIT

Ask: What have you eaten so far today? How do you think the foods you've eaten—and the foods you haven't eaten—affect how you feel? What exercise have you had, or are you planning to have later today? How does that activity make you feel? After volunteers have shared their experiences and ideas, let students form cooperative learning groups. Ask the members of each group to work together in writing a meal plan—including snacks—and an exercise plan for one average day in the life of a student their own age. Provide an opportunity for groups to share their ideas with the rest of the class. Explain that students will learn more about healthy eating and exercising as they read and discuss the chapters in Unit 2. At the end of the unit, let students meet in the same groups to review and revise their diet and exercise plans.

VIDEODISC/VHS

Teen Health Course 2

You may wish to use video segment 6, "Sports and Conditioning," in which members of a swim team share secrets of their success in setting and achieving performance goals.

Videodisc Side 1, Chapter 6
Sports and Conditioning

Search Chapter 6, Play to 7

DEALING WITH SENSITIVE ISSUES

A General Approach As a health teacher, you have the opportunity to help students deal with sensitive issues in several ways. One way is by sharing accurate information and professional resources about the issues with students and doing so in ways that won't embarrass them, make them feel uncomfortable, or jeopardize their self-esteem. Knowledge alone is not enough, however. Experience shows that just conveying information is less effective in changing behaviors than is teaching *skills*. For example, telling students to "just say no" to drugs is a less effective deterrent to drug use than teaching them *how* to say no.

Planning Guide

Chapter 4
Food and Nutrition

	Features	Classroom Resources
Lesson 1 **The Food Guide Pyramid** *pages 92–95*		Concept Map 15 Enrichment Activity 15 Lesson Plan 1 Lesson 1 Quiz Reteaching Activity 15 Transparency 14
Lesson 2 **Nutrients for Health and Wellness** *pages 96–101*	Life Skills: Making Healthful Choices at Fast-Food Restaurants *page 100*	Concept Map 16 Cross-Curriculum Activity 7 Decision-Making Activity 7 Enrichment Activity 16 Lesson Plan 2 Lesson 2 Quiz Reteaching Activity 16 Transparencies 15, 16
Lesson 3 **Healthful Meal Planning** *pages 102–105* TEEN HEALTH DIGEST *pages 106–107*	Making Healthy Decisions: Making Breakfast Important *page 104*	Concept Map 17 Cross-Curriculum Activity 8 Decision-Making Activity 8 Enrichment Activity 17 Lesson Plan 3 Lesson 3 Quiz Reteaching Activity 17 Transparency 17
Lesson 4 **The Digestive and Excretory Systems** *pages 108–113*	Health Lab: The Digestion of Protein *pages 110–111*	Concept Map 18 Enrichment Activity 18 Health Lab 4 Lesson Plan 4 Lesson 4 Quiz Reteaching Activity 18

National Health Education Standards

One of the main goals of the National Health Education Standards is to move students toward "health literacy." Health literacy is the capacity of individuals to obtain, interpret, and understand basic health information and services and the competence to use such information and services in ways which promote health. The health standards were developed by applying the characteristics of a well-educated, literate person within the context of health. The health literate person is:

- ▶ a critical thinker and problem solver
- ▶ a self-directed learner
- ▶ a responsible, productive citizen
- ▶ an effective communicator

Listed below are the Health Education Standards Performance Indicators addressed in each lesson of this chapter.

Lesson	Health Standards Performance Indicators
1	(6.1, 6.3)
2	(1.6, 3.1, 6.1)
3	(3.4, 6.1, 6.6)
4	(1.1, 1.7, 3.1)

ABCNEWS InterActive Videodisc Series

You may wish to use the videodisc *Food and Nutrition* with this chapter. See side one, video segments 3, 4, 6, 8, 9, 13, 14; side two, video segments 4, 5, 6, 12, 13. Use the *ABCNews InterActive™ Correlation Bar Code Guide* for title reference. Also available in VHS format.

Chapter Resources

Teacher's Classroom Resources
- Chapter 4 Test
- Parent Letter and Activities 4
- Performance Assessment 4
- Testmaker Software

Student Activities Workbook
- Study Guide 4
- Applying Health Skills 15–18
- Health Inventory 4

Student Diversity Strategies
- Audiocassette Program (English)
- Audiocassette Program (Spanish)
- Spanish Parent Letters
- Spanish Summaries, Quizzes, and Activities

Multimedia Components
- English Audiocassette Program
- Spanish Audiocassette Program
- *Teen Health* Videodisc/VHS Series

Other Resources

Readings for the Teacher
Duyff, Roberta Larson, R.D. *The American Dietetic Association's Complete Food and Nutrition Guide.* Minneapolis, MN: Chronimed Publishing, 1996.

Napier, Kristine M. *How Nutrition Works.* Emeryville, CA: Ziff-Davis Press, 1995.

Snyder, S. H., M.D., ed. *Nutrition and the Brain.* New York: Chelsea House, 1992.

Readings for the Student
Denny, Sharon. "Food News: What Can You Believe?" Current Health 2, December 1995, pp. 6–12.

Jacobson, M. F., et al. *Safe Food: Eating Wisely in a Risky World.* Venice, CA: Living Plant Press, 1991.

The Nutrition-Fitness Link: How Diet Can Help Your Body and Mind. Brookfield, CT: Millbrook Press, Inc., 1993.

Out of Time?

If time does not permit teaching this chapter, you may use these features: Life Skills on page 100; Making Healthy Decisions on page 104; Teen Health Digest on pages 106–107; Health Lab on pages 110–111; and the Chapter Summary on page 114.

Chapter 4
Food and Nutrition

CHAPTER OVERVIEW

Chapter 4 helps students understand their own nutritional needs and provides guidelines for planning a healthful, balanced diet.

Lesson 1 explains how the Food Guide Pyramid can be used to make healthful daily food choices.

Lesson 2 identifies the six categories of nutrients and explains how RDAs can be used as guidelines for including necessary nutrients in a daily diet.

Lesson 3 discusses the importance of eating a nutritious breakfast and explains how students can make healthful food choices throughout the day.

Lesson 4 identifies the major organs of the digestive and excretory systems and explains how those systems function.

PATHWAYS THROUGH THE PROGRAM

As their bodies change and grow, most teens are highly conscious of their own appearance. At the same time, teens are likely to ignore the connections between diet and health, between eating well and feeling and looking good. The lessons in this chapter help students recognize how they can make daily food choices that provide proper nutrition and enhance all three aspects of their own health—physical, mental/emotional, and social. As students begin studying this chapter, help them recognize that by taking responsibility for their own eating habits, they can gain an important measure of control over their health, both now and in the future.

Chapter 4
Food and Nutrition

Student Expectations
After reading this chapter, you should be able to:
1. Identify ways you can use the Food Guide Pyramid to plan healthful food choices.
2. Recognize the nutrients that you need each day for energy, growth, and health.
3. Explain healthful choices for breakfast, other meals, and snacks.
4. List the steps in the digestive process and describe ways to take care of your digestive and excretory systems.

Key to Ability Levels

Teaching strategies that appear throughout the chapter have been identified by one of three codes to give you an idea of their suitability for students of varying learning styles and abilities.

L1 **Level 1** strategies should be within the ability range of all students. Full class participation is often required. Teacher direction is usually needed.

L2 **Level 2** strategies are for average to above-average students or for small groups. Some teacher direction is necessary.

L3 **Level 3** strategies are designed for students able and willing to work independently. Minimal teacher direction is necessary.

Teen Chat Group

LaShawna: I don't know what I want. What are you two getting?

Jessica: I'm getting the salad.

Tiffany: Oh, no! Now *I* have to get something healthy, too.

LaShawna: I don't even know what's healthy anymore. One week some food is really bad for you, and the next week it's okay. I'm just going to eat what I like.

Tiffany: Sounds good to me!

Jessica: I know what you mean, LaShawna, but it's not really that hard. I try to eat fruits and vegetables whenever I can. I like them anyway. You know I'm not a fanatic about what I eat—I still eat dessert.

Tiffany: Then *I'm* getting a burger. It's what I want. What about you, LaShawna?

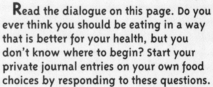

Read the dialogue on this page. Do you ever think you should be eating in a way that is better for your health, but you don't know where to begin? Start your private journal entries on your own food choices by responding to these questions.

▶ What would you say if you were eating lunch with these girls? What might you order?

▶ What influences the choices you make about the foods you eat?

▶ How might your eating habits affect the way you look and feel? How much thought do you give to meal planning?

When you reach the end of the chapter, you will use your journal entries to make an action plan.

Chapter 4: Food and Nutrition 91

Chapter 4

INTRODUCING THE CHAPTER

▶ Divide the class into groups of three or four, and give each group a menu from a local restaurant. Have group members decide on what to order for one meal—breakfast, lunch, or dinner. After the groups have shared their orders with the rest of the class, lead a brief discussion of what was ordered. Ask: What did you take into consideration in making your choices? How healthful do you think this meal is? Why? Conclude the discussion by telling students that they will learn more about food and food choices as they study the lessons of Chapter 4.

▶ Write the word *food* on the board. Then give students three or four minutes to write all the words and phrases that come to mind in response to that word. Let volunteers share some of their responses. Without intruding on any student's privacy, help the class recognize that eating is an activity that relates to physical, emotional, and social health and that can evoke a wide range of both positive and negative responses.

Cooperative Learning Project

International Food Fair

The Teen Health Digest on pages 106 and 107 provides students with high-interest articles related to the content of this chapter.

The material in the Teacher's Wraparound Edition presents suggestions for a class project in which students prepare a variety of healthful foods from around the world to be shared with their families, the student body, or other members of the community.

Making their own food choices is one of the most significant ways in which teens assume responsibility for their own health. Selecting food in the school cafeteria, in restaurants, or in the family kitchen can be seen as first steps toward adult responsibilities such as planning and preparing daily meals. Remind students that, in order to make meaningful decisions, they must be honest about their own preferences and habits. The students' journals, in which they can record personal entries with assurances of privacy, offer excellent opportunities for exploring the food choices students make now, the reasons for those choices, and the changes they want to make in those choices.

91

Lesson 1
The Food Guide Pyramid

Focus

LESSON OBJECTIVES

After studying this lesson, students should be able to
- discuss factors that influence their food choices.
- explain how they can use the Food Guide Pyramid to make healthy food choices.
- discuss the importance of a varied diet.

MOTIVATOR

Ask students to list all the foods they have eaten in the past 24 hours. On their lists, have students note how much of each food they ate, the approximate time they ate each food, and whether it was a snack or part of a meal.

INTRODUCING THE LESSON

Have volunteers read aloud the lists they wrote as part of the Motivator activity. Select certain foods from each volunteer's list and ask: How many of the rest of you ate that same food? Why do you think you chose to eat that food? Should you have eaten less or more of it? How healthy was your choice?

INTRODUCING WORDS TO KNOW

Go over the Words to Know with the class, and let volunteers offer an informal definition of each term. Then ask students to read the formal definitions in the Glossary. Have students work in pairs to prepare their own sets of flashcards, one consisting of the Words to Know and the other of formal definitions. Encourage students to play matching games with these flashcards as an aid to learning the word definitions.

Lesson 1
The Food Guide Pyramid

This lesson will help you find answers to questions that teens often ask about their daily diets. For example:
- What influences my diet?
- How can I use the Food Guide Pyramid to plan a healthful diet?

Words to Know

nutrition
diet
Food Guide Pyramid
nutrients
calories

Food in Your Life

Food provides your body with the fuel it needs to grow and work properly. After you eat, your body breaks down the food and uses it to build and repair body cells and provide you with energy. **Nutrition** (noo·TRI·shuhn) is the term for *the process of taking in food and using it for growth and good health*. This is the main reason why people need to eat—to fulfill the body's physical needs. Eating satisfies social needs, too, however. Mealtimes can provide a chance to enjoy the company of family members and friends. Eating also has an emotional element. Babies, for example, learn to associate feeding with feelings of closeness. The choices you make about food, therefore, have an impact on every area of your life.

What Influences Your Food Choices?

Now that you are a teen, you make more of the decisions about your **diet,** *all the things you regularly eat and drink.* Many factors influence these choices, including:

- **Sensory appeal.** The ways that foods look, taste, smell, and feel influence what you choose to eat.

Traditions and ethnic background have an influence on food choices. How do your family's traditions and background affect your food choices?

- **Geography.** The land, climate, and agricultural products of your region influence your food choices and food availability.

- **Your Cultural and Family Background.** You may make certain food choices because they are part of your family's cultural or ethnic background.

- **Advertising.** Food advertisements are designed to influence you to choose one food over another.

Chapter 4: Food and Nutrition

Lesson 1 Resources

Teacher's Classroom Resources
- Concept Map 15
- Enrichment Activity 15
- Lesson Plan 1
- Lesson 1 Quiz
- Reteaching Activity 15
- Transparency 14

Student Activities Workbook
- Study Guide 4
- Applying Health Skills 15

The Food Guide Pyramid

There are many ways to combine different foods to create a wholesome and delicious diet. **Figure 4.1** shows the **Food Guide Pyramid,** *a guide for making healthful daily food choices.* The Food Guide Pyramid was developed by the U.S. Department of Agriculture. It puts foods into groups on a diagram, based on the substances they provide to the body. One of the benefits of such a guide is that you can see at a glance the variety of foods you need for good health.

interNET CONNECTION
Research the fat in fast foods, or find simple but delicious-and-nutritious recipes by weighing on to the Web!
http://www.glencoe.com/sec/health

Figure 4.1
The Food Guide Pyramid: A Guide to Daily Food Choices

The Food Guide Pyramid is an excellent tool for planning your food choices each day. At the narrowest part of the pyramid are foods you should eat only in small amounts. As the pyramid gets wider, the suggested number of servings from each food group increases.

KEY
- ○ Shows the amount of naturally occurring and added fats in each group
- ▼ Shows the amount of added sugars in each group

These symbols show that fats and added sugars come mostly from foods in the pyramid tip. The substances can, however, be part of or added to foods in the other food groups as well.

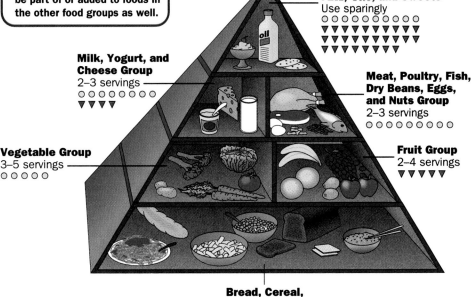

Fats, Oils, and Sweets
Use sparingly

Milk, Yogurt, and Cheese Group
2–3 servings

Meat, Poultry, Fish, Dry Beans, Eggs, and Nuts Group
2–3 servings

Vegetable Group
3–5 servings

Fruit Group
2–4 servings

Bread, Cereal, Rice, and Pasta Group
6–11 servings

More About...

The Food Guide Pyramid The Food Guide Pyramid represents a clear, useful depiction of current recommendations for healthful eating.

▶ A diet that is balanced according to the pyramid guidelines results in increased energy and improved outlook.

▶ Following the pyramid guidelines can help children, teens, and adults achieve or maintain healthful body weight and physical fitness.

▶ A diet based on the Food Guide Pyramid offers protection against five of the ten leading causes of death, including stroke, coronary heart disease, certain cancers, diabetes, and atherosclerosis.

Chapter 4 Lesson 1

Teach

Teacher Talk
Dealing with Sensitive Issues

Students who are—or who feel they are—overweight may be uncomfortable discussing their eating habits with or in front of their peers. Try to keep discussions of diet as general as possible, and ask questions about personal food choices only of students who volunteer.

L1 Recalling
Ask students to recall the three components of health. (*physical, mental/emotional, and social*) How do food and eating relate to all three of these components?

L2 Journal Writing
Ask students to write journal entries in response to these questions: How do you think your peers affect your diet choices? In what sense is their influence positive? Negative?

Visual Learning
Using the Food Guide Pyramid as a visual learning tool, present questions for class discussion. Some examples: Why is the food section at the base of the pyramid so large? Why is the section at the point or apex so small? How does the size of each section relate to the number of servings from that section? Do you think the pyramid helps people recognize that they need a variety of foods? The U.S. government spent nearly $2,000,000 to develop this pyramid—do you think the money was well spent?

Chapter 4 Lesson 1

L1 Analyzing
Discuss with students the differences in the number of recommended servings for foods from the major groups: How do you think age, gender, and activity levels affect the recommended number of servings? Do you think all 12-year-old boys should eat more than all 12-year-old girls? Why or why not? Do you think all the members of a soccer team should have the same number of servings from each food group? Why or why not?

L2 Demonstrating
Let a group of volunteers work together to prepare a set of food cards. Each card should show a picture of a single food, either drawn or cut out from a magazine. The cards should include a wide variety of specific foods from each group in the Food Guide Pyramid. Display the complete set of cards for the class. Ask students to choose cards to plan a healthy meal based on their own age, gender, activity level, and food preferences.

L3 Researching
What is the U.S. Department of Agriculture? When and why was it established? Who is in charge of the U.S.D.A.? Besides publishing the Food Guide Pyramid, what does this agency do? Have interested students find answers to these questions and write brief reports summarizing their answers.

VIDEODISC/VHS

Teen Health Course 2

You may wish to use video segment 4, "Food and Nutrition," to show how a balanced diet and healthful snacking can give teens the energy and nutrients they need for growth and an active life.

Videodisc Side 1, Chapter 4
Food and Nutrition

Search Chapter 4, Play to 5

Getting What You Need

The Food Guide Pyramid shows a suggested range of servings for foods within each of the five major food groups. The number of servings you need depends on your energy needs. Your energy needs depend on your age, gender, and activity level. **Figure 4.2** shows the recommended daily servings in each group for teen girls and for teen boys. **Figure 4.3** shows what this means in terms of an actual day's menu.

Figure 4.2
Recommended Daily Servings
Daily servings should always take into account your energy needs, and should be based on *calorie intake* and *activity level*.

Food Group	Sample Serving Sizes	Teen Girls	Teen Boys
Milk, Yogurt, and Cheese	1 cup (8 oz.) milk or yogurt 1½ oz. natural cheese 2 oz. processed cheese	3 or more servings	3 or more servings
Meat, Poultry, Fish, Dry Beans, Eggs, and Nuts	2–3 oz. cooked lean meat, poultry, or fish The following are equivalent to 1 oz. meat: 1 egg, ½ cup cooked dry beans, 2 Tbs. peanut butter, ⅓ cup nuts	2 servings	3 servings
Vegetables	½ cup cooked or raw chopped vegetables 1 cup raw leafy vegetables ¾ cup vegetable juice	4 servings	5 servings
Fruits	1 medium apple, banana, orange ½ cup chopped, cooked, or canned fruit ¾ cup fruit juice	3 servings	4 servings
Breads, Cereals, Rice, and Pasta	1 slice of bread or 1 muffin 1 oz. ready-to-eat cereal ½ cup cooked cereal, rice, or pasta	9 servings	11 servings

Figure 4.3
A Day's Menu
This menu fulfills the recommended daily servings from the Food Guide Pyramid for a teen girl.

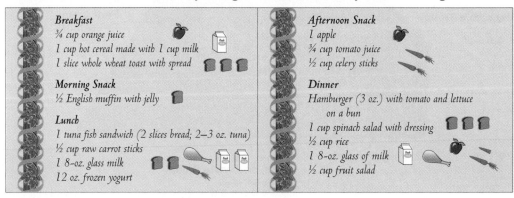

94   Chapter 4: Food and Nutrition

Cultural Awareness
In some cases, students' dietary choices may be guided by religious precepts. Eating pork, for example, is prohibited by Orthodox Judaism, Islam, and Hinduism. For some teens, avoiding specific foods is an important part of practicing their religious faith.

Occasionally fasting from food altogether is another practice of certain faiths. For Muslims, the celebration of Ramadan involves abstaining from food and drink from sunrise to sunset. Many Jews fast during Yom Kippur, and some Christians participate in modified fasts (avoiding only certain foods) during Lent.

Eating a Variety of Foods

Using the Food Guide Pyramid can help you get enough **nutrients** (NOO·tree·ents)—*the substances in foods that your body needs in order to grow, have energy, and stay healthy.* Of course, many of the foods you eat will be made from several food groups—beef stew, for example, will have both meat and vegetables. Even within a food group, different foods have different amounts of nutrients. It's a good idea, therefore, to eat many different types of foods over time from every group. Your meals will also be more interesting that way.

Foods that are high in sugars and fats—represented by the tiny tip of the Food Guide Pyramid—are generally low in other nutrients. Additionally, these foods are high in **calories** (KA·luh·reez), which are *units of heat that measure the energy available in foods.* A diet that is high in calories often leads to weight gain and other types of health problems. Most 11- to 14-year-olds need about 2,200 to 2,500 calories per day.

in Your journal

List the foods that you ate yesterday—breakfast, lunch, snacks, and dinner. Evaluate your daily food choices, using the Food Guide Pyramid and the Recommended Daily Servings chart on page 94. Give specific suggestions for improving your daily diet, based on your findings.

What food groups are represented in this meal?

Review — Lesson 1

Using complete sentences, answer the following questions on a separate sheet of paper.

Reviewing Terms and Facts

1. **Vocabulary** Define the term *Food Guide Pyramid,* and explain how it can be used to plan a healthful diet.
2. **List** Identify the five major food groups in the Food Guide Pyramid, and list the range of servings recommended daily for each one.
3. **Restate** Foods provide *nutrients* and *calories.* Define each term.

Thinking Critically

4. **Differentiate** List two factors that may influence a person's food choices. Describe how these influences could be negative as well as positive.
5. **Explain** Reread the menu in Figure 4.3 on page 94. How might this menu be adapted to the recommended daily number of servings for an active teen boy? Describe additions you could make to the menu.

Applying Health Concepts

6. **Personal Health** Create a three-day menu for yourself that satisfies recommendations and is both varied and appetizing. Share your menu with a classmate.

Lesson 1: The Food Guide Pyramid 95

Personal Health

Food Allergies Many people avoid foods that they consider themselves allergic to, but true food allergies are relatively uncommon. About 6 percent of children suffer from food allergies, and many outgrow their allergies by adulthood. The symptoms of a food allergy may appear immediately after eating the food or several hours later. These are some of the varied indications of a food allergy:

▶ skin rash or red skin
▶ swollen lips, tongue, or face
▶ difficulty in breathing
▶ nausea

Lesson 2
Nutrients for Health and Wellness

Focus

LESSON OBJECTIVES

After studying this lesson, students should be able to
- identify the six major categories of nutrients.
- discuss the most important guidelines to follow in planning a healthful diet.
- read and interpret Nutrition Facts labels.

MOTIVATOR

Give students five minutes to draw pictures of an appealing and healthful meal.

INTRODUCING THE LESSON

Let students share and discuss the pictures they drew for the Motivator activity. Select several pictured meals and help students consider the combination of foods suggested in them: What nutrients does each food provide? How do the foods work toward meeting your daily needs? What foods should be included in the other meals and snacks of the day to balance your daily food choices? Conclude the discussion by explaining that students will learn more about specific nutrients and about making wise food choices as they study the lesson.

INTRODUCING WORDS TO KNOW

Guide students in reading aloud the list of Words to Know. Then have pairs of students work together to skim the lesson, find the definition given for each term, and write an original sentence using each word that demonstrates an understanding of the definition.

Lesson 2
Nutrients for Health and Wellness

This lesson will help you find answers to questions that teens often ask about the nutrients in the foods they eat. For example:

▶ What nutrients do I need in order to be healthy?
▶ What substances should I limit in my diet?
▶ How can I use nutrition labels to make wise food choices?

Words to Know

- carbohydrates
- proteins
- vitamins
- minerals
- fats
- fiber
- saturated fats
- cholesterol
- unsaturated fats

The Six Major Nutrients

Using the Food Guide Pyramid to plan your daily diet is the best way to make sure that you are getting all the nutrients you need in order to build and repair your body and to provide enough energy to help you feel fit and well. **Figure 4.4** shows the six categories of nutrients. Each plays an important role in maintaining your health and wellness.

Figure 4.4
The Six Types of Nutrients
These are the major nutrient catagories.

Carbohydrates (kar·boh·HY·drayts) *are the starches and sugars that provide the body with most of its energy.*
   Simple carbohydrates are sugars. Some sugars are found in fruits, milk, and honey. Table sugar is processed from sugar cane or sugar beets.
   Complex carbohydrates include the starches found in breads, pasta, rice, and such starchy vegetables as potatoes and corn. Your body breaks down starch and converts it to sugar.

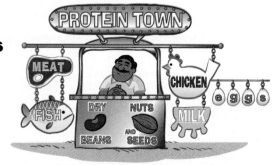

Proteins (PROH·teenz) *are essential nutrients used to repair body cells and tissues.* Proteins are made up of a group of 22 chemicals called amino (uh·MEE·noh) acids. Your body can make many of them on its own, but essential amino acids can be obtained only from some foods.
   Complete proteins are found in foods—such as milk, eggs, and fish—that come from animals. They contain all 9 of the essential amino acids.
   Incomplete proteins are found in foods—such as dry beans and grains—from plants. Such foods do not contain all the essential amino acids. People who do not eat foods that come from animals can get all the amino acids that they need by eating a variety of plant-based foods.

96

Lesson 2 Resources

Teacher's Classroom Resources
- 📁 Concept Map 16
- 📁 Cross-Curriculum Activity 7
- 📁 Decision-Making Activity 7
- 📁 Enrichment Activity 16
- 📁 Lesson Plan 2
- 📁 Lesson 2 Quiz
- 📁 Reteaching Activity 16
- Transparencies 15, 16

Student Activities Workbook
- 📁 Study Guide 4
- 📁 Applying Health Skills 16

Vitamins are *substances that help to regulate the body's functions*. Vitamins also help the body use other nutrients. Some help fight infection. Fruits, vegetables, and whole-grain breads are good vitamin sources.

Water-soluble vitamins, including vitamin C and many B vitamins, dissolve in water. Because the body gets rid of extra amounts of these vitamins in urine, water-soluble vitamins must be replaced each day.

Fat-soluble vitamins, including vitamins A, D, E, and K, dissolve in fat. The body can therefore store these vitamins until they are needed. Fruits and vegetables, whole-grain breads and cereals, and fortified milk are the best sources of vitamins.

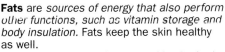

Minerals are *substances that strengthen the muscles, bones, and teeth; enrich the blood; and keep the heart and other organs operating properly.*

The minerals calcium and fluoride are particularly important for strong bones and teeth. Iron is needed for building and strengthening red blood cells. Potassium, sodium, and chloride help regulate the water balance in body tissues. Milk, meat, fish, spinach, fruits, and dry beans are rich sources of minerals.

Fats are *sources of energy that also perform other functions, such as vitamin storage and body insulation.* Fats keep the skin healthy as well.

Food energy that is not used by the body, however, is stored as body fat. Too much body fat puts stress on the skeleton and the heart. This is one reason why fats should be eaten in small quantities. Fats are found in meat, butter, margarine, cooking oil, whole milk, egg yolks, cheese, and most salad dressings.

Water is essential to survival. At birth, water makes up 75 percent of your total body weight; by adulthood, it makes up 60 percent. Water helps break down food, carries nutrients throughout the body, removes wastes from the body, and keeps the body at a comfortable temperature.

You should drink six to eight glasses of water each day. You can obtain additional water from many foods, including fruits and fruit juices, celery, milk and cabbage.

Lesson 2: Nutrients for Health and Wellness   **97**

Chapter 4 Lesson 2

Teach

L1 Recalling

Guide students in recalling the Food Guide Pyramid and the suggestions it presents for choosing a healthful diet.

Teacher Talk
Dealing with Sensitive Issues

Ask students to study and discuss the pictures and labels in Figure 4.4: What foods are important sources of each kind of nutrient? Why is that nutrient important in your daily diet? Which foods high in that nutrient do you like? Which are easy for you to include in your diet?

L3 Gathering Information

Assign interested students to research vegetarian diets (no meat, poultry, or fish) and vegan diets (no animal products at all, including eggs and milk). Have them investigate the practice of combining foods to include complete proteins in these diets. Ask the students to collect or create recipes and/or menu plans that provide complete proteins while avoiding animal products. Encourage students to begin a file or notebook to which additions can be made throughout the year.

L3 Reporting

Some students may be interested in expanding their study of nutrients by learning about malnutrition: What is malnutrition? What causes it? Where and why are people most likely to suffer from malnutrition? What efforts are being made to prevent malnutrition?

More About...

Carbohydrate Loading For athletes and nonathletes alike, carbohydrates should provide the primary source of energy in a daily diet. Many athletes, especially those who have long daily training sessions and those who participate in athletic events that last more than 90 minutes, make "carbohydrate loading" an important part of their meal plans. Rather than relying on carbohydrates to supply 60 to 65 percent of their energy needs, they boost their intake to supply 70 percent. Even athletes who train and compete at full strength have to be careful to avoid "overloading." The muscles can store only a limited amount of glycogen; the remaining carbohydrates are stored as fat.

97

Chapter 4 Lesson 2

L1 Discussing

Ask students to consider the best sources of vitamins and minerals: How can you include the necessary vitamins and minerals in your daily diet? Why do you think so many people take vitamin and mineral supplements? Which makes better sense—eating a balanced diet or taking supplements? Are there individuals for whom supplements *do* make sense? Who and why?

L3 Reporting

Various claims have been made about the special benefits of Vitamin C, from fighting off colds to preventing diabetes and cancer. Ask interested students to learn more about these claims and present a short report to the rest of the class.

L2 Interviewing

Have students interview family members who are responsible for planning and preparing meals. Suggest that they ask questions such as these: What are your most important considerations in planning meals? Why? In recent years, how have you changed the meals you prepare? Why? Would you like to make other changes? If so, what changes and why? If not, why not? Let students compare and discuss the responses to their interview questions.

L2 Creating Charts

Ask: What foods can you include in your daily diet to be sure you get 25–30 grams of fiber? Let students work in groups to find answers. Then have group members cooperate in putting their answers into charts to share with their fellow students.

Your Total Health

Your Bones Need Calcium!

The disabling bone disease called osteoporosis (ahs·tee·oh·puh·ROH·suhs) affects about 25 million Americans. Osteoporosis causes bones to become brittle, leading to fractures and spinal injuries. Recent studies indicate that Caucasian and Asian-American women are at the highest risk for developing this disease. An adequate supply of calcium, especially in the teen years when so much of the skeleton is developing, is very important. Start protecting yourself today by eating more calcium-rich foods, such as milk, yogurt, broccoli, and calcium-fortified cereals and fruit juices. In addition, make sure that you engage in regular physical activity.

Guidelines for Healthful Choices

To help you plan a healthful diet, the United States government has created a series of dietary guidelines. These guidelines are designed to help you to meet the Recommended Dietary Allowances, often abbreviated as RDAs, a set of measured amounts of particular nutrients that will meet the needs of most healthy people. RDAs are figured based on a person's gender, age, and activity level. By following dietary guidelines and using the Food Guide Pyramid, you will get the nutrients you need. The major points covered by the guidelines are explained on these pages. How well do you follow these guidelines in your diet?

Vary Your Diet

You already know that the most healthful diet consists of many different foods, chosen from the five basic groups. By varying your diet and following Food Guide Pyramid recommendations for food choices and servings, you can get the right amounts of the nutrients you need. No one food or food group can supply all that your body needs for growth and good health.

Manage Your Weight Through Exercise

Physical activity has many benefits. One of these benefits is to burn off the calories you consume every day. (Chapter 5 discusses physical activity and weight management.) Try to get at least 30 minutes of moderate physical activity most days of the week. Everything from brisk walking to in-line skating and dance—even cleaning your room—will provide you with exercise.

Eat Plenty of Grains, Vegetables, and Fruits

Foods in the three groups at the bottom of the pyramid provide energy as well as vitamins and minerals for healthy eyes, skin, bones, and blood. Additionally, many of these foods are good sources of **fiber,** *the part of grains, fruits, and vegetables that the body cannot break down.* Examples of fibrous parts of foods include apple and potato skins and the tough outer coating on kernels of corn and wheat. Many types of beans are also high in fiber. Although your body does not use fiber for energy, fiber helps you move food and wastes through your system. It also helps prevent some diseases, such as cancer. You should eat at least 20 grams of fiber each day to keep your digestive system healthy.

Getting enough exercise and eating foods high in fiber will help keep you healthy.

98 Chapter 4: Food and Nutrition

Health of Others

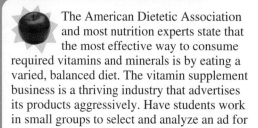

The American Dietetic Association and most nutrition experts state that the most effective way to consume required vitamins and minerals is by eating a varied, balanced diet. The vitamin supplement business is a thriving industry that advertises its products aggressively. Have students work in small groups to select and analyze an ad for vitamins: What is the product? For what age group or activity-level group is the product intended? How does the advertisement appeal to that special group (or to the parents of that group)? What foods provide the vitamins highlighted by the ad? Ask each group to display their ad and discuss their analysis with the rest of the class.

Cut Down on Fats

Fats are an essential nutrient, and you need them in your diet. However, no more than 30 percent of your daily calories should come from fat. No more than 10 percent of your total calories should come from saturated fats. **Saturated fats** are *fats found mostly in meats and dairy products.* They include the fats in meats, poultry, and eggs, as well as those in milk and butter. Physicians suggest that people go easy on saturated fats because they tend to raise the body's level of **cholesterol** (kuh·LES·tuh·rawl), *a waxlike substance used by the body to build cells and make other substances.* The body produces all the cholesterol it needs, so including large amounts of cholesterol in your diet can contribute to heart disease and stroke. **Unsaturated fats,** *liquid fats that come mainly from plants,* are found in olive oil and canola oil. You can cut down on cholesterol by choosing foods that contain unsaturated fats. A high-fiber and low-fat diet is linked to a lower risk for some types of cancer.

Avoid Too Much Added Sugar

Most people enjoy the sweet taste that sugar naturally gives to such foods as apples and strawberries. Sugar is a simple carbohydrate that the body uses for energy. Consuming too much added sugar (or carbohydrates in any form), however, can lead to tooth decay. Foods at the tip of the pyramid are high in fats and added sugars and low in other nutrients. They should be eaten sparingly.

Cut down on your sugar intake by not adding table sugar to fruit desserts or breakfast cereal and by avoiding sugary snacks such as cookies and candy. Remember to check canned and processed foods for hidden sugars. **Figure 4.5** shows some of the names of various sugars as they appear on food labels.

Science Connection

Keep Those Vitamins In!

Follow these important rules to preserve the vitamins in fruits and vegetables.

▸ Eat fruits and vegetables raw or lightly steamed. Heat and cooking in water cause loss of some vitamins.

▸ Eat fresh fruits and vegetables soon after buying them.

▸ Make sure that you use the outer leaves of lettuce or cabbage. They are especially rich in vitamins.

Figure 4.5
Other Names for Sugar

If a food has one of these substances early in the ingredients list, or if several appear within the list, the food is likely to be high in added sugars.

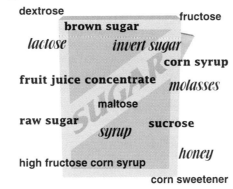

Watch Sodium and Salt

Sodium is a mineral that the body needs in small amounts to help regulate fluids. In the diet, its most common form is sodium chloride, commonly called table salt. Although each person's daily intake of salt should be no more than a teaspoonful, most people consume far more than that. Like sugar, salt or other high-sodium ingredients are often added to processed foods.

Cut down on sodium and salt by seasoning your food with herbs and spices rather than with table salt. Read nutrition labels on packaged foods to choose those with less sodium. In addition, go easy on salty snacks such as potato chips and salted peanuts.

Lesson 2: Nutrients for Health and Wellness

Chapter 4 Lesson 2

Health Update

Here is one reason for sticking to a low-fat diet: Fats may cause headaches. In one study, migraine sufferers reported fewer and less severe headaches when they limited their intake of fats to 20 grams per day.

L1 Listing Examples

Divide the class into two groups. Assign one group to list as many high-fat foods as they can think of in five minutes. Assign the other group to list as many low-fat foods as possible in the same period. Which group generated the longer list?

L3 Science

Have interested students investigate the differences among these sugars: dextrose, fructose, lactose, maltose, and sucrose. Ask the students to make a chart or other display explaining the differences.

L1 Discussing

Lead a class discussion of the recommended daily intake of salt: How much salt do you think you and other teens consume each day? Do you think you consume more or less salt than adults do? Why is it important to cut back on salt now, before you reach adulthood? What are some substitutes for salt? When and how can you use them? How can you recognize salt listed on food labels? (any form of the word *sodium*, including *sodium chloride*, *monosodium glutamate*, *disodium phosphate*, and so on)

L2 Brainstorming

Potato chips are popular snacks—and even meals—with many teens. Have small groups of students brainstorm a list of other favorite snack foods. Then let group members rate each snack food on the list according to its appeal and health value.

Growth and Development

 Drinking plenty of water before, during, and after participating in sports or other activities is important for people of all ages. However, specific fluid needs vary depending on several factors, including age. Teens in general share the water needs of adults, but young teens might require extra water during sports activities.

Students may have heard that drinking water—especially cold water—during a workout or a competition can cause stomach cramps. For most people, this isn't true. In fact, an athlete is much more likely to suffer stomach cramps as a result of dehydration than as a result of drinking cold water.

Chapter 4 Lesson 2

L3 Analyzing

Let students work independently or with partners to select a diet plan—for maintaining, losing, or gaining weight—from a health brochure or a popular diet book. Then have students apply the guidelines for healthful choices presented in this lesson to analyze the diet. How healthful is it? How likely is it to be effective? Ask students to write short reports summarizing their findings and conclusions.

L2 Math

Explain that carbohydrates provide 4 calories per gram, proteins 4 calories per gram, and fats 9 calories per gram. Ask students to calculate the calories provided in one serving of food that has 24 grams of carbohydrates (24 × 4 = 96), 3 grams of protein (3 × 4 = 12), and 2 grams of fat (2 × 9 = 18). What is the total number of calories? (96 + 12 + 18 = 126)

Assess

EVALUATING THE LESSON

Assign Reviewing Terms and Facts and Thinking Critically on page 101; then assign the Lesson 2 Quiz in the TCR to evaluate students' understanding.

 **in your journal**

Look back over the guidelines in this lesson. Which do you think you follow well? Which do you need to work on? In your journal, make two lists: Things I Do and Things I Need to Work On.

Reading a Nutrition Facts Label

The Nutrition Facts label found on most food packages is another tool to help you make smart food choices. Learn to use this resource to evaluate packaged foods before you buy them. **Figure 4.6** shows the information that appears on a Nutrition Facts label. The illustration also shows how you can use this label to make wise choices.

The percentages shown on the label are based on amounts called Daily Reference Values. Daily Reference Values are guidelines concerning the amounts of certain nutrients that are part of a healthy diet.

LIFE SKILLS
Making Healthful Choices at Fast-Food Restaurants

In the United States in 1995, on any given day, 57 percent of Americans ate at least one meal or snack away from home. The same study found that when children and teens ate out, they did so in fast-food restaurants more than 80 percent of the time. Fast-food restaurants can offer benefits to teens: enjoyable food at reasonable prices, quick service, and an informal and convenient place to gather with friends. However, many fast-food meals are high in saturated fat, cholesterol, sodium, sugar, and calories.

How can you make healthful choices at fast-food restaurants? Follow these tips.

▶ **Have food you like, but have it in a different way.** Ask the server to hold (leave out) the creamy spread on your sandwich. Pass up the french fries for a plain baked potato, or have a small order of fries.

▶ **Take advantage of items on the menu with lower amounts of fat.** Try the salad bar, and use the low-fat dressing. Instead of having a sandwich with meat that has been fried, choose meat that has been grilled or baked.

▶ **Try a new drink.** Many fast-food restaurants offer fruit juices and low-fat or skim milk.

▶ **Eat your vegetables.** If you order a pizza, top it with green peppers, mushrooms, onions, or broccoli.

▶ **Be creative about dessert.** For less fat, try frozen yogurt rather than ice cream. Or choose fresh fruit from the salad bar.

Follow-up Activity

Think about a fast-food restaurant that you enjoy. What do you usually order? Write down your order, and then analyze its nutritional value. What foods might you substitute, or what cooking methods might you avoid?

100 Chapter 4: Food and Nutrition

LIFE SKILLS
Making Healthful Choices at Fast-Food Restaurants

Focus on Life Skills

Ask students to identify local fast-food restaurants they enjoy. Write the names of those places on the board. Help students discuss their eating experiences at those restaurants by posing questions such as these: What meal or snack do you like to eat at each restaurant? With whom do you go there most frequently? What influences your choices? How much time do you spend considering the nutrient value of each food choice before you order? If you gave more thought to nutrient values and the Food Guide Pyramid, how would your order change?

Making Health Connections

Health has three essential components: physical, mental/emotional, and social. For most students, a stop at a local fast-food restaurant with friends or family members aids the development of good emotional health and good social health. However, most

Figure 4.6
Reading a Nutrition Facts Label
Most processed foods now carry this label. You may also see such labels on or displayed near fruits, vegetables, meats, poultry, and fish.

Ⓐ The amount of calories, nutrients, and other substances is based on one serving of the package's contents. Also shown is the serving size. Measures are shown in both standard and metric units.

Ⓑ One serving contains this many calories. Next to this amount is the number of calories from fat.

Ⓒ Major nutrients are measured in grams (g) or milligrams (mg) and the percentage of daily values. On labels from larger containers, like this one, additional information is provided at the bottom of the label.

Ⓓ This shows the amount of total fat, as well as of saturated fat, in one serving. Less than 0.5 grams is rounded to 0.

Ⓔ This shows the amounts of these substances in one serving of the food.

Ⓕ This shows the percentage of daily values for selected vitamins and minerals in one serving of the product.

Review Lesson 2

Using complete sentences, answer the following questions on a separate sheet of paper.

Reviewing Terms and Facts
1. **List** Identify the six major types of nutrients.
2. **Restate** What are RDAs?
3. **Vocabulary** Explain what is meant by the term *fiber,* and describe its value to the body.

Thinking Critically
4. **Compare and Contrast** State the difference between saturated and unsaturated fats, including the foods each come from.
5. **Explain** How can the Food Guide Pyramid help you avoid having too much fat and sugar in your diet?

Applying Health Concepts
6. **Consumer Health** Study the Nutrition Facts panel on a package or can of food. Which of the dietary guidelines apply to this food? Write a short report in which you list your findings and explain the product's nutritional value.
7. **Growth and Development** Find out more about vitamin D. Discover which foods contain moderate to high amounts of this vitamin and how it affects growth and development. Then find out what conditions might result from having too little vitamin D in the diet. Write a paragraph summarizing your findings.

Lesson 2: Nutrients for Health and Wellness

Lesson 3
Healthful Meal Planning

Focus

LESSON OBJECTIVES

After studying this lesson, students should be able to
- explain the importance of eating a healthful breakfast.
- identify ground rules that can help in planning enjoyable, nutritious meals.
- discuss healthful snack choices.

MOTIVATOR

Write the following questions on the board: Is it OK to snack between meals? Why or why not? Give students five minutes in which to write their responses.

INTRODUCING THE LESSON

Have students meet in groups to compare and discuss their responses to the Motivator questions. Ask the members of each group to consider everyone's ideas and then compose a brief statement expressing the group's opinion of snacking. Let each group share its opinion with the rest of the class. Then explain to students that they will learn more about snacks—and meals—as they read Lesson 3.

INTRODUCING WORDS TO KNOW

Ask students to share what they already know about the two words in the term *nutrient density*. (Nutrients, as students know from Lesson 1, are the substances in foods that the body needs to grow and stay healthy; *density* means the degree to which something is filled.) Have them reach their own conclusions about the definition of *nutrient density;* then have them skim the lesson to find the formal definition.

Lesson 3 Healthful Meal Planning

This lesson will help you find answers to questions that teens often ask about planning meals. For example:

▶ Why is breakfast so important?
▶ How can I plan nutritious meals?
▶ What snacks will provide important nutrients?

Words to Know

glucose
nutrient density

Cultural Connections

A European Breakfast ACTIVITY

To add some fun and variety to breakfast, try these foods, which are traditional parts of breakfast in many parts of Europe:

▶ Crusty rye or French bread
▶ Thin slices of low-fat ham and cheese
▶ Granola mixed with yogurt and honey
▶ Chunks of fresh fruit topped with a sprinkling of cinnamon or nutmeg

Find out about breakfast foods from other parts of the world. Report on your findings to your class.

If you get into the habit of eating a good breakfast now, you'll probably continue that habit into adulthood.

The Importance of Breakfast

When you get up in the morning, at least 10 or 12 hours may have passed since you last ate. You need a new supply of nutrients to build and repair cells. Your body also needs to be resupplied with energy. When the body breaks down carbohydrates, it converts them to a sugar called **glucose** (GLOO·kohs). This sugar is *the body's main source of energy.* Because your body cannot store extra reserves of glucose, you need to resupply it in the morning.

Several 1995 nutritional studies clearly showed that eating a healthful breakfast helps teens to think and work better. Students who eat breakfast have a longer attention span and a more positive attitude toward school. What's a good breakfast? Fortunately, you have a number of delicious choices.

■ The best breakfasts are high in complex carbohydrates. Toast, bagels, English muffins, and waffles are good choices. To cut down on fat, top them with yogurt, fruit spreads, cottage cheese, or applesauce rather than with butter or margarine.

■ If time is short, choose easy-to-prepare items, such as fresh or canned fruits, yogurt, and ready-to-eat cereals. Look for cereals made from whole grains or fortified with vitamins and minerals. You can top ready-to-eat cereals with dried fruits or chopped nuts, too.

■ Take something with you. Celery stuffed with peanut butter or a lowfat cheese stick along with fruit or vegetable juice are good choices.

Lesson 3 Resources

Teacher's Classroom Resources
- 📁 Concept Map 17
- 📁 Cross-Curriculum Activity 8
- 📁 Decision-Making Activity 8
- 📁 Enrichment Activity 17
- 📁 Lesson Plan 3
- 📁 Lesson 3 Quiz

- 📁 Reteaching Activity 17
- 🖨 Transparency 17

Student Activities Workbook
- 📁 Study Guide 4
- 📁 Applying Health Skills 17

- If you aren't hungry, at least drink a glass of milk or juice. Remember, however, that cereal with milk or a piece of fruit will provide more lasting energy than just milk or juice.
- Finally, you don't have to eat traditional breakfast foods. A bagel pizza or a reheated plate of spaghetti will provide you with an excellent start to the day.

Planning Your Other Meals

A good breakfast sets the pattern for good nutrition all day. The Food Guide Pyramid can help you make healthful meal and snack choices. What kinds of ground rules will help you get the nutrients you need and enjoy your food, too? Here are a few suggestions.

- **Eat regular meals.** Avoid skipping meals. People who skip meals often overeat at another meal.

- **Watch portion sizes.** Suggested portion sizes may be smaller than you think. For example, one serving of meat is only 2-3 ounces, which is about the size of a deck of cards. (The palm of your hand is also a good, easy guide for meat serving size). Many people eat a cup of pasta at one sitting. Keep in mind that this represents *two* servings—not one—from the bread, cereals, rice, and pasta group.

- **Eat less of the "pyramid tip" foods, but don't feel forced to cut them out entirely.** After all, you eat for enjoyment as well as for good nutrition. It's okay to have a soft drink or candy occasionally.

- **Plan and make healthful lunches.** If you pack your own lunch to take to school, try to make nutritious food choices. For example, using whole-wheat bread instead of white bread for sandwiches will provide more fiber. If you use mayonnaise or butter on a sandwich, use a small amount. Add moisture with lettuce leaves and tomato slices. If you like luncheon meats, try turkey, chicken, or low-fat ham to cut down on fat.

- **Achieve balance over time.** Balance your nutrients and food groups over several days or a week. If you eat a lunch that is high in fats one day, make sure that dinner and perhaps the next day's choices are low in fats.

Q & A

No Fat, No Problem?

Q: I've seen fat-free snacks at the store, but I've heard that they can make people sick. Is that true?

A: Some of these fat-free snacks may contain the chemical olestra. It is used in snack foods such as chips or crackers because it creates a food that tastes almost exactly like the high-fat variety, but has half the calories and no fat. This chemical can cause digestive problems in some people, however, so foods containing olestra must carry a warning label.

Many teens enjoy planning and making some of their own meals.

Lesson 3: Healthful Meal Planning

Home and Community Connection

Food Banks As students read about the importance of a balanced, healthy diet, encourage them to consider people who do not have the resources to provide that kind of diet for themselves and their families. Which local organizations offer food for those people? Why is that kind of aid important? How does providing nutritious food help meet people's physical, emotional, and social health needs? You may want to help students organize a food drive to help supply a local food bank. You might also encourage interested students to volunteer at a food bank or similar organization. Discuss with students how volunteering in this way can meet their own health needs.

Chapter 4 Lesson 3

Teach

L1 Discussing
Ask students to share their own breakfast preferences: What do you like to eat in the morning? Why? What other breakfast options appeal to you? How soon after breakfast do you feel hungry? What do you usually do in response to that hunger?

L1 Recalling
Review with students what complex carbohydrates are and which foods are good sources of complex carbohydrates. (Refer students to page 96 if necessary.)

L2 Family and Consumer Sciences
Ask a group of volunteers to learn more about frozen breakfasts available in local markets: What does the Nutrition Facts label reveal about each breakfast? How much does each cost? How much time is a consumer likely to save by preparing each? Which of the frozen breakfasts—if any—might be a good meal choice? Let members of the group share their findings with the rest of the class.

L2 Demonstrating
Ask volunteers to prepare and present demonstrations of serving sizes of foods from each of the sections of the Food Guide Pyramid. Let students use measuring cups, shaped dough or other food-sized objects, or actual food samples to show the serving sizes.

L1 Discussing
Guide students in considering this statement from the text: *It's okay to have a soft drink or candy occasionally.* Ask questions such as these: How often is "occasionally"? What tasty, healthful alternatives might you consider? (*fruit juice, fresh or dried fruit, low-fat sweets*)

Chapter 4 Lesson 3

L2 Applying Knowledge
Ask students to record all the food choices available in the school lunch program for a given day. Then have them discuss which of those foods could be combined to make the most healthful and appealing meal.

L1 Discussing
Let students discuss where and when they are most likely to snack. Ask: What prompts you to snack? Is it hunger? social activities? boredom? What techniques can you use to avoid snacks with low nutrient density?

L3 Applying Life Skills
Have students write a one-week meal plan for their own families. The plan should take into account the ages and activity levels of the family members, particular food preferences, and the family's cultural and ethnic background. Snacks for individuals and/or the whole family should also be included. Encourage students to share their completed meal plans with adults at home. Suggest that students might offer to plan and prepare some of the meals on the plan.

Assess

EVALUATING THE LESSON
Assign Reviewing Terms and Facts and Thinking Critically on page 105 to review the lesson; then assign the Lesson 3 Quiz in the TCR to evaluate students' understanding.

Smart Snacking

Between meals during the day, many people like to have a snack. Unfortunately, they may reach for snack foods that are high in calories, fat, and sugar, and low in nutrients. A healthful diet includes snacks with high **nutrient density,** *the amount of nutrients in a food relative to the number of calories.* **Figure 4.7** shows a comparison of various snack foods, based on nutrient density.

Figure 4.7
Snacks and Nutrient Density
It is fairly easy to substitute snacks with high nutrient density for those with lower nutrient density.

Snacks with LOW Nutrient Density (high in calories, low in nutrients)	Snacks with HIGH Nutrient Density (low in calories, high in nutrients)
potato chips	pretzels
candy bars	raisins
soft drinks	mixed vegetable juice
ice cream	low-fat frozen yogurt
chocolate chip cookies	graham crackers
dips made with sour cream	dips made with salsa

MAKING HEALTHY DECISIONS
Making Breakfast Important

Jameel has begun to take more of an interest in his diet since he entered middle school. One area in which he has seen improvement is breakfast. He and his parents used to have daily arguments about his refusal to eat in the morning. Now, although his choices aren't always standard breakfast foods, he eats a nutritious breakfast every day.

Jameel's best friend, Raymond, has the same arguments with his parents that Jameel used to have. He either eats nothing at all or grabs a doughnut or a toaster pastry. Jameel would like to encourage his friend to eat more healthful breakfasts, but he isn't sure how to introduce the subject. Then he remembers the six-step decision-making process he learned in school.

1. **State the situation**
2. **List the options**
3. **Weigh the possible outcomes**
4. **Consider your values**
5. **Make a decision and act**
6. **Evaluate the decision**

Follow-up Activities
1. Apply the six steps of the decision-making process to Jameel's situation.
2. With your classmates, role-play a number of different decisions that Jameel might make. Discuss these role plays.

MAKING HEALTHY DECISIONS
Making Breakfast Important

Focus on Healthy Decisions
Lead students in discussing positive changes they have made in their own lives: What was the change? What motivated you to make that change? How did you feel about making it? Then, after students have read the Making Healthy Decisions feature, ask them about Jameel's positive decision: How does the change he has made affect his decision-making process?

Activity
Use the following suggestions to discuss Follow-up Activity 1:
Step 1 Jameel has to decide how and when to introduce the subject of eating healthful breakfasts with his friend Raymond.
Step 2 Jameel might wait until Raymond says something about feeling tired or hungry during the school day. He also might introduce the subject by talking about his own healthful breakfast or by telling

Think of your total daily snacks as a fourth meal, a fourth chance to provide your body with the nutrients it needs. If you often look forward to snack time for an energy boost, what should you reach for? Studies indicate that the best energy snacks combine complex carbohydrates with small amounts of protein. Be creative by combining foods that you enjoy from these two food groups. Here are a few examples:

- Chunks or slices of unpeeled fresh fruit or vegetables dipped in low-fat yogurt
- A bowl of whole-grain cereal with skim milk
- Popcorn topped with Parmesan cheese
- Whole-wheat crackers spread with peanut butter
- A microwaveable soft pretzel and a stick of string cheese
- Half of a pita pocket stuffed with low-fat turkey

in your journal

From the snacks you have read about in this lesson, or from your own knowledge of healthful snacks, write a weekly "snack plan" for yourself, with the goal of making this part of your diet more healthful.

These snacks are both nutritious and delicious.

Review — Lesson 3

Using complete sentences, answer the following questions on a separate sheet of paper.

Reviewing Terms and Facts

1. **Recall** Why is breakfast so important?
2. **List** Suggest two tips for planning meals other than breakfast.
3. **Vocabulary** Define the term *nutrient density*.
4. **Explain** Which two nutrient categories can be combined for a high-energy snack?

Thinking Critically

5. **Infer** Why do you think that people do not need to cut out foods from the pyramid tip entirely?
6. **Evaluate** How does technology affect the way people eat? Describe how the technology available today affects your meal planning choices.

Applying Health Concepts

7. **Health of Others** Using what you have learned, write an advertisement for a healthful breakfast or snack. Combine words and pictures to persuade readers that your featured foods are healthful, easy to prepare, and delicious.

Lesson 3: Healthful Meal Planning

Focus

The Teen Health Digest articles can be used in two ways: as an individual activity for reflection and enrichment or as a cooperative learning activity as described below.

Motivator

Draw the Food Guide Pyramid on the board, and ask students to identify and label the sections. As students read each article in the Teen Health Digest, ask: Which section (or sections) of the pyramid does this article relate to? What additional information does it provide about that group of foods? Which article (or articles) relate to the entire pyramid?

Cooperative Learning Project

International Food Fair

Setting Goals

Ask students to name foods from other cultures or countries that they have tried and enjoyed. Tell students they will plan and put on an International Food Fair. Together, they will select and prepare a variety of healthful foods from other countries and cultures.

Delegating Tasks

Direct students to work together to make general plans for their International Food Fair: When and where will it be held? Who will be invited? What countries will be represented? Then let students choose which of these committees they want to participate in: Food Preparation (needs the greatest number of students), Publicity, Decorations.

TEEN HEALTH DIGEST

Teens Making a Difference

Strength in Numbers

Along with their regular classes, students at the William Penn Charter School in Philadelphia can take a course in helping others. As part of a program led by James Ballengee, Director of Service Learning, teens spend at least 40 hours each year helping people in need.

Once a month, a group of students spends four to five hours on a Saturday helping out at a shelter and soup kitchen for homeless men. Another group volunteers for Philabundance, a nonprofit agency that picks up surplus foods from restaurants, caterers, and hotels and delivers it to shelters and other resources for needy people.

Jeff Riddle, a staff member at Philabundance, knows that he can count on the Penn Charter students to help out. "Jim Ballengee's students define the term *youth in action!*" he said. One week, Philabundance learned that someone wanted to donate 10,000 pounds of sweet potatoes. The potatoes would have spoiled if they had been left on the truck for very long. A call to Mr. Ballengee brought a team from Penn Charter. They spent most of a day unloading, sorting, and repacking the potatoes for distribution.

Try This: Find out about a group in your community that fights hunger. Ask what opportunities there are for teens to help.

106 Chapter 4: Food and Nutrition

Myths and Realities

The Truth About Sugar

Q: I hear parents talk about how sugar affects their kids and makes them act wild. Is that really true?

A: Despite what many people believe, sugar does not seem to cause hyperactive behavior. One study of 23 children conducted over nine weeks, for example, found no link between sugar consumption and boisterous behavior. Eating a lot of sugar can cause other problems for children, however. It promotes tooth decay and can lead to excessive weight gain. Children who are severely overweight may experience health problems such as diabetes or high blood pressure when they reach adulthood.

TECHNOLOGY OPTIONS

On-line Services The Internet may be a good source of recipes and other information that students can use in planning and preparing their contributions to the Food Fair. Let partners work together, using search engines to find instructions for making foods from specific countries or cultures.

Computer Students may be able to use banner-creating programs to make signs and other decorations for the Food Fair. They might also use computer programs to make advertising posters or individual invitations. In addition, they may want to input their recipes and make neat copies to be distributed to guests at the fair.

HEALTH UPDATE

Safe Fruits and Vegetables

Fruits and vegetables are an important part of a healthful diet. However, you should know that some of today's farming and food production methods may expose foods to dangerous germs that can make people ill.

To protect yourself, rinse produce thoroughly, especially if you plan to eat it without cooking it. Raw meat, poultry, seafood, and eggs may also contain harmful germs. If you are preparing those foods, therefore, wash your hands and any utensils thoroughly before working with other foods.

Personal Trainer

Make a Commitment with Calcium!

Getting enough calcium is crucial for strong bones. Just consuming calcium alone, however, does not make bones as dense as they could be. You also need physical activity.

What kind of activity will help you use the calcium you consume? Weight-bearing exercise, which puts stress on your bones, makes them grow stronger. There are many forms of weight-bearing exercise, including

- brisk walking.
- dancing.
- stair climbing.
- jumping rope.

CON$UMER FOCU$

Good Bargains, Good Nutrition

Smart shoppers save money by buying foods that are on sale. However, you should be careful *not* to buy perishables—foods that may spoil, such as meats, dairy products, and fresh fruits and vegetables—in such large quantities that you can't use them up while they are still fresh. Make sure you read the expiration, or "Use by," date on packaged sale items. Such dates tell you the *last* date that the product is considered fresh.

Try This:

Help a family member with a shopping trip by checking store advertisements and coupons ahead of time and suggesting bargains.

Teen Health Digest

Chapter 4

The Work Students Do

Following are suggestions for implementing the project:

1. If students are learning about particular countries or regions in their social studies classes, you may want to have them research and prepare foods from those countries to be included in the Food Fair.
2. Be sure students working on decorations coordinate their efforts with those who are preparing foods, so that the appropriate countries and cultures will be represented in the signs, banners, tablecloths, and other items.
3. Students should use the Food Guide Pyramid and other nutritional information in the text to evaluate the healthful attributes of each food. Guide students in avoiding high-fat or high-sugar recipes.
4. Suggest that students who prepare food for sampling provide copies of the recipes as well. Then guests who enjoy a particular food will be able to make it at home.
5. Remind students that they are all part of the Clean-Up Committee. They should work cooperatively to restore their Food Fair site to a clean, tidy condition.

Assessing the Work Students Do

Ask students to evaluate their own participation in the Food Fair. Meet with students individually to discuss their responses to these questions, or let them respond in writing: What did you enjoy most about the project? What did you find most difficult? most rewarding? What is the most important thing you learned from helping present the International Food Fair?

Meeting Student Diversity

Physically Disabled As students consider possible sites for their Food Fair, remind them to keep issues of accessibility in mind. Students with physical disabilities should be able to get to the site and move around it with ease. In addition, some of the Food Fair guests may have special needs that can be accommodated by selecting an appropriate site and by placing tables and signs thoughtfully.

Language Diversity Encourage students who are still learning English to work closely with native speakers. In addition, help students consider the family members or other guests who will be attending the Food Fair. If appropriate, have students write some of the invitations, signs, and/or recipes in the language familiar to each population.

Lesson 4

The Digestive and Excretory Systems

Focus

LESSON OBJECTIVES

After studying this lesson, students should be able to
- describe how digestion works.
- explain how wastes are removed from the body.
- discuss health habits that are essential to taking care of the digestive and excretory systems.

MOTIVATOR

Write this question on the board: Which parts of your body help digest the food you eat? Give students a set period of time in which to write their answers in list form.

INTRODUCING THE LESSON

Let students read aloud their lists from the Motivator activity, and have a volunteer record all the responses on the board. With input from other students, the volunteer should write the body parts in anatomical order, so that the beginning of digestion in the mouth is shown at the top of the list. Point out to students that they already know quite a bit about digestion and that they will learn more as they study Lesson 4.

INTRODUCING WORDS TO KNOW

Help students read the Words to Know, along with the terms' definitions in the Glossary. Then have students work with partners to create their own word-search puzzles using at least seven of the Words to Know. Let pairs of students exchange papers and solve each other's puzzles.

108

Lesson 4

The Digestive and Excretory Systems

This lesson will help you find answers to questions that teens often ask about how their bodies break down and use food. For example:

▶ Why does food have to be digested?
▶ How does digestion work?
▶ How are waste products removed from the body?

Words to Know

- digestion
- digestive system
- saliva
- stomach
- small intestine
- liver
- pancreas
- excretory system
- excretion
- kidney
- colon

Energy for Life

Food provides nutrients that the body needs in order to grow, develop, and stay healthy. Before the body can use the nutrients from food, it must break food down into smaller parts. **Digestion** (dy·JES·chuhn) is *the process by which the body breaks food down into smaller components that can be absorbed by the bloodstream and sent to each cell in your body.* The process of digestion is accomplished by your **digestive** (dy·JES·tiv) **system**, *a series of organs that work together to break down foods into substances that your cells can use.*

How Digestion Works

Think for a moment about what happens in a typical factory. Raw materials, such as logs or iron ore, enter the factory. Inside, processes occur that change the material into another form. For example, logs might be ground into pieces to make paper, or iron ore might be melted to make steel. At some point, usually, chemicals are added. The end product is stored for later use, and waste materials are thrown away.

Digestion is similar in some ways. It involves both physical changes, such as the crushing of food by the teeth, and chemical changes, such as the transforming of food by body chemicals called *enzymes*.

It takes anywhere from 16 to 24 hours for your body to process food into energy and get rid of waste products.

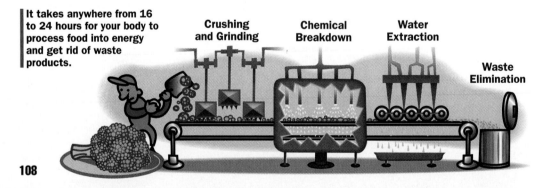

Crushing and Grinding · Chemical Breakdown · Water Extraction · Waste Elimination

108

Lesson 4 Resources

Teacher's Classroom Resources
- Concept Map 18
- Enrichment Activity 18
- Health Lab 4
- Lesson Plan 4
- Lesson 4 Quiz
- Reteaching Activity 18

Student Activities Workbook
- Study Guide 4
- Applying Health Skills 18
- Health Inventory 4

The Mouth and Throat

Does your mouth ever water when you sit down to eat, or when you smell something delicious roasting in the oven? That "water" is **saliva** (suh·LY·vuh), *a digestive juice produced by the salivary glands in your mouth.* It starts to flow as a physical signal from your body that it is ready to begin the digestive process. **Figure 4.8** shows the first steps in digestion.

Figure 4.8
The Beginning of Digestion
The process of digestion begins in the mouth.

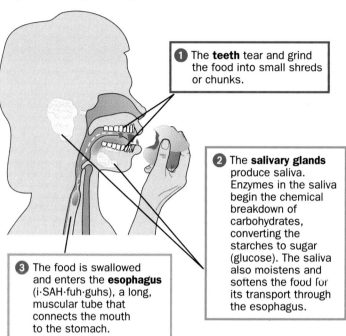

❶ The **teeth** tear and grind the food into small shreds or chunks.

❷ The **salivary glands** produce saliva. Enzymes in the saliva begin the chemical breakdown of carbohydrates, converting the starches to sugar (glucose). The saliva also moistens and softens the food for its transport through the esophagus.

❸ The food is swallowed and enters the **esophagus** (i·SAH·fuh·guhs), a long, muscular tube that connects the mouth to the stomach.

Teen Issues

Bulimia Hurts ACTIVITY!

People who suffer from the eating disorder called bulimia nervosa (boo•LEE• mee•uh ner•VOH•suh) make themselves vomit after eating in order to avoid weight gain. This dangerous eating disorder can lead to malnutrition. Frequent vomiting is harmful to the body in other ways as well. Gastric juices, powerful enough to dissolve food particles, come up the esophagus and into the mouth with the purged (vomited) food. There they often do terrible damage to the teeth, the gums, and the tender tissues of the throat.

Research an eating disorder and its health effects. Write several paragraphs describing what you have found.

The esophagus is located right next to the trachea (TRAY·kee·uh), or windpipe. The trachea is the passageway through which air gets to your lungs. The passages up to the nose are at the back of your throat. Getting food into your air passages could cause you to choke. For that reason, two flaps of skin close airways when you swallow. The uvula (YOO·vyuh·luh) closes the airway to the nose. The epiglottis (e·puh·GLAH·tis) closes the airway to the lungs, as shown in **Figure 4.9**.

Figure 4.9
When You Swallow
Before you swallow, the passages to the nose and lungs are open. When you swallow, these passages are automatically closed.

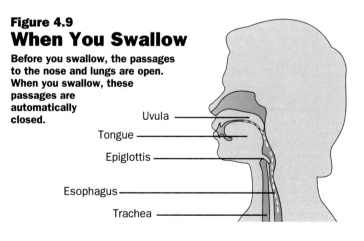

- Uvula
- Tongue
- Epiglottis
- Esophagus
- Trachea

Lesson 4: The Digestive and Excretory Systems **109**

Visually Impaired
Students with limited visual ability may have trouble seeing and understanding the diagrams that identify the organs and functions of the digestive and excretory systems. Let these students work in pairs or small groups with visually able students to carefully describe and discuss each aspect of the diagrams. Encourage visually impaired students to ask questions and add comments and facts during these discussions.

If your school has models of the digestive and excretory systems or of some of the organs from these systems, make the models available to students with visual impairment. Encourage them to handle, describe, and discuss the models, while their partners or other group members refer to the labeled diagrams in the text.

Chapter 4 Lesson 4

Teach

L1 Recalling
Have students recall the most important reasons for eating. (Refer them to page 92 if necessary.) Ask: Which purpose of eating does digestion serve?

L1 Discussing
Guide students in discussing the comparison between digestion and a factory: What raw materials are used in the "digestion factory"? What is the essential product of this factory? If a factory that pulps logs is called a paper mill and a factory that melts iron ore is called a steel mill, what would you name the "digestion factory"? (Be sure students recognize that the product of digestion is the energy a body needs to grow, to keep itself healthy, and to fuel activity.)

L2 Demonstrating
Show the class a clear, realistic picture of an appetizing food, and describe the food's taste, smell, and texture in rich detail. (If possible, ask a capable volunteer to prepare and present this demonstration.) Then ask: Does this picture and description make your mouth water? Why?

L3 Investigating
Ask a group of volunteers to learn about the different kinds of salivary glands and their location in the mouth. Have the volunteers draw a cartoon or diagram to share this information with the rest of the class.

L1 Discussing
Almost everyone has been told, "Slow down—chew your food!" Ask students why that reminder is important.

Would You Believe?

Nearly 4,000 Americans die from accidental choking every year.

109

Chapter 4 Lesson 4

L2 Language Arts
Have students read about gastritis (a common stomach ailment that affects the stomach lining). Ask each student to write a paragraph explaining how gastritis occurs.

L1 Discussing
Guide students in discussing the digestive process inside the stomach: In what form does food reach the stomach? How does the stomach change the form of that food? How long does food usually stay in the stomach? What effect does this have on how often you are likely to feel hungry?

Would You Believe?
 A normal stomach can hold as much as three cups of pureed food.

L2 Investigating
Assign students to find and discuss the answers to these questions: What is acid indigestion? How is it related to the function of the stomach in digesting food?

L3 Family and Consumer Sciences
Consumers purchase an amazing array of remedies for upset stomach, nausea, and indigestion. Ask a group of volunteers to select five or more such products and to compare their ingredients, their intended uses, and their costs. Have them summarize the results of their investigation in a chart or graph.

Q & A
Baby Talk

Q: Why do babies sometimes spit up milk after feeding?

A: Between the stomach and the esophagus there is a muscle designed to keep the food in the stomach. In young babies, this muscle is not mature. The gases that normally build up in the stomach during feeding create pressure. That pressure—a burp—forces some of the milk in the baby's stomach past the muscle and up the esophagus. Over the first year of life this muscle gradually strengthens. Spitting up does not usually happen after the age of one.

The Stomach and Small Intestine

Figure 4.10 shows the path that food takes during the next part of the digestive process. Once food enters the esophagus, muscles lining the walls of the esophagus contract and relax, moving the food along until it reaches the **stomach,** *a muscular organ in which some digestion occurs.* The stomach holds and processes the food for up to four hours. During this time the stomach's strong muscular walls churn and mix the food. When this part of the digestive process is complete, the food resembles a fairly thin soup, called chyme (KYM).

Figure 4.10
The Digestive System
Most of the process of digestion goes on in the small intestine.

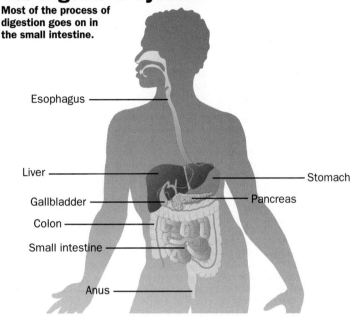

HEALTH LAB
The Digestion of Protein

Introduction: During digestion, organs of the digestive system secrete enzymes to break down proteins. Once proteins are broken down, their amino acids can flow into the bloodstream and be used by the body.

All plants and animals contain enzymes. In this experiment, you will use the enzyme *papain* (puh-PAY-uhn), a protein-splitting enzyme that comes from the papaya fruit. You will also use gelatin, a substance made from a strong protein called *collagen* (KAH-luh-juhn). In meat, collagen is a tough material found in tendons and ligaments.

Objective: Observe how enzymes break down proteins during digestion.

Materials and Method: Work with a partner or small group. You will need a package of flavored or unflavored gelatin; a measuring cup; a spoon for stirring; water; two petri dishes or other shallow, nonmetallic containers; measuring spoons; and unflavored meat tenderizer made from papain.

Follow the package directions to make the gelatin. Pour the gelatin into each shallow container, to a depth of about ½ inch. Label one container A and the

110   Chapter 4: Food and Nutrition

HEALTH LAB
The Digestion of Protein

Time Needed
2 class periods

Supplies
► gelatin (unflavored or flavored)
► measuring cups
► stirring spoons
► water
► petri dishes (or other shallow, nonmetallic containers)
► measuring teaspoons
► meat tenderizer (made from papain)

Focus on the Health Lab
Show students a small chain of pop-beads or similar linking units. Explain that the chain represents a protein molecule, and each bead represents an amino acid within that molecule. During digestion,

Next the food moves to the **small intestine,** *a coiled tube, about 20 feet long, where most of the digestive process takes place.* Three other organs help the small intestine in the digestive process:

- **Liver.** The **liver** is *the body's largest gland, which secretes a liquid called bile that helps to digest fats.* The liver also helps to regulate the level of sugar in the blood, breaks down harmful substances such as alcohol, and stores some vitamins.
- **Gallbladder.** After the liver produces bile, it sends it to the gallbladder (GAWL·bla·der), which stores this bile until it is needed.
- **Pancreas.** The **pancreas** (PAN·kree·uhs) is *a gland that helps the small intestine by producing pancreatic juice, a blend of enzymes that breaks down proteins, carbohydrates, and fats.*

When food has been completely broken down, nutrients are absorbed through the walls of the small intestine into the bloodstream. Blood carries the nutrients to body cells. **Figure 4.11** shows this absorption process.

Your Total Health

Liver Damage

One of the functions of the liver is to act as a "detox center" for poisons that enter the body. For this reason, overconsumption of alcohol can damage the liver over time. A condition called cirrhosis (suh·ROH·sis) can occur, in which liver cells are destroyed and replaced by scar tissue. If excessive drinking continues, the condition can cause death.

Figure 4.11
Nutrient Absorption
The walls of the small intestine are covered with fingerlike projections called villi (VI·ly). Digested material is absorbed into the villi, which allow it to enter the bloodstream.

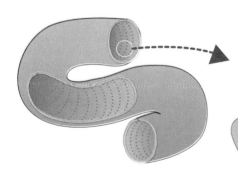
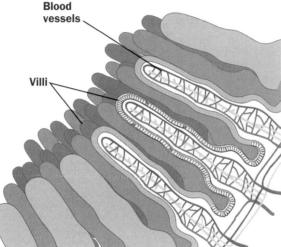

other container B. Allow the gelatin to set until firm. With the measuring spoons, sprinkle about 1 teaspoon of the meat tenderizer on the top of the gelatin in dish A. Wash your hands thoroughly. (**NOTE:** Do not eat the gelatin in either dish.) Wait one-half hour to make your observations.

Observations and Analysis:
Observe dish A and dish B. Compare the appearance of the gelatin. What was the effect of the papain on the gelatin? What conclusions can you draw regarding the effect of human digestive enzymes on meat and poultry?

Note that papain is an ingredient in meat tenderizer, a substance that chefs sometimes sprinkle on tough meats before cooking them. How do you think papain makes meat tender?

Lesson 4: The Digestive and Excretory Systems

Chapter 4 Lesson 4

Visual Learning
Refer students to Figure 4.12 to study and discuss the process of excretion. Let volunteers identify the location and the function of the kidneys, the bladder, and the urethra.

L1 Discussing
Have the class consider and discuss the importance of each guideline on page 113:

▶ How can eating a balanced diet help keep your digestive and excretory systems healthy? What are other benefits of eating a balanced diet?

▶ Which foods are high in fat? What problems can they cause your digestive and excretory systems? Which foods are low in fat? How can they help keep you healthy? Why is it important to eat high-fiber foods?

▶ How does drinking plenty of water help protect the health of your digestive and excretory systems?

▶ Why do you think it's important to eat three meals a day at regular times? How do you think snacks should fit in to your meal schedule?

▶ What problems are you likely to have if you don't chew your food carefully?

▶ How does taking care of your teeth help keep your digestive system healthy? What are other benefits of keeping your teeth clean?

▶ Which parts of the digestive process benefit when you exercise regularly? In what other ways does regular exercise help you stay healthy?

L2 Presenting Skits
How can you encourage friends or family members to follow the guidelines for caring for their digestive and excretory systems? How effective are examples? reminders? direct instructions? Let small groups of students plan and present short skits demonstrating effective encouragement.

Did You Know?
Drink Cranberry Juice!

Many people find cranberry juice a refreshing drink, and it is high in vitamin C. Several studies indicate that cranberry juice may be helpful in keeping the kidneys and bladder working properly. Drinking cranberry juice has been linked to a lower incidence of infections of the excretory system.

Removing Wastes

Although digestion is complete, the body still has tasks to perform. As food is being broken down, the body separates out the parts that it cannot use. Those wastes must then be removed from the body.

The body produces three kinds of wastes. Solid wastes are made up of foods that the body could not break down, including fiber. Liquid wastes and carbon dioxide gas are wastes produced by cell activity. Your **excretory** (EK·skruh·tor·ee) **system** is *the system that removes wastes from your body and controls water balance.* Your lungs perform some functions of the excretory system by getting rid of carbon dioxide gas when you exhale. Your skin also gets rid of some wastes through sweat. The major organs of the excretory system, however, are the kidneys, bladder, and colon. These remove most of the liquid and solid wastes from your system.

The Kidneys, Bladder, and Colon

Approximately 60 percent of your body is water, and most waste materials are dissolved in it. **Excretion** (ek·SKREE·shuhn) is *the process by which the body gets rid of liquid waste materials.* Figure 4.12 shows the process of filtering, storing, and excreting liquid wastes.

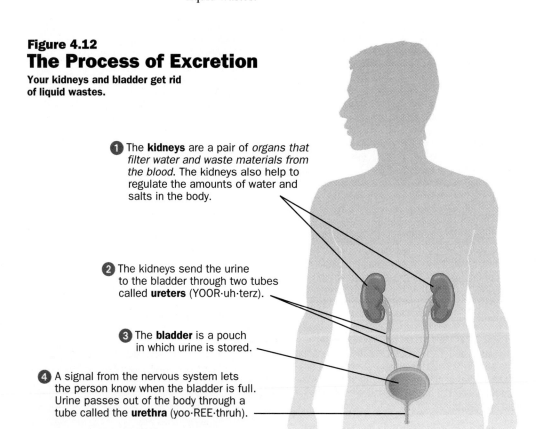

Figure 4.12
The Process of Excretion
Your kidneys and bladder get rid of liquid wastes.

① The **kidneys** are a pair of *organs that filter water and waste materials from the blood.* The kidneys also help to regulate the amounts of water and salts in the body.

② The kidneys send the urine to the bladder through two tubes called **ureters** (YOOR·uh·terz).

③ The **bladder** is a pouch in which urine is stored.

④ A signal from the nervous system lets the person know when the bladder is full. Urine passes out of the body through a tube called the **urethra** (yoo·REE·thruh).

Cooperative Learning

Informative Display Have the entire class work together to plan a mural, a series of cartoons, or a set of posters that explain the digestive process and emphasize the importance of taking good care of the digestive and excretory systems. Direct students by asking them what information they want to communicate, to whom, and how they can communicate that information most effectively. Then let students plan how many groups they need and what responsibilities each group should have; provide help as appropriate in forming the groups. Guide group members in working cooperatively to complete their section of the project, and let students display their completed mural or set of illustrations in a public area of the school.

The body sends a mixture of water and undigested solid wastes, into the **colon** (KOH·luhn), or large intestine, *a storage tube for solid wastes.* Most of the water is absorbed by the colon and returned to the body. The remaining solid wastes become material called feces. When the colon becomes full, strong muscles in its walls contract. This movement pushes the feces out of the body through an opening called the anus.

Caring for Your Digestive and Excretory Systems

Good health habits are essential to the care of the digestive and excretory systems. Follow these guidelines.

- Eat a balanced diet, based on the Food Guide Pyramid.
- Eat plenty of foods that are low in fat and high in fiber.
- Drink six to eight glasses of water every day.
- Eat at regular times each day.
- Always chew your food slowly and carefully.
- Brush and floss your teeth at least twice a day and have regular dental checkups. Your teeth are important to the digestive process.
- Be active! Participate in moderate physical activity most, if not all, days of the week.

List in your journal two actions you can take that will provide you with better nutrition and at the same time keep your digestive and excretory systems healthy.

Fiber-rich foods are particularly important to the health of your digestive and excretory systems.

Review — Lesson 4

Using complete sentences, answer the following questions on a separate sheet of paper.

Reviewing Terms and Facts

1. **Vocabulary** Define the terms *digestive system* and *excretory system.*
2. **Recall** How do the liver, gallbladder, and pancreas aid in the digestive process?
3. **Restate** What does the colon do?

Thinking Critically

4. **Explain** Describe possible consequences for the rest of the body of a digestive system that is not working properly.
5. **Hypothesize** Choose two of the tips for caring for your digestive system, and explain why you think each might be beneficial.

Applying Health Concepts

6. **Personal Health** Make a diagram or flow chart to show the passage of food from the time it enters the mouth to the time that its undigested parts are excreted from the body.

Lesson 4: The Digestive and Excretory Systems   **113**

Chapter 4 Review

CHECKING COMPREHENSION

Use the Chapter Summary and the Chapter 4 Review to help students go over the most important ideas presented in Chapter 4. Encourage students to ask questions and add details as appropriate.

CHAPTER 4 REVIEW ANSWERS

Reviewing Key Terms and Concepts

1. Your diet is all the things you regularly eat and drink. Nutrition is the process of taking in food and using it for growth and good health. By planning a diet based on the Food Guide Pyramid, a person can achieve good nutrition.
2. Bread, Cereal, Rice, and Pasta Group should be the source of the largest number of servings.
3. Cholesterol is a waxlike substance used by the body to build cells and make other substances. Cholesterol should be limited in the diet because the body produces all the cholesterol it needs; high levels of cholesterol can contribute to heart disease and other health problems.
4. Any three pieces of information from Figure 4.6 on page 101.
5. Glucose is a form of sugar that is the body's main source of energy.
6. Achieving balance in your diet over time means getting all necessary nutrients and eating foods in correct proportion from the Food Guide Pyramid over a period of several days or even a week.
7. Saliva, a digestive juice produced by salivary glands in the mouth, begins the chemical breakdown of carbohydrates. It also moistens and softens food.

Chapter 4 Review

Chapter Summary

▶ **Lesson 1** Basing your food choices on the Food Guide Pyramid helps you plan meals that provide the nutrients you need each day.

▶ **Lesson 2** The body needs a daily supply of the six major nutrients: carbohydrates, proteins, vitamins, minerals, fats, and water. Although fats are needed for overall health, they should be eaten sparingly.

▶ **Lesson 3** Breakfast is an important meal because it refuels the body with energy. Wise planning decisions such as eating regular meals and achieving balance over time in your meal and snack choices will help you get the nutrients you need.

▶ **Lesson 4** In order to use food, the body must digest it—break it down into tiny particles that can be absorbed by the blood—and excrete waste products.

Reviewing Key Terms and Concepts

Using complete sentences, answer the following questions on a separate sheet of paper.

Lesson 1

1. Define the terms *diet* and *nutrition,* and explain the relationship between them.
2. Which food group should be the source of the largest number of servings each day?

Lesson 2

3. Define *cholesterol*. What is the role of cholesterol in your health?

Thinking Critically

Using complete sentences, answer the following questions on a separate sheet of paper.

9. **Explain** Use what you have learned about nutrients to show why the sections of the Food Guide Pyramid vary in size.
10. **Evaluate** Analyze the nutritional value of each of the following breakfast menus. Then suggest several foods that might be

4. List three pieces of information listed on a Nutrition Facts panel of a food label.

Lesson 3

5. Define *glucose*.
6. What does it mean to achieve balance in your food choices over time?

Lesson 4

7. Explain the role of saliva in digestion.
8. Briefly describe the processes of digestion and excretion after the food leaves the stomach.

substituted to make each menu more nutritious and lower in fat.

Breakfast #1: orange juice, two fried eggs with bacon, whole-wheat toast

Breakfast #2: whole-grain cereal with whole milk and sliced peaches, white toast with butter and jelly, water

11. **Describe** Explain the value of a high-fiber, low-fat diet.

Chapter 4: Food and Nutrition

Meeting Student Diversity

Language Diversity Use the following suggestions to help students who have difficulty with English:

▶ Pair those learners with native speakers of English who can restate the Chapter Summary in language that helps students comprehend important concepts.

▶ Direct auditory learners or those students with language diversity to the Teen Health Audiocassette Program. Available in English and Spanish, this component provides an audio and written summary of the chapter.

Chapter 4 Review

Your Action Plan

The decisions you make about your diet will affect your health and energy level, both now and in the future.

Step 1 Look over your journal entries from Chapter 4. What do you need to do to improve your nutrition? Use your journal entries to set a long-term goal. You might decide to follow the Food Guide Pyramid in making food choices or to increase or decrease the amount of a particular nutrient in your diet.

Step 2 Describe steps you can take to reach your long-term goal. For example, you might decide to begin eating breakfast or making your lunch every day instead of buying it.

Set specific dates to assess your progress. On those dates, write a paragraph in your journal that describes how you feel about achieving your goals.

In Your Home and Community

1. **Health of Others** Offer to make dinner for your family one night at home. Plan a healthful menu that you think everyone will enjoy. Go over it with a parent or guardian to make sure that person approves. Then make the meal and serve it to your family. If you like the results, you may decide to do this on a regular basis, monthly or even weekly.

2. **Growth and Development** Talk with a preschool teacher in your area about the concepts the school teaches to young children about nutrition. Offer to make some colorful posters, shoot a video, or take some photographs to enliven his or her next lesson.

Building Your Portfolio

1. **Illustration of the Food Guide Pyramid** Create your own version of the Food Guide Pyramid. Draw the outline on a large piece of poster paper, and use pictures from magazines or pictures you draw to illustrate the sections. Do additional research to find out some of the main nutrients that are found within each food group (for example, calcium in the milk, yogurt, and cheese group).
2. **Nutrient Advertisement** Create an advertisement for a food or a particular nutrient. Show how you would promote the value of that food or nutrient. Add the advertisement or its description to your portfolio.
3. **Personal Assessment** Look through all the activities and projects you did for this chapter. Choose one or two that you would like to include in your portfolio.

Performance Assessment

▶ **Self-evaluation** Direct students to review the activities that are provided throughout the chapter. Encourage each student to select one finished product or activity that demonstrates his or her best work for the chapter. Have students explain what they learned and how the examples they selected show their progress.

▶ **Teacher's Classroom Resources** Assign Performance Assessment 4, "Food and Nutrition Survey," in the TCR.

Planning Guide

Chapter 5
Physical Activity and Weight Management

	Features	Classroom Resources
Lesson 1 **Physical Fitness and You** *pages 118–122*	Life Skills: Everyday Fitness *page 120*	Concept Map 19 Cross-Curriculum Activity 9 Enrichment Activity 19 Lesson Plan 1 Lesson 1 Quiz Reteaching Activity 19
Lesson 2 **The Circulatory System** *pages 123–127*		Concept Map 20 Enrichment Activity 20 Lesson Plan 2 Lesson 2 Quiz Reteaching Activity 20 Transparency 18
Lesson 3 **The Skeletal and Muscular Systems** *pages 128–133* TEEN HEALTH DIGEST *pages 134–135*	Health Lab: Pulling Pairs *page 132*	Concept Map 21 Decision-Making Activity 9 Enrichment Activity 21 Health Lab 5 Lesson Plan 3 Lesson 3 Quiz Reteaching Activity 21 Transparencies 19, 20
Lesson 4 **Planning a Fitness Program** *pages 136–140*		Concept Map 22 Decision-Making Activity 10 Enrichment Activity 22 Lesson Plan 4 Lesson 4 Quiz Reteaching Activity 22 Transparency 21
Lesson 5 **Weight Management** *pages 141–145*	Making Healthy Decisions: Helping a Friend with an Eating Disorder *page 144*	Concept Map 23 Cross-Curriculum Activity 10 Enrichment Activity 23 Lesson Plan 5 Lesson 5 Quiz Reteaching Activity 23

National Health Education Standards

One of the main goals of the National Health Education Standards is to move students toward "health literacy." Health literacy is the capacity of individuals to obtain, interpret, and understand basic health information and services and the competence to use such information and services in ways which promote health. The health standards were developed by applying the characteristics of a well-educated, literate person within the context of health. The health literate person is:

- a critical thinker and problem solver
- a self-directed learner
- a responsible, productive citizen
- an effective communicator

Listed below are the Health Education Standards Performance Indicators addressed in each lesson of this chapter.

Lesson	Health Standards Performance Indicators
1	(1.1, 1.2, 1.3, 3.4)
2	(1.1, 1.6, 3.1, 3.7)
3	(1.1, 1.6, 3.1)
4	(1.1, 1.6, 6.4, 6.5, 6.6)
5	(1.1, 1.6, 3.1, 7.4)

ABCNEWS InterActive Videodisc Series

You may wish to use the following videodiscs with this chapter: *Making Responsible Decisions,* side one, video segment 4; *Food and Nutrition,* side one, video segments 5 and 11. Use the *ABCNews InterActive™ Correlation Bar Code Guide* for title reference. Also available in VHS format.

Chapter Resources

Teacher's Classroom Resources
- Chapter 5 Test
- Parent Letter and Activities 5
- Performance Assessment 5
- Testmaker Software

Student Activities Workbook
- Study Guide 5
- Applying Health Skills 19–23
- Health Inventory 5

Student Diversity Strategies
- Audiocassette Program (English)
- Audiocassette Program (Spanish)
- Spanish Parent Letters
- Spanish Summaries, Quizzes, and Activities

Multimedia Components
- English Audiocassette Program
- Spanish Audiocassette Program
- *Teen Health* Videodisc/VHS Series
- *Teen Health* Video Kit:
  *Teenage Nutrition:
  Prevention of Obesity*

Other Resources

Readings for the Teacher

Berg, Frances M. *Afraid to Eat: Children and Teens in Weight Crisis.* Hettinger, ND: Healthy Weight Journal, 1997.

Feinberg, Brian. *The Musculoskeletal System.* New York: Chelsea House, 1994.

Kolodny, Nancy J. *When Food's a Foe: How to Confront and Conquer Eating Disorders.* Boston: Little, Brown, 1992 (revised edition).

Readings for the Student

Reef, Catherine. *Stay Fit: Build a Strong Body.* New York: Twenty-First Century Books, 1993.

Silverstein, Alvin, et al. *The Circulatory System.* New York: Twenty-First Century Books, 1994.

Silverstein, Alvin, et al. *The Muscular System.* New York: Twenty-First Century Books, 1994.

Out of Time?

If time does not permit teaching this chapter, you may use these features: Life Skills on page 120; Health Lab on page 132; Teen Health Digest on pages 134–135; Making Healthy Decisions on page 144; and the Chapter Summary on page 146.

Chapter 5
Physical Activity and Weight Management

CHAPTER OVERVIEW
Chapter 5 focuses on the advantages of being physically fit. It explains how physical fitness helps maintain healthy circulatory, skeletal, and muscular systems and promotes healthy weight management.

Lesson 1 discusses the meaning and importance of physical fitness.

Lesson 2 identifies the parts of the circulatory system, explains how they function, and explores the relationship between fitness activities and a healthy circulatory system.

Lesson 3 explains how the parts of the skeletal and muscular systems work to facilitate movement; it also discusses practices that help maintain the health of these two body systems.

Lesson 4 explains how students can plan and implement their own personal fitness programs.

Lesson 5 helps students understand how they can recognize, achieve, and maintain a weight that is healthy for them.

PATHWAYS THROUGH THE PROGRAM
Achieving and maintaining physical fitness is a cornerstone of good health. Teens who establish healthy habits of physical activity and nutrition will be taking important steps toward improving their physical, mental/emotional, and social health. They will also be making a worthwhile investment in their ability to maintain good health as they mature.

Chapter 5
Physical Activity and Weight Management

Student Expectations
After reading this chapter, you should be able to:
1. Define what it means to be fit.
2. Describe how the circulatory system works.
3. Explain how the skeletal and muscular systems work.
4. Describe how to plan a fitness program.
5. Identify healthful attitudes for and principles of weight management.

Key to Ability Levels

Teaching strategies that appear throughout the chapter have been identified by one of three codes to give you an idea of their suitability for students of varying learning styles and abilities.

L1 Level 1 strategies should be within the ability range of all students. Full class participation is often required. Teacher direction is usually needed.

L2 Level 2 strategies are for average to above-average students or for small groups. Some teacher direction is necessary.

L3 Level 3 strategies are designed for students able and willing to work independently. Minimal teacher direction is necessary.

Teen Chat Group

Stephan: James! Are you here for the cancer research benefit, too?

James: Yeah. It's my first. I hope I can do the whole five kilometers.

Krystina: It's not that hard. Just remember to pace yourself. Are you going to walk or jog?

James: I'm going to jog until I get tired and then walk for a while. I jog every morning now, so I want to see how far I can go.

Stephan: This is my first time, too. I decided to get involved because my grandfather was diagnosed with cancer earlier this year. I thought maybe this could help him and be good for me at the same time.

Krystina: You won't regret it. I did it last year. It's great to know there's something you can do to make a difference.

James: Come on, let's head for the starting line.

in your journal

Read the dialogue on this page. Have you ever thought about how your personal fitness could make a difference to others? Start your private journal entries on fitness by responding to these questions:

- What does being fit mean to you?
- What activities could help you stay fit?
- How do you think fitness is related to weight management?

When you reach the end of the chapter, you will use your journal entries to make an action plan.

Chapter 5: Physical Activity and Weight Management **117**

Chapter 5

INTRODUCING THE CHAPTER

▶ To introduce this chapter, encourage students to share and discuss their own understanding of physical activity: What types of exercise give you the best workout? Which activities make your heart rate increase? Which types of exercise do you enjoy most? Conclude the discussion by telling students that Chapter 5 will teach them more about how and why they can move their bodies.

▶ Display several magazine ads featuring teen and young adult models. Ask students to describe and discuss the models' appearance: Do they look physically fit? From these photographs, what conclusions can you draw about the models' physical, mental/emotional, and social health? Explain that students will learn more about fitness, weight, and health as they read Chapter 5.

▶ Ask students to use their left hands to "explore" their right hands and wrists. Why are the hands made up of so many small, short bones? Do you think there are many muscles in the hands? Why? Let students share their ideas, then explain that they will learn more about the skeletal and muscular systems as they study Chapter 5.

Cooperative Learning Project

Fun-Times Fitness Day

The Teen Health Digest on pages 134 and 135 provides students with high-interest articles related to the content of this chapter.

The material in the Teacher's Wraparound Edition presents suggestions for a class project in which students plan and put on a day of fun fitness activities for elementary school students.

Most teens enjoy imagining themselves as physically fit—with extra energy and an appealing appearance. Still, it's difficult for many to make a connection between sports or other forms of exercise and physical fitness. A consideration of the teens in this situation may help them make that connection—and begin applying it in their own lives.

Let students meet in small groups to discuss the photograph and dramatize the Teen Chat Group dialogue. Encourage group members to add their own comments to the dialogue. Then let the group members read the journal questions and discuss their ideas before they begin writing independently.

117

Lesson 1
Physical Fitness and You

Focus

LESSON OBJECTIVES

After studying this lesson, students should be able to
- explain what it means to be fit.
- describe the varied benefits of fitness.
- identify and discuss the elements of fitness.

MOTIVATOR

Display photographs (from magazines or other sources) showing active people of different ages and various physiques. Ask students: Which of the people in these pictures are physically fit? How can you tell? Give students a few minutes in which to record their ideas.

INTRODUCING THE LESSON

Ask several volunteers to share their responses to the Motivator questions. Then ask: What do you think fitness is? How can you judge whether you are fit? How can you judge whether other people are fit? Conclude by explaining that Lesson 1 will help students develop a more complete and accurate understanding of physical fitness.

INTRODUCING WORDS TO KNOW

Let students work in small groups to read the Words to Know and to find the definition of each term within the lesson. Have group members work together to choose criteria for categorizing the terms. Using those criteria, students will organize the terms into two or more lists. Let each group share its lists with the rest of the class and explain the system of categorizing.

Lesson 1 Physical Fitness and You

This lesson will help you find answers to questions that teens often ask about physical fitness. For example:
▶ What does it mean to be fit?
▶ How will fitness help me?
▶ What kinds of activities will help me be fit?

Words to Know

fitness
exercise
body composition
strength
endurance
aerobic exercise
anaerobic exercise
flexibility

 Your Total Health

Fitness for Others *ACTIVITY*

When you are fit, it will not only improve every area of your life, but it can also help the people around you. A fit person can respond effectively to emergencies. If, for example, you are with a friend who experiences a sudden injury, you will be able to run for help. Knowing first aid is another way to be prepared for an emergency. Find out where you can learn first-aid procedures in your area.

Fitness for Life

What does it mean to be fit? **Fitness** is *the ability to handle the physical work and play of everyday life without becoming overly tired.* Being fit also means that you have energy in reserve to meet unexpected demands. Think about all the different things you do each day: going to school, taking part in after-school activities, doing chores. You're fit if you can meet all these demands—as well as extra challenges, such as having to run for the bus or spend an evening waiting on tables for a school fundraising dinner.

Do you think that you are fit? How might you raise your level of fitness? **Exercise** is *physical activity that develops fitness.* By exercising, you become more fit. When you exercise regularly, you find that you are able to get through each day with energy to spare. The more demanding your activities are, the more you must exercise to stay fit. A soccer player, for example, needs to do wind sprints to be fit to play in games. Developing the habit of exercising and keeping fit can help you remain healthy throughout your life.

Physical activity is one way for people of different ages to stay fit and enjoy one another's company.

Chapter 5: Physical Activity and Weight Management

Lesson 1 Resources

Teacher's Classroom Resources
- Concept Map 19
- Cross-Curriculum Activity 9
- Enrichment Activity 19
- Lesson Plan 1
- Lesson 1 Quiz
- Reteaching Activity 19

Student Activities Workbook
- Study Guide 5
- Applying Health Skills 19

What Fitness Does For You

The benefits of fitness are much more than just physical well-being. Being fit also helps you mentally and emotionally as well as socially. Fitness is an excellent way to keep your health triangle in balance. Take a look at **Figure 5.1** below. Do you see how the benefits of fitness in one area of health lead to benefits in the other areas as well?

Figure 5.1
The Benefits of Fitness
Staying fit is a way of life.

interNET CONNECTION
Shape up with healthy ideas on the Web, like customized meal and fitness plans to help achieve your goals.
http://www.glencoe.com/sec/health

Benefits to Physical Health
Higher energy level
Decreased risk of getting certain diseases
Better performance of heart and lungs
Improved strength and endurance
Improved posture
Greater freedom of movement
Better coordination
Better sleep
Better weight control

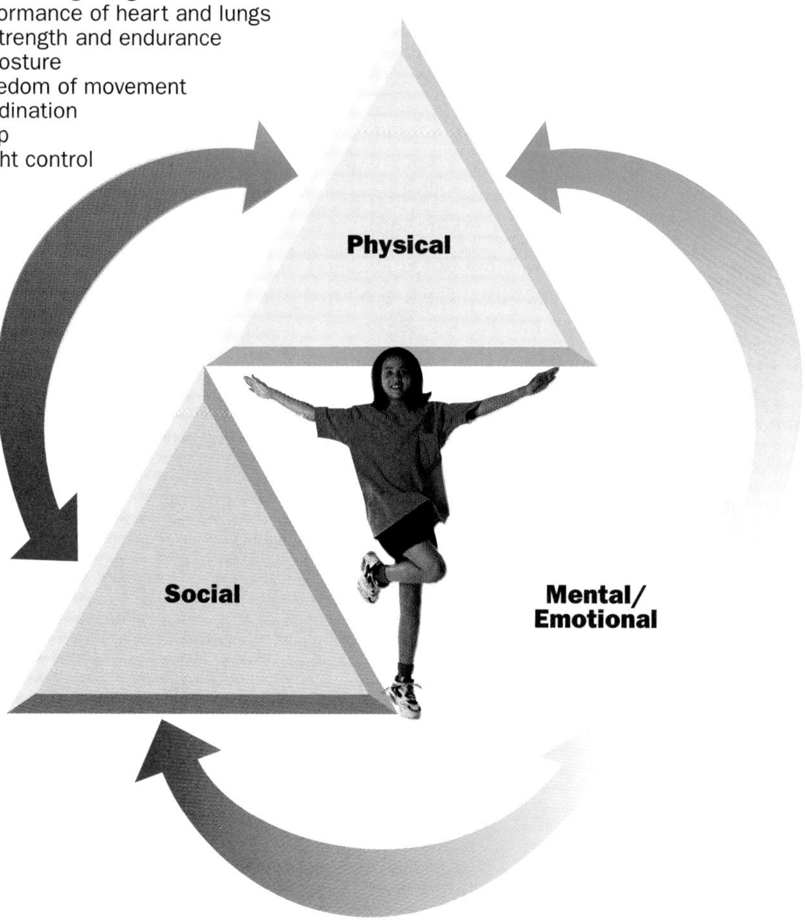

Physical

Social

Mental/Emotional

Benefits to Social Health
Additional chances to meet new people
Opportunities to share common goals with others
Greater ability to interact and cooperate with others
Opportunities to use talents to help others
Ability to work more efficiently

Benefits to Mental/Emotional Health
Enhanced self-esteem
Sharpened mental alertness
Reduced stress
More relaxed attitude
More enjoyment of leisure time

Lesson 1: Physical Fitness and You   119

Chapter 5 Lesson 1

Teach

L1 Identifying Examples
Ask students what it might mean to "have energy in reserve to meet unexpected demands." (Having to run up a flight of stairs or being able to recall what is the best response in a safety emergency, for example.) Inquire: What do you think it means to be "excessively tired"? What activities are easy or comfortable for you when you are physically fit? How do you feel about those activities when you are not fit?

L1 Discussing
Encourage students to describe and discuss the activity in the photograph at the bottom of page 118. Then ask: What activities do you enjoy? How do these activities help you become or stay physically fit? With whom can you participate in those activities?

L1 Critical Thinking
Ask students to consider how fitness activities benefit social health: What effect do you think a solitary activity—lifting weights, for example—might have on a teen's social health? Why? How do you think partcipating in competitive sports can affect a teen's social and emotional health?

Visual Learning
Guide students in describing and discussing the health triangle in Figure 5.1. Let the class select one specific kind of fitness activity—jogging, for example— and explain how each of the listed benefits might result from that activity.

Meeting Student Diversity

Physically Challenged
Physical fitness is for everyone. It's important for all your students to understand that they have certain special physical abilities and that they will find certain other kinds of physical activities more difficult. If your class includes teens with specific physical challenges, it is essential for students to recognize that the challenged students should also work to establish and maintain physical fitness. Understanding one's own capabilities and setting individual goals are essential parts of planning a fitness program—for everyone.

Some students (not necessarily those who are physically challenged) may be interested in learning about sporting activities and competitions for physically challenged athletes, such as wheelchair basketball or ski racing for amputees.

Chapter 5 Lesson 1

L3 Researching

Assign volunteers to learn more about healthy body composition: What is the recommended range of percentage of body fat for male teens? for female teens? for adult men? for adult women? Why are they different? Have the volunteers prepare a chart and present their findings to the class.

L1 Identifying Examples

How does increased strength help you in your daily life? Let groups of students work together to list specific activities that they can do more easily or more skillfully as they increase their strength.

L2 Demonstrating

Ask three volunteers or three pairs of volunteers to demonstrate the three standard strength-training exercises. Have these volunteers research potential problems with each exercise and then demonstrate the correct form and technique while explaining mistakes that should be avoided.

L1 Discussing

Discuss with students the definition of *endurance*: How can you tell whether you are "overly tired"? How hard do you think you should push yourself in continuing a tiring activity? Why? Against what standard do you think you should judge your own endurance?

! Would You Believe?

At rest, an average person takes about 14 or 15 breaths each minute. Vigorous exercise can raise that rate to 100 breaths per minute.

Q & A
Rope Tricks

Q: I haven't jumped rope since I was little. How do I find a rope the right size for me?

A: Hold the handles of the rope together and step with one of your feet on the middle of the rope. The handles should reach to about the middle of your chest.

To build your strength, gradually increase the number of repetitions you do at each exercise session.

The Elements of Fitness

When you exercise, you develop three elements of fitness: strength, endurance, and flexibility. Another aspect of fitness is **body composition**—*the proportions of fat, bones, muscle, and fluid that make up body weight.* By eating a nutritious diet that is low in fats and getting plenty of exercise, you will keep your body composition within healthful limits.

Strength

Strength is *the ability of your muscles to exert a force.* In fitness, strength is measured by the most work your muscles can do at a given time. By exercising to build strength, you can more easily lift heavy objects without injury or strain. You will also be more skilled in the sports and other activities in which you participate. Here are three standard strength-training exercises:

- **Curl-ups.** Curl-ups, also called crunches, strengthen your abdominal muscles. Strong abdominal muscles take some of the strain off the muscles of your back. To do curl-ups, lie on your back, bend your knees, and put your hands on the floor at your sides. (For a slightly easier exercise, cross your hands over your chest.) Roll yourself up far enough so that your shoulder blades clear the floor, then lower yourself to the floor again. Make sure that your heels do not come up off the floor. When you do this exercise, be careful to avoid pulling forcefully on your neck.

- **Push-ups.** Push-ups strengthen the muscles of your upper arms and chest. Lie on the floor face down, with arms bent and hands flat on the floor under your shoulders. Press your whole body upward until your arms are straight, lower your body to the floor again, and repeat. Keep your legs and back straight throughout the exercise. For a somewhat easier exercise, bend your knees and rest them on the floor. Push up only your hips and upper body. Be sure to keep your back and upper legs straight.

- **Step-ups.** This exercise strengthens leg muscles. Step up onto a step with your left foot, bring the right foot up, step down with the right foot, then bring the left foot down. Repeat, alternating legs.

LIFE SKILLS
Everyday Fitness

It's good to have a regular exercise routine to improve your strength, endurance, and flexibility. However, you can also include activities in your day-to-day life that will help you become fit. These activities shouldn't replace vigorous activities such as swimming laps or jogging with a friend, but they can help you to develop a more active lifestyle. Try some of the following ideas:

▶ Use your bicycle for transportation, making sure that you ride in safe areas. Wear your backpack and take your bike for local errands.

▶ When you get together with your friends, take a walk instead of sitting around someone's house.

▶ Plan family outings that include walking, riding bikes, or playing active games.

LIFE SKILLS
Everyday Fitness

Focus on Life Skills

Introduce the Life Skills feature by asking students how they got to school this morning. Record their responses on the board in four columns: *Walked, Biked, Rode in Car, Rode on Bus.* Add more columns if appropriate (such as *Skateboarded* or *Rollerbladed.*) Then ask: If you rode to school (either in a car or on a bus), could you have walked or ridden a bike instead? What difference would that morning exercise have made in your day? Encourage students to share their ideas about using the trip to school as an opportunity for added physical activity.

Making Health Connections

Exercise and physical activity are essential to all aspects of health—physical, mental/emotional, and social. Teens who establish the habits of an active lifestyle will be able to improve their fitness levels and maintain good health. They will further benefit from the habits of exercise and activity, which they can carry on into their adult years.

Endurance

Endurance (in·DUR·uhnts) is *your ability to perform vigorous physical activity without getting overly tired.* There are two basic types of endurance. Muscle endurance is the measure of how long a group of muscles can exert a force without tiring. Heart and lung endurance is the measure of how effectively your heart and lungs work during exercise and how quickly they return to normal.

The best way to build endurance is through **aerobic** (e·ROH·bik) **exercise**—*nonstop, rhythmic, vigorous activity that increases breathing and heartbeat rates.* Aerobic exercises include running, bicycling, and swimming. (The other type of exercise is **anaerobic** (an·e·ROH·bik) **exercise,** *intense physical activity that requires short bursts of energy,* such as weight lifting or sprinting.)

To build your endurance, it is best to perform aerobic exercises for at least 20 minutes, at least three times a week. Doing this will raise your heart and breathing rates enough to benefit your cardiovascular system. The following exercises will build endurance:

- **Walking/Jogging.** Begin by walking briskly, working up to a 20-minute walk. Then alternate jogging with walking until you can jog or run the entire time.

- **Jumping Rope.** Alternate jumping rope with running in place. When you jump, protect your joints from strain by raising your feet only enough to clear the rope—no more than two inches off the floor.

- **Swimming.** Swim near the surface of the water. The lower you are, the harder you have to work. Gradually increase your swimming time to 20 minutes of sustained laps at a steady pace.

Swimming is a total body workout that is easy on the joints.

Did You Know?

The Government Reports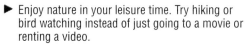

In July of 1996 the Centers for Disease Control and Prevention (CDC) released "Physical Activity and Health: A Report of the Surgeon General." The report showed that nearly half of American youths aged 12 to 21 are not physically active on a regular basis.

The report detailed the benefits of physical activity as well as ways people might increase their overall fitness. Locate a copy of the summary of this report, and share it with your class.

- ▶ Enjoy nature in your leisure time. Try hiking or bird watching instead of just going to a movie or renting a video.
- ▶ Go window shopping. Even during bad weather, you can get a lot of walking done in the mall.
- ▶ Take the stairs whenever possible instead of riding the elevator or the escalator. For your own safety, however, stay out of empty stairwells in buildings.

Follow-up Activity

Think of other ways to make your lifestyle more active. Make entries in your journal or calendar any time you try one of these physical activities. Write about how you feel. Do you feel stronger or better about yourself? Do you feel as if you have more energy? Share what you've been doing with your friends and family members.

Chapter 5   Lesson 1

L1 Identifying Examples

Ask students to name as many sports or fitness activities as they can. Record their ideas on the board. Then have students consider each sport or activity on the list: Is this an aerobic exercise or an anaerobic exercise? How can you tell?

L1 Critical Thinking

Ask the class to consider the relationship between strength and flexibility: Do you think there is a direct relationship, an inverse relationship, or no consistent relationship? Encourage students to cite specific examples to support their ideas.

VIDEODISC/VHS

Teen Health Course 2

You may wish to use video segment 5, "Physical Fitness," which outlines the goals set by the President's Challenge for teens to become physically fit.

Videodisc Side 1, Chapter 5
Physical Fitness

Search Chapter 5, Play to 6

Health Minute

Regular stretching can reduce muscle tension and encourage relaxation. An effective stretch produces a mild tension that diminishes as the stretch is held. If a stretch causes pain or interrupts your natural, relaxed breathing pattern, you've pushed your body too hard.

Meeting Student Diversity

Most everyday fitness activities are more enjoyable—and sometimes safer—when done with others. Teens who are shy or who are new to the community may have special trouble asking an acquaintance to go along on a bike ride, a walk to school, or a quick stroll around the mall. To provide special help, have all students participate in brief role-plays with one teen asking another to join him or her in an enjoyable fitness activity. Be sure that the role-plays include positive responses. If possible, arrange several role-playing opportunities for students who seem to need additional encouragement.

Assessing the Work Students Do

Have students work together to create their own ads for everyday fitness activities, using the medium of their choice (in the style of a radio ad, a magazine ad, or a TV ad, for example). Arrange for them to share their ads with other members of the student body. Assign credit based on participation, effort, and attitude.

Chapter 5 Lesson 1

Assess

EVALUATING THE LESSON

Assign Reviewing Terms and Facts and Thinking Critically on page 122 to review the lesson; then assign the Lesson 1 Quiz in the TCR to evaluate students' understanding.

LESSON 1 REVIEW

Answers to Reviewing Terms and Facts

1. The three elements of fitness are strength, endurance, and flexibility.
2. Body composition is the ratio of body weight that is fat, bone, muscle, or fluid.
3. Endurance is the ability to engage in a vigorous physical activity without getting overly tired. The two basic types of endurance are muscle endurance and heart and lung endurance.
4. Aerobic exercise is non-stop, rhythmic, vigorous activity that increases breathing and heartbeat rates. Examples will vary; possible examples include running and bicycling.

Answers to Thinking Critically

5. Responses may vary.
6. Responses may vary.
7. Responses will vary.

RETEACHING

▶ Assign Concept Map 19 in the TCR.
▶ Have students complete Reteaching Activity 19 in the TCR.

ENRICHMENT

Assign Enrichment Activity 19 in the TCR.

Close

Have students identify an element of fitness that they would like to incorporate into their physical fitness routine. Ask: How will you accomplish this?

Music Connection
Easy Listening

You may like to listen to music with a strong beat when you're jumping rope or riding a stationary bike, but don't play it when you stretch. When you stretch to music, try listening to something that will help you move smoothly, without bouncing.

Flexibility

Flexibility is the *ability to move joints fully and easily.* Some people are more naturally flexible than others, but everyone benefits from increased flexibility. By doing regular, gentle stretching, bending, and twisting, you will feel more comfortable, improve your posture, and reduce your risk of injury during strength or endurance training. **Figure 5.2** offers tips for safe stretching.

Figure 5.2
Safety in Stretching
Stretching should be gentle, not forced.

ⓐ Wear loose-fitting, comfortable clothing.

ⓑ Stretch to a point where you feel a gentle pull. Do not force your muscles to stretch too far. Hold there for a count of 15 or 20.

ⓒ Do not bounce or jerk to try to get more of a stretch. This can cause an injury.

ⓓ Stretch both sides of your body equally.

Lesson 1 Review

Using complete sentences, answer the following questions on a separate sheet of paper.

Reviewing Terms and Facts

1. **List** Identify the three elements of fitness.
2. **Explain** What is body composition?
3. **Vocabulary** Define *endurance*, and identify the two types.
4. **Vocabulary** Describe *aerobic exercise*, and give two examples.

Thinking Critically

5. **Describe** Explain how the three parts of the health triangle may be linked through fitness.
6. **Compare and Contrast** Reread the descriptions of the strength and endurance exercises. In what ways are they similar? How are they different?
7. **Design** Create a workout that includes exercises that develop each of the elements of fitness.

Applying Health Concepts

8. **Consumer Health** Interview a health club employee or another exercise professional. Ask questions about exercise equipment for use at home. What products are helpful, and which may be harmful? Present your report to the whole class.

Cooperative Learning

Making Presentations Point out that certain stretches are often recommended as part of the preparation and/or cool-down for specific sports or other activities. Let students form cooperative groups, based on a common interest in a particular sport or other activity. Then have group members work together to research the best forms of stretching to be used in combination with their chosen activity. Ask them to prepare a presentation that explains the usefulness of the stretches and demonstrates the correct way to perform those stretches. Allow time for the members of each group to share their presentation with the rest of the class.

The Circulatory System

Lesson 2

This lesson will help you find answers to questions that teens often ask about the heart and about blood circulation. For example:

▶ What does the circulatory system do for the body?
▶ How does blood circulate through the body?
▶ How can I keep my circulatory system healthy?

Transporting Materials Through the Body

Part of being fit involves having a healthy circulatory system. The **circulatory system** is *the group of organs and tissues that transport essential materials to body cells and remove their waste products.* This system consists of the heart, the blood vessels, and the blood itself. The **cardiovascular** (KAR·dee·oh·VAS·kyoo·ler) **system** is *another name for the circulatory system.* Cardio- refers to the heart and -vascular means having to do with vessels.

The heart is a muscle that pumps blood throughout the network of blood vessels. Blood flows through three types of blood vessels, as shown in **Figure 5.3.**

Words to Know

circulatory system
cardiovascular system
artery
capillary
vein
pulmonary circulation
systemic circulation
plasma
blood pressure

Figure 5.3
The Circulatory System
There are three types of blood vessels: arteries, capillaries, and veins (VAYNZ).

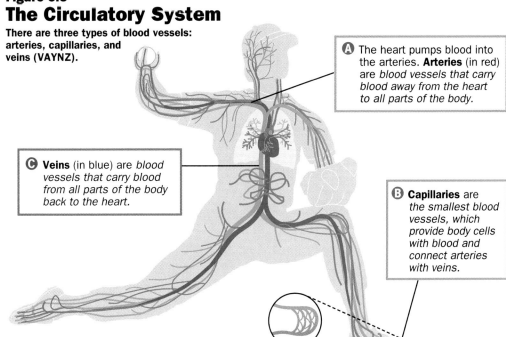

A The heart pumps blood into the arteries. **Arteries** (in red) are *blood vessels that carry blood away from the heart to all parts of the body.*

C **Veins** (in blue) are *blood vessels that carry blood from all parts of the body back to the heart.*

B **Capillaries** are *the smallest blood vessels, which provide body cells with blood and connect arteries with veins.*

Lesson 2: The Circulatory System 123

Chapter 5   Lesson 2

Teach

L1 Discussing
Help students examine the relationship between being fit and having a healthy circulatory system: Which kinds of exercise target the health of your circulatory and respiratory systems? What place should those exercises have in a physical fitness program? How do you think having a healthy circulatory system affects your ability to participate in fitness activities?

L2 Language Arts
Point out that, in poetry and other forms of literature, the heart is depicted as the seat of deep emotions, including love, compassion, and empathy. Let students work in groups to find and share poems in which the word *heart* is used to represent such emotions.

L1 Comprehending
Consider with students the importance of understanding that the heart is a muscle: What do we do to keep other muscles strong and healthy? What should we do to keep our hearts strong and healthy? Why?

Visual Learning
Guide the class in describing and discussing the vessels shown in Figure 5.3. Have students point to the heart, to any artery, to any vein, and to the capillaries in the enlarged picture. Tell students to use their fingers to trace the flow of blood from the heart to a foot and back again.

Would You Believe?
If all your blood vessels could be laid end to end, they would be long enough to wrap around the earth's equator 2½ times.

---

in your journal
Constant stress can harm the heart and blood vessels. You can reduce stress by using effective time management skills and having people you can talk with when you have problems.

In your private journal, write about your feelings when you experience stress. Imagine how these feelings might affect your circulatory system. Then make a list of actions you can take to keep stress to a minimum in your life.

Circulation Through the System

The cardiovascular system transports blood along a pathway that includes two types of circulation. **Pulmonary** (PUL·muh·nehr·ee) **circulation** *carries blood from the heart, through the lungs, and back to the heart.* This stage of circulation allows the blood to become enriched with oxygen before it is sent throughout the body. **Systemic** (sis·TE·mik) **circulation** *sends oxygen-rich blood to all the body tissues except the lungs.* Take a look at **Figure 5.4**. It shows how these two types of circulation work together to keep your body cells supplied with nutrients and free of waste products.

Figure 5.4
How Circulation Works
Oxygen-rich blood coming from the lungs is circulated through the heart and pumped to body tissues. This blood returns to the heart depleted of oxygen and is pumped to the lungs.

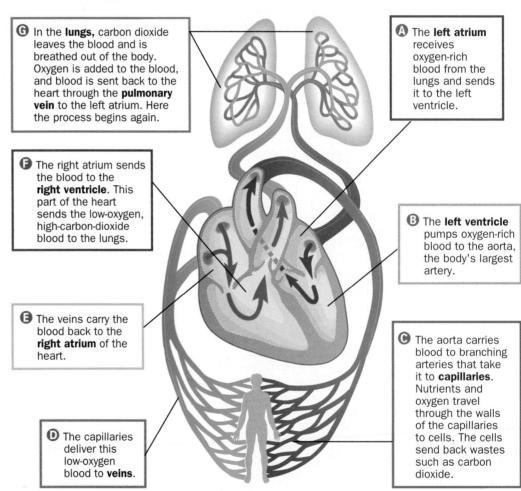

G In the **lungs,** carbon dioxide leaves the blood and is breathed out of the body. Oxygen is added to the blood, and blood is sent back to the heart through the **pulmonary vein** to the left atrium. Here the process begins again.

F The right atrium sends the blood to the **right ventricle**. This part of the heart sends the low-oxygen, high-carbon-dioxide blood to the lungs.

E The veins carry the blood back to the **right atrium** of the heart.

D The capillaries deliver this low-oxygen blood to **veins**.

A The **left atrium** receives oxygen-rich blood from the lungs and sends it to the left ventricle.

B The **left ventricle** pumps oxygen-rich blood to the aorta, the body's largest artery.

C The aorta carries blood to branching arteries that take it to **capillaries**. Nutrients and oxygen travel through the walls of the capillaries to cells. The cells send back wastes such as carbon dioxide.

124   Chapter 5: Physical Activity and Weight Management

Meeting Student Diversity

Kinesthetic Learners
To help students comprehend the circulation process as shown in Figure 5.4, provide an opportunity for them to "act it out." Let students help you set up areas of the room to serve as the left atrium, the left ventricle, the aorta, a system of capillaries, the right atrium, the right ventricle, the lungs, and the pulmonary vein. Students may want to make and post labels for each of these areas. Also, agree with students on an article—perhaps a sheet of red paper—to represent oxygen. Then let individual students move through the system, beginning in the left atrium with "oxygen" in hand, moving to the capillaries and depositing "oxygen" there, and then returning to the heart and collecting more "oxygen" from the lungs.

The Blood

Blood performs many important functions in the body. Many of the functions of the blood have to do with transporting substances through the system or protecting the body against harm. The blood does all of the following.

- Blood delivers oxygen from the lungs to all body parts.
- Blood transports carbon dioxide to the lungs for removal.
- Blood transports other wastes to the kidneys for removal.
- Blood delivers nutrients, such as vitamins and sugars, to cells.
- Blood carries special cells that fight germs in the body.
- Blood carries hormones, the messenger chemicals that regulate body processes.
- Blood promotes healing by clotting at wounds.

Parts of the Blood

Over half of the volume of blood is made up of plasma. **Plasma** (PLAZ·muh) is *a yellowish fluid, the watery portion of blood.* The rest of the volume of blood is made up of three kinds of cells: red blood cells, white blood cells, and cell fragments called platelets (PLAYT·luhts). Each of these cell types is shown in **Figure 5.5**.

Figure 5.5
Plasma and Blood Cells
Each element of the blood helps the body in a different way.

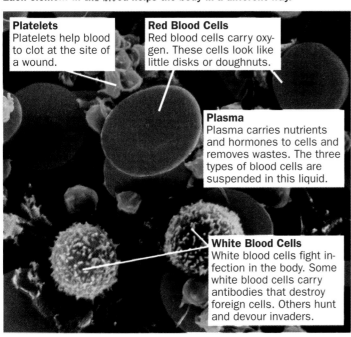

Platelets
Platelets help blood to clot at the site of a wound.

Red Blood Cells
Red blood cells carry oxygen. These cells look like little disks or doughnuts.

Plasma
Plasma carries nutrients and hormones to cells and removes wastes. The three types of blood cells are suspended in this liquid.

White Blood Cells
White blood cells fight infection in the body. Some white blood cells carry antibodies that destroy foreign cells. Others hunt and devour invaders.

Q & A

Tough Guys

Q: How do platelets make blood clot?

A: When a break occurs in the skin, platelets immediately begin to clump together. They also release chemicals to create a sticky mesh that traps even more blood cells. Eventually the clot hardens into a scab. This gives the skin and blood vessel wall time to heal.

Science Connection

Ups and Downs *ACTIVITY*

Blood has to flow against gravity in some parts of the body. How do you think this affects you? To find out, try this experiment.

Stretch one arm up so that your hand is held high above your head. Let the other arm hang down so that your hand is held low by your side. Hold this position for a minute, then examine your hands. What differences can you detect?

Lesson 2: The Circulatory System **125**

Chapter 5 Lesson 2

Visual Learning
Let volunteers read aloud and then restate in their own words the labels in Figure 5.4. Ask students to explain the difference between the red areas *(oxygenated blood)* and the blue areas *(de-oxygenated blood).*

L1 Discussing
Ask volunteers to read aloud each of the listed functions blood performs. Encourage discussion by asking of each function: Why is this important? What would happen to you if your blood could not perform this function?

L3 Researching
A thrombus is a clot that occurs inside a blood vessel. What causes a thrombus? What can result from a thrombus? How can it be prevented and treated? Let interested volunteers research this topic and prepare written reports of their findings.

L3 Reporting
Interested students can work in groups to learn about specific blood disorders—anemia, sickle-cell anemia, infectious mononucleosis, leukemia, and hemophilia. Ask group members to work together in selecting a disorder, researching it, and then planning and presenting a short report to the rest of the class.

L3 Researching
Let volunteers research the most recent advances in producing and using "artificial blood." Have them summarize their findings in written reports.

L1 Critical Thinking
To help students understand blood pressure, let them discuss responses to these questions: What is the difference between your pulse and your blood pressure? How is each measured? Why are they both important?

Personal Health

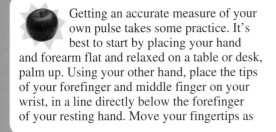

Getting an accurate measure of your own pulse takes some practice. It's best to start by placing your hand and forearm flat and relaxed on a table or desk, palm up. Using your other hand, place the tips of your forefinger and middle finger on your wrist, in a line directly below the forefinger of your resting hand. Move your fingertips as necessary until you feel the steady beat of a pulse in your wrist. Don't try using your thumb to find your pulse, because your thumb itself has a pulse. Once you can recognize the pulse, use a watch or clock with a second hand to measure exactly one minute. Count each pulse you feel during that minute.

Chapter 5 Lesson 2

L2 Guest Speaker
Invite a nurse or other health professional to demonstrate the use of a sphygmomanometer and stethoscope for measuring blood pressure.

L3 Reporting
Have a group of interested volunteers research answers to these questions: What is shock? How is it related to blood pressure? When and why does shock usually occur? How is it treated? Why is it dangerous? Have the volunteers plan and present a short skit or demonstration.

L1 Critical Thinking
Explain to the class that when they feel stressed, their blood pressure is likely to rise. When they exercise vigorously, their blood pressure is likely to rise still higher. Do these rises constitute high blood pressure, or hypertension? Why or why not? Are these rises necessarily healthy? Encourage students to explain their responses.

L3 Social Studies
Have students research early attempts in various cultures to transfuse blood from a donor to a patient. Ask them to read about Dr. Karl Landsteiner, who discovered the four main blood types and who used that new understanding to make routine blood transfusions feasible. Encourage students to compare and discuss their findings.

Assess

EVALUATING THE LESSON
Assign Reviewing Terms and Facts and Thinking Critically on page 127 to review the lesson; then assign the Lesson 2 Quiz in the TCR.

126

Blood Pressure

Blood pressure is *the force of blood pushing against the walls of the blood vessels.* A blood pressure reading consists of two numbers, usually written in this way: 120/80. The first number is the pressure at its highest point, when the heart contracts and forces blood into the arteries. The second number is the lowest point of pressure, when the heart relaxes to refill with blood.

A blood pressure reading can reveal how your circulatory system is working. A high reading means that your heart is working harder than it should be. High blood pressure may also indicate that the blood vessels are not as elastic as they should be.

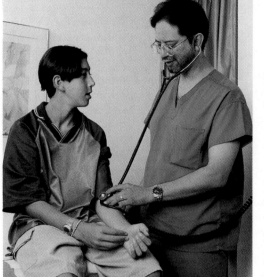

A normal blood pressure reading for teens is about 110/70.

Blood Types

Not all blood is the same. The four types—A, B, AB, and O—are classified according to the type of red blood cells they contain. Some blood types can be mixed, while others cannot. When blood types that are not compatible are mixed, the chemicals in one type of blood clump together and block the blood vessels. That is why hospitals make sure of a person's blood type before allowing one person to receive blood from another. In most cases, the blood given should be the same type as the recipient's blood. A person can, however, receive a different, compatible blood type in an emergency. **Figure 5.6** shows compatible blood types.

Blood may also contain a chemical called an Rh factor. Most people are Rh-positive, meaning that their blood has this chemical. Some are Rh-negative, meaning that their blood does not contain the chemical. People with Rh-positive blood can receive blood from people who are either Rh-positive or Rh-negative. People who are Rh-negative, however, can accept blood only from others who are Rh-negative.

Did You Know?

Where Do Blood Cells Come From?

Many red blood cells and most white blood cells are made in the bones. In the center of many bones there is fatty tissue called bone marrow. New blood cells are manufactured there and then released into the bloodstream.

Figure 5.6
Blood Type Compatibility
Type A blood has substances called A antigens (AN·ti·jenz). Type B blood has B antigens. Type AB has both A and B antigens, and type O has no A or B antigens.

Type	Can Receive	Can Donate To
A	O, A	A, AB
B	O, B	B, AB
AB	all	AB
O	O	all

126   Chapter 5: Physical Activity and Weight Management

Growth and Development

Unlike your heartbeat, which is likely to slow gradually as you mature and age, your blood pressure typically rises during your lifetime. A newborn baby may have a systolic pressure of about 40, which rises to an average of about 100 by the age of 10. In older adults, systolic blood pressure may rise to an average of 140 by the age of 60 and go up to 160 by the age of 80.

The range of blood pressure for teens is considered to be 100–120/60–80; this is the range of reading expected at rest. Under stress, teens' pressure is likely to rise to 140/90; after vigorous exercise, it may go up to 180/100.

Caring for Your Circulatory System

You can take care of your circulatory system in several ways. First, you can eat a balanced diet that is low in fats. You should also try to avoid tension, because tension can put strain on your heart and blood vessels. Two conditions that can harm your circulatory system are being overweight and smoking. Excess weight makes your heart work harder. Cigarette smoke contains chemicals that keep your blood from carrying oxygen effectively.

Staying fit is an important way to keep your circulatory system healthy. When you exercise vigorously, blood is pumped around your body faster than when you are at rest. This supplies extra oxygen and nutrients to cells. How else does fitness help this system?

- Exercise makes the muscle fibers in your heart stronger and thicker. This makes your heart more powerful and able to work more efficiently. Your heart actually has to beat less often.
- When you are exercising vigorously, blood flushes through your arteries. This may help reduce clogging by fatty materials.
- Fitness helps you stay at your ideal weight. Being overweight increases your risk for heart disease.

Handball is ranked high, along with jogging and swimming, in terms of its cardiovascular benefits.

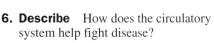

Review — Lesson 1

Using complete sentences, answer the following questions on a separate sheet of paper.

Reviewing Terms and Facts

1. **Describe** What does the circulatory system do?
2. **Vocabulary** List and define the three kinds of blood vessels.
3. **Identify** Name the two types of circulation.
4. **List** What are the components of blood?

Thinking Critically

5. **Analyze** Describe ways that the circulatory system and the digestive and excretory systems work together.

6. **Describe** How does the circulatory system help fight disease?
7. **Explain** Why is it important for hospitals to know the blood types of patients?

Applying Health Concepts

8. **Health of Others** With a group, create an advertisement or commercial for physical fitness. Do research that will help you to promote the idea that being physically fit protects and improves the circulatory system. Try to convince others that exercise will offer them long-term health benefits.

Lesson 2: The Circulatory System

Home and Community Connection

Blood Donor Centers Essential to the life-saving techniques of blood transfusions are the blood donor centers within the local community. If possible, arrange for the class to visit a blood donor center at a local hospital or Red Cross facility. Ask a technician or other professional there to explain to students the process of drawing, typing, and storing blood and to define the qualifications for becoming a blood donor. The speaker may also be able to discuss the processes and purposes of autologous blood donations and pheresis donations. If a class trip cannot be arranged, consider inviting a speaker to discuss these matters with students in the classroom.

Chapter 5 Lesson 2

LESSON 2 REVIEW

Answers to Reviewing Terms and Facts

1. Transports essential substances to body cells and removes their waste products.
2. Arteries carry blood away from the heart. Capillaries are tiny blood vessels that provide body cells with blood and that connect arteries with veins. Veins carry blood back to the heart.
3. Pulmonary circulation and systemic circulation.
4. Plasma, red blood cells, white blood cells, and platelets.

Answers to Thinking Critically

5. The digestive system provides the nutrients and vitamins that the circulatory system distributes to body cells. The circulatory system also carries waste products from cells to the excretory system.
6. White blood cells fight infection in the body. When wounds occur, clotting promotes healing and helps prevent infection.
7. In case patients require blood transfusions.

RETEACHING

▶ Assign Concept Map 20 in the TCR.
▶ Have students complete Reteaching Activity 20 in the TCR.

ENRICHMENT

▶ Assign Enrichment Activity 20 in the TCR.

Have each student complete the following sentence, being as specific as possible: I can help keep my circulatory system healthy by _____.

Lesson 3
Skeletal and Muscular Systems

Focus

LESSON OBJECTIVES

After studying this lesson, students should be able to
- identify the major parts of the skeletal system and discuss the function of the system within the body.
- identify the major parts of the muscular system and discuss the function of the system within the body.
- describe what they can do to keep their skeletal and muscular systems healthy.

MOTIVATOR

Ask students to imagine that they have no skeletal system and no muscular system. What would they look like? Have them sketch their ideas.

INTRODUCING THE LESSON

Let volunteers tell what they think their skeletal and muscular systems are, and explain their function in the body. Then have students share and discuss their sketches from the Motivator activity.

INTRODUCING WORDS TO KNOW

Have students work in pairs to read the Words to Know and to find the definition of each in the Glossary. Instruct each pair to make two sets of vocabulary cards for these words. The front of each card should be divided in half; have students write a Word to Know on one half of each card and the definition of a different Word to Know on the other half. Let the pairs use their word cards to play Health Dominoes, matching either term and term, term and definition, or definition and definition.

Lesson 3
The Skeletal and Muscular Systems

This lesson will help you find answers to questions that teens often ask about how the body supports itself and moves. For example:

▶ What do the skeletal and muscular systems do for the body?
▶ How do bones and muscles work together to allow movement?
▶ How can I keep my skeletal and muscular systems healthy?

Words to Know

skeletal system
muscular system
joint
cartilage
ligament
tendon
contract
extend

in your journal

Write down three exercises that you do or that you would like to begin doing. Describe the effects that these exercises might have on your bones, joints, or muscles.

Support and Movement

Whether you are taking a leisurely walk or playing a game of baseball, you depend on your skeletal and muscular systems to support you and help you move. The **skeletal system** is the *framework of bones and other tissues that support the body.* Your skeletal system supports your body, protects major organs and other soft parts of your body, and helps to make it possible for you to move.

Muscles provide the power to move the body. The **muscular system** *consists of tissues that move parts of the body and operate internal organs.* For example, your leg muscles help you run and your stomach muscles help you digest food.

There are 206 bones—of many shapes and sizes—and more than 600 muscles in an adult human body. The skeletal system consists of bones, joints, and connecting tissues.

The Bones and the Joints

Bones are made of living tissue and, like other organs, are supplied by blood vessels. They do more than just support and protect your body. Bones store calcium and other minerals. In addition, the inner part of bones, called the marrow, makes new blood cells.

Joints are *places where two or more bones meet.* Some joints are immovable, such as those in the skull. Others allow a wide range of movement. **Figure 5.7** on the next page identifies some of the major bones in the skeletal system and describes several types of joints.

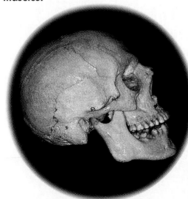

In adults the joints in the skull are fused, protecting the brain.

128 Chapter 5: Physical Activity and Weight Management

Lesson 3 Resources

Teacher's Classroom Resources
- 📁 Concept Map 21
- 📁 Decision-Making Activity 9
- 📁 Enrichment Activity 21
- 📁 Health Lab 5
- 📁 Lesson Plan 3
- 📁 Lesson 3 Quiz

- 📁 Reteaching Activity 21
- 🖨 Transparencies 19, 20

Student Activities Workbook
- 📁 Study Guide 5
- 📁 Applying Health Skills 21

128

Figure 5.7
The Skeletal System
Here are some of the major bones and joints of the skeletal system.

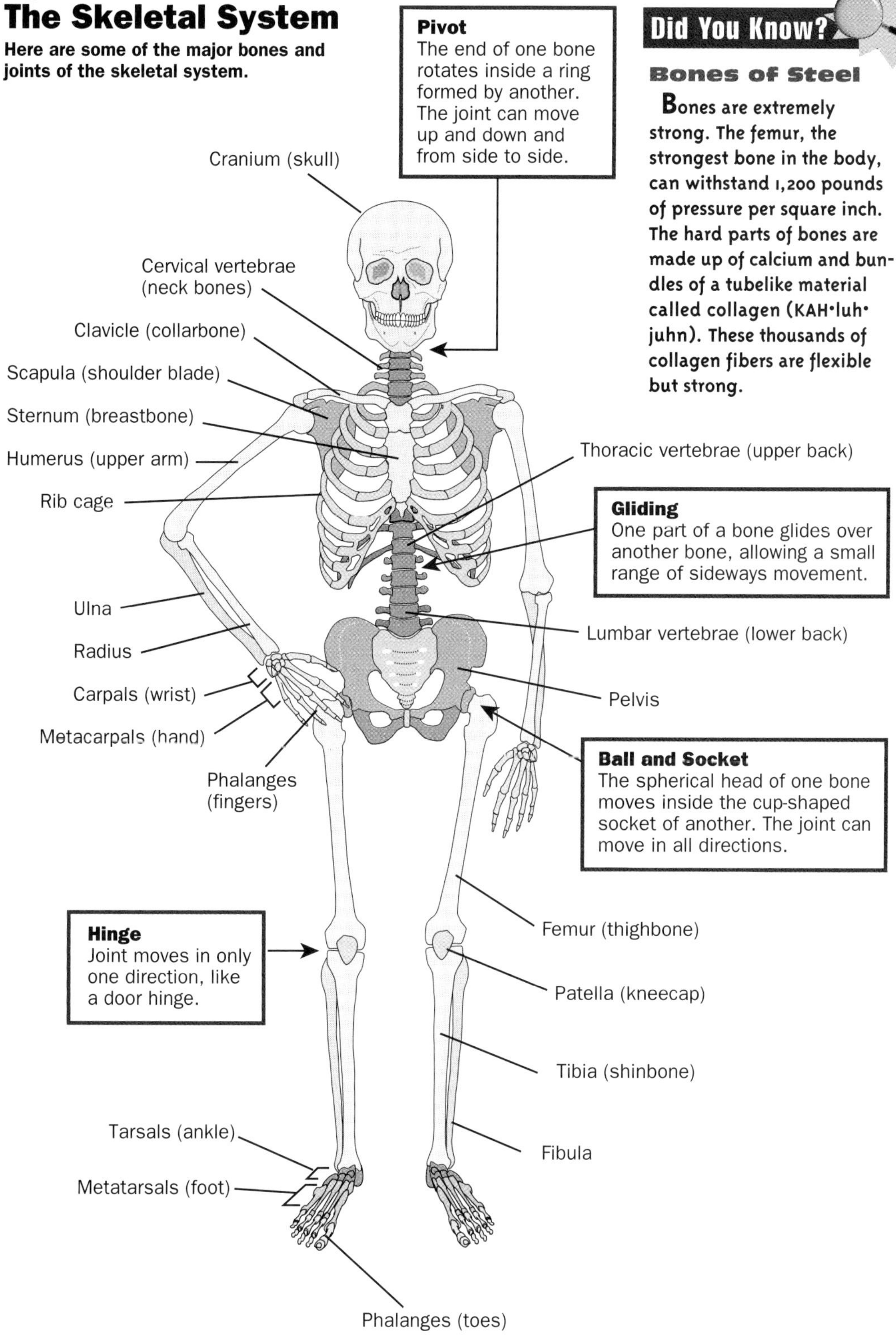

Pivot
The end of one bone rotates inside a ring formed by another. The joint can move up and down and from side to side.

Bones labeled:
- Cranium (skull)
- Cervical vertebrae (neck bones)
- Clavicle (collarbone)
- Scapula (shoulder blade)
- Sternum (breastbone)
- Humerus (upper arm)
- Rib cage
- Thoracic vertebrae (upper back)
- Ulna
- Radius
- Carpals (wrist)
- Metacarpals (hand)
- Phalanges (fingers)
- Lumbar vertebrae (lower back)
- Pelvis
- Femur (thighbone)
- Patella (kneecap)
- Tibia (shinbone)
- Fibula
- Tarsals (ankle)
- Metatarsals (foot)
- Phalanges (toes)

Gliding
One part of a bone glides over another bone, allowing a small range of sideways movement.

Ball and Socket
The spherical head of one bone moves inside the cup-shaped socket of another. The joint can move in all directions.

Hinge
Joint moves in only one direction, like a door hinge.

Did You Know?
Bones of Steel

Bones are extremely strong. The femur, the strongest bone in the body, can withstand 1,200 pounds of pressure per square inch. The hard parts of bones are made up of calcium and bundles of a tubelike material called collagen (KAH·luh·juhn). These thousands of collagen fibers are flexible but strong.

Lesson 3: The Skeletal and Muscular Systems

Chapter 5 Lesson 3

L1 Discussing
Guide students in discussing the skeletal system and the muscular system: Which bones are you aware of? Which can you see and identify in your own body? Which can you not see at all? How do you think the muscles and bones in your body depend on each other?

Visual Learning
Examine and discuss with students the bones and joints shown in Figure 5.7. If your classroom has a large model of a skeleton, display the model and let individual students come forward to identify each named bone from Figure 5.7 in the model. If you have a smaller model, allow students to examine it in small groups, and have them point to each of the identified bones.

Would You Believe?
The smallest bone in the body is only one fifth of an inch—or 5 mm—long. It's the stirrup bone, or stapes, inside the middle ear.

L3 Science
Encourage interested students to use blocks of wood and popsicle sticks (or similar craft sticks) to construct a lever system similar to the skeletal system. Let these students demonstrate and share their models with the rest of the class.

L3 Reporting
Ask volunteers to research surgical joint replacements: What kinds of joints can be replaced? What disease or injury might necessitate replacement surgery? From what material is the replacement joint usually made? How is the surgery performed? Let the volunteers present their findings to the rest of the class.

Health of Others

The most common bone disease afflicts older people, primarily women. It's called osteoporosis, which means "porous bones." As a person ages, bones may lose some of their minerals and thus become less dense. As this process continues, bones become brittle and more likely to break; they also become less capable of healing quickly. In people with advanced osteoporosis, a bone may fracture simply in response to daily, routine activity. Although osteoporosis develops later in life, healthy habits early in life can help prevent this disease. These healthy habits include avoiding smoking, limiting consumption of alcohol, getting plenty of exercise, and consuming calcium-rich foods.

Chapter 5 Lesson 3

L1 Comprehending

After students have studied Figure 5.8, ask: What does cartilage do? How are ligaments and tendons alike? How are they different? Ask volunteers to draw simple pictures showing the connections made by ligaments and tendons.

L2 Researching

Ask students to read about common injuries to bones (fractures) and to ligaments (sprains). How are the two kinds of injuries alike? How are they different? Let students meet in groups to compare and discuss their findings.

L2 Demonstrating

Have students work with partners to draw pictures that show the relationships between bones, muscles, cartilage, and tendons. Suggest that students use colored pencils or markers to draw the different kinds of tissues.

L1 Critical Thinking

Challenge students to sit so still that they don't move a single muscle. Wait for about one minute, and then ask who succeeded. (Be sure students recognize that their smooth muscles and cardiac muscle continued to move.)

  Would You Believe?

In general, large animals have slower heart rates than small animals. A gray whale's heart beats about 9 times per minute, and an elephant's beats about 30 times. The heart rate of a small bat is about 660 beats per minute, unless the animal is hibernating—then its heart beats only 30 times per minute. The tiny hummingbird has an amazing heart rate of 1,200 beats per minute.

 Q&A

Can't Lift A Finger? ACTIVITY!

Q: How come I can't move some of my fingers without moving others?

A: Each finger is connected to muscles in your forearm with tendons. However, there is also a connection between the tendons of your middle finger and those of your ring finger. This connection makes it hard to move those two fingers separately. You can demonstrate this by placing your hand palm down on a table with the middle finger curled under. Try to lift each of your fingers. You probably won't be able to lift your ring finger.

Connecting Tissues

Cartilage (KAR·tuhl·ij), ligaments (LI·guh·ments), and tendons (TEN·duhns) are connecting tissues. They link bones and muscles so that the two can work together to move parts of the body. Figure 5.8 describes the types of connecting tissue.

Figure 5.8
Connecting Tissues
Connecting tissues join bones to muscles and other bones.

Tissue Type	Description	Job
Cartilage	Strong, flexible tissue that covers the ends of bones; also supports some structures	Allows joints to move easily, cushions bones, supports soft tissues (in nose and ear)
Ligaments	Strong bands of tissue that hold bones in place at the joints	Hold bones in place, support joints
Tendons	Strong, flexible, fibrous tissue that joins muscle to muscle or muscle to bone	Move bones when muscles contract

Muscles

Muscle is tissue that responds to messages from the brain and **contracts** (kuhn·TRAKTS), or *shortens,* to cause movement. Some muscles are voluntary, or under your control, such as the muscles in your arms and legs. Others are involuntary and move without your being aware of it. These include the muscles of internal organs and blood vessels. There are three main types of muscle tissue, as shown in **Figure 5.9**. **Figure 5.10** shows some of the major muscles of the body.

Figure 5.9
Types of Muscles
Here are the three types of muscles, shown as magnified under a microscope.

Skeletal Muscle
Muscles of this type are voluntary and are attached to the bones.

Smooth Muscle
Muscles of this type are involuntary and are found in internal organs such as the stomach.

Cardiac Muscle
This type of muscle, found only in the heart, is involuntary. It contracts and relaxes about 70 times per minute, pumping blood to all parts of your body.

130 Chapter 5: Physical Activity and Weight Management

More About...

Studying Bones The first scientist to use a microscope in studying the structure of bones was Clopton Havers, an English anatomist who worked during the last part of the seventeenth century. Among the parts of the bone that he identified are the tiny channels that bear his name, the haversian canals. A system of these canals runs through the compact bone; this outer bone layer, which lies just below the periosteum, or tough membrane covering, is harder than any other body material except enamel. Each haversian canal within the compact bone carries blood vessels, lymph vessels, connective tissue, and nerves.

Figure 5.10
The Muscular System
These are some of the main skeletal muscles of your body.

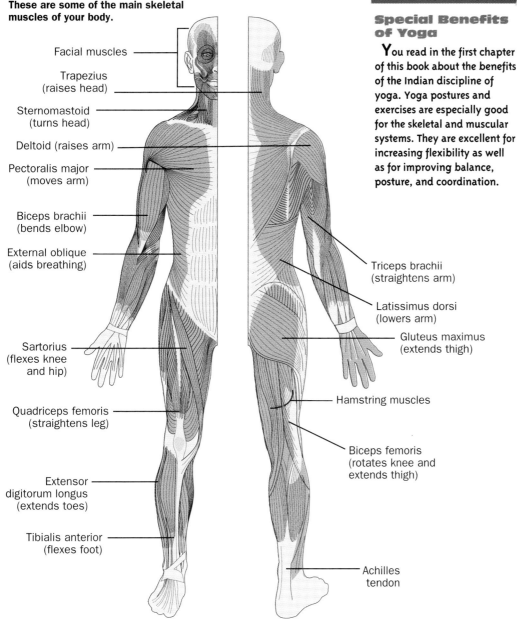

- Facial muscles
- Trapezius (raises head)
- Sternomastoid (turns head)
- Deltoid (raises arm)
- Pectoralis major (moves arm)
- Biceps brachii (bends elbow)
- External oblique (aids breathing)
- Sartorius (flexes knee and hip)
- Quadriceps femoris (straightens leg)
- Extensor digitorum longus (extends toes)
- Tibialis anterior (flexes foot)
- Triceps brachii (straightens arm)
- Latissimus dorsi (lowers arm)
- Gluteus maximus (extends thigh)
- Hamstring muscles
- Biceps femoris (rotates knee and extends thigh)
- Achilles tendon

Cultural Connections
Special Benefits of Yoga
You read in the first chapter of this book about the benefits of the Indian discipline of yoga. Yoga postures and exercises are especially good for the skeletal and muscular systems. They are excellent for increasing flexibility as well as for improving balance, posture, and coordination.

Working Together for Movement

Skeletal muscles work in pairs to move bones. Each member of the pair is connected to the bone that is to be moved. When one muscle contracts, the other muscle **extends,** or *lengthens*. The muscles switch roles to move the bone back to its original position. **Figure 5.11** shows this process in the arm.

Lesson 3: The Skeletal and Muscular Systems   **131**

Figure 5.11
Muscles and Bones Work Together
Muscle pairs are said to work in opposition. This means that to create movement, two muscles must do opposite things.

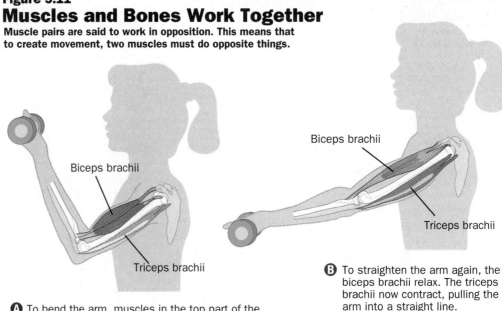

A To bend the arm, muscles in the top part of the arm—especially the biceps brachii—contract, or shorten, pulling the bone of the lower arm upward. At the same time, the triceps brachii—on the other side of the arm—must relax and extend, or lengthen.

B To straighten the arm again, the biceps brachii relax. The triceps brachii now contract, pulling the arm into a straight line.

HEALTH LAB
Pulling Pairs

Introduction: What makes your arm bend? Bend your elbow, bringing your hand close to your shoulder. With your other hand, feel your biceps, the muscle in the front of your upper arm. Then straighten your arm, feeling your triceps, the muscle at the back of your upper arm. Can you feel how the muscles are working?

Objective: With a partner, create a model to explore how paired muscles cause bones to move.

Materials and Method: You'll need two cardboard rectangles, each about 7 inches by 3 inches; a hole punch, a brad, scissors, and two large rubber bands. Overlap the ends of the rectangles and connect them with the brad. Punch two holes in each piece of cardboard, as shown in the diagram. Then cut the rubber bands, run them through the holes as shown, and retie them.

Pull on each rubber band to make the "arm" bend. Try pulling on one rubber band, then the other. What happens?

Observations and Analysis: Observe how the model works. Then think about how this applies to the more complex muscles in your arm. Could you have made the model move by pushing on the rubber bands instead of pulling them? How is this also true of the way muscles allow movement?

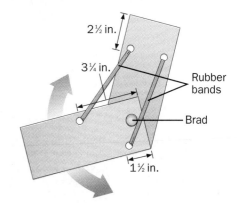

132 Chapter 5: Physical Activity and Weight Management

Caring for Your Skeletal and Muscular Systems

Follow these tips to keep your skeletal and muscular systems healthy.

- **Exercise regularly.** Exercises that build strength will make your bones and muscles stronger. Exercises that build endurance will make muscles more efficient and will also strengthen the muscles in your heart. Exercises that increase flexibility will make it easier for you to move and may prevent injuries.
- **Eat a nutritious diet.** Include foods that are rich in calcium and vitamin D for bone growth and strength. Carbohydrates will give your muscles energy, and proteins will build muscle tissue.
- **Watch your posture.** Sit and stand in a correct but relaxed manner, so that bones, joints, and muscles maintain proper placement.
- **Treat injuries promptly.** If you hurt yourself, see a physician. Avoid putting stress on an injured body part.

in your journal

Look back at what you wrote in your journal entry at the beginning of this lesson. In what ways were you correct? Now that you know more about how your muscles and bones work, describe what is happening in these systems when you engage in a physical activity.

Many teens enjoy gymnastics, which has many fitness benefits.

Review — Lesson 3

Using complete sentences, answer the following questions on a separate sheet of paper.

Reviewing Terms and Facts

1. **Vocabulary** Describe the skeletal system and the muscular system, and tell how they work together.
2. **List** Name three types of joints, giving an example of each.
3. **Identify** What are the three types of connecting tissue?
4. **Vocabulary** Use the words *contract* and *extend* in a sentence together.

Thinking Critically

5. **Analyze** How might a muscle injury affect a bone, and how might a bone injury affect a joint or muscle?
6. **Hypothesize** Why do you think there are more muscles than bones in the body?

Applying Health Concepts

7. **Health of Others** With a small group, create a skit in which you show ways to care for the skeletal and muscular systems. You may want to use pantomime to show good exercise and diet habits, with a narrator providing an explanation.
8. **Growth and Development** Children's bones are softer and more flexible than those of adults. They are less likely to break because they are able to bend slightly. Do some research to find out what causes bones to grow harder and more brittle over time. Present your findings to the class.

Lesson 3: The Skeletal and Muscular Systems **133**

Chapter 5 Lesson 3

LESSON 3 REVIEW

Answers to Reviewing Terms and Facts

1. The skeletal system is the body's framework of bones and other tissues that supports the body; the muscular system consists of tissues that move parts of the body and operate internal organs. Muscles and bones are attached to each other by tendons and work together to provide support and allow movement.
2. Any three: pivot joint, the joint between the neck and the head; gliding joint, joints in the backbone; ball and socket joint, the joint in the hip; hinge joint, the joint in the knee.
3. The three types of connecting tissue are cartilage, ligaments, and tendons.
4. Sentences will vary.

Answers to Thinking Critically

5. Responses will vary.
6. Responses will vary.

RETEACHING

▶ Assign Concept Map 21 in the TCR.
▶ Have students complete Reteaching Activity 21 in the TCR.

ENRICHMENT

Assign Enrichment Activity 21 in the TCR.

Divide the class into small groups to review what they have learned. Have them discuss their responses to the three Objectives questions posed at the beginning of the lesson.

After students have read the Health Lab Introduction, ask: When does your triceps muscle contract? When does it relax? How is it related to your biceps muscle?

Understanding Objectives

As they build and move their models, students should understand the extend-and-contract relationship between pairs of muscles and should recognize how muscles and bones work together.

Observation and Analysis

Be sure both students in each pair participate in building and manipulating the model. Then let the entire class discuss responses to the Observation and Analysis questions.

Further Investigation

Ask students to identify an important difference between the rubber bands in their models and the muscles in their arms: How does repeated use affect each?

Assessing the Work Students Do

Ask students to evaluate their own work on this Health Lab activity.

133

TEEN HEALTH DIGEST

Focus
The Teen Health Digest articles can be used in two ways: as an individual activity for reflection and enrichment or as a cooperative learning activity as described below.

Motivator
Ask students to list a number of fitness activities they participate in. Inquire: Which do you enjoy most? How does your enjoyment of a fitness activity relate to your participation in it? Guide students in recognizing that the best activities are those they enjoy and thus continue. Ask volunteers to summarize the Teen Health Digest articles and explain how they relate to the topic of the chapter.

Cooperative Learning Project

Fun-Times Fitness Day

Setting Goals
Tell students that they will plan and put on a day of fun fitness activities specially designed for elementary school children. They will have the opportunity to help young children develop positive health habits by emphasizing the fun of fitness.

Delegating Tasks
Let students divide into small groups, and make each group responsible for a fitness activity. Group members should select an appropriate activity, plan how to share that activity with individual children or groups, collect or create any equipment required, and make signs identifying their activity.

TEEN HEALTH DIGEST

CON$UMER FOCU$
Don't Fall for These Ads

"You don't have to be a doctor or a scientist to spot a phony diet ad," says Richard Cleland, a lawyer with the Federal Trade Commission's division of advertising practices. "Ads that use testimonials where people have lost 15 pounds in the first week or claims that a product works by a secret formula or is the result of some new scientific breakthrough are simply not credible."

When a product claim sounds too good to be true, it probably is. That goes especially for diet plans that say that you can eat unlimited amounts of certain foods. If you take in more calories than you use up, you'll gain weight—no matter where the calories came from.

Remember: you didn't gain weight overnight, and you aren't going to lose it that way, either. But you can damage your future health through poor choices. Make sure that your weight-loss program is good for your total health.

Try This: List at least three other claims diet ads might make that should tip off a consumer.

Personal Trainer
Water Workout

This water exercise builds strength in your abdominal and back muscles. Don't forget to warm up with easy swimming before trying it!

In deep water, face the pool side and hold onto it with both hands. Raise both legs out to your sides, then bring them together, crossing them over each other. Continue raising and crossing, alternating which leg is in front.

You could also try walking through waist-high water, taking giant steps. This exercise builds abdominal, back, and leg muscles.

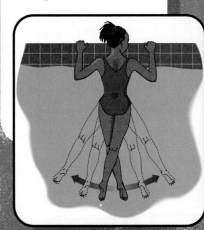

TECHNOLOGY OPTIONS

Video If the class has access to a video camera, students might take turns videotaping the activities at the various stations during the Fun-Times Fitness Day. This tape will allow students to review and evaluate their own work, and to share it with other students and family members.

If the video camera has an instant-viewing feature, students can let the children watch their own participation in the fitness activities. Encourage students to ask questions that can help the children improve their performance. ("Where should you be looking when you throw the ball?" for example.) However, remind students that the main objective is to make the activities fun for the children.

Myths and Realities

Realistic Expectations

"If I buy that fitness video and do all the exercises, will I look like the model on the box?" Some people think so. When the results don't measure up, disappointment can rob them of their enthusiasm for fitness.

Peg Jordan, a California researcher, explains that unrealistic goals can chip away at self-esteem and actually prevent people from forming healthful habits. Exercising to improve your total fitness is a healthy, realistic goal. Hoping to look like a professional fitness trainer who works out for a living is not.

Balance and common sense are the keys to setting your fitness goals. You don't need to look like a model to have a fit, healthy body.

Try This: Ask friends what fitness videos they have used or like. Find out the reasons for their comments.

HEALTH UPDATE

Time to Eat

According to Roberta Schwartz Wennick, a registered dietitian, eating the right foods at the right time is an important part of your exercise schedule. You can make the most of your fitness routines by following a few of Wennick's guidelines.

- If you exercise early in the morning—eat a light breakfast *about a half hour* before exercising.
- If you exercise in the middle of the day—have a light lunch or snack *about one hour* before exercising.
- If you exercise at night—eat dinner *about one to two hours* before exercising.

Sports and Recreation

Maine's Main Woman

Cindy Blodgett is so popular at the University of Maine that boys and girls all over the state wear basketball jerseys with her number—14—on them.

"She's a great role model for kids because of her personality and the fact that she's an excellent student," says her coach, Joanne Palombo-McCallie.

Kids aren't her only fans. Two of her team's most prominent supporters are the horror writer Stephen King and his wife, Tabitha King, who wrote a biographical book about Cindy.

The 5'9" point guard led the nation in scoring in the 1996 season. As a high school player in Clinton, Maine, Cindy led her team to four state championships. She was recruited by many colleges but chose to stay in her home state. The people of Maine support her in return.

Teen Health Digest 135

Chapter 5

The Work Students Do

Following are suggestions for implementing the project:

1. Early in the project, provide time for a class meeting in which each group identifies the activity it has selected. Guide students in offering constructive suggestions and in making sure the groups' activities are sufficiently different from each other.
2. Ask two or three students to contact a local elementary school and make arrangements for putting on the Fun-Times Fitness Day there. Be sure the time and location are confirmed, and procure any parental permissions necessary for your students' participation.
3. Guide students in considering the age, abilities, and interests of the children who will be participating in the Fitness Day. Students with younger siblings may be able to offer helpful suggestions; an elementary school physical education teacher might also be invited to speak to the class.
4. Help students consider the setting in which young children will participate in their fitness activities. Will the Fitness Day be held indoors or out? How much space will be available? What kind of ground surface will the children be playing on?

Assessing the Work Students Do

After the Fun-Times Fitness Day, let students discuss the project: How did the children react to the activities? What do you think they learned about fitness? Then have group members work together to evaluate their own contribution to the Fitness Day. Ask group members to identify the best aspects of their activity and of their interaction with the children. Also, ask them to explain what they would change if they were to organize a second Fun-Times Fitness Day.

Ability Levels You may want to guide the formation of the project groups to assure a good mix of ability levels and learning styles within each group. Whenever necessary, emphasize that every group member has important contributions to make and that all group members bear equal responsibility for a successful project.

Physically Disabled Students with any form of physical disability should be encouraged to participate fully in this project. All the students will demonstrate, by their example, that individual fitness is possible for everyone. In addition, students should make arrangements to accommodate and include any physically disabled children who will attend the Fun-Times Fitness Day.

135

Lesson 4
Planning a Fitness Program

Focus

LESSON OBJECTIVES

After studying this lesson, students should be able to
- identify factors they should consider in planning a fitness program.
- explain how they can conduct exercise sessions that will help them meet their fitness goals.
- discuss how they can assess their fitness progress.

MOTIVATOR

Ask students to list at least five reasons people give for starting a fitness program and five reasons people give for abandoning a fitness program.

INTRODUCING THE LESSON

Ask volunteers to read aloud their lists from the Motivator activity. Note students' ideas on the board. Guide students in discussing each of the reasons on the board: Of the reasons listed, which do you consider the best for starting a fitness program? Which if any are valid reasons for quitting? Conclude by telling students they will learn more about fitness programs as they read Lesson 4.

INTRODUCING WORDS TO KNOW

Write the listed Words to Know on the board, and have volunteers read the terms aloud. Then, from the lesson, read to the class the formal definitions of the terms, in random order. After each definition, tell students to write down the Word to Know they think it matches. Let students use their texts to check and correct their responses.

136

Lesson 4
Planning a Fitness Program

This lesson will help you find answers to questions that teens often ask about setting up their own fitness programs. For example:

▶ What do I need to think about when I plan a fitness program?
▶ How should I plan my workouts?
▶ How can I assess my progress?

Words to Know
warm-up
cool-down
exercise frequency
exercise intensity
target heart rate

in your journal

List three reasons why you might want to increase your level of fitness. If you are happy with your fitness level, list three reasons why you are satisfied.

First Things First

Perhaps you've decided that a fitness program will benefit your circulatory, skeletal, and muscular systems. What's next? First of all, think about your goals. What do you want to achieve? You may want to increase your endurance or develop specific skills for a team sport. You may just want to feel healthier. Having a specific goal to work toward can inspire you to stick with your fitness program—and help you know when you've accomplished something.

Selecting the Right Exercises

You also need to consider the practical aspects of your workout. What kind of exercises do you want to do? Do you need to buy equipment? When and where will you exercise? You need to plan a program that is convenient, affordable, and enjoyable for you.

There are many different activities from which you can choose. Ideally, you will want to work on strength, endurance, and flexibility. What types of exercise will help you most? **Figure 5.12** shows how some common forms of exercise measure up in these three categories.

Figure 5.12
Types of Physical Activity
Which type of exercise offers the highest level of benefit in all three categories of fitness?

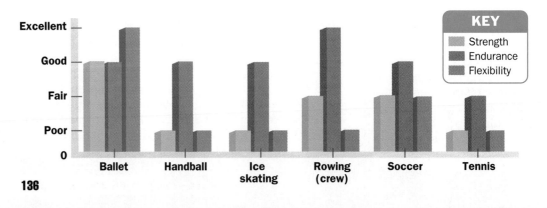

136

Lesson 4 Resources

Teacher's Classroom Resources
- Concept Map 22
- Decision-Making Activity 10
- Enrichment Activity 22
- Lesson Plan 4
- Lesson 4 Quiz

- Reteaching Activity 22
- Transparency 21

Student Activities Workbook
- Study Guide 5
- Applying Health Skills 22

Thinking About Safety

You must also think about your safety when you work out. You can exercise safely by taking sensible precautions. **Figure 5.13** lists some of the most important safety issues. Consider these points as you plan your program.

Figure 5.13
Safety Issues
Think about your total health when you plan for fitness.

Choose Safe Places and Times
Soft, even surfaces will be easier on your bones and muscles. Be careful about exercising outdoors alone, especially in a deserted place or at night. If possible, take along a couple of friends.

Dress Appropriately
Loose-fitting clothing is best for exercising. If you will be exercising outdoors at night, wear light-colored clothing and reflective coverings. This will make you more visible to others.

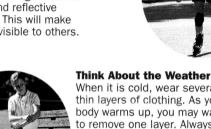

Choose Good Equipment
Good equipment is not necessarily the most expensive. Take special care, however, that your shoes or skates provide good support and are comfortable. Wear appropriate equipment for your activity.

Think About the Weather
When it is cold, wear several thin layers of clothing. As your body warms up, you may want to remove one layer. Always protect your skin from freezing temperatures. On extremely hot days, shorten your outdoor workout, and remember to apply sunscreen. Always drink plenty of fluids.

Listen to Your Body
A little discomfort when beginning new exercises or increasing intensity is normal. Pain is *not* normal, however. If you feel pain, stop exercising. If the pain continues, see a doctor. Also, don't exercise if you feel sick. You'll get well sooner if your body uses its energy for healing.

Making a Schedule

Making a schedule will help you exercise consistently. Write out a weekly plan that includes your school physical education classes and your activities before and after school. Then make a wall chart or calendar to remind yourself of what you'll do and when you'll exercise each day. Keep track of the amount of time you spend exercising each day and each week.

Be flexible, however. You don't have to write a plan for six months in advance. Your goals and needs may change as your fitness develops. Be willing to try new activities. In fact, varying your workout will help prevent boredom.

Lesson 4: Planning a Fitness Program

More About...

Exercise and Mental Activity

Exercisers have long maintained that physical activity sharpens mental acuity and stimulates productive thinking. Now research appears to confirm this connection. Scientists studied the brain waves of research subjects as they rode stationary bicycles. During a 15-minute exercise session, the subjects showed a measurable increase in brain wave activity. Later, the brain waves of the same subjects were recorded while they sat and watched television; their brain-wave activity showed measurable decreases during a 15-minute session in front of the TV.

Chapter 5 Lesson 4

Teach

L1 Discussing
Encourage individual students to share their own specific reasons for starting, expanding, or maintaining fitness programs. If necessary, remind students of the benefits of exercise to their mental/emotional and social health as well as their physical health.

L1 Recall
Review with students the importance of developing all three forms of fitness—strength, endurance, and flexibility.

L1 Discussing
Help students consider the information presented in Figure 5.12: What fitness benefits are associated with each form of exercise? Do you think your benefits would be increased if you combined that exercise with another fitness activity? What activity? Why? Ask students to come up with several other forms of exercise. Have them make educated guesses about the ratings of those forms of exercise in each of the three categories.

L1 Listing Examples
Have students work together to list specific places in their community where they can exercise safely. For example, they may start by listing the school gym, school tracks, public parks, a Boys and Girls Club, and community basketball courts. Then help students discuss the hours during which these facilities are available and the times of day when they consider themselves to be safe at those facilities.

L2 Applying Knowledge
Let students bring to class photographs (from magazine ads or newspaper articles) of specialized exercise clothing: What—if anything—makes each outfit appropriate for a particular sport or other form of exercise? Do you think this kind of outfit is worth purchasing? Why or why not?

Chapter 5 Lesson 4

L1 Listing Examples

Let students suggest several of their favorite sports or other exercise activities. Then have them list the equipment necessary for each activity. Guide them as needed in recalling appropriate safety equipment. Why is it important to wear safety gear, such as helmets, *every* time you participate in an activity?

L1 Discussing

Lead a class discussion of particular weather conditions in your area: How does the local weather affect outdoor activities? What special accommodations should students make in summer? in winter?

L1 Applying Knowledge

Discuss with students how they can learn to listen to their own bodies: What is the difference between discomfort and pain? If you are just starting to exercise, what can help you distinguish between the two? Why is it important to know the difference?

L1 Critical Thinking

Ask students to consider the exercise maxim "No pain, no gain," which has now been largely discredited: What does the saying mean? What dangers does it pose? Why do you think those who exercise—and their coaches and teachers—believed it?

L1 Discussing

Help students focus on the importance of maintaining a balance within their lives and in their exercise schedule: What problems are you likely to have if you plan to begin your fitness program by spending too much time exercising? How does keeping some free time in your schedule contribute to your good health?

L1 Comprehending

Let students respond to these questions: How do warm-up activities prepare your body for a workout? How do you think they help prepare your mind and attitude?

Science Connection

Exercise in Space

When astronauts are in space for long periods of time, they lose bone and muscle mass. The problem is caused by the weightlessness of space, where gravity has little or no effect. Scientists are working on exercises especially for astronauts that emphasize aerobics and endurance training as well as strength-building.

Planning Your Workout

Plan your exercise session to include a beginning warm-up, the workout activity, and a cool-down period. You may choose an aerobic workout to develop overall fitness and endurance or do strength-building exercises. If you do both, do the strength training after the aerobic work. After aerobic exercise, your muscles will work more smoothly.

Warm-Up and Cool-Down

A **warm-up** routine is *gentle exercise you do to prepare your muscles for vigorous activity.* Warm-up exercises bring blood into muscles, supplying them with nutrients and oxygen. Your muscles and tendons actually become warmer, so that they are more flexible. Warming up also allows you to increase your heart rate gradually and safely. Spend about ten minutes on your warm-up, which should consist of easy aerobic exercise—a brisk walk or some gentle body exercises.

Stretching should be done *after* you are warmed up. **Figure 5.14** shows two stretching exercises. You will also want to do some stretching after your **cool-down.** The cool-down stage involves *gentle exercises that let your body adjust to ending a workout.* If you stop exercising abruptly, your muscles may tighten, or you may feel faint. Continue the movements of your workout in a slower fashion. After running, for example, jog and then walk for about ten minutes to cool down. Cooling down brings your blood circulation back to normal and lowers your body temperature.

Figure 5.14
Stretching Exercises
These exercises will stretch the muscles of your upper body and your legs.

A Lean against a wall for support. Keep your arms straight while moving your upper body downward. Keep your knees slightly bent, and keep your hips over your feet.

B Stand close to a wall and lean toward it, placing your palms flat against it. Keep one leg bent and the other extended. Keeping the heel of the extended leg on the ground, move your hips forward until you feel a stretch in the calf muscle.

138 Chapter 5: Physical Activity and Weight Management

Promoting Comprehensive School Health

Community Influences In teaching students about the importance of physical fitness and a healthful diet, it is important to keep in mind the many external influences that affect teens' activities and attitudes. How fit are adults within the community? What part does family exercise play in most local households? How many of the families eat meals together on a regular basis? What facilities are available for physical fitness activities? Are local adults available to supervise activities and offer training suggestions? What programs might help improve physical fitness among all the members of the community? For more information, consult *Promoting a Comprehensive School Health Program* in the TCR.

Your Workout

To meet your workout goals, you need to pay attention to the frequency, intensity, and duration of your workout sessions. All of these factors should increase over time.

- **Exercise frequency** means *how often you work out each week.*
- **Exercise intensity** means *how much energy you use when you work out.*
- Exercise time, or duration, is how long each workout lasts.

Figure 5.15 shows how you can work with these factors to reach your fitness goals.

Figure 5.15
Meeting Your Goals
Over time, frequency, intensity, and time should all progress.

Frequency. Gradually increase your exercise frequency. Begin by working out two or three times a week. Work up to exercising every day. Remember that you will probably want to vary your routine from day to day.

Intensity. You can increase your intensity by working harder. You may try to bicycle 3 miles in less time than you did the week before. You might also include more hills in your route.

Time. Begin by exercising in 10- to 15-minute sessions. Increase the duration of individual workouts. Gradually, work up to 30 or 45 minutes of exercise at each session.

Your Target Heart Rate

You can measure your exercise intensity, or find out how hard you're really working, by figuring out your target heart rate. Your **target heart rate** is the *number of heartbeats per minute you should aim for during vigorous exercise for cardiovascular benefit.* Your target heart rate is usually a range, not a single number. If you are just starting to exercise, you should try to reach the lower part of the range. As you become more fit, try to attain higher levels. Do not, however, go beyond the top of your range. To find your target heart rate, do the following:

- Subtract your age from 220. This is your maximum heart rate.
- Multiply by 0.6 for the low end of your heart rate range.
- Multiply by 0.8 for the high end of your heart rate range.

To check your actual heart rate, take your pulse for 15 seconds in the middle of an exercise session, then multiply the number of pulse beats by four. Where are you within your range?

Lesson 4: Planning a Fitness Program **139**

Teen Issues
Do What You Like

The exercise plan that will help you the most is the one you can stick to and enjoy. Talk to people you know who are physically active. Ask them what motivates them to exercise. Share your findings with the class.

Chapter 5 Lesson 4

L2 Applying Life Skills

Do students have trouble finding time for exercise? Challenge them to give up three or four television programs next week and to use the saved time for fitness activities. At the end of the week, encourage students to discuss the results of their schedule changes.

L2 Guest Speaker

Invite a school coach or a personal trainer to speak to students about planning and starting a fitness program. Ask the speaker to discuss how he or she takes individual needs, abilities, and preferences into account, and what kinds of goals he or she helps individuals set for themselves.

L1 Math

Have students take their resting pulse. Let them calculate their pulse rate as a percentage of the typical resting rate of 70 beats per minute.

L2 Promoting Health

Have students make posters encouraging other teens to begin—or stick with—fitness programs. Display these posters in the school hallways or in other classrooms.

VIDEODISC/VHS

Teen Health Course 2

You may wish to use video segment 6, "Sports and Conditioning," in which members of a swim team share secrets of their success in setting and achieving performance goals.

Videodisc Side 1, Chapter 6
Sports and Conditioning

Search Chapter 6, Play to 7

Cooperative Learning

Pulse Comparisons Let students form small groups in which to check, compare, and graph their pulse rates. Have each group member take and record his or her own pulse three times: while sitting, relaxed, in a chair; while standing; and then again after running in place for two minutes. Encourage students to discuss how their pulse rates compare in each situation. What do the differences indicate? Then have group members work together to plan and draw an appropriate graph (such as a multiline graph) that shows the pulse rates of all group members in all three situations.

Chapter 5 Lesson 4

Assess

EVALUATING THE LESSON
Assign Reviewing Terms and Facts and Thinking Critically on page 140 to review the lesson; then assign the Lesson 4 Quiz in the TCR to evaluate students' understanding.

LESSON 4 REVIEW

Answers to Reviewing Terms and Facts
1. Exercise at safe places and times. Dress appropriately. Choose good equipment. Consider the weather. Listen to your body.
2. Warm-up exercises prepare the muscles for vigorous activity; warm-ups bring blood to the muscles and make them more flexible.
3. Cool-down activities are gentle exercises that let the body adjust to the end of a workout. Cooling down helps bring blood circulation back to normal and lowers the body temperature.
4. Exercise frequency means how often you work out each week. Exercise intensity means how much energy you use each time you work out.

Answers to Thinking Critically
5. Responses will vary.
6. Responses will vary.

RETEACHING
▶ Assign Concept Map 22 in the TCR.
▶ Have students complete Reteaching Activity 22 in the TCR.

ENRICHMENT
Assign Enrichment Activity 22 in the TCR.

Close
Have each student answer the following question, either orally or in writing: What is your next step in planning, implementing, or maintaining your own fitness program?

Q & A
RICE is Nice

Q: I pulled a muscle in my leg while exercising. A friend of mine told me to use RICE. What is it?

A: RICE is a way to help people remember four rules for treating an injury: Rest, Ice, Compression, and Elevation. The first letter of each word, put together, spells RICE. First, rest by stopping your exercise program. Use ice to keep swelling down and reduce blood flow to the area. Compression means putting pressure on the injured area to reduce swelling, as with a wrapped bandage. Elevation involves raising the injured part, also to reduce swelling. Check with a doctor or trainer before trying this method.

Checking Your Progress
You've established your fitness program and followed your weekly schedule. How are you doing? How do you feel? Look back at your original goals. Are you achieving them? The following tips can help you assess your program and your progress.

■ If you've been working out for four to eight weeks, you should see some results. You may feel stronger, be more flexible, or have more endurance. Keeping a fitness log as you go will help you see how far you have come.

■ If you feel that you're not any closer to your goal, think about whether you have been keeping to your schedule. If not, how can you make sure that you do? If you have stuck to your schedule, you may need to reevaluate your goal. Is it realistic? Maybe you need more time than you thought.

■ If you've already achieved your goal, it's time for you to set a new, more challenging one.

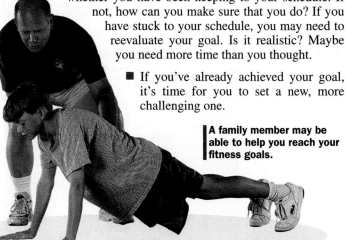
A family member may be able to help you reach your fitness goals.

Lesson 4 Review

Using complete sentences, answer the following questions on a separate sheet of paper.

Reviewing Terms and Facts
1. **List** Identify five safety precautions you should take when exercising.
2. **Describe** What is the purpose of a warm-up?
3. **Explain** Define *cool-down*, and identify its purpose.
4. **Vocabulary** Define *exercise frequency* and *exercise intensity*.

Thinking Critically
5. **Apply** What are some factors you would need to consider when planning your exercise program? Explain why each factor is important.
6. **Hypothesize** List three factors that you think might make the difference between a successful exercise program and one that a person tries and then abandons.

Applying Health Concepts
7. **Personal Health** Create a fitness log that you could use for an exercise program. Be ready to explain how you would use such a log to monitor your fitness program.

140   Chapter 5: Physical Activity and Weight Management

Home and Community Connection

Fitness in the Schools As students understand the importance of fitness activities, they may become interested in exploring the role of physical education and fitness activities in their community's schools. What physical education opportunities are offered in the district's elementary schools? middle schools or junior highs? high schools? How do the physical education requirements differ from the opportunities? Encourage students to find the answers to these questions, or invite representatives from various physical education departments to speak with the class. If students feel that physical education programs need to be expanded, help them compose, revise, and send letters to the appropriate local officials.

Weight Management

Lesson 5

This lesson will help you find answers to questions that teens often ask about managing their weight. For example:

- How do I know what is a healthy weight for me?
- How can I manage my weight through diet and exercise?
- Why are eating disorders dangerous, and how can I recognize them?

Wellness and Weight Management

Knowing and maintaining the weight that is right for you is an important part of your total health. To some extent, your ideal weight—the weight that is right for you—is determined by your height, age, and gender. Heredity also plays a part. You have a particular body type, and the weight that is right for you may not be ideal for your friend of the same height and age.

Your ideal weight is a range, not a single number. While your weight is one way to measure your health, many other factors—such as good nutrition, avoidance of risks, and overall fitness—also affect your level of wellness.

Your Body Mass Index

One way to determine your ideal weight is the Body Mass Index, or BMI. The **Body Mass Index** is *a measure of weight based on comparing body weight to height.* Calculating your BMI can help you see whether you are in a healthy range. **Figure 5.16** on the next page shows you how to calculate this number.

Words to Know

Body Mass Index (BMI)
obesity
eating disorder
anorexia nervosa
bulimia nervosa

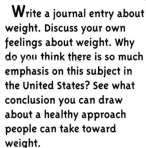

in your journal

Write a journal entry about weight. Discuss your own feelings about weight. Why do you think there is so much emphasis on this subject in the United States? See what conclusion you can draw about a healthy approach people can take toward weight.

If you are not sure whether your weight is right for you, talk with a health professional.

Chapter 5 Lesson 5

Teacher Talk
Dealing with Sensitive Issues

Many teens—both boys and girls—may be uncomfortable with any discussion that even seems to refer to their own weight. It is important to keep discussions of weight management as impersonal—but as helpful—as possible. These discussions may also help you recognize signs of undue anxiety in individual students. Make an effort to talk with such students privately, or ask the school nurse or counselor to speak with them.

L1 Discussing

Guide students in considering how they can identify and deal with their own desired weight: How good a judge can you be of your own weight? How much should you trust the judgments of your friends? Why is advice from adults probably more useful? Which adults do you think can help you recognize whether your current weight is right for you?

L1 Analyzing

Present the following situation to students: One teen, whose weight is healthy, weighs 20 pounds more than another teen, who is the same height and whose weight is also healthy. What factors could account for this difference?

L1 Applying Knowledge

Guide students in studying and discussing the Body Mass Index in Figure 5.16. Present several sample cases—for example, one teen who is 5'4" tall and weighs 125 pounds, another teen who is 5'7" tall and weighs 120 pounds—and let students find the BMI for each. Have students use the chart to find their own BMI at home or in another private setting. Encourage students who are concerned about their BMI to discuss their concerns with you, the school nurse, or another health professional.

142

Figure 5.16
Body Mass Index

To find your Body Mass Index, find your height in column A. Then find your weight in column B. Using a ruler, place it on the illustration so that the edge of the ruler lines up with both your height and your weight. The point at which the ruler intersects the scale in column C is your BMI. Generally, a BMI of over 25 indicates that you are overweight. This is only an approximation, however. A doctor can perform more precise tests of body composition.

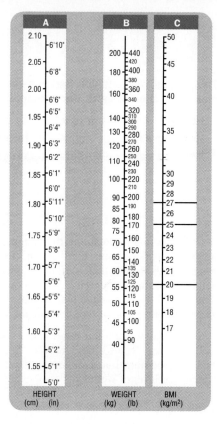

Did You Know?
Another Health Risk ACTIVITY!

Another risk related to being overweight is the risk of developing diabetes. This disease is characterized by difficulties in converting food into energy. Do some research to find out about this link, and write a one-page report displaying your findings.

Weight and Good Health

Many people think about their weight only in terms of how they look. Being underweight or overweight, however, can also create health risks and problems. If you weigh too little, you may experience fatigue, irritability, sleeping problems, or dry skin. Women who are too thin may experience problems with their reproductive systems.

Being overweight has other consequences. Obesity is especially dangerous. **Obesity** is *weighing 20 percent more than your ideal weight.* Excess weight creates a number of health risks.

- Your skeletal and muscular systems have to bear more weight than they should. Muscle and joint problems may result.

- Your heart must work harder. People who are obese have a higher risk of heart disease and stroke.

- Higher rates of certain cancers are associated with a diet high in fats. Such a diet may lead to obesity.

- People who are obese have a lower life expectancy than people who maintain their ideal weight.

142   Chapter 5: Physical Activity and Weight Management

Health of Others

Students who have become enthusiastic about physical fitness and healthy weight management may be eager to share their ideas with older family members. In one study, more than half the parents of elementary school students reported that they didn't engage in any regular, vigorous exercise. Still, teens have to proceed cautiously if they want to succeed. Discuss approaches that are likely to succeed with family members, and help students recognize the importance of setting a good example and of encouraging any improvements they recognize in the eating and exercise behaviors of their relatives.

Nutrition and Exercise

Staying at your ideal weight involves both good nutrition *and* adequate exercise. The food you eat affects your weight, but so does your activity level. By developing healthy eating and exercise habits, you can control your weight.

Figuring Out Calories

Your body runs on energy from food, much as a car runs on energy from gasoline. A calorie is a unit of heat that measures the energy available in foods. Your body needs a certain amount of energy just to stay warm and to build and repair tissues. It needs additional energy for the activities you engage in each day.

The calories in food are converted by your body to a type of energy that your cells can use. If more of this energy is produced than your cells require, the body converts the calories into body fat. Fat is stored energy, ready for use at a later time. (Every 3,500 calories that are stored rather than used become one pound of body fat.) To achieve your ideal weight, keep the following points in mind.

- To maintain your weight, you must take in the same number of calories each day that you use for energy.
- To gain weight, you must take in more calories than you use.
- To lose weight, you must take in fewer calories than you use. Your body will then turn to reserves of fat for the extra calories it needs to function. As this fat is used up, it will disappear.

You can see that if losing weight is your goal, you can approach it by consuming fewer calories *and* increasing your activity level. That way you will use more calories every day than you take in, and excess body fat will be burned up. **Figure 5.17** shows how many calories are burned during some common activities.

Q & A

Fat Versus Calories ACTIVITY!

Q: Will low-fat products help me control my weight?

A: Not by themselves. Weight management involves eating sensibly and exercising regularly. Many people believe that if they eat low-fat foods, they will lose weight no matter how much they eat. However, food labeled "low-fat" may still be high in calories.

With a partner, check out the nutrition labels on packages of five low-fat snacks. Make a chart showing the actual calorie count for each snack serving, and share your results with the class.

Figure 5.17
Calories Burned in Selected Activities
The number of calories burned is calculated for a 100-pound person performing the activity for one hour. The longer you perform an activity, the more calories you burn.

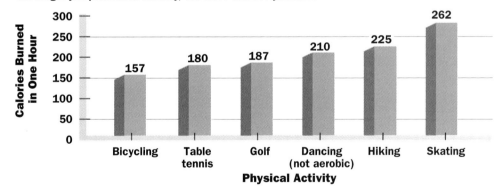

Lesson 5: Weight Management 143

Consumer Health

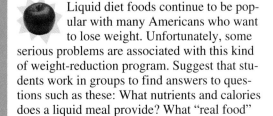

Liquid diet foods continue to be popular with many Americans who want to lose weight. Unfortunately, some serious problems are associated with this kind of weight-reduction program. Suggest that students work in groups to find answers to questions such as these: What nutrients and calories does a liquid meal provide? What "real food" meal could provide the same—or more—nutrients and the same number of calories? What specific diseases and disorders have been associated with the regular use of liquid meals? What percentage of consumers who have lost weight on liquid diets have been able to maintain their goal weights? Let students share their findings and opinions.

Chapter 5 Lesson 5

L1 Math
Have students team up with partners to devise two math problems that involve calculating whether an unnamed person is obese. (In most cases, the problems will involve a current weight and a desirable weight for a given individual, but students should feel free to be creative in developing their own problems.) Tell each pair of students to exchange problems with another pair; let the partners work together to solve the math problems they have been given.

L1 Critical Thinking
Being overweight may cause problems for an individual's mental/emotional health and social health as well as physical health. Ask students what those problems are. In addition to working toward achieving a healthy weight, what could an overweight teen do to avoid or minimize those problems?

L2 Listing Examples
Let students work in groups to come up with two lists of healthy snacks: one for teens who want to lose weight, and one for teens who want to gain weight. Then have them find the number of calories provided by each snack.

VIDEODISC/VHS

Teen Health Course 2
You may wish to use video segment 4, "Food and Nutrition," to show how a balanced diet and healthful snacking can give teens the energy and nutrients they need for growth and an active life.

Videodisc Side 1, Chapter 4
Food and Nutrition

Search Chapter 4, Play to 5

Chapter 5 Lesson 5

L1 Discussing
Lead a discussion of the Tips for Weight Management, Figure 5.18. Of each tip, ask: How do you think this can help a person lose (or gain) weight? If you wanted to lose (or gain) weight, what practices and habits would you need to change to follow this tip?

L2 Presenting Skits
Working in pairs, students will plan short two-person skits in which one teen suggests going on an extreme diet to lose weight quickly. The other teen should present convincing arguments for avoiding extreme eating patterns, regardless of weight concerns. Allow time for each pair of students to present their skit to the rest of the class.

L3 Language Arts
Ask each of several volunteers to read a short story or novel in which a central character suffers from an eating disorder. (Examples include *Nell's Quilt*, *Please Don't Go*, and *Picture Perfect*.) Let each volunteer give a short oral report, summarizing the book or story and sharing her or his opinion of it.

Assess

EVALUATING THE LESSON

Assign Reviewing Terms and Facts and Thinking Critically on page 145 to review the lesson; then assign the Lesson 5 Quiz in the TCR to evaluate students' understanding.

Managing Your Weight

How can you manage your weight effectively? You will need to keep track of the number of calories you take in and the number you use up. **Figure 5.18** gives tips for gaining or losing weight in healthful ways.

Figure 5.18
Tips for Weight Management
These tips can help you lose or gain weight in a healthful way.

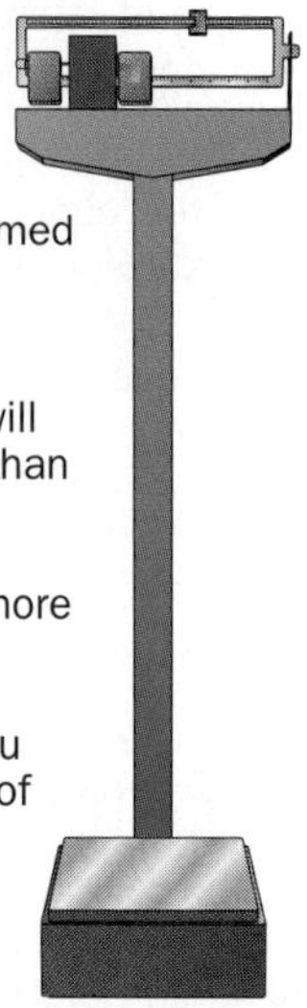

To Lose Weight

A Do not try to lose more than 1–2 pounds per week.

B Eat smaller servings of food with fewer calories. Follow Food Guide Pyramid guidelines, choosing foods high in complex carbohydrates and fiber.

C Avoid fried foods. Choose broiled, baked, or steamed foods instead.

D Increase the amount of exercise you get. Include exercises that build muscle, because your body will burn more energy supporting a pound of muscle than it will supporting a pound of fat.

E Eat slowly. Less food will satisfy you if you take more time to eat.

F Don't skip meals. Always eat breakfast to give you energy for the day. Avoid eating within two hours of bedtime, however.

G Avoid extreme diets. They can seriously damage your health.

To Gain Weight

A Eat larger servings of nutritious foods.

B When choosing foods for weight gain, concentrate on complex carbohydrates and some dairy products, but not items high in fat.

C Continue to exercise so that the weight you gain is muscle not fat. Exercises that build strength and endurance are good choices.

D Eat healthful snacks between meals. Avoid eating too close to mealtimes, however.

MAKING HEALTHY DECISIONS
Helping a Friend with an Eating Disorder

Chantal and Tamara have been taking ballet classes together since first grade. In the past three months, Chantal has noticed that Tamara is steadily losing weight. Tamara is always tired, too. As Chantal thinks back, she realizes that Tamara has been making constant comments about being too fat. In fact, Chantal realizes that she hasn't seen Tamara eat anything for a long time.

When Chantal asks Tamara about her weight, Tamara gets angry and tells Chantal that everything is fine. Chantal has to make a decision. Should she do something, and if so, what? Chantal uses the six steps of the decision-making process.

① **State the situation**
② **List the options**
③ **Weigh the possible outcomes**
④ **Consider your values**
⑤ **Make a decision and act**
⑥ **Evaluate the decision**

Follow-up Activities

1. Apply the six steps of the decision-making process to Chantal's situation.
2. With a classmate, role-play Chantal talking the problem over with Tamara or with an adult.
3. Discuss the situation in a small group. Should Chantal try to talk to Tamara again before going to an adult? What do you think a teen should do when a friend's health seems threatened?

144 Chapter 5: Physical Activity and Weight Management

MAKING HEALTHY DECISIONS

Helping a Friend with an Eating Disorder

Focus on Healthy Decisions
Allow students to discuss their responses to these questions: What would you do if you thought your friend was getting the flu? What would you do if you thought your friend had a problem that affected both her or his physical and mental health? After a brief discussion, let students read the Making Healthy Decisions feature.

Activity
Use the following suggestions to discuss Follow-up Activity 1:

Step 1 Chantal has to decide what to do in response to her suspicion that her friend Tamara has an eating disorder.

Step 2 Chantal can discuss the problem with Tamara or with a trusted adult, or she can ignore the problem.

144

Eating Disorders

Many teens are concerned about their weight. Some, however, develop **eating disorders,** which are *extreme eating behaviors that can lead to serious illness or even death.* These disorders are dangerous to mental and emotional health as well as to physical health.

- **Anorexia nervosa** (an·uh·REK·see·uh ner·VOH·suh) is *an eating disorder in which a person has an intense fear of weight gain and starves himself or herself.* People who have anorexia nervosa often become dangerously thin. They may see themselves as overweight no matter how thin they are.

- **Bulimia nervosa** (boo·LEE·mee·uh ner·VOH·suh) is *an eating disorder in which a person repeatedly eats large amounts of food and then purges.*

Why some teens and adults develop eating disorders is not clear. Several factors may be at work, however, including life stresses, a desire for control or perfection, and depression. Our society's emphasis on appearance, especially for girls and women, may also play a part. No matter what the cause of the disorder, professional help is almost always needed to help a person recover.

Recovering from an eating disorder can be a long and difficult process. Seeking professional help is an important step.

Teen Issues

Your Body Image

When you watch television or read a magazine, you see lots of thin people. You may feel that your goal should be to look like them. Remember, however, that you cannot base your goals on someone else's body. Learn to love the body you have, and work to keep it in the best possible shape for you.

Review — Lesson 5

Using complete sentences, answer the following questions on a separate sheet of paper.

Reviewing Terms and Facts

1. **Vocabulary** What is the *Body Mass Index?*
2. **Describe** List three harmful effects of obesity.
3. **Restate** What is the relationship between calories and weight gain?
4. **Identify** What are two types of eating disorders?

Thinking Critically

5. **Explain** How can exercise be important for both losing and gaining weight?
6. **Recommend** Suppose that you have a friend who does not have an eating disorder, but who seems to have unrealistic expectations of how thin she should be. What could you say to her to help her avoid future problems?

Applying Health Concepts

7. **Consumer Health** Choose a weight-loss product or service (such as an exercise machine, diet program, or diet clinic) to research. Find out how much it would cost to use the product or service for a year. Then explain how a person might achieve the same or better results for free with good nutrition and regular exercise. Present your findings to the class.

Lesson 5: Weight Management 145

Chapter 5 Review

CHECKING COMPREHENSION

Use the Chapter Summary and the Chapter 5 Review to help students go over the most important ideas presented in Chapter 5. Encourage students to ask questions and add details as appropriate.

CHAPTER 5 REVIEW ANSWERS

Reviewing Key Terms and Concepts

1. Exercise is physical activity that develops fitness.
2. Aerobic exercise is non-stop, rhythmic, vigorous activity that increases breathing and heartbeat rates. Anaerobic exercises are not rhythmic or continuous; they require short, intense bursts of energy.
3. Another name for the circulatory system is the cardiovascular system.
4. Blood pressure is the force of blood pushing against the walls of the blood vessels.
5. The bones of the skeletal system surround and protect the major organs and soft parts of the body.
6. The three types of muscle tissue are skeletal muscle, smooth muscle, and cardiac muscle.
7. A typical workout session has three stages: warm-up, the workout activity, and cool-down.
8. To determine your target heart rate, subtract your age from 220 and multiply the difference in two ways: by 0.6 and by 0.8. The range between those two numbers is your target heart rate.
9. Obesity occurs when a person weighs 20 percent over his or her ideal weight.
10. Eat larger servings of nutritious food; get extra calories from carbohydrates, not from fats; exercise to gain muscle; eat healthy snacks between meals.

Chapter 5 Review

Chapter Summary

▶ **Lesson 1** Fitness is the ability to handle the physical work and play of everyday life. A person can become fit by exercising regularly to develop strength, endurance, and flexibility.

▶ **Lesson 2** The circulatory system transports blood throughout the body to carry nutrients, oxygen, and other important materials to body cells. The heart pumps the blood through a system of blood vessels.

▶ **Lesson 3** The skeletal and muscular systems provide a supportive framework for your body and enable it to move. Regular exercise and a balanced diet help maintain a healthy skeleton and muscles.

▶ **Lesson 4** To plan a personal fitness program, you need to set goals, select the right exercises, and think about safety. Your workout should include warming up and cooling down.

▶ **Lesson 5** Achieving and maintaining your ideal weight will help you remain healthy. Weight management principles involve good nutrition and adequate exercise.

Reviewing Key Terms and Concepts

Using complete sentences, answer the following questions on a separate sheet of paper.

Lesson 1
1. What is the meaning of the term *exercise?*
2. Distinguish between *aerobic exercise* and *anaerobic exercise.*

Lesson 2
3. What is another name for the circulatory system?
4. What is *blood pressure?*

Lesson 3
5. In what way does the skeletal system protect the body?

6. What are the three types of muscle tissue?

Lesson 4
7. What are the stages in a typical workout session?
8. Describe how you can determine your target heart rate.

Lesson 5
9. What does the term *obesity* mean?
10. List four tips for people trying to gain weight.

Thinking Critically

Using complete sentences, answer the following questions on a separate sheet of paper.

11. **Predict** How could physical fitness and remaining active help an elderly person remain healthy?
12. **Integrate** Describe how a program of regular exercise designed to improve strength, endurance, and flexibility could help the skeletal and muscular systems.
13. **Analyze** Why might you want to vary your workout over time?
14. **Distinguish** How could you tell the difference between a healthful weight management program and an eating disorder?

146 Chapter 5: Physical Activity and Weight Management

Meeting Student

Language Diversity Use the following suggestions to help students who have difficulty with English:

▶ Pair those learners with native speakers of English who can restate the Chapter Summary in language that helps students comprehend important concepts.

▶ Direct auditory learners or those students with language diversity to the Teen Health Audiocassette Program. Available in English and Spanish, this component provides an audio and written summary of the chapter. 🎧

Chapter 5 Review

Your Action Plan

Making a commitment to fitness involves dedication, not only to exercise, but to maintaining a balanced diet and having a healthful attitude about weight.

Step 1 Look over your journal entries and come up with at least one long-term goal and several short-term goals. For example, if your long-term goal is to exercise for half an hour four times a week, your short-term goals might include walking to school every day or riding your bike every weekend.

Step 2 Think about these goals in terms of your overall health—physical, mental/emotional, and social. Would you add or change any goals in order to increase gains in any of these areas? You might, for example, have a goal of learning a new physical activity. You may also feel, however, that you need more social interaction. You might refine your original goal to include an activity that you can take part in with others.

Decide on dates for reaching your goals. As you reach every short-term goal, check it off and give yourself a reward.

In Your Home and Community

1. **Community Resources** Propose that your school or community hold a fitness fair. Work with interested adults to promote fitness and good nutrition. Fitness and health professionals and local merchants could be invited to share information.

2. **Health of Others** Offer to be an exercise buddy for a younger neighbor or family member. You can encourage your buddy with your support and increase your own fitness program at the same time.

Building Your Portfolio

1. **Personal Fitness Program** Design a personal fitness program. Include a list of long-term and short-term goals, your schedule, and descriptions of your activities in your portfolio.

2. **Editorial** Write an editorial on the subject of body image in American culture. Discuss how both obesity and extreme thinness can be dangerous to health. Suggest actions that might bring about a more realistic view of weight and health. Include the editorial in your portfolio.

3. **Personal Assessment** Look through all the activities and projects you did for this chapter. Choose one or two that you would like to include in your portfolio.

Chapter 5: Chapter Review **147**

Performance Assessment

▶ **Self-evaluation** Direct students to review the activities that are provided throughout the chapter. Encourage each student to select one finished product or activity that demonstrates his or her best work for the chapter. Have students explain what they learned and how the examples they selected show their progress.

▶ **Teacher's Classroom Resources** Assign Performance Assessment 5, in the TCR.

Chapter 5 Review

Thinking Critically

11. People who have stayed physically fit throughout their lives are likely to have healthy circulatory, skeletal, and muscular systems. This may reduce the chances of some of the typical diseases of older age, including heart disease, osteoporosis, and joint problems.

12. Strength-building exercises are good for strong muscles and bones. Exercises for endurance help make all types of muscle tissue, including the heart, stronger. Exercises for flexibility keep the joints and muscles healthy.

13. Varying your workout can help prevent boredom, exercise different muscle groups, build strength in different areas, and may allow you to exercise outdoors in good weather and indoors in bad.

14. A healthful weight management program involves a balanced diet and exercise. An eating disorder involves an unnatural focus on food. A person with an eating disorder often loses weight too quickly.

RETEACHING

Assign Study Guide 5 in the Student Activities Workbook.

EVALUATE

▶ Use the reproducible Chapter 5 Test in the TCR, or construct your own test using the Testmaker Software.
▶ Use Performance Assessment 5 in the TCR.

EXTENSION

Let students work together to compile information on local programs that promote physical fitness. Have them find out about the time, location, sponsoring organization and cost of each program. Then let students publish the information they have gathered in a brochure or flier that can be distributed to other members of the student body.

Planning Guide

Chapter 6
Sports and Conditioning

Lesson		Features	Classroom Resources
1	**Individual and Team Sports** *pages 150–153*	Life Skills: Teamwork *page 153*	Concept Map 24 Decision-Making Activity 11 Enrichment Activity 24 Health Lab 6 Lesson Plan 1 Lesson 1 Quiz Reteaching Activity 24 Transparencies 22, 23
2	**Sports and Physical Wellness** *pages 154–158*		Concept Map 25 Decision-Making Activity 12 Enrichment Activity 25 Lesson Plan 2 Lesson 2 Quiz Reteaching Activity 25 Transparency 24
3	**Conditioning Goals and Techniques** *pages 159–163* TEEN HEALTH DIGEST *pages 164–165*	Health Lab: Getting into Condition *pages 162–163*	Concept Map 26 Cross-Curriculum Activity 11 Enrichment Activity 26 Lesson Plan 3 Lesson 3 Quiz Reteaching Activity 26 Transparency 25
4	**Balancing School, Sports, and Home Life** *pages 166–169*	Making Healthy Decisions: Balancing a Life *page 169*	Concept Map 27 Cross-Curriculum Activity 12 Enrichment Activity 27 Lesson Plan 4 Lesson 4 Quiz Reteaching Activity 27

National Health Education Standards

One of the main goals of the National Health Education Standards is to move students toward "health literacy." Health literacy is the capacity of individuals to obtain, interpret, and understand basic health information and services and the competence to use such information and services in ways which promote health. The health standards were developed by applying the characteristics of a well-educated, literate person within the context of health. The health literate person is:

- ▶ a critical thinker and problem solver
- ▶ a self-directed learner
- ▶ a responsible, productive citizen
- ▶ an effective communicator

Listed below are the Health Education Standards Performance Indicators addressed in each lesson of this chapter.

Lesson	Health Standards Performance Indicators
1	(1.1, 5.1, 5.5, 6.1)
2	(1.1, 3.1, 3.5)
3	(6.4, 6.5, 6.6)
4	(1.2, 6.1, 6.2, 6.3)

ABCNEWS InterActive VIDEODISC SERIES

You may wish to use the videodisc *Food and Nutrition* with this chapter. See side one, video segments 5 and 11. Use the *ABCNews InterActive™ Correlation Bar Code Guide* for title reference. Also available in VHS format.

Chapter Resources

Teacher's Classroom Resources
- Chapter 6 Test
- Parent Letter and Activities 6
- Performance Assessment 6
- Testmaker Software

Student Activities Workbook
- Study Guide 6
- Applying Health Skills 24–27
- Health Inventory 6

Student Diversity Strategies
- Audiocassette Program (English)
- Audiocassette Program (Spanish)
- Spanish Parent Letters
- Spanish Summaries, Quizzes, and Activities

Multimedia Components
- English Audiocassette Program
- Spanish Audiocassette Program
- *Teen Health* Videodisc/VHS Series

Other Resources

Readings for the Teacher
Miller, Mary. *Opportunities in Fitness Careers.* Lincolnwood, IL: VGM Career Horizons, 1997.

Simon, N. *Good Sports: Plain Talk About Health and Fitness for Teens.* New York: Harper & Row, 1990.

Zumerchik, John, ed. *Encyclopedia of Sports Science.* New York: Macmillan Library Reference, 1997.

Readings for the Student
Dinn, Sheila. *Hearts of Gold: A Celebration of Special Olympics and Its Heroes.* Woodbridge, CT: Blackbirch Press, 1996.

Farrington, Jan. "What's the World's Best Exercise? It's Your Move!" Current Health 2, May 1995, pp. 6–12.

Gard, Carolyn. "Get with the Program That's Right for You," Current Health 2, February 1997, pp. 22–24.

Out of Time?

If time does not permit teaching this chapter, you may use these features: Life Skills on page 153; Health Lab on pages 162–163; Teen Health Digest on pages 164–165; Making Healthy Decisions on page 169; and the Chapter Summary on page 170.

Chapter 6
Sports and Conditioning

CHAPTER OVERVIEW

Chapter 6 helps students understand the interrelationship between fitness and sports. It also emphasizes the importance of maintaining a healthy balance among the elements of teen life.

Lesson 1 focuses on individual and team sports, and discusses some of the benefits of sports competition.

Lesson 2 explains the importance of a nutritious, balanced diet for athletes. It also discusses tips for avoiding sports injuries and emphasizes the importance of avoiding anabolic steroids.

Lesson 3 presents information about conditioning for teen athletes and describes cross-training as a healthy means of developing whole-body fitness.

Lesson 4 helps students understand how they can evaluate the demands of school, sports, and family to achieve a healthy balance in their lives.

PATHWAYS THROUGH THE PROGRAM

The lessons in Chapter 6 guide students in developing skills that will help them achieve and maintain good health throughout their lives. Involvement in sports can help improve physical health, with added benefits to both mental/emotional health and social health. In addition, an ability to recognize and deal with competing demands, such as those of schoolwork, sports, and home life, can help individuals of every age maintain a high level of mental/emotional health.

Chapter 6
Sports and Conditioning

Student Expectations
After reading this chapter, you should be able to:
1. Describe the role sports play in a healthful lifestyle.
2. Recognize the importance of good nutrition and prevention of injuries in playing sports.
3. Explain how to use conditioning techniques effectively.
4. Identify ways you can balance school, sports, and home life.

Key to Ability Levels

Teaching strategies that appear throughout the chapter have been identified by one of three codes to give you an idea of their suitability for students of varying learning styles and abilities.

L1 **Level 1** strategies should be within the ability range of all students. Full class participation is often required. Teacher direction is usually needed.

L2 **Level 2** strategies are for average to above-average students or for small groups. Some teacher direction is necessary.

L3 **Level 3** strategies are designed for students able and willing to work independently. Minimal teacher direction is necessary.

Teen Chat Group

Courtney: Well, I did it. I put my name on the list for field hockey tryouts.

Julie: Go, girl! Just remember, the school team is a lot tougher than our community league was.

Ian: What are you saying—that Courtney shouldn't try out?

Julie: No, it's just that *I'm* scared to. Hardly any seventh graders make the team.

Courtney: I know that already. I talked it over with my dad last night. He said if I didn't make it this year, I could work on my skills and try out again next year.

Ian: Right. And you'll be even better by then.

Julie: I love playing hockey, but I'm not ready for the pressure of trying out. We can still practice together, Courtney, right?

Courtney: Sure—whether I make the team or not!

in your journal

Read the dialogue on this page. Have you ever wanted to try a sport but not felt sure that you could succeed? Start your private journal entries on sports and conditioning by responding to these questions:

- What would you say to Courtney if you were part of this conversation?
- Would you be willing to try out for a team if you weren't sure that you could make it? Why or why not?
- What do you consider to be a healthy attitude toward playing sports? Describe your feelings in your journal.

When you reach the end of the chapter, you will use your journal entries to make an action plan.

Chapter 6: Sports and Conditioning **149**

Chapter 6

INTRODUCING THE CHAPTER

▶ Ask students to identify and discuss the various sports opportunities in their school and community, including both team and individual sports for boys, for girls, and for boys and girls together: In what sports can teens compete? How are the teams or competitors selected? How, where, and when do they practice? Where and when do they compete? Encourage students to share their own experiences with, and reactions to, these sports activities. Conclude this discussion by telling students they will learn more about sports and fitness as they study Chapter 6.

▶ Show a short videotape of athletic competition. After they have watched the videotape, let students discuss the athletes and events. Encourage students to share their ideas about the athletes' training and mental conditioning: How long do you think these athletes have been preparing for these events? What have they probably given up to be able to compete at this level? What do you think they have gained? Conclude by telling students that, as they read Chapter 6, they will learn more about sports conditioning and competition.

Cooperative Learning Project

Working Out: An Exercise Manual

The Teen Health Digest on pages 164 and 165 provides students with high-interest articles related to the content of this chapter.

The material in the Teacher's Wraparound Edition presents suggestions for a class project in which students plan, write, illustrate, and publish a short guidebook of appropriate exercises for teens.

---

in your journal

Introduce the Journal activity with a brief class discussion: How many of you have tried out for a sports team? What motivated you to try out? How did you feel before the tryouts? during the tryouts? How did you feel afterward, when you found out whether or not you had made the team? What makes it hard to try out for a team? What are some of the benefits of trying out, whether or not you make the team? Then let three volunteers read the Teen Chat Group aloud, dramatizing the dialogue and adding comments of their own. Encourage students to share and discuss their responses to the three In Your Journal questions before they begin writing independently.

Lesson 1
Individual and Team Sports

Focus

LESSON OBJECTIVES

After studying this lesson, students should be able to
▶ explain the advantages of participating in sports.
▶ identify the most important differences between individual sports and team sports.
▶ discuss the value of competition in sports.

MOTIVATOR

Give students two minutes in which to list as many different sports as they can.

INTRODUCING THE LESSON

Ask volunteers to read aloud the lists they wrote in response to the Motivator activity. Then help students discuss some of the sports they have named: Which sports do you enjoy watching? Which do you enjoy playing? Which of these sports have you only heard about? How could you learn more about those sports? Close by telling students that they will have an opportunity to explore sports further as they study Lesson 1.

INTRODUCING WORDS TO KNOW

Let volunteers read aloud the four Words to Know and explain what they think each term means. Help students consider the terms in pairs: How do you think individual sports and team sports are alike? How do you think they are different? Which two Words to Know start with the same prefix? What meaning is associated with that prefix? Then let students skim the lesson to find the definition of each Word to Know.

150

Lesson 1 Individual and Team Sports

This lesson will help you find answers to questions that teens often ask about playing sports. For example:

▶ What role do sports play in a healthy lifestyle?
▶ How can I choose which sports to play?
▶ What are some benefits of competition?

Words to Know

commitment
individual sports
team sports
competition

Language Arts Connection

Writing About Sports

If you're interested in your home team, you probably read the sports pages in the newspaper. Even if you're not a big fan, you may enjoy human interest sports stories, such as the kind that tells how a player became successful.

Attend a school or community sports event, and try "covering" it as a sportswriter would. Report on one game, or interview a player or a coach for your article.

Sports and You

One way to achieve fitness is to start and follow a well-thought-out exercise program. One way to remain fit is by playing sports. Some people play sports just for fun. Others make a serious **commitment,** *a pledge or promise,* to one or more sports. Commitment involves dedicating yourself to something over a period of time. If you're committed to a sport, you learn all you can about it, practice regularly, and work to develop the necessary skills.

Becoming involved in sports can result in a lifetime of fitness. Activities such as tennis, golf, and swimming are good lifetime sports. They can be enjoyed by people of almost any age.

Choosing Sports Activities

What do you like about sports? Do you like staying fit, the excitement of a close game, or the satisfaction of mastering a new skill? It is important to think about these questions as you choose a sports activity. You'll be most likely to get the greatest benefit out of a sport if you enjoy it.

Golf can provide excellent exercise for people of varied ages.

150   Chapter 6: Sports and Conditioning

Lesson 1 Resources

Teacher's Classroom Resources
📁 Concept Map 24
📁 Decision-Making Activity 11
📁 Enrichment Activity 24
📁 Health Lab 6
📁 Lesson Plan 1
📁 Lesson 1 Quiz

📁 Reteaching Activity 24
🎞 Transparencies 22, 23

Student Activities Workbook
📁 Study Guide 6
📁 Applying Health Skills 24

Individual Sports

Individual sports are *physical activities you can take part in by yourself or with a friend, without being part of a team.* There are many individual sports to choose from, including the following:

- Biking
- Hiking
- Swimming
- Running
- Horseback riding
- Skateboarding
- Surfing
- Skating

Individual sports have several advantages. For example, you can set your own schedule and determine your own level of commitment. You don't have to be compared to anyone else, and you can set the pace of the activity. On the other hand, you miss some of the social, mental, and emotional benefits of playing on a team.

interNET CONNECTION

Team up with the World Wide Web to set fitness goals, avoid sports injuries, and focus your concentration.
http://www.glencoe.com/sec/health

When you are involved in an individual sport, you need self-discipline to follow your regimen.

Team Sports

Team sports are defined as *organized physical activities with specific rules, played by opposing groups of people.* Baseball, basketball, football, soccer, and volleyball are some of the most popular team sports. Team sports have many advantages.

- You have the companionship and encouragement of teammates and coaches.
- Playing against another team may push you to excel.
- Keeping up with regularly scheduled practice can help you become more responsible.

Some teens, however, find team sports too restricting. Individual sports are a better choice for these teens.

Many teens find that they form close friendships with their teammates.

Lesson 1: Individual and Team Sports   151

Growth and Development

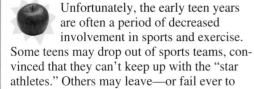

Unfortunately, the early teen years are often a period of decreased involvement in sports and exercise. Some teens may drop out of sports teams, convinced that they can't keep up with the "star athletes." Others may leave—or fail ever to join—because their friends are not involved in athletics. Not surprisingly, many teens are actually less physically fit than they were during their elementary-school years. Fitness researchers found that about three-quarters of the 9-year-olds tested met a minimum standard in running one mile. Among 14-year-olds, however, less than 60 percent of the boys and less than 50 percent of the girls achieved that same standard.

Chapter 6 Lesson 1

Teach

L1 Discussing

Let students indicate by a show of hands whether they would rather achieve fitness by following an exercise program or by participating in sports. Have several students from each side of the question explain their responses. Help students recognize benefits to both plans.

Visual Learning

Help students discuss the photograph at the bottom of page 150. Who is participating in this sport? How do you think this activity benefits each person's physical health? mental/emotional health? social health? What activities might offer *you* those benefits?

L1 Identifying Examples

Ask students to extend the list of individual sports. Write their ideas on the board, and emphasize that some individual sports may involve team practice and competition. Encourage students to share specific information about community opportunities for participating in these sports.

L1 Discussing

Let students share their ideas in response to this question: What are some of the social, mental, and emotional benefits of committing yourself to an individual sport?

L2 Creating Displays

Have students organize themselves into two or more groups. Ask the members of each group to work together to plan and make a bulletin board (or other display) of teens involved in individual sports or in team sports. Encourage students to incorporate words as well as images in their display.

151

Chapter 6 Lesson 1

L2 Current Events
Suggest that students look through the sports section of a local newspaper every day for a week. Provide a few minutes of class time in which students can recall and discuss the articles they looked at. Let interested volunteers select, clip out, and carefully read articles they find interesting. Have the volunteers summarize these articles for the rest of the class and then post them on a bulletin board.

L1 Discussing
Encourage students to share their own experiences in competitive sports: At what age did you begin competing? Did you enjoy it then? Why or why not? Do you still compete? Why or why not?

L2 Critical Thinking
Help students consider some of the "stars" of professional sports teams: How do they help their teams compete successfully? Is there a point at which a star's performance detracts from the performance of the team? If so, what is that point?

VIDEODISC/VHS

Teen Health Course 2
You may wish to use video segment 6, "Sports and Conditioning," in which members of a swim team share secrets of their success in setting and achieving performance goals.

Videodisc Side 1, Chapter 6
Sports and Conditioning

Search Chapter 6, Play to 7

---

in your journal

How do you feel about competition? Do you enjoy it, or would you rather play just for fun or to become more skilled? Do you enjoy competition only in some sports, or with some people but not with others? Write down your feelings in your journal.

Sports and Competition

Most sports can be played on several levels: to increase your degree of fitness, to improve your skills, or with the goal of competition. **Competition** is *rivalry between two or more individuals or groups trying to reach the same goal.* When you play team sports competitively, your goal is to help your team win the game. In an individual competitive sport, your goal is to run faster, jump higher, or score more points than another person. You can even compete against yourself by trying to improve on your previous performance.

Some people thrive on competition. They enjoy the challenge of working to be the best. When they don't win, the experience makes them feel even more motivated to win next time. Others don't like competing. They find losing very painful. Still others find it hard to enjoy winning if their friends have to lose.

Competition can be both enjoyable and valuable, however, as **Figure 6.1** shows. Competitive sports can have value far beyond winning one game or even one season. They're a way to have fun, to build skills and fitness, and to work together with others toward a common goal.

Figure 6.1
The Value of Competition
Playing competitive sports can offer you opportunities to grow.

A Competition gives you a reason to work hard, practice, and make your best effort.

B Competition helps you develop mental focus, a skill that will be useful in other areas of your life.

C Competition allows you to learn the value of encouraging others and receiving their encouragement.

D Competition offers you a chance to improve your skills and feel good about your accomplishments.

152 Chapter 6: Sports and Conditioning

LIFE SKILLS
Teamwork

Focus on Life Skills
Let students share their ideas in response to questions such as these: What do you like best about being part of a sports team? What do you think is the hardest part of being a team member? What are the most important differences between playing a team sport and playing an individual sport? How is a sports team similar to a group of students working together on a school assignment? To a group of employees working together to achieve a specific goal at work?

After the introductory discussion, have students read the Life Skills feature, restate each suggested teamwork skill in their own words, and describe its importance.

Making Health Connections
Understanding and practicing teamwork skills can improve all aspect of teens' health. As teens put these skills into use within the context of sports teams, they gain important physical exercise benefits and establish enjoyable exercise habits that can be

Review

Using complete sentences, answer the following questions on a separate sheet of paper.

Reviewing Terms and Facts
1. **Vocabulary** Describe what it means to make a *commitment* to a sport.
2. **List** Identify three individual sports and three team sports.
3. **Vocabulary** What is the meaning of *competition*?

Thinking Critically
4. **Recommend** Write a short paragraph that explains how sports can be helpful to people throughout their lives.
5. **Evaluate** Describe your own thoughts and feelings about the benefits of individual sports as compared to the benefits of team sports.

Applying Health Concepts
6. **Personal Health** Intramural sports are team sports that allow different teams from the same school to compete against each other. As a class, hold a brainstorming session about the advantages and difficulties of having such a program. Questions to consider may include how to organize teams, where to hold practices, and how to get a faculty sponsor. If your school does not have an intramural program already, consider presenting a formal plan for such a program to the administration.

LIFE SKILLS
Teamwork

Working as a productive member of a team is a lifetime skill. Teamwork skills can help you get along in school, in your community, and at a job. What skills will help you be a good team player?

- **Cooperate.** Use your playing skills to help your team. Don't try to win the game alone. Players need to work together to be successful.
- **Communicate.** Let others know when you need help or when you have a good idea. Also, listen to your coach and your teammates.
- **Be sensitive.** Instead of criticizing a teammate for making a mistake, offer support. Everybody makes mistakes—even you.
- **Be generous.** Pay attention to the game when you're on the bench. Encourage your teammates.
- **Be responsible.** Show up for practices on time, and work at home on your skills whenever you can. Come to practice ready to work.

Practice these skills. If you aren't on a team, you can apply similar skills to a team project at school or to interactions within your family.

Follow-up Activity
With a group of four to six people, prepare a skit about teamwork. You may use the setting of a sports team or of some other type of team. Illustrate the benefits of good teamwork or the problems caused by not working together. Perform your skit for the class.

Lesson 1: Individual and Team Sports

Lesson 2
Sports and Physical Wellness

Focus

LESSON OBJECTIVES

After studying this lesson, students should be able to
- discuss the relationship between calorie intake and energy output.
- explain how nutrition affects sports performance.
- identify practices that help prevent sports injuries.
- discuss the importance of avoiding performance-enhancing drugs.

MOTIVATOR

Present the following situation to the class: You are a professional athlete, competing in the sport of your choice. Write a list of everything you eat and drink on the day before a big game or an important competition.

INTRODUCING THE LESSON

Let volunteers share their lists with the rest of the class. Ask: How do your ideas of an athlete's diet differ from your diet? Why? How do you think a gymnast's diet differs from a football player's diet? Why? Conclude by telling students that they will learn more about sports nutrition, as well as other aspects of sports training, as they study Lesson 2.

INTRODUCING WORDS TO KNOW

Have students write down the Words to Know, along with their definitions as presented either in the lesson or in the Glossary. Then ask students to sketch simple pictures that will help them remember the meanings of these terms.

154

Lesson 2 Sports and Physical Wellness

This lesson will help you find answers to questions that teens often ask about sports nutrition and sports injuries. For example:
▶ How can nutrition affect my level of performance?
▶ How can I avoid sports injuries?
▶ Why is it harmful to take drugs to improve sports performance?

Words to Know

dehydration
anabolic steroids

Sports and Energy

You already know that to maintain your weight, you need to use the same number of calories as you put into your body by eating. If you play sports, you will use more energy. Therefore, to maintain your weight, you will need additional calories.

Not all sports require the same amount of energy. Chapter 5 gave information on how many calories are burned during various activities. **Figure 6.2** divides sports into three categories: low-energy, moderate-energy, and high-energy. Knowing the energy requirements of the sports you play can help you plan your diet.

Figure 6.2
Sports and Energy Use
The number of calories burned, as shown in this figure, is calculated for a 100-pound person playing for one hour.

High-energy Sports (over 350)
- Cross-country skiing
- Handball
- Racquetball
- Rope jumping
- Running
- Soccer

Moderate-energy Sports (250–350)
- Aerobic dancing
- Badminton
- In-line and ice-skating
- Swimming
- Tennis
- Volleyball

Low-energy Sports (less than 250)
- Baseball and softball
- Basketball (half-court)
- Bowling
- Golf
- Gymnastics
- Hiking
- Judo and karate

154   Chapter 6: Sports and Conditioning

Lesson 2 Resources

Teacher's Classroom Resources
- 📁 Concept Map 25
- 📁 Decision-Making Activity 12
- 📁 Enrichment Activity 25
- 📁 Lesson Plan 2
- 📁 Lesson 2 Quiz

- 📁 Reteaching Activity 25
- 🖨 Transparency 24

Student Activities Workbook
- 📁 Study Guide 6
- 📁 Applying Health Skills 25

Sports Nutrition

Whether or not you are active in sports, you should have a balanced, nutritious diet. Getting enough complex carbohydrates, proteins, vitamins, and minerals—and not consuming too much fat—is important for anyone. If you play sports, however, you will need to keep your body supplied with additional energy. This may require some changes in what you eat as well as when you eat.

What to Eat

If you play sports, make sure that you get enough of the six major types of nutrients. You can do this most effectively by eating a variety of foods from the five major food groups. Here are a few pointers.

- **Eat plenty of complex carbohydrates.** Fruits and vegetables, pasta, and whole-grain breads will provide your body with carbohydrates, a very efficient source of energy, as well as fiber. If you need to increase your calorie intake because of sports activity, eat larger amounts of foods high in complex carbohydrates.

- **Get enough vitamins and minerals.** A balanced diet should provide the vitamins and minerals that you need. Two minerals that are especially important are iron and calcium. Iron helps to supply your muscles with oxygen while you are exercising. Good sources are red meat and green leafy vegetables. Calcium strengthens bones and helps muscles work properly. Dairy products, salmon, and collard greens are good sources of calcium.

- **Don't load up on protein or fats.** An athlete does not need more protein or fats than a nonathlete. Although protein is used in building muscle tissue, muscles become developed only through exercise and training. Eating two to three servings from the meat, poultry, fish, dry beans, eggs, and nuts group daily will provide all the protein you need. A diet high in fats could put you at risk for several kinds of diseases later in life.

- **Eat breakfast.** If you have a game in the morning, make sure that you eat breakfast. You'll perform better if you've given your body fuel to go on. Eat lightly, however, if your sport requires running.

- **Drink plenty of water.** Make sure that you get enough water. Eight glasses of water are recommended each day for most individuals. If you are playing a sport, raise that to 9–13 glasses of water. While engaging in sports you lose more water because you perspire. **Dehydration** (dee·hy·DRAY·shuhn) is *excessive loss of water from the body.* It can cause muscle cramps and heatstroke, and it is also harmful to other body systems.

Your Total Health

Eat Up

Here are some snacks and meals that provide 100 grams of carbohydrates, as well as proteins, vitamins, and minerals. They're good for you and good for your game.

- 1 bagel with peanut butter and 2/3 cup of raisins
- 1 cup of low-fat yogurt, 1 banana, and 1 cup of orange juice
- 1 turkey sandwich (without mayonnaise) and 1 cup of applesauce
- 2 cups of spaghetti with meat sauce and 1 piece of bread

Athletes need access to liquids while participating in any sport.

Lesson 2: Sports and Physical Wellness

Chapter 6 Lesson 2

Teach

L1 Recalling

Pose these questions to be sure students understand the relationships between food intake and exertion: What is a calorie? (a unit for measuring energy) How does your body get calories? (by eating, and from fat stored in the body) How does your body use calories? (by maintaining itself and by moving)

L1 Discussing

Guide students in considering the sports listed in Figure 6.2, Sports and Energy. Have volunteers read the sports in each column. Why do you think these are low- (or medium- or high-) energy sports? What other sports do you think might fit into this category? Why?

L2 Comprehending

Let students work in small groups to consider changes that should be made in serving sizes to accommodate an active athlete who wants to take in an extra 800 calories per day. Have group members create a sample menu for one day's meals and snacks, indicating the size of each serving.

L2 Interviewing

Have students talk with high school athletes or adults who train and participate in sports activities: How do the athletes change their eating habits when they are in training? when they are preparing for a game or other competition? What do they eat before, during, and after a game or competition? How much water (or other liquids) do they drink? Let the students share and discuss the results of their interviews.

More About...

Eating Breakfast Teens may acknowledge that eating a healthy breakfast can help improve an athlete's performance. It is important to realize that breakfast can help improve a student's performance as well. Studies show that people who eat breakfast are more alert and productive than people who skip this meal. Doctors and nutritionists consider breakfast the most important meal of the day; it gets body systems functioning well and reduces the likelihood of consuming empty calories later in the day. However, the National Adolescent Student Health Survey found that one-third of teenage boys and half of all teenage girls do not eat breakfast regularly.

Chapter 6 Lesson 2

L1 Discussing

As students read each pointer for *What to Eat*, ask questions such as these: How is this guideline different from the general guidelines for healthy eating? What specific foods does this pointer suggest to you? When do you need to increase your intake of those foods?

Health Minute

It's important for athletes to drink water or other healthy liquids, even if they don't feel thirsty. Dehydration can lead to heat exhaustion, which causes more than 400 deaths in the United States each year.

L2 Family and Consumer Sciences

Ask students to investigate one of the special sports-related foods or drinks now on the market; these include high-energy bars and drinks intended to replace lost sodium and potassium: What benefits do these special products offer? In which cases—if any—do they offer advantages over natural snack foods or plain water or juice?

L2 Demonstrating

Let students with particular interests in specific sports demonstrate safe, effective movements used in those activities. For example, one student might show how to move the arm for a forehand tennis stroke; another might demonstrate the correct leg movements for race-walking.

L1 Brainstorming

Let students work in groups to brainstorm a list of problems that might result when an athlete loses control of his or her emotions during a game.

L2 Researching

Have students use newspaper or periodical indexes to find recent articles about sports injuries. Instruct each student to choose and read one article; then let students meet in groups to compare and discuss what they have read.

Q & A

Better than Water?

Q: Do I need to drink sports drinks when I exercise or play sports?

A: Water alone is enough if your workout or game lasts less than an hour, or if you are not playing continuously. Sports drinks do replenish fluids, however, as well as carbohydrates that your body can use for energy. These drinks can be helpful for long, intense games.

When to Eat

Although you may not have to change *what* you eat very much when you begin to play a sport, deciding *when* to eat may require more thought. Important factors include how much energy you will need and when, and how your sport will affect your digestion. **Figure 6.3** provides tips on nutrition before, during, and after playing sports.

Figure 6.3
Knowing When to Eat

If you don't eat enough before or during a strenuous game, you could feel faint. To play your best, follow these tips.

① Before You Play
Good foods to eat before a game are those rich in complex carbohydrates. Bananas, bagels, and fruit juices are good choices. Eat about one to two hours before your game. Drink plenty of fluids as well. Some athletes find that they achieve a better energy level by eating a regular meal a couple of hours before a game and then having a light snack about a half hour before they play.

② During Your Game
Drink plenty of cool water, at least ½ cup every 20 minutes. Usually it is not necessary to replace carbohydrates, even sugars, unless your game will last more than an hour and you will be playing almost nonstop. Some coaches, however, provide orange or apple slices, rice cakes, or fruit juice on the sidelines, along with water.

③ After Your Game
After your game you're likely to be hungry. At this point, eat a balanced meal with plenty of carbohydrates but also some protein and fats. Make sure that the meal also replenishes vitamins and minerals. Continue to drink water and other liquids.

156 Chapter 6: Sports and Conditioning

Personal Health

Both diet and exercise play important roles in maintaining healthy cholesterol levels. Elevated cholesterol is associated with heart disease in adults, and elevated cholesterol during childhood and teen years typically leads to high-cholesterol problems during adulthood.

The National Cholesterol Education Program suggests that teenage boys with cholesterol readings of 183 to 202 mg/dl, and teenage girls with readings of 192 to 216 mg/dl, are at moderate risk of heart disease. Higher readings indicate even higher risks. Teens with readings in these risky ranges should work to reduce their intake of fats, especially saturated fats, and to increase their aerobic activities.

Preventing Injuries

When you are playing a sport, you are using your body to run, jump, throw, and kick. You are often thinking and moving fast—and so are the people around you. Here are some tips on how to avoid sports injuries.

- Always warm up, stretch adequately, and cool down.
- Learn the proper techniques for your sport. For example, when you throw a ball, avoid overextending your elbow. When you land from a jump, bend your knees.
- If you get injured during a game, don't return to play until you've been checked out by the coach, a trainer, or a doctor. To be safe, don't return to the game at all unless the injury was a minor one *and* you have been cleared by your coach.
- Control your emotions. Getting angry may cloud your judgment and cause you to get hurt or to hurt someone else.

Having the right equipment and using it properly are the best ways to protect yourself against injury. **Figure 6.4** describes some of the equipment that can help protect you.

Did You Know?

Danger Zone

In 1994 the three team sports with the highest number of injuries for children aged 5 to 14 were basketball, football, and baseball. However, well over half of the injuries in each of these sports occurred during informal activities, not organized ones. Being part of an organized team with a qualified coach in charge reduces your chance of injury in any sport.

Figure 6.4
Protective Equipment
Different sports require different kinds of protective gear.

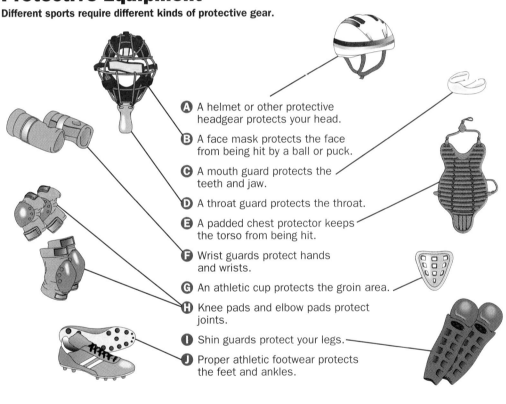

- **A** A helmet or other protective headgear protects your head.
- **B** A face mask protects the face from being hit by a ball or puck.
- **C** A mouth guard protects the teeth and jaw.
- **D** A throat guard protects the throat.
- **E** A padded chest protector keeps the torso from being hit.
- **F** Wrist guards protect hands and wrists.
- **G** An athletic cup protects the groin area.
- **H** Knee pads and elbow pads protect joints.
- **I** Shin guards protect your legs.
- **J** Proper athletic footwear protects the feet and ankles.

Visual Learning

Help students identify and discuss each kind of protective equipment shown in Figure 6.4: In what kind of sport is this protective equipment used? What kinds of injuries do you think it helps prevent? Should it be worn during practice? during games or other forms of competition? Encourage students to discuss their own experiences with wearing various kinds of protective equipment.

L3 Reporting

Assign interested students to research specific incidents of drug testing during the Olympic Games: Why was testing conducted? Who was tested? With what results? Have these students share their findings in a brief oral report.

L3 Researching

What are the legal, medical uses of anabolic steroids? Let interested students research the answers to this question and summarize their findings in written reports.

L1 Discussing

Guide students in describing the photograph and reading the caption on page 158: What sport have these teens committed themselves to? Do you think they consider themselves successful? Do you think they would feel equally successful if they were using performance enhancers? Why not?

Health Minute

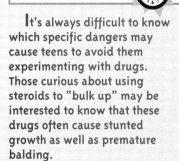

It's always difficult to know which specific dangers may cause teens to avoid them experimenting with drugs. Those curious about using steroids to "bulk up" may be interested to know that these drugs often cause stunted growth as well as premature balding.

Meeting Student Diversity

Learning Styles It is important for students to recognize each kind of protective equipment and to understand its proper use. For some students, this learning can be facilitated by concrete experience. Make available some or all of the pieces of protective equipment shown in Figure 6.4. Let students form small groups in which to handle and discuss each piece, trying the equipment on when feasible. Encourage group members to identify specific kinds of injuries that could be prevented by this equipment.

Chapter 6 Lesson 2

Assess

EVALUATING THE LESSON
Assign Reviewing Terms and Facts and Thinking Critically on page 158 to review the lesson; then assign Lesson 2 Quiz in the TCR to evaluate students' understanding.

LESSON 2 REVIEW

Answers to Reviewing Terms and Facts
1. Responses will vary. Possible response: eat plenty of complex carbohydrates; don't load up on proteins or fats; drink plenty of water.
2. Dehydration is excessive or abnormal loss of water from the body.
3. Responses will vary. See Figure 6.4 on page 157.

Answers to Thinking Critically

4. Responses will vary. Possible response: Calm down by breathing deeply or focusing on a peaceful thought. If you can't control your anger, ask to be removed from play.
5. Responses will vary. Possible responses: Personal and external pressures to excel may cloud a young athlete's judgment. Many teens believe that they can handle any unpleasant side effects or that they will not have any problems with the drugs.

RETEACHING
- Assign Concept Map 25 in the TCR.
- Have students complete Reteaching Activity 25 in the TCR.

ENRICHMENT
Assign Enrichment Activity 25 in the TCR.

Close

Let students work in small groups to list the three most important things they learned from Lesson 2.

158

 **in your journal**

What would you do if a friend told you that he or she was considering using steroids or some other performance-enhancing drug? In your journal, describe how you would feel, and what you would say or do to handle the situation.

Avoiding Harmful Substances

You can build strength and endurance safely with regular workouts and good nutrition. Some people, however, use performance-enhancing drugs such as anabolic steroids. **Anabolic steroids** (a·nuh·BAH·lik STIR·oydz) are *synthetic compounds that cause muscle tissue to develop at an abnormally high rate.*

Anabolic steroids, often referred to as *steroids,* have legitimate medical uses, such as treating some types of cancer. It is illegal, however, to use them to improve athletic performance. They are also very dangerous because of their side effects, which include

- development of acne.
- weakening of tendons, possibly leading to joint or tendon injury.
- damage to the cardiovascular system, affecting heart rate and blood pressure and increasing the risk of heart attack.
- bone damage, since bones can become more brittle.
- harmful effects on sexual characteristics, including growth of facial hair in females and breast development in males.
- mental and emotional effects, such as irritability, anxiety, suspicion, or sudden rage.
- liver and brain cancers.

Avoid anabolic steroids and all other performance-enhancing drugs. They can harm your health and ruin your athletic career.

You don't need performance enhancers to achieve success in a sport.

Lesson 2 Review

Using complete sentences, answer the following questions on a separate sheet of paper.

Reviewing Terms and Facts
1. **List** Identify three tips for good sports nutrition.
2. **Vocabulary** What is *dehydration?*
3. **Recall** List three body areas that might be injured while you are playing sports, and describe a piece of equipment for protecting each one.

Thinking Critically
4. **Recommend** What advice would you give to a player who begins to feel very angry during a game?
5. **Hypothesize** Why do you think people take anabolic steroids and other performance-enhancing drugs, despite the dangers of these substances?

Applying Health Concepts
6. **Personal Health** Create a list of ways to avoid injuries in the sports you enjoy most. You may include information from this lesson as well as from other sources.
7. **Health of Others** Make a poster to highlight the dangers of anabolic steroids and other performance-enhancing drugs. Describe the serious side effects of various substances.

158 Chapter 6: Sports and Conditioning

 **Home and Community Connection**

Is there a "Bike Rodeo" or "Bike Safety Day" in your area? Typically, law enforcement officers and other volunteers show young children how to ride their bikes safely, emphasizing the importance of always wearing a helmet when riding. Ask students to find out about this kind of safety fair in their community. When and where is it held? Whom does it serve? Encourage students to volunteer to help in advertising the event or in working with children on the day of the event.

If your community does not have any bike safety events for children, suggest that interested students work together to enlist the help of community adults in planning and presenting this kind of safety program.

Conditioning Goals and Techniques

Lesson 3

This lesson will help you find answers to questions that teens often ask about conditioning goals and techniques. For example:

- What is conditioning?
- What techniques will help me get into condition without doing more than I should?
- How can keeping records help me to reach my goals?

Conditioning and Your Goals

If you play a sport regularly, you'll need to devote time to getting into shape for it. *Training to get into shape* is called **conditioning**. As in other aspects of life, you'll need to set goals for your conditioning if you expect to accomplish anything. The goals you set, however, will depend on what you want to get out of the sports you play. Ask yourself questions like these.

- **What are my priorities?** What is most important to me? Fitness? Fun? Friendship? Improving specific skills? Excelling at the sport I have chosen?
- **What are the demands of this sport?** What do my coach and my teammates expect of me? How serious do I need to be?
- **What will happen in the future?** What will be expected of me? How might my other responsibilities increase? Do I want to keep on playing this sport?

Words to Know

conditioning
cross-training
overtraining

Did You Know?

Hooray for Hoops

A recent survey of more than 25,000 15- to 18-year-olds in 41 countries asked teens what sports they liked to play and to watch. Basketball was listed by 71 percent of the respondents. Teens in Taiwan, Greece, and Korea showed the most enthusiasm for the game. Soccer came in second with 67 percent of the teens placing it on their lists.

Different sports require different types of conditioning.

Lesson 3: Conditioning Goals and Techniques 159

Lesson 3 Resources

Teacher's Classroom Resources
- Concept Map 26
- Cross-Curriculum Activity 11
- Enrichment Activity 26
- Lesson Plan 3
- Lesson 3 Quiz
- Reteaching Activity 26
- Transparency 25

Student Activities Workbook
- Study Guide 6
- Applying Health Skills 26

Lesson 3
Conditioning Goals and Techniques

FOCUS

LESSON OBJECTIVES

After studying this lesson, students should be able to

- explain how to set conditioning goals.
- define cross-training and identify some of the benefits of cross-training.
- discuss the dangers of over-training.

MOTIVATOR

Display photographs of three athletes: a figure skater, a football player, and a sprinter. Write these questions on the board: Which of these athletes might take ballet lessons as part of his or her training? Why? Give students five minutes in which to write their responses.

INTRODUCING THE LESSON

Encourage students to share and discuss their responses to the Motivator questions. If appropriate, direct students' thinking with questions such as these: What kinds of fitness do you think ballet promotes? Which of these athletes is likely to benefit from increased flexibility? from increased strength? Conclude by telling students they will learn more about various kinds of athletic training as they read Lesson 3.

INTRODUCING WORDS TO KNOW

Divide the class into three groups, and assign each group one of the Words to Know. Let group members work together to plan and present a skit that helps explain the definition of their assigned term.

Chapter 6 Lesson 3

Teach

L1 Comprehending

Ask students to cite specific examples of conditioning: Have you ever trained to get into shape? How did you train? What do you know about the conditioning of high school and college athletes? of professional athletes?

L1 Discussing

Let students organize their own groups based on their interest in specific sports. Have group members discuss the three goal-setting questions on page 159. Help students recognize that their responses will vary according to their own situations and interests.

L2 Language Arts

Have students select and read biographies or autobiographies of athletes. Provide time for students to discuss the subjects of these books, either with the whole class or in small groups: When did the athlete begin serious conditioning? What kinds of conditioning did he or she use? With what results?

L1 Identifying Examples

Ask students to identify specific people within the school and the community to whom they can turn for advice on conditioning. Write their suggestions on the board.

L1 Comparing

Help students compare single-sport training with cross-training: How can single-sport training lead to boredom? What happens when a teen gets burned out on a specific sport? In terms of avoiding injuries, what are the most important differences between single-sport training and cross-training? How do you think cross-training benefits one's mental/emotional health and social health?

160

Q&A

What a Pain

Q: Why are my muscles stiff and sore after I've exercised strenuously?

A: Muscles become stiff and sore because of microscopic damage to muscle fibers. The small tears in the muscles fill up with fluids and waste products. As the fibers heal, your muscles feel better. The pain is often at its worst about two days after exercising. For this reason, the pain is known as delayed-onset muscle soreness. If you continue to exercise your muscles regularly, soreness will no longer be a problem.

Getting Good Advice

To set appropriate goals, you may want to ask someone for advice. Your physical education teacher, a coach, or possibly a family physician can give you good advice on how to get into condition for a particular sport.

Different sports demand different levels of strength, endurance, and flexibility. For example:

- A gymnast needs strength, flexibility, and endurance to perform a variety of gymnastic routines.
- A basketball player needs endurance, speed, and agility to handle the ball, as well as strength for shooting.
- A wrestler needs strength, agility, flexibility, and balance.

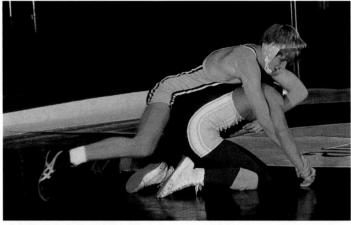

What kinds of exercises do you think wrestlers might do to condition their bodies?

Cross-Training

Cross-training is *any fitness program that includes a variety of physical activities to promote balanced fitness.* If you play several sports, cross-training is essential. It is also helpful for people who do not play team sports. What are the benefits of cross-training?

Figure 6.5
A Sample Cross-Training Program

Here's a sample weekly program that would promote conditioning for several different sports and be interesting as well. Notice that this teen combines outdoor and indoor activities as well as solitary activities with activities she does with other people.

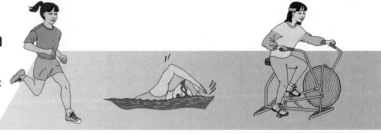

On day 1, run two miles on the school track.

On day 2, take an easy swim in an indoor pool.

On day 3, go for a bicycle ride or ride a stationary bike.

160 Chapter 6: Sports and Conditioning

Health of Others

Flexibility varies from individual to individual. However, most people lose flexibility as they grow older. In a given family, the young children are likely to be more flexible than the teenage children, who are in turn probably more flexible than the parents. Students may want to compare their own flexibility with that of other family members. Suggest that students have their family members (or friends of various ages) sit on the floor, with their legs straight out in front of them. Then have them bend forward to touch their toes. How far beyond their toes can they reach? Students can measure each person's reach beyond the toes and compare the results.

- If you combine a variety of activities you enjoy, you are less likely to become bored or "burned out."
- You avoid injury by strengthening and stretching complementary muscles, such as the hamstrings in the back of the thighs and the quadriceps in the front of the thighs. You also exercise different muscle groups.
- Your total health, including mental/emotional and social health as well as physical health, is improved by achieving overall fitness of your whole body.

Figure 6.5 illustrates the type of variety you might see in a cross-training program.

Avoiding Overtraining

It's great to be enthusiastic about getting into condition, but don't overdo it. Too much exercise without enough rest can be harmful. **Overtraining** is *exercising too hard or too often, without enough rest in between.* What are the signs of overtraining? They include

- an abnormally high heart rate when you are not exercising.
- feeling sore or tired all the time.
- frequent illness.
- disturbed sleeping habits.
- irritability or inability to concentrate.
- frequent muscle strain or injury.

How can you avoid overtraining? First, take at least one day off every week. During the rest of the week, alternate tough workout days with easy workout days. For example, the day before a major event, exercise lightly. (Take another look at Figure 6.5 to see how this might be done.) Finally, every two months, reduce your exercise intensity for a week.

Teen Issues

Trying Too Hard

Some teen athletes train so hard that it delays their normal growth and development. Others develop eating disorders in trying to stay at an ideal size for their sport. This is a frequent problem for female athletes in such sports as gymnastics, figure skating, and distance running, in which small size and low body weight are advantages. If you think that you or someone you know may be overtraining, talk to a trusted adult.

On day 4, take a rest. Don't exercise at all on this day.

On day 5, take a half-hour walk with a friend. Carry light arm weights to strengthen your upper body.

On day 6, spend twenty minutes doing toning exercises.

On day 7, play a game of tennis with a friend.

Consumer Health

Private exercise classes, gyms, and fitness centers attract adults and teens—and in some cases even children—in many communities. You may want to have students investigate some of these exercise opportunities in their neighborhoods: What exercise equipment is provided? What guidance is offered in using the equipment properly and safely? What kinds of classes are offered? What levels of individual attention and correction are available? What is the charge for using the facilities? Schedule class time during which students can compare and discuss their findings. Guide them in evaluating whether these facilities might be "good buys."

Chapter 6 Lesson 3

L1 Comprehending
Let students share their responses to these questions: Why do you think taking a day off of training each week is important? How can a day off affect your attitude toward training? Are there any problems that might be associated with taking a day off from training? How could you avoid or overcome those problems?

L1 Discussing
Guide students in discussing the importance of keeping records of their conditioning activities: How can your own record encourage you to continue and improve? How could a therapist or physician use your training records if you sustain a sports injury?

L1 Journal Writing
Have students write journal entries about their own conditioning goals. Encourage them to distinguish between their short- and long-term goals.

L1 Discussing
Help students consider the idea of rewarding themselves for reaching their long-term conditioning goals: How long will it probably take to reach a long-term goal? What reward would be most beneficial? Why? What other kinds of rewards are you likely to receive as you move toward your long-term goal? (Be sure students recognize their own improved fitness as one of these rewards.)

in your journal

Set some long-term goals for your exercise or conditioning program. Consider such factors as weather changes and vacations. Will you need to change your routine in the middle of winter or when your family goes on a trip? In your journal, list some of your long-term goals as well as personal circumstances that might affect your ability to reach them.

Keeping Records

Keeping track of your conditioning program helps you to stay aware of your progress. As you improve, you can take pride in your accomplishments. If you experience muscle soreness, you can change activities or seek help before the problem gets worse. **Figure 6.6** shows a useful format for charting your progress.

When you begin your conditioning program, set some short-term and long-term goals. As you reach each short-term goal—for example, riding your bike for 2 miles—record it on your conditioning chart. When you reach a long-term goal, such as finishing in a major race, give yourself a reward.

Figure 6.6
Conditioning Chart
This chart might have been kept by the teen whose cross-training program was presented in Figure 6.5.

Day of the Week	Exercise	Time Spent/Distance Covered	Comments
Monday (5/16)	Run	3 miles	Good run.
Tuesday (5/17)	Swim	30 minutes	Varied my strokes today.
Wednesday (5/18)	Cycle	3 miles	Had to ride inside—bad weather.
Thursday (5/19)	none	none	————
Friday (5/20)	Walk	1.5 miles	Reached my goal of walking 1.5 miles carrying two 2-pound weights.
Saturday (5/21)	Toning exercises	20 minutes	Next week, work harder on abdominal muscles.
Sunday (5/22)	Tennis	1 hour	I won! Try basketball for a few weeks?

HEALTH LAB
Getting into Condition

Introduction: Fitness testing is one way to find out how fit you are and where you need to improve. Fitness tests can also be a way to measure your progress in a conditioning program.

Objective: With a partner, measure your strength, endurance, and flexibility.

Materials and Method: Perform each of the following tests for strength, endurance, and flexibility. For the strength test you will need a horizontal bar that is anchored firmly enough to support your weight. For the endurance test you will need a stopwatch as well as a sturdy step about 8 inches high.

▶ **Strength Test: Pull-Ups.** This is a test for upper body strength that requires a horizontal bar. Grasp the bar with your palms facing away. Hang with your arms straight. Pull yourself up until your chin clears the bar. You have completed a pull-up when you have lowered yourself to the starting position. Count the number of pull-ups you can do without resting. (Average rating: females 2; males 6–7)

162   Chapter 6: Sports and Conditioning

HEALTH LAB
Getting into Condition

Time Needed
30 minutes

Supplies
▶ horizontal bar for pull-ups
▶ stopwatch
▶ sturdy step about 8 inches high

Focus on the Health Lab
Review with students the three elements of fitness developed through exercise: strength, endurance, and flexibility. Ask volunteers to define each element and to suggest activities that target each. Let students rate themselves, privately, on their own strength, endurance, and flexibility fitness: below average, average, or above average.

Have students choose partners for the Health Lab activity, and ask the partners to read the Health Lab feature together.

Understanding Objectives
Students should get an objective measure of their

Chapter 6 Review

Your Action Plan

Many people do not take the time to think about the role that sports play in their lives. This activity will give you a chance to do that.

Step 1 Take a moment to look over your private journal entries for this chapter. Are you satisfied with the role that sports play in your life?

Step 2 Is there anything you would like to change? Do you want to start a new program or increase your amount of activity?

Step 3 Once you have answered these questions, formulate a long-term goal statement. Then create three to five short-term goals to help you reach that long-term goal. For example, if your long-term goal were to jog a mile in 7 minutes, a short-term goal might be to jog half a mile in 3½ minutes and then finish the mile at a comfortable speed.

Keep track of your progress. When you reach a goal, write one or two sentences about how it makes you feel. Share your thoughts with a friend.

In Your Home and Community

1. With family members, discuss the subject of balancing school, sports, and home responsibilities. Ask members to contribute to a list of healthful ways in which young athletes might respond to the demands that sports make on them. Discuss your list in class.

2. Do some research—either at the library or by interviewing preschool teachers—about noncompetitive games for young children. Find out about the value of group games that involve cooperation rather than competition. Write a short report, and present it to your class.

Building Your Portfolio

1. Look in magazines and newspapers for a sports story involving a teen. The story might be about a teen who has overcome obstacles to succeed in a sport or who has used a sport to help others. Write a summary of the article, and place it in your portfolio. Be sure to identify the source of the article.

2. Conduct library research to find out about a sport that you have never played. Write a one-page report about the sport, telling whether you think that you would enjoy it and why. Add the report to your portfolio.

3. Look through all the activities and projects you did for this chapter. Choose one or two that you would like to include in your portfolio.

Chapter 6: Chapter Review **171**

Performance Assessment

▶ **Self-evaluation** Direct students to review the activities that are provided throughout the chapter. Encourage each student to select one finished product or activity that demonstrates his or her best work for the chapter. Have students explain what they learned and how the examples they selected show their progress.

▶ **Teacher's Classroom Resources** Assign Performance Assessment 6, "Get Active!" in the TCR.

Chapter 6 Review

8. Responses will vary. Possible responses (any four): choose sports with seasons that don't overlap; eliminate one sport; join a different team or league; balance the sports you play by using cross-training techniques.

Thinking Critically
Responses will vary. Possible responses are given.

9. Discuss the fun and the fitness activities involved in being on the team; explain that he could help the team and that the two of you could play together.

10. Excitement could prevent a player from being careful. Depression over a loss or mistake could prevent a player from maintaining his or her focus.

11. A conditioning program would help you stay in shape for the next season. You might choose to modify some activities or exercise less frequently or less strenuously.

12. The principles teens use to balance sports, schoolwork, and home life can be used later to balance the elements of adult life, including work, family, fitness, and social activities.

RETEACHING
Assign Study Guide 6 in the Student Activities Workbook.

EVALUATE
▶ Use the reproducible Chapter 6 Test in the TCR, or construct your own test using the Testmaker Software.
▶ Use Performance Assessment 6 in the TCR.

EXTENSION
Let students work together to plan a Sports and Fitness Day for the rest of their grade or for the entire school. Suggest that students learn about new games and unfamiliar sports—perhaps from other cultures—to teach and play during the Sports and Fitness Day.

Unit 3
Understanding Yourself and Others

UNIT OBJECTIVES
Students will gain a clearer understanding of themselves as growing, developing individuals. They will learn about important influences on their own mental/emotional and social health.

UNIT OVERVIEW

Chapter 7 Growth and Development
Chapter 7 provides information about the changes of adolescence. Students become familiar with the endocrine system and with both the male and female reproductive systems. They also consider adolescence in the context of all the stages of life.

Chapter 8 Mental and Emotional Health
Chapter 8 helps students examine the central issues of mental health, including personality, self-concept, and self-esteem. Students learn about the indicators of positive mental health, as well as symptoms of mental disorders. The chapter also presents information on coping with stress and on finding help for mental/health problems.

Chapter 9 Social Health: Families and Friends
Chapter 9 provides students an opportunity to examine the importance of social relationships. Students learn about communication skills, family relationships, and friendships. They also discuss the importance of abstinence, develop refusal skills, and learn positive approaches to disagreements and conflict.

Bulletin Board Suggestion

Getting to Know You Ask each student to bring a photograph of himself or herself, to be posted on the bulletin board. (As alternatives, let students draw self-portraits, take individual candid photographs of each student, or get copies of students' school photos.) Let students put their pictures, along with name labels, onto the bulletin board. As they study the chapters in this unit, have students add words and phrases around their own pictures, identifying some of their own positive traits and abilities, their healthy interests, and other indicators of their mental/emotional and social health.

Unit 3
Understanding Yourself and Others

Chapter 7	Growth and Development
Chapter 8	Mental and Emotional Health
Chapter 9	Social Health: Families and Friends

DEALING WITH SENSITIVE ISSUES

The Importance of Listening A crucial part of communicating with students is listening to their thoughts and feelings. A teacher who is a good listener provides an environment where students feel they are being heard without being judged, corrected, or interrupted. This feeling of acceptance helps them speak more openly and confidently. A teacher who is a good listener also shows empathy for students. This ability to put yourself in another's place helps you understand not only what your students are *saying* but also what they are *feeling*. As a result, if students can't express how they feel in words, but you can empathize with them, you will be able to share their feelings and show you really *hear* what they are saying.

Unit 3

INTRODUCING THE UNIT

Present the following situations to the class. For each situation, have students sketch a quick picture or write a few descriptive words or phrases indicating their feelings.

- Three of your best friends walk right past you in the school hall, without saying hello or even waving.
- Your family members are planning to spend the evening together, but no plans have been made yet. One of the adults asks, "What would *you* like to do?"
- You've spent hours working on a skill you want to develop—a long jump shot or an overhead smash, for example. Finally, everything "clicks" and you know you've mastered the skill.
- You see a group of teens in a park, all smoking cigarettes. They act surprisingly friendly and say, "Come on over and have a smoke. What do you mean you don't smoke? Everyone does! Give it a try."

Allow volunteers to share some of their responses, and then ask: Which aspects of your health triangle do these situations deal with most directly? Which situations reflect your increasing independence and responsibility? Explain that students will consider these and other, related issues as they study the chapters in Unit 3.

VIDEODISC/VHS

Teen Health Course 2

You may wish to use video segment 9, "Communication Skills," to demonstrate how effective communication can help resolve conflicts.

Videodisc Side 2, Chapter 9
Communication Skills

Search Chapter 9, Play to 10

173

Planning Guide

Chapter 7
Growth and Development

Lesson	Features	Classroom Resources
1 Adolescence *pages 176–181*	Life Skills: Dealing with Conflicting Feelings *page 179*	Concept Map 28 Decision-Making Activity 13 Enrichment Activity 28 Health Lab 7 Lesson Plan 1 Lesson 1 Quiz Reteaching Activity 28 Transparencies 26, 27
2 The Endocrine System *pages 182–185*		Concept Map 29 Enrichment Activity 29 Lesson Plan 2 Lesson 2 Quiz Reteaching Activity 29 Transparency 28
3 The Male Reproductive System *pages 186–189*		Concept Map 30 Enrichment Activity 30 Lesson Plan 3 Lesson 3 Quiz Reteaching Activity 30
4 The Female Reproductive System *pages 190–193*	TEEN HEALTH DIGEST *pages 194–195*	Concept Map 31 Enrichment Activity 31 Lesson Plan 4 Lesson 4 Quiz Reteaching Activity 31
5 Human Development *pages 196–200*	Health Lab: Your Unique Fingerprints *page 198*	Concept Map 32 Cross-Curriculum Activity 13 Enrichment Activity 32 Lesson Plan 5 Lesson 5 Quiz Reteaching Activity 32 Transparency 29

Lesson 6

Life Stages
pages 201–205

Features

Making Healthy Decisions: Helping a Friend Deal with Loss
page 204

Classroom Resources

- Concept Map 33
- Cross-Curriculum Activity 14
- Decision-Making Activity 14
- Enrichment Activity 33
- Lesson Plan 6
- Lesson 6 Quiz
- Reteaching Activity 33

National Health Education Standards

Listed below are the Health Education Standards Performance Indicators addressed in each lesson of this chapter.

Lesson	Health Standards Performance Indicators
1	(1.2, 4.4, 5.3, 6.1)
2	(1.3, 1.7)
3, 4	(1.1, 1.7)
5	(1.5, 1.8)
6	(1.2, 3.1, 6.1)

ABCNEWS InterActive™ Videodisc Series

You may wish to use the videodisc *Teenage Sexuality* with this chapter. See side one, video segments 3, 4, 5, 6, 7, 9, 15. Use the *ABCNews InterActive™ Correlation Bar Code Guide* for title reference. Also available in VHS format.

Chapter Resources

Teacher's Classroom Resources
- Chapter 7 Test
- Parent Letter and Activities 7
- Performance Assessment 7
- Testmaker Software

Student Activities Workbook
- Study Guide 7
- Applying Health Skills 28–33
- Health Inventory 7

Student Diversity Strategies
- Audiocassette Program (English)
- Audiocassette Program (Spanish)
- Spanish Parent Letters
- Spanish Summaries, Quizzes, and Activities

Multimedia Components
- English Audiocassette Program
- Spanish Audiocassette Program
- *Teen Health* Videodisc/VHS Series

Other Resources

Readings for the Teacher
DiGiulio, Robert, and Rachel Kranz. *Straight Talk about Death and Dying.* New York: Facts on File, 1995.

Feldman, Robert S., and Joel Feinman. *Who You Are: Personality and Its Development.* New York: Franklin Watts, 1992.

Petrikin, Jonathan S., ed. *Male/Female Roles: Opposing Viewpoints.* San Diego, CA: Greenhaven Press, 1995.

Readings for the Student
Bode, Janet. *Death Is Hard to Live with: Teenagers Talk about How They Cope with Loss.* New York: Delacorte Press, 1993.

Garell, H. C., M.D., et al., eds. *Adulthood.* New York: Chelsea House, 1993.

Monroe, Judy. "The Great Debate: Gender Differences," Current Health 2, December 1995, pp. 22–25.

Out of Time?

If time does not permit teaching this chapter, you may use these features: Life Skills on page 179; Teen Health Digest on pages 194–195; Health Lab on page 198; Making Healthy Decisions on page 204; and the Chapter Summary on page 206.

Chapter 7
Growth and Development

CHAPTER OVERVIEW

Chapter 7 helps students understand the changes of adolescence, including the development of the male and female reproductive systems, within the context of a sequence of life stages.

Lesson 1 explains what adolescence is and discusses the physical, mental/emotional, and social changes that take place during this time of life.

Lesson 2 identifies the major glands in the endocrine system and explains their functions.

Lesson 3 identifies the parts of the male reproductive system and discusses their role in producing, storing, and releasing sperm.

Lesson 4 identifies the parts of the female reproductive system and explains how the system produces egg cells, allows fertilization to occur, and nourishes the fertilized egg.

Lesson 5 explains how a human individual develops before birth and discusses the importance of both heredity and environment in prenatal development.

Lesson 6 helps students understand the stages of life through which each individual passes, from birth through adulthood.

PATHWAYS THROUGH THE PROGRAM

Your students are probably already dealing with the various changes associated with adolescence. The more teens understand these changes and their causes, the better equipped they are to deal successfully with the changes, as well as with themselves, their peers, their parents and teachers, and others in their lives.

Chapter 7
Growth and Development

Student Expectations
After reading this chapter, you should be able to:
1. Describe the development and changes that occur during adolescence.
2. Identify the parts of the endocrine system and explain how the system works.
3. List the parts of the male reproductive system and describe their function.
4. List the parts of the female reproductive system and describe their function.
5. Describe human development after fertilization and identify factors that affect development.
6. Describe the stages of life from infancy to adulthood.

Key to Ability Levels

Teaching strategies that appear throughout the chapter have been identified by one of three codes to give you an idea of their suitability for students of varying learning styles and abilities.

L1 Level 1 strategies should be within the ability range of all students. Full class participation is often required. Teacher direction is usually needed.

L2 Level 2 strategies are for average to above-average students or for small groups. Some teacher direction is necessary.

L3 Level 3 strategies are designed for students able and willing to work independently. Minimal teacher direction is necessary.

Teen Chat Group

Dan: I can't believe it! I've been taller than you since second grade. Now you've got 2 inches on me.

Kyle: It's about time.

Dan: It feels weird. I'm not used to reaching up to block your shots.

Kyle: Well, both my parents are tall, so I always figured I'd be tall too.

Dan: How tall is your dad?

Kyle: He's 6 foot 3. My older brother is 6 foot 2. I hope I reach that size.

Dan: I'll *never* block your shots then! Anyway, it's your ball. Let's play.

in your journal

Read the dialogue on this page. Have you had feelings or experiences like these? Start your private journal entries by answering these questions:

- What would you say to Dan and Kyle if you were part of this discussion?
- In what ways have you changed during the past year or two?
- Which recent changes in your life have been positive? Which have not?
- What have you learned from the changes in your life?

When you reach the end of the chapter, you will use your journal entries to make an action plan.

Chapter 7: Growth and Development

Chapter 7

INTRODUCING THE CHAPTER

▶ Display several photographs (as from magazines) of groups of adolescents. Be sure the pictures show individuals of various body types and in various stages of physical development. Have students describe the teens in the pictures: What do all these individuals have in common? What makes each individual different from the others? Conclude the discussion by telling students that they will learn more about this stage of life—adolescence—and other stages as they study Chapter 7.

▶ You may choose to encourage students to bring in photographs of their own family groups or of their family's set of close friends, showing as many different generations as possible. Let students meet in groups to share and discuss their relatives and friends as shown in the photo. Guide the discussion to include identification of babies, children, adolescents, and adults. After all the students have had a chance to share their photos in groups, tell the class that Chapter 7 presents information about people at various stages of life and explains aspects of the process of maturing, from the beginning of life to late adulthood.

Cooperative Learning Project

On the Road: A Display Presenting the Stages of Life

The Teen Health Digest on pages 194 and 195 provides students with high-interest articles related to the content of this chapter.

The material in the Teacher's Wraparound Edition presents suggestions for a class project in which students will plan, prepare, and present a detailed display showing the stages of human life.

in your journal

A journal provides a comfortable, convenient setting in which students can explore the changes they are experiencing, evaluate the developments they have achieved, and anticipate the new opportunities and emotions that lie ahead. This journal activity is one that students may want to reuse several times in the coming months.

Let students work in small groups to read the Teen Chat Group aloud and to extend the conversation, inserting their own personalities and ideas. Then encourage group members to discuss their responses to the In Your Journal questions in preparation for writing their own, private journal entries.

Lesson 1
Adolescence

Focus

LESSON OBJECTIVES

After studying this lesson, students should be able to
- explain what adolescence is.
- identify the physical changes that take place in females and males during adolescence.
- discuss the mental and emotional growth that takes place during adolescence.

MOTIVATOR

Write the word *adolescence* on the board. Give students two or three minutes to list all the words and phrases that come to mind in response to that word.

INTRODUCING THE LESSON

Let students begin by sharing their understanding of the terms *adolescence* and *adolescent*. Then have several volunteers read aloud the lists they wrote in response to the Motivator activity. Promote discussion by asking questions such as these: How do you think adolescents are usually portrayed on TV? How is that portrayal similar to and different from most adolescents you know? What do you think the best parts of being an adolescent are? What are the worst parts? Conclude by telling students that they will learn more about adolescence as they study Lesson 1.

INTRODUCING WORDS TO KNOW

Let volunteers find and read aloud the definition of each Word to Know as presented in the lesson. Ask other volunteers to restate these definitions in their own words. Then guide students in discussing the relationships between adolescence, hormones, and puberty.

176

Lesson 1 Adolescence

This lesson will help you find answers to questions that teens often ask about changes that take place during the teen years. For example:

▶ What physical changes take place during adolescence?
▶ What mental and emotional growth occurs during adolescence?
▶ How will I grow socially during adolescence?

Words to Know

adolescence
hormones
puberty

A Time of Change

As you enter your teen years, you'll experience many changes. Some will be physical: your body will grow and develop. Other changes may be less obvious. You'll begin to look at life differently from the way you did when you were younger. You'll also experience new and different thoughts and feelings. Some of these changes will be pleasant. Others, at least at first, may make you feel a little uncomfortable.

This *time of life, between childhood and adulthood,* is **adolescence** (a·duhl·E·suhns). It is a time of growth and change, usually starting somewhere between ages 11 and 15. Some of your friends will begin to develop physically before you do, while others may develop later. This is normal. Each individual goes through the changes of adolescence at his or her own rate. Typically, however, girls experience these changes about two years earlier than boys.

Many of the physical and emotional changes of adolescence are brought about by hormones. **Hormones** (HOR·mohnz) are *chemical substances, produced by glands, which help to regulate the way your body functions.* The changes that hormones cause during adolescence prepare you for adulthood.

Hormones cause physical and emotional changes in adolescents.

176 Chapter 7: Growth and Development

Lesson 1 Resources

Teacher's Classroom Resources
- 📁 Concept Map 28
- 📁 Decision-Making Activity 13
- 📁 Enrichment Activity 28
- 📁 Health Lab 7
- 📁 Lesson Plan 1
- 📁 Lesson 1 Quiz

- 📁 Reteaching Activity 28
- 📽 Transparencies 26, 27

Student Activities Workbook
- 📁 Study Guide 7
- 📁 Applying Health Skills 28

Physical Changes

Adolescence begins with the physical changes of puberty. **Puberty** (PYOO·ber·tee) is *the time when you start to develop certain physical characteristics of adults of your own gender.* **Figure 7.1** lists some of these characteristics.

Physical growth during puberty generally occurs at a rapid pace. Many girls grow 3 inches taller between ages 11 and 14. Boys usually begin their growth spurt later than girls. Boys may grow 6 or 7 inches between ages 13 and 16. Both girls and boys often keep growing for several more years after their initial growth spurt.

The rate of physical growth and development varies greatly from person to person and often causes teens to feel uncomfortable or self-conscious. Some teens worry that they are developing too quickly. Others are concerned that they are not developing quickly enough. Such concerns are a normal part of adolescence.

interNET CONNECTION
Take a journey to self-understanding by exploring adolescent development on the Internet.
http://www.glencoe.com/sec/health

Figure 7.1
What Are the Physical Changes of Adolescence?
Many types of physical changes occur during adolescence.

Females
- Sudden, rapid growth occurs.
- All permanent teeth come into place.
- Acne may appear.
- Underarm hair appears.
- Pubic hair appears.
- Perspiration increases.
- External genitals enlarge.
- Breasts develop.
- Hips get wider.
- Waistline gets narrower.
- Ovulation occurs.
- Menstruation starts.
- Uterus and ovaries enlarge.

Males
- Sudden, rapid growth occurs.
- All permanent teeth come into place.
- Acne may appear.
- Underarm hair appears.
- Pubic hair appears.
- Perspiration increases.
- External genitals enlarge.
- Breasts may enlarge somewhat.
- Shoulders get broader.
- Muscles develop.
- Sperm production starts.
- Facial hair appears.
- Larynx gets larger and voice deepens.
- Hairline begins to recede.

Lesson 1: Adolescence   **177**

Home and Community Connection

Volunteer Opportunities Most communities offer many different situations in which teens can volunteer, for example, working with young children, with older adults, with disabled individuals, with homeless individuals and families, and with other students who need tutoring. Some teens might prefer volunteering in soup kitchens, community offices, thrift warehouses, or parks, doing chores that support other people without involving direct personal contact.

Guide students in compiling a list of appropriate volunteer opportunities in their community. Let several volunteers organize the list into a useful format and publish it for the use of other students in the school.

Chapter 7 Lesson 1

Teach

Teacher Talk
Dealing with Sensitive Issues
Young teens may be uncomfortable talking about the specific changes of adolescence and puberty; these discussions can be more productive if they are not allowed to become personal. If possible, keep on display photos (as from posters or magazines) of teens in various stages of development. You may even want to name the teens in these pictures. You and the students can then refer to these pictures when talking about the physical and emotional changes of adolescence.

L1 Discussing
Encourage students to discuss important changes they have experienced, such as leaving a familiar neighborhood, or moving from elementary school to middle school or junior high: What changes have been fun for you? What changes have been especially difficult? Is any change—even a pleasant change—ever problem-free? Why not? Then help students relate these experiences to the changes of adolescence: In what ways is moving from childhood to adulthood an exciting transition? Is it reasonable to expect that there will be no problems or worries associated with these changes? Why or why not?

L2 Journal Writing
Suggest that students use their journals to record their personal responses to the changes of adolescence: Do you think you are developing more slowly or more quickly than your friends? How do you feel about your own schedule of development? What could you do to make your personal schedule of development easier to accept?

177

Chapter 7 Lesson 1

L1 Comprehending

Focus students' attention on the definition of puberty. Let them explain their responses to these questions: Can you be in adolescence without being in puberty? Can you be in puberty without being in adolescence?

L2 Applying Knowledge

To help students observe the variations in rates of physical growth and development, ask them to watch a TV show featuring teenage characters, or tape a segment of such a show and play it for the class. Then guide a discussion or ask students to write descriptions of their observations: What differences did you notice among teen characters? What signs did those characters give that they felt uncomfortable or self-conscious about their own growth and development?

Visual Learning

Let students begin their examination of Figure 7.1 by describing the teens in the photograph behind the lists: What similarities can you identify? How is each teen different from the others in the group? Then rea aloud the changes in each list, and encourage students to demonstrate their understanding by restating each list item in their own words.

! Would You Believe?

In a recent survey, more than 25 percent of seventh- and eighth-graders expressed anxiety that their bodies might not be growing normally.

---

in your journal

How do events in your life affect your emotions? In your journal, make two columns. In one column, write down the emotions you remember feeling yesterday. In the other, write what you think was the reason you felt each emotion. If you don't think that there was a reason for a particular emotional change, write "no reason." What conclusions can you draw? For example, what events cause you to feel positive emotions? Do you often experience emotions for what seems to be no reason at all?

Mental and Emotional Growth

During adolescence you develop more complex thinking skills. For example, you learn to analyze and solve more complicated problems. You come to understand that many questions do not have simple right or wrong answers. At the same time you begin to recognize points of view that are different from your own. You will use all of these thinking skills as an adult.

You also realize that you have the power to make choices. You recognize that your actions have consequences, and that you should take responsibility for these consequences. All of these aspects of mental growth may feel overwhelming, but they are a natural part of becoming a mature individual.

Emotional Needs

Adolescence is a time of emotional growth as well. Although people have emotional needs throughout their lives, these needs change over time. Basic emotional needs include the need to care for others and to feel that others care for you. Another basic need is to feel accepted for who you are. You also need to feel that your actions have meaning and value.

As you grow, your emotional needs change. For example, as a child you probably tried hard to win the approval of your parents. As you enter your teen years, parental approval is still important. However, the approval and acceptance of your peers also begins to matter greatly.

Many of the actions people take are ways to meet their own emotional needs. They have to satisfy these needs in order to feel good about themselves and to be content with their lives. Learning to meet your emotional needs in positive and realistic ways is an important task of adolescence.

Doing volunteer work is a way to help others while meeting your own emotional needs.

178 Chapter 7: Growth and Development

LIFE SKILLS
Dealing with Conflicting Feelings

Focus on Life Skills

Present students with a hypothetical situation such as this: Congratulations! You have just been selected to represent our school district at a special convention in Washington, D.C. How do you feel?

Encourage students to identify the wide range of feelings they might experience in response to this situation. Help them explore each feeling: Which aspect of the honor or of the trip evokes that feeling in you? Why? Ask: How many different responses do you have to this situation? How is this mixture of feelings similar to reactions you have had to real-life events? Then have students read the Life Skills feature to learn more about coping with mixed feelings.

Making Health Connections

Young teens are still likely to expect to have a single emotional response to a given situation; some

Emotional Changes

Many teens find that adolescence brings on powerful emotions. Perhaps one day you feel excited and happy. You want to be out having fun with your friends. The next day you may feel gloomy enough to cry. All you want to do is stay in your room alone. In a single day you may experience a confusing up-and-down mix of emotions. Such swings in mood, like the changes in your body, are often caused by hormones. You need to realize that changing moods are a normal part of adolescence.

It is also important to understand, however, that these changes are also signs of emotional growth. During adolescence, your relationships with other people can deepen. You come to recognize that others have needs, just as you do. For example, when you listen to a friend's problems and offer advice, you are helping to meet some of these needs.

During your teen years, you may also have a greater desire to interact with members of the opposite sex. Feelings of attraction may confuse or even scare you, but they are part of growing up, too.

Q & A

Time Out

Q: Sometimes I feel so stressed out, I just want to be by myself. What's the matter with me?

A: Everyone feels the need to get away from other people at times. Having some time to yourself may help. Many people find that listening to music, reading a book, or just resting is a good way to unwind after a hard day. Setting aside some quiet time will help you keep your life in balance.

LIFE SKILLS
Dealing with Conflicting Feelings

Have you ever experienced two different feelings at the same time, pulling you in opposite directions? When you start a new year in a new school, for example, part of you may feel excited, while another part feels nervous and uncertain. Here are a few tips for dealing with conflicting feelings.

- Keep in mind that you're not alone in feeling this way. Everyone experiences emotional conflicts. Learning to deal with them is part of becoming a mature person.

- Remember also that having several opposing emotions is a sign of maturity. After all, few situations are so clear-cut that they could bring forth only a single emotion.

- Try to sort out your feelings. Be honest with yourself. Ask yourself: What's the *real* reason behind my feelings?

- Take time to *think* before you speak or act. Consider the consequences—short-term and long-term—of what you are about to do.

- People often experience conflicting feelings when they have to make decisions. If you face a difficult choice, weigh your options carefully.

Using the six-step decision-making process will help. Seek advice from others, if necessary, but remember that the final decision is yours.

Follow-up Activity

Think about a recent situation in which making a decision caused you to have conflicting feelings. How did it turn out? How might you have made the decision-making process less stressful? Write down one piece of advice that you would give to a friend in a similar situation.

Lesson 1: Adolescence

Chapter 7 Lesson 1

L1 Identifying Examples

Ask students to identify some of the choices they can now make and to note the consequences of those choices. You may want to encourage students to focus on helping others as a way to enhance their own growth in understanding and compassion.

L2 Presenting Skits

An adolescent's growing interest in his or her peers often causes conflict with parents, siblings, or other family members. What kinds of conflicts may arise? How can those conflicts be settled? Let students work in pairs or small groups to plan and present short skits illustrating answers to those questions.

L3 Writing Stories

Have students write a short story in which an adolescent makes an effort to meet the needs of another person.

VIDEODISC/VHS

Teen Health Course 2

You may wish to use video segment 7, "Growth and Development," to show how teens can use compromise and effective communication skills to solve problems.

Videodisc Side 1, Chapter 7
Growth and Development

Search Chapter 7, Play to 8

---

may even try to feel what they think they *should* feel. The ability to recognize, accept, and deal successfully with conflicting feelings in themselves and others will have a positive impact on all aspects of their health—physical, mental/emotional, and social—throughout their lives.

Meeting Student Diversity

Students with certain learning disabilities and those who are acquiring English skills may have trouble reading and comprehending the Life Skills tips. To provide assistance, let all students work with partners to read and discuss the information in the feature; carefully pair each student who has reading difficulties with a capable and helpful reader.

Assessing the Work Students Do

Have students work in small groups to plan and rehearse skits that illustrate the Life Skills presented in this feature. Provide time for the members of each group to perform their skit for the rest of the class. Assign credit based on students' participation and attitude.

Chapter 7 Lesson 1

L1 Discussing

Ask students to name five or six familiar emotions. Record their suggestions on the board. Guide students in discussing each emotion: Is this a feeling that needs to be controlled? Why or why not? What are some appropriate ways to express this feeling? What are some inappropriate ways? How can you avoid those inappropriate expressions?

L2 Creating Collages

Have students make individual collages to show "the true me." Suggest that they use photographs of themselves, drawings, pictures cut from magazines, and/or words as part of their projects. Display students' completed collages in the classroom.

L2 Math

Which of the listed aspects of advancing toward adulthood do students consider most important? Ask a small group of volunteers to survey classmates and other young teens to find the answer to that question. Have the volunteers summarize their findings in a large graph to be presented and discussed with the rest of the class.

> *Visual Learning*
>
> Let students describe and discuss the teen in Figure 7.2: What different identities does he imagine for himself? Do you think he has to select just one? When and how will he make his choice?

L1 Journal Writing

Let students describe the photograph and read the caption on page 181. Then encourage them to write journal entries in response to this question: What do your own interests and activities say about you?

180

Teen Issues

Social Life   ACTIVITY!

Adolescence is a time of growing independence and more socialization with friends. Make a list of five activities you enjoy with friends now. Then list five activities you and your friends liked when you were in fifth grade. How do the two lists compare? Do you think that your list will change again by the time you reach ninth grade? How?

Emotional Expression

Whatever your feelings may be at a given time—joy, anger, fear—it's important to express them in healthy and appropriate ways. For instance, if you're having a disagreement with a friend or parent, discussing the problem calmly will more quickly lead to a solution than either shouting or keeping silent. Learning how to control and express your feelings is another way that you grow during adolescence.

Social Development

As you move from childhood to adulthood, you're constantly learning about yourself, about others, and about how you fit into the world. Sometimes you may feel confused. You may wonder, "What is the true *me* really like?"

Discovering your "self" will take some time, as **Figure 7.2** shows. At first you may look to your friends to help you figure out your identity. Being part of a group may seem very important. As you get older, however, you'll come to see yourself as a separate individual. You'll still want to be accepted by your friends, of course, but you'll also want to be unique, with your own personal views. In a similar way, you'll still want to be close to your family, but you'll also want to feel independent.

Figure 7.2
Who Am I?
During your teen years, you'll start to get a clearer image of who you are.

180   Chapter 7: Growth and Development

More About...

The Concerns of Adolescents
In a recent survey, junior high school students disclosed some of the wide-ranging issues about which they feel concerned. Here are some of their responses:

▶ 56 percent worry about their performance in school.

▶ 53 percent feel concerned about their own appearance.

▶ 38 percent are worried about hunger and poverty in this country.

▶ 30 percent are already concerned that they might not get good jobs as adults.

Moving Toward the Future

Your growth and development during adolescence will take many paths. You will advance toward adulthood in important, exciting, and challenging ways. In the coming years, you will

- learn to accept your body and its characteristics.
- gain a masculine or feminine image of yourself.
- become more independent in your thoughts and feelings.
- discover who you are and what makes you unique.
- develop your own set of values.
- learn to solve problems and make decisions in a mature way.
- learn to accept responsibility for your own actions.
- establish more mature relationships with people of both sexes.
- develop a greater awareness of, and concern for, your community and the world.

Your teen years open the door to your future. The experiences you have and the knowledge you gain at this time will prepare you for life as an adult.

Your interests and activities tell a great deal about you as an individual and about what is important to you.

Review Lesson 1

Using complete sentences, answer the following questions on a separate sheet of paper.

Reviewing Terms and Facts

1. **Vocabulary** What is the difference between *adolescence* and *puberty?*
2. **List** Identify six physical changes that occur in females during adolescence and six changes that occur in males during adolescence.
3. **Recall** Give three examples of mental or emotional growth that occur during adolescence.
4. **Recall** Describe two ways in which hormones affect adolescents.

Thinking Critically

5. **Compare** In what ways are you more independent now than you were a few years ago?
6. **Analyze** Which of the many changes of adolescence do you think are most challenging for teens to cope with? Why did you choose these changes?

Applying Health Concepts

7. **Personal Health** Make a montage entitled "The Real Me." Include snapshots, magazine pictures, and other items that highlight various aspects of your life and offer clues to what you are really like.

Lesson 1: Adolescence 181

Health of Others

Students can help both themselves and others by sharing their understanding of adolescence with parents or guardians. Suggest that each student write a private letter explaining what adolescence is and discussing the kinds of growth—physical, mental/emotional, and social—that adolescents experience. Let students work on their own to plan and draft their letters, referring to the list of changes on page 181 as appropriate. Encourage students to personalize their letters as appropriate.

Lesson 2
The Endocrine System

Focus

LESSON OBJECTIVES

After studying this lesson, students should be able to
- describe the role of hormones in regulating body functions.
- identify the major glands of the endocrine system.
- explain how the endocrine system works.
- identify the most common endocrine disorders.

MOTIVATOR

Have students jot down their responses to these questions: At what age do you think you will reach your full height? What do you think that height will be? What determines your adult height and when you will reach it?

INTRODUCING THE LESSON

Let students share and discuss their responses to the Motivator questions. Introduce or reinforce the concept that although height is largely determined by heredity, the timing and the extent of growth are controlled by a specific gland within the endocrine system. Tell students that they will learn more about the endocrine system and its functions as they study Lesson 2.

INTRODUCING WORDS TO KNOW

Have students work with partners to find and discuss the definitions of the three Words to Know. Then ask each pair of students to draw a simple chart that demonstrates the relationships between the three terms.

182

Lesson 2
The Endocrine System

This lesson will help you find answers to questions that teens often ask about the endocrine system. For example:

▶ What is the endocrine system?
▶ How does the endocrine system work?
▶ What problems can occur in the endocrine system?

Words to Know

endocrine system
gland
pituitary gland

Did You Know?

In a Heartbeat

The hormone epinephrine (e•puh•NE•fruhn)—more commonly known as adrenaline—is a powerful heart stimulant. It is so powerful, in fact, that an injection of epinephrine directly into the heart muscle is sometimes used to restart a heart that has stopped beating.

Regulating Body Functions

The hormones that cause physical and emotional changes during adolescence come from your endocrine system. The **endocrine** (EN·duh·krin) **system** consists of *glands throughout your body that regulate body functions.* A **gland** is *a group of cells, or an organ, that produces a chemical substance.*

Hormones produced by the endocrine glands pass directly into your bloodstream. They are then carried to different parts of your body, where they control various functions. For example, some hormones regulate growth. Others aid digestion. Your body produces some hormones continually and others only at certain times. Hormones are involved in some way in nearly every body function.

Excitement or fear can cause your glands to release adrenaline. One effect of this hormone is to make your heart pound.

182   Chapter 7: Growth and Development

Lesson 2 Resources

Teacher's Classroom Resources
- Concept Map 29
- Enrichment Activity 29
- Lesson Plan 2
- Lesson 2 Quiz
- Reteaching Activity 29
- Transparency 28

Student Activities Workbook
- Study Guide 7
- Applying Health Skills 29

The Glands of the Endocrine System

Figure 7.3 describes the glands of the endocrine system and the body functions they regulate. Signals from the brain or from other glands keep the endocrine system working. For example, when the brain detects a low level of thyroid hormone in the blood, it sends a signal to the pituitary gland. The pituitary then signals the thyroid, which adds more thyroid hormone to the bloodstream.

Figure 7.3
The Endocrine System
This table describes some of the major glands of the endocrine system.

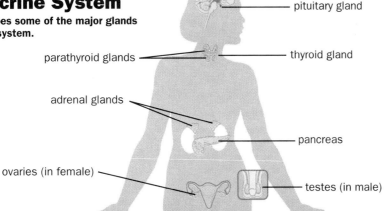

Gland	Description
pituitary (pi-TOO-i-tehr-ee) **gland**	Located at the base of the brain, the pituitary gland *produces several hormones that control other glands.* For example, pituitary hormones regulate the thyroid gland, adrenal glands, and kidneys. Pituitary gland hormones also regulate your body's growth and development. The pituitary is sometimes called the "master gland."
thyroid (THY-royd) **gland**	The hormone produced by the thyroid gland regulates body growth and the rate of metabolism. The thyroid is located alongside the trachea (windpipe).
parathyroid (par-uh-THY-royd) **glands**	Located within the thyroid gland, the small parathyroid glands regulate the levels of calcium and phosphorous in the blood.
pancreas	The pancreas, located behind the stomach, is part of both the endocrine system and the digestive system. The pancreas controls the level of sugar in the blood and provides the small intestine with digestive juice.
adrenal (uh-DREE-nuhl) **glands**	Hormones produced by the adrenal glands help regulate the balance of salt and water in the body. They also aid in digestion and control the body's response to emergencies. The adrenal glands are located on the kidneys.
ovaries (OH-vuh-reez)	The ovaries are the female reproductive glands. Hormones produced in the ovaries control sexual development and the production of eggs.
testes (TES-teez)	The testes are the male reproductive glands. The hormone produced in the testes controls sexual development and the production of sperm.

Lesson 2: The Endocrine System **183**

Chapter 7 Lesson 2

Visual Learning
Have students describe the teens in the photo on page 182: What are they doing? What emotions are they experiencing? How are their bodies responding to this emotion? Encourage students to describe specific situations in which they have experienced similar emotions and responses: In addition to a pounding heart, what other physical reactions did you experience?

Would You Believe?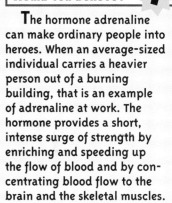
The hormone adrenaline can make ordinary people into heroes. When an average-sized individual carries a heavier person out of a burning building, that is an example of adrenaline at work. The hormone provides a short, intense surge of strength by enriching and speeding up the flow of blood and by concentrating blood flow to the brain and the skeletal muscles.

L1 Discussing
Let students work as a class or in several smaller groups to read and discuss the information in Figure 7.3, The Endocrine System. Encourage them to use their own words to identify the location of each gland in the body and to restate its function.

L1 Comprehending
Ask one or two volunteers to explain why the pituitary gland is called the master gland.

L2 Researching
Point out that the pancreas is part of two different systems, the endocrine system and the digestive system. Ask students to find out how unusual this is: Which other organs function within two different body systems? Let students compare and discuss their findings.

Meeting Student Diversity

Learning Styles Many students may find it easier to understand and recall the location of the major glands if they have an opportunity to work with anatomical figures or on dolls. As students discuss Figure 7.3 in groups, provide each group with a small human figure. Let them work together to point out the correct location of each gland on that figure.

Language Diversity If the class includes students who are acquiring English skills and who share a common first language, let them work together to study and discuss Figure 7.3. Encourage these students to use their native language to name and/or describe each gland shown in the figure. Have them use both English and their native language to identify the functions of the glands.

Chapter 7   Lesson 2

Visual Learning
Guide the class in studying and discussing the information presented in Figure 7.4. Let students begin by describing the illustration of each step in the sequence and explaining what they think it illustrates. Then have students read the label identifying each step. To reinforce learning, you may want to have volunteers act out the sequence, with one student portraying the tissue, another the gland, and a group portraying the hormone.

L3 Reporting
Let an interested volunteer use library and Internet resources to find answers to these questions: What are artificial growth hormones? How are they manufactured? How can they be used to treat growth disorders? Ask the volunteer to summarize his or her findings in an oral report to the rest of the class.

L1 Analyzing
Explain that, a century or so ago, goiter was a much more common problem among Americans than it is today. Ask: Are you related to or acquainted with anyone who has a goiter? anyone who has a thyroid disorder? Let students share their ideas about why goiters have become so uncommon in this country.

L1 Critical Thinking
Have the class discuss the scene in the photograph on page 185 and then read the caption. Ask: What benefits do you think specialized summer camps can provide for children with diabetes? Do you think there are any problems associated with attending this kind of specialized camp? If so, what are those problems?

in your journal

List in your private journal some of the physical changes that you have experienced during adolescence. Then explain what role your endocrine system has played in those changes.

How the Endocrine System Works

You may wonder how the body, with so many different glands, keeps track of how much of each hormone to produce and when to produce it. The endocrine system depends on continuous feedback to meet the body's changing needs. That is, the endocrine glands make adjustments based on information that comes back to them from the body. **Figure 7.4** shows how this feedback system works.

Figure 7.4
How the Body Controls Hormone Levels
Hormones are powerful chemicals that regulate body functions.

1. Hormones affect specific tissues in various parts of the body. When the level of a particular hormone is lower than it should be, the tissue affected by that hormone sends a chemical signal.
2. The gland that produces the needed hormone receives the signal and increases production of the hormone.
3. The level of the hormone in the bloodstream rises. When the necessary level is reached, another signal is sent.
4. The gland receives the signal and responds by reducing hormone production.

More About...

Hormones and Teens The same hormone that promotes the growth of body hair also stimulates oil glands in the skin. Often the oil is produced more quickly than it can be released, and a pore becomes plugged up with oil. Then dead skin cells can no longer be pushed out through that pore, and a skin blemish develops.

What's the best way to treat acne? For most teens, simply waiting is the best solution. Eventually, the body reduces its hormone production, and problems with acne gradually subside. Surprisingly, vigorous scrubbing and lots of soap are not solutions to acne. Instead, they may actually stimulate additional oil production.

Problems of the Endocrine System

If the endocrine system produces too little or too much of a hormone, problems will result. Symptoms of endocrine system problems vary greatly because hormones affect so many different body functions. Doctors can usually use a blood test to detect endocrine disorders. For treatment, doctors may prescribe a medication or a synthetic version of a hormone. The following list shows several of the most common endocrine disorders.

- Diabetes, or diabetes mellitus (dy·uh·BEE·teez ME·luh·tuhs) occurs when the pancreas does not produce enough insulin, a hormone that regulates chemicals in the blood. This lack of insulin keeps the body from using the sugars and starches in food for energy. Chapter 13 discusses this disorder in detail.

- Growth disorders occur when the endocrine system releases too much or too little of the hormones that regulate growth.

- Goiter (GOY·ter) is the name for an enlargement of the thyroid gland in the neck. The gland enlarges in an attempt to produce more thyroid hormone.

Did You Know?
World's Tallest Man

The tallest person who ever lived was Robert Wadlow of Illinois. He was 8 feet 11 inches tall and still growing when he died at the age of 22. He suffered from giantism, or abnormal growth caused by a pituitary disorder. Most people who have this disorder die before they reach the age of 30.

Specialized summer camps can help children who have diabetes learn to manage their disorder.

Review — Lesson 2

Using complete sentences, answer the following questions on a separate sheet of paper.

Reviewing Terms and Facts

1. **Vocabulary** What is the *endocrine system?*
2. **Vocabulary** What is a *gland?*
3. **Describe** How does the body use feedback to regulate hormone levels?

Thinking Critically

4. **Compare and Contrast** How are the functions of the ovaries and the testes alike? How do they differ?

5. **Analyze** How could knowing about the role of hormones in triggering strong emotions help teens learn to cope with them?

Applying Health Concepts

6. **Growth and Development** Work with a partner to create an illustrated "Guide to the Endocrine System." Include information about the significance of the endocrine system during the teen years.

Lesson 2: The Endocrine System **185**

Cooperative Learning

Hormone Research Scientists have already identified more than a hundred different hormones in mammals, and the list is extended each year. Our understanding and appreciation of hormones is constantly refined by research. Let students work in cooperative groups to learn what the most recent research indicates about human hormones and the glands of the endocrine system.

Have the members of each group select a specific question or topic they want to investigate. Remind group members to work together in using library and Internet resources to find answers to their question or current information on their topic. Ask each group to plan and prepare a poster or other display that publishes the results of their research.

Chapter 7 Lesson 2

Assess

EVALUATING THE LESSON

Assign Reviewing Terms and Facts and Thinking Critically on this page to review the lesson; then assign the Lesson 2 Quiz in the TCR to evaluate students' understanding.

LESSON 2 REVIEW

Answers to Reviewing Terms and Facts

1. The endocrine system consists of glands that regulate body functions.
2. A group of cells, or an organ, that produces a chemical substance.
3. When the level of a hormone is too low, the tissue affected by that hormone sends out a chemical signal. The gland that produces the hormone receives the signal and increases production of the hormone. When the necessary level is reached, the tissue sends out another signal, causing the gland to reduce hormone production.

Answers to Thinking Critically

4. Both control sexual development. The ovaries are female glands. The testes are the male glands.
5. Responses will vary.

RETEACHING

▶ Assign Concept Map 29 in the TCR.

▶ Have students complete Reteaching Activity 29 in the TCR.

ENRICHMENT

Assign Enrichment Activity 29 in the TCR.

Close

In a closing discussion, let each student cite a fact about the endocrine system that he or she found especially interesting.

Lesson 3

The Male Reproductive System

Focus

LESSON OBJECTIVES

After studying this lesson, students should be able to
- define the terms *reproduction* and *reproductive system*.
- identify the parts of the male reproductive system.
- discuss proper care of the male reproductive system.
- identify the most common disorders that can affect the male reproductive system.

MOTIVATOR

Have students draw on their prior knowledge to sketch the approximate shape of a sperm. In addition, have them consider the size of a sperm and write the number of sperm cells they think would fit on the head of a pin.

INTRODUCING THE LESSON

- Help the class recall that sperm are the male reproductive cells. Ask several students to share their sketches with the rest of the class.
- Let other students share their ideas about the size of sperm cells. Explain that scientists estimate that 400 million sperm cells could fit on the head of a pin (or in a single drop of semen).

INTRODUCING WORDS TO KNOW

Let students work with partners to find the meanings of the Words to Know in the lesson and to make a flashcard for each, with the term on one side and the definition on the other side. Have the partners use their flashcards to quiz each other.

Lesson 3 — The Male Reproductive System

This lesson will help you find answers to questions that teens often ask about the male reproductive system. For example:

▶ What parts make up the male reproductive system?
▶ How does the male reproductive system work?
▶ What care does the male reproductive system require?

Words to Know

reproduction
reproductive system
sperm
semen

in your journal

Why is it sometimes difficult for teens to ask questions or talk about problems involving the reproductive system? How do you think teens can best obtain valid information about their concerns? Respond to these questions in your private journal.

The Human Reproductive System

Reproduction (ree·pruh·DUHK·shuhn) is the name for *the process by which living organisms produce new individuals of their kind.* Reproduction is essential to all living things. Without reproduction, groups of organisms would disappear over time.

Human life results from the union of two cells: one from the female and one from the male. These cells are produced in the reproductive system. The human **reproductive** (ree·pruh·DUHK·tiv) **system** *consists of body organs and structures that make possible the production of offspring.* Unlike most other human body systems, the parts of the male and female reproductive systems are not alike. Each of the two systems is specially suited to perform its role in reproduction. This lesson will discuss the male reproductive system.

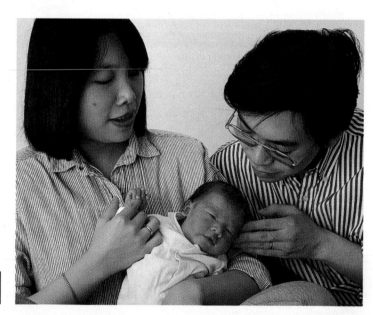

The arrival of a baby is a happy event in a couple's life.

Chapter 7: Growth and Development

Lesson 4 Resources

Teacher's Classroom Resources
- Concept Map 30
- Enrichment Activity 30
- Lesson Plan 3
- Lesson 3 Quiz
- Reteaching Activity 30

Student Activities Workbook
- Study Guide 7
- Applying Health Skills 30

Parts of the Male Reproductive System

Figure 7.5 shows the external and internal parts of the male reproductive system. During puberty, the testes start to produce **sperm**, *the male reproductive cells*. Sperm production usually begins between ages 12 and 15. The creation of new life takes place when male reproductive cells unite with female reproductive cells.

Various organs and structures are responsible for producing, storing, and releasing sperm. After the testes produce sperm, the sperm travel through tubes to the urethra. Along the way, the sperm mix with fluids. This *mixture of sperm and fluids* is called **semen** (SEE·muhn). Semen is released from the urethra through the penis (PEE·nuhs). The muscular action that forces semen through the urethra and out of the penis is called ejaculation (i·ja·kyuh·LAY·shuhn).

Q & A

How Many?

Q: How many sperm are released during a typical ejaculation?

A: Normally 300 to 500 million sperm are released during one ejaculation.

Figure 7.5
The Male Reproductive System
These illustrations show the internal and external parts of the male reproductive system.

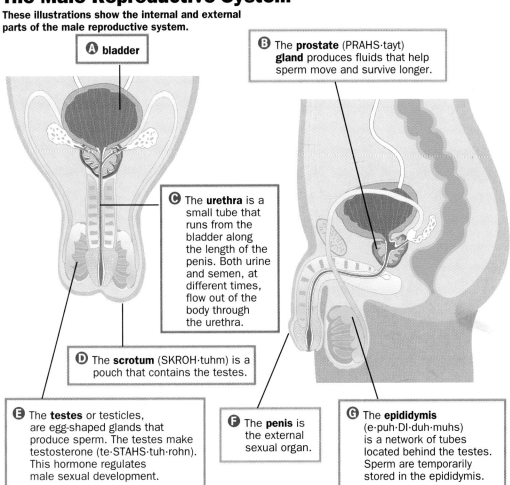

A bladder

B The **prostate** (PRAHS·tayt) **gland** produces fluids that help sperm move and survive longer.

C The **urethra** is a small tube that runs from the bladder along the length of the penis. Both urine and semen, at different times, flow out of the body through the urethra.

D The **scrotum** (SKROH·tuhm) is a pouch that contains the testes.

E The **testes** or testicles, are egg-shaped glands that produce sperm. The testes make testosterone (te·STAHS·tuh·rohn). This hormone regulates male sexual development.

F The **penis** is the external sexual organ.

G The **epididymis** (e·puh·DI·duh·muhs) is a network of tubes located behind the testes. Sperm are temporarily stored in the epididymis.

Lesson 3: The Male Reproductive System

More About...

Testosterone The anabolic steroids that some athletes unwisely take to bulk up their muscles are actually artificial forms of testosterone, the hormone that regulates male sexual development. This fact helps explain some of the unpleasant side effects associated with illegal steroid use: mood swings, deepening of the voice, increased facial hair and body hair, coarser skin, baldness, and aggressive and even violent outbursts. These side effects occur in both men and women; women who take steroids typically become "masculinized" in appearance. Steroid use also frequently results in sterility for both men and women.

Chapter 7 Lesson 3

Teach

L1 Language Arts
Ask students to identify the verb from which the noun *reproduction* is formed. What is the prefix in that verb? What meaning does it convey? To what verb has the prefix been added? How does an understanding of these word parts help explain the meaning of *reproduction*?

L1 Discussing
Encourage students to describe and discuss the family in the photograph on page 186: How do the parents appear to feel about their new baby?

L1 Comprehending
Ask students why the bladder is included in Figure 7.5: Is it a part of the male reproductive system? Which body system does the bladder belong to? Be sure students understand that either urine (from the bladder) or semen can flow out of the body through the urethra, but the two cannot leave the body at the same time.

Visual Learning
Let students describe each of the organs identified in Figure 7.5. Have volunteers explain how each organ relates to others in the system.

L2 Researching
Ask students to find the answers to these questions: Why are the testes located away from the abdominal cavity, in the scrotum? How does the scrotum protect the testes? Let students compare and discuss their findings.

Would You Believe?
If the epididymis behind each testicle were uncoiled, it would stretch to a length of about 20 feet.

Chapter 7 Lesson 3

L1 Identifying Examples
Let students name as many contact sports as they can. Then ask: Why is it important to wear protective gear when participating in these sports?

L2 Guest Speaker
Invite a local coach, either from the school or from a community team, to speak to the class about the importance of wearing protective gear. Ask the coach to bring and display samples of protective equipment, including gear that protects the male reproductive organs.

Health Minute
Testicular cancer is the most common form of cancer among young men (15-24 years). However, survival rates are very high among patients in whom testicular cancer is diagnosed in its early stages. These facts make it especially important for teens to learn and practice testicular self-examination.

L1 Critical Thinking
Tell students that many teens and adults feel reluctant to consult a health professional about a possibly "minor" problem, including, for example, soreness in the testes. Ask: What do you think causes this reluctance? Why is it important to overcome these feelings and seek professional help? What changes in actions and/or attitudes do you think might help people feel more willing to discuss these problems with a health professional?

L3 Reporting
Let several volunteers work together to learn more about hernias: What are the different kinds of hernias? What are the symptoms? What dangers do hernias pose? How are hernias treated? Ask the volunteers to give an oral presentation summarizing their findings for the rest of the class.

Your Total Health

Exams That Save Lives ACTIVITY!

If you are a male, find out how to perform a testicular self-examination (TSE). If you are a female, find out how to perform a breast self-exam (BSE). Write down the instructions on an index card so that you can use them to perform regular self-exams.

Caring for the Male Reproductive System

Maintaining a healthy reproductive system requires basic care and common sense. Here are some guidelines for males.

- Take a shower or bath daily to keep your external reproductive organs clean.
- Always wear protective gear when you are participating in contact sports.
- Examine your testes monthly for lumps, swelling, soreness, or other problems. Discuss any concerns with your doctor.
- Have regular physical checkups.

Wearing protective gear can help prevent injuries to the reproductive organs.

Problems of the Male Reproductive System

A number of disorders can affect the male reproductive system. The following are some of the most common:

- **Hernia.** An inguinal hernia (IN·gwuh·nuhl HER·nee·uh) occurs when a part of the intestine pushes into the groin. This occurs because of a weakness in the muscle wall. Hernias can be repaired by surgery.

188   Chapter 7: Growth and Development

Health of Others

Enlarged Prostate Gland As men age, it is not uncommon for the prostate gland to become enlarged. Assign a group of students to learn more about this condition: How common is it? At what age is it most likely to occur? What are the symptoms? What forms of surgery are used to treat enlarged prostate glands? What new kinds of surgery or alternative treatments are being developed? What is the relationship between an enlarged prostate and cancer of the prostate?

Have these students design and make an illustrated brochure that explains the condition. Encourage students to share copies of their booklet with older family members.

- **Sterility.** Being unable to produce enough healthy sperm to reproduce is a condition called sterility. Causes of sterility include certain diseases and exposure to certain drugs.
- **Testicular cancer.** Cancer of the testes is rare, but it is most common in males between the ages of 15 and 34. Symptoms may include a lump or swelling in the scrotum, pain or tenderness in one of the testicles, or a heavy feeling in a testicle. Surgery is necessary to prevent the spread of the disease. Regular self-examinations as well as checkups can help males to discover this disease in its early stages.

Did You Know?

Mumps and Sterility

One of the diseases that children are regularly immunized against is mumps. Mumps can cause inflammation and swelling of structures within the testes if the disease is caught after puberty, and this condition can lead to sterility.

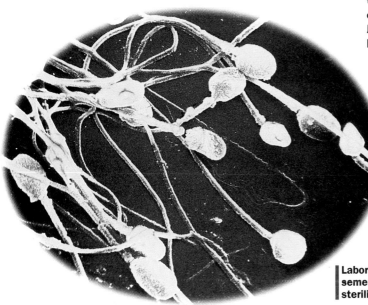

Laboratory analysis of semen can help diagnose sterility.

Review — Lesson 3

Using complete sentences, answer the following questions on a separate sheet of paper.

Reviewing Terms and Facts

1. **Vocabulary** Describe the function of the human *reproductive system.*
2. **Vocabulary** What is the difference between *sperm* and *semen?*
3. **Recall** What can males do to keep their reproductive systems healthy?
4. **List** Identify three disorders of the male reproductive system.

Thinking Critically

5. **Explain** Why is regular self-examination of the testes important?
6. **Describe** How are the male reproductive system and the endocrine system related?

Applying Health Concepts

7. **Consumer Health** Urologists treat many diseases and disorders of the male reproductive system. Do research about this medical specialty, and prepare a short brochure that might explain to a male what kinds of problems this doctor treats.

Lesson 3: The Male Reproductive System

Lesson 4
The Female Reproductive System

Lesson 4

The Female Reproductive System

This lesson will help you find answers to questions that teens often ask about the female reproductive system. For example:

- How does the female reproductive system work?
- What is menstruation?
- What care does the female reproductive system require?

Words to Know

- fertilization
- ovaries
- uterus
- menstruation
- gynecologist

The Female Reproductive System

The female reproductive system has three main functions. First, it stores and releases female reproductive cells, called egg cells. Second, this system allows fertilization to take place. **Fertilization** (fer·til·i·ZAY·shuhn) is *the joining of a male sperm cell and a female egg cell to form a new human life*. Third, the female reproductive system nourishes and protects the developing child until it is able to survive outside the female's body. **Figure 7.6** shows the female reproductive system.

Figure 7.6
The Female Reproductive System
Fertilization and pregnancy occur in the female reproductive system.

A At puberty the two *female reproductive glands*, the **ovaries**, start to release eggs, or ova. The ovaries also produce estrogen (ES·truh·jen), a hormone that regulates female sexual development.

B From the ovaries, eggs travel down the **fallopian** (fuh·LOH·pee·uhn) **tubes** to the uterus. Fertilization takes place in the fallopian tube.

C A fertilized egg becomes implanted in the **uterus** (YOO·tuh·ruhs), or womb (WOOM). This is a *pear-shaped organ in which a developing child is nourished*.

D The lining of the uterus is called the **endometrium** (en·doh·MEE·tree·uhm).

E The **cervix** (SER·viks) is the opening at the bottom of the uterus.

F The **vagina** (vuh·JY·nuh) is a muscular passageway leading from the uterus to the outside of the body.

Focus

LESSON OBJECTIVES

After studying this lesson, students should be able to

- discuss the functions of the female reproductive system.
- identify the parts of the female reproductive system.
- explain the cycle of menstruation.
- discuss proper care of the female reproductive system.
- identify the most common disorders that can affect the female reproductive system.

MOTIVATOR

Write the following questions on the board, and let students write phrases or short sentences in response to each: What is one important similarity between the male reproductive system and the female reproductive system? What is one important difference between the two systems?

INTRODUCING THE LESSON

Let students share and discuss their responses to the Motivator questions. Explain that they will learn more about the female reproductive system as they study Lesson 4.

INTRODUCING WORDS TO KNOW

In small groups, allow students to discuss the five Words to Know. Have group members share their understanding of each term before checking its formal definition in the Glossary. Then ask group members to find the words in a dictionary: From what language does each word derive? What other related words derive from the same root?

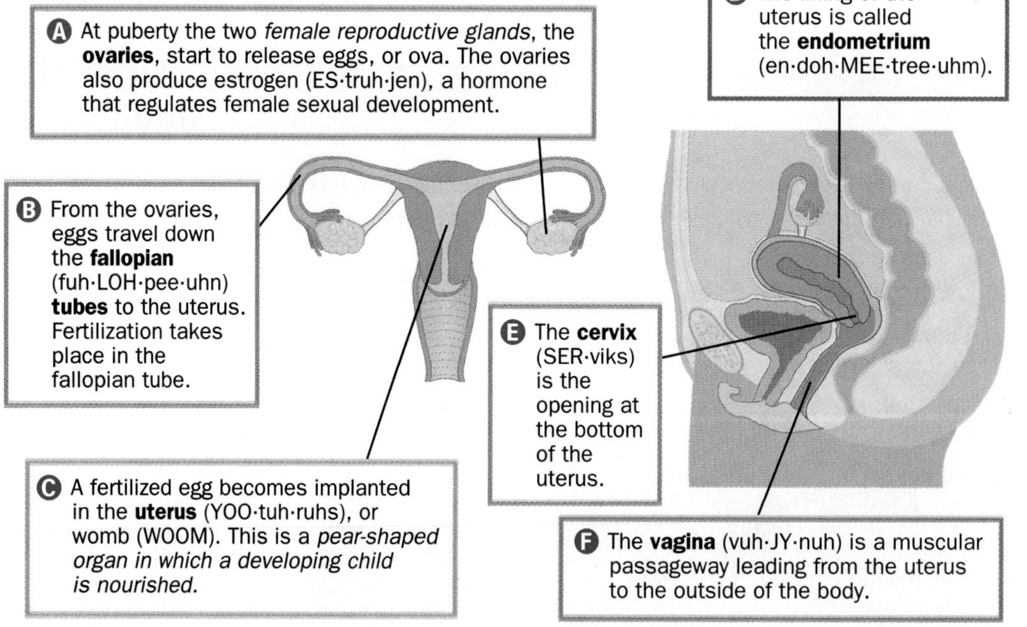

Lesson 4 Resources

Teacher's Classroom Resources
- Concept Map 31
- Enrichment Activity 31
- Lesson Plan 4
- Lesson 4 Quiz
- Reteaching Activity 31

Student Activities Workbook
- Study Guide 7
- Applying Health Skills 31

Menstruation

Once puberty begins, the ovaries start to release a single mature egg cell each month. This is the process known as ovulation (ahv·yuh·LAY·shuhn). At the same time, the lining of the uterus thickens. If fertilization occurs, this lining will nourish the fertilized egg. If fertilization does not occur, the egg is shed along with this thickened lining. *The flow of the uterine lining material from the body* is called **menstruation** (men·stroo·WAY·shuhn). For most girls, menstruation starts between ages 9 and 16.

Menstruation usually happens about every 28 days and lasts, on average, from 5 to 7 days. However, the timing of this menstrual (MEN·struhl) cycle may vary widely from one person to another. An adolescent female may find that the length of her menstrual cycle varies from month to month. This variation is normal and should not be a cause for concern unless menstruation stops for months at a time. **Figure 7.7** describes a typical menstrual cycle.

Did You Know?
Eggs Galore

At the time of a girl's birth, her ovaries contain between 200,000 and 400,000 egg cells. However, only about 400 to 500 of these develop into mature eggs. One egg is released each month, starting at puberty.

Figure 7.7
The Menstrual Cycle
Menstrual cycles vary. This is a typical cycle of 28 days.

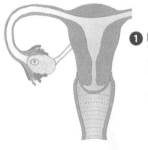
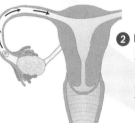
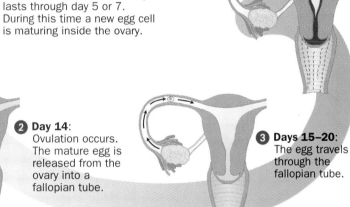

❶ **Days 1–13**: The cycle begins with the menstrual flow, which usually lasts through day 5 or 7. During this time a new egg cell is maturing inside the ovary.

❷ **Day 14**: Ovulation occurs. The mature egg is released from the ovary into a fallopian tube.

❸ **Days 15–20**: The egg travels through the fallopian tube.

❹ **Day 21**: The egg enters the uterus. If the egg has not been fertilized, the uterine lining begins to break down. Menstruation begins about 7 days later, on day 28.

Fertilization

Sperm entering the vagina travel to the fallopian tubes. Fertilization takes place in the fallopian tube when a male sperm cell and a female egg cell unite. The fertilized egg then moves through the fallopian tube to the uterus. In the uterus, the fertilized egg gradually develops into a baby.

Lesson 4: The Female Reproductive System

Personal Health

Healthy girls may have their first menstrual period as early as age nine or as late as age sixteen. It is important for teens to understand that there is nothing unusual or worrisome in beginning to menstruate either "early" or "late." Similarly, teens should understand that their second menstrual period is unlikely to begin exactly 28 days after their first period. An individual's menstrual cycle may take years to establish itself in a predictable pattern.

It is also important for teens to understand that the 28-day menstrual cycle is merely an average. Some girls and women may have shorter cycles; others may have longer cycles.

Chapter 7 Lesson 4

L1 Comparing
Ask students: How is the release of egg cells (ova) similar to the production of sperm? How are the two different? If appropriate, point out that a female has all her egg cells at birth; mature eggs are released, one each month, once she begins to menstruate. By contrast, the mature male's reproductive system produces new sperm nearly constantly.

L2 Guest Speaker
Invite the school nurse or another health professional to speak with the class about the beginning of menstruation in adolescent females. Ask the speaker to emphasize the variety of ages at which healthy girls may begin to menstruate and to explain that the menstrual cycle is typically not regular at first. Encourage students to think of and ask approriate questions after the speaker's presentation.

L3 Social Studies
Have interested students research coming-of-age ceremonies in various cultures. Let students compare and discuss their findings.

L1 Critical Thinking
Let students share their responses to these questions: Do you think all women experience PMS symptoms? How do you think the experiences and attitudes of friends and family members might affect a teen's experience of PMS symptoms?

L2 Researching
Have each student find and read a magazine article about a woman's experience with infertility. Then let students meet in groups to discuss the articles.

L1 Comprehending
Guide students in discussing the importance of regular checkups in detecting all kinds of cancers.

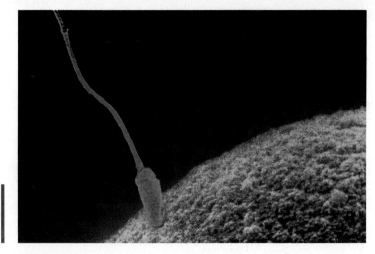

Fertilization occurs when a male sperm cell penetrates a female egg cell. Only one sperm cell can enter an egg cell.

Q & A
An Important Test

Q: I've heard people refer to something called a "Pap test," but I'm not sure what it is.

A: A Pap test (named after George Papanicolaou, the doctor who developed it) is a test for early detection of uterine cancer. It is quick, easy, and reliable. Gynecologists include a Pap test as part of the routine examination. They use a cotton swab and a small wooden paddle to collect cells from the cervix.

192 Chapter 7: Growth and Development

Caring for the Female Reproductive System

For females, just as for males, keeping the reproductive system healthy requires a combination of basic care and common sense. Here are some guidelines for females.

- Shower or bathe daily to keep external reproductive organs clean. Keeping clean is particularly important when menstruation is occurring.

- Be sure to schedule regular physical checkups by a **gynecologist** (gy·nuh·KAH·luh·jist), *a doctor who specializes in the female reproductive system,* beginning at the age recommended by your regular physician.

- Examine your breasts monthly for any unusual lumps, thickening, or discharge. Ask your doctor to explain the self-examination procedure.

- Keep a record of your menstrual cycle. If your cycle becomes irregular, check with a doctor. Check also if you experience severe or unusual pain or excessive bleeding during menstruation.

Problems of the Female Reproductive System

Various disorders can affect the female reproductive system. They include the following:

- **Premenstrual syndrome (PMS).** To different degrees, women may experience physical and emotional changes before menstruation. Symptoms of premenstrual (pree·MEN·struhl) syndrome, such as headache, breast tenderness, and irritability, may range from mild to severe. Regular exercise and dietary changes can often help. If you are troubled by PMS, talk with your doctor, especially if your symptoms make you very uncomfortable.

Health of Others

Several different over-the-counter medications are marketed specifically to reduce the symptoms of acne. Divide the class into groups, assigning one product to each. Have group members work together to learn more about their assigned product:

▶ What are its main ingredients?

▶ What symptoms does it claim to treat?

▶ What is the cost per dose?

▶ What information is available in reputable journals or other medical resources regarding the effectiveness of the product?

Let the members of each group share their findings with the rest of the class.

- **Toxic shock syndrome.** This bacterial infection is rare, but it has been linked to tampon use. Symptoms include high fever, a rash, and vomiting. It can be serious and even fatal if not treated. To protect yourself, follow directions for tampon use carefully. Change tampons at least every 4 hours, and avoid using superabsorbent tampons. Check with your doctor if you have questions.
- **Infertility.** Being unable to produce children is called infertility (in·fer·TIL·i·tee). There are a number of possible causes for infertility. Surgery or hormone treatment can overcome some types of infertility.
- **Vaginitis.** Pain, itching, and discharge are symptoms of this infection of the vagina. Doctors treat vaginitis (va·juh·NY·tis) with medication.
- **Cancer.** Cancer can affect the breasts, ovaries, uterus, or cervix. Self-examinations and regular checkups can detect cancer early and increase the likelihood of cure.

in your journal

Do you have any questions about your reproductive system that you would like answered? With whom would you feel comfortable talking about your questions? Write your answers in your private journal.

A physician can answer your questions about the female reproductive system.

Review — Lesson 4

Using complete sentences, answer the following questions on a separate sheet of paper.

Reviewing Terms and Facts

1. **Recall** What are the three most important functions of the female reproductive system?
2. **Vocabulary** Briefly define the term *fertilization*.
3. **Vocabulary** What is *menstruation*?
4. **Recall** What actions can females take to keep their reproductive systems healthy?

Thinking Critically

5. **Analyze** Why are the fallopian tubes important to reproduction?
6. **Explain** Provide three sources of valid information about the reproductive system.

Applying Health Concepts

7. **Growth and Development** Make a chart or poster showing what happens to an egg after it is formed in an ovary. Include what happens if the egg is fertilized and what happens if it is not.

Lesson 4: The Female Reproductive System 193

More About...

Menstrual Cramps About one-third of all girls and women experience uncomfortable menstrual cramps, usually during the first day or two of menstruation. They are not, as many people assume, the result of the uterine lining detaching from the interior wall of the uterus. Rather, the cramps are caused by contractions in the uterus itself, as the organ squeezes the lining away and out.

Why do some women have cramps with nearly every menstrual period, whereas other women never experience cramps? Hormones make the difference. The contracting of the uterus is stimulated by a hormone called prostaglandin.

Chapter 7 Lesson 4

Assess

EVALUATING THE LESSON

Assign Reviewing Terms and Facts and Thinking Critically on this page to review the lesson; then assign the Lesson 4 Quiz in the TCR to evaluate students' understanding.

LESSON 4 REVIEW

Answers to Reviewing Terms and Facts

1. Stores egg cells, allows fertilization to take place, and nourishes and protects the developing child until it can survive outside the female's body.
2. The joining of a male sperm cell and a female egg cell, which is necessary to form a new human life.
3. Menstruation is the flow of the uterine lining material, blood, and other fluids from the body.
4. Females should shower or bathe daily, have regular checkups, examine their breasts monthly, and keep a record of their menstrual cycle.

Answers to Thinking Critically

5. Eggs travel down the fallopian tubes to the uterus. Fertilization occurs in the fallopian tubes.
6. Responses will vary.

RETEACHING

▶ Assign Concept Map 31 in the TCR.
▶ Have students complete Reteaching Activity 31 in the TCR.

ENRICHMENT

Assign Enrichment Activity 31 in the TCR.

Close

Lead a brief class discussion in which students share their responses to this question: Why is it important for all teens to understand the female reproductive system?

193

Focus
The Teen Health Digest articles can be used in two ways: as an individual activity for reflection and enrichment or as a cooperative learning activity as described below.

Motivator
Ask students to turn to pages 201, 202, and 203 and review the stages shown in "The Stages of Life." Have them refer to these stages as they read and discuss the Teen Health Digest articles: Which stage or stages does each article highlight? What facts and attitudes does the article present?

Cooperative Learning Project

On the Road: A Display Presenting the Stages of Life

Setting Goals
Tell students that they will plan and mount a detailed display with information about, and illustrations of, the stages of life. Class guests, including family members and other students, will be invited to view their display.

Delegating Tasks
Using the stages in Figures 7.11 and 7.12 as a guide, have students identify the specific life stages they want to present in their display. (For example, students may want to treat early childhood, middle childhood, and later childhood as separate stages, or they may prefer to present them all in the single stage of childhood.) Help students divide themselves into equal groups, with one group for each of the selected stages.

TEEN HEALTH DIGEST

People at Work
Providing Care and Comfort

Job: Hospice worker
Responsibilities: To comfort and care for people who are terminally ill; to help families cope with the loss of a loved one; to administer medication; to help with practical matters
Education and Training: Hospice workers include nurses, doctors, counselors, health aides, and others. Besides training in their specialty area, workers receive special instruction in caring for the terminally ill.
Workplace: Care may be provided at the patient's home or at a residential facility.
Positives: The satisfaction that comes with helping and comforting others
Negatives: The work is emotionally demanding; it may include working nights or weekends and being on call.

Try This: Interview a worker at a hospice. Ask the person about the challenges and rewards of the work.

Teens Making a Difference
Bringing People Together

People have a basic need to care about others and to feel that others care about them. While working at a home for abused children, Connecticut teen Lauren Garsten learned what this need could mean to two very diverse groups of people.

Lauren's idea was to pair up children who desperately needed to feel loved with senior citizens who were seeking ways to make a difference in someone else's life. With the help of other students, Lauren brought the children and the seniors together once a month. The initial purpose was simple: just to have fun. But over time, children who had never trusted an adult before discovered that there were people they could count on. Lauren also discovered that no matter how different people are in age or experience, they can find something to talk about and share.

194 Chapter 7: Growth and Development

TECHNOLOGY UPDATE

Slide Projector Some or all of the groups may want to integrate photographic slides into their displays. They may use a projector to show one or two slides continuously, perhaps as part of the display background. If a more advanced projector is available, group members can set up a slide show with frequently changing photographs as a central part of their display.

On-Line Computer If group members want to find facts and/or photographs on the Internet, review with them the precautions for choosing and using Web sites and for protecting privacy on the Internet. Suggest that students access the Internet with partners.

HEALTH UPDATE
Senior Power

One of the fastest-growing population groups around the world is what demographers (di·MAH·gruh·fuhrz), people who study the characteristics of human populations, call the "oldest old." This group is made up of people over 80 years of age.

Why is the percentage of older people increasing? Throughout the world the birth rate is slowing down, and medical advances allow people to live longer. Adjusting to this aging population will require new kinds of thinking in countries all over the world. The invention and availability of devices to make life easier for an aging population and the extension of the retirement age past the current norm of 65 are two possible adjustments that could be made.

Myths and Realities
Determined at Birth?

Q: Is a baby's brain "wired" at birth—are all the pathways there just waiting to be used?

A: Not at all. Brain research has shown that although the brain at birth has nearly all the nerve cells it will ever have, the way these cells are linked together develops in extraordinary ways after birth. The stimulation a baby receives from its environment can affect greatly the potential that baby will have as a child and as an adult.

CON$UMER FOCU$
Eating to Stay Young

Q: I've seen ads for products that are supposed to keep a person young. I know that most of these products really don't do much, but I was wondering if there is anything that *does* keep the body from aging.

A: Although nothing can stop the aging process, certain foods may affect how quickly or slowly you age. For example, substances known as antioxidants (an·tee·AHK·suh·dunhtz) may work to prevent some of the damage done to the body over time by environmental pollutants and harmful foods. Fruits and vegetables high in vitamin C, such as broccoli, cantaloupe, carrots, collard greens, and peppers, are rich in antioxidants. Whole-grain breads and cereals are also good sources.

Try This: Find out more about the claims made for antioxidants. Report to your class on your findings.

Teen Health Digest 195

Chapter 7

The Work Students Do

Following are suggestions for implementing the project:

1. Students should begin by discussing, as a class, the kinds of displays they want to prepare. Possibilities include but are not limited to photographs, brief essays, drawings, object collections, brief skits or tableaux, or any combination. Encourage students to be as creative as possible and to select displays that can be produced successfully.
2. With students' input, decide on a specific date and time for presenting the completed displays. Encourage students to invite friends and family members, as well as other members of the student body, to the presentation.
3. Work with students to select a site for presenting their displays. Help students keep this planned location in mind as they develop their displays. Will each display fit easily into the available space? Will visitors view the display from only one side, or will they walk around the display?
4. Suggest that groups review one another's progress at least once. What constructive suggestions can students make about other groups' ideas? About the implementation of those ideas?

Assessing the Work Students Do

Confer regularly with the members of each group to be sure their display will be ready on time. Offer suggestions if necessary to be sure the displays will be appropriate, informative, and engaging.

After the displays have been presented, let group members meet again to discuss their contributions. What were the best aspects of the group's display? What would students do differently if they had the opportunity?

Meeting Student Diversity

Visually Impaired Students and/or guests with visual impairments will benefit if the groups' displays are not solely visual. Encourage students to include oral presentations, either live or taped, in their displays. You may also want to make oral interpreters available to accompany guests with visual impairments.

Learning Styles As students form groups, make sure each group includes students that provide a good mix of learning styles. Encourage group members to contribute in ways that are comfortable but also challenging for themselves, and that make a positive difference for the group.

195

Lesson 5
Human Development

Focus

LESSON OBJECTIVES

After studying this lesson, students should be able to
- discuss how the body is organized from its smallest parts—cells—to complex body systems.
- describe the stages in human development before birth.
- explain the influences of both heredity and environment on early development.

MOTIVATOR

Ask students to a write short paragraph describing their own earliest development.

INTRODUCING THE LESSON

Let three or four volunteers read aloud their Motivator paragraphs. Do any describe development before birth? Explain that students will learn more about this pre-birth development as they study Lesson 5.

INTRODUCING WORDS TO KNOW

Guide students in reading aloud the Words to Know and in skimming the lesson to find the definition of each term. Then have each student prepare a personal "bingo" card by dividing a plain sheet of paper into nine boxes, marking the center box *free*, and writing each Word to Know in one of the remaining boxes. Let one student select a Word to Know and read its definition aloud; the other students should cover or mark the correct word on their game cards. The first player to cover or mark three words in a row is the winner and may read the definitions aloud for the next game.

Lesson 5
Human Development

This lesson will help you find answers to questions that teens often ask about the earliest stage of human development. For example:

- What is the basic unit of life?
- How does a baby develop before birth?
- What factors affect a baby's development?

Words to Know

cell
tissue
organ
body system
embryo
fetus
chromosome
gene

Parts of a Whole

Your body is made up of more than 50 trillion microscopic cells. A **cell** is *the basic unit, or building block, of life.* The cells in your body are grouped to form tissues, organs, and body systems (see **Figure 7.8**).

Figure 7.8
From Cell to System
The body is organized from its smallest parts, cells, to complex body systems.

Cells
Cells come in many different forms and shapes. Each type of cell has a particular function. This is a cell from the lining of the stomach.

Tissues
Tissues are *groups of similar cells that do a particular job.* Each kind of tissue is designed to perform one function. These cells from the lining of the stomach protect the stomach from acidic stomach fluids.

Organs
Organs are *body parts made up of different tissues joined together to perform a function.* For example, the stomach is an organ made up of muscle, mucous membranes, and other types of tissue. These tissues work together to store food and prepare it for digestion.

Systems
A group of organs working together to carry out related tasks forms a **body system**. Examples include the digestive system, shown here; the endocrine system; and the reproductive system.

196 Chapter 7: Growth and Development

Lesson 5 Resources

Teacher's Classroom Resources
- Concept Map 32
- Cross-Curriculum Activity 13
- Enrichment Activity 32
- Lesson Plan 5
- Lesson 5 Quiz

- Reteaching Activity 32
- Transparency 29

Student Activities Workbook
- Study Guide 7
- Applying Health Skills 32

Development After Fertilization

Although the body is made up of countless cells, every human being begins as a single cell. This cell is formed as a result of fertilization. After a sperm cell and an egg cell join, the fertilized egg starts to divide. One cell becomes two, two become four, four become eight, and so on. These cells are referred to as an **embryo** (EM·bree·oh), *the name for the organism from fertilization to about the eighth week.* After the eighth week it is called a **fetus** (FEE·tuhs), *the name for the developing child from about the ninth week until the time of birth.* From fertilization until birth, the growing number of cells develop into the tissues, organs, and systems of the newborn's body (see **Figure 7.9**).

Science Connection

Brain Facts

At six weeks the brain of an embryo is almost as big as the rest of its body. At birth the baby's brain has 100 billion nerve cells, just about the same number of nerve cells as there are stars in the Milky Way.

Figure 7.9

Development Before Birth

It takes only a little over nine months for a baby to develop from two microscopic cells into a child ready to be born.

Time Passed	Approximate Length and Weight	Development
1 month	less than $\frac{1}{3}$ inch long	Major internal organs are forming. Heart starts to beat.
2 months	1 inch long	Organs continue to develop. Tiny arms, legs, fingers, and toes have started to form.
3 months	3 inches; 1 ounce	Fetus starts to move; can open and close mouth; fingers are visible; heartbeat can be heard by using a special instrument.
4 months	5 inches; 6 ounces	Facial features are taking shape; mother can feel movement of fetus.
5 months	almost 10 inches long; 1 pound	Eyes, nose, and mouth are well-developed; movements are stronger.
6 months	12.5 inches; 1.5 pounds	Fetus can kick and cry; footprints appear; fetus can hear sounds.
7 months	14.5 inches; 2 pounds	Eyes open; limbs are able to move freely.
8 months	18 inches long; 4 pounds	Hair grows; skin becomes smoother.
9 months	18-21 inches; 6-9 pounds	Fingers are able to grasp; body organs and systems can function on their own.

Cooperative Learning

Making Models To help students comprehend the amazing growth and development that take place before birth, have them work in groups to create a series of models showing the length and weight of an embryo or fetus at each month of pregnancy. Group members should work together to decide on answers to these questions:

- What materials should we use for our models?
- How will we display our models?
- How will we divide the responsibilities for making and displaying our group's models?

Once group members have resolved these questions, they will need time during or outside of class to prepare their models. Provide an opportunity for each group to discuss its display with the rest of the class.

Teach

L1 Discussing

Help students consider the number of cells in their bodies. Let a volunteer write the number *50,000,000,000,000* (50 trillion) on the board. Ask: What else might there be 50 trillion of? What does this number tell you about the size of each cell?

Visual Learning

Guide students in studying and describing the illustrations in Figure 7.8. Let volunteers read aloud the labels, and ask other volunteers to restate each label in their own words.

L2 Demonstrating

Suggest that a group of interested students use building blocks or connecting blocks to demonstrate the relationships between cells, tissues, organs, and systems. Have these students work together to plan an oral presentation to support their demonstration.

L2 Guest Speaker

Invite an obstetrician to speak with students about the development of an embryo and fetus. How does the doctor monitor the development of the baby and the health of the mother? What specific recommendations does the doctor usually make to help the mother protect the health and development of her baby and to protect her own health?

L2 Discussing

Discuss with students the rapid growth and development of a baby before birth. Suggest that students use rulers to note the length of the fetus or embryo at each stage: How many inches does the developing baby grow each month? During which month of pregnancy do you think the most important development takes place? Why do you think that?

Chapter 7 Lesson 5

Would You Believe?

Pre-birth development proceeds at an amazing rate. A baby who continued to double its size every month after birth would be more than a mile tall by its first birthday!

L2 Guest Speakers

Invite several new mothers and/or pregnant women to visit the class together. Ask them to share with students their own experiences of their babies' pre-birth development, including feeling the baby move, listening to its heartbeat, and perhaps seeing the developing baby in a sonogram.

L3 Researching

Ask interested students to read about the kinds of tests currently used to check the development of unborn babies (such as ultrasound examination, fetal stress test, and chorionic villus sampling): What specific forms of development does each test check? How is the test administered? For which mothers is the test generally recommended? When and how was the test developed? Let these students discuss their findings with the rest of the class.

L1 Recalling

Let students recall the definitions of *heredity* and *environment*, or refer them to the definitions given on page 9.

Cultural Connections

Baby Talk

Researchers have found that parents in many different cultures all tend to speak to their infants in similar ways. They put their faces close to the baby and speak in a singsong tone. Babies actually seem to respond more quickly and learn language more effectively when it is delivered this way than when parents speak as they would to each other.

Birth

The fertilized egg attaches itself to the lining of the uterus. There it grows and develops into a baby, receiving nourishment and oxygen from its mother through a tube attached to its abdomen. This period of pregnancy lasts a little over nine months. When the baby is ready to be born, muscles in the wall of the uterus begin contracting, causing the baby to be pushed out of the mother's body through the vagina.

Factors that Influence Early Development

In Chapter 1, you read about how heredity and environment shape your development as a unique individual. These factors affect you even before birth and continue to affect your development throughout your life.

HEALTH LAB

Your Unique Fingerprints

Introduction: The ridges and grooves on the pads of your fingers help you grip and feel objects. They also form fingerprints. Each person's fingerprints are different from everyone else's. That's why fingerprints can be used to identify people.

Objective: Identify and compare fingerprint patterns.

Materials and Method: You'll need a washable-ink pad, some white paper, and a magnifying glass. You should also have access to soap and water.

Make clear prints of the five fingers of one hand. You can do this by firmly pressing your finger into the ink pad and then pressing it onto the white paper. Also take prints from two classmates. Wash your hands.

Observations and Analysis: Study your own fingerprints. Look for the three basic fingerprint patterns: *arches, loops,* and *whorls.* How are the patterns of your fingerprints different from one another? How are they alike?

Compare your fingerprints with those of your classmates. What similarities can you find? What differences? How do people use the uniqueness of fingerprints for identification?

arch

loop

whorl

198 Chapter 7: Growth and Development

HEALTH LAB

Your Unique Fingerprints

Time Needed
1 class period

Supplies
- ink pad
- plain white paper
- magnifying glass
- access to soap and water

Focus on the Health Lab

Ask students: What makes you unique? Encourage a wide variety of responses. Next, help students focus on the physical characteristics that set them apart from everyone else. Ask: What do you have that no one else in the world has or has ever had? If necessary, help students recognize one response: "my fingerprints." Then have students read the Health Lab feature.

Heredity

Structures within cells influence heredity. **Chromosomes** (KROH·muh·sohmz) are *threadlike structures that carry the codes for inherited traits.* There are 46 chromosomes—23 pairs—in almost all human cells. Each chromosome is divided into many thousands of small parts called genes. **Genes** are *the basic units of heredity.* They determine the traits that you inherit, such as height, facial features, and the color of skin, hair, and eyes. Children inherit two genes for each trait—one from each parent. Children of the same two parents, however, inherit different combinations of chromosomes and genes.

Only sperm cells and egg cells do not have 46 chromosomes. Each has 23 chromosomes. When the sperm and egg cell join during fertilization, they produce a fertilized cell with 46 chromosomes. Chromosomes determine whether a baby will be a boy or a girl. Each egg cell contains one X (female) chromosome. Each sperm contains either an X (female) or a Y (male) chromosome. An XX chromosome combination will produce a female child; an XY combination will produce a male.

Occasionally, however, genetic problems occur that cause a baby to be born with a disorder. Such disorders may affect the baby's physical or mental development, or both. Some genetic disorders are mild; others are severe. A few rare disorders are fatal.

in your journal

Have you ever known anyone with an inherited disorder? How did it affect the person's life and the life of the rest of the family? Write your answers in your private journal.

Down syndrome is a genetic disorder in which a person's cells have 47 chromosomes instead of the usual 46. A person with Down syndrome has characteristic physical features and some degree of mental retardation, which may be very mild.

Fetal Environment

In addition to heredity, a key factor that affects development is environment. For a developing fetus the environment is the mother's uterus. The health of the baby is directly affected by the actions and health of the mother (see **Figure 7.10**).

Lesson 5: Human Development

Chapter 7 Lesson 5

L3 Science
Suggest that students interested in heredity read about the experiments of botanist Gregor Mendel. Have them prepare a chart illustrating his findings, and let them discuss their chart with the rest of the class.

L3 Career Education
Encourage students to learn about genetics as a career field: What kinds of jobs are available to people interested in genetics? What education and training are required? Ask students to present their findings in brief, illustrated reports.

L1 Identifying Examples
Have students identify genetic disorders they have heard of or know about. Record those examples on the board. Ask: Which do you think are mild? Which are severe?

L1 Discussing
Guide students in reading and discussing the guidelines for Healthy Mother, Healthy Baby: Why is each important? What is likely to happen if a woman becomes pregnant without planning to? if she fails to recognize or acknowledge her pregnancy?

L2 Creating Posters
Have students work in groups to plan and make posters about the importance of good prenatal care. Tell group members to select one guideline from Figure 7.10, find more information supporting that suggestion, and include the new information, along with a persuasive picture or slogan, in their poster.

Understanding Objectives
Students will make prints of their own fingers and observe the patterns of their fingerprints. This activity will help them appreciate the incredible diversity among humans and see themselves as unique individuals.

Observation and Analysis
For best results, let students conduct the lab in groups of three. Have the members of each group to discuss the patterns of arches, loops, and whorls they can identify in each set of prints. Ask each group to write a short description of the similarities and differences they have identified.

Further Investigation
Encourage students to learn more about fingerprints and their uses. Have the members of each group work together to research the uses of fingerprints in solving crimes, or the most recent developments in using computers to recognize and categorize fingerprints. Let each group share its findings with the class.

Assessing the Work Students Do
Have the members of each group write a brief assessment of their own efforts and a summary of what they have learned.

Chapter 7 Lesson 5

Assess

EVALUATING THE LESSON

Assign Reviewing Terms and Facts and Thinking Critically on this page to review the lesson; then assign the Lesson 5 Quiz in the TCR to evaluate students' understanding.

LESSON 5 REVIEW

Answers to Reviewing Terms and Facts

1. In its early stage of development, a fertilized egg is called an embryo. After about two months, it is called a fetus.
2. A chromosome is one of the threadlike structures that carry the codes for inherited traits, and a gene is the basic unit of heredity. More than 1,000 genes are found on each chromosome.
3. A pregnant woman should eat healthful foods, avoid alcohol, avoid medications and drugs, beware of infections, and avoid smoking and secondhand smoke.

Answers to Thinking Critically

4. Cells grouped together to perform particular functions form tissues. Tissues join together to form organs. Organs work together in body systems.
5. Responses will vary.

RETEACHING

▶ Assign Concept Map 32 in the TCR.
▶ Have students complete Reteaching Activity 32 in the TCR.

ENRICHMENT

Assign Enrichment Activity 32 in the TCR.

Close

Ask students to write short answers to this question: What do you consider the most important thing teens should know about a baby's development after fertilization?

Figure 7.10
Healthy Mother, Healthy Baby
During her pregnancy, there are several important ways for a mother to help ensure her baby's good health.

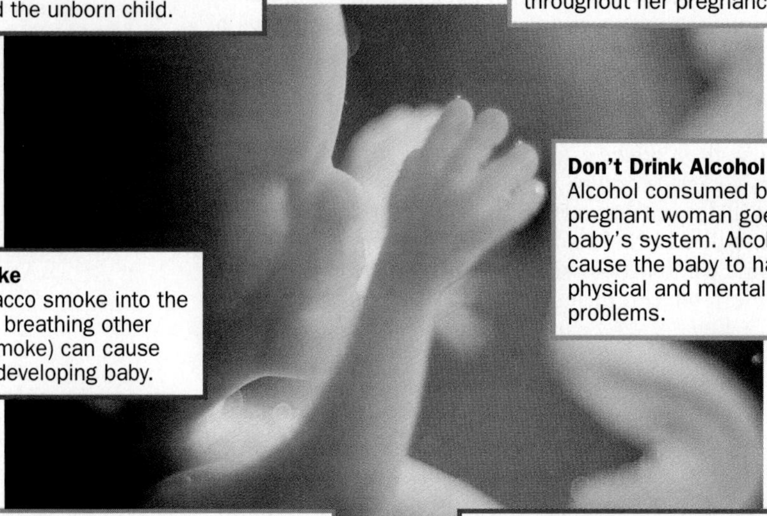

Eat Healthful Foods
The baby's nourishment comes from the mother. Nutritious foods are essential to the health of both the mother and the unborn child.

Get Regular Checkups
A woman should see her doctor as soon as she suspects that she is pregnant. After that, she should keep all regular appointments throughout her pregnancy.

Don't Smoke
Taking tobacco smoke into the body (even breathing other people's smoke) can cause harm to a developing baby.

Don't Drink Alcohol
Alcohol consumed by a pregnant woman goes into the baby's system. Alcohol can cause the baby to have both physical and mental problems.

Beware of Infections
Certain diseases, such as rubella (German measles) and some sexually transmitted diseases, pose severe danger to an unborn baby. A vaccine is available to protect against rubella. A physician can explain how best to avoid other dangerous infections.

Avoid Medications and Drugs
All drugs, even those available without a prescription, can affect a developing baby. A pregnant woman should take medication only if absolutely necessary, and only with a doctor's approval.

Lesson 5 Review

Using complete sentences, answer the following questions on a separate sheet of paper.

Reviewing Terms and Facts

1. **Vocabulary** What is the difference between an *embryo* and a *fetus*?
2. **Vocabulary** Describe the relationship between *chromosomes* and *genes*.
3. **Recall** List five ways in which a pregnant woman can help ensure her unborn baby's good health.

Thinking Critically

4. **Explain** Describe how cells, tissues, organs, and systems are related.
5. **Infer** Why would regular checkups during pregnancy be important?

Applying Health Concepts

6. **Growth and Development** Work with a partner to make a poster showing the relationships among cells, tissues, organs, and systems. Display your poster in the classroom.

Home and Community Connection

Prenatal Clinics Careful monitoring and thoughtful counseling from a doctor or other health professional are an essential part of the care a pregnant woman should seek. Some women receive this kind of prenatal care from a private obstetrician or midwife; others go to health clinics supported by private or public agencies in their community.

What are the options for pregnant women in your community? Let students work together to compile information about local prenatal clinics and other sources of help and support for pregnant women. Have them publish this information in an illustrated flier.

Life Stages

This lesson will help you find answers to questions that teens often ask about growing up. For example:

▶ What are the stages of life?
▶ What does it mean to be an adult?
▶ How do people deal with death and their feelings about death?

Birth and Growth

The years between infancy and adulthood are a period of huge change and development. During this time, a child's abilities and confidence grow. As the years pass, the once helpless baby gradually develops into an independent adult.

Infancy

The time of life from birth to the teen years is full of changes. **Infancy** (IN·fuhn·see) *the first year of a baby's life,* is a time of amazing growth. The baby's weight triples, and the baby's size increases by half. Infants learn to sit, crawl, and stand. They reach for objects. They smile and laugh. They start to explore their surroundings. They imitate sounds and may say a few words. During this first year, babies also develop mentally, emotionally, and socially. Their growth and development at this early stage prepares them for the stages of life to come (see **Figure 7.11**).

Words to Know

infancy
toddler
preschooler
grief

in your journal

Talk with a parent or relative about your childhood. In your private journal, write a paragraph explaining how your childhood experiences have helped to shape you as a person.

Figure 7.11
The Stages of Life
Throughout life, people continue to learn, change, and develop. Each period of life prepares you for what follows.

Birth Even before birth, babies begin to sense and react to their surroundings. As they adjust to life outside the uterus, newborns are totally dependent on others for their care.

Lesson 6: Life Stages   **201**

Chapter 7 Lesson 6

Teach

L1 Discussing

Tell students that, among developed nations, the United States ranks thirteenth in infant mortality. Ask: Why do you think our country does so poorly in promoting the health of babies?

L1 Critical Thinking

Help students consider the importance of good care for infants: How do you think a lack of attention might affect a baby's physical health? Mental/emotional health? social health? How would a lack of attention affect the baby's health in later stages of life?

> **Visual Learning**
>
> Let students describe and discuss Figure 7.11, "The Stages of Life": What items are being used at different stages? Why is each item associated with that stage? How do you think each stage is different from the one before it and the one that follows it?

L1 Comparing

Encourage students to draw on their own experiences as well as on information in the lesson as they respond to these questions: What are the most important differences between toddlers, preschoolers, and children in later childhood? How are children alike in all three of these stages?

L3 Language Arts

Ask interested students to select several books written for toddlers and several written for preschoolers. Let them respond to these questions, either in a short written report or in an oral presentation: What makes each book unique? Do you think each book will appeal to children in its intended age group? What are the differences between the books for toddlers and those for preschoolers?

Teen Issues

Be Your Own Person ACTIVITY

During adolescence, teens try to fit in with their peer group. However, adolescence is also a time to learn to accept responsibility for your own actions. This includes not doing what peers do when you think their actions are wrong. On a sheet of paper, jot down some situations in which you followed your own values. What gave you the strength to do this?

Childhood

Children between the ages of one and three are called **toddlers.** At this early childhood stage, children learn to walk, talk, and use the bathroom. They actively explore their surroundings. They begin to become less dependent on their parents. They take pride in their accomplishments and are eager to master new tasks. Help and encouragement from parents build a toddler's self-confidence.

Children between ages three and five are called **preschoolers.** Physical and mental skills develop rapidly in this period of middle childhood. Preschoolers enjoy singing, sharing stories, and playing make-believe. They like to imitate parents and older siblings. By supporting and encouraging a preschooler's efforts, parents can help develop the child's self-esteem.

Between six and eleven, a period sometimes called late childhood, children continue to develop. At this stage children are going to school and expanding their knowledge and social skills. They can read on their own and perform a growing number of activities independently. Their ability to express themselves increases, as do their problem-solving skills. Games and activities become more challenging. At this age children often like to build or create things. Interaction with friends becomes more and more important.

Adolescence and Adulthood

As you have read in Lesson 1 of this chapter, adolescence is a time of important physical, emotional, mental, and social development. Except for infancy, at no time during your life do you grow and change as much as you do during adolescence. The personal development that occurs during your teen years serves as a bridge from childhood to young adulthood.

Figure 7.11 (continued)

Infancy (to age 1) During this period, infants develop trust.

Early Childhood (ages 1–3) In this toddler period, children learn that they are independent beings.

Middle Childhood (ages 3–5) In this preschool period, children begin to learn about rules and consequences.

202 Chapter 7: Growth and Development

Growth and Development

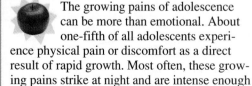

The growing pains of adolescence can be more than emotional. About one-fifth of all adolescents experience physical pain or discomfort as a direct result of rapid growth. Most often, these growing pains strike at night and are intense enough to awaken the adolescent. The young person may experience aching in the shins, calves, and/or thighs. Usually, a short massage or the application of heat brings relief. Intermittent growing pains of this kind may last for as long as a year without being a cause for concern. However, an adolescent who suffers severe leg pain or frequent leg pain for more than a year should be urged to consult a physician.

The adult years bring responsibility and satisfaction, joy and disappointment, challenge and accomplishment. During adulthood, people develop close and lasting relationships. They raise families, work toward personal goals, and make contributions to society.

Like childhood, adulthood can be seen as a series of stages. **Figure 7.12** shows three stages of adulthood.

Figure 7.12
The Adult Years

During young adulthood, from age 19 to about age 30, people work on developing a career and forming close personal relationships. In middle adulthood, approximately age 31 to 60, people work to achieve goals and to contribute to their families or community. In late adulthood, starting after age 60, individuals reflect on their lives and what they have accomplished.

Late Childhood (ages 6–11)
In this period, children seek the approval of their parents and begin to establish themselves among their peers. They gain mastery over objects and activities.

Adolescence (ages 12–18)
In this period, teens search for their identity and take on greater responsibility.

Adulthood (age 19 and onward)
During adulthood, people work to develop relationships, to achieve goals, and to understand the meaning of their lives.

Lesson 6: Life Stages

Chapter 7 Lesson 6

L2 Interviewing
Have students interview caregivers or teachers who work with infants, toddlers, preschoolers, or children in later childhood. Suggest they begin by working in groups to come up with interview questions regarding the characteristics, needs, and special qualities of children in each stage of life. After they have conducted their interviews, let students meet in groups again to discuss their findings.

L3 Reporting
Ask one or two students to read more about self-esteem: What is it? Why is it considered important? What suggestions do psychologists make to help parents and teachers enhance children's self-esteem? Let these students present their findings in an oral report to the rest of the class.

L2 Language Arts
Have students discuss novels or short stories they have recently read in which the main characters are adolescents: What particular problems do the characters face in dealing with peers? with family members? with teachers and other adults? Encourage students to relate the experiences of the adolescent characters to their own experiences.

L3 Social Studies
Ask a group of interested students to do research on adolescence and childhood as distinct stages of life: During which periods of history did these stages begin to be viewed as separate from adulthood? What have been some of the effects of considering childhood and adolescence as separate stages? Have these students discuss their findings with the rest of the class.

Meeting Student Diversity

Cultural Perspectives
American culture is largely youth-oriented. Many students may be unfamiliar with the respect and even veneration that is shown to older adults in other cultures. Members of Asian, Native American, and traditional European cultures can help students gain a new perspective on aging. If possible, invite several speakers to discuss the place elders hold in their culture and/or community. If speakers are not available, ask students to do library research. How is respect shown to older adults? What special responsibilities do older adults have? What is the relationship between older adults and children? Between older adults and young adults? In a class discussion, have the students share the results of their research.

Chapter 7 Lesson 6

L2 Interviewing
Ask each student to interview five adults. Have them ask about each adult's experience with adolescence: What was the most interesting part of your own adolescence? What was the most difficult part? Let students write short essays summarizing and discussing their interviews.

L1 Discussing
Guide students in discussing the life cycle: At what stage in life do you think people begin to understand that both birth and death are part of the life cycle? When do you think people are likely to accept death as part of the cycle of their own lives? How does living with family members and friends who are in various stages of life help people understand and accept the cycle of life?

L2 Guest Speaker
If a school counselor or psychologist conducts grief work with students, either individually or in groups, invite that staff member to speak with the class about available sessions and about teens' reactions to the death of a loved one.

L1 Comparing
Let students share their ideas in response to these questions: How is the end of a marriage similar to a death of a family member or close friend? How are the two events different? What special problems do you think teens face when their parents divorce? when a parent dies? Where can teens go to find help in facing those problems?

Q&A

How Can I Be Angry?

Q: My favorite uncle died in a car accident a few months ago. Sometimes I find myself feeling angry at him for dying. Is that normal?

A: Anger is a very normal response to the death of someone you love. Your anger comes in part because someone important was taken away from you by circumstances beyond your control. Don't feel guilty about these feelings. They will pass in time. If they really bother you, talk with your parents, a counselor, or another trusted adult.

The Life Cycle

The people around us, and we ourselves, move continuously through the cycle of life. Each day people are born and people die. Eventually people we love will die, and we ourselves will die, too. Part of growing up is learning to face this reality and accept it as part of the cycle of life.

Accepting Death

People deal with the reality of death in individual ways. When faced with the prospect of dying, however, most people go through five stages.

- **Stage 1: Denial.** The person refuses to believe that he or she is really going to die.
- **Stage 2: Anger.** The person becomes angry about the unfairness of his or her death.
- **Stage 3: Bargaining.** The person tries to prolong life through bargaining, perhaps promising to live a better life in exchange for escaping death.
- **Stage 4: Depression.** The person feels an intense sadness.
- **Stage 5: Acceptance.** The person accepts the reality of death and makes peace with both self and others.

MAKING HEALTHY DECISIONS

Helping a Friend Deal with Loss

Binh and David have been close friends for years. Binh has become worried about David lately. Ever since David's grandmother died two months ago, David has been quiet and withdrawn. He rarely participates in after-school activities anymore. One night David confides to Binh that he's feeling guilty. David wishes that he'd spent more time with his grandmother in the past year instead of hanging out so much with friends. Now, having fun with friends makes him feel somehow disloyal to the memory of his grandmother.

Binh is not sure what to do. Should he just keep quiet and wait for David to work things out on his own? David seems to have become more and more isolated. To make up his mind, Binh uses the decision-making process:

1. **State the situation**
2. **List the options**
3. **Weigh the possible outcomes**
4. **Consider your values**
5. **Make a decision and act**
6. **Evaluate the decision**

Follow-up Activities

1. Apply the six steps of the decision-making process to Binh's concern.
2. With a partner, role-play a conversation between Binh and David, in which Binh offers constructive suggestions to help his friend.

204 Chapter 7: Growth and Development

MAKING HEALTHY DECISIONS

Helping a Friend Deal with Loss

Focus on Healthy Decisions
Ask students to name examples of what they do to help their friends. When is it easy to help a friend? What circumstances sometimes make it difficult to help a friend? How can a friendship be affected if one of the friends suffers a loss? After this introductory discussion, have students read the Making Healthy Decisions feature.

Activity
Use the following suggestions to discuss Follow-up Activity 1:
Step 1 Binh has to decide what—if anything—he should try to do to help his friend David deal with the death of David's grandmother.
Step 2 Binh might decide to do nothing and wait for David to feel better, try to help David himself, or share his concerns about David with a parent or another trusted adult.

Dealing with Grief

The *sorrow caused by the loss of something precious* is known as **grief.** Grief, whether over the death of a relative, a friend, or a pet, can be intensely painful. People also grieve over other types of loss, including the loss of a job or the end of a marriage.

Everyone experiences grief in his or her own way. Some people feel numb at first, then intensely sad. Others feel anger. Still others are troubled by guilt. They believe that they should have done or said something differently while the person was alive. Most people find that their grief comes in waves—just when they think they are feeling better, they are struck by sadness again.

No matter how deep a person's grief, however, the ache of loss gradually lessens over time. People usually find the grieving process easier if they share their feelings with other people. If you are grieving over a loss or death, don't be afraid to talk about how you feel with a trusted friend or adult.

Funeral services and other ceremonies help people deal with the loss of a loved one.

Review

Using complete sentences, answer the following questions on a separate sheet of paper.

Reviewing Terms and Facts

1. **Recall** List the stages of life from infancy through adulthood.
2. **List** What stages do people go through when faced with death?
3. **Vocabulary** What is *grief?*

Thinking Critically

4. **Analyze** How can parents help their children move successfully through the stages of childhood?
5. **Explain** Why is it helpful for people who are grieving to talk about their feelings with others?

Applying Health Concepts

6. **Health of Others** Find out about support groups in your community for people who have suffered a loss. What kinds of groups are there? Who sponsors these groups? How do they help people cope with their grief? Write a short report about your findings.

Lesson 6: Life Stages 205

Chapter 7 Review

CHECKING COMPREHENSION

Use the Chapter Summary and the Chapter 7 Review to help students go over the most important ideas presented in Chapter 7. Encourage students to ask questions and add details as appropriate.

CHAPTER 7 REVIEW ANSWERS

Reviewing Key Terms and Concepts

1. Hormones are chemical substances that are produced in the glands and that help to regulate the way the body functions.
2. Responses will vary.
3. The pituitary gland produces hormones that control other glands.
4. Responses will vary.
5. Reproduction is the process by which living organisms produce new individuals of their kind.
6. The testes produce sperm and make testosterone.
7. The ovaries release eggs. The uterus is the organ in which a fertilized egg is nourished and allowed to develop.
8. A gynecologist is a doctor who specializes in the female reproductive system.
9. Genes are the basic units of heredity.
10. Responses will vary. Possible response: Alcohol and tobacco can harm a fetus.
11. Infants are babies from birth to one year; toddlers are children between the ages of one and three; preschoolers are children between the ages of three and five.
12. Responses will vary.

Chapter 7 Review

Chapter Summary

- **Lesson 1** Adolescence brings physical, mental, emotional, and social changes that prepare teens for adulthood.
- **Lesson 2** The endocrine system produces hormones and regulates body functions.
- **Lesson 3** The male reproductive system produces, stores, and releases sperm, the male reproductive cells.
- **Lesson 4** The female reproductive system stores female reproductive cells (egg cells), allows fertilization to occur, and nourishes the fertilized egg.
- **Lesson 5** Humans develop in predictable patterns before birth. Both hereditary and environmental factors have an effect on development.
- **Lesson 6** People move through stages of development beginning at birth and continuing through adulthood.

Reviewing Key Terms and Concepts

Using complete sentences, answer the following questions on a separate sheet of paper.

Lesson 1
1. What are *hormones*?
2. What mental and emotional growth takes place during adolescence?

Lesson 2
3. Why is the pituitary gland important to proper body functioning?
4. List three glands other than the pituitary gland, and describe their functions.

Lesson 3
5. What is *reproduction*?
6. List two functions of the testes.

Lesson 4
7. Describe the functions of the ovaries and the uterus in female reproduction.
8. What is a *gynecologist*?

Lesson 5
9. What are *genes*?
10. List two substances that a woman might take into her body that could harm a fetus.

Lesson 6
11. What are the age ranges for infants, toddlers, and preschoolers?
12. What are typical tasks in each of the three stages of adulthood?

Thinking Critically

Using complete sentences, answer the following questions on a separate sheet of paper.

13. **Explain** How do the changes of adolescence prepare you for adulthood?
14. **Summarize** Why is the endocrine system important during adolescence?
15. **Compare and Contrast** Compare and contrast the functions of the male and female reproductive systems.
16. **Select** Choose a particular stage of life, and describe how it prepares a person for a future stage.

206 Chapter 7: Growth and Development

Meeting Student Diversity

Language Diversity Use the following suggestions to help students who have difficulty with English:

- Pair those learners with native speakers of English who can restate the Chapter Summary in language that helps students comprehend important concepts.
- Direct auditory learners or those students with language diversity to the Teen Health Audio-cassette Program. Available in English and Spanish, this component provides an audio and written summary of the chapter.

Chapter 7 Review

Your Action Plan

From the day we are born until the day we die, change is a part of our lives. Some changes are part of the life cycle—happy changes, such as the birth of a child, and sad changes, such as the death of a loved one. Throughout your life, you will want to use these changes to help you grow as a person.

Step 1 Review your journal entries to answer these questions: What changes have you already experienced? Which have you dealt with successfully? Which changes have caused you problems?

Step 2 Set a short-term goal to help you deal with a change that you expect to occur soon. If your family expects to move, you might write: *I will talk with my parents about visiting my new school before the year begins.*

Step 3 Set a long-term goal to help you deal with unexpected changes. For example: *I will work on coping with my emotions so that it is easier to adjust to change.*

Periodically review your goals and check your progress. Evaluate how you have grown as a person, and think about how you would like to continue to grow.

In Your Home and Community

1. **Growth and Development** Work with classmates to create a guidebook to adolescence entitled *What's Normal?* Include descriptions of the typical changes teens go through. Obtain permission to share the guidebook with other classes.

2. **Health of Others** Create a poster suggesting ways for pregnant women to help ensure that their babies will be healthy. Ask if you can display your poster in the office of a gynecologist or pediatrician, or at a local hospital.

Building Your Portfolio

1. **Illustrated Chart** Create an illustrated chart on the stages of life. Use pictures from magazines to show people in various life stages. Add some written information about each stage, or provide quotations from people talking about that stage of life. Place your chart in your portfolio.

2. **Interview** Interview two senior citizens. Ask them to describe the various stages of their lives. Which events were most memorable? Which experiences did they learn the most from? Summarize the interviews in a few paragraphs. Add the summary to your portfolio.

3. **Personal Assessment** Look through all the activities and projects you did for this chapter. Choose one or two that you would like to include in your portfolio.

Chapter 7: Chapter Review **207**

Chapter 7 Review

Thinking Critically
Responses will vary. Possible responses are given.

13. Physical changes make reproduction and the establishment of new families possible; mental/emotional changes make mature thought possible; social changes create more independence.

14. The endocrine system controls growth and sexual development.

15. Both systems contribute to reproduction. They release different kinds of cells: sperm in males and eggs, or ova, in females. Fertilization and pregnancy occur within the female reproductive system.

16. During late childhood, the development of the mind and a greater interest in peers prepare individuals to enter adolescence.

RETEACHING
Assign Study Guide 7 in the Student Activities Workbook.

EVALUATE
▶ Use the reproducible Chapter 7 Test in the TCR, or construct your own test using the Testmaker Software.

▶ Use Performance Assessment 7 in the TCR.

EXTENSION
Invite a local mental health provider to speak with students about the changes individuals experience as they move through life's stages. Ask the speaker to focus on adolescence. Encourage students to prepare appropriate questions to ask after the presentation.

Performance Assessment

▶ **Self-evaluation** Direct students to review the activities that are provided throughout the chapter. Encourage each student to select one finished product or activity that demonstrates his or her best work for the chapter. Have students explain what they learned and how the examples they selected show their progress.

▶ **Teacher's Classroom Resources** Assign Performance Assessment 7, "A Song About Changes," in the TCR.

Planning Guide

Chapter 8
Mental and Emotional Health

	Features	**Classroom Resources**

Lesson 1: What Is Mental Health?
pages 210–213

- Concept Map 34
- Enrichment Activity 34
- Lesson Plan 1
- Lesson 1 Quiz
- Reteaching Activity 34

Lesson 2: Building Positive Self-Esteem
pages 214–218

Health Lab: The Qualities You Admire — page 215

- Concept Map 35
- Cross-Curriculum Activity 15
- Decision-Making Activity 15
- Enrichment Activity 35
- Lesson Plan 2
- Lesson 2 Quiz
- Reteaching Activity 35
- Transparencies 30, 31

Lesson 3: Managing Stress
pages 219–225

Life Skills: Time Management — page 221

Personal Inventory: Life Changes and Stress — page 223

TEEN HEALTH DIGEST
pages 226–227

- Concept Map 36
- Cross-Curriculum Activity 16
- Decision-Making Activity 16
- Enrichment Activity 36
- Health Lab 8
- Lesson Plan 3
- Lesson 3 Quiz
- Reteaching Activity 36
- Transparencies 32, 33

Lesson 4: Mental Disorders
pages 228–231

- Concept Map 37
- Enrichment Activity 37
- Lesson Plan 4
- Lesson 4 Quiz
- Reteaching Activity 37

Lesson 5: Sources of Help
pages 232–235

Making Healthy Decisions: Deciding Whether to Tell — pages 234–235

- Concept Map 38
- Enrichment Activity 38
- Lesson Plan 5
- Lesson 5 Quiz
- Reteaching Activity 38

National Health Education Standards

One of the main goals of the National Health Education Standards is to move students toward "health literacy." Health literacy is the capacity of individuals to obtain, interpret, and understand basic health information and services and the competence to use such information and services in ways which promote health. The health standards were developed by applying the characteristics of a well-educated, literate person within the context of health. The health literate person is:

- a critical thinker and problem solver
- a self-directed learner
- a responsible, productive citizen
- an effective communicator

Listed below are the Health Education Standards Performance Indicators addressed in each lesson of this chapter.

Lesson	Health Standards Performance Indicators
1	(1.2, 3.1, 6.2)
2	(1.4, 3.1, 3.4, 6.4)
3	(1.1, 3.2, 3.7, 6.1)
4	(1.2, 2.6, 6.1, 7.4)
5	(2.2, 2.4, 2.6, 6.1)

ABC NEWS INTERACTIVE VIDEODISC SERIES

You may wish to use the videodisc *Making Responsible Decisions*. Use the *ABC-News InterActive™ Correlation Bar Code Guide* for title reference. Also available in VHS format.

Chapter Resources

Teacher's Classroom Resources
- Chapter 8 Test
- Parent Letter and Activities 8
- Performance Assessment 8
- Testmaker Software

Student Activities Workbook
- Study Guide 8
- Applying Health Skills 34–38
- Health Inventory 8

Student Diversity Strategies
- Audiocassette Program (English)
- Audiocassette Program (Spanish)
- Spanish Parent Letters
- Spanish Summaries, Quizzes, and Activities

Multimedia Components
- English Audiocassette Program
- Spanish Audiocassette Program
- *Teen Health* Videodisc/VHS Series

Other Resources

Readings for the Teacher
Barbour, William, ed. *Mental Illness: Opposing Viewpoints.* San Diego, CA: Greenhaven Press, 1995.

Goldman, M. Nikki. *Emotional Disorders.* New York: M. Cavendish, 1994.

McFarland, Rhoda. *Coping through Self-Esteem.* New York: Rosen Publishing Group, 1993.

Readings for the Student
Chiles, J. *Teenage Depression and Suicide.* New York: Chelsea House, 1992.

McFarland, Rhoda. *Coping through Self-Esteem.* New York: Rosen Publishing Group, 1993.

Maloney, M., and R. Kranz. *Straight Talk about Anxiety and Depression.* New York: Dell Publishing, 1993.

Out of Time?
If time does not permit teaching this chapter, you may use these features: Health Lab on page 215; Life Skills on page 221; Personal Inventory on page 223; Teen Health Digest on pages 226–227; Making Healthy Decisions on pages 234–235; and the Chapter Summary on page 236.

Chapter 8
Mental and Emotional Health

CHAPTER OVERVIEW

Chapter 8 helps students focus on the mental/emotional aspect of good health. Students learn the importance of developing positive self-concept, coping with stress, recognizing mental disorders, and knowing where and how to seek help with mental/emotional problems.

Lesson 1 defines mental health and details the factors that shape an individual's personality.

Lesson 2 examines self-concept and self-esteem and lists ways of developing a positive self-concept and high self-esteem.

Lesson 3 offers definitions of positive and negative stress and presents coping strategies.

Lesson 4 highlights key types of mental disorders and common warning signs of suicide.

Lesson 5 helps students identify sources of help when an individual needs assistance with an emotional problem.

PATHWAYS THROUGH THE PROGRAM

Chapter 8 focuses on the characteristics of mental health. The chapter offers an opportunity to emphasize that emotional turmoil is typical during the teen years. Teens—especially young teens—are often frightened by their own sudden and extreme shifts in emotions, and it may be difficult for them to see the proverbial light at the end of the tunnel. Reassure students that any attempt to seek help with emotional and mental distress is a sign of maturity and responsibility.

Chapter 8
Mental and Emotional Health

Student Expectations
After reading this chapter, you should be able to:
1. Identify the traits of good mental health.
2. Explain the benefits of a positive self-concept and high self-esteem.
3. Describe ways of dealing with stress.
4. Identify major mental disorders.
5. List sources of help to solve problems.

Key to Ability Levels

Teaching strategies that appear throughout the chapter have been identified by one of three codes to give you an idea of their suitability for students of varying learning styles and abilities.

L1 **Level 1** strategies should be within the ability range of all students. Full class participation is often required. Teacher direction is usually needed.

L2 **Level 2** strategies are for average to above-average students or for small groups. Some teacher direction is necessary.

L3 **Level 3** strategies are designed for students able and willing to work independently. Minimal teacher direction is necessary.

Teen Chat Group

Katie: Try out? Are you kidding? You know I'm shy. I can't try out for the play with you. It'll never work out.

Jerome: Just make yourself do it. It might not be as hard as you think. It could even be fun.

Carmen: Jerome, don't tell her to try out if she's too shy. Anyway, they'll need people to do makeup and build scenery, too. What about doing that, Katie?

Katie: Well, maybe. . . I wish I felt more comfortable up in front of people, though.

Jerome: That's what I mean. If you don't like being shy, you should try to change it. Take a chance. What do you have to lose?

Katie: I could look stupid, for one thing. I'm just not sure of what I really want to do.

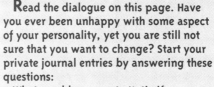

Read the dialogue on this page. Have you ever been unhappy with some aspect of your personality, yet you are still not sure that you want to change? Start your private journal entries by answering these questions:
▶ What would you say to Katie if you were part of this conversation?
▶ Are there aspects of your personality that you would like to change?
▶ How do you think a person can stay mentally and emotionally healthy?

When you reach the end of the chapter, you will use your journal entries to make an action plan.

Chapter 8: Mental and Emotional Health   209

Chapter 8

INTRODUCING THE CHAPTER

▶ Ask volunteers to name adjectives to describe their normal waking moods. (Examples include *happy, lively, stressed, sleepy*.) Then guide students in discussing how the rest of the day is affected by their moods at the start of the day.

▶ Ask students to describe their most common states of mind as if they were amusement park rides: Are the rides exciting ones in which you can expect sharp changes in direction? Or are they slow, calm rides that seldom change pace or direction? What are the advantages and disadvantages of each state of mind? Explain that students will learn more about these and related issues as they study the lessons in Chapter 8.

Cooperative Learning Project

Mind-Body Newsletter

The Teen Health Digest on pages 226 and 227 provides students with high-interest articles related to the content of this chapter.

The material in the Teacher's Wraparound Edition presents suggestions for a class project in which students plan and publish a newsletter for teens on the mind-body connection.

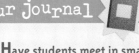

Have students meet in small groups to read the Teen Chat Group dialogue out loud and to discuss their responses to it: What do you think of Katie's reason's for not trying out for the play? How convincing are her friends' encouragements? Why? If you were Katie's friend, would you urge her to try out? If so, why? If not, why not? Do you think Katie and her friends are "normal"? Do you think that all three of them are mentally healthy? What signs do you see to support your opinion? After these group discussions, provide time for students to write private journal entries.

Lesson 1
What Is Mental Health?

Focus

LESSON OBJECTIVES

After studying this lesson, students should be able to
- recognize the signs of good mental health.
- discuss tips for developing good mental health.
- identify the factors that shape a personality.

MOTIVATOR

Ask students: Do you think you can recognize a person who has good mental health? If so, what signs do you look for? If not, why not? Have students write their responses in sentences or list form.

INTRODUCING THE LESSON

Let several volunteers share their responses to the Motivator activity. Ask why some people seem to enjoy good mental health and others do not: Are certain individuals just lucky, or does good health require effort? Assure students that good mental health can be fostered, and explain that Lesson 1 presents more information about good health and how it can be achieved.

INTRODUCING WORDS TO KNOW

Write the lesson's Words to Know on the board. Have students record the terms in their notebooks, write a definition for each word, and then use the word in an original sentence. Ask the class to share their definitions and sentences. Correct any misconceptions the students' answers reveal.

Lesson 1 — What Is Mental Health?

This lesson will help you find answers to questions that teens often ask about mental health. For example:

▶ How can I tell if I am in good mental health?
▶ How can I develop good mental health habits?
▶ What shapes my personality?

Words to Know

mental health
personality

in your journal

Review the signs of good mental health. For the next three days watch for examples of these signs in the behavior of yourself and others. Write these examples in your journal.

Mental Health

The three important parts of good health rely on one another for total balance and strength. Two sides of your health triangle are physical health and social health. The remaining side is your mental and emotional health. **Mental health** is *your ability to deal in a reasonable way with the stresses and changes of everyday life.* When people are in good mental health, they usually

■ have a positive outlook on life and welcome challenges. (As **Figure 8.1** shows, a positive outlook has many benefits.)

■ accept their limitations and set realistic goals for themselves.

■ feel good about themselves and about others.

Figure 8.1
The Benefits of a Positive Outlook
How can a positive outlook benefit you?

A Having a positive outlook leads to making an effort.

B Making an effort leads to encouragement from others.

210   Chapter 8: Mental and Emotional Health

Lesson 1 Resources

Teacher's Classroom Resources
- Concept Map 34
- Enrichment Activity 34
- Lesson Plan 1
- Lesson 1 Quiz
- Reteaching Activity 34

Student Activities Workbook
- Study Guide 8
- Applying Health Skills 34

- can accept disappointment without overreacting.
- act responsibly at work and in relationships.
- are aware of their feelings and are able to express those feelings in healthful ways.
- accept honest criticism without anger.

If you showed all of the qualities of mental health all the time, you would have perfect mental health. Of course, no one does. These qualities are goals to work toward. As with physical health, there are many different levels of mental health, and everyone's mental health has its ups and downs. Don't worry if you aren't showing all these qualities at all times. You can, however, improve your overall mental health by practicing good mental health habits. Look at the following tips for achieving good mental health. The first three have to do with the way you view others. The last five have to do with how you view yourself. Which of them do you find the easiest? In which areas do you need to improve?

- Accept other people as they are. It isn't fair to judge everyone by your own background and behavior.
- Focus on other people's strengths, not on their weaknesses.
- Consider other people's feelings.
- Focus on your own strengths.
- Accept the parts of yourself that you cannot change.
- Work to improve what you can change.
- Don't dwell on failures and disappointments.
- Learn from your mistakes.

interNET CONNECTION
Discover on the Web how capable you are of managing stress and achieving balance.
http://www.glencoe.com/sec/health

C Encouragement from others leads to success.

D Success strengthens your positive outlook and leads to efforts in new areas.

Lesson 1: What Is Mental Health?   211

More About...

Developing Good Mental Health Remind students that one of the keys to good mental health is accepting that they have no real control over how others feel and act; they have control only over how they themselves act. The next time students feel someone has not lived up to their expectations, encourage them to remember the following:

▶ Everyone makes mistakes.
▶ Everyone has value, even if they do not behave as others would like them to.
▶ Showing compassion will help the other person feel better about his or her mistake.

Chapter 8 Lesson 1

Teach

L1 Discussing

Have students identify and discuss famous people (actors, singers, athletes, and politicians, for example) they consider to be in good mental health. Then have them identify and discuss others they consider to be in poor mental health. On what do they base their opinions? Guide students in considering the subjects' probable feelings of self-worth and accomplishment.

L1 Analyzing

Encourage students to share their responses to these questions: Why is it important to recognize and accept your own limitations? What are some specific examples of limitations? How does knowing your limitations help you set goals?

Visual Learning

Let students describe each photograph in Figure 8.1: What is the teen doing? How does that reflect her positive outlook? What are some other activities that might be included to illustrate the benefits of having a positive outlook?

L1 Discussing

Encourage students to identify some of the problems that can be caused by trying to be perfect. What other, related problems are caused by expecting perfection in others?

Health Minute

Perfectionists—people who feel they must be perfect, and perfectly in control, at all times—feel that they will let others down if they make even one mistake. In fact, the opposite is true: Healthy individuals enjoy spending time with people who can forgive themselves and others for their mistakes, and who can learn from those mistakes.

211

Chapter 8 Lesson 1

L3 Analyzing

Have each student select a fictional character (as from television or a book) and make a list of that character's personality traits. Then let volunteers introduce their characters to the class and describe the characters' mental health. Ask: Do you find the characters realistic? Why or why not? Conclude by helping students discuss how many of the fictional characters benefit from a positive outlook.

L1 Writing Descriptions

Tell students to print their names in a vertical line on a sheet of paper. Instruct them to list an adjective that begins with each letter of their name and that closely describes their personality traits. Afterward, encourage students to discuss how they feel about their individual lists of adjectives.

L2 Language Arts

Have the students work in groups to write one-paragraph descriptions of a character named Terry. Say that Terry, who may be either male or female, is the picture of good health. Paragraphs should tell what Terry looks like, how he or she acts, and so on. Have a member from each group read aloud the group's description, and encourage students to discuss whether they would like to meet Terry: Why or why not?

VIDEODISC/VHS

Teen Health Course 2

You may wish to use video segment 1, "A Healthy Foundation," to show how overall wellness requires balancing the three aspects of health.

Videodisc Side 1, Chapter 1
A Healthy Foundation

Search Chapter 1, Play to 2

Teen Issues

Snap Judgments ACTIVITY!

We all give and receive impressions of one another at first meetings. Some people rely too much on first impressions in their opinions of others. They make snap judgments from visible personality traits about people they just met. These judgments usually are based on a bias about certain physical or cultural characteristics.

The next time you meet someone new, take the time to get to know the person before you decide if you like him or her. Isn't that the way you want to be judged?

Personality and Mental Health

Just as no two snowflakes are alike, no two people are exactly the same. Each person is an individual with a **personality,** which is a *special mix of traits, feelings, attitudes, and habits.* You may have heard a person described as having no personality. This can never be. Everyone has a personality. Your personality is everything about you that makes you the person you are.

Factors That Shape Your Personality

Many factors influence the development of your personality. The three most important factors are heredity (the passing on of traits from your parents), environment (all of your surroundings), and behavior (the way you act in the many different situations and events in your life). Behavior is the factor over which you have the most control. Study the examples in **Figure 8.2;** then think of one specific example of each factor that describes you.

To some extent, each factor will continue to shape your personality throughout your life. Some factors, such as heredity, are beyond your control. No one gets to choose her or his inherited traits. Much of your environment is also beyond your control. Most young people live where the adults in their family choose to live.

Figure 8.2
What Shapes Your Personality?

Heredity	Environment	Behavior
Height	Community	Caring for yourself
Skin, hair, and eye color	Family and friends	Caring for others
Body type	Experiences	Reflecting your values

Your behavior toward other people is an important factor in shaping your personality.

212

Personal Health

Inform students that the foods they eat influence the moods they experience. Low-fat proteins tend to stimulate, while carbohydrates tend to have a calming effect. Among the brain-booster foods are low-fat milk, skinned chicken, and dried beans. Mood-soother foods include bread, pasta, and cereals. Help students discuss the implications of these influences: What specific foods might make a good study snack? an appropriate lunch during an especially busy day? a healthful snack during an evening of quiet relaxation?

What you do have control over, however, is how you will deal with your inherited traits and your environment. The final decision about how you act is yours. You decide if you will focus on your strengths or weaknesses. You decide if you will work to improve those situations that can be changed. Your behavior is your choice and your responsibility. A big part of who you are is up to you.

The Factor You Control

Behavior is the way you act in the various situations of your life. It is the factor of your personality over which you have the most control. Your behavior is based on your values, which are beliefs and ideas about what is important in your life. Most of your values are learned from your family, and to a lesser extent from your friends. How you behave reflects those values. If good health is important in your life, then you will behave in a manner that reflects this. You will actively take good care of your mind and your body. You will choose behaviors that promote good health. You will not take risks that endanger your health or the health of others.

Your heredity, environment, and behavior have combined to shape the person you are today. How you handle these factors will define your personality as you grow and mature.

Your family teaches you about values through their actions and behavior.

Review — Lesson 1

Using complete sentences, answer the following questions on a separate sheet of paper.

Reviewing Terms and Facts

1. **Vocabulary** Define the term *mental health*. Use it in an original sentence.
2. **Recall** What are the three factors that shape personality?

Thinking Critically

3. **Apply** List at least five specific examples of good mental health qualities.
4. **Describe** Give examples of two times in the last week that your personal behavior demonstrated good mental health habits.

Applying Health Concepts

5. **Growth and Development** Role-play the following situation. A friend on your sports team has seemed unhappy lately. Your friend makes an error that causes the team to lose. You confront your friend angrily. Discuss with your classmates alternative responses to the loss of the game. Replay the situation, using responses that are more healthful.
6. **Personal Health** Examine an aspect of your physical surroundings, such as your bedroom, the place where you study, or the outside area around your home. How does this place affect your mental well-being? What can you do to improve the space? Share your ideas with your classmates.

Lesson 1: What Is Mental Health? 213

Cooperative Learning

Clustering Also known as the "think-pair-share" strategy, clustering is an arrangement in which randomly chosen individuals switch group affiliations in the middle of class brainstorming sessions. Best implemented with small groups, clustering stimulates thought and discussion through the importing of ideas from a fellow group.

Academically, the clustering approach helps students master material and aids content reinforcement. Socially, clustering promotes openness to new points of view. For more information about group strategies, consult the Cooperative Learning booklet in the TCR.

Chapter 8 Lesson 1

Assess

EVALUATING THE LESSON

Assign Reviewing Terms and Facts and Thinking Critically on this page to review the lesson; then assign Lesson 1 Quiz in the TCR to evaluate students' understanding.

LESSON 1 REVIEW

Answers to Reviewing Terms and Facts

1. Mental health is the ability to deal with life's daily stresses and changes. Sentences will vary.
2. Heredity, environment, and behavior shape personality.

Answers to Thinking Critically

3. Responses will vary.
4. Responses will vary.

RETEACHING

▶ Assign Concept Map 34 in the TCR.
▶ Have students complete Reteaching Activity 34 in the TCR.
▶ Have each student give one positive example and one negative example of the role each of the following can play in the development of a person's personality: heredity, environment, and behavior.

ENRICHMENT

▶ Assign and distribute Enrichment Activity 34 in the TCR.
▶ Present the following situation for students' consideration: Pedro wishes he could look more muscular, like other boys in class. He has tried everything but nothing has worked. Ask: What mental health habit does Pedro need to work on? What solutions would you suggest for Pedro? Why?

Close

Ask each student to identify one lifestyle behavior that promotes good mental health.

213

Lesson 2

Building Positive Self-Esteem

This lesson will help you find answers to questions that teens often ask about liking themselves. For example:

- Why do I feel the way I do about myself?
- How does self-concept differ from self-esteem?
- How can I feel better about myself?

Word to Know

self-concept

Write a description of yourself in your journal. How do you think you usually feel, think, look, and act? Mark your calendar to reread what you have written in three weeks. Do you still think it is accurate? Add your answer to your journal.

How Do You See Yourself?

Imagine for a moment that you are about to enter a room to meet some people for the first time. You ready yourself at the doorway, pull open the door, and walk in. Describe the person (you) that these people are about to meet. Are you confident, fun to be with, intelligent, honest, well groomed, happy, healthy, organized, or creative? Are you sincerely interested in other people? Your description paints a word picture of how you view yourself and how you believe others view you. This *view that you have of yourself* is called your **self-concept**.

Your self-concept could be realistic, which means that you have a pretty accurate awareness of the strengths and weaknesses of your personality. Some teens, however, have an unrealistic self-concept. They dismiss their strengths and focus only on their "faults," usually exaggerating them. Think back to the tips for developing good mental health on page 211. Which of the points listed would you associate with having a realistic self-concept?

Focusing only on what you perceive as your "faults" makes them seem greater than they actually are.

214 Chapter 8: Mental and Emotional Health

How Self-Concept Develops

The self-concept that you have today has been in the making for a long time. It was built gradually from your experiences with other people. In general, people who are given support, encouragement, and love tend to develop a positive self-concept. People who are neglected, discouraged, criticized often, or spoken to harshly tend to develop a negative self-concept.

Your self-concept began in your early years, but it keeps developing as you grow. Remarks that your family, friends, and teachers make and ways in which they act toward you can reinforce, or strengthen, the view you have of yourself. **Figure 8.3** illustrates how various experiences contribute to your self-concept.

Thanks for mowing the lawn, Tom. That was a great help.

Figure 8.3
Shaping Your Self-Concept
What kind of reinforcement—positive or negative—does each of these messages represent?

- Your older sister calls you clumsy when you spill the juice.
- You join two friends at lunch, but they ignore you.
- Your teacher frowns during your oral report.
- The coach gives you a thumbs-up sign.
- Your friends wait for you to catch up and walk with them.
- Your teacher smiles encouragement when you give your report.

HEALTH LAB
The Qualities You Admire

Introduction: "I wish I had Rosa's confidence." "Everyone likes Garrett. I wish I was more like him." Have you ever made statements like these? Most people admire certain qualities in others. Do you look for and admire qualities in others that you also have, or do you most admire qualities that you do not have? You can find out by carrying out the following experiment.

Objective: Look for evidence during the next week that tells you if you have any of the six qualities that you most admire in others.

Materials and Method: List eight to ten qualities that you admire in others. Examples could include being a good listener, keeping secrets, and being reliable.

When you complete your list, circle the six qualities that you most admire. Clip three sheets of notebook paper together, and fold them in half. On each half page, write one of the six qualities. During the next week, look for signs that you have, or don't have, each of the qualities. Every time you have any evidence, either way, write it down under the appropriate quality heading.

Observations and Analysis: At the end of the week, ask a classmate to help you analyze your observations. Remember to weigh the evidence fairly. For example, returning a library book a day late does not mean that you are not reliable. However, breaking your word might. Do you have the qualities that you most admire in others?

Lesson 2: Building Positive Self-Esteem **215**

Chapter 8 Lesson 2

Teach

L1 Critical Thinking
Let students share their ideas in response to these questions: Is there one true, realistic self-concept that you *should* have? Why or why not? How do you think you can tell whether your self-concept is realistic?

L3 Researching
How does the care and attention given to an infant affect that individual's self-concept through the years? What research has been conducted? What specific recommendations do psychologists make to parents of infants? Why? Ask interested students to do library research to find answers to these questions and then summarize their findings in written reports.

L1 Identifying
Remind students that nonverbal messages sometimes affect a person's self-concept. For example, a person's turning his or her back on another person might have a negative impact on that person's self-concept. Let students identify and demonstrate similar nonverbal messages, some positive, some negative.

L2 Language Arts
Ask students to think of a person who has had a positive influence on their self-concept. Have each student write a paragraph about that person, describing how he or she has influenced the student's self-concept.

HEALTH LAB
The Qualities You Admire

Time Needed
1 week

Supplies
▶ paper
▶ pen or pencil

Focus on the Health Lab
After students have read the Health Lab feature, review "Shaping Your Self-Concept" on this page. Brainstorm other positive reinforcements that teens might give to family members and friends.

Observation and Analysis
Have students share and discuss their findings in small groups. Later, ask each group to choose one finding to share with the class.

Assessing the Work Students Do
Have students create a brochure or flier "advertising" one or more qualities they admire in themselves.

215

Chapter 8 Lesson 2

Teacher Talk
Dealing with Sensitive Issues

Recognizing their own low self-esteem may make some students feel worse about themselves. Be particularly sensitive to the needs of these students. To make sure they don't feel alone, point out that many people have low self-esteem and that everyone who makes an effort can improve upon it. Tell students that the most important thing they can learn from this lesson is how to improve their self-esteem.

L1 Brainstorming

Divide the class into four or five groups. Give the members of each group five minutes or so to brainstorm a list of experiences that would be likely to raise a teen's self-esteem. Then have the same groups brainstorm lists of experiences likely to have a negative impact on a teen's self-esteem.

L2 Presenting Skits

Instruct groups of students to work together to imagine a typical day in the life of a fictitious teen who has both positive and negative experiences with family members, friends, and teachers. Have group members present short skits showing events from that day, dramatizing some experiences likely to improve the teen's self-esteem and others likely to diminish the teen's self-esteem.

L1 Critical Thinking

Encourage students to explore their responses to these questions: How do the actions and attitudes of parents affect the self-esteem of children? How important are the actions and attitudes of peers to the self-esteem of children? How does the balance between parents and peers shift for adolescents? What effect does this change have on teens and their families?

Teen Issues
The Power of Your Words

 ACTIVITY

As a teen, you know the effect others have on your self-concept. At the same time, remember that you are reinforcing the self-concepts of others. Make an effort to say and do things that help friends and family members improve their self-concepts. If you see people doing something kind, compliment their behavior. If you like a meal at home, say so and thank the person who made it. Keep a list for a week of all the times you have said or done something positive to reinforce someone's self-concept. Remember that your words have power.

Self-Concept and Self-Esteem

Closely tied to your self-concept is your self-esteem. Self-esteem is the confidence and pride you have in yourself. The way you feel about your body, your mind, your emotions, and your interactions with others are all part of your self-esteem. **Figure 8.4** lists the types of behavior that indicate self-esteem.

Figure 8.4
Behaviors That Indicate Self-Esteem
Where do you fall on the self-esteem continuum?

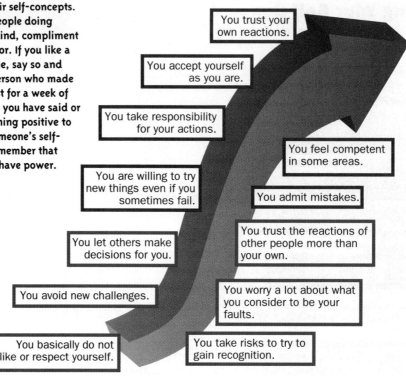

Low self-esteem is often linked to an unrealistic self-concept. For example, people who overlook their strengths and exaggerate their weaknesses in their own minds will probably not feel very good about themselves. In other words, people who do not like or respect themselves have low self-esteem.

Like your self-concept, your self-esteem is formed in part by messages, both positive and negative, that you receive from other people. Your self-esteem is also formed by messages that you send

More About...

Self-esteem The importance of high self-esteem for a healthy, happy life cannot be overemphasized. Students should be made aware of the influence of self-esteem on every aspect of their lives. Tell them that self-esteem affects their success in school, their relationships, the decisions they make, and their self-respect. Then have the class brainstorm examples of ways that self-esteem influences important areas of life. For example, people with low self-esteem may not believe that they can do well in school even if they try, so they do not bother to try; others with low self-esteem may not think they are likable, so they are shy and withdrawn around others.

yourself. The messages you give yourself are sometimes powerful enough to change the meaning of messages from others. Have you ever received a compliment that you were unable to accept? Perhaps the more the person praised you, the more you rejected the praise. If you have had this experience, you have let your own negative messages override the positive ones from someone else. What incident can you recall when your own positive messages lessened the sting of negative messages from another person?

Self-Esteem and Your Health

Probably no single factor has a greater impact on your total health than your self-esteem. People with high self-esteem are more likely to practice good health habits than people with low self-esteem. People with high self-esteem are also more likely to avoid harmful behaviors, such as overeating, abusing alcohol and drugs, and refusing to take personal safety measures.

Because people with high self-esteem like themselves, they take good care of themselves. Their health, safety, and appearance are important. They try to accept and learn from fair criticism, and they are able to ignore mean remarks. High self-esteem lets you accept the negative incidents in your life as exceptions, not the rule. People with high self-esteem generally act responsibly toward themselves and others. People with low self-esteem do just the opposite. They tend to believe that all experiences are negative.

Were you ever feeling down when suddenly something really positive happened that made you feel great for the rest of the day? Everyone's level of self-esteem changes from day to day and sometimes from hour to hour. Although your self-esteem goes up and down, it usually falls within a specific range on the self-esteem continuum. Which of the signs of high and low self-esteem in **Figure 8.4** do you think describe you most of the time?

Improving Your Self-Esteem

When your self-esteem is high, you are usually happy with yourself and get along well with others. You are able to bounce back quickly after a loss, and your total health seems to get a boost. For these reasons, having high self-esteem is a worthy goal. There are several actions you can take to help you reach the goal of high self-esteem. **Figure 8.5** on the next page describes some of these actions. What other items would you add to the list?

in your journal

Your behavior can sometimes trigger just the response you do not want. For example, if you want more freedom at home, acting irresponsibly will convince your parents that you are not ready for more personal freedom. Think of two recent incidents: one that resulted in the desired response and one that did not. Describe both incidents in your journal.

Wearing a helmet when riding a bicycle is an example of a personal safety measure.

Personal Health

Tell students that if they were working in a store, they might be asked to take inventory of the available stock—that is, tally up lists of what goods are on hand. In a similar way, to assess their self-esteem they might periodically make "inventory" lists of the following qualities or assets that they have "in stock":

- personality qualities
- skills
- natural talents
- friendships
- value to their families
- goals
- values
- achievements

Chapter 8 Lesson 2

L1 Identifying Examples

Have students suggest specific comments or greetings teens can make to encourage positive feelings in others. Record their suggestions on the board. Then ask: How does encouraging others affect your own self-esteem? Why?

Teacher Talk

Journal Writing

Have students use their private journals to examine their level of self-esteem and gain insight into ways to adjust it.

▶ I feel good about myself.
▶ I can face problems.
▶ I get involved in activities.
▶ I make friends easily.
▶ I take good care of myself.
▶ I try to do my best.

Visual Learning

Guide students in discussing the music student in Figure 8.5: How does each of her thoughts indicate a high level of self-esteem? Suppose she struck a series of wrong notes. What response on her part might indicate a high level of self-esteem? What might be the response of a music student with low self-esteem?

VIDEODISC/VHS

Teen Health Course 2

You may wish to use video segment 8, "Building Positive Self-Esteem," to show how having a positive mental outlook lays the foundation for esteem-building accomplishments.

Videodisc Side 2, Chapter 8
Building Positive Self-Esteem

Search Chapter 8, Play to 9

Chapter 8 Lesson 2

Assess

EVALUATING THE LESSON
Assign Reviewing Terms and Facts and Thinking Critically on this page to review the lesson; then assign Lesson 2 Quiz in the TCR to evaluate students' understanding.

LESSON 2 REVIEW

Answers to Reviewing Terms and Facts
1. Self-concept is the view that a person has of himself or herself; self-esteem is the confidence and pride a person feels.
2. You can improve self-esteem by (any three): focusing on successes; giving yourself positive messages; setting realistic goals; asking for help when you need it; learning from mistakes.

Answers to Thinking Critically
3. Responses will vary.
4. Even people who generally have high self-esteem have days when they feel sad.

RETEACHING
▶ Assign Concept Map 35 in the TCR.
▶ Have students complete Reteaching Activity 35 in the TCR.
▶ Have students define self-concept and identify three ways of building their own self-concept.

ENRICHMENT
▶ Assign and distribute Enrichment Activity 35 in the TCR.

Close

Have each student name one way that the school environment helps students develop positive self-concept.

Keeping a realistic self-concept and raising your self-esteem takes effort. Although some people always appear confident and happy, everyone has occasional self-doubts and personal concerns. The teen years are usually times of changing self-esteem levels. Temporary periods of unrealistic self-concept and low self-esteem seem to happen frequently, but usually they do not last long. Teens who feel unloved, unimportant, and unworthy most of the time, however, should talk about their feelings with an adult family member, clergy member, school counselor, or coach.

Figure 8.5
Working Toward High Self-Esteem
Here are some suggestions for raising your self-esteem.

A Focus on your successes.
B Send yourself positive messages.
C Set realistic goals for yourself.
D Ask people for help when you need it.
E Learn from your mistakes.

Lesson 2 Review

Using complete sentences, answer the following questions on a separate sheet of paper.

Reviewing Terms and Facts
1. **Vocabulary** What is the difference between *self-concept* and *self-esteem?*
2. **Give Examples** List three ways to improve self-esteem.

Thinking Critically
3. **Explain** How and when does a person's self-concept develop?
4. **Analyze** Why do people with generally high self-esteem have times when their self-esteem is low?

Applying Health Concepts
5. **Growth and Development** Think of a common problem that affects a teen's self-esteem. Draw a cartoon strip that shows what teens with this problem do to feel better about themselves.
6. **Health of Others** Write down the number of times during the day that you reinforced someone else's self-concept. Perhaps you smiled at someone, complimented a friend on a job well done, or invited someone to join an activity. Include any negative reinforcements you may have given.

218   Chapter 8: Mental and Emotional Health

Cooperative Learning

Interviewing Divide the class into pairs. Ask one student in each pair to interview the other for two minutes. Sample interview questions: What do you enjoy doing? When do you feel you are at your best? When the time is up, have students reverse roles and give them another two minutes. Then ask each interviewer to list as many positive attributes as possible about the person he or she interviewed. Ask volunteers to read aloud their lists. After each list is read, ask the person whom the report was about to share how he or she felt during the reading.

Conclude by discussing the following questions: Why are we often uncomfortable when someone says nice things about us? How is self-concept related to what people say about us?

Managing Stress

Lesson 3

This lesson will help you find answers to questions that teens often ask about stress. For example:

▶ How does stress affect me?
▶ What circumstances in my life might be causing me stress?
▶ How can I cope with the stress in my life?

Stress in Your Life

Stress is a familiar term. You probably hear it mentioned almost every day. Most often stress is mentioned in a negative way, as something to avoid. **Stress,** however, is only *your body's response to changes around you.* Those responses can certainly have a negative effect. Feeling nervous and maybe having an upset stomach because you are worried about an upcoming event are definitely not pleasant experiences. *Stress that keeps you from doing the things you need to do or that causes you discomfort* is called negative stress, or **distress.**

Some stress is considered positive. Positive stress helps you to accomplish and reach goals. The stress that makes you feel excited or challenged by an activity is an example of positive stress.

Like stress, changes in your life may also be labeled either positive or negative. Winning a race is usually considered positive, and losing a race most often seems negative. Your body, however, cannot tell the difference. Because your body responds to every change, any change—positive or negative—causes stress. The Personal Inventory feature on page 223 lists some life changes that cause stress. How many of these events have you experienced lately?

Words to Know

stress
distress
stressor
adrenaline
fatigue
physical fatigue
psychological fatigue
defense mechanism

in your journal

Consider the sources of stress in your life. List them in your journal. Write why you think each one is causing you stress. Write down the signs of stress you notice in yourself.

Positive stress can challenge athletes and motivate them to work hard to meet a goal.

219

Chapter 8 Lesson 3

Teach

L1 Comprehending
Let volunteers explain how stress can help you reach your goals. Then ask: Do you think there is a clear-cut distinction between positive stress and negative stress? What are some examples of positive stress? Can you name examples of situations that involve both positive and negative stress?

L2 Journal Writing
Ask students to think of the most stressful day they can remember. In their journals, have them list their activities that day, from the time they awoke to the time they went to bed. Instruct them to label any activity that was crucial (such as being on time for school) with the letter A, any activity that was not crucial (such as watching television) with a C, and any activity that was in-between (such as helping a friend study for a test) with a B. Then instruct them to draw bar graphs or circle graphs showing the amount of time spent on the activities in each category: What do the graphs reveal about stressful activities? about time management?

L3 Art
Assign interested students to research the effects that various colors and images have on stress levels. How have the results of this research been used in designing and decorating offices and other public spaces? Let these students discuss their findings with the rest of the class.

Did You Know?
Amazing Acts

At times of extreme stress, adrenaline helps the body perform amazing acts, such as lifting automobiles to rescue people trapped under them. Normally, such acts of strength are not possible, but the body's response to stress sometimes supplies it with amazing power.

Stress and Stressors

Stress is a natural part of everyday life. The *triggers of stress* are called **stressors.** It isn't just major events that cause stress. Stress is caused by everyday irritations and pleasures.

Your body responds to most stressors by getting ready to act. This response is called the "fight-or-flight" response because your body prepares to fight the stressor or flee from it (see **Figure 8.6**). One part of this response is the release of **adrenaline** (uh·DRE·nuhl·in). This hormone *increases the level of sugar in your blood, which gives your body extra energy.*

Figure 8.6
The Fight-or-Flight Response
When your body responds to a stressor, certain physical reactions occur.

A More blood is directed to your muscles and brain.

B Your heart beats faster.

C Your muscles tighten up and are ready for action.

D Your senses sharpen. You become more alert.

E Your air passages widen so that you can take in more air.

F The level of sugar in your blood increases, which gives you extra energy.

LIFE SKILLS
Time Management

Focus on Life Skills
Brainstorm a list of school and local community activities in which teens participate. Have students read the feature and then discuss the difficulty of balancing schoolwork, family life, and other activities.

Making Health Connections
Time management is one of the most effective ways of reducing stress. Emphasize the importance of keeping priorities in mind when determining how to spend time, and remind students that they should assume responsibility for finding effective solutions to their time problems. Encourage students to set realistic time goals, to identify time wasters, and to make daily and weekly lists of chores and commitments. Also emphasize the importance of making time for exercise and relaxation. Help students discuss how their ability to reduce stress can affect not only their own health but the health of others.

Stress and Fatigue

Once the stressor is gone, your body's response usually stops. However, if the stress is great, or if it lasts long enough, the response may continue. After a time, your body can become exhausted. **Fatigue,** or *extreme tiredness,* then sets in.

There are actually two types of fatigue. **Physical fatigue** is *extreme tiredness of the whole body.* It usually occurs after vigorous activity. Muscles may be overworked and sore, and your body feels tired all over. When this happens, you need rest.

The other type is **psychological** (sy·kuh·LAH·ji·kuhl) **fatigue,** or *extreme tiredness caused by your mental state.* This type is brought on by stress, worry, boredom, or depression. Activity, such as exercise or doing a project, can help relieve this kind of fatigue.

Teen Issues

Reducing Stress ACTIVITY

Which day of the week is most stressful for you? Try to analyze why you feel more stress on that day. Write your thoughts on a sheet of paper. Examine what you have written to determine if there are ways to reduce the stress of that day.

LIFE SKILLS
Time Management

People who manage their time well are better able to control this major source of stress in their lives.

To rate your time management skills, answer the following questions with *yes* or *no.*

- Are you almost always in a hurry?
- Do you leave tasks or chores incomplete?
- Do you feel as if you are working hard but not accomplishing much?
- Do you not have enough time for rest or for personal relationships?
- Are you regularly late with assignments and for appointments?
- Are you overwhelmed by demands?
- Do you often try to do several things at once?
- Do you have trouble deciding what to do next?

If most of your answers are yes, your time management skills need improvement. Try putting the following tips to work in your life.

1. **Set priorities.** Decide which activities are obligations and which are choices. Of the activities that are choices, decide which are most important to you.
2. **Make a schedule** (see **Figure 8.7**). Decide when you will do each activity. Try to be realistic about how much you can do, and leave time for relaxing. If you need to drop some activities, check your priority list.
3. **Learn to say no.** Most important, know your limitations and when to say no.

Follow-up Activity
After three weeks, take the time management quiz again. Have any of your answers changed? What else can you do to manage your time better?

Figure 8.7
Managing Your Time

Sunday	Monday	Tuesday	Wednesday	Thursday	Friday	Saturday
29	30	31	1	2	3	4
Picnic with friends	Photography club after school	Baby-sitting for neighbor from 6:00 to 8:00 p.m.	Marching band tryouts	Trumpet lesson after school	Movie night with Aunt Celia	Grandma's 80th birthday party

Lesson 3: Managing Stress

Chapter 8 Lesson 3

L2 Debating
Have students generate a list of common stressors (car horn, oversleeping, winning a contest), and record their suggestions on the board. Divide the class into two teams. Let the teams debate whether each of the listed items is a positive or a negative stressor. Encourage both teams to present persuasive arguments in support of their positions.

Visual Learning
Ask volunteers to identify the stressor in Figure 8.6 and to describe the teen's reaction. Have students read each lettered explanation of the fight-or-flight response and discuss how it will help the teen deal with his stressor. Then ask: Suppose that, instead of a snake on the trail, the stressor was a request from a teacher to stand up and present an oral report that the teen had not prepared: Would his body respond in the same ways?

L1 Discussing
Guide the class in a discussion of the relationship between physical fatigue and psychological fatigue: In what ways do they feel the same? different? How do you think one might lead to the other? Why does one require rest and the other activity?

Meeting Student Diversity
Learning to set priorities might be difficult for a student who has not identified some of his or her long-term goals (such as being with family, making friends, doing well in school, maintaining or improving health, getting a part-time job, and so on). Suggest that such a student make a list of personal, long-term goals, and beside each one write one or more short-term obligations and/or choices that relate to the long-term goal.

Assessing the Work Students Do
Ask students to make personal lists of obligations and choices for three days. After three days, have them review their lists to see how many activities they completed. Encourage them to think of ways they can better manage their time so that they can do the activities on their list. Assign credit based on the completeness of their work.

Chapter 8 Lesson 3

L1 Comparing

Ask students to compare fight-or-flight responses with defense mechanisms: How are they similar? How different?

L1 Discussing

Allow volunteers to read aloud the explanations of specific defense mechanisms, and guide students in suggesting other specific examples of each mechanism at work. Ask: How can this mechanism be a helpful short-term solution to a problem? What is likely to happen if a person continues to use this mechanism rather than face the problem?

L1 Science

Have students discuss the types of protection possessed by various animals. Examples include a porcupine's quills and a chameleon's ability to change colors. What are the similarities between those forms of defense and human defense mechanisms? What are the differences?

Visual Learning

Ask students to describe and discuss each scene in Figure 8.8: What defense mechanism is the teen in each scene using? With what effect? How do you think the father in the first scene should respond? Why? What do you think the friend in the second scene should say? Why?

L2 Researching

Suggest that interested students learn more about repression or another of the listed defense mechanisms: What are some of the symptoms indicating that the use of this mechanism has become unhealthy? How does unconscious reliance on the mechanism affect a person's mental health? What are appropriate treatments for that health problem? Let these students discuss their findings with the rest of the class.

Defense Mechanisms

The fight-or-flight response is your body's reaction to stress. Your mind also reacts to stress. These reactions, called **defense mechanisms** (di·FENS MEK·uh·nizms), are *short-term ways of dealing with stress*. Defense mechanisms allow you to set aside a certain amount of stress until you are better able to face the problem and deal with it (see **Figure 8.8**). When used as a temporary solution to problems, defense mechanisms can actually help people. However, if you use them as a permanent substitute for facing the problem, your mental health could be harmed. There are several types of defense mechanisms. Which of the following defense mechanisms are familiar to you?

- **Denial** is a refusal to accept reality. Ted's parents are getting a divorce. Ted refuses to accept that his dad is moving out and acts as if everything is fine with the family.

- **Rationalization** is justifying behavior, ideas, or feelings to avoid guilt or to obtain approval or acceptance. Marlene says she did not finish her homework because she was busy helping her grandmother with her food shopping.

- **Repression** is blocking out unpleasant thoughts. Kahlil has a lot of homework to do tonight, but he does not even think of it when he agrees to go out with his friends.

- **Displacement** is having bad feelings toward someone not really related to the cause of the problem. Wynona had an argument with her friend that left her hurt and angry. Later, at home, Wynona yelled at her sister for no reason.

- **Projection** is blaming someone else for your problem. Will got up late, spilled his juice on his shirt, could not find his homework, and finally missed his bus. While these were his problems, he blamed his mother for everything.

Did You Know?
Automatic Relief

Defense mechanisms can kick in automatically to bring you relief from stress. In other words, you may be using defense mechanisms without being aware of it.

Figure 8.8
Examples of Defense Mechanisms
Identify the defense mechanism or mechanisms that may be at work in each example.

Pete's more popular than I am because you give him everything.

You have to tell Lisa that you were hurt by what she did.

No, no. It's not a problem.

Cooperative Learning

Skits Divide the class into five groups. Assign each group a defense mechanism discussed in this lesson and give them 10 minutes to plan a skit demonstrating it. Have groups present their skits to the class. Ask other class members to identify which mechanism was enacted.

Follow up by asking each participant how he or she felt while acting out the part. Guide students in discussing the positive and negative aspects of the use of defense mechanisms: At what point does the use of a defense mechanism indicate that a person is having more problems than he or she may admit?

Personal Inventory
LIFE CHANGES AND STRESS

Stress and the effect it will have on people is difficult to measure. What causes one person a great deal of stress may hardly affect another person at all. The following chart gives values in "stress points" to certain life changes. Accumulating between 150 and 299 stress points in one year increases a person's chance of getting sick. Whether sickness will actually occur depends on the person. It is not the amount of stress that is important; it is how you respond to it.

Rank	Event	Stress points
1.	Death of a parent	98
2.	Death of a sister or brother	95
3.	Death of a friend	92
4.	Divorce or separation of parents	86
5.	Failure in one or more school subjects	86
6.	Getting arrested	85
7.	Repeating a grade in school	84
8.	Family member's alcohol or drug problem	79
9.	Starting to use alcohol or drugs	77
10.	Loss or death of a pet	77
11.	Family member's serious illness	77
12.	Making choices about sexual relationships	75
13.	Losing money you've saved	74
14.	Breaking up with girlfriend or boyfriend	74
15.	Quitting or being suspended from school	73
16.	Pregnancy of a close friend	69
17.	Father or mother losing a job	69
18.	Being seriously sick or hurt	64
19.	Arguing with parents	64
20.	School troubles with teacher or principal	63
21.	Discomfort and concern about weight, height, acne	63
22.	Going to a new school	57
23.	Moving to a new home	51
24.	Change in physical appearance due to braces, glasses	47
25.	Arguing with sister or brother	46
26.	Beginning to menstruate (girls)	45
27.	Making a decision about smoking	45
28.	Having someone, such as a grandparent, move in	35
29.	Mother's pregnancy	31
30.	Beginning to go out on dates	31
31.	Making new friends	27
32.	Marriage of a sister or brother	26

Lesson 3: Managing Stress 223

Cooperative Learning

Activity After dividing students into groups of four, ask them to review the Personal Inventory on this page. Ask them to work as a group to assign stress points to the following situations on a scale of 1 to 100, with 100 indicating the most stressful situation: being late to class; getting a poor test grade; getting ready to go to a party; arguing with your best friend; arguing with your parents; being compared with your brother or sister; not being allowed to stay out late; getting a big "zit" on your nose. Have groups share their results and discuss their findings.

Chapter 8 Lesson 3

L1 Journal Writing
Provide quiet class time in which students can read the Personal Inventory and tally their own total stress points. (Or, ask students to complete this activity at home.) Then suggest that students record their responses to the stress inventory in their journals. Remind students that they can and should discuss with a trusted adult any concerns about the levels of stress in their lives.

L1 Comprehending
Help students identify and discuss the effects of stress on all three aspects of their health: physical, mental/emotional, and social.

Visual Learning
Guide students in discussing the photograph on page 224: What are the two teens doing? Why do you think they are doing it? What choices about timing and work habits have they probably made? What steps do you think they have taken to reduce stress associated with what they are working on? Have they completely avoided stress? If so, how? If not, why not?

L1 Discussing
Ask students to skim the items in the Personal Inventory on page 223. Which inventory items involve confronting the unknown? Which involve starting over? Which involve feeling that you have to prove yourself? Why do you think such events are sources of stress, especially for teens?

L1 Identifying Examples
Ask: What are some specific stressors that teens can completely avoid by changing their behavior? Record their suggestions on the board.

Chapter 8 Lesson 3

Visual Learning
Let students describe the character in Figure 8.9: How accurately do you think that character represents the way people feel when they are dealing with too much stress? Have volunteers read aloud the tips in the boxes of Figure 8.9: Why is each important? What specific action can you take to put each tip into effect?

L2 Creating Collages
Have the class work in groups to create collages illustrating stress management for teens. Suggest that students include magazine pictures, words, phrases, and original drawings in their collages. Display the completed collages in a section of the school where other students will be able to view them.

Would You Believe?
Talking to a pet or even just watching fish in an aquarium can lower your blood pressure.

Assess

EVALUATING THE LESSON
Assign Reviewing Terms and Facts and Thinking Critically on page 225 to review the lesson; then assign Lesson 3 Quiz in the TCR to evaluate students' understanding.

RETEACHING
► Assign Concept Map 36 in the TCR.
► Have students complete Reteaching Activity 36 in the TCR.
► Have students write paragraphs using the Words to Know on page 219. The paragraphs should define the words and explain how they relate to each other and to the concepts covered in this lesson.

224

Planning for Health
Serious and long-lasting stress can cause illness. By being ready for changes and planning for them whenever possible, you can reduce both the stress and your chances of becoming sick.

Coping with Stress

Stress can have a strong effect on your total health. Stress can alter your emotions and behavior. You may be so worried about a relative's health that you don't eat or sleep well. Being tired may cause you to snap angrily at a friend or family member.

No matter how you feel pushed and pulled by stress, it's important to keep your head. Defense mechanisms are a temporary way to deal with the effects of stress. They are not, however, a long-term solution to the problem. Coping with stress involves avoiding some stressors entirely by changing your behavior. For those stressors that you cannot avoid, you can use management techniques to keep yourself from suffering negative effects.

Giving yourself plenty of time to complete a project will help you avoid stress.

Avoiding Stress
Fortunately, you can get a head start on coping with stress by keeping certain types of stress out of your life. Think about situations that make you uncomfortable or anxious. Are there ways for you to change your behavior to avoid these sources of stress? For example, you are likely to experience distress if you must stay up late completing an assignment. By using good time management and study skills, however, you could avoid this distress by starting the project in plenty of time to finish it comfortably. Common sense will also help you to avoid certain types of stress. If you know, for example, that there will be alcohol or drugs at a party, you'll save yourself from having to deal with negative peer pressure by making other plans.

High self-esteem and a positive attitude will also help you cope with stress. If you are confident in your abilities, you won't be as likely to feel overwhelmed when things go wrong. Your aim should not be to get rid of stress entirely. That just isn't possible. Changes will always occur in your life, and learning how to adapt effectively to them is an important skill to learn.

224   Chapter 8: Mental and Emotional Health

TECHNOLOGY UPDATE

Biofeedback Biofeedback trains patients to relax certain muscles and even to slow their heart rates. A patient is connected to a computer and polygraph using external sensors placed on the body. The sensors measure specific physiological responses such as muscle tension, perspiration, and heart rate. The responses are then translated into signals—tones, light, or meters—that the patient can see or hear. To alter the signals, the patient must learn to relax. Combined with other stress management techniques, such as progressive relaxation, deep breathing, and visualization, biofeedback has proven effective as a method to control pain and manage chronic illness, sometimes eliminating the need for drugs.

Managing Stress Effectively

Stress is a factor in everyone's life. The key to managing it is to identify sources of stress and learn how to handle them in ways that promote good mental and emotional health. **Figure 8.9** lists some tips for handling stress.

Figure 8.9
Tips for Effective Stress Management
Which of these tips do you use?

Stay Healthy During periods of stress, be sure to eat right and get enough sleep. Stress management takes energy.

Redirect Your body reacts to all stress by producing adrenaline, which raises your energy level. Rechannel all that extra energy into something worthwhile.

Relax Try imagining yourself in a quiet, peaceful place, such as under a tree by a lake. As you relax, try to empty your mind of troubling thoughts.

Talk Just talking things out with another person can relieve stress. People who aren't directly involved can often see solutions to your problems that you cannot.

Laugh Spend time with people who enjoy a good laugh. See a funny movie after an especially stressful day. Laughter relieves stress.

Review — Lesson 3

Using complete sentences, answer the following questions on a separate sheet of paper.

Reviewing Terms and Facts

1. **Vocabulary** What is the difference between *stress* and *distress?*
2. **Define** What is a *stressor?* Give three examples.
3. **Recall** What are the two types of *fatigue?*

Thinking Critically

4. **Hypothesize** How could defense mechanisms harm your health? Why do you think people use them?
5. **Analyze** Your friend tells you that she is feeling very tired. She says that she is so worried about some problems at home that she can't think of anything else. What kind of stress is she feeling? What can she do right away to relieve her stress?

Applying Health Concepts

6. **Health of Others** Interview five people about how they handle stress. You can interview family members, teachers, classmates, and friends. Ask each person to suggest at least one tip for coping with stress. Share the tips with your classmates.

Lesson 3: Managing Stress

Home and Community Connection

Trusted adults are always the recommended resource for teens who want to discuss the effects of stress on their lives—or who need help with other problems. Many teens may not feel comfortable turning to their parents to discuss issues they consider personal. For this reason, it is important to help students identify other people who may meet the definition of "trusted adult." Remind them that parents of friends, adult neighbors, and other family members can be helpful. Also, discuss the fact that, the adult does not have to be the same age as their parents; a young adult friend or a grandparent can also be a good source of help.

Focus

The Teen Health Digest articles can be used in two ways: as an individual activity for reflection and enrichment or as a cooperative learning activity as described below.

Motivator

Remind students of the health triangle. Ask: What is the relationship between physical health and mental/emotional health? Then point out that the articles on these pages address aspects of mental health and the effects a healthy mind can have on an individual's physical well-being. Ask volunteers to read the Teen Health Digest articles aloud as other students follow along in their texts.

Cooperative Learning Project

Mind-Body Newsletter

Setting Goals

Ask students to imagine that they are the publishers of a teen newsletter on the mind-body connection. What information would you want to include? What opinions would you want to express? What regular features would your publication contain? Tell students that they will use ideas they have generated as they publish one issue of their newsletter, the *Mind-Body Connection*.

Delegating Tasks

Divide a sheet of paper into four columns labeled *Research, Interviewing, Writing,* and *Art*. Pass the sheet around the class, and ask students to sign up for the task or tasks they want to perform.

TEEN HEALTH DIGEST

Personal Trainer

Take It Easy

Simple breathing exercises can help you deal with stress—right where you are, in ten minutes or less. Here's how.

1. **Choose a quiet place.** Sit in a chair with your feet flat on the floor.
2. **Close your eyes.** Feel how the chair supports you. Let your shoulders relax.
3. **Breathe slowly and deeply.** Count as you breathe, taking twice as long to breathe out as to breathe in. Continue for about five minutes. Stop, however, if you begin to feel dizzy.
4. **As you breathe, clear your mind of all thoughts.** Another method is to think of the most peaceful scene you can imagine.

Do these exercises whenever you feel stressed—or plan to do them twice every day. You'll be surprised how much better you feel.

226 Chapter 8: Mental and Emotional Health

People at Work

Music Therapist

Interview with Sal Marino, a music therapist.

Q. What made you choose this job?

A. Music has always been part of my life. When I started college, I was torn between studying music and psychology. Then a teacher told me how to combine them into a very satisfying career.

Q. What makes your career such a rewarding one?

A. I help people to understand that playing music can bring relief from mental and emotional problems. For example, I have several patients who have trouble expressing anger in a healthful way. I often teach people with this problem to play the drums. How good the music sounds isn't really important. How good the patient feels is.

Try This:

Try expressing your own emotions through your voice or a musical instrument. Describe the experience in a paragraph.

TECHNOLOGY OPTIONS

Computer The art group can use a computer to design and lay out the pages of the newsletter. If desktop publishing software is not available, students can use a word processing program. The finished letter may be printed on a school computer printer or photocopied from a single printed master.

On-Line Computer Members of the research group can send letters via E-mail to researchers, requesting specific advice that might help teens find hope at critical times. In their queries, students should explain that they are creating a newsletter and should ask for permission to quote the researchers.

Sports and Recreation
Very Special Athletes

In 1963, Eunice Kennedy Shriver started a summer camp for children and adults with mental retardation. She especially encouraged the campers to explore sports and physical activities. Five years later, Shriver's summer camp became the First International Special Olympic Games. One thousand athletes with mental retardation came to Chicago from 26 states and Canada to compete in track and field, floor hockey, and swimming.

Over time the Special Olympics grew, drawing coverage from television networks and gaining support from many different people. In 1995 over 7,000 athletes from 143 countries competed in 21 sports at the Ninth Special Olympics World Games. What had started as a small idea had become a hugely successful and very inspiring international event.

CON$UMER FOCU$
Self-Image Bait

Advertisements aimed at teens imply that if they buy certain products, they will feel good about themselves and be popular.

By encouraging self-doubt, the messages in these ads create a problem and then offer a solution. Many teens have a shaky self-image to start with. Ads like these take advantage of this lack of confidence to undermine the consumer's self-image even more. The ad hints that your looks are in need of improvement and that you do not have enough friends. Don't let ads hurt your self-image or self-confidence.

Teens Making a Difference
Choosing Life

You might say that Tonya Smith's eighth-grade year was both her worst and her best ever. That year several of her friends began talking about suicide. One attempted to end his life. Shaken by the experience, Tonya turned to a program called SAIL—Self Acceptance Is Life.

SAIL helps teens like Tonya get help for troubled friends. SAIL emphasizes that teens cannot become responsible for their friends' lives. SAIL members encourage teens to guide their friends to trusted school or religious counselors, teachers, or parents. That's exactly what has happened with Tonya's friends. Today Tonya speaks at schools, churches, and other community groups so that others can benefit from what she has learned. Tonya emphasizes this message: "With someone to talk to, you can get through troubled times."

Try This: *Bring several ads of this type to class and discuss how their messages could be handled by teens.*

Teen Health Digest 227

Chapter 8
The Work Students Do

Following are suggestions for implementing the project.

1. Members of the research group can contact the school guidance counselor or a community mental health clinic for information about teen mentor programs. Information about the programs can be published in a "News Briefs" section of the newsletter.
2. As a follow-up to an article on the therapeutic nature of music, interviewers can ask friends and family members to respond to the saying "Music hath charms to soothe the savage beast." (respond to the quotation "Music hath charms to soothe a savage breast, to soften rocks or bend a knotted oak.")
3. Writers can create a feature called "Teen Self-Image Watch." They can cite advertisements and other come-ons, like those addressed in the Consumer Focus article, that exploit the shaky self-image of teens.
4. Writers may want to include at least one editorial in the newsletter. Encourage students from all groups to suggest appropriate topics and outline the opinions they would like to see expressed.

Assessing the Work Students Do

Have student groups hold regular staff meetings in which they record the progress they have made and address any problems. Ask the groups to turn in these written reports; include them as factors in the overall assessment of each group's contribution to the project.

Meeting Student Diversity

Language Diversity Encourage interviewers to arrange meetings with community members who are not native speakers of English, as well as with those who are. Suggest that students fluent in the non-English languages common to your community accompany the interviewers and act as translators or assistants, as appropriate.

Physically Challenged Suggest that the newsletter include a "Perspectives" article addressing the concerns of the teen years from the vantage point of those with physical challenges. Encourage interested students, both disabled and non-disabled, to work together on this article.

Lesson 4 Mental Disorders

Lesson 4: Mental Disorders

This lesson will help you find answers to questions that teens often ask about mental health problems and mental disorders. For example:

- What is the best way to deal with my problems?
- How can I tell if my mental problems are serious?
- What causes mental illness?
- What should I do if I suspect that a friend is thinking of suicide?

Words to Know

anxiety disorder
mood disorder
depression
schizophrenia
suicide

Facing Problems

Life is filled with events that create problems and cause stress. Some problems can be solved easily. A misplaced notebook might cause stress for only a few minutes. Other problems, such as a death in the family, have a much more lasting effect on your life.

Regardless of the problem, the healthy way to deal with it is first to face up to it. If you are having trouble dealing with a problem on your own, ask for help. **Figure 8.10** offers information on how to handle problems.

Figure 8.10
How to Handle a Problem
Some problems require professional help.

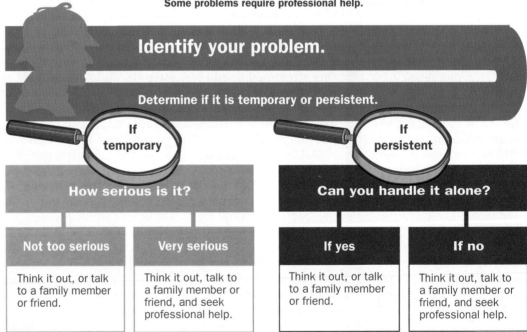

Focus

LESSON OBJECTIVES

After studying this lesson, students should be able to

- discuss the importance of facing and dealing with problems.
- identify categories of mental disorders and discuss their treatment.
- list the warning signs of suicide.
- explain the importance of seeking help for themselves or for friends who are considering suicide.

MOTIVATOR

Display photographs or realistic drawings of individuals from various backgrounds and of various ages. Have students write short responses to these questions: Which of these people has a mental disorder? How can you tell?

INTRODUCING THE LESSON

Let volunteers share their responses to the Motivator questions. Guide students in recognizing that anyone can suffer from a mental disorder. Encourage students to share what they know about mental disorders, focusing on available help and treatment if possible. Explain that Lesson 4 presents more information about mental health problems and mental disorders.

INTRODUCING WORDS TO KNOW

Have students work with partners to read the Words to Know and to find the definitions presented in the lesson. Then ask the partners to create their own hidden word puzzle, using all five terms.

Lesson 4 Resources

Teacher's Classroom Resources
- Concept Map 37
- Enrichment Activity 37
- Lesson Plan 4
- Lesson 4 Quiz
- Reteaching Activity 37

Student Activities Workbook
- Study Guide 8
- Applying Health Skills 37

Understanding Mental Disorders

Feeling anxious, fearful, or sad from time to time is natural. When these feelings continue for a long time or make a person feel out of control or unable to deal with life, they could signal a mental disorder. The causes of mental disorders cannot always be identified. Sometimes a mental disorder has a physical cause, such as inherited genetic traits, injury to the brain, or the effects of certain drugs. Emotional causes are harder to pinpoint. A person may, for example, develop a mental disorder in an attempt to avoid a repeated stressor, or in response to a negative experience that is overwhelming.

Types of Disorders

Mental health experts recognize several different groups of mental disorders. They include anxiety disorders, mood disorders, and schizophrenia.

Everyone has, at some time, felt fear or been anxious about something. In some people, however, this nervousness or fear takes the form of an **anxiety disorder**, *a disorder in which real or imagined fears keep a person from functioning normally.* Examples of anxiety disorders follow.

- Phobias are inappropriate or exaggerated fears of something specific. There are many kinds of phobias, including acrophobia (ak·ruh·FOH·bee·uh), or the fear of high places, and social phobia, the fear of group situations.

- Obsessive-compulsive disorder is a condition in which a person cannot keep certain thoughts or images out of his or her mind. The person may then develop repetitive behaviors as a way of relieving the anxiety.

- Various stress disorders may affect people who have been through overwhelming experiences, such as a violent attack, or a conflict, such as a war.

Another category of disorders is made up of mood disorders. A **mood disorder** is *a disorder in which a person undergoes changes in mood that seem inappropriate or extreme.* This is different from the normal mood swings experienced by teens or the emotions that all people go through in response to stressful life events. The mood changes of a person who has a mood disorder are not necessarily related to any event. **Depression** is a *mood disorder involving feelings of hopelessness, helplessness, worthlessness, guilt, and extreme sadness that continue for periods of weeks.* It's common to feel some of these emotions at times, but depression is a very serious condition that may leave a person completely unwilling or unable to function. Severely depressed people may even think about ending their own lives. Another mood disorder is bipolar disorder, in which a person has extreme mood swings for no apparent reason. The person usually veers between periods of hyperactivity and depression, often taking dangerous risks during the hyperactive periods.

Science Connection

The Winter Blues

If you've ever felt sluggish, tired, or irritable on gray winter days, you're not alone. Scientists think that millions of Americans suffer from seasonal affective disorder (SAD). Some researchers theorize that during the short, dark days of winter, people don't get enough sunlight. This lack can affect a hormone called melatonin, which is related to sleep patterns and overall mood. If you feel down in winter, try to get outside for walks and exercise, and sit near windows with the curtains open. Special lamps are available for people who are strongly affected by seasonal affective disorder.

Lesson 4: Mental Disorders 229

More About...

Phobias A phobia is a fear disorder. Severe and repeated fears triggered by specific objects or experiences are very common. Among recognized phobias are the following:

- ailurophobia, a fear of cats
- agoraphobia, a fear of open places or of leaving the house
- arachnophobia, a fear of spiders
- cynophobia, a fear of dogs
- mysophobia, a fear of dirt or germs
- nyctophobia, a fear of night
- ophidiophobia, a fear of snakes
- xenophobia, a fear of strangers

Chapter 8 Lesson 4

Teach

Visual Learning
Guide the class in describing and discussing the chart in Figure 8.10. Let volunteers offer synonyms or informal definitions for *temporary* and *persistent*. Have all the students use their fingers or pencil erasers to trace the steps for handling temporary and persistent problems. Ask students to suggest specific problems that might fit into each category: temporary and not too serious, temporary and very serious, persistent and okay to handle alone, persistent and not okay to handle alone.

L1 Analyzing
Encourage students to share their responses to these questions: Why does mental illness carry such a stigma? What could be done to help people understand it better? How can you and other teens help overcome that stigma?

Would You Believe?
Experts estimate that more than 20 percent of Americans under the age of 18 have some form of mental disorder.

Health Update
Sleepwalking, sleep talking, nightmares, night terrors, and other abnormal events that occur during sleep are called parasomnias. Sleepwalking (somnambulism) is a fairly common disorder among children. Although an estimated 15 percent of all children between the ages of 5 and 12 have walked in their sleep, most of them outgrow this sleep disorder. Sleepwalking in adults, however, is more serious. Frequently, it is attributable to alcohol or drug abuse, stress, or sleep deprivation.

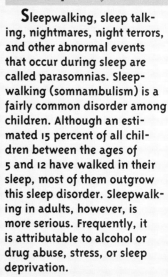

Chapter 8 Lesson 4

L2 Creating Posters

Divide the class into small groups. Have groups list general techniques for overcoming momentary bouts of sadness. Then let group members work together to plan and make a large, colorful poster encouraging other teens to use one of those techniques.

L1 Discussing

Let students suggest specific examples of the problems teens face. What is the best way to cope with the stress brought on by such problems?

L2 Guest Speaker

Invite a speaker from a local crisis center or community mental health center to speak with students about suicide, particularly among teens.

L2 Writing Lists

Have students work in small groups to list problems that teens might give as reasons for attempting suicide. Then let group members select one of the problems and discuss healthy ways to cope with it. Let each group share its ideas with the rest of the class.

Assess

EVALUATING THE LESSON

Assign Reviewing Terms and Facts and Thinking Critically on page 231 to review the lesson; then assign Lesson 4 Quiz in the TCR to evaluate students' understanding.

RETEACHING

▶ Assign Concept Map 37 in the TCR.
▶ Have students complete Reteaching Activity 37 in the TCR.
▶ Have students use each of the following terms in an original sentence that explains the meaning of the term: *anxiety disorder, mood disorder, depression, schizophrenia, suicide.*

Q & A

Eating Disorders

Q: My friend hates food. She practically never eats. Even when she has food, she seems to just play with it. She is getting thinner and thinner, yet she says she needs to lose weight. What's going on with her? Does she have a mental disorder or an eating disorder?

A: Both. Eating disorders such as anorexia nervosa, in which a person refuses to eat, have long been regarded as emotionally based mental disorders. Eating disorders are dangerous and lead to serious physical illness and even death. Get some help for your friend now. Tell an adult why you are worried about her.

Another category of mental disorder includes **schizophrenia** (skit·zoh·FREE·nee·uh), which is *a serious disorder in which a person's perceptions lose their connection to reality.* A person with schizophrenia may hear voices or see images that are not really there. He or she may suffer from false beliefs—for example, that he or she has unusual powers or is on a special mission. Schizophrenia may be so disruptive that it is often considered the most serious mental disorder.

Treating Mental Disorders

Mental disorders are often very difficult to treat and cure. They are linked to a complex network of causes and effects throughout a person's life. Many mental disorders are treated with a combination of strong medicines and long-term therapy under the care of a psychiatrist. If you believe that you or someone you know may be suffering from a mental disorder, seek professional help as soon as possible.

Seriously Troubled Teens

The teen years are full of changes—in the body, in feelings, in relationships, and in responsibilities. These changes cause stress, which sometimes becomes more than a person can cope with alone. If that happens to you, ask for help. There is nothing wrong with needing help with mental problems. Most people need help sometime. People who do not get help suffer more than they need to.

Suicide

Each year thousands of teenagers attempt suicide. **Suicide,** *intentionally killing oneself,* has been the second-leading cause of death for people between the ages of 15 and 19 since 1986. Suicide is a serious matter, and threats of suicide should never be ignored. The best thing to do is to get help immediately.

Sometimes there are warnings that a person is thinking about suicide. If you notice any of the following signs in someone, try to get the person to talk to someone who can help. If he or she will not, tell an adult why you are worried about that person. It is important to get help before it is too late.

If a friend has problems that are overwhelming her, urge her to talk to a concerned adult or professional counselor.

230 Chapter 8: Mental and Emotional Health

Meeting Student Diversity

Culture Shock Students who come to a new country or even to a new area of their native country may experience culture shock. They may be confused by a strange language, by the food, and by the way people act. Culture shock may cause lack of appetite, inability to sleep, and stress. Fortunately, it is usually a temporary problem, unlike the more serious problems mentioned in this lesson. Humans are very flexible, and if they make the effort, they can learn to understand and adapt to new cultures and customs. Humans can—and do—learn to function in many different cultural settings.

The Warning Signs of Suicide

- Statements such as "They'll be sorry when I'm gone" or "I wish I could sleep forever"
- Avoiding activities involving friends or family
- Low level of energy
- Taking greater risks than usual
- Loss of interest in hobbies, sports, job, or school
- Giving away prized personal possessions
- A past history of suicide attempts—80 percent of suicides have attempted suicide before

What the Numbers Mean

Every year, more than 5,000 teens and young adults commit suicide, and many more make an attempt. Most teens who talk about suicide or attempt it are really pleading for help. They do not really want to die—they just want their troubles to go away. When anyone attempts suicide, he or she is choosing a permanent solution to a temporary problem.

Although most teens who attempt suicide do not really want to die, many do want to end their lives. That is why it is important to seek help for yourself or a friend when problems seem overwhelming. Remember that you are never alone. With the help of a concerned adult or a professional counselor, you can find solutions to your problems. You can prevent suicide.

in your journal

Imagine that you write an advice column for teens. You receive a letter from a teen who says he is beginning to think that suicide is the only solution to the stress he feels. In your journal, write an answer to that person.

Review Lesson 4

Using complete sentences, answer the following questions on a separate sheet of paper.

Reviewing Terms and Facts

1. **Vocabulary** What are some major characteristics of *anxiety disorders?* How are they different from *mood disorders?*
2. **Explain** What is *depression?*
3. **Define** Explain the meaning of the term *schizophrenia*.

Thinking Critically

4. **Explain** Why do you think some people avoid seeking help for a mental or emotional problem?
5. **Compare** Explain how the behavior of someone with normal anxiety is similar to and different from that of someone who has an anxiety disorder.

Applying Health Concepts

6. **Health of Others** Imagine that you have a friend who is having mental/emotional problems, but who does not want to seek help. Write a note explaining to that person why he or she should seek help.

Lesson 4: Mental Disorders

TECHNOLOGY UPDATE

A Visit to the Mayo Clinic The current CD-ROM version of the *Mayo Clinic Family Health Book* offers consumers access to health information with the speed and ease of computer technology. Recently, new avenues for the Mayo Clinic's multimedia products include CD-ROMs ranging from consumer publications to on-line services and continuing education programs for physicians. What doctors once learned only by attending conferences and reading through medical publications will be available at the touch of the keyboard. With the reality of these capabilities, talking on-line to Mayo's doctors cannot be far away.

Chapter 8 Lesson 4

LESSON 4 REVIEW

Answers to Reviewing Terms and Facts

1. Anxiety disorders are characterized by real or imagined fears that keep a person from functioning normally. Mood disorders involve extremes of mood, such as hyperactivity or sadness.
2. Depression is a mood disorder involving feelings of extreme sadness or hopelessness that continue for periods of weeks.
3. Schizophrenia is a serious disorder in which a person's perceptions lose their connection to reality.

Answers to Thinking Critically

4. Responses will vary. Students may suggest that people are afraid of what others will think or do not know where to go for help.
5. Both people may behave in ways that show their anxiety, or fear. A person with an anxiety disorder, however, has anxieties that prevent him or her from functioning normally.

ENRICHMENT

▶ Assign and distribute Enrichment Activity 37 in the TCR.

▶ Have students imagine that they are hosting a talk show on a local radio station. A caller says that he is having trouble sleeping, is eating much more than usual, and suddenly dislikes being with people. What conclusion would you draw about the caller? What would you say to him?

Close

Use the lesson objectives presented on page 228 of the Teacher's Wraparound Edition to help students review the main concepts of this lesson.

Lesson 5
Sources of Help

Focus

LESSON OBJECTIVES

After studying this lesson, students should be able to
- name some signs that point to a need for help with emotional problems.
- list people to whom teens with emotional problems can turn for help.
- discuss how to help someone who talks about suicide.

MOTIVATOR

Ask students to identify the person or people they could go to if they had difficulty with the following: finding a specific room or building on the school campus; locating a medical specialist; learning the pronunciation of a foreign word.

INTRODUCING THE LESSON

Have students share their responses to the Motivator activity. Then ask: What do all the people you have named have in common? If necessary, help students recognize that they are sources of help. Tell students that when people need help with their mental and emotional health, they can turn to trained professionals. In this lesson, they will learn more about how a person goes about finding such help.

INTRODUCING WORDS TO KNOW

Write the Words to Know on the board. Ask the class to identify the words that make up the compound word in both terms. Remind students that sometimes the meaning of a compound word can be understood by looking at the shorter terms that make up the word. Is that true of these words? Have students verify their answers by looking the terms up in the Glossary.

Lesson 5
Sources of Help

This lesson will help you find answers to questions that teens often ask about where to get help for mental health problems. For example:

▶ How do I know when the situation is serious enough to ask for help?
▶ Who can really help?
▶ What responsibility do I have for friends in trouble?

Words to Know

support system
teen hot line

in your journal

Is it easy for you to ask for help with problems? Why or why not? Write your answer in your private journal.

Knowing When to Go for Help

Everyone needs help in solving problems at one time or another. Being able to ask for help is a sign that you are growing up. It shows that you are capable of deciding which problems or parts of a problem you need help with and which you can solve yourself.

Figure 8.11
Warning Signs
The following signs may signal a serious problem.

■ Suspecting that everyone is against you	■ Aches and pains that seem to have no medical cause
■ Continually feeling sad	■ Feelings of hopelessness
■ Sudden or extreme changes in mood	■ Trouble sleeping or frequent nightmares
■ Trouble concentrating or making decisions	■ Taking extreme or unusual risks
■ Not taking care of yourself	■ Loss of appetite

Teens who are mature are not afraid to ask for help in solving problems.

232 Chapter 8: Mental and Emotional Health

Lesson 5 Resources

Teacher's Classroom Resources
- Concept Map 38
- Enrichment Activity 38
- Lesson Plan 5
- Lesson 5 Quiz
- Reteaching Activity 38

Student Activities Workbook
- Study Guide 8
- Applying Health Skills 38
- Health Inventory 8

Chapter 8 Review

Your Action Plan

Most teens experience periods of stress and low self-esteem. Perhaps there are aspects of your mental and emotional health that you would like to improve. An action plan can help you do this.

Step 1 Review your private journal entries for this chapter to decide on a long-term goal. Once you've established what your long-term goal is, write it down. Perhaps, for example, you want to be more assertive in your behavior.

Step 2 Now think of a series of short-term goals—actions you can take to reach your long-term goal. If your goal is to be more assertive, possible short-term goals could involve behaving more assertively with particular people. You will have reached a short-term goal when you are comfortable with the way you relate to that person.

Plan a schedule for checking your progress. You might ask a family member or friend to help you evaluate how you are doing.

In Your Home and Community

1. **Health of Others** Choose a family member and give him or her a specific day to be special. On that day, go out of your way to build up that family member's self-esteem just by being nice to him or her. Spend time with the person, listen, and show that you care. Choose a different family member each week.

2. **Community Resources** Helping other people is one way to build self-confidence and self-esteem. Talk to a person who does volunteer work in your community. Ask the person about how this volunteer work affects his or her self-concept and self-esteem. Write a short account of your interview.

Building Your Portfolio

1. **Interview** Interview other students about how they manage stress. Ask about situations that produce stress and how the students deal with them. Take notes and write a short summary of any of the ideas that you could use to help you manage stress. Keep these ideas in your portfolio.

2. **Script** Write a script involving a conversation between two teens, one of whom is very depressed. Show how the other teen can help his or her friend to find help. Put the finished script in your portfolio.

3. **Personal Assessment** Look through all the activities and projects you did for this chapter. Choose one or two that you would like to include in your portfolio.

Chapter 8: Chapter Review **237**

Performance Assessment

▶ **Self-evaluation** Direct students to review the activities that are provided throughout the chapter. Encourage each student to select one finished product or activity that demonstrates his or her best work for the chapter. Have students explain what they learned and how the examples they selected show their progress.

▶ **Teacher's Classroom Resources** Assign Performance Assessment 8, "Mental and Emotional Health Video," in the TCR.

Chapter 8 Review

Thinking Critically

Responses will vary. Possible responses are given.

11. They are similar because having high self-esteem, or feeling confident about yourself, helps you maintain good mental health and deal with the problems you face every day.

12. Friends and family can help you talk about your stress and plan ways to adapt to it. Their support and encouragement also helps you deal with stress.

13. A phobia is an anxiety disorder characterized by unreasonable or exaggerated fears of something specific. Depression is a mood disorder. A person experiencing depression has feelings of hopelessness and extreme sadness.

14. Students may mention discussing problems frequently, asking about sources of help another person has used, or making vague references to pressures or problems.

RETEACHING

Assign Study Guide 8 in the Student Activities Workbook.

EVALUATE

▶ Use the reproducible Chapter 8 Test in the TCR, or construct your own test using the Testmaker Software.

▶ Use Performance Assessment 8 in the TCR.

EXTENSION

Direct interested students to learn more about traditional and contemporary theories of mental health treatment. They might consult library resources, including general-interest nonfiction works and journal articles. If possible, they might also be encouraged to interview mental health professionals and/or psychology teachers or professors. Ask students to summarize the information they discover and present their findings to the class.

Planning Guide

Chapter 9
Social Health: Families and Friends

Features   **Classroom Resources**

Lesson 1: Communication Skills
pages 240–244

Life Skills: Giving and Taking Criticism — pages 242–243

- Concept Map 39
- Cross-Curriculum Activity 17
- Enrichment Activity 39
- Lesson Plan 1
- Lesson 1 Quiz
- Reteaching Activity 39
- Transparency 34

Lesson 2: Understanding Family Relationships
pages 245–248

- Concept Map 40
- Enrichment Activity 40
- Lesson Plan 2
- Lesson 2 Quiz
- Reteaching Activity 40
- Transparencies 35, 36

Lesson 3: Friendships and Peer Pressure
pages 249–253

Personal Inventory: Are You a Good Friend? — page 250

Making Healthy Decisions: Facing Peer Pressure — pages 252–253

- Concept Map 41
- Decision-Making Activity 17
- Enrichment Activity 41
- Health Lab 9
- Lesson Plan 3
- Lesson 3 Quiz
- Reteaching Activity 41
- Transparency 37

Lesson 4: Abstinence and Refusal Skills
pages 254–257

TEEN HEALTH DIGEST *pages 258–259*

- Concept Map 42
- Enrichment Activity 42
- Lesson Plan 4
- Lesson 4 Quiz
- Reteaching Activity 42

Lesson 5: Resolving Conflicts at Home and at School
pages 260–265

Health Lab: Mediating a Conflict — pages 262–263

- Concept Map 43
- Cross-Curriculum Activity 18
- Decision-Making Activity 18
- Enrichment Activity 43
- Lesson Plan 5
- Lesson 5 Quiz
- Reteaching Activity 43

National Health Education Standards

One of the main goals of the National Health Education Standards is to move students toward "health literacy." Health literacy is the capacity of individuals to obtain, interpret, and understand basic health information and services and the competence to use such information and services in ways which promote health. The health standards were developed by applying the characteristics of a well-educated, literate person within the context of health. The health literate person is:

- a critical thinker and problem solver
- a self-directed learner
- a responsible, productive citizen
- an effective communicator

Listed below are the Health Education Standards Performance Indicators addressed in each lesson of this chapter.

Lesson	Health Standards Performance Indicators
1	(5.1, 5.4, 5.5, 6.1)
2	(1.2, 1.4, 5.2, 6.2)
3	(3.2, 5.2, 5.5, 5.6)
4	(1.6, 3.1, 5.3, 5.6)
5	(5.3, 5.7, 5.8, 7.4)

ABCNEWS InterActive™ VIDEODISC SERIES

You may wish to use the following videodiscs with this chapter: *Teenage Sexuality*, side one, video segments 14, 16, 17; side two, video segments 4, 5, 6, 7, 8, 9, 10, 11, 13, 14, 15, 16; *Making Responsible Decisions;* and *Violence Prevention.* Use the ABCNews InterActive™ Correlation Bar Code Guide for title reference. Also available in VHS format.

Chapter Resources

Teacher's Classroom Resources
- Chapter 9 Test
- Parent Letter and Activities 9
- Performance Assessment 9
- Testmaker Software

Student Activities Workbook
- Study Guide 9
- Applying Health Skills 39–43
- Health Inventory 9

Student Diversity Strategies
- Audiocassette Program (English)
- Audiocassette Program (Spanish)
- Spanish Parent Letters
- Spanish Summaries, Quizzes, and Activities

Multimedia Components
- English Audiocassette Program
- Spanish Audiocassette Program
- *Teen Health* Videodisc/VHS Series
- *Teen Health* Video Kit:
  *Choose to Refuse:
  Saying No and Keeping Your Friends*

Other Resources

Readings for the Teacher
Brandon, R. N. *Social Accountability: A Conceptual Framework for Measuring Child and Family Well-Being.* Washington, DC: Center for the Study of Social Policy, 1992.

Eager, George. *All about Peer Pressure.* Valdosta, GA: Mailbox Club Books, 1994.

Scarf, Maggie. *Intimate Worlds: Life Inside the Family.* New York: Random House, 1995.

Readings for the Student
Dreher, Nancy. "Divorce and the American Family," Current Health 2, November 1996, pp. 6–12.

Kreiner, Anna. *Everything You Need to Know about Creating Your Own Support System.* New York: Rosen Publishing Group, 1996.

Lang, Denise V. *But Everyone Else Looks So Sure of Themselves: A Guide to Surviving the Teen Years.* White Hall, VA: Shoetree Press, 1991.

Out of Time?

If time does not permit teaching this chapter, you may use these features: Life Skills on pages 242–243; Personal Inventory on page 250; Making Healthy Decisions on pages 252–253; Teen Health Digest on pages 258–259; Health Lab on pages 262–263; and the Chapter Summary on page 266.

Chapter 9
Social Health: Family and Friends

CHAPTER OVERVIEW

Chapter 9 focuses on teens' social relationships. It presents information that will help students communicate and get along with friends and family members.

Lesson 1 explains the communication process. It also presents tips that will help students express themselves clearly and listen effectively.

Lesson 2 discusses different kinds of families, identifies the basic needs that families meet, and explains what teens can do to strengthen family relationships.

Lesson 3 deals with the importance of friendships and helps students understand how they can identify and deal with peer pressure, both positive and negative.

Lesson 4 explains what abstinence is, discusses the importance of abstaining from risk behaviors, and helps students develop and use refusal skills.

Lesson 5 presents information on dealing with disagreements in healthy ways and resolving conflicts through neutrality, mediation, and negotiation.

PATHWAYS THROUGH THE PROGRAM

The lessons of Chapter 9 concentrate on the social side of the health triangle. This aspect of health is especially important to young teens, and good social relations enhance both mental/emotional health and physical health. As students work with these lessons, stress the importance of developing social relations by making choices based on positive personal values.

Chapter 9
Social Health: Families and Friends

Student Expectations
After reading this chapter, you should be able to:
1. Identify and apply effective communication skills.
2. Explain the benefits of healthy family relationships.
3. Describe the importance of friendship and the effects of peer pressure.
4. Practice effective refusal skills to avoid risk behaviors.
5. Recognize appropriate methods for preventing and resolving conflicts at home and at school.

Key to Ability Levels

Teaching strategies that appear throughout the chapter have been identified by one of three codes to give you an idea of their suitability for students of varying learning styles and abilities.

L1 Level 1 strategies should be within the ability range of all students. Full class participation is often required. Teacher direction is usually needed.

L2 Level 2 strategies are for average to above-average students or for small groups. Some teacher direction is necessary.

L3 Level 3 strategies are designed for students able and willing to work independently. Minimal teacher direction is necessary.

Teen Chat Group

Lisa: Before anybody else gets here, I want to talk to you two.

Sharon: What's up?

Lisa: I wanted to tell you that I invited Miyoko to the cookout. You know her—she was in my pre-algebra class this year.

Andy: She's the girl who moved from Japan this year, right?

Lisa: Right. The problem is, Carl is coming, too. Sharon, you remember how he kept making fun of her accent and getting other people to laugh at her, too.

Sharon: Carl never seems to think about other people's feelings. He's really popular, though, and people go along with him.

Lisa: Well, I just don't want that to happen today. If you hear Carl getting started, find a way to change the subject. I'm depending on you guys!

in your journal

Read the dialogue on this page. Have you ever been in a situation like this one? Start your private journal entries about your own relationships with friends and other classmates by responding to these questions:

- If you were at this cookout and Carl began making fun of Miyoko, what would you do?
- Why is it hard to speak up when someone is encouraging others to do wrong, even if you strongly disagree?
- How can good communication skills help you in your relationships with friends and family?

When you reach the end of the chapter, you will use your journal entries to make an action plan.

Chapter 9: Social Health: Families and Friends

Chapter 9

INTRODUCING THE CHAPTER

▶ Pose these questions for students to think about: Who are the most important people in your life? (Students are most likely to name members of their immediate and extended families, as well as same-age friends.) Which of those people are family members? Which are friends? What makes you feel close to those people? (spending time together; sharing feelings; helping and/or being helped) What happens when you feel angry with one or more of them? How could you add more good friends to this group? Explain that these are the kinds of issues students will be considering as they study the lessons of Chapter 9.

▶ Display several photographs of active teens. Each picture should show two or more friends together or a teen with one or more other family members. Let students describe the teens in these pictures, focusing on the social relationships: What are they doing together? How do they appear to feel? What kind of relationship do you think they have with each other? What makes you think that? Explain to students that participating in activities together is one way to enjoy and strengthen relationships with family members and friends.

Cooperative Learning Project

Celebrating Family and Friends

The Teen Health Digest on pages 258 and 259 provides students with high-interest articles related to the content of this chapter.

The material in the Teacher's Wraparound Edition presents suggestions for a class project in which students will plan and hold a party celebrating their own family members and friends.

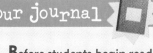

Before students begin reading the Teen Chat Group dialogue, ask three volunteers to read aloud the teens' roles, stopping after Lisa says, "...he kept making fun of her accent and getting other people to laugh at her too." Ask the rest of the class: Why do you think Lisa is bringing this up with her friends? What do you think she wants them to do? Why? If you were one of Lisa's friends, what would you say? Why?

Let the volunteers finish reading the dialogue aloud, and encourage students to discuss the devlopment of the dialogue and their own responses to the In Your Journal questions. Finally, provide time for students to write private journal entries.

239

Lesson 1
Communication Skills

Focus

LESSON OBJECTIVES

After studying this lesson, students should be able to
- explain what communication is.
- identify different types of communication.
- discuss how to develop speaking skills, listening skills, and the ability to compromise.

MOTIVATOR

Have students think of—or make up—a situation in which a problem developed because one person misunderstood another. Let students draw cartoons or write short narratives to depict those situations.

INTRODUCING THE LESSON

Encourage students to share and discuss the situations they depicted in response to the Motivator activity: What message did one person intend to communicate? What message did the other person receive? What problem resulted? How could the problem have been avoided? Explain that Lesson 1 presents information about communication and about improving communication skills.

INTRODUCING WORDS TO KNOW

Have students work in groups to discuss the Words to Know. Then ask them to skim the lesson, finding and reading the definition of each term. Have group members work together to brainstorm a list of words and phrases each Word to Know brings to mind.

Lesson 1
Communication Skills

This lesson will help you find answers to questions that teens often ask about communication. For example:
- How do people communicate?
- What are the advantages of expressing myself clearly?
- How can I improve my speaking skills?
- What can I do to become a better listener?

Words to Know

communication
verbal communication
nonverbal communication
body language
tact
empathy
compromise

in your journal

Think about the communication habits of people you know. Which people do you think are especially good communicators? What types of skills do they use effectively? How might you work to improve your own communication skills? Respond to these questions in your journal.

What Is Communication?

Communication is *the exchange of thoughts, feelings, and beliefs among people.* Communication has three basic components:

- Sender
- Receiver
- Message

Communication is not always simple, however. The sender must express the message clearly, and the receiver or receivers need to listen attentively and thoughtfully. In many cases, people don't say what they really intend to say, and listeners often hear only what they want to hear. The result may be a failure to communicate.

When sender and receiver do communicate effectively, everyone benefits. Effective communication builds positive relationships between people. It creates a healthy social environment in which people can not only be themselves but also understand and appreciate each other.

Effective communication leads to positive and fulfilling relationships.

240 Chapter 9: Social Health: Families and Friends

Lesson 1 Resources

Teacher's Classroom Resources
- Concept Map 39
- Cross-Curriculum Activity 17
- Enrichment Activity 39
- Health Lab 9
- Lesson Plan 1
- Lesson 1 Quiz

- Reteaching Activity 39
- Transparency 34

Student Activities Workbook
- Study Guide 9
- Applying Health Skills 39

Types of Communication

Communication can be verbal (using words) and nonverbal (without words). Most communication is a mixture of both types. **Figure 9.1** shows several types of communication.

Communicating Effectively

Speaking and listening skills play an important role in a healthy life. Expressing your thoughts and feelings clearly allows you to make your experiences and opinions known to others. In order to maintain relationships, you must also understand the messages that are sent to you. When conflicts arise, your speaking and listening skills go into action to allow you to reach the best possible solutions.

interNET CONNECTION
Visit our site to improve your skills at getting along with family and friends.
http://www.glencoe.com/sec/health

Figure 9.1
Types of Communication

How many types of communication can you find in this scene?

Verbal Communication means *using words to express yourself, either in speaking or in writing.*

Nonverbal Communication includes *all the ways in which you can get a message across without using words.* The most common form of nonverbal communication is body language. **Body Language** includes *posture, gestures, and facial expressions.*

Lesson 1: Communication Skills   241

Personal Health

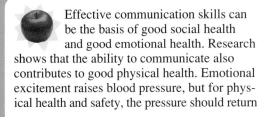

Effective communication skills can be the basis of good social health and good emotional health. Research shows that the ability to communicate also contributes to good physical health. Emotional excitement raises blood pressure, but for physical health and safety, the pressure should return to normal as soon as possible. Researchers found that individuals who were able to identify and express their feelings experienced a return to normal blood pressure more quickly than did people who had trouble labeling and talking about their emotions.

Chapter 9 Lesson 1

Teach

◻ Discussing

Guide the students in discussing the role of the sender and the receiver in communication: Why is the role of the receiver so important? How is the role of the receiver during a lecture different from the role of the receiver in a two-person conversation? In a personal conversation, how does each person function as a receiver and, simultaneously, as a sender?

Visual Learning

Guide students in discussing the examples of communication in Figure 9.1. For each identified message, ask: Is this communication verbal, nonverbal, or mixed? What are some others ways you might communicate the same message?

◻ Comprehending

Let students identify several ways of using verbal communication without nonverbal communication; most examples will involve written messages, including those sent on a computer. Help them recognize recordings and telephone conversations as strictly verbal, and then point out that different uses of the voice (such as inflection and tone) can affect the message communicated by the words. Let volunteers say the same basic sentence (such as *I told you I would do it*) several ways, modifying the tone of their voice and placing emphasis on different words to vary the meaning.

◻ Demonstrating

To help students recognize body language, let them play a variation of charades. Give each student a turn to communicate an emotional message, such as joy, surprise, or fear, using only posture, facial expressions, and gestures. How quickly can the other students identify the message?

Chapter 9 · Lesson 1

Creating Collages

Ask students: What does your appearance communicate about you? How do you change your appearance to communicate different messages to different audiences? How do teens use their appearance to define themselves as different from adults? After they have discussed their ideas, let students work in groups to create collages that convey their ideas about the messages a person's appearance can communicate. Suggest that they use original photographs, drawings, and/or images cut from magazines.

Career Education

Point out that some college students major in communications. Ask interested class members to research that major: What does the study involve? For what careers are students prepared? Let volunteers share their findings with the rest of the class.

VIDEODISC/VHS

Teen Health Course 2
You may wish to use video segment 9, "Communication Skills," to demonstrate how good communication skills can help resolve conflicts.

Videodisc Side 2, Chapter 9
Communication Skills

Search Chapter 9, Play to 10

Art Connection

Communicating Through Colors ACTIVITY

Russian artist Wassily Kandinsky (1866-1944) assigned feelings to certain colors in his paintings. To Kandinsky, orange meant health, green meant happiness, red meant the presence of strong feelings, and gray meant sadness. Use paints or markers in colors of your own choice to make a drawing that nonverbally communicates a thought or feeling.

Expressing Yourself Clearly

With the words you speak, you can share your likes and dislikes, your interests, and your hopes for the future. In addition, you can express the care and consideration you feel for the people in your life. Here are a few tips for expressing yourself clearly.

- Think before you speak. A few seconds of thought can make it much easier to find the right words.
- Be direct, but avoid being rude or insulting.
- Control your tone of voice. In most cases, a friendly and polite tone will encourage a listener to pay attention.
- Avoid speaking too quickly or too slowly.
- Make eye contact, and use gestures as needed to clarify your meaning.
- Let your listener respond. Then you will see what you need to make clearer.
- Use "I" messages. An "I" message keeps the focus on your own thoughts and feelings instead of blaming or criticizing someone else. For example, saying "You're wrong!" may make your listener feel defensive and can lead to an argument. Saying "I disagree because. . ." makes your opinions clear without starting an argument.

Although honesty is important, expressing your exact feelings may at times hurt other people. Therefore, it is crucial to use **tact**—*the quality of knowing what to say to avoid offending others.* Tact is essential in developing mature relationships. Ask yourself: *Would I be hurt if someone said this to me?* If the answer is yes, find another way to make your point.

LIFE SKILLS
Giving and Taking Criticism

Criticism—both receiving it and giving it—is a part of daily life. Criticism can be helpful, revealing weaknesses in a plan or a way of behaving. However, criticism can also make a person feel foolish. Whether criticism has a positive or a negative effect depends on how it is given and how it is received. Follow these tips when you give and take criticism.

When Giving Criticism

▶ **Choose the right time and place.** Be considerate of the other person's feelings. Don't criticize anyone in front of other people.

▶ **Don't criticize actions that can't be changed.** Saying "You should never have said that about Elise" doesn't change what has been done. It would be more helpful to say "Elise is pretty upset. Do you think you should call her and explain what you meant when you said . . . ?"

▶ **Be as positive as possible.** Praise the good before pointing out problems. For example, "I like your plan for getting a sponsor for the team. A couple of us have some ideas, too. In future meetings, we'd also like the chance to present other plans and discuss them."

▶ **Be specific and helpful.** Saying to a teammate "You let us down" provides only negative information. Instead, you might say "I know it's tough to shoot free throws. Do you want me to show you how I learned to do it?"

242   Chapter 9: Social Health: Families and Friends

LIFE SKILLS
Giving and Taking Criticism

Focus on Life Skills

Write the word *criticism* on the board, and let several students explain what it is. Then ask: Who is most likely to criticize you? How do you feel about being criticized? Whom are you most likely to criticize? Why? Do you think you usually end up helping the person you criticize or just hurting that person's feelings? Why?

After students have read the Life Skills feature, encourage them to discuss what they have learned: Are there individuals you think you should avoid criticizing? If so, who and why? If not, why not? What do you think you can learn from criticism? How can you learn, even from criticism that isn't given in the best possible way? How do you think you should respond to criticism that you believe is inappropriate?

242

Listening to Others

You are an essential part of a conversation even when you are not speaking. Listening should be an active process in which your thoughts and feelings become involved in what someone else is saying. In fact, skilled listening can actually help you to express your own thoughts and feelings more clearly.

Effective listeners try to understand not only what people are saying but also what they are feeling. **Empathy,** *the ability to identify and share another person's feelings,* is a crucial part of effective communication. You will need to pay attention both to what a speaker says and to the way he or she says it. A speaker's facial expression, gestures, and tone of voice will help you identify the feelings behind the message. Effective listeners also let the speaker know they are listening with techniques like these.

- Look at the speaker when he or she is talking.
- Use appropriate facial expressions or gestures, or nod your head to show that you are listening.
- Avoid interrupting the speaker, but ask questions when they are appropriate.
- Don't jump to conclusions. Keep an open mind about what the speaker is saying.

What is this teen doing to show that he is actively listening?

When Receiving Criticism

▶ **Consider why the person is criticizing you.** Someone who risks criticizing you is often someone who also cares about you. Use this knowledge to keep the criticism in perspective.

▶ **Focus on the message.** Your critic is trying to help you improve. Remember that it is your performance on a task or your behavior, not you as a person, that is being criticized.

▶ **Don't let someone else's poor communication skills cause conflict.** Even helpful messages can be delivered badly. When people are nervous or upset, they may communicate poorly. If you don't understand what the person is trying to tell you, ask for clarification.

▶ **Evaluate the message.** Listen carefully to the message and be as objective as possible. Is the criticism helpful? Is some part of it helpful? If the criticism is not helpful, ignore it—then try to forget about it.

Follow-up Activity

Perform a skit with several other students. The skit should show how criticism could be given or received for a positive result. In your skit, describe a scene in which someone is criticizing a friend's health habit, such as smoking or skipping meals. The skit should show both helpful criticism and its positive results.

Chapter 9 Lesson 1

L2 Presenting Skits
Let pairs of volunteers plan and present short skits in which teens exhibit tact without telling any lies. Encourage the rest of the class to identify and comment on the tactful responses in the skits.

L1 Discussing
Let students share their responses to questions such as these: What do you think is the difference between hearing and listening? What special listening skills do you need to use during a lecture? during a telephone conversation?

L1 Demonstrating
Let volunteers demonstrate appropriate and inappropriate facial expressions for attentive listening.

L1 Journal Writing
Ask students to write a journal entry about a teen or an adult they consider to be a good listener. To help students identify such a person, ask: When you have a problem, whom do you want to talk with? When something great happens, whom do you want to tell?

L1 Listing Examples
Let students work in groups to write two lists of situations: one in which teens should compromise and the other in which teens should not compromise. Ask the members of each group to share some of their examples with the rest of the class: What compromise would be appropriate in each situation in the first group? Why is compromise not appropriate for the situations in the second group?

Making Health Connections
A critical remark can save a person's life—and can certainly help protect a person's physical health. Giving and receiving criticism can also be a comfortable part of everyday life; handling criticism with confidence can support a person's mental/emotional health and enhance his or her social health.

Meeting Student Diversity
Participating in story writing or skit planning can be especially challenging for students with poor organizational and/or verbal skills. To help all students experience success with this project, form mixed-ability groups. Remind the class that all group members have something to contribute to the planning and production of the story or skit.

Assessing the Work Students Do
Suggest that students identify criticism given by characters on their favorite television programs. Guide them in discussing how TV characters give and receive criticism: What suggestions would you make? Assign credit based on students' contributions and participation to their group's story or skit.

Chapter 9 Lesson 1

Assess

EVALUATING THE LESSON

Assign Reviewing Terms and Facts and Thinking Critically to review the lesson; then assign the Lesson 1 Quiz in the TCR.

LESSON 1 REVIEW

Answers to Reviewing Terms and Facts

1. Tact is the quality of knowing what to say to avoid offending others. Examples will vary.
2. Responses will vary. Students may mention any five of the tips for expressing yourself clearly (page 242) or the tips for listening (page 243).
3. Empathy is the ability to identify and share another person's feelings. Sentences will vary.

Answers to Thinking Critically

4. Examples will vary.
5. Situations will vary. Responses should indicate that compromise is appropriate if values and rules are not jeopardized and if some give-and-take can result in a workable solution.

RETEACHING

▶ Assign Concept Map 39 in the TCR.
▶ Have students complete Reteaching Activity 39 in the TCR.

ENRICHMENT

Assign Enrichment Activity 39 in the TCR.

Close

Have students identify, either orally or in writing, at least one change they want to make to improve their listening skills.

---

in your journal

Write a summary of a situation in which you and a friend, or you and a sibling, compromised to solve a disagreement. Do you feel that the compromise was successful? Explain your answer.

Knowing When to Compromise

People put speaking and listening skills to work every day in solving disagreements. Some disagreements, however, can be solved only by **compromise,** *a method in which each person gives up something in order to reach a solution that satisfies everyone.* Compromise is the give-and-take at the heart of healthy and mature relationships.

It is important to know the difference between situations that require compromise and those in which you should stand your ground and say no. Never compromise your deepest values; your sense of right and wrong; or the rules of your parents, school, and community. You will, however, need to make compromises in everyday situations in which you and a family member or friend disagree on common, nonthreatening matters. For example, you can easily compromise on which movie to see or who plays a game first.

Compromise can solve many everyday conflicts. What do you think these teens are disagreeing about? What compromise would you suggest to them?

Lesson 1 Review

Using complete sentences, answer the following questions on a separate sheet of paper.

Reviewing Terms and Facts

1. **Vocabulary** Define *tact,* and give an example of how a person might display this quality.
2. **List** Identify five tips for communicating effectively.
3. **Restate** Define *empathy,* and use the word in an original sentence.

Thinking Critically

4. **Apply** Give at least three examples of body language, and explain the message each might send.
5. **Compare and Contrast** Give three examples of situations in which a person could safely compromise and three in which a person should stand her or his ground.

Applying Health Concepts

6. **Personal Health** With a small group of classmates, write and perform a skit that illustrates effective or ineffective communication skills. Make the situation as realistic as possible, involving a familiar issue that might arise between friends or family members.

244 Chapter 9: Social Health: Families and Friends

Meeting Student Diversity

Cultural Awareness Students with roots in other cultures may find some of the conventions of polite conversation different and even difficult to adjust to.

▶ Some students may be reluctant to look at a speaker or to maintain eye contact during a conversation. This kind of behavior is considered disrespectful in some cultures.

▶ Silence is a more important part of conversation in many other cultures than it is in ours. A student who is patiently waiting for another person to complete a thought may be considered inattentive in modern American culture.

▶ Gestures may play a more important role in communication for some students; for others, they may be considered inappropriate.

Understanding Family Relationships

Lesson 2

This lesson will help you find answers to questions that teens often ask about family relationships and family issues. For example:

▶ What does a family do for its members?
▶ How can family members deal with life's changes and challenges?
▶ What can family members do to strengthen their relationships?

Families and Society

Throughout your life you will be a member of many different groups—teams, classes, clubs, businesses, circles of friends. These groups are held together by common goals and shared interests. You are already a member of your first and most important group: your family.

The **family** is *the basic unit of society*. Through your family you develop values and social skills that enable you to participate in society. In your family you also develop into the unique individual you are. As **Figure 9.2** shows, there are many different kinds of families.

Words to Know
family
nurture

Figure 9.2
Kinds of Families
Many different types of families are common in the United States.

Family Structure	Description
Couple Family	Two adults with no children
Nuclear Family	A mother, father, and one or more children
Single-Parent Family	One parent and one or more children
Blended Family	A parent, a stepparent, and children of one or both parents
Extended Family	A nuclear, single-parent, or blended family, with other relatives—such as grandparents, aunts, uncles, or cousins—living in the same dwelling

Lesson 2: Understanding Family Relationships   **245**

Lesson 2 Resources

Teacher's Classroom Resources
- 📁 Concept Map 40
- 📁 Enrichment Activity 40
- 📁 Lesson Plan 2
- 📁 Lesson 2 Quiz
- 📁 Reteaching Activity 40
- 📁 Transparencies 35, 36

Student Activities Workbook
- 📁 Study Guide 9
- 📁 Applying Health Skills 40

Lesson 2
Understanding Family Relationships

Focus

LESSON OBJECTIVES
After studying this lesson, students should be able to
▶ identify different kinds of families.
▶ explain how families nurture their members.
▶ identify and discuss some of the changes and challenges families may face.
▶ discuss what family members can do to strengthen family relationships.

MOTIVATOR
Ask students: What are the most important benefits of belonging to a family? Let students write their responses in short paragraphs or in lists of words and phrases.

INTRODUCING THE LESSON
Encourage the students to compare and discuss their responses to the Motivator question. Direct the discussion as necessary to help students identify the physical safety, the emotional support, and the social enjoyment that families can provide. Then ask: What responsibilities do you have as a family member? How do you help others enjoy family life? Tell students they will learn more about family life in Lesson 2.

INTRODUCING WORDS TO KNOW
Let volunteers read aloud the Words to Know and their definitions, as given in the Glossary. Then have each student write two original sentences using the vocabulary terms; suggest that students sketch illustrations for their sentences.

245

Chapter 9 Lesson 2

Teach

Teacher Talk

Dealing With Sensitive Issues

Keep in mind that some students may have difficult family situations; others may feel that they have no family. Avoid asking students any direct questions about their own families; instead, keep class discussions as general as possible.

L1 Discussing

Guide the students in discussing the different kinds of families listed on page 245. Which do you consider more important: how families are alike, or how they are different? Why?

L2 Researching

Let interested volunteers work in small groups to research the role of the extended family in other cultures. Ask the members of each group to share their findings with the class.

L1 Critical Thinking

Ask students to consider the four categories of needs that families meet, as shown in Figure 9.3: Do you consider any of these needs more important than the others? If so, which one and why? If not, why not?

L1 Discussing

Help students recognize that, in family nurturing, parents or other adult family members meet all the needs of young children. Ask: How is your role in the system of family nurturing changing? How will it continue to change?

What Families Are For

Families throughout the world share a common goal: to **nurture,** or *provide for the physical, emotional, mental, and social needs* of each of their members. **Figure 9.3** shows ways in which families accomplish these essential tasks.

Figure 9.3
How Families Nurture Their Members
Families take care of their members in many ways.

Meeting Social Needs
Families are the first learning centers for the skills their members need to get along with other people and be productive members of a group. Family members learn to take turns, to share, to accept responsibilities, and to respect one another's rights and individuality. Families are miniature societies, and within them people develop the social awareness that keeps the larger society healthy.

Meeting Emotional Needs
Families provide a base of emotional security so that each member can feel accepted, supported, and loved. Families help their members face life's challenges, deal with disappointments, and celebrate successes.

Meeting Physical Needs
Families provide the basic needs for survival: food, clothing, and shelter. They care for family members during illness and protect each other against accidents and injuries.

Meeting Mental Needs
Families are the first teachers, teaching everyday tasks from tying shoes to speaking languages. As children develop, families promote intellectual development through reading, helping with schoolwork, and answering questions about the world and how it works. Eventually children and adults share their knowledge, skills, and experience equally, enriching the lives of all family members.

A family also shares values with its members. These values are based on many different influences, including family history, traditions, and religious beliefs. Values such as justice, honesty, and compassion usually reflect the personal beliefs of individual family members as well as a culture's highest goals for itself.

Your values, of course, are much more than ideas or words. They are the basis of your decision making in the real world. The values learned through family life have an enormous influence on the lives of all the members of a society.

Chapter 9: Social Health: Families and Friends

Personal Health

Adolescence is a time of examining and shaping identity. For many adopted teens, this includes curiosity and often concern about their birth parents. Unless an "open adoption" situation has been maintained, adopted teens may have little or no access to information about biological family members. The best choice for these teens is to talk with their parents about their questions. However, many teens are reluctant to raise this issue at home, fearing that this interest may hurt their parents' feelings or make their parents angry. In these cases, teens should be encouraged to talk with a school counselor or other mental health professional.

Changes and Challenges That Families Face

Family life involves changes and challenges that are sometimes hard to handle. In a healthy family, each member tries to help the other members adjust to new circumstances, make difficult decisions, and deal with painful situations. **Figure 9.4** lists some of the most common changes and challenges that families face, along with suggested ways to cope with them. When challenges are truly serious—substance abuse or physical abuse, for example—help from outside the family is usually needed. Families in such situations should get help from professional counselors, medical or law enforcement workers, or religious leaders.

Q & A

Families on the Move

Q: My family is moving to a new city next month. How can I find out what it's like?

A: Try calling or writing the local Chamber of Commerce for information. You could also check a library, a bookstore, or the Internet. Many cities publish brochures or set up Web pages to help newcomers.

Figure 9.4
Facing Family Changes and Challenges
Family members can support each other in a crisis.

Change or Challenge	Suggestions for Action
Moving	Learn as much as possible about your new neighborhood and school before you move. Find out what clubs and other activities are available that match your interests. Get involved as soon as you can. Keep a positive attitude. Pull together as a family and help other family members adjust. Make plans to spend more time with parents and siblings until you've made new friends.
Job Loss	Don't blame or criticize. Make the situation a family project. Think about ways in which you might help out with the family finances, such as cutting down on some expenses or contributing baby-sitting money.
Separation and Divorce	Try to think positively. The future may bring new peace of mind to individuals in your family. Share your feelings. If possible, tell your parents what you're feeling. If your parents are unable to give you the support you need, seek help from a counselor or another trusted adult. Show both parents that you care for them. Try to help younger brothers and sisters understand and cope.
Illness and Accidents	Share your thoughts, fears, and feelings. Don't hold them in. Provide emotional support to other family members. Listen to them, and give them a chance to express their feelings. Take responsibility for tasks that you can accomplish, both in the care of the sick or injured family member and in everyday family chores.
Death	Accept your emotions, even unexpected ones such as anger or fear. Understand that your grief may last for quite a while. Do not be critical of family members who show their grief in ways that are different from yours. Many people hide sadness behind unusual cheerfulness or what seems like a lack of emotion. Pay special attention to younger members of the family, offering as much comfort and security as you can and helping them look toward the future.

Lesson 2: Understanding Family Relationships

Chapter 3 Lesson 2

L2 Interviewing

Suggest that each student interview a parent or other adult family member. Ask students to find out what values that adult hopes to pass along to children in the family. Then ask students: Were you able to predict what the adult in your family would say about values? Do you feel those values are being shown and shared in your family? Let students record their responses in journal entries.

L1 Discussing

Let students share their ideas about each of the changes or challenges that families may face. Help students recognize that these situations are usually difficult for all family members; teens who feel they aren't getting support from family members should turn to counselors or other trusted adults for help. Ask: How can talking to someone outside your family be especially helpful? What kinds of problems and emotions do you think teens want to discuss with their friends? How can you find an adult with whom to talk?

L2 Guest Speaker

Ask a librarian from the school or community library to speak with the class about books on family problems, especially divorce and death. If possible, check out a selection of these books and make them available in your classroom library. Encourage students to browse through the books and discuss their contents.

L1 Discussing

Help students discuss each of the tips for strengthening family relationships. Suggest that students describe or demonstrate examples of each. Ask: How do these tips serve as suggestions for getting along with any other group of people? Why is it so important for all family members to do what they can to strengthen family relationships?

Cooperative Learning

Family Bulletin Board Let students work together, either as a class or in several groups, to create a family bulletin board. Encourage students to begin by discussing the various groupings that make up families. Be sure they at least note all the kinds of family groups to which class members belong. Then have group members plan and make a bulletin board display that shows the diversity of families. They may want to use a collage technique, gathering magazine photographs and headlines or even using their own family pictures; they may want to use drawings, cartoons, or a combination of family depictions.

Chapter 9 Lesson 2

Assess

EVALUATING THE LESSON
Assign Reviewing Terms and Facts and Thinking Critically to review the lesson; then assign the Lesson 2 Quiz in the TCR to evaluate students' understanding.

LESSON 2 REVIEW

Answers to Reviewing Terms and Facts
1. To nurture is to provide for the physical, emotional, mental, and social needs of another person.
2. Changes and challenges that families face include (any three): moving, job loss, separation and divorce, illness and accidents, and death. Tips for dealing with any one of those changes or challenges will vary.
3. Families can remain strong by (any three): showing appreciation; communicating ideas, information, and feelings; spending time together; getting to know grandparents.

Answers to Thinking Critically
4. Responses will vary.
5. Responses will vary.

RETEACHING
▶ Assign Concept Map 40 in the TCR.
▶ Have students complete Reteaching Activity 40 in the TCR.

ENRICHMENT
Assign Enrichment Activity 40 in the TCR.

Close
Let students discuss their ideas in response to these questions: How is being part of a family like being a member of a circle of friends? How are the two groups different?

---

in your journal

Summarize the strengths and positive qualities of your family. Point out specific contributions made by each family member. Consider physical, mental, emotional, and social contributions.

Strengthening Family Relationships

Every family member plays several different roles. A woman, for example, may be a mother, a wife, a daughter, and a sister. Multiple roles mean that everyone is busy, and juggling roles and responsibilities can lead to stress. Members of your family can work together to relieve some of this stress. By strengthening the bonds that hold your family together, everyone will find it easier to get through difficult times. Empathy, respect, humor, and understanding can help. Here are a few additional tips.

- **Show appreciation.** Don't take the members of your family for granted. Give compliments and encouragement to each member. Stay on the lookout for signs of stress, and offer help.

- **Communicate ideas, information, and feelings.** Be willing to share your thoughts and feelings. Ask for help when you need it. Talk with family members on a regular basis, not only when a problem arises. Be a good listener, too.

- **Spend time together.** As often as possible, eat dinner together, and use the time to talk about everyone's experiences that day. Plan family outings—and don't be afraid to make compromises so that everyone can have a good time.

- **Get to know your grandparents and other extended family members.** Their experiences may point you in new directions and help you make decisions about your own future.

A close relationship with a grandparent can be valuable for both of you.

Lesson 2 Review

Using complete sentences, answer the following questions on a separate sheet of paper.

Reviewing Terms and Facts
1. **Vocabulary** Define the term *nurture*.
2. **Restate** List three changes and challenges that may occur in families. Suggest two ideas for dealing with one of these changes or challenges.
3. **Explain** Describe at least three ways for families to remain strong.

Thinking Critically
4. **Compare and Contrast** State the similarities and differences between two types of families described in **Figure 9.2**.
5. **Infer** Members of strong and nurturing families make communities more stable by becoming involved in the life of the community. In what other ways do you think strong families benefit society?

Applying Health Concepts
6. **Growth and Development** Pretend that you are moving to a new town. Using the ideas in this lesson, write a letter to yourself that would help you accept this major change and adjust to your new surroundings.

248   Chapter 9: Social Health: Families and Friends

Health of Others

Students' parents or other adult family members may come to you for information on family-support resources. The following national organizations can be helpful:
▶ Family Resource Coalition
▶ Parental Stress Services
▶ Parents Without Partners
▶ Grandparents Raising Grandchildren

Also, suggest that parents look under such headings as "Social Service Organizations," "Marriage and Family Counselors," and "Social Workers" in the local Yellow Pages. In addition, a school counselor may have brochures or listings of recommended sources of family support.

Friendships and Peer Pressure

Lesson 3

This lesson will help you find answers to questions that teens often ask about their relationships with friends. For example:

- What are the qualities of a good friend?
- What is peer pressure?
- How can I deal with negative peer pressure?

The Importance of Friends

Everyone forms relationships with people beyond her or his own family. As social experience grows, people form **friendships,** *relationships between people who like each other and who have similar interests and values.* Friends become especially important during the teen years. A circle of friends adds to the enjoyment of life and plays a major role in physical, mental/emotional, and social health.

Friendships usually develop over time, based on shared experiences, values, and goals. The old saying, "To have a friend you must be a friend," still holds true. Friendship grows out of each person's willingness to reach out to the other person, to listen, support, and care, in order to create a healthy, growing relationship.

Good friends are a reliable source of companionship, shared fun, and mutual support.

Words to Know

friendship
peers
peer pressure

Did You Know?

Something to Say *ACTIVITY*

A friend who needs you to listen may not know how to begin. Watch for clues that a person is having a problem, such as

- questions like "Are you busy right now?"
- being unusually quiet.
- a drop in grades.
- irritation or a quick temper.
- any behavior that is abnormal for this person.

With a partner, role-play a scene in which someone discovers that a friend needs to talk about a problem. Make the dialogue as natural as possible, and have the friends display an honest desire to communicate and to listen.

Lesson 3: Friendships and Peer Pressure 249

Lesson 3 Resources

Teacher's Classroom Resources
- Concept Map 41
- Decision-Making Activity 17
- Enrichment Activity 41
- Health Lab 9
- Lesson Plan 3
- Lesson 3 Quiz
- Reteaching Activity 41
- Transparency 37

Student Activities Workbook
- Study Guide 9
- Applying Health Skills 41

Lesson 3
Friendships and Peer Pressure

Focus

LESSON OBJECTIVES

After studying this lesson, students should be able to

- discuss the importance of having friends.
- identify some of the qualities of true friends.
- explain techniques for standing up to negative peer pressure.

MOTIVATOR

Ask: What makes a good friend? Have students list at least four qualities they look for in a friend.

INTRODUCING THE LESSON

Encourage the class to compare and discuss their lists of the characteristics of a good friend. Which qualities are included on all (or most) students' lists? Why do you think so many teens consider those qualities important in a friend? Encourage students to apply their lists of qualities to themselves: How good a friend do you think you are? Why? Tell students they will learn more about friends and friendships as they read Lesson 3.

INTRODUCING WORDS TO KNOW

Guide students in sharing what they already know about the Words to Know. What is friendship? How are friends and peers alike? How are they different? Who can exert peer pressure? What kinds of effects can peer pressure have? Have students skim the lesson to find the formal definition of each term. Suggest that they record the Words to Know and their definitions in their notebooks.

249

Chapter 9 Lesson 3

Teach

L1 Discussing

Help students consider how their friendships have grown since early childhood: Who were your friends when you were very young? How has the balance changed between time spent with your family and time spent with your friends?

Teacher Talk

Dealing with Sensitive Issues

Your class may include one or more students who are uncomfortable with discussions of friendship. In some cases, personality issues make a teen feel isolated; in others, differences in ability (physical or mental) may make a teen feel separated from peers. In addition, serious problems at home (including abuse) may make a teen withdraw from peers.

L1 Applying Knowledge

After students have read and discussed the qualities of true friends, present each of the following situations. Ask students how a true friend might respond.

- It's the night before an important math test, and you need to study. A friend says to you, "I don't feel like studying. Let's go shoot baskets."
- A friend tells you, "Ever since my mom got sick, I don't feel like doing much of anything. You'd better go to the game without me."
- A friend whispers, "Aunt Cleo told me the most amazing thing, but she said I shouldn't tell a soul! Want to know what it is?"
- A friend suggests, "It probably wouldn't hurt to try smoking just a little marijuana. What do you think?"

What Is a Good Friend?

A friend is much more than an acquaintance, someone you see occasionally or know casually. Your relationship with a friend is deeper and means more to you. Although there is no accepted test for friendship, most people you call friends will have the following qualities:

- **Sympathy and empathy.** True friends understand how you feel. In fact, close friends even identify with strong feelings such as joy, sadness, or disappointment. They appreciate your talents and strengths and help you overcome your weaknesses.

- **Loyalty.** True friends remain on your side and don't desert you, even when the going gets tough. They are available to you when you need support.

- **Reliability.** A true friend is someone you can depend on to keep promises and to live up to realistic expectations. You feel confident that your true friends will make sacrifices for you.

- **Shared values.** True friends will not ask you to do things that are wrong or dangerous. They respect your beliefs and help you hold true to your values.

Personal Inventory
ARE YOU A GOOD FRIEND?

To have good friends, you must be willing to be a good friend. Use this self-assessment form to examine the qualities that make you a true friend. On a separate sheet of paper, write yes or no to tell whether each statement describes you.

1. I do what I can to be a positive influence and help my friends to make healthy decisions.
2. I listen carefully when my friends want to talk about serious issues or problems.
3. My friends know that they can trust me to keep promises and secrets.
4. I respect my friends' right to have opinions that are opposed to mine.
5. I am willing to compromise if my friends and I disagree, but my friends know that I won't compromise my values.
6. I recognize my friends' strengths and talents and tell my friends about them.
7. I accept my friends' feelings, even if they are different from mine.
8. My friends know that I will accept their weaknesses, but that I will also try to help them overcome them.
9. I am honest with my friends.
10. I am loyal to my friends.

Give yourself 1 point for each yes. A score of 9–10 is very good, 7–8 is good, and 5–6 is fair. If you scored 0–4, you need to brush up on your friendship skills.

Cooperative Learning

Social Opportunities Brochure Many young teens, including those who spend time with several close friends, complain of being or feeling friendless. In most cases, these teens are expressing a different lack: they feel they do not belong to a special, defined—and defining—group. This is one of the needs that teens satisfy when they join a sports team, club, or other organization. Usually, these organizations also meet teens' needs for physical or mental activity, for caring but limited adult supervision, and for positive peer support. To help teens find the best groups for themselves, let students work together to compile a complete list of possibilities in the community. Have students select the best form in which to publish their list.

Recognizing Peer Pressure

As you move through the teen years, your relationships with your **peers**—*people close to your age who are similar to you in many ways*—become more and more important. What your peers think of you, how they react to you, and how they accept you can affect your decisions about how you should act and think. This *influence that your friends have on you to believe and act like them* is called **peer pressure**. Peer pressure can be either positive or negative, as **Figure 9.5** shows.

Figure 9.5
Positive and Negative Peer Pressure

There are two types of peer pressure: positive and negative.

List at least two examples of positive peer pressure that you have experienced. Then select one example and describe how the peer pressure helped you grow as an individual.

With positive pressure, peers can
- Ⓐ challenge you to work hard on a team or project.
- Ⓑ encourage you to do your best in school.
- Ⓒ persuade you to work to help others.
- Ⓓ inspire you to be fit, healthy, and safe.
- Ⓔ urge you to work toward your goals.

Self-Improvement

Higher Self-Esteem

Health, Safety, and Fitness

With negative pressure, peers can
- Ⓐ urge you to use tobacco, alcohol, or other drugs.
- Ⓑ dare you to commit an unlawful act or take unnecessary risks.
- Ⓒ coax you to break the rules of your family, school, or community.
- Ⓓ encourage you to betray your own values.
- Ⓔ expect you to fear or dislike someone who is different.

Dangerous or Unlawful Situations

Lower Self-Esteem

Health and Safety Risks

Handling Negative Peer Pressure

Standing up for what you know is right can be difficult, particularly when friends are very persuasive. After all, everyone wants to be liked and accepted. You may worry that if you don't go along with your peers, you will be unpopular, or people will laugh at you.

As a teen, however, you have developed the ability to think for yourself. Unlike a small child, you make many of your own decisions. As you become more independent, you gain more strength as an individual. You can use that strength to resist negative peer pressure. **Figure 9.6** on the following page provides some practical tips for handling negative peer pressure.

More About...

Shy Teens An amazing 40 percent of American adults consider themselves shy. That statistic makes it likely that far more than half of all young teens are—or feel—shy. For most teens, shyness is primarily a sense of being different (which is shared by most other teens) and of being uncertain (which is shared by *everyone*, at least occasionally). Talking about shyness, especially in small discussion groups, can be helpful. Teens who think all other students look so sure of themselves will be surprised and reassured to find out otherwise. Directed practice with basic social skills can also be helpful; let students role-play scenes in which they initiate friendly conversations.

Chapter 9 Lesson 3

L1 Discussing
Guide the students in discussing each statement in the Personal Inventory: Are You a Good Friend? How does this kind of behavior strengthen a friendship? What happens to a friendship if this quality is missing or unreliable? After the discussion, provide time for students to respond to the inventory in private. Ask them to record their reactions to the inventory in their journals.

L1 Creating Displays
What do most teen's peers, acquaintances, and friends have in common? How are they different? Let students work in cooperative groups to plan and draw graphic organizers or other displays to show the relationship between the terms *peer*, *acquaintance*, and *friend*.

Visual Learning
Encourage students to describe the staircases in figure 9.5: Why does the positive peer pressure staircase lead up? What specific things can teens say and do to exert positive peer pressure? Why does the negative staircase lead down? What are some of the words and actions that exert negative peer pressure on teens?

L1 Discussing
Ask students to share their responses to these questions: Why do you think people so often think of negative pressure when they hear the term *peer pressure*? What can you do to choose the kind of peer pressure you are exposed to? How can you exert positive peer pressure on your friends? on other students at school?

L1 Critical Thinking
Encourage students to identify specific ways in which many teens at school dress alike. Then ask: Is this similarity of dress a form of peer pressure? If so, what kind? If not, why not?

Chapter 9 Lesson 3

L2 Current Events

Have students scan newspapers and/or magazines for articles about teens in the news. The teens may be involved in positive activities, such as community clean-up, or in negative—or even criminal—activities. Let students meet in groups to summarize and discuss what kind of peer pressure was involved in each activity?

L2 Language Arts

Ask each student to recall or read a novel and to identify the peer pressure exerted on or by one of the main characters. Have each student write a paragraph about the peer pressure in that novel and draw an illustration suggesting part of the novel's plot. Post these "peer pressure book reports" on a classroom bulletin board.

L1 Demonstrating

As students discuss the peer pressure situations in Figure 9.6, ask volunteers to role-play specific examples of each situation.

Assess

EVALUATING THE LESSON

Assign Reviewing Terms and Facts and Thinking Critically on page 253 to review the lesson; then assign Lesson 3 Quiz in the TCR to evaluate students' understanding.

RETEACHING

▶ Assign Concept Map 41 in the TCR.
▶ Have students complete Reteaching Activity 41 in the TCR.

Figure 9.6
Standing Up to Negative Peer Pressure
Here are some ways to handle several types of negative peer pressure situations.

Types of Situations	What to Do
Situations that have a strong potential for danger. For example, a friend asks you to go to an unsupervised party, or someone who has been drinking alcohol offers to drive you home.	**Say no.** You don't have to defend your position. If, however, you *want* to say more, state your reasons clearly. If possible, suggest an alternative plan that is safer, or join forces with another peer who agrees with you.
Situations that are clearly unsafe or illegal. For example, a peer urges you to play a dangerous practical joke on someone or to shoplift.	**Rely on your values.** If the situation is dangerous, point out what might happen. If the situation involves breaking the law, say that you don't want to do something illegal.
Situations in which compromise is clearly inappropriate. For example, someone offers you a cigarette. When you refuse, he or she says, "Oh, come on. Just take one little puff. It won't hurt you."	**Stand your ground.** Be firm, and don't compromise. If the person won't take no for an answer, make up an excuse, change the subject, or simply walk away.
Situations in which your refusal leads a peer to make fun of you. For example, you refuse to go swimming in an unsafe place, and a peer says, "What are you, a baby?"	**Focus on the issue.** Tell the other person why you are saying no. Don't exchange insults, however.
Situations in which your refusal causes a peer to get angry or abusive. For example, you refuse to go along with a peer's hurtful verbal abuse of someone. The peer then turns on you and begins insulting you, too.	**Walk away.** The most dangerous kind of negative peer pressure involves threats or other forms of abuse. Don't let yourself be hurt by a bully. Report abuse to a parent, teacher, counselor, or other trusted adult. Then avoid contact with the abusive peer.

MAKING HEALTHY DECISIONS
Facing Peer Pressure

Julie is looking at the items displayed in a downtown store window when she notices Marcie and Tina, two of her friends, at the counter inside. Marcie and Tina don't see Julie. In fact, they think that no one is watching them as they slip several pairs of earrings into their pockets. As they turn to leave the store, they spot Julie. They realize from the look on her face that she has seen them shoplifting.

Marcie and Tina continue out of the store and walk over to Julie. After greeting her, they whisper: "Want some free earrings? It's easy to take the ones on that counter. They never watch at this store. Take a few. They'll never miss them."

Julie feels very uncomfortable. When she doesn't respond, Marcie and Tina just look at each other, laughing, and walk away.

Later that night, Julie is thinking about the incident. She knows that she would never shoplift, but she has to think about what to do the next time she sees her friends. She tries the six-step decision-making process.

252   Chapter 9: Social Health: Families and Friends

MAKING HEALTHY DECISIONS
Facing Peer Pressure

Focus on Healthy Decisions

Tell students that peer pressure is not always direct. For example, rather than pressuring you to join in an activity you disapprove of, your friends might encourage you to ignore their involvement in that activity. Ask: What makes this kind of peer pressure especially hard to resist? How do you think you would react if an acquaintance pressured you in this way? What if your best friend pressured you?

Activity

Use the following suggestions to discuss Follow-up Activity 1:

Step 1 Julie has to decide what she will say to her friends, whom she saw shoplifting.

Step 2 Julie can ignore what happened and never mention the shoplifting. She can tell her friends that shoplifting is wrong and that she will never shoplift with them. Or, she can tell her friends that shoplifting is wrong and that they should return the earrings.

Review — Lesson 3

Using complete sentences, answer the following questions on a separate sheet of paper.

Reviewing Terms and Facts

1. **Vocabulary** Define the terms *friendship* and *peers*.
2. **List** Identify three qualities of a true friend, and provide an example of each quality.
3. **Describe** Explain the difference between the two types of peer pressure. Support your description by providing one example of each.

Thinking Critically

4. **Compare and Contrast** Explain what makes some people your friends and some just acquaintances. How are they alike? How are they different?
5. **Hypothesize** How might one teen who resists negative peer pressure be helpful to other teens? Support your response with a specific example.

Applying Health Concepts

6. **Health of Others** Make a poster explaining to young children the qualities to look for in a true friend and how to *be* a true friend. Be sure to use language and examples that younger children will be able to understand.

1. State the situation
2. List the options
3. Weigh the possible outcomes
4. Consider your values
5. Make a decision and act
6. Evaluate the decision

Follow-up Activities

1. Apply the six steps of the decision-making process to Julie's situation.
2. Role-play the scene that takes place as Julie meets her friends the next time.
3. In small groups, discuss options that Julie might choose. What would you have done in her situation? Do you think that she has a responsibility to talk with the store owner? Explain your answer.

Lesson 3: Friendships and Peer Pressure 253

Lesson 4

Abstinence and Refusal Skills

Focus

LESSON OBJECTIVES

After studying this lesson, students should be able to

- discuss their own increasing responsibility for protecting and guiding themselves.
- explain what abstinence is and how it can protect teens' health.
- list tips for developing and using refusal skills effectively.

MOTIVATOR

How many ways can you say no? Give students four minutes in which to list as many words, phrases, and sentences as possible in response to this question.

INTRODUCING THE LESSON

Let students share words, phrases, and sentences from their Motivator lists; record these ways of saying no on the board. Ask: What are some risk behaviors you should avoid? Explain to students that Lesson 4 helps teens avoid risk behaviors and gives them tips for effective ways of saying no.

INTRODUCING WORDS TO KNOW

Have students use dictionaries to find the verb from which the Word to Know, *abstinence*, is formed. What is the definition of that verb? Have them find the definition of *abstinence* given in the lesson. How does knowing what the verb *abstain* means help you understand the meaning of *abstinence*? Let students record the Word to Know, along with its definition and an original sentence using the word, in their notebooks.

Lesson 4 Abstinence and Refusal Skills

This lesson will help you find answers to questions that teens often ask about protecting themselves from dangerous situations. For example:

▶ Why should I avoid high-risk activities and situations?

▶ How can I make good choices?

▶ What should I say and do if someone pressures me to engage in risk behaviors?

Word to Know

abstinence

Acting Responsibly

Parents establish rules in order to protect and guide their children. As you grow up, an important part of becoming responsible is to accept the task of protecting and guiding yourself. You can make wise choices and decisions right now that will protect your health and happiness in the years ahead.

The teen years are very challenging. You and your peers are no longer young children—you are adolescents. You may feel like travelers crossing an ocean to a new country. You are naturally eager to get to that new country, adulthood, as quickly as possible. You feel that you are becoming ready to make your own decisions, set your own rules, and be responsible for your own actions.

Avoiding Risk Behaviors

When you were a small child, you learned to avoid certain risk behaviors—not to run out into the street and not to accept rides from strangers, for example. Now your world has become larger and more complex than it was when you were younger. Teens make decisions about many issues, including the use of alcohol and tobacco and whether they will engage in sexual activity.

It is important to keep a balanced outlook—positive but cautious, confident but careful. Be honest with yourself. Consider the consequences of your actions, and protect yourself. Why take chances with all that you have going for you?

In your teen years you have more freedom and more responsibility.

Chapter 9: Social Health: Families and Friends

Lesson 4 Resources

Teacher's Classroom Resources
- 📁 Concept Map 42
- 📁 Enrichment Activity 42
- 📁 Lesson Plan 4
- 📁 Lesson 4 Quiz
- 📁 Reteaching Activity 42

Student Activities Workbook
- 📁 Study Guide 9
- 📁 Applying Health Skills 42

What Is Abstinence?

In today's world, the pressures on teens to engage in risk behaviors can come from peers, from adult role models, from advertising, from movies and television, and from many other sources. In obvious ways and not-so-obvious ways, teens are urged to take risks, to experiment, to engage in activities that make them seem more mature. Many of these activities are very dangerous.

Refusing to participate in unsafe behaviors or activities is called **abstinence.** To abstain means to say no to sexual activity before marriage, to the use of tobacco, alcohol, and drugs, and to other risk behaviors. Abstinence is the only sure way to protect yourself against the potentially dangerous consequences of risk behaviors.

Figure 9.7 lists some of the ways in which abstinence can help you stay physically, emotionally, and socially healthy.

Did You Know?
Abstinence Saves Lives

Studies by the Guttmacher Institute, released in February 1997, show that the risk of death from childbearing is two to four times higher for girls under the age of 17 than for women aged 20 and older. In addition, babies born to mothers between the ages of 15 and 19 have a 30 percent higher risk of dying in their first year of life.

Figure 9.7
Practicing Abstinence
Abstaining from certain behaviors can save your life.

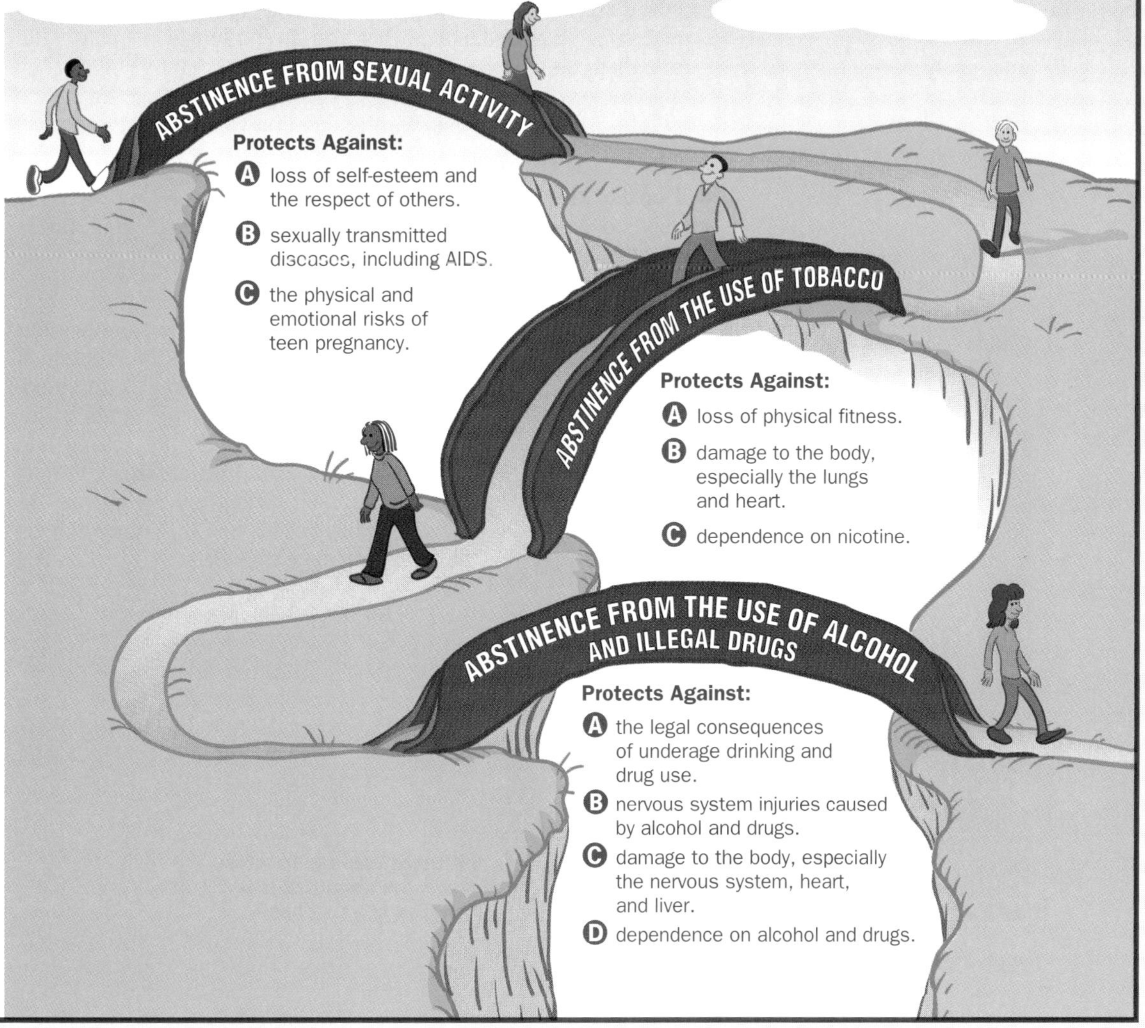

ABSTINENCE FROM SEXUAL ACTIVITY

Protects Against:
- Ⓐ loss of self-esteem and the respect of others.
- Ⓑ sexually transmitted diseases, including AIDS.
- Ⓒ the physical and emotional risks of teen pregnancy.

ABSTINENCE FROM THE USE OF TOBACCO

Protects Against:
- Ⓐ loss of physical fitness.
- Ⓑ damage to the body, especially the lungs and heart.
- Ⓒ dependence on nicotine.

ABSTINENCE FROM THE USE OF ALCOHOL AND ILLEGAL DRUGS

Protects Against:
- Ⓐ the legal consequences of underage drinking and drug use.
- Ⓑ nervous system injuries caused by alcohol and drugs.
- Ⓒ damage to the body, especially the nervous system, heart, and liver.
- Ⓓ dependence on alcohol and drugs.

Growth and Development

One of the purposes of abstaining from sexual activity is the prevention of teen pregnancies. In January 1996, President Clinton delivered a "Statement on Teen Pregnancy" from the White House. He announced the formation of the National Campaign to Reduce Teen Pregnancy and said: "This morning we want to talk about teen pregnancy, because it is a moral and a personal problem and a challenge that individual young people should face, and because it has reached such proportions that it is a very significant economic and social problem for the United States…Teen parents often don't have the education they need, don't have the self-awareness they need, don't have the self-confidence they need to make the most of their own lives in the workforce or to succeed themselves as parents."

Chapter 9   Lesson 4

Teach

L1 Discussing
Ask students: What are some decisions you make for yourself now that you weren't able or allowed to make a few years ago? What other decisions do you want to make for yourself now? Why do you think you're not permitted to make those decisions yet?

L1 Identifying Examples
Help the class identify some of the not-so-obvious ways in which teens are urged to take risks. What are some examples from advertising? from movies? from TV programs? from music?

L1 Comprehending
Guide students in understanding that it is never too late to practice abstinence. Many teens feel that once a person has experimented with sex or tobacco or alcohol and other drugs, it is too late to go back. Ask: Why is it harder to say no to a risk behavior after you have tried it? Why is it especially important to make that extra effort to say no?

Visual Learning
Lead a class discussion of each of the risk behaviors identified in Figure 9.7: What threats does that behavior pose to your physical health? your mental/emotional health? your social health? What are some of the pressures that encourage you to try or engage in that kind of behavior? What can you do to avoid that pressure?

L2 Investigating
Have students find and read magazine articles or news stories about teen pregnancy. Then let students meet in discussion groups to share what they have read: What are the particular problems that pregnant teens have to face? How does abstinence allow teens to avoid those problems.

Chapter 9 Lesson 4

L2 Presenting Skits

Divide the class into five groups, and assign each group one of the tips for developing and using refusal skills. Have the members of each group work together to plan a skit that shows the effectiveness of their assigned skill. After each skit has been presented, guide the rest of the class in discussing how refusal skills were used.

L1 Discussing

Help students consider how they can choose adults with whom to discuss difficult situations and decisions: Why is it helpful to talk with your parents? Why is it sometimes helpful to talk with adults other than your parents? Why is it important to know the values of an adult with whom you want to discuss problems? How can you tell what an adult values?

Visual Learning

Let volunteers describe the situation shown in the photograph near the bottom of page 256: What do you think has just happened? What pressure do you think this teen has refused? How do you think she feels now? How do you think she will feel later about her decision? Why?

VIDEODISC/VHS

Teen Health Course 2

You may wish to use video segment 11, "Alcohol and Decision Making," in which a teen uses decision-making skills to avoid being pressured to use alcohol.

Videodisc Side 2, Chapter 11
Alcohol and Decision Making

Search Chapter 11, Play to 12

---

in your journal

Reread the possible responses to the pressure statements in Figure 9.8. Suggest another response for each pressure statement. Explain why you feel that your response would be effective.

Effective Refusal Skills

Standing up to negative peer pressure requires the confidence and strength to say no. You learned how to use some specific refusal skills—ways to say no effectively—in Lesson 3 as you looked at negative peer pressure situations. Like other skills, refusal skills take time and commitment to develop. Refusal skills involve being honest and polite, but also being firm. The tips below will help you develop and use refusal skills effectively.

- **Choose your friends carefully.** Sometimes teens take dangerous risks in order to be accepted by a group or to become popular. True friends, however, won't pressure you to put your health or your life at risk. Instead, they'll look out for your welfare, and they'll rely on you to help keep them safe, too.

- **Choose your situations carefully.** Avoid situations in which you might be pressured to engage in risk behaviors. A party at which you know there will not be adult supervision is an example of a dangerous situation. Evaluate every situation carefully. Does it sound dangerous? Suspicious? Just plain wrong? Say no.

- **Talk to adults you trust.** Parents, teachers, and other trusted adults *want* you to be happy. They also want you to be healthy and safe. They understand that you are no longer a small child, and that you have developed many decision-making skills. They also know, however, that their knowledge and experience can be helpful. If you're under pressure and don't know what to do, get some good advice.

- **Look to the future.** Always try to focus on what you want and need to do to fulfill your goals. Don't jeopardize your future freedom, independence, and success by taking a dangerous action today.

- **Be prepared to leave a situation—or end a relationship—if the pressure becomes too strong.** If someone threatens you or urges you to do something that makes you uncomfortable, walk away. Walking away may be difficult, but getting into a high-risk situation could be far worse. Protect your pride and self-esteem by giving reasons for your refusal. The tips in **Figure 9.8** on the following page may help.

It takes courage to refuse pressure from friends to engage in a risk behavior.

256   Chapter 9: Social Health: Families and Friends

Home and Community Connection

Abstinence Programs President Clinton's 1996 "Statement on Teen Pregnancy" included the following: "Ultimately, I believe what is needed on this issue is a revolution of the heart. We have to work to instill within every young man and woman a sense of personal responsibility, a sense of self-respect, and a sense of possibility. Having a child is the greatest responsibility anyone can assume…and it is not the right choice for a teenager to make before she or he is ready. This message has to be constantly enforced and reinforced by community organizations and by other groups who are in a position to help our children make good choices."

What organizations in your community are involved in this effort? Invite representatives to speak to the class about their programs.

Figure 9.8
Pressure Statements and Possible Responses

On the left are some common statements peers may use to persuade you to engage in risk behaviors. On the right are possible responses you might make. Can you think of others?

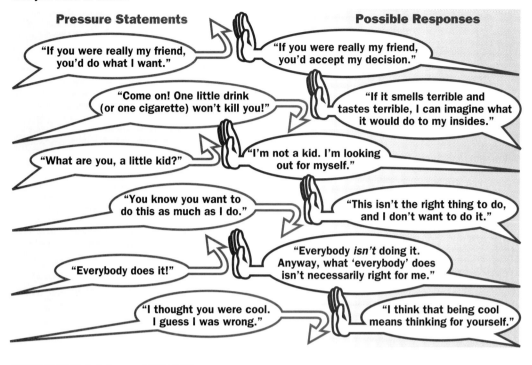

Review — Lesson 4

Using complete sentences, answer the following questions on a separate sheet of paper.

Reviewing Terms and Facts

1. **Vocabulary** Define *abstinence*, and explain why it is the most effective way to avoid risk behaviors.
2. **List** What serious problems can a teen avoid by abstaining from drugs and alcohol?
3. **Identify** List five tips that will help teens avoid risk behaviors.

Thinking Critically

4. **Infer** How can choosing your friends carefully help you avoid risk behaviors? Explain your answer.

5. **Apply** Give an example of a risk behavior that a teen might engage in that could have long-term negative effects.

Applying Health Concepts

6. **Health of Others** With a small group, present a panel discussion about the effectiveness of refusal skills in avoiding risk behaviors or unsafe activities. You might open the discussion with the question: "How well do refusal skills work?"

Lesson 4: Abstinence and Refusal Skills   257

Chapter 9 Lesson 4

Assess

EVALUATING THE LESSON
Assign Reviewing Terms and Facts and Thinking Critically on this page to review the lesson; then assign the Lesson 4 Quiz in the TCR.

RETEACHING
- Assign Concept Map 42 in the TCR.
- Have students complete Reteaching Activity 42 in the TCR.

LESSON 4 REVIEW

Answers to Reviewing Terms and Facts
1. Abstinence is avoiding a dangerous behavior or activity. It removes the teen from the risk situation.
2. Responses will vary.
3. To avoid risk behaviors, teens should choose friends carefully, choose situations carefully, talk to adults they trust, look to the future, and be prepared to leave a situation or end a relationship to avoid negative peer pressure. Students may also mention that teens should have good refusal skills.

Answers to Thinking Critically
4. Responses will vary. Possible response: If your friends are engaged in safe behaviors, you will not be faced with as many high-risk situations.
5. Examples will vary.

ENRICHMENT
Assign Enrichment Activity 42 in the TCR.

Close

Let students take turns presenting one another with pressure statements (such as those listed in Figure 9.8). Encourage other students to make original, effective responses.

More About...

Saying No Teens who are committed to abstaining from risk behaviors need to be encouraged to say no clearly and convincingly. These tips can help:

- Use the word *no*. It's much more convincing than saying "I don't think so" or "That doesn't sound like a good idea."
- Don't be shy about repeating yourself. If your friends fail to understand *no* the first time, say it again.
- Back up your verbal communication with a strong nonverbal message—look like you mean it.
- Suggest another activity, one that is both fun and safe.

TEEN HEALTH DIGEST

Focus
The Teen Health Digest articles can be used in two ways: as an individual activity for reflection and enrichment or as a cooperative learning activity as described below.

Motivator
Write the words *family* and *friends* on the board. Ask students to suggest how the two are related. Why are both important to an individual's mental/emotional, social, and even physical health? Then let students read the Teen Health Digest articles and discuss how each relates to the key words on the board.

Cooperative Learning Project

Celebrating Family and Friends

Setting Goals
Help students briefly review the importance of family and friends in the lives of all teens. Then tell students that they are going to plan and put on a party to celebrate their families and friends.

Delegating Tasks
Engage students in a class discussion about the party or social event they are going to hold: When and where should the party be held? Who will be invited? How will invitations be issued? What special party elements—decorations, games, refreshments, announcements, presentations, and so on will be included? After students have made general plans, let them form groups in which they can work cooperatively on specific aspects of the celebration.

258

TEEN HEALTH DIGEST

Sports and Recreation
Big-Time Players, Lifetime Friends

Hockey stars Mark Messier and Wayne Gretzky have known each other since childhood and have been best friends for 20 years. They have also been able to play as teammates for some of that time.

From 1980 to 1987, Mess and Gretz, as they are called by their teammates, played for the Edmonton Oilers and led the team to four Stanley Cup victories. Then Messier moved to the New York Rangers and Gretzky to the Los Angeles Kings. In 1996, when Gretzky had the chance to play for the Rangers, both men were pleased. Unfortunately, they were able to play together for only one year. In 1997 Messier joined the Vancouver Canucks. Trying to see the bright side, Messier pointed out that at least he would be in a different conference—not in danger of opposing his friend on the ice.

Try This: What are the advantages of teammates being friends? What challenges might friends face as teammates? Write a short paragraph explaining your ideas.

258   Chapter 9: Social Heath: Families and Friends

HEALTH UPDATE
New Phones for the Hearing Impaired

Technology can have a great impact on the ways in which people communicate. In the 1970s, scientists developed special phones called TDDs—telecommunication devices for the deaf. Each TDD displayed conversations line by line, much like today's Internet conversations. TDDs were not portable, however, and they could transmit messages only from one TDD to another.

In 1996 a marketing manager named Mark Elderkin developed a device that makes communication easier—a cellular phone for the deaf. Consisting of a tiny modem and a portable computer the size of a small book, the phone is easy to carry and can transmit calls to and from anyone with a computer and modem. A battery allows the computer to stay switched on at all times—just like a regular telephone.

TECHNOLOGY OPTIONS

Computer Students will probably have several suggestions for using computers to support their work on this cooperative project. For example, they may want to use computer programs as part of creating invitations, making memento books and/or award certificates, and designing party decorations.

Video Camera Because this party is a special celebration of family and friends, students may want to make a video recording of the event. Alternately, they might want several students to take candid photographs during the event and later to compile a celebration album.

People at Work

A Therapist for Children

Psychologist Becky Talmadge is a therapist for foster children at Foster Care Connection, an agency in Colorado. Becky's clients, who range in age from 4 to 17, have all been separated from their parents because of severe abuse or neglect.

Talmadge explains that the goal of therapy is to make children feel listened to and safe. "It's impossible to undo what abused or neglected children have been through," Becky says, "but we can help them to learn to cope and to develop skills that will help them protect themselves."

Becky first became interested in this type of work as a teen. In high school she worked as a peer mediator. After four years of college, she went on to get a master's degree in psychology.

To be a therapist, Becky says, "You must have empathy. You must be patient, too. These kids have been through such crises that their progress is often in baby steps."

Try This: Make a list of activities or volunteer jobs a teen might get involved in to prepare for a career like Talmadge's.

Teens Making a Difference

A Friend in Need

While working on her paper route, 15-year-old Rachael Nadeau of Connecticut made friends with a 92-year-old customer. Upon reaching her friend's house one day, however, Rachael noticed that newspapers had begun to pile up out front. When she knocked on the door and got no response, Rachael went home and tried to reach the woman by phone. When that failed, Rachael went back to the house. Peering through a window, she saw her friend lying on the floor. Rachael called 911. The woman had fallen two days earlier and had not been able to get up by herself. Rachael's friendly concern probably saved the woman's life.

Myths and Realities

The Family Balancing Act

Teens are old enough to share in family responsibilities. Parents who are both bringing in paychecks need to stay flexible and keep lines of communication open. Here are some tips for success offered by families in this situation.

- **Communicate.** Everybody needs to know what needs to be done during a particular day or week and who is available to help.

- **Do your share.** Helping your family with household chores will make the day run more smoothly. For example, you could prepare part of the evening meal, sort laundry, or vacuum your bedroom.

- **Make time for the family.** Some families select one or two nights each week to spend together. This aids communication and also strengthens relationships.

Teen Health Digest 259

Chapter 9

The Work Students Do

Following are suggestions for implementing the project:

1. Encourage students to keep the purpose of their celebration in mind: They should plan each aspect of the party to honor the family members and friends who contribute to their health and happiness.
2. Students may want to write personal invitations to their own family members and friends. In this case, members of an invitations group can design, decorate, and duplicate invitation forms in which students can write personal notes.
3. Suggest that students create a special guest book to be placed in a central spot during the celebration. Encourage all the guests to sign the book and to add remarks and responses.
4. Students may also want to create memento books—with quotations and photographs, for example—to be presented to the guests.
5. Guide students in considering appropriate decorations for the party area; photographs, sayings, autographs, and/or students' artwork might be included.
6. If students want to present awards, certificates, or other honors to their guests, suggest that these be distributed to all guests (or to all guests in a particular category, such as mothers). Help students recognize that this action should not encourage feelings of competition or winning and losing.

Assessing the Work Students Do

Ask students to draw or write descriptions of their favorite aspect of the class celebration. In brief individual conferences, let students share their descriptions and assess their own contributions to the party.

Meeting Student Diversity

Language Diversity Help students consider the languages with which their celebration guests will be most comfortable. Encourage small groups of students to prepare signs, awards, decorations, and/or other party elements in those languages.

Physically Challenged As students select a party location and then set up tables, chairs, and decorations, encourage them to keep in mind the particular needs of physically challenged guests and/or students. Ramps, an absence of steps, and plenty of open space make access easier for everyone.

Lesson 5
Resolving Conflicts at Home and at School

Focus

LESSON OBJECTIVES
After studying this lesson, students should be able to
- discuss the most causes of conflict.
- identify ways to prevent conflict.
- describe ways to help others avoid conflicts.

MOTIVATOR
Ask: What can happen when one teen "disses" another? Let students respond to this question by writing paragraphs or sketching cartoons.

INTRODUCING THE LESSON
Let volunteers read their paragraphs aloud and share their cartoons with the rest of the class. Ask: How many of you described a conflict that developed in response to one teen's "diss"? What other outcome could there be? After students have shared their ideas, explain that Lesson 5 presents information about what causes conflicts and how conflicts can be peacefully resolved.

INTRODUCING WORDS TO KNOW
In pairs, have students discuss the Words to Know and read the definitions presented in the Glossary. Then have the partners divide the terms into two or more meaningful groups, according to criteria they develop together. Ask each pair to share their groupings with the rest of the class: Can the other students recognize the criteria used to group the Words to Know?

260

Lesson 5
Resolving Conflicts at Home and at School

This lesson will help you find answers to questions that teens often ask about conflict resolution. For example:

▶ How can I recognize when a situation may be building toward a conflict?
▶ How can I stop a conflict before it gets serious?
▶ How can I help others avoid conflicts?

Words to Know
conflict
prejudice
nonviolent confrontation
neutrality
mediation
negotiation

How Conflicts Begin

A **conflict** is *a disagreement between people with opposing viewpoints.* Conflicts occur for all kinds of reasons. For example, your brother borrows your baseball mitt without asking first. A teen accuses a classmate of stealing her athletic shoes.

Sometimes the reason for a conflict may seem unimportant. Even so, a small quarrel may result in a nasty, even deadly, fight. When fights occur, it is usually because the people involved do not know healthy ways to settle their differences. Disagreements do not have to end in violence. In this lesson, we will examine causes of fights and nonviolent ways to deal with conflicts.

Teen Issues
Fight Stoppers

If one approach fails to stop a disagreement, try another.

▶ Do something creative and unexpected. For example, write a friendly note on a lunch bag.
▶ Offer the other person a way out of the conflict by proposing a truce or a compromise.
▶ Apologize for your part in the conflict.

Conflicts are a natural part of life. The way you handle them, however, determines whether or not you learn from them.

260 Chapter 9: Social Health: Families and Friends

Lesson 5 Resources

Teacher's Classroom Resources
- 📁 Concept Map 43
- 📁 Cross-Curriculum Activity 18
- 📁 Decision-Making Activity 18
- 📁 Enrichment Activity 43
- 📁 Lesson Plan 5

- 📁 Lesson 5 Quiz
- 📁 Reteaching Activity 43

Student Activities Workbook
- 📁 Study Guide 9
- 📁 Applying Health Skills 43
- 📁 Health Inventory 9

Arguments

Arguments occur when people are not communicating well or when they are disrespectful of one another. When arguments get out of hand, fights may result. Here are some of the most common reasons for teen arguments:

- **Property.** Teens may not respect one another's property. They may use items that belong to someone else without getting permission ahead of time.
- **Jealousy.** Young people may feel jealous when they are not included in certain activities, or when a boyfriend or girlfriend pays attention to someone else.
- **Territory.** Teens may not want others to cross the boundaries that make up their neighborhood.
- **Values.** Teens may refuse to do something that goes against their values, such as lying, cheating, or stealing.

Hurt Pride

Sometimes fights begin because someone's pride has been hurt. A teen may do something hurtful such as insult a member of someone else's family, spread a rumor, or ridicule another in public. Often the injured party feels hurt or angry and responds by fighting back.

Peer Pressure

Fights also begin when teens encourage others to "fight it out." They may stand on the sidelines, heckling and cheering the fighters. This behavior only worsens the situation. Once a crowd has gathered, the chances for settling the problem peacefully decrease greatly.

Revenge

One mean act or insult can start a chain of events in which the victim wants to get even. He or she may recruit family or friends to get involved in the conflict. As the need for revenge grows, the fighting may become more intense and more dangerous.

Acts of revenge are common among rival gangs when one of their own members has been harmed. Because gangs frequently use weapons, a minor misunderstanding can result in violence, such as a stabbing or a shoot-out.

Prejudice

Sometimes people refuse to accept others who are different from them. Their feelings may be based on **prejudice** (PRE·juh·dis), *a negative and unjustly formed opinion, usually against people of a different racial, religious, or cultural group.*

People who are prejudiced may single out a person and harass, intimidate, or threaten him or her. That person, in turn, may strike back, either alone or with supporters. As a result, fights or even gang warfare may occur.

Art Connection

Peace Mural

A mural is a scene painted on a very large space such as the side of a building. Get a large roll of paper and some friends to help. Work together to create a mural of peace, showing many different kinds of people cooperating or having fun. Ask for permission to display your mural at school.

Teen Issues

An Eye for an Eye

In ancient times, the law often required revenge—"an eye for an eye." Today revenge is discouraged. Instead, a show of respect to someone during a nonviolent confrontation increases the chances that a resolution will be reached and that no one will be injured. Remember to choose your words carefully, and look the other person in the eye. This will let him or her know that you are paying attention and showing respect.

Chapter 9 Lesson 5

Teach

L1 Discussing
Let the class suggest some of the constructive results of disagreements. Guide students in recognizing that disagreements call attention to problems that need resolution. Disagreements can also bring about change, result in better decisions, and help develop healthy relationships.

L1 Critical Thinking
Point out that arguments often develop when teens refuse to do something that goes against their values: What are the alternatives to these kinds of arguments? Do you think teens should just give in, rather than arguing? Why or why not?

L1 Comprehending
Instead of fighting back, what can teens do when their pride has been hurt? Encourage students to suggest a variety of healthy, safe responses.

L1 Analyzing
Encourage students to share their ideas about prejudice: How do people become prejudiced? How do you think people can overcome their own prejudices? Why do you think it is easier to recognize prejudiced attitudes in other people than to recognize them in yourself? Why is it important to identify your own prejudices?

VIDEODISC/VHS

Teen Health Course 2
You may wish to use video segment 7, "Growth and Development," to show how teens can use compromise and effective communication skills to solve problems.

Videodisc Side 1, Chapter 7
Growth and Development

Search Chapter 7, Play to 8

Lesson 5: Resolving Conflicts at Home and at School   **261**

Cooperative Learning

Avoiding Arguments Ask students to form cooperative groups of three or four. Distribute to the groups slips of paper, each with a word naming a common cause of arguments—*property*, *jealousy*, *territory*, or *values*. Let a member of each group draw one of the slips at random. Then have group members work together to write a dialogue between two or more teens who begin to argue over that issue. The dialogue should show teens failing to communicate well and/or being disrespectful of one another, and should express anger on both sides. Once the dialogues are complete, let each group share its "argument" with the rest of the class. Ask: How could these teens have avoided this argument? What could they have done to disagree fairly and to avoid anger?

261

Chapter 9 Lesson 5

Teacher Talk
Fighting Fair
Sometimes verbal arguments cannot be prevented. Encourage students to keep their arguments fair by following these rules:

- Focus on behaviors, not personalities.
- Don't bring up past hurt or anger.
- Don't resort to name-calling or insults.
- Listen to the other person's side without interrupting and without walking away.
- Don't argue with someone who is under the influence of alcohol or other drugs.

L1 Speculating
Ask students: Why do you think people sometimes become violent with the people they love most—their spouses, children, or parents? What are some of the effects of such violence?

L1 Discussing
Encourage students to share their ideas about expressing anger: Why do you think people so often keep their anger in, rather than communicating their feelings to the person they are angry with? How is keeping your anger hidden likely to make you feel? Do you think it can be helpful to talk with someone about your feelings? Why or why not?

Preventing Conflicts

It is not always easy to avoid conflicts, but it is possible. Like a balloon that is inflated too much, anger can build up inside you until the pressure makes it explode. However, you can learn healthy ways to keep conflicts from reaching the explosion stage. The best way to prevent problems is to recognize conflict early, control your anger, and ignore some conflicts. When you are unable to avoid a conflict, nonviolent methods can help you resolve the problem in a peaceful way.

There are many ways to prevent fights from erupting. Which ones can you recognize in this situation?

Recognize Conflict Early

There are usually signs that a problem exists. For example, there may be name-calling, insults, threats, or shoves. The key to preventing fights is to recognize the signs early and deal with them before they reach the danger stage. It is easier to resolve a conflict peacefully when you are still in control of your emotions.

HEALTH LAB
Mediating a Conflict

Introduction: In some schools, when students have a conflict they cannot resolve on their own, they sign up to meet with a student mediator, or a neutral third person. Usually adults are not present at this meeting. The mediator takes the students through the following steps to resolve the conflict.

1. Emphasize neutrality, and assure the participants that everyone will cooperate to reach a satisfactory solution.
2. Set guidelines for the meeting. For example, there should be no name-calling, insults, swearing, or interrupting.
3. Allow each person to give her or his side of the situation without interruptions. A mediator should listen carefully but does not react. Ask people to repeat or clarify their points when necessary.
4. Help the participants brainstorm solutions that will feel right to both sides.
5. Have both sides promise to abide by the agreement.

Objective: During the next week, observe conflicts around you that would benefit from mediation. Consider examples of these conflicts in your own family, among your friends, or in the news.

262 Chapter 9: Social Health: Families and Friends

HEALTH LAB
Mediating a Conflict

Time Needed
1 week

Supplies
- index cards
- long strips of paper
- pen or pencil

Focus on the Health Lab
Ask students to give examples of conflicts they have observed among teens on television sitcoms or news programs. Which were successful in resolving their conflict without violence? Which were unsuccessful? Why? After students have read the Health Lab feature, ask them to suggest examples of typical conflicts that might be resolved through mediation.

Control Your Anger

The first step in managing your anger is recognizing its early signs so you can stay in control. The body usually reacts to anger with physical changes such as increased heart rate and breathing, sweaty palms, flushed face, stuttering, or a high-pitched voice. By being alert for these signs, you can try to resolve the conflict peacefully or ignore it altogether.

To manage your anger, find a way to relieve pent-up feelings that works for you. Some suggestions follow.

- Walk, jog, swim, or shoot some baskets.
- Listen to quiet music.
- Take a long bath or shower.
- Pound a pillow.
- Talk it out with a good friend.
- Have a good cry.
- Sit quietly for half an hour or so.

If, however, you feel really angry, find someone such as a school counselor or health care professional who can help you sort it out.

Ignore Some Conflicts

Some issues are not worth your time and effort. For instance, if the other person is a stranger or someone you will never see again, it is probably best just to walk away. If the other person is someone you care about, you need to communicate your feelings in a calm and reasonable manner.

Chapter 9   Lesson 5

L2 Creating Collages

Have each student create an individual anger collage, including photographs, drawings, words, and phrases that represent his or her anger triggers. Let students meet in groups to share and discuss their collages.

L1 Identifying Examples

Ask the class to name specific situations in which it makes sense to ignore and walk away from conflicts.

Teacher Talk

Journal Writing

Suggest that students use their private journals as an aid in identifying the factors that underlie their own interpersonal conflicts. The next time they sense an important conflict developing, they can write private journal entries answering these questions:

- Is this conflict really necessary and healthy?
- What are my goals in this situation?
- What are my needs in this situation?
- What are my boundaries or limits?
- What prejudices or stereotypes may I be holding against the other person?
- How much can I compromise without giving up my safety, my values, or my basic rights?

Materials and Method: You will need one 3-by-5-inch card for each conflict you are observing and a long strip of paper. On each card, write *Observations* on one side and *Analysis* on the other. On the *Observations* side, write the facts—the words, gestures, facial expressions, and actions of the people you are observing. On the *Analysis* side, write how these behaviors make it more or less likely that the conflict will be resolved. Select one of the cards, and imagine that you are the mediator in that conflict. Divide the strip of paper into frames, and create a mediation cartoon for this conflict. In a series of drawings with dialogue in cartoon balloons, take the argument through the mediation process.

Observations and Analysis: At the end of the week, share your observations, analyses, and cartoon with your classmates. Find out what other solutions they might suggest.

Understanding Objectives

In this activity, students begin to realize that conflicts are a normal part of life and that many conflicts can be resolved through mediation if both sides are willing.

Observation and Analysis

If there is not sufficient time for a thorough discussion of all examples, have students discuss their observations, analyses, and cartoons in small groups. Afterward, ask each group to choose one example from their discussions to present to the entire class.

Further Investigation

Suggest that students investigate the use of mediation to resolve conflicts among government entities, such as cities, counties, states, and nations.

Assessing the Work Students Do

Ask students to write one-paragraph responses to this question: What did you learn about mediation from participating in this Health Lab activity? Assign credit based on the thoroughness of their responses.

263

Chapter 9 Lesson 5

Visual Learning
Let students describe the two different scenes in Figure 9.9: What do you think the teen in a hurry said and did in each situation? In most cases, what are the consequences of insisting on having your way? of politely requesting what you want?

L2 Applying Knowledge
Let students form small cooperative groups. Instruct the group members to work together in planning and writing a short narrative about a conflict between two or more teens. Have group members conclude their narrative with a nonviolent resolution of the conflict.

L2 Creating Posters
Ask: What impact do fear and concern for personal safety have on students' daily lives? Have students make posters promoting safe schools. Display the posters in hallways and other public areas of the school.

L2 Guest Speaker
Invite a student/peer or professional mediator to speak to the class about his or her work: What kinds of problems does the mediator help to resolve? What attitudes do those in conflict often bring to the mediation process? What approaches does the mediator find most effective?

Assess

EVALUATING THE LESSON
Assign Reviewing Terms and Facts and Thinking Critically on page 265 to review the lesson; then assign the Lesson 5 Quiz in the TCR to evaluate students' understanding.

Q & A

Oh, Brother!

Q: My brother is having a big feud with another guy at school. They always fight in the locker room. Things are really heating up. Now my brother wants me to back him up when they meet after school to fight it out. What should I do?

A: Don't get caught in the middle of someone else's battle, even your brother's. Tell him that fighting will only make it worse. Urge him to talk with a peer counselor or mediator at school. Suggest that he use some of the skills you've learned for nonviolent confrontation.

Use Nonviolent Confrontation

Nonviolent confrontation means *resolving your conflict by peaceful methods.* In nonviolent confrontation, you settle matters without angry words or looks, threats, punches, or weapons. The benefit of nonviolent confrontation is that the argument is likely to be settled so that both parties are satisfied. See **Figure 9.9** and try some of the following guidelines.

- Carefully plan what to say, stay calm, and stick to the subject.
- Pick the right time and place.
- Talk to the other person when he or she is alone.
- Be a good listener; do not interrupt.
- Be sensitive to body language, or any nonverbal communication.
- Be positive; avoid insults, blame, sarcasm, accusations, and threats.
- Be willing to compromise.
- Leave the area if there is a weapon present.

Helping Others Avoid Fights

Friends can help one another to avoid fights by showing disapproval of fighting. For instance, they can refuse to spread rumors, and they can ignore people who talk badly about others. Advising your friends to act in these ways can help them stay safe.

When people you know and care about are starting to argue, you can help without getting hurt. You can assist by sharing what you know about neutrality, mediation, and negotiation.

Figure 9.9
Two Approaches to Conflict
Look at the pictures on this page and the next one. The teens are reacting differently in each picture. Why does one response promote settling the conflict while the other makes it worse?

264

Cooperative Learning

Role-Playing Have a small group of students role-play a conflict involving teens. Tell them to approach the conflict as though it were an athletic competition. Each side should be shown trying to win by defeating the other side. After students have presented their skit, ask the rest of the class to suggest ways in which communication and cooperation, instead of competition, could have been used to arrive at a win-win, rather than a win-lose, solution.

- **Neutrality** (noo·TRA·luh·tee) is *not taking sides when others are arguing.* Avoid fights and urge others to do the same.
- **Mediation** (mee·dee·AY·shuhn) is *resolving conflicts by using a neutral person to help reach a solution that is acceptable to both sides.* Many people use mediation to resolve disagreements.
- **Negotiation** (ni·goh·shee·AY·shuhn) is *the process of discussing problems face-to-face in order to reach a solution.* Negotiation involves talking, listening, considering the other point of view, and compromising.

Toward a Win-Win World

People often think of situations in terms of winning and losing. In relationships, win-lose thinking can leave people feeling angry or cheated.

Mediation is a way to turn a win-lose situation into a win-win situation. Many families, schools, and communities are using mediation as a way to prevent violence. When people resolve their differences peacefully, both sides, as well as society, benefit.

in your journal

Think of a recent argument you had at home or at school. Take the conflict through the mediation steps presented in this lesson. How might the conflict have been resolved? Write your answer in your private journal.

Review — Lesson 5

Using complete sentences, answer the following questions on a separate sheet of paper.

Reviewing Terms and Facts

1. **Give Examples** List three causes of conflicts.
2. **Vocabulary** Compare and contrast the following terms: *neutrality, negotiation.*

Thinking Critically

3. **Suggest** How can you manage your anger before it gets out of control?
4. **Recall** How does *mediation* result in a positive outcome for all people involved in a conflict?

Applying Health Concepts

5. **Health of Others** Do some research on Mohandas Gandhi and Martin Luther King, Jr., and their commitment to nonviolence. Share your findings in class.
6. **Personal Health** List your own signs and symptoms of anger, and decide at what point you are likely to use poor judgment. What do you need to do before you reach this point of anger?

In a hurry? Sure, you can go ahead of me.

Lesson 5: Resolving Conflicts at Home and at School

Home and Community Connection

Mediation in Practice Have students identify and read about situations in which real-life conflicts were resolved through mediation. Possible subjects for research include labor-management conflicts in the workplace and international trade or military conflicts mediated by presidents or other statespeople. Ask students to investigate how mediation was used to resolve the conflict. Also, have them speculate what might have happened if the conflict had not been resolved through mediation. Let students summarize their findings in writing and then share them with the rest of the class.

Chapter 9 Lesson 5

LESSON 5 REVIEW

Answers to Reviewing Terms and Facts

1. Conflicts are usually caused by (any three) arguments, hurt pride, peer pressure, revenge, prejudice.
2. Both neutrality and negotiation are ways to help others avoid fights. Neutrality involves not taking sides when others are arguing; negotiation is the process of discussing problems face-to-face in order to reach a solution.

Answers to Thinking Critically

3. Recognize anger's early signs in order to stay in control, and find a positive way to relieve pent-up feelings.
4. Mediation uses a neutral person to help reach a solution that is acceptable to all people involved.

RETEACHING

▶ Assign Concept Map 43 in the TCR.
▶ Have students complete Reteaching Activity 43 in the TCR.

ENRICHMENT

▶ Assign Enrichment Activity 43 in the TCR.
▶ Have students learn more about peer mediation. Suggest that they read articles in magazines and/or selections from nonfiction books on the topic. In addition, if your school has a peer mediation program, ask students to talk with peer mediators about the types of conflicts they help resolve, the methods they use, and the training they receive.

Close

Let students respond to the following question, either orally or in writing: What do you consider the most important concepts presented in this lesson?

Chapter 9 Review

CHECKING COMPREHENSION

Use the Chapter Summary and the Chapter 9 Review to help students go over the most important ideas presented in Chapter 9. Encourage students to ask questions and add details as appropriate.

CHAPTER 9 REVIEW ANSWERS

Reviewing Key Terms and Concepts

1. Communication is the exchange of thoughts, feelings, and beliefs between two or more people. The two basic types are verbal and nonverbal.
2. Compromise is a method of solving disagreements in which each person gives up something in order to reach a solution that satisfies everyone.
3. The family is the basic unit of society. A blended family includes a parent, a stepparent, and children of one or both parents. An extended family includes a nuclear or single-parent family with other relatives.
4. A family meets its members' physical needs by providing food, clothing, shelter, and care during injury or illness. Meeting emotional needs involves providing acceptance, support, and love.
5. Peer pressure is the influence that your friends have on you to believe and act as they do. Positive peer pressure encourages self-improvement and wellness. Negative peer pressure encourages health and safety risks.
6. Compromise is not appropriate, if negative peer pressure urges you to go against your deepest values, or the rules of your family, school, or community.

Chapter 9 Review

Chapter Summary

▶ **Lesson 1** Communication is the exchange of thoughts, feelings, and beliefs. Effective verbal and nonverbal communication skills help people express themselves clearly and listen actively to others.

▶ **Lesson 2** Families nurture their members by meeting physical, emotional, mental/intellectual, and social needs, and by sharing values.

▶ **Lesson 3** Friends become especially important in the teen years. True friends display sympathy, empathy, loyalty, and reliability, and usually share similar values. Peers can exert both positive and negative peer pressure.

▶ **Lesson 4** Abstinence is the only sure way to protect yourself from the results of risk behaviors, such as engaging in sexual activity and using tobacco, alcohol, or other drugs.

▶ **Lesson 5** Conflicts can be resolved through neutrality, mediation, or negotiation.

Reviewing Key Terms and Concepts

Using complete sentences, answer the following questions on a separate sheet of paper.

Lesson 1
1. Define *communication*, and identify the two basic types.
2. What does it mean to *compromise*?

Lesson 2
3. Define the term *family*. Then describe a blended family and an extended family.
4. Explain how a family can meet its members' physical and emotional needs.

Lesson 3
5. Define *peer pressure*. Then explain how it can be either negative or positive.

6. Why is compromise a poor way to handle negative peer pressure?

Lesson 4
7. List three sources of pressure for teens to engage in risk behaviors.
8. What problems can a teen avoid by abstaining from tobacco?

Lesson 5
9. Define the term *prejudice*, and explain how prejudice can cause a conflict.
10. What is the benefit of nonviolent confrontation?

Thinking Critically

Using complete sentences, answer the following questions on a separate sheet of paper.

11. **Explain** How might a person use good communication to strengthen relationships?
12. **Compare** How are your relationships with family members like your relationships with your friends? In what ways are they different?
13. **Compose** Create a list of three pressure statements (in addition to those provided in the text) and an effective response to each.
14. **Predict** How might poor communication skills lead to conflict?

Chapter 9: Social Health: Families and Friends

Meeting Student Diversity

Language Diversity Use the following suggestions to help students who have difficulty with English:

▶ Pair those learners with native speakers of English who can restate the Chapter Summary in language that helps students comprehend important concepts.

▶ Direct auditory learners or those students with language diversity to the Teen Health Audiocassette Program. Available in English and Spanish, this component provides an audio and written summary of the chapter.

Chapter 9 Review

Your Action Plan

Throughout this chapter you have reflected on your relationships with family members and friends. You have also examined ways to resolve conflicts. You can apply what you have learned in your everyday life.

Step 1 Review your journal entries from this chapter. Think about the communication skills that are your strongest. Then think about the areas in which you might want to strengthen your skills.

Step 2 Summarize your thinking in a two-column chart. Use the chart to set a long-term goal. For example, you might want to get into fewer conflicts.

Step 3 Having decided on a long-term goal, set some short-term goals that will help you reach it. Examples of short-term goals might be finding ways to relieve stress and walking away when a situation is getting out of hand.

Write a contract with yourself in which you commit to reaching your long-term goal in a specific length of time. Decide which of your skills will help you in reaching your long-term goal. Your reward for reaching your goal will be closer, happier relationships with family and friends.

In Your Home and Community

1. **Personal Health** For one week, concentrate on communicating effectively with each member of your family. After the week has passed, write a paragraph that describes the positive results of your improved communication.

2. **Community Resources** Search out groups in your community that work to strengthen families and family relationships. What techniques do they use to help and encourage families? Write a short report on what you have found out.

Building Your Portfolio

1. **News Analysis** Find one or more articles in newspapers and magazines about a conflict between political opponents, workers and managers, or athletes and team owners. Report on how negotiation was used in resolving the conflict. If negotiations did not resolve the conflict, suggest a possible solution.
2. **Short Story** Write a short story in which a teen effectively uses refusal skills to stand up to negative pressure. Remember that pressure can come from anyone. Make the situation and dialogue as realistic as possible.
3. **Personal Assessment** Look through all the activities and projects you did for this chapter. Choose one or two that you would like to include in your portfolio.

Chapter 9: Chapter Review **267**

Performance Assessment

- ▶ **Self-evaluation** Direct students to review the activities that are provided throughout the chapter. Encourage each student to select one finished product or activity that demonstrates his or her best work for the chapter. Have students explain what they learned and how the examples they selected show their progress.

- ▶ **Teacher's Classroom Resources** Assign Performance Assessment 9, "Communication Pamphlets," in the TCR.

Chapter 9 Review

7. Responses will vary. Possible response: Pressure to engage in risk behaviors can come from (any three) peer, adult role models, advertising, movies, television.
8. Abstinence from tobacco protects against loss of physical fitness, damage to the lungs and heart, and dependence on nicotine.
9. Prejudice is an opinion about people formed without having facts about those people. Those who show prejudice may single out a person to harass, intimidate, or threaten. That person may strike back, either alone or with supporters, and fights may occur.
10. Nonviolent confrontation allows conflict to be clarified and resolved peacefully.

Thinking Critically
11. Possible response: Expressing yourself clearly allows you to let friends and family members know what you need and feel; listening skills help you understand friends and family. These skills deepen relationships.
12. Possible response: Both family members and friends give and receive caring, support, and encouragement. Family members are provided, and friends are chosen.
13. Responses will vary.
14. Possible response: Poor communication skills could lead someone to say something in a way that would offend someone else; it might also lead someone to misunderstand what another person has said. These situations can lead to conflicts.

RETEACHING
Assign Study Guide 9 in the Student Activities Workbook.

EVALUATE
- ▶ Use the reproducible Chapter 9 Test in the TCR, or construct your own test using the Testmaker Software.
- ▶ Use Performance Assessment 9 in the TCR.

267

Unit 4
Protecting Your Health

UNIT OBJECTIVES

Students will become familiar with the dangers posed by tobacco, alcohol, and other drugs, and they will learn how to make a commitment to being tobacco-, alcohol-, and drug-free. Students will also learn about recognizing, avoiding, and treating common communicable and noncommunicable diseases.

UNIT OVERVIEW

Chapter 10 Tobacco

As they read Chapter 10, students learn about the specific health risks of tobacco and tobacco smoke, particularly the risks to the human respiratory system. They also examine the problems of tobacco dependency and addiction.

Chapter 11 Alcohol and Drugs

Chapter 11 helps students recognize and avoid the dangers of alcohol and other drugs. Students learn about the human nervous system and how it can be affected by drugs, including alcohol.

Chapter 12 Understanding Communicable Diseases

In Chapter 12, students learn about common communicable diseases and the role of the human immune system in resisting and fighting these diseases. In the final lesson, students learn about sexually transmitted diseases and HIV/AIDS.

Chapter 13 Understanding Noncommunicable Diseases

Chapter 13 makes students familiar with major noncommunicable diseases, including heart disease, cancer, allergies, and asthma.

Bulletin Board Suggestion

Putting a STOP to Unhealthy Behaviors Make a STOP sign out of red construction paper, and post it in the middle of a bulletin board. Around the sign, post pictures or sentences identifying behaviors that represent risks to students' health. For example, you might post pictures of a hand offering an open pack of cigarettes, a teen sneezing without covering his or her mouth and nose, and a can of beer. Let students discuss the health risks posed by each picture or description on the board. You may want to let volunteers make copies of the "no" symbol (a circle with one diagonal line) to put over the pictures. Then, as the class studies the chapters in Unit 4, encourage students to watch for other unhealthy behaviors, and have volunteers draw pictures, cut out photographs, or write brief descriptions to be added to the display.

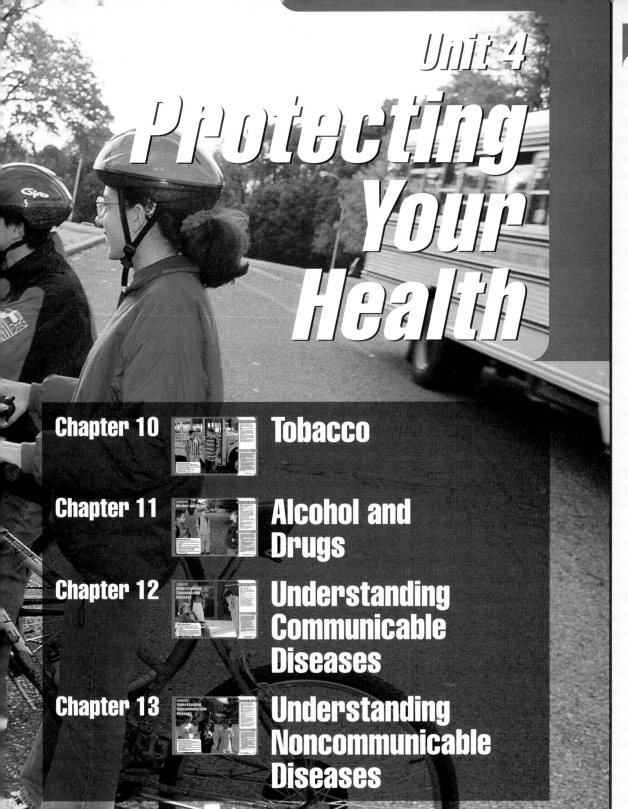

Unit 4

Protecting Your Health

- **Chapter 10** Tobacco
- **Chapter 11** Alcohol and Drugs
- **Chapter 12** Understanding Communicable Diseases
- **Chapter 13** Understanding Noncommunicable Diseases

Unit 4

INTRODUCING THE UNIT

Write the unit title, *Protecting Your Health*, on the board. Let students form small groups, and have the members of each group work together, listing as many ways to "protect your health" as possible. Suggest that one group member assume the job of recorder, writing down each idea. Set a time limit (six or seven minutes, for example) and challenge students to see which group can generate the longest list. At the end of the assigned time, let groups count their recorded precautions and compare their totals. Then ask groups to share some of their ideas. Ask: How does this behavior or attitude help protect your health? Help students determine which precautions were most commonly listed by the groups. Also help them identify the least commonly listed precautions. Finally, let students point out the precautions that deal with tobacco, alcohol and other drugs, communicable diseases, and noncommunicable diseases, and tell students that they will learn more about these issues in Unit 4.

VIDEODISC/VHS

Teen Health Course 2

You may wish to use video segment 12, "Understanding Communicable Diseases," to show how germs get spread and how good hygiene can keep them in check.

Videodisc Side 2, Chapter 12
Understanding
Communicable Diseases

Search Chapter 12, Play to 13

DEALING WITH SENSITIVE ISSUES

Sharing Knowledge Sharing knowledge with students about sensitive issues will give them the information they need to make informed decisions and to solve problems wisely. Unless students are fully aware that eating disorders are life-threatening conditions, for example, they may not know how they can help a friend with anorexia. Similarly, unless students know that heterosexual activities can lead to the transmission of HIV, they may not appreciate the vital importance of sexual abstinence. Sharing accurate information about sensitive issues will also help make students aware that many other youths have similar problems and concerns. As a result, students feel less isolated and less different from their peers.

Planning Guide

Chapter 10
Tobacco

	Features	Classroom Resources
Lesson 1 **What Tobacco Does to Your Body** *pages 272–275*	Health Lab: Smoking and Breathing *page 275*	Concept Map 44 Cross-Curriculum Activity 19 Enrichment Activity 44 Lesson Plan 1 Lesson 1 Quiz Reteaching Activity 44 Transparencies 38, 39
Lesson 2 **The Respiratory System** *pages 276–279*  *pages 280–281*		Concept Map 45 Enrichment Activity 45 Lesson Plan 2 Lesson 2 Quiz Reteaching Activity 45 Transparency 40
Lesson 3 **Tobacco Addiction** *pages 282–286*	Life Skills: Analyzing a Media Message *page 285*	Concept Map 46 Decision-Making Activity 19 Enrichment Activity 46 Lesson Plan 3 Lesson 3 Quiz Reteaching Activity 46 Transparency 41
Lesson 4 **Choosing to Be Tobacco Free** *pages 287–291*	Making Healthy Decisions: Overcoming the Pressure to Smoke *pages 288–289*	Concept Map 47 Cross-Curriculum Activity 20 Decision-Making Activity 20 Enrichment Activity 47 Health Lab 10 Lesson Plan 4 Lesson 4 Quiz Reteaching Activity 47

National Health Education Standards

One of the main goals of the National Health Education Standards is to move students toward "health literacy." Health literacy is the capacity of individuals to obtain, interpret, and understand basic health information and services and the competence to use such information and services in ways which promote health. The health standards were developed by applying the characteristics of a well-educated, literate person within the context of health. The health literate person is:

- ▶ a critical thinker and problem solver
- ▶ a self-directed learner
- ▶ a responsible, productive citizen
- ▶ an effective communicator

Listed below are the Health Education Standards Performance Indicators addressed in each lesson of this chapter.

Lesson	Health Standards Performance Indicators
1	(1.1, 1.3, 1.8)
2	(1.1, 1.7, 3.1)
3	(1.3, 1.4, 1.6, 4.2)
4	(5.6, 6.1, 6.3)

ABCNEWS INTERACTIVE VIDEODISC SERIES

You may wish to use the following videodiscs with this chapter: *Tobacco,* side one, video segments 3, 10, 11, 12; side two, video segments 3, 6, 7, 9, 10, 11, 14, 15, 16, 17, 18, 20; *Making Responsible Decisions.* Use the *ABCNews InterActive™ Correlation Bar Code Guide* for title reference. Also available in VHS format.

Chapter Resources

Teacher's Classroom Resources
- Chapter 10 Test
- Parent Letter and Activities 10
- Performance Assessment 10
- Testmaker Software

Student Activities Workbook
- Study Guide 10
- Applying Health Skills 44–47
- Health Inventory 10

Student Diversity Strategies
- Audiocassette Program (English)
- Audiocassette Program (Spanish)
- Spanish Parent Letters
- Spanish Summaries, Quizzes, and Activities

Multimedia Components
- English Audiocassette Program
- Spanish Audiocassette Program
- *Teen Health* Videodisc/VHS Series

Other Resources

Readings for the Teacher
Glantz, Stanton A., et al. *The Cigarette Papers.* Berkeley: University of California Press, 1996.

Monroe, Judy. *Nicotine.* Springfield, NJ: Enslow, 1995.

Sullivan, M. *Peer Pressure: A Parent/Child Manual.* New York: TOR Press, 1991.

Readings for the Student
Lang, Susan, and Beth Marks. *Teens and Tobacco: A Fatal Attraction.* New York: Twenty-First Century Books, 1996.

Pringle, Laurence P. *Smoking: A Risky Business.* New York: Morrow Junior Books, 1996.

Silverstein, Alvin, et al. *The Respiratory System.* New York: Twenty-First Century Books, 1994.

Out of Time?
If time does not permit teaching this chapter, you may use these features: Health Lab on page 275; Teen Health Digest on pages 280–281; Life Skills on page 285; Making Healthy Decisions on pages 288–289; and the Chapter Summary on page 292.

Chapter 10
Tobacco

CHAPTER OVERVIEW

Chapter 10 introduces students to the dangers of using tobacco products. It explains the effects of tobacco on the respiratory and other body systems and offers strategies for avoiding—or, if necessary, breaking—the tobacco habit.

Lesson 1 identifies the toxic substances in tobacco and tobacco smoke and explains how these substances harm the body.

Lesson 2 explains the parts and functions of the respiratory system and helps students understand how they can keep this body system healthy.

Lesson 3 focuses on the social pressures and advertising techniques that often lead teens into dependency and addiction.

Lesson 4 examines reasons teens try tobacco and offers them strategies for saying no when offered tobacco.

PATHWAYS THROUGH THE PROGRAM

Some of the most compelling, yet controversial, research in the area of health behavior and total wellness deals with the impact of tobacco on the health of smokers and on the health of nonsmokers who breathe tobacco smoke. The connection between tobacco use and life-threatening health problems is well documented. Teens who begin smoking seldom realize that addiction to nicotine is as strong as addiction to other drugs. Although cigarettes, cigars, and tobacco products carry warning labels, they are not banned substances; therefore, teens may be tempted to view them as less damaging than unlawful drugs. Throughout this chapter, emphasize to students the damaging effects tobacco has on physical, mental/emotional, and social health.

Chapter 10
Tobacco

Student Expectations

After reading this chapter, you should be able to:

1. Identify the harmful effects of tobacco on the body.
2. Describe how your respiratory system works and how it may be affected by tobacco.
3. Discuss how tobacco addiction develops.
4. Identify ways to say no to tobacco and to learn skills that can help break the tobacco habit.

Key to Ability Levels

Teaching strategies that appear throughout the chapter have been identified by one of three codes to give you an idea of their suitability for students of varying learning styles and abilities.

L1 Level 1 strategies should be within the ability range of all students. Full class participation is often required. Teacher direction is usually needed.

L2 Level 2 strategies are for average to above-average students or for small groups. Some teacher direction is necessary.

L3 Level 3 strategies are designed for students able and willing to work independently. Minimal teacher direction is necessary.

Teen Chat Group

Carlos: What's wrong, Rob?

Rob: My mom caught my brother smoking last night. They had a big fight.

Carlos: I thought you said he was quitting.

Rob: He tried to, but it was really hard. My mom said he just wasn't trying hard enough. But my brother says she doesn't realize how hard it is to quit.

Carlos: Well, think about how she feels. I don't like it when he smokes around me either. The smoke makes my clothes stink.

Rob: I know. It's so bad for him. I'm scared he might get cancer or something. I wish he'd never started.

in your journal

Read the dialogue on this page. Do you know about people who have experienced problems in their lives because of a tobacco habit? Start your private journal entries on tobacco use by answering these questions:

▶ If you were Rob's brother, how would you solve the problems caused by your cigarette habit?
▶ Why do you think teens start smoking?
▶ How does the decision to use tobacco affect a person's physical, mental/emotional, and social health?
▶ How does a person's use of tobacco affect the lives of other people?

When you reach the end of the chapter, you will use your journal entries to make an action plan.

Chapter 10: Tobacco

Chapter 10

INTRODUCING THE CHAPTER

▶ The American humorist and writer Mark Twain is famous for this statement: "To quit smoking is the easiest thing I ever did; I ought to know because I did it hundreds of times." Help students appreciate and discuss Twain's statement: What makes it a fitting description of tobacco's influence? How has the influence of tobacco changed, if at all, since Twain's time? (Twain died in 1910.) What do you think modern teens can learn from Twain's humorous remark?

▶ Invite a former smoker who has recently given up tobacco to speak to the class about the experience. Encourage the speaker to discuss with students the best and worst aspects of becoming tobacco free.

Cooperative Learning Project

Talent Show

The Teen Health Digest on pages 280 and 281 provides students with high-interest articles related to the content of this chapter.

The material in the Teacher's Wraparound Edition presents suggestions for a class project in which students will plan and perform a talent show emphasizing the importance of avoiding tobacco.

---

in your journal

Despite widespread knowledge of the health risks associated with tobacco consumption, tobacco use has not decreased among adolescents. At greatest risk are teens whose parents smoke or condone tobacco use. Preventing chronic use and addiction is a pressing public health concern. Experimentation with tobacco should never be sanctioned; however, the majority of teens who use tobacco in response to peer pressure or in isolated acts of rebellion should be encouraged to evaluate the long-term negative consequences of their behavior.

Lesson 1

What Tobacco Does to Your Body

Focus

LESSON OBJECTIVES

After studying this lesson, students should be able to

- discuss the harmful substances found in tobacco products.
- name several forms of tobacco.
- list the harmful effects tobacco has on various body systems.
- analyze the high cost of tobacco use on the health of society.

MOTIVATOR

Ask students to list reasons they have heard people give for smoking or chewing tobacco.

INTRODUCING THE LESSON

Have students share their responses to the Motivator activity. Then ask students what reasons they have heard people give for *not* using tobacco. Ask: Based on what you know about smoking now, is there any reason to run the risk of using tobacco? Let students vote, and have volunteers explain their responses. Tell students that they will learn more about the contents of cigarettes and the risks of using tobacco products as they study Lesson 1.

INTRODUCING WORDS TO KNOW

Write the Words to Know on the board. For each word, have a volunteer supply the correct definition and two false definitions (in any order). Let students vote for the definition they think is the correct one and verify their responses by checking the Glossary.

272

Lesson 1

What Tobacco Does to Your Body

This lesson will help you find answers to questions that teens often ask about using tobacco. For example:

▶ What is in tobacco that causes health problems?
▶ Is tobacco harmful in all of its forms?
▶ What parts of my body would be affected by tobacco?

Words to Know

nicotine
tar
carbon monoxide

in Your journal

In your journal, write some questions you have about smoking. If any of the questions have not been answered by the end of this chapter, ask a parent, teacher, or health care provider to answer them. Then write the answers in your journal.

The Facts About Tobacco

A single puff of tobacco smoke exposes the body to more than 4,000 chemicals. Almost all of these make the body unable to work as it should. At least 43 of the chemicals in tobacco smoke are known to cause cancer in smokers. Smoke also harms the health of nonsmokers. Even smokeless tobacco causes health problems, including cancer, in its users.

What Is in Tobacco?

Three substances in tobacco smoke are especially harmful to your health. **Nicotine** (NIK·uh·teen) is *an addictive drug found in tobacco*. This drug makes tobacco users crave more nicotine. **Tar** is *a thick, dark liquid that forms when tobacco burns*. This liquid covers the lining of the lungs, causing disease. **Carbon monoxide** (KAR·buhn muh·NAHK·syd) is *a colorless, odorless, poisonous gas produced when tobacco burns*. All of these dangerous substances are in tobacco smoke whether the user is smoking a cigarette, a cigar, or a pipe (see **Figure 10.1** on the next page).

Tobacco in Many Forms

Tobacco products come in several forms, which are smoked or chewed. The most commonly used form is cigarettes.

In this smoker's lung, chemicals from tobacco smoke have coated the inner structures. This makes the lungs work less efficiently and can cause serious diseases, such as cancer and emphysema.

272 Chapter 10: Tobacco

Lesson 1 Resources

Teacher's Classroom Resources
- Concept Map 44
- Cross-Curriculum Activity 19
- Enrichment Activity 44
- Lesson Plan 1
- Lesson 1 Quiz

- Reteaching Activity 44
- Transparencies 38, 39

Student Activities Workbook
- Study Guide 10
- Applying Health Skills 44

Cigarettes

Cigarettes are made from shredded tobacco leaves. Although filters can reduce the amount of nicotine and tar in cigarette smoke, they do not help to decrease the amount of carbon monoxide and other disease-causing chemicals. Some kinds of cigarettes contain other ingredients, such as spices, that make the cigarettes taste and smell sweet.

Smokeless Tobacco

Two forms of tobacco are placed in the mouth instead of being smoked. Chewing tobacco is made from compressed, coarsely ground leaves. A wad is placed between the cheek and gum, where it is sucked or chewed. Snuff is a finely ground, powdery substance. It is placed between the lower lip and gum, where it mixes with saliva and is absorbed.

Smokeless tobacco is not a safe alternative to smoking tobacco. The nicotine in smokeless tobacco is just as harmful and addictive as that in cigarette smoke. In addition, smokeless tobacco is linked to an increased incidence of mouth cancer and cancers of the esophagus, larynx, and pancreas. It also causes inflamed gums, bad breath, yellowed teeth, and stomach ulcers.

interNET CONNECTION

Learn on-line how to keep your lungs healthy, including tips on how to resist peer pressure and stay tobacco free.

http://www.glencoe.com/sec/health

Figure 10.1
Some Harmful Substances in Tobacco Smoke
Tobacco smoke contains many dangerous chemicals.

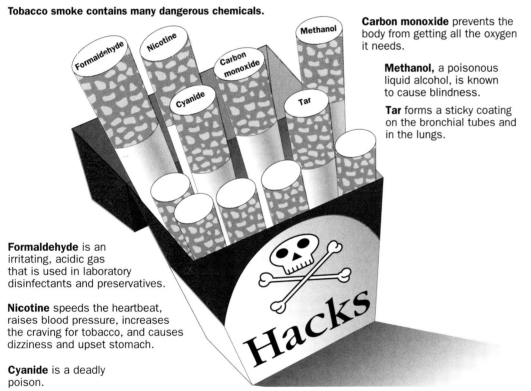

Carbon monoxide prevents the body from getting all the oxygen it needs.

Methanol, a poisonous liquid alcohol, is known to cause blindness.

Tar forms a sticky coating on the bronchial tubes and in the lungs.

Formaldehyde is an irritating, acidic gas that is used in laboratory disinfectants and preservatives.

Nicotine speeds the heartbeat, raises blood pressure, increases the craving for tobacco, and causes dizziness and upset stomach.

Cyanide is a deadly poison.

Lesson 1: What Tobacco Does to Your Body   **273**

Cooperative Learning

News about Tobacco Divide the class into several learning groups. Have the members of each group work together to draw a life-size outline of a teen's body on a strip of butcher paper. Then instruct group members to locate, clip, and read articles (as from newspapers and magazines) about the health effects of tobacco. For each article, have students draw a corresponding illustration inside the body outline, then post the article outside the body outline. Let members of each group show their completed drawing and summarize their news articles for the rest of the class. Display their completed work on walls in the classroom or in school hallways.

Chapter 10 Lesson 1

Teach

Teacher Talk
Dealing with Sensitive Issues

Many students may resist recognizing the negative qualities of tobacco use. Their responses may convey this attitude: Tobacco *really can't be that dangerous—just look at all the people who smoke or chew regularly.* Bring in guest speakers who suffer from chronic diseases attributed to tobacco use. (A local chapter of the American Cancer Society can recommend speakers.) Usually, such speakers are eager to tell students how they themselves ignored, or were ignorant of, the long-term consequences when they first started using tobacco.

Visual Learning
Use Figure 10.1 to help students practice the pronunciation and spelling of the harmful substances in tobacco smoke. Have students create ways to use the illustration as a learning aid. For example, a rap rhythm might help:
   nicotine, cyanide,
   methanol, formaldehyde,
   carbon monoxide, tar.

L3 Social Studies
Have students research current local, state, and/or federal laws regulating the use and sale of tobacco products. Let students meet in groups to discuss their findings.

Would You Believe?

It's against the law to use formaldehyde to preserve dissection specimens for school use—but the chemical is still a legal component of cigarettes.

Chapter 10 Lesson 1

Health Minute

The tobacco industry continues to research new, "safer" cigarettes. Although the gases given off by the so-called smokeless cigarettes may be invisible, the burning tobacco still produces the same dangerous poisons as other cigarettes. Tobacco, regardless of the type of cigarette, exposes both smokers and nonsmokers to harmful effects.

L1 Demonstrating

To demonstrate the harmful effects of the chemicals in tobacco, find a plant infested with aphids and cut off a twig or leaf cluster that has many insects on it. Remove the paper from several cigarettes. Boil the tobacco in a cup of water for 15 minutes. Strain the solution through filter paper or a paper towel into a glass jar. Add a drop of detergent to this tobacco solution so it will stick to the aphids and plant clipping. Pour the solution into an atomizer or other small spray can. Spray the twig or leaves on top and bottom. (Do not breathe in the spray.) Observe what happens to the aphids. (The solution kills the aphids because nicotine, a chemical in tobacco, is poisonous.)

EVALUATING THE LESSON

Assign Reviewing Terms and Facts and Thinking Critically on page 275 to review the lesson; then assign the Lesson 1 Quiz in the TCR.

Math Connection

The Price of Smoking ACTIVITY!

Smokers not only damage their bodies; they spend money on the products that hurt them. Calculate the cost of cigarettes over the course of one year, two years, five years, and ten years, at $2.50 per pack. Assume that a smoker buys one pack of cigarettes a day.

Pipes and Cigars

Pipes and cigars also use shredded tobacco leaves, some of which may be flavored. Pipe and cigar smokers develop lung cancer less often than cigarette smokers because they usually inhale less smoke. However, pipe and cigar smokers are more likely to develop cancers of the lip, mouth, tongue, and throat.

Tobacco and Your Body

Tobacco damages the body in many ways. **Figure 10.2** illustrates tobacco's effect on five body systems.

Figure 10.2
What Tobacco Does to the Body
Tobacco has harmful effects on nearly every body system.

Body System	Effects of Tobacco Use
Nervous system	Smoking reduces the flow of oxygen to the brain, possibly leading to a stroke.
Respiratory system	Tar and other chemicals leave a sticky residue that destroys structures in the lungs. Smoking also damages the alveoli (al·VEE·uh·ly), the tiny air sacs in the lungs. This damage causes difficulty in breathing and prevents oxygen from getting to the rest of the body. Smokers are ten times more likely than nonsmokers to develop lung cancer.
Circulatory system	Smoking weakens the blood vessels. Smoking also causes a fatty buildup that clogs the blood vessels, increasing the risk of heart attack or a stroke.
Digestive system	Tobacco causes bad breath. It stains the teeth and makes them susceptible to cavities. Tobacco dulls the taste buds and can cause cancer of the mouth and throat. It is also a cause of stomach ulcers.
Excretory system	Smoking increases the danger of bladder cancer. Smokers have twice the risk of bladder cancer that nonsmokers have.

The Costs to Society

Smokers pay a high price for their tobacco habit, including the price of tobacco products and the cost of health care. Nonsmokers who are exposed to the tobacco smoke of others also pay a price. They are at increased risk for lung cancer and other respiratory diseases. The developing babies of pregnant women who smoke can also suffer serious effects. To help reduce all these costs to society, the government has regulated the tobacco industry. In 1965 health warnings began to appear on cigarette packs. In 1971 cigarette advertisements were banned from radio and television. New regulations, approved in 1996, limit the access persons under 18 have to tobacco. Store owners must verify the age of a person who is purchasing tobacco products, and cigarette vending machines are allowed only in places where people under 18 are not admitted.

HEALTH LAB
Smoking and Breathing

Time Needed
2 or 3 class periods

Supplies
▶ squeezable rubber bulbs
▶ rubber or glass tubing
▶ clean white handkerchiefs
▶ cigarettes
▶ matches

Focus on the Health Lab
Let volunteers name some of the most harmful substances in tobacco smoke. Ask: What is intended to go into your lungs when you breathe? What goes into your lungs when you smoke a cigarette? Explain that students will have a chance to see some of the effects of cigarette smoke as they complete the Health Lab activity.

Understanding Objectives
Students should become more aware of the respiratory system and the harmful effects of the chemicals in tobacco.

Review — Lesson 1

Using complete sentences, answer the following questions on a separate sheet of paper.

Reviewing Terms and Facts

1. **Vocabulary** Describe the following substances in tobacco smoke: *nicotine, tar,* and *carbon monoxide.*
2. **List** Name three forms of tobacco.
3. **Give Examples** List four ways in which smoking or chewing tobacco harms the body.

Thinking Critically

4. **Explain** If you had a friend who chewed tobacco, how could you persuade him to quit using it?
5. **Hypothesize** Why do you think it is difficult to quit smoking?

Applying Health Concepts

6. **Health of Others** Collect newspaper and magazine articles about the harmful effects of using tobacco. Use the articles and draw illustrations to make a bulletin board display.

HEALTH LAB
Smoking and Breathing

Introduction: For people who smoke, breathing deeply can be difficult. Tar covers their airways and lungs, paralyzing or destroying the hairs that trap dirt from air. Without these hairs, smoke that is inhaled into the lungs deposits harmful gases and particles in the bronchi and alveoli. Taking deep breaths of fresh air causes irritation, causing "smoker's cough."

Objective: Work with a partner to recognize that chemicals produced by cigarette smoke stay inside the smoker's body.

Materials and Method: You will need the following materials: a squeezable rubber bulb, rubber or glass tubing, a clean white handkerchief, cigarettes, and matches. You will also need a sheet of paper to record your results. Divide the sheet into two columns: *Observations* and *Analysis.* In the observations column, write what occurred during the experiment. In the analysis column, write your interpretation of the experiment.

Attach the bulb to the tubing, placing a handkerchief between the bulb and the tubing. Compress the bulb, and hold a lighted cigarette at the other end of the tube. Release the bulb, drawing the smoke through the tube and handkerchief. Repeat this procedure three or four times, until tar accumulates on the handkerchief. Put out the cigarette safely.

Observations and Analysis: Do you think that what happened to the handkerchief happens to the smoker's body, too? Where do you think the chemicals are deposited in the body? What harm might that do? Share your analysis of the experiment with your classmates. Discuss the effects of smoking on breathing.

Lesson 1: What Tobacco Does to Your Body

Chapter 10 Lesson 1

LESSON 1 REVIEW

Answers to Reviewing Terms and Facts

1. Nicotine is an addictive drug. Tar is a thick, dark liquid. Carbon monoxide is a colorless, odorless, poisonous gas.
2. Forms of tobacco include (any three): cigarettes, chewing tobacco, snuff, pipe tobacco, cigars.
3. Any four: Smoking causes emphysema; cancer of the lungs, mouth, throat, and bladder; increases risk of heart disease, stroke, and stomach ulcers.

Answers to Thinking Critically

4. You could tell your friend that chewing tobacco damages the teeth and gums and causes oral cancer and several other types of cancer. In addition, the nicotine in chewing tobacco is addictive.
5. Nicotine is addictive.

RETEACHING

▶ Assign Concept Map 44 in the TCR.
▶ Have students complete Reteaching Activity 44 in the TCR.

ENRICHMENT

▶ Assign Enrichment Activity 44 in the TCR.
▶ Ask students to research the influence of tobacco on the U.S. economy.

Ask students to explain how smoking endangers their physical health.

Observation and Analysis

Ask: How do you think a heavy smoker might feel when conducting or observing this experiment? How do you think an occasional smoker might respond? What excuses might smokers give for ignoring the observations they make during the activity?

Further Investigation

Ask students to create and practice a variety of tactful yet effective requests they might use to ask a smoker not to smoke in their presence; these requests should focus on the harmful effects of secondhand smoke. Hold a contest to select the most convincing and/or creative request.

Assessing the Work Students Do

Ask students to imagine that their respiratory systems can talk: Have students draw cartoons or write paragraphs showing or describing the reaction of their lungs to cigarettes. Then ask students to explain how they can apply what they have learned in this Health Lab to future situations.

Lesson 2
The Respiratory System

Focus

LESSON OBJECTIVES
After studying this lesson, students should be able to
- identify the parts of the respiratory system.
- explain how the breathing process works.
- identify problems of the respiratory system.
- explain what they can do to keep their respiratory system healthy.

MOTIVATOR
Pose these questions: What happens when you breathe onto a mirror or a pane of glass? What do you think causes that effect? Ask students to sketch and/or write short responses.

INTRODUCING THE LESSON
Let several volunteers share their responses to the Motivator questions. Guide students as necessary in recognizing that the "fog" on the mirror or pane of glass is water vapor (a gas) that condenses (becomes liquid). If you have a mirror available, you may want to let several volunteers demonstrate. Then ask: What do you take into your body when you breathe in? (oxygen) In addition to the water vapor, what do you breathe out? (carbon dioxide) Tell students that Lesson 2 will help them learn more about how and why they breathe.

INTRODUCING WORDS TO KNOW
Guide students in reading aloud the Words to Know, and ask volunteers to find and read the definitions. Have students work with partners to draw simple diagrams labeled with the Words to Know and their definitions.

276

Lesson 2
The Respiratory System

This lesson will help you find answers to questions that teens often ask about the respiratory system. For example:
- What are the parts of the respiratory system?
- What is my body doing when I breathe?
- How can I take care of my lungs?

Words to Know
respiratory system
trachea
bronchi
diaphragm
alveoli

in your journal
Most of the time, you breathe without even thinking about it. Sit down and concentrate on your breathing for one minute. Then write a description in your journal of how breathing makes your body feel. Do you think that someone who smokes tobacco breathes differently? How and why?

Oxygen for Life
Take a deep breath. Now let the air out. These two motions, inhaling and exhaling, are the basic actions of your **respiratory system,** *the set of organs that supply your body with oxygen and rid your body of carbon dioxide.* This process is crucial because you can only survive for a few minutes without oxygen.

Parts of the Respiratory System
Many body parts work together to help you breathe. Air goes in and out through your nose and mouth. The lungs exchange oxygen and carbon dioxide. Muscles in the chest allow the lungs to expand. The table in **Figure 10.3** lists and describes many of the body parts that make breathing possible.

Figure 10.3
Parts of the Respiratory System
All of these structures work together to help you breathe. Damage to any one of them can make breathing difficult.

Body Part	Description
Nose/Mouth	Passages for air; nose lined with cilia (SIH·lee·uh), fine hairs that trap dirt from air
Trachea (TRAY·kee·uh)	Tube in throat that takes air to and from lungs (also called the windpipe)
Epiglottis	Flap of tissue in back of mouth that covers the trachea to prevent food from entering
Bronchi (BRAHNG·ky)	Two tubes that branch from the trachea, one to each lung
Lungs	Two large organs that exchange oxygen and carbon dioxide
Diaphragm	Large dome-shaped muscle below the lungs that draws air in and pushes air out

Chapter 10: Tobacco

Lesson 2 Resources

Teacher's Classroom Resources
- Concept Map 45
- Enrichment Activity 45
- Lesson Plan 2
- Lesson 2 Quiz
- Reteaching Activity 45
- Transparency 40

Student Activities Workbook
- Study Guide 10
- Applying Health Skills 45

How Breathing Works

Breathing consists of three main stages. When you inhale, you take air and oxygen into your lungs. There oxygen enters the bloodstream, replacing the carbon dioxide that must leave the body. Then you exhale, breathing out the carbon dioxide. These three stages repeat in a cycle. Look again at the parts of the respiratory system listed on the previous page. Now find them in **Figure 10.4**, which shows the steps in the breathing cycle.

Figure 10.4
The Breathing Process
Every day you take about 25,000 breaths of air.

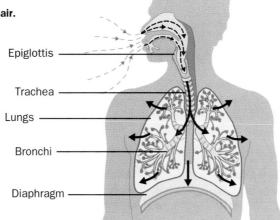

❶ Inhaling
Your diaphragm (DY·uh·fram) moves down, and your rib cage expands, creating more room in your chest. This expansion causes air to rush in. The air enters the nose and mouth, then moves past the epiglottis into the trachea and bronchi.

Epiglottis
Trachea
Lungs
Bronchi
Diaphragm

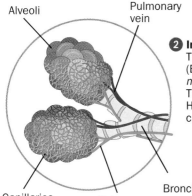

Alveoli
Pulmonary vein
Capillaries
Pulmonary artery
Bronchiole

❷ Inside the Lungs
The bronchi divide into smaller passages called bronchioles (BRAHN·kee·ohlz). Air passes through to the **alveoli** (al·VEE·uh·ly), *microscopic air sacs in the lungs where gases are exchanged.* Tiny blood vessels called capillaries surround the alveoli. Here the oxygen in the air moves into your bloodstream, and carbon dioxide enters the alveoli.

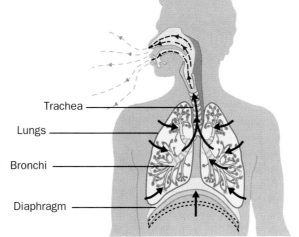

Trachea
Lungs
Bronchi
Diaphragm

❸ Exhaling
Your diaphragm pushes up, and your ribs move in and down, forcing air out of your lungs. The air, now containing carbon dioxide, moves back through the bronchioles and bronchi, up the trachea, and out through the nose and mouth.

Lesson 2: The Respiratory System   **277**

More About...

Bronchitis Acute bronchitis comes on suddenly and is usually caused by a virus entering the respiratory system and settling in the bronchial tubes. Treatment for acute bronchitis is medication to halt the infection. Chronic bronchitis is usually caused by smoking. A person suffering from chronic bronchitis has a continuing heavy, barklike cough. The lungs secrete an excessive amount of sticky mucus as a means of dealing with the irritating smoke. The "smoker's cough" is a warning that the body may not be able to counteract the irritation of continual smoke and that the lung's passageways are narrowing. This narrowing causes pressure on the alveoli, which sometimes collapse, leading to emphysema.

Chapter 10   Lesson 2

Teach

L1 Discussing
As students begin reading the lesson, encourage them to follow the instructions: Take a deep breath and let it out. What changes do you notice? How did the action make you feel? Ask students to share any experiences they may have had in holding their breath, perhaps for longer than was comfortable.

Visual Learning
Have students describe each part of the respiratory system using Figures 10.3 and 10.4: Where is it? What does it do? Then let students use their fingers to trace the path of an inhalation and an exhalation.

Health Minute
Cilia in the nose and trachea play an important role in preserving the health of the respiratory system. These fine hairs trap dirt and prevent it from entering the lungs. (Tar from tobacco products impairs the functioning of the cilia.) In addition, the body produces mucus to trap dirt. When you blow your nose or cough up mucus, you are ridding yourself of dirt that your body has prevented from entering your lungs.

L2 Guest Speaker
Ask a singer or voice teacher to discuss and demonstrate the process of diaphragm breathing. Suggest that students try this form of breathing. Ask: How is it different from normal, automatic breathing?

L2 Math
Ask the class to calculate responses to these questions: About how many breaths do you take in a year? How many breaths have you taken since you were born? How many more breaths will you take before finishing high school?

277

Chapter 10 Lesson 2

L1 Discussing
Encourage students to discuss their own experiences with having and treating colds and flu: What are the most familiar symptoms of these infections? How do you usually treat those symptoms? Under what circumstances are you likely to stay home from school with a cold or the flu? Why? How might tobacco smoke impact a cold or flu?

Would You Believe?
Young children who live in families with at least one smoking adult are three times as likely to suffer from asthma as children in smoke-free homes.

L2 Demonstrating
To help others understand the sensation of an asthma attack, one asthma sufferer has described it as running in place for two minutes, then trying to breathe through a straw, with the nose and the rest of the mouth blocked off. Suggest that students act out this description.

L2 Reporting
Let volunteers read more about the two respiratory problems directly related to smoking, emphysema and lung cancer. How many cases are diagnosed each year in this country? How many of those patients are or were smokers? How many live or lived with smokers? What treatment is recommended for each disease? What is the outlook for patients? Provide time for these volunteers to discuss their findings with the rest of the class.

L3 Social Studies
Let interested students research answers to these questions: What current laws regulate acceptable levels of ozone and particulate matter? What goals for smog and soot have been set, either for the nation or for your state or community? How are those goals being met? Have the students present their findings in a chart or poster.

Problems of the Respiratory System

As **Figure 10.4** shows, the respiratory system is made up of many complex and delicate parts. Damage to the system may occur in several ways. For example, germs may enter through the nose and mouth. Tobacco smoking, inhaled chemicals, or environmental pollution may also cause damage. Look at **Figure 10.5** at the bottom of this page. It describes some of the illnesses that can harm your respiratory system. Which ones can be most easily avoided?

Caring for Your Respiratory System

If you've ever held your breath too long in the swimming pool or gagged on fumes from a passing truck, then you know what it's like when you can't breathe freely. Your whole body depends on your respiratory system and on the air you breathe. **Figure 10.6** on the next page shows several ways to take good care of your respiratory system.

Figure 10.5
Problems of the Respiratory System
Some respiratory problems are common; others are rare. All of them can be dangerous if they are not treated properly and at an early stage.

Disease or Disorder	Description	Treatment
Colds/Flu	Infection caused by virus; symptoms include runny nose, cough, fever, aches	Bed rest and fluids; flu vaccine can prevent some types
Pneumonia	Infection of lungs by virus or bacteria; symptoms include fever, chest pain, breathing difficulty	Antibiotics for bacterial type; bed rest for viral type
Asthma	Disease in which airways narrow; symptoms include wheezing, shortness of breath, coughing	Medication can relieve symptoms; individual can also avoid substances that trigger the attacks
Tuberculosis	Communicable disease in which lungs become infected by bacteria; symptoms include cough, fatigue; can be fatal	Antibiotics
Emphysema	Disease in which alveoli are destroyed; symptoms include extreme difficulty in breathing; often caused by smoking	No known cure; pure oxygen can make breathing easier
Lung cancer	Uncontrolled growth of tumors in lungs; often caused by smoking	Surgery, radiation, and medications; survival rates are very low

278 Chapter 10: Tobacco

Cooperative Learning

Demonstrating the Hazards of Smoking Show students one demonstration of the effects of smoking on the respiratory system: Pour a cup of dark Karo syrup into a clear glass jar. Display the syrup in the jar and explain that this is what tar would look like in the lungs of a teen who had smoked half a pack of cigarettes a day for just one year. Encourage students to share their reactions to the demonstration. Then ask students to work in groups to create other demonstrations of the effects smoking has on the respiratory system. Suggest that group members use library or Internet resources to gather facts and to stimulate their thinking. Provide time for each group to share its demonstration with the rest of the class.

Figure 10.6
Keeping Your Respiratory System Healthy

These teens are taking action to keep their breathing strong and healthy. How can you make these tips work for you?

A **Exercise and breathe deeply.** Giving your lungs a good workout makes them stronger and more efficient.

B **Stay away from smoke and air pollution.** Don't start smoking, and stay away from people who do. Avoid heavily polluted air whenever possible. For example, don't jog on streets where there is heavy traffic. If you know that you are allergic to a substance, such as pollen, do all you can to avoid it.

C **Take care of illnesses.** See a doctor right away if you have any problems with breathing. Preventing serious illness will keep your lungs healthy for life.

Review — Lesson 2

Using complete sentences, answer the following questions on a separate sheet of paper.

Reviewing Terms and Facts

1. **Vocabulary** List the parts of the respiratory system through which air passes when you inhale.
2. **Recall** What is the name of the large muscle below your lungs? How does it help you breathe?
3. **Identify** What two gases are exchanged in your lungs when you breathe?

Thinking Critically

4. **Explain** Why are the alveoli important? What happens if they are damaged by smoking?
5. **Describe** List some ways in which the respiratory system protects itself from damage and infection.

Applying Health Concepts

6. **Growth and Development** Read at least two recent articles on the effects of smoking on an unborn baby. Write a letter from the baby to its mother, asking her to stop smoking. Use facts from the articles in your letter.

Lesson 2: The Respiratory System   **279**

Promoting Comprehensive School Health

The Cost of Prevention Citizens are becoming increasingly aware that their communities cannot afford the costs of problems stemming from student difficulties. The costs include: low self-esteem; dropping out of school; tobacco and other drug use; sexually transmitted diseases; teen pregnancy; poor or even nonexistent prenatal care; ineffective parenting practices; malnutrition; and poor mental health. Most schools recognize that it costs a great deal more to remedy such problems than to prevent them. Consequently, a comprehensive school health program takes into account the human costs suffered by those who do poorly academically, socially, and economically. For more information, consult *Promoting a Comprehensive School Health Program* in the TCR.

Chapter 10 Lesson 2

Assess

EVALUATING THE LESSON
Assign Reviewing Terms and Facts and Thinking Critically on this page to review the lesson; then assign the Lesson 2 Quiz in the TCR to evaluate students' understanding.

LESSON 2 REVIEW

Answers to Reviewing Terms and Facts
1. When you inhale, air passes through the nose or mouth, trachea, and bronchi.
2. The diaphragm is the muscle below the lungs. It moves down to pull air into the lungs and moves up to push air out.
3. Oxygen and carbon dioxide are exchanged in the lungs.

Answers to Thinking Critically
4. The alveoli are important because they are the sites at which oxygen and carbon dioxide are exchanged. If the alveoli are damaged, it is very difficult to supply oxygen to the body and to remove waste carbon dioxide.
5. Responses will vary. Possible response: Cilia trap dirt from the air; coughing expels foreign substances.

RETEACHING
▶ Assign Concept Map 45 in the TCR.
▶ Have students complete Reteaching Activity 45 in the TCR.

ENRICHMENT
Assign Enrichment Activity 45 in the TCR.

Close

Have students write a list of the steps they plan to take to protect the health of their respiratory system.

Focus

The Teen Health Digest articles can be used in two ways: as an individual activity for reflection and enrichment or as a cooperative activity as described below.

Motivator

Review with students the three components of good health—physical, mental/emotional, and social—and remind them that tobacco use can threaten all these aspects. Let students read the Teen Health Digest articles independently. Then guide them in discussing the ideas and information presented: Which component or components of health does the article discuss? How can you apply the information in this article to your own life?

Cooperative Learning Project

Talent Show

Setting Goals

Ask: What are the most convincing reasons for teens to avoid tobacco? List volunteers' responses on the board. Tell students they can communicate these ideas to other teens, using music, dance, skits, and other performance skills. Explain that they will work together to plan and put on a "Steer Clear of Tobacco" talent show.

Delegating Tasks

Let students work together to share and discuss ideas for their talent show. Then have them develop their own list of working groups (such as Staging, Publicity, Music/Dance, Skits, Other Performances). Have students join one or more groups.

TEEN HEALTH DIGEST

Personal Trainer

The Recovery Workout

When smokers quit, they can experience unpleasant withdrawal symptoms, such as fatigue, stomach problems, and nervousness. Many of these symptoms can be relieved by doing the following.

- Exercise, or just take a brisk walk.
- Avoid sweets, spicy foods, and foods that are high in fat. Add more fiber to your diet gradually. Drink plenty of water and other fluids.
- Relax. Take slow, deep breaths.
- Think positively. Withdrawal symptoms are temporary, but the benefits of quitting smoking last a lifetime.

Myths and Realities

It's Smart to Be a (Cold) Turkey

Myth: I want to quit smoking, but I'm worried about the withdrawal symptoms. I've heard that if I taper off gradually, I'll have an easier time adjusting.

Reality: In fact, this method can actually make withdrawal symptoms worse. If you are in the habit of smoking many cigarettes each day, your body is used to a constant high level of nicotine. Any time your nicotine level drops below that point, you experience withdrawal symptoms. When you gradually reduce the number of cigarettes smoked, your body is in a constant state of withdrawal. On the other hand, if you quit suddenly—go "cold turkey"—your body rids itself of nicotine much more quickly. After that, your body adjusts to being free of nicotine.

Try This:

Ask several people who successfully quit smoking or using other tobacco products how they did it. Make a list of their suggestions.

TECHNOLOGY OPTIONS

Video Camera If the class has access to a video camera, have one of the students (or an adult volunteer) record a rehearsal of the talent show. View and discuss the tape with students, focusing on constructive comments that can improve the performance. In addition, students may want to arrange for a taping of the final show.

Slide Projector Encourage members of the staging group to consider using slides to project background images for some or all of the talent show performances. Slides can be selected to underscore the message of each act.

CON$UMER FOCU$
Smoking Out the Beauty Myths

Have you ever seen an unattractive person in a cigarette advertisement? Tobacco companies appeal to young people's insecurities about their bodies with images of physically appealing smokers. Cigarettes that are marketed with women in mind include a special emphasis. Often these brands of cigarettes are called "slims" or "lights" to give the impression that smoking them will make you slender.

In fact, smoking is likely to damage a person's appearance. Nicotine reduces the amount of oxygen in the skin, and so causes wrinkles. Cigarette smoke also stains the teeth and fingers and causes bad breath. These unattractive qualities are never shown in the advertisements.

Try This: Look through a magazine that prints cigarette ads. Compare the people in these ads with the people in ads for other products. Are they noticeably different? What messages are these companies sending, and why?

HEALTH UPDATE
Smoke Less, See More

Smoking is known to harm the lungs, heart, and stomach. Now scientists are finding that it can also damage the eyes. Macular degeneration (MA·kyuh·ler di·je·nuh·RAY·shuhn), the most common cause of blindness in people over the age of 65, is more than twice as common among smokers. It occurs when membranes in the eye break down, causing damage to the macula, the center of the retina. More than one and a half million Americans have lost their vision due to macular degeneration.

Sports and Recreation
Getting Tobacco Off the Field

In the past, tobacco companies often attempted to promote their brands by sponsoring sports events. Seeing the brand name posted everywhere led sports fans to have positive feelings about that brand of tobacco product. Many of these fans were teens.

In the future, tobacco companies may not have this opportunity. Tobacco companies and some antismoking groups are engaged in a legal battle over regulation of tobacco advertising. Antismoking groups want to make it illegal for tobacco brands to be sponsors of sports events. Once these brand names have been removed from the sports arena, antismoking groups hope that teens may get the message that sports and tobacco don't mix.

Teen Health Digest 281

Meeting Student Diversity

Language Diversity Help students consider whether they should present a bilingual or even multilingual talent show. Perhaps several students can write and perform a song in a language other than English, and another group can write and act out a skit in a third language. This diversity may broaden the appeal and the effectiveness of the talent show.

Physically Challenged It is important that all students participate in this talent show. Encourage physically challenged students to perform as well as to make behind-the-scenes contributions. These performances and contributions (which may be adapted to meet particular needs) can help communicate the talent show's message effectively, both to participants and to audience members.

Chapter 10

The Work Students Do

Following are suggestions for implementing the project:
1. Help students schedule a convenient time and place for their talent show; they will want as many friends and family members as possible to attend.
2. Remind students that those working with the performance groups will not necessarily appear on stage. Writing, composing, choreographing, directing, and other behind-the-scenes responsibilities are as important as performing responsibilities; group members should work cooperatively to prepare effective contributions to the show.
3. Encourage groups to coordinate their efforts. Students working on staging will, of course, need to know about all the performances other groups are planning. In addition, various performance groups should compare plans to ensure a variety of ideas and approaches. Also, students working on publicity should be familiar with all acts so they can decide how to highlight them in posters and other forms of advertising.
4. Guide students in keeping the purpose of their talent show in mind: Each act should emphasize the importance of teens' avoiding all forms of tobacco use.
5. Provide time in which students can have several rehearsals and, if possible, a full-dress run-through before they perform their talent show for an audience.

Assessing the Work Students Do

Observe group meetings whenever possible, and note the cooperation and participation of group members. After the talent show, have each student write a paragraph in response to this question: How did you contribute to the success of the talent show?

Lesson 3
Tobacco Addiction

Focus

LESSON OBJECTIVES

After studying this lesson, students should be able to
- list reasons teens use tobacco.
- analyze media messages about tobacco.
- discuss physical and psychological addiction to tobacco.

MOTIVATOR

Display several photographs (as from magazine ads) showing young adults smoking. Let students select one of the smokers and make up a short history of that person: What led him or her to try cigarettes? Does he or she want to quit smoking? Suggest that students record their ideas in notes or short sentences.

INTRODUCING THE LESSON

Allow several volunteers to share the histories they have created for the pictured smokers. Guide students as necessary to identify addiction as the reason that at least some of their characters continue to smoke. Tell students that, as they read Lesson 3, they will learn more about the risks of becoming addicted to tobacco.

INTRODUCING WORDS TO KNOW

Instruct students to find definitions in the lesson for the terms listed in Words to Know. In the process, they might locate other words that they are not familiar with. Encourage students to learn the definitions of those new words as well. Ask them to explain how the words they have learned relate to the use of tobacco.

Lesson 3: Tobacco Addiction

This lesson will help you find answers to questions that teens often ask about tobacco dependency. For example:
- Why do teens begin using tobacco?
- How do people become addicted to tobacco?

Words to Know

addiction
withdrawal
physiological dependence
psychological dependence

Did You Know?

Serious Stuff
- Nicotine addiction is the most common form of drug addiction. It causes more death and disease than all other addictions combined.
- Nicotine is as addictive as alcohol, cocaine, or heroin.
- The earlier in life people start smoking cigarettes, the more likely they are to become strongly addicted to nicotine.

Why Teens Begin Using Tobacco

Why do teens start to use tobacco? One of the main reasons is that their friends smoke, although nationally only about one in four teens does smoke. Many teens think smoking will give them confidence in social situations or will make them appear sophisticated and cool. Some teens use tobacco because their parents and other adults smoke. They become curious and want to experiment. Unfortunately, before they know it, they are hooked.

Other teens smoke because tobacco advertising on billboards and in magazines makes smoking look attractive. Teens also may think that the bad effects of smoking happen only after many years of smoking or when people are older. They do not realize that the negative effects on their health begin with the first cigarette.

Most young smokers believe they can quit at any time. Some do not realize how addictive smoking is. Others know about the addictive effects of nicotine but believe that they can avoid this problem.

Parents can serve as positive role models for their children by avoiding all tobacco products.

282 Chapter 10: Tobacco

Lesson 3 Resources

Teacher's Classroom Resources
- Concept Map 46
- Decision-Making Activity 19
- Enrichment Activity 46
- Lesson Plan 3
- Lesson 3 Quiz
- Reteaching Activity 46
- Transparency 41

Student Activities Workbook
- Study Guide 10
- Applying Health Skills 46

Reasons for Tobacco Use Among Teens

Every day about 3,000 American teens begin using tobacco. Even though schools and the media send messages to warn them about the health hazards of tobacco use, teens continue to smoke and chew tobacco. Why? Here are some of the reasons.

- Teens smoke because of peer pressure. As one teen put it, "All my friends smoke, so I smoke, too."

- Some teens blame their smoking on their parents' habit. "Both my mom and dad smoke. I wanted to find out what it was like," explains a teen smoker.

- "I tried cigarettes because I thought it would make me be more grown up and in control of my life," explained another teen.

- "I wanted to look cool like the models I see in the cigarette advertisements," admitted still another teen.

- One teen explained how his chewing tobacco habit began: "I wanted to be just like my favorite pro baseball player."

- After just a few cigarettes, teens may find it difficult to quit. "It wasn't hard to get hooked," said one.

Tobacco Addiction

Teens often wonder why people get hooked on tobacco products. The answer is nicotine. Nicotine has a powerful effect on the brain and nervous system. The tobacco user forms an **addiction** (uh·DIK·shuhn), *a physical or psychological need for a drug or other substance.* For a person addicted to nicotine, it is extremely difficult to stop using tobacco. When tobacco users try to quit or reduce their use of tobacco, they go through what is known as **withdrawal**—*unpleasant symptoms that occur when someone stops using an addictive substance.* People experiencing withdrawal usually become anxious, depressed, irritable, and tired. As an addictive drug, nicotine causes two kinds of dependence: physiological (fi·zee·uh·LAH·ji·kuhl) and psychological.

All stores that sell tobacco products are required to ask any customer who looks younger than 27 for a photo ID. If customers cannot prove that they are 18 or older, they cannot buy tobacco products.

Did You Know?

An Early Decision

The decision to smoke is nearly always made during the teen years. Studies have shown that if people do not begin to smoke as children or teens, it is unlikely that they ever will. Therefore, saying no to tobacco now could save your life many years down the road.

Cultural Connections

Making a Difference

Thailand has some of the strongest antismoking laws in the world. Its leadership has played an important role in encouraging the antismoking movements that have taken root in many parts of Asia, including China, Japan, Mongolia, and South Korea.

Lesson 3: Tobacco Addiction   283

Consumer Health

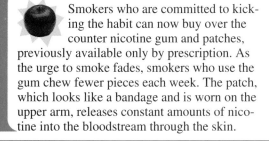

Smokers who are committed to kicking the habit can now buy over the counter nicotine gum and patches, previously available only by prescription. As the urge to smoke fades, smokers who use the gum chew fewer pieces each week. The patch, which looks like a bandage and is worn on the upper arm, releases constant amounts of nicotine into the bloodstream through the skin. Over time, a recovering smoker can apply smaller patches, which release decreasing amounts of nicotine. Health educators warn that this kind of treatment is simply an aid, not a complete solution to addiction. Smokers must also be motivated to learn new ways to manage stress and should seek counseling about methods of avoiding weight gain and resisting the temptation to resume smoking.

Chapter 10   Lesson 3

Teach

L1 Listing Examples
Ask students to write down at least ten healthy activities or attitudes that help them feel confident in social situations. Encourage students to share and discuss some of the items on their lists. Then ask: How can smoking actually make people feel less self-confident?

L3 Interviewing
Let the class work together to prepare a questionnaire to be administered both to adults who now smoke and to adults who never smoked or who have quit. Help students, as appropriate, to frame questions about when and why the respondent began smoking; how much the respondent smokes now; what tobacco-related health problems the respondent has had; whether, how, and with what success the respondent tried to stop smoking; and how old the respondent is now. Have each student administer the questionnaire to four or more adults of various ages. Then let students work together to compile and publish their findings.

L1 Analyzing
Have students analyze ads for different brands of cigarettes. What words are used to describe the cigarettes? To whom is each ad designed to appeal? What warnings appear on each brand? Does the warning weaken the message the advertiser is trying to convey? Why or why not?

Visual Learning
Let students describe or act out the scene in the photograph near the bottom of page 282. Ask: In addition to parents, what other adults can serve as positive role models for teens? In what ways can teens serve as positive role models for their own parents? What can you do to serve as a positive role model for younger children in your family or community?

Chapter 10 Lesson 3

Would You Believe?

Evidence from Harvard University shows that the risk of colon cancer is significantly higher in smokers, and the risk remains high even after smokers quit smoking. In contrast, the risk of lung cancer declines quickly after smokers quit.

Health Update

In recent years, smoking has been a deciding factor in several child custody cases. In each case, a parent lost custody of a child or children because he or she smoked cigarettes. Some antismoking advocates even say that exposing children to second-hand smoke is a form of child abuse. Ask students how they feel about this issue.

Visual Learning

While students study Figure 10.7, explain that by causing blood vessels to narrow, nicotine reduces the oxygen supply to the heart, which increases the risk of heart attack. In addition, narrowed blood vessels may lead to blood clots that can block an artery to the brain, causing stroke.

Health Minute

Smokers are much more likely to develop cataracts than are nonsmokers. Cataracts, or the clouding of the eyes' lenses, are the leading cause of blindness worldwide, and smoking contributes to a substantial percentage of cases.

Q & A

Signs of Addiction

Q: How can you tell if someone is addicted to tobacco?

A: If a person experiences a craving for tobacco or cannot stop using it without becoming anxious and irritable, then he or she is probably addicted.

Physiological Dependence

Physiological (fi·zee·uh·LAH·ji·kuhl) **dependence** is *a type of addiction in which the body itself feels a direct need for a drug.* In tobacco, the drug to which a person becomes addicted is nicotine. Nicotine affects many parts of the body, as shown in **Figure 10.7**.

Nicotine addiction is very strong. The tobacco user does not feel normal until he or she has another dose of the drug. Only by chewing tobacco or smoking is this need met. The smoker feels better after smoking, but the feeling does not last long. Soon the smoker feels a need to smoke again. As the smoker's body becomes more accustomed to the drug, he or she needs it more often to feel its effect. This same dependence is caused by the nicotine in other forms of tobacco, such as chewing tobacco.

Figure 10.7
Nicotine's Negative Effects on the Body

Nicotine causes serious problems for many of the body's important organs.

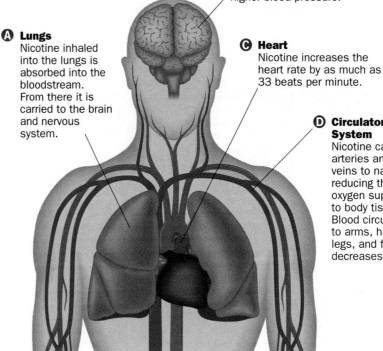

A Lungs
Nicotine inhaled into the lungs is absorbed into the bloodstream. From there it is carried to the brain and nervous system.

B Brain
Nicotine inhaled from a cigarette reaches the brain in about 20 seconds, causing the adrenal glands to produce adrenaline. This, in turn, leads to a faster heart rate and higher blood pressure.

C Heart
Nicotine increases the heart rate by as much as 33 beats per minute.

D Circulatory System
Nicotine causes arteries and veins to narrow, reducing the oxygen supply to body tissue. Blood circulation to arms, hands, legs, and feet decreases.

LIFE SKILLS

Analyzing the Media's Message

Focus on Life Skills

Ask students to describe recent newspaper or magazine ads they have seen for tobacco products: What—if anything—made each ad appealing?

When students finish reading the Life Skills feature, ask for their reactions to cigarette ads: Why do you think so many ads appear to be directed toward young people, even though the sale of cigarettes to minors is illegal? Guide students in discussing the importance of analyzing cigarette ads: How can understanding the hidden messages in an ad help you resist its appeal?

Making Health Connections

Cigarette ads send powerful messages that encourage people—especially teens—to smoke. As informed consumers, students should be able to recognize hidden messages and look beyond them for the truth about advertised products. Encourage

Psychological Dependence

Psychological (sy·kuh·LAH·ji·kuhl) **dependence** is *an addiction in which the mind sends the body the message that it needs more of a substance.* This kind of dependence can be caused by physiological dependence—the effect of nicotine, for example—as well as by other factors. Psychological dependence can be created by the pleasurable experiences or rewards that smokers may associate with smoking. Tobacco use is often linked with daily routines. A smoker may develop the habit of always smoking a cigarette after a meal, while reading the newspaper, or during a work break. Some people reward themselves with a cigarette after they have completed a difficult task. These habits, along with the physiological dependence on nicotine, make it harder for a person to quit using tobacco.

Psychological dependence may form as a result of tobacco's perceived effects. Some smokers believe that cigarettes provide them with extra energy. Others think that cigarettes help calm them down when they are tense. Still others feel that they need tobacco to keep their weight under control because smoking reduces their appetites. Although these benefits are imaginary, smokers may become convinced that they're real.

in your journal

For a week, take note of every tobacco ad you see. In your journal, record where the ads appeared, and include descriptions of them. What are the messages in the ads? Why do you think the ads were placed where they were? Write your answers in your journal.

LIFE SKILLS
Analyzing a Media Message

What flashes through your mind when you see a cigarette ad that shows smokers having fun? Do you believe that smoking is a way to make friends and share a good time? This is the hidden message that the cigarette makers want you to receive. They send this message to influence your decision about smoking. They show appealing pictures to make you feel good about the idea of smoking. They try to persuade you to smoke. The ad is concerned with image, or the way people or objects appear. This image is the ad's message, but it does not tell the truth about smoking.

You can avoid being confused by this false advertising message. Just analyze any ad to see the hidden messages in its pictures and words. Look carefully at the ad on this page. Then, to analyze its message, ask yourself the following questions:

1. What is the hidden message in the ad's picture? What is the hidden message in the ad's words?
2. After analyzing this ad, what decision about smoking would you make? Explain your choice.

**Be Cool
Have Fun
Smoke *Flavor***

SURGEON GENERAL'S WARNING: Quitting Smoking Now Greatly Reduces Serious Risks to Your Health.

Lesson 3: Tobacco Addiction

Chapter 10 Lesson 3

L1 Guest Speaker
Invite an addiction specialist, such as a drug counselor, to speak to the class about addiction to nicotine. Suggest that the speaker address issues such as physiological and psychological dependence on nicotine and how nicotine compares with other addictive drugs in its health impact, economic costs, and addictive potential.

L1 Examining the Issue
Explain to the class that physiological dependence on nicotine can be treated with nicotine patches or chewing gum. How would using one of these aids make quitting easier? What problems might be associated with using nicotine patches or gum? Also explain that these patches and gum can be toxic to young children, and ask: What special precautions should users take?

Assess

EVALUATING THE LESSON
Assign Reviewing Terms and Facts and Thinking Critically on page 286 to review the lesson; then assign the Lesson 3 Quiz in the TCR.

RETEACHING
► Assign Concept Map 46 in the TCR.
► Have students complete Reteaching Activity 46 in the TCR.
► Ask volunteers to explain how psychological dependence differs from physiological dependence. Have other students help correct any misconceptions.

---

students to recall the real dangers of tobacco whenever they see advertisements for tobacco products. Besides dry mouth and bad breath, the reality of smoking includes addiction, lung cancer, heart disease, and emphysema.

Meeting Student Diversity
Class members with special abilities or interests might be encouraged to do further research, such as contrasting cigarette ads in magazines from previous years with those of today. You may want to have these students make an exhibit to highlight the changes. They might also research the changing demographics of smokers and nonsmokers during the twentieth century, including information about gender, age, race, and ethnic group. A further possible topic for research is the changing global market for cigarettes.

Assessing the Work Students Do
Have students write letters to government representatives, describing the results of the activity and requesting greater truth in the advertising of tobacco products. Ask students to share any responses they receive to their letters. Assign credit based on a predetermined level of participation.

Chapter 10 Lesson 3

LESSON 3 REVIEW

Answers to Reviewing Terms and Facts

1. Responses may vary. Possible response: Factors that influence teens to smoke include having friends who smoke; thinking it will boost self-confidence; rebelling against adults; believing ads that make smoking look attractive; believing they can quit anytime.
2. Addiction is a physical or mental need for a drug or other substance. Withdrawal is the unpleasant symptoms that occur when someone stops using an addictive substance. Sentences will vary.
3. Nicotine causes arteries and veins to narrow, reducing oxygen supply to body tissues.

Answers to Thinking Critically

4. Responses will vary. Possible response: Ads usually show healthy-looking, active, young adults. The picture encourages smoking by implying that smokers look great, feel great, and are popular. The words encourage smoking by giving the impression that smoking refreshes your mouth.
5. Students who answer *physiological dependence* will point to the body's need for the drug. Those who answer *psychological dependence* will stress the pleasurable experiences, daily routines, perceived effects, and rituals that are part of the mind's attachment to smoking.

ENRICHMENT

- Assign Enrichment Activity 46 in the TCR.
- Display a tobacco ad. Have students write descriptions of the ad, identifying the ad's hidden message and explaining what information about tobacco the ad avoids.

Close

Ask students to summarize the negative effects of nicotine.

Your Total Health

Bad Habits Multiply ACTIVITY

A person who has one bad habit is likely to have others. A survey conducted by a fitness research center found that people who smoke are more likely to have other unhealthful habits. Examine your own habits. List them on a separate sheet of paper. If there are any habits on your list that you think are unhealthful, make a plan to change them. Predict what might happen to the status of your health if you choose not to change them.

Replacing smoking with a different ritual is one way to break the cigarette habit. For example, if a person smokes after a meal, he or she could try chewing sugarless gum instead.

Smokers also may develop rituals that create psychological dependence. For example, some smokers reach for a cigarette when they begin certain activities. Others talk with an unlit cigarette in their mouths before they light it. These rituals can become as difficult to stop as the actual habit of smoking. One key to breaking the habit is to change the ritual. Instead of reaching for a cigarette, think of something else to do—take a walk, call a friend, play on the computer, or work on a hobby.

Lesson 3 Review

Using complete sentences, answer the following questions on a separate sheet of paper.

Reviewing Terms and Facts

1. **Give Examples** List five factors that have an influence on teens' decisions to use tobacco.
2. **Vocabulary** Define the terms *addiction* and *withdrawal*. Use each word in an original sentence.
3. **Recall** How does nicotine affect the circulatory system?

Thinking Critically

4. **Apply** Select a magazine advertisement for cigarettes. Describe how the picture and the words of the advertisement encourage tobacco use.
5. **Analyze** Which part of tobacco addiction do you think is more powerful—the physiological dependence or the psychological dependence? Give reasons for your answer.

Applying Health Concepts

6. **Consumer Health** Create an advertisement to discourage teens from smoking. Use the same kinds of hidden messages in your advertisement that the media use to encourage teens to smoke.
7. **Personal Health** With a classmate, write a short play about breaking a psychological dependence on tobacco.

286 Chapter 10: Tobacco

Home and Community Connection

Smokers in the Family What can teens do to discourage smoking by other members of their family? How can teens help other young family members avoid experimenting with cigarettes? How should teens respond when parents or other adult family members smoke? Why? Under what circumstances is it appropriate to encourage an adult—especially an adult addicted to tobacco—to quit smoking? You may want to have students form small groups in which to explore their responses to these difficult questions. Then let the members of each group share their ideas with the rest of the class.

Choosing to Be Tobacco Free

Lesson 4

This lesson will help you find answers to questions that teens often ask about remaining tobacco free or giving up tobacco use. For example:

▶ How can I remain tobacco free?
▶ What are some ways of breaking the tobacco habit?
▶ How can I defend my rights as a nonsmoker?

Saying No to Tobacco

Saying no when friends pressure you to use tobacco can be difficult. There are ways to resist, however. The first line of defense is to be prepared for the pressure. Practice your refusal skills. Know ahead of time what you can say to respond to the pressure to use tobacco.

If the pressure is light, you can simply refuse. Say that you're not interested or that you don't like the taste or smell of cigarettes. If you think you will need more help, refer to **Figure 10.8**. It will provide you with 12 very good reasons to remain tobacco free.

Sometimes the pressure to use tobacco can be very strong. If you are an athlete, you can say that you want to keep your breathing healthy for your sport. You can simply say no and ask people to respect your right to make your own decisions. If the pressure continues, leave the scene. One way to resist the pressure to use tobacco is to choose friends who do not use it.

Words to Know

secondhand smoke
sidestream smoke
mainstream smoke
passive smoker

in your journal

Look at the reasons for quitting tobacco use. Review the list again. Are there any reasons you can think of that are not included in the list? If so, write them down in your journal.

Figure 10.8
The Top 12 Reasons to Be Tobacco Free
There are plenty of good reasons not to smoke.

1. You will be healthier.
2. Your breath will smell better.
3. You can save money.
4. Your senses of taste and smell will be sharper.
5. You will have fewer allergies.
6. You will not be confined to smoking areas.
7. You will have more energy and stamina for sports.
8. Your skin will be healthier and look better.
9. Your hair and clothes will not smell like smoke.
10. You will not be forcing others to breathe smoke.
11. You will not have to lie to parents, teachers, or friends.
12. You will not be breaking the law.

Lesson 4: Choosing to Be Tobacco Free   287

Lesson 4 Resources

Teacher's Classroom Resources
- Concept Map 47
- Cross-Curriculum Activity 20
- Decision-Making Activity 20
- Enrichment Activity 47
- Health Lab 10
- Lesson Plan 4
- Lesson 4 Quiz
- Reteaching Activity 47

Student Activities Workbook
- Study Guide 10
- Applying Health Skills 47
- Health Inventory 10

Lesson 4
Choosing to Be Tobacco Free

Focus

LESSON OBJECTIVES
After studying this lesson, students should be able to
▶ list ways to refuse nicotine.
▶ discuss ways to kick the tobacco habit.
▶ explain how tobacco affects nonsmokers.

MOTIVATOR
Write the following on the board for students to work on: *Write five ways to say no to something you would rather not do.*

INTRODUCING THE LESSON
Have volunteers share their responses to the Motivator activity. Help students recognize that deciding to say no—and saying no convincingly—can be hard. This is especially true when you feel you might lose something by saying no. You may worry about disappointing friends, losing the respect of peers, and so on. Explain that this lesson will help students learn how to say no to tobacco.

INTRODUCING WORDS TO KNOW
Write the Words to Know on the board. Ask students to identify the individual words that make up the compound words. (For example, *side* and *stream* form one compound word, and *sidestream* and *smoke* form another.) Have students use the meanings of the three shorter words to decide what the longer term means. Then have them verify their responses by checking the Glossary. Ask them to do the same for the other Words to Know. Have students use the terms in original sentences.

287

Chapter 10 Lesson 4

Teach

L1 Discussing
Peer pressure is often considered a negative influence. Ask students to name ways in which peer pressure can be positive. How can peer pressure be used to prevent teens from starting to use tobacco?

L1 Applying Life Skills
Divide the class into groups of three. Have each group act out a short scene in which two teens try to persuade a third to use tobacco. Repeat the activity, but this time have just one teen try to persuade the other two to use tobacco. After students have had turns at each part, encourage them to compare the two scenes: Why is it easier to say no when a friend is there to back you up?

L1 Critical Thinking
Let students discuss and explain their responses to these questions: Can a person who smokes or chews tobacco be your friend? Can a person who pressures you to try tobacco be your friend? If this person won't take no for an answer, can she or he be your friend?

Visual Learning
Let students meet in small groups to read and discuss the reasons listed in Figure 10.8: Which reasons are most compelling for you? Why? Have group members rephrase each reason as a phrase or sentence they could say when peers urge them to use tobacco.

Friends who do not smoke can provide support and encouragement for people who are trying to quit.

Did You Know?
Television Taboo

Television networks are doing their part to change the public's ideas about cigarette smoking. Talk-show hosts, actors, and actresses rarely smoke today.

Kicking the Habit

Many people who used tobacco in the past are kicking the habit. It is not easy to do. However, there are many ways to stop. There are also people and places to go to for help in breaking the habit.

Since the nicotine in tobacco is addictive, people who quit using it experience withdrawal symptoms. These symptoms do not last long, but they include nervousness, moodiness, and difficulty in sleeping. If you or someone you know is trying to quit smoking, here are some tips that can help.

- Make a list of reasons why you want to quit smoking. Read the list whenever you get the urge to smoke.
- Set small goals. Try quitting one day at a time. Every year the American Cancer Society sponsors the "Great American Smokeout," a campaign that calls for all smokers to avoid smoking for one day.
- If possible, avoid being with people who smoke. Stay away from places where lots of smokers hang out.
- Change any habits you have that are linked to smoking. For example, if you smoke after a meal, take a short walk instead. Learn other methods of relaxing and relieving stress.
- Exercise when you feel the urge to reach for tobacco. Stretch, take deep breaths, go for a walk, or take a ride on your bike.
- Think positively. Seek encouragement from your nonsmoking friends.
- Eat healthful snacks instead of reaching for tobacco.

MAKING HEALTHY DECISIONS
Overcoming the Pressure to Smoke

All the science classes in seventh grade made a trip to Cape Henry every spring for four days. Tracy was looking forward to being with her friends away from home and at the beach.

Every evening after dinner, the students had half an hour of free time. The first evening, Tracy was returning to her cabin when a friend called her to come around back. Tracy was surprised to find two of her friends and two girls she didn't know smoking cigarettes. After Jill introduced Tracy to Sally and Liz, she offered Tracy a cigarette and said,

"Have a few puffs. It's a great way to unwind after following teachers' orders all day."

Tracy declined Jill's offer and said that she had to leave to help set up for that night's special program. The next evening she noticed that the four girls were behind the cabin again. Tracy could hear them laughing and thought about joining in.

The next morning, one of the girls, Sally, confronted Tracy. "Why don't you have a smoke with us after dinner? Do you think you're too good for us?"

288   Chapter 10: Tobacco

MAKING HEALTHY DECISIONS
Overcoming the Pressure to Smoke

Focus on Healthy Decisions
Display a colorful newspaper ad for cigarettes. Then let a small group of students role-play a conversation in which several teens urge another to smoke. Ask: Which kind of pressure is harder to deal with—hidden messages in advertising, or peer pressure? Why?

After students have read the Making Healthy Decisions feature, guide them in discussing Tracy's situation: How does her friendship with two of the girls make it more difficult to resist their pressure?

Activity
Use the following suggestions to discuss Follow-up Activity 1:

Step 1 Tracy must decide whether to remain tobacco free or to smoke with her peers.

Step 2 Tracy can join the four girls but not smoke, she can join them for a cigarette, or she can spend time with nonsmokers.

Step 3 Tracy might join the four girls and ignore their pressure; she may be able to enjoy their company,

Programs That Help

The "cold turkey" method of quitting tobacco is popular, and it is recommended by many experts. In this method, the smoker simply stops using all forms of tobacco. Some people need support or assistance to quit. They should contact a group that offers a program to help people quit. The American Lung Association, the American Heart Association, and the American Cancer Society are just a few of the many groups that offer such programs. You can contact these groups by telephone or by mail to get more information about quitting.

Books and recordings can help people quit smoking on their own. Many can be borrowed from local libraries. There are also several products available to help lessen withdrawal symptoms. They include over-the-counter drugs (such as chewing gum that contains nicotine and the nicotine patch) and sets of graduated filters designed to reduce tar.

People who are trying to quit need support and encouragement from family and friends. If you are close to someone who wants to kick the habit, do what you can to help. Praise the person for each day that he or she avoids smoking.

The support of friends is a powerful tool in fighting nicotine addiction.

Teen Issues

Tips for Supporters

Helping someone quit using tobacco is not an easy task. You must be patient, caring, and supportive. Help the ex-tobacco user avoid situations in which he or she might be tempted to reach for tobacco. Offer to join him or her in healthy activities such as walking or bike riding. Reassure the person that his or her irritability, lack of energy, or other withdrawal symptoms will pass in time, and that the health benefits will make the effort of quitting worthwhile.

Tracy was embarrassed by Sally's comments. She was torn between her unwillingness to smoke and her desire to be part of the group. Tracy decided to use the six-step decision-making process to make up her mind about smoking:

1. **State the situation**
2. **List the options**
3. **Weigh the possible outcomes**
4. **Consider your values**
5. **Make a decision and act**
6. **Evaluate the decision**

Follow-up Activities

1. Apply the six steps of the decision-making process to Tracy's story.
2. With a classmate, role-play a scene in which Tracy resists the peer pressure to use tobacco.

Lesson 4: Choosing to Be Tobacco Free   289

Chapter 10   Lesson 4

L3 Life Skills

Ask students to take a poll of people who have kicked the tobacco habit: What methods did they use to give up tobacco? How long have they have been tobacco free? Encourage students to publish their results in the school's newspaper.

L1 Math

Besides being a health risk, smoking is very costly. Have students find out how much a pack of cigarettes costs in your community. Ask: If a pack-a-day smoker kicks the habit, how many days will it take before he or she can afford a computer game that sells for $35? A jacket that sells for $60?

L2 Creating Bumper Stickers or Buttons

"If you don't start smoking—you'll never have to quit" sounds like it could be a slogan for a bumper sticker or a button. Ask the class to suggest other antismoking messages; write their suggestions on the board. Then have each student design a bumper sticker or button featuring an antismoking message.

Health Minute

Many young people are getting involved in the fight against the use of tobacco. Students in one Texas city started a group called Students L.E.A.D. The letters stand for "Loving Everyone Against Drugs." Group members encourage parents and their children to talk openly about all drugs, including tobacco.

---

but she runs the risk of giving in and smoking. She might join the four girls and smoke with them; this choice would jeopardize her physical health. She might spend her time with other students who do not smoke; this choice would protect her physical health and keep her away from negative peer pressure.

Step 4 Tracy must consider whether she values being tobacco free more than being part of a specific group.

Step 5 If Tracy values remaining tobacco free, she is most likely to spend her time with other students; she might also choose to spend time with the four girls but resist their pressure.

Step 6 Encourage students to evaluate both the short- and long-term consequences of their decisions.

Assessing the Work Students Do

Have students work with partners or in small groups to role-play the scene in Follow-up Activity 2. Assign credit based on students' ability to use the decision-making process.

How Tobacco Affects Nonsmokers

Even if you are not lighting up and smoking, you may be breathing *air that has been contaminated by tobacco smoke,* or **secondhand smoke.** Each time a smoker lights a cigarette, smoke fills the air from two sources. *The smoke coming from the burning tip of the cigarette is* called **sidestream smoke.** It contains twice as much tar and nicotine as *the smoke that the smoker exhales,* or **mainstream smoke.** This is because sidestream smoke has not passed through the cigarette filter or the smoker's lungs.

Nonsmokers who breathe secondhand smoke become **passive smokers.** Passive smoking is harmful to your health because it contributes to respiratory problems. Passive smoking irritates your nose and throat, and it also causes itchy and watery eyes, headaches, and coughing. Some people are much more sensitive to secondhand smoke than others.

A smoke-filled room has high levels of nicotine, carbon monoxide, and other pollutants. In such a room, a nonsmoker can inhale as much nicotine and carbon monoxide in one hour as if he or she had smoked a whole cigarette. Long-term exposure to secondhand smoke poses the same risk of serious illness for passive smokers as smoking does for active smokers. These risks include heart and lung diseases and respiratory problems. According to the U.S. Environmental Protection Agency, secondhand smoke is a human carcinogen, or cancer-causing substance, that is responsible for 3,000 lung cancer deaths each year.

Your Total Health
Substitute Healthy Snacks

Some people who quit smoking gain weight because they eat more snacks as a substitute for smoking. Choosing healthy snacks such as fruits and vegetables will make the former smoker less likely to gain weight. In addition, a diet rich in fruits and vegetables has been found to reduce the risk of heart disease and cancer in smokers and nonsmokers.

Most restaurants provide separate nonsmoking sections for their customers.

Personal Health

Because chemicals in cigarette smoke use up the body's supply of vitamin C, smokers need to take in more vitamin C than nonsmokers. A California study found that the same is true for people exposed to secondhand smoke. Study participants who were exposed to secondhand smoke for 20 hours per week on average had 24 percent less vitamin C than those not exposed to smoke. Researchers believe the shortfall of vitamin C could increase the risk of heart disease and cancer in those exposed to secondhand smoke because vitamin C seems to protect against both diseases. They recommend that people who can't avoid secondhand smoke eat more foods rich in vitamin C and consider taking supplements.

Chapter 10 Lesson 4

L2 Applying Life Skills
Have a group of students work together to brainstorm a list of reasons that teens might give for starting to smoke cigarettes. Then, for each reason on their list, have the group members think of at least one related reason for choosing not to smoke. Finally, have the group members use their list to write a public service announcement suitable for radio or television broadcast that would help teens make healthy decisions about smoking.

L3 Researching
Ask volunteers to research the effects of smoking on fetal development. Have them prepare an illustrated report to present to the rest of the class. The report should include topics such as the path by which smoke reaches the fetus, birth defects and other prenatal problems associated with smoking while pregnant, and number of births affected.

L1 Critical Thinking
Encourage students to discuss their responses to these questions: When you enter the home of adults who smoke, what happens to your right to breathe air that is free of harmful tobacco smoke? What happens to that right when adult members of your own family smoke? What do you think is the best way to handle these situations?

VIDEODISC/VHS
Teen Health Course 2
You may wish to use video segment 10, "Choosing to Be Tobacco Free," to show how teens can take action, alone or together with others, to get people to stop smoking.

Videodisc Side 2, Chapter 10
Choosing to Be Tobacco Free

Search Chapter 10, Play to 11

Unborn Babies and Children

A woman who smokes during pregnancy seriously endangers the health of her unborn child. Cigarette smoking during pregnancy is associated with increased chances of miscarriage, stillbirth, and low birth weight. The lower a baby's birth weight, the higher the risk of complications in the baby's development. The effects of tobacco may affect the growth, mental development, and behavior of children for up to 11 years after their birth. In addition, infants whose mothers smoke during and after pregnancy are three times more likely to die from Sudden Infant Death Syndrome (SIDS) than infants whose mothers do not smoke. If their mothers stop smoking during pregnancy but resume immediately after the birth, infants are twice as likely to die from SIDS as those born to nonsmoking mothers.

Rights of Nonsmokers

The fact is clear that inhaling secondhand smoke is as dangerous as smoking. Laws continue to be passed to restrict smoking in public places. In 1989 smoking was banned on all domestic airplane flights. In 1994 more than 600 state and local ordinances restricted smoking. Employers now have a legal right to restrict smoking in the workplace, and some have banned smoking.

What does all of this mean to you? As a nonsmoker, you have the right to breathe air that is free of harmful tobacco smoke. You have the right to express your preference that people not smoke around you. You also have the right to work for the passage of laws that prevent nonsmokers from being exposed to tobacco smoke in public places.

Because of the potential damage to children, both before birth and during their later development, women should not smoke while they are pregnant.

Review — Lesson 4

Using complete sentences, answer the following questions on a separate sheet of paper.

Reviewing Terms and Facts

1. **Recall** List four of the top 12 reasons to be tobacco free.
2. **List** Name three methods that can help smokers to quit.
3. **Vocabulary** How are *secondhand smoke, sidestream smoke,* and *mainstream smoke* different from each other?

Thinking Critically

4. **Suggest** Miguel is at a party with his friend Sam. Another friend offers them cigarettes, but Miguel does not want to smoke. How can Miguel say no to tobacco and still keep his friends?
5. **Apply** A friend of yours has asked you to help her to quit smoking. What could you do?
6. **Analyze** Although you enjoy bowling, the local bowling alley has no restrictions on smoking. You would like to exercise your rights as a nonsmoker. What could you do in this situation?

Applying Health Concepts

7. **Health of Others** Organize a support group for students who want to stop using tobacco. Ask the school nurse to train some students as peer supporters. Have the students who are former smokers keep a record of the days they have been tobacco free.

Lesson 4: Choosing to Be Tobacco Free **291**

The Health of Others

Data from the National Center for Health Statistics show that women who smoke during pregnancy face a greater risk of giving birth to low-birth-weight babies than do women who don't smoke. Women who smoke are almost twice as likely to have babies under 5 pounds—the cutoff point for normal birth weight—as women who don't smoke. Low birth weight is a risk factor for many other perinatal problems. Babies who are low in weight at birth are at higher risk of infections, birth defects, mental retardation, developmental delays, and even death than babies of higher birth weight.

Chapter 10 Review

CHECKING COMPREHENSION

Use the Chapter Summary and the Chapter 10 Review to help students go over the most important ideas presented in Chapter 10. Encourage students to ask questions and add details as appropriate.

CHAPTER 10 REVIEW ANSWERS

Reviewing Key Terms and Concepts

1. Harmful substances in tobacco smoke include (any three): formaldehyde, nicotine, cyanide, carbon monoxide, methanol, and tar.
2. Smokeless tobacco contains nicotine, which is addictive. It is linked to an increased incidence of mouth cancer and cancers of the esophagus, larynx, and pancreas. It also causes red, inflamed gums and an increased incidence of dental cavities.
3. Problems that can affect your respiratory system include (any three): colds, flu, pneumonia, asthma, tuberculosis, emphysema, and lung cancer.
4. Responses will vary. Possible response: You can help keep your respiratory system healthy be getting exercise, avoiding cigarette smoke, and taking care of illnesses.
5. Responses will vary. Possible response: Nicotine causes veins and arteries to narrow; it also affects the heart, increasing the heart rate.
6. It may begin with physiological dependence but then deepens as pleasurable experiences are associated with smoking.
7. Effects of nicotine withdrawal include nervousness, moodiness, and difficulty sleeping.

Chapter 10 Review

Chapter Summary

▶ **Lesson 1** There are many harmful substances in tobacco. Tobacco comes in several forms, which are either placed in the mouth or smoked.

▶ **Lesson 2** Your respiratory system supplies your body with oxygen and gets rid of carbon dioxide. Breathing is a cycle that includes inhaling, an exchange of oxygen for carbon dioxide, and exhaling.

▶ **Lesson 3** Many teens start smoking because they are influenced by friends, advertisements, or other factors. Smokers can easily become addicted because nicotine causes physiological dependence. Smokers can also develop a psychological dependence on tobacco.

▶ **Lesson 4** Resisting the pressure to smoke is an important decision for your health. There are many ways to kick the tobacco habit and many individuals and groups who are willing to help a user quit.

Reviewing Key Terms and Concepts

Using complete sentences, answer the following questions on a separate sheet of paper.

Lesson 1

1. List three harmful substances found in tobacco smoke.
2. How is using smokeless tobacco harmful to the body?

Lesson 2

3. Identify three problems that can affect your respiratory system.
4. Name three actions you can take to keep your respiratory system healthy.

Lesson 3

5. List two parts of the body that are negatively affected by nicotine, and tell how they are affected.
6. How does psychological dependence develop?

Lesson 4

7. What are some of the effects of nicotine withdrawal?
8. Describe how secondhand smoke affects passive smokers.

Thinking Critically

Using complete sentences, answer the following questions on a separate sheet of paper.

9. **Assess** Some brands of cigarettes contain very little tar. Do you think these brands are safe to smoke? Why or why not?
10. **Hypothesize** If your respiratory system were damaged by smoking or disease, how might the rest of your body be affected?
11. **Compare and Contrast** Do you think that adults who start smoking do so for the same reasons as teens? What reasons do they have in common, and how might their reasons differ? For which group do you think that it is harder to quit?
12. **Recommend** You are in a restaurant, enjoying your dinner, when a person at the next table lights a cigarette. You cannot stand the smell of smoke, especially while you are eating. How could you handle this situation?

292   Chapter 10: Tobacco

Meeting Student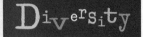

Language Diversity Use the following suggestions to help students who have difficulty with English:

▶ Pair those learners with native speakers of English who can restate the Chapter Summary in language that helps students comprehend important concepts.

▶ Direct auditory learners or those students with language diversity to the Teen Health Audiocassette Program. Available in English and Spanish, this component provides an audio and written summary of the chapter.

Chapter 10 Review

Your Action Plan

Tobacco can have a negative impact on your life whether you use it or not. Protecting yourself from the health risks of tobacco is very important.

Step 1 Look back through your journal entries. What do they tell you about your feelings toward tobacco? What changes would you like to make to improve your health or the health of others?

Step 2 Decide on a long-term goal for yourself, and write it down. Your long-term goal might be to help a friend kick the tobacco habit.

Step 3 Next, write down a series of short-term goals to help you achieve your long-term goal. For example, you might try to persuade your friend to join a support group.

Plan a schedule for accomplishing each short-term goal. When you reach your long-term goal, reward yourself. If your goal involves a friend, include her or him in the reward.

In Your Home and Community

1. **Community Resources** Action on Smoking and Health (ASH) is a national organization working for a smoke-free America. Find out whether your community has a local chapter of ASH. If not, write to: Action on Smoking and Health, 2013 H Street NW, Washington, DC 20006. Ask how you can start a chapter in your community.

2. **Health of Others** Write or call the American Lung Association or the American Cancer Society, and ask if they have any training programs in which you might learn how to help your peers stop smoking. Once you are trained, ask an adult who does such work to help you begin a program in your school or through your community recreation department.

Building Your Portfolio

1. **Poster** Interview an administrator in your school to find out what the penalties are for teens who buy or use tobacco. Use this information to create a "Wanted" poster for an underage tobacco user. Include the person's "crimes" and the penalties he or she can expect to pay. Put the poster in your portfolio.

2. **Script** Write a script about teens being pressured to use tobacco. Have some of your friends or classmates act out the script. Place a copy of the script in your portfolio.

3. **Personal Assessment** Look through all the activities and projects you did for this chapter. Choose one or two that you would like to include in your portfolio.

Chapter 10: Chapter Review **293**

Performance Assessment

▶ **Self-evaluation** Direct students to review the activities that are provided throughout the chapter. Encourage each student to select one finished product or activity that demonstrates his or her best work for the chapter. Have students explain what they learned and how the examples they selected show their progress.

▶ **Teacher's Classroom Resources** Assign Performance Assessment 10, "Antismoking Cartoon," in the TCR.

Chapter 10 Review

8. Secondhand smoke causes respiratory problems; irritates the nose and throat; and causes itchy and watery eyes, headaches, and coughing.

Thinking Critically

9. These brands are not safe because they contain other harmful substances, such as carbon monoxide.
10. The rest of the body would not get as much oxygen. You could be easily tired and suffer damage to other organs.
11. Reasons common to teens and adults are the influences of advertising and the belief that they can quit whenever they want. Teens are more likely than adults to start smoking in response to pressure from peers. The longer and the more people smoke, the harder it is to quit.
12. The nonsmoker could ask the smoker to put the cigarette out, ask to move to another table, or complain to the management and suggest separate nonsmoking areas for the future.

RETEACHING

Assign Study Guide 10 in the Student Activities Workbook.

EVALUATE

▶ Use the reproducible Chapter 10 Test in the TCR, or construct your own test using the Testmaker Software.

▶ Use Performance Assessment 10 in the TCR.

EXTENSION

Ask two or more volunteers to debate the right of smokers to smoke versus the right of nonsmokers to be protected from exposure to secondhand smoke. The students should first research the topic so that they are familiar with the most important legal, ethical, and health issues.

293

Planning Guide

Chapter 11
Alcohol and Drugs

Lesson		Features	Classroom Resources
1	**Use and Abuse of Alcohol** *pages 296–300*	Personal Inventory: Attitudes About Alcohol *page 298* Life Skills: Helping Someone with a Drinking Problem *page 299*	Concept Map 48 Decision-Making Activity 21 Enrichment Activity 48 Lesson Plan 1 Lesson 1 Quiz Reteaching Activity 48 Transparency 42
2	**Use and Abuse of Drugs** *pages 301–306*		Concept Map 49 Cross-Curriculum Activity 21 Decision-Making Activity 22 Enrichment Activity 49 Lesson Plan 2 Lesson 2 Quiz Reteaching Activity 49 Transparencies 43, 44
3	**The Nervous System** *pages 307–311* TEEN HEALTH DIGEST *pages 312–313*	Health Lab: Nervous System Tricks *pages 310–311*	Concept Map 50 Enrichment Activity 50 Health Lab 11 Lesson Plan 3 Lesson 3 Quiz Reteaching Activity 50 Transparency 45
4	**Risks Involved with Alcohol and Drug Use** *pages 314–318*		Concept Map 51 Enrichment Activity 51 Lesson Plan 4 Lesson 4 Quiz Reteaching Activity 51 Transparency 46
5	**Avoiding Substance Abuse** *pages 319–323*	Making Healthy Decisions: Offering Alternatives *pages 322–323*	Concept Map 52 Cross-Curriculum Activity 22 Enrichment Activity 52 Lesson Plan 5 Lesson 5 Quiz Reteaching Activity 52

Lesson 6

Addiction and Recovery
pages 324–327

Features

Classroom Resources
- Concept Map 53
- Enrichment Activity 53
- Lesson Plan 6
- Lesson 6 Quiz
- Reteaching Activity 53

National Health Education Standards

Listed below are the Health Education Standards Performance Indicators addressed in each lesson of this chapter.

Lesson	Health Standards Performance Indicators
1	(1.6, 1.8, 3.1, 3.2)
2	(1.6, 1.8, 2.2, 3.1)
3	(1.1, 1.8, 3.1)
4	(1.2, 1.6, 1.8, 3.1)
5	(3.1, 5.6, 6.1, 6.4)
6	(2.4, 2.6)

ABCNEWS InterActive™ VIDEODISC SERIES

You may wish to use the following videodiscs with this chapter: *Alcohol,* side one, video segments 3, 4, 5, 6, 7, 8, 9, 10, 11; side two, video segments 3, 4, 5, 6, 13, 14; *Drugs and Substance Abuse,* side one, video segments 3, 6, 7, 8, 9, 10, 11, 12, 13, 14; side two, video segments 11, 12, 18; *Making Responsible Decisions;* and *Violence Prevention.* Use the *ABCNews InterActive™ Correlation Bar Code Guide* for title reference. Also available in VHS format.

Chapter Resources

Teacher's Classroom Resources
- Chapter 11 Test
- Parent Letter and Activities 11
- Performance Assessment 11
- Testmaker Software

Student Activities Workbook
- Study Guide 11
- Applying Health Skills 48–53
- Health Inventory 11

Student Diversity Strategies
- Audiocassette Program (English)
- Audiocassette Program (Spanish)
- Spanish Parent Letters
- Spanish Summaries, Quizzes, and Activities

Multimedia Components
- English Audiocassette Program
- Spanish Audiocassette Program
- *Teen Health* Videodisc/VHS Series

Other Resources

Readings for the Teacher

Landau, Elaine. *Teenage Drinking.* Springfield, NJ: Enslow Publishers, 1994.

Monroe, Judy. *Antidepressants: The Drug Library.* Springfield, NJ: Enslow Publishers, 1996.

Weatherly, Myra. *Inhalants.* Springfield, NJ: Enslow Publishers, 1996.

Readings for the Student

Berger, Gilda. *Alcoholism and the Family.* New York: Franklin Watts, 1993.

McLaughlin, Miriam, and Sandra P. Hazouri. *Addiction: The High That Brings You Down.* Springfield, NJ: Enslow Publishers, 1997.

Sherry, Clifford J. *Inhalants.* New York: Rosen Publishing Group, 1994.

Out of Time?

If time does not permit teaching this chapter, you may use these features: Personal Inventory on page 298; Life Skills on page 299; Health Lab on pages 310–311; Teen Health Digest on pages 312–313; Making Healthy Decisions on pages 322–323; and the Chapter Summary on page 328.

Chapter 11
Alcohol and Drugs

CHAPTER OVERVIEW

Chapter 11 presents important facts about alcohol and a wide variety of drugs, guides students in understanding the dangers of trying or using these substances, and explains the process of recovering from alcohol or drug addiction.

Lesson 1 describes the effects of alcohol on the body and discusses alcoholism and its treatment.

Lesson 2 explains the distinction between drugs and medicines, discusses the appropriate use of medicines, and identifies groups of dangerous and illegal drugs.

Lesson 3 presents information on the nervous system and its functions, and identifies safety measures that help keep the nervous system healthy.

Lesson 4 identifies the risks to physical, mental/emotional, and social health associated with the use of alcohol and drugs.

Lesson 5 discusses attitudes and approaches teens can use to avoid alcohol and drugs.

Lesson 6 explains the process of recovering from drug addiction and discusses how teens can find help for those dealing with addiction.

PATHWAYS THROUGH THE PROGRAM

Choosing to be alcohol- and drug-free is an essential part of safeguarding one's health. Teens who understand the serious dangers posed by alcohol and other drugs are better equipped to make a strong personal decision to avoid those substances. Knowing how to resist alcohol and drugs and how to stand up to peer pressure will help students protect and maintain their good health now and in the future.

Chapter 11
Alcohol and Drugs

Student Expectations
After reading this chapter, you should be able to:
1. Describe the effects of the use and abuse of alcohol on the body.
2. Identify how misuse and abuse of drugs can affect health.
3. Explain how to keep the nervous system healthy.
4. Recognize the risks of alcohol and drug use to physical, mental/emotional, and social health.
5. Identify ways to avoid substance abuse.
6. Describe the process of addiction to and withdrawal from alcohol or drugs.

Key to Ability Levels

Teaching strategies that appear throughout the chapter have been identified by one of three codes to give you an idea of their suitability for students of varying learning styles and abilities.

L1 Level 1 strategies should be within the ability range of all students. Full class participation is often required. Teacher direction is usually needed.

L2 Level 2 strategies are for average to above-average students or for small groups. Some teacher direction is necessary.

L3 Level 3 strategies are designed for students able and willing to work independently. Minimal teacher direction is necessary.

Teen Chat Group

Robert: I'm having a party at my house on Saturday night. Do you guys want to come?

Eduardo: Are your parents going to be home?

Robert: Yeah, they'll be there. Why?

Eduardo: Last weekend I went to a party at Heather's house. I didn't know that her parents were away for the weekend. Some of her brother's friends started drinking, and they wanted me to drink, too.

Erin: What did you do?

Eduardo: At first I just said no, but they kept pressuring me, so I left the party.

Erin: That was a good idea. You could get into a lot of trouble just being around people who are drinking.

Robert: Well, you won't have to worry about that on Saturday. Just get there around eight.

in Your Journal

Read the dialogue on this page. Have you ever felt pressured to try illegal substances? Start your private journal entries on alcohol and drugs by answering these questions:

▶ Do you think that Eduardo's decision to leave the party was a wise one? Explain your answer.

▶ How might knowing the effects of alcohol and drugs on your body help you make healthful choices?

▶ What are some risks involved in alcohol and drug use?

▶ What are some ways in which you could resist pressure to try alcohol or drugs?

When you reach the end of the chapter, you will use your journal entries to make an action plan.

Chapter 11: Alcohol and Drugs

Chapter 11

INTRODUCING THE CHAPTER

▶ Prepare a bulletin board display with colorful magazine ads for various types of alcohol, including beer and wine, and for various medicines. Encourage students to examine and discuss the ads. Then ask: What attitude does our society seem to have toward the use of alcohol? Toward the use of medicinal drugs? What problems do you think this attitude toward alcohol and drugs creates for teens who are trying to make decisions about alcohol and drug use? Tell students that they will learn more about alcohol and other drugs and about ways to resist using these substances in Chapter 11.

▶ Point out that aspirin, cold medicine, and other common pain- and symptom-relievers are drugs. Encourage discussion by posing questions such as these: Are these drugs always good for you, no matter how you use them? Why not? What are some dangerous drugs? What effects can those drugs have on your body and your mind? What drug is legal for adults but not for teens? What are some of the dangers of using that drug—that is, of drinking alcohol? Tell students that, as they read Chapter 11, they will learn more about these issues.

Cooperative Learning Project

Television Ads

The Teen Health Digest on pages 312 and 313 provides students with high-interest articles related to the content of this chapter.

The material in the Teacher's Wraparound Edition presents suggestions for a class project in which students will plan and produce television ads encouraging other teens to avoid alcohol and other drugs.

---

in Your Journal

Let volunteers do a dramatic reading of the Teen Chat Group dialogue, and encourage students to share their responses to the situation: What difference does it make if parents are home during a party? How does being at a party at which people are smoking marijuana affect teens who don't want to smoke marijuana? What response would you expect if you asked whether a friend's parents would be home? How can asking a question like that help you avoid high-risk situations? Then read aloud the In Your Journal questions, and ask several volunteers to share their responses to each. Finally, provide time for students to write private journal entries in response to the activity.

Lesson 1
Use and Abuse of Alcohol

Focus

LESSON OBJECTIVES

After studying this lesson, students should be able to
- explain what alcohol is and discuss why some teens begin drinking alcohol.
- discuss the effects alcohol has on the body.
- explain what alcoholism is and identify sources of help, both for alcoholics and for those living with alcoholics.

MOTIVATOR

Write this word on the board: *PARTY!* Give students time to list words and phrases that the word brings to mind.

INTRODUCING THE LESSON

Ask volunteers to read aloud their lists from the Motivator activity. Comment on whether references have been made to drinking alcohol. Ask students: Why do you think some teens associate drinking alcohol with partying? What are the best ways to avoid alcohol at parties? Explain that students will learn more about alcohol and its effects in Lesson 1.

INTRODUCING WORDS TO KNOW

Have students work with partners to read the Words to Know and to find the definition of each term presented in the lesson. Tell the partners to write three statements using the three Words to Know, but have them leave a blank space in place of the vocabulary word. Then let each pair of students exchange sentences with another pair. Can students complete the sentences they have been given by filling in the correct Words to Know?

Lesson 1 Use and Abuse of Alcohol

This lesson will help you find answers to questions that teens often ask about alcohol. For example:

▶ How does alcohol affect the body?
▶ Why do different people react differently to alcohol?
▶ What is alcoholism, and what can be done about it?

Words to Know

alcohol
cirrhosis
alcoholism

Science Connection

Deadly Alcohol

The word alcohol actually refers to several related chemicals. The type found in alcoholic drinks is ethanol (E•thuh•nawl), also called ethyl (E•thuhl) alcohol and grain alcohol. Other types of alcohol include methanol (ME•thuh•nawl), also called wood alcohol, and isopropyl (eye•suh•PROH•puhl) alcohol, the main ingredient in rubbing alcohol. Both methanol and isopropyl alcohol are highly poisonous.

What Is Alcohol?

Alcohol is found in beer, wine, whiskey, and other beverages. **Alcohol** is *a drug created by a chemical reaction in some foods, especially fruits and grains.* This drug can have strong physical and mental effects on the drinker.

Alcohol is the most commonly abused drug in this country. It is illegal for people under the age of 21 to drink alcohol. Even so, the average age for taking a first drink is 13. Nearly 90 percent of older teens feel that alcohol abuse is a critical problem in their schools.

Why do some teens begin drinking alcohol? Some are trying to escape from their problems or relieve stress. Others may feel peer pressure to drink at parties and on other social occasions. Learning about the effects of alcohol on the body, however, will help you make the decision not to drink.

Staying away from alcohol helps keep you physically and mentally healthy. What healthy activities do you enjoy doing with your friends?

296 Chapter 11: Alcohol and Drugs

Lesson 1 Resources

Teacher's Classroom Resources
- Concept Map 48
- Decision-Making Activity 21
- Enrichment Activity 48
- Lesson Plan 1
- Lesson 1 Quiz

- Reteaching Activity 48
- Transparency 42

Student Activities Workbook
- Study Guide 11
- Applying Health Skills 48

How Alcohol Affects the Body

Alcohol has both short-term and long-term effects on the body. Even one drink can impair judgment; large quantities of alcohol can cause death. **Figure 11.1** shows how alcohol affects various parts of the body.

Figure 11.1
Effects of Alcohol on the Body

The short-term effects of alcohol occur within a few minutes of drinking. The long-term effects may develop in a person who drinks heavily over an extended length of time.

interNET CONNECTION
Control your own destiny by knowing how to avoid substance abuse. You have the power!
http://www.glencoe.com/sec/health

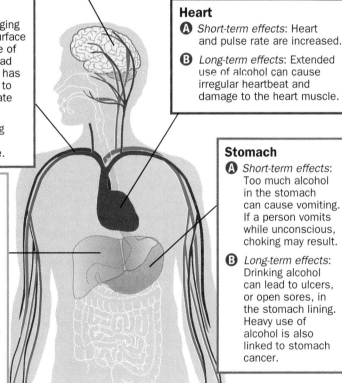

Brain
- **Ⓐ** *Short-term effects*: Alcohol reaches the brain soon after it is swallowed. It impairs such functions as judgment, reasoning, memory, and concentration. Reaction time is slowed, coordination decreases, and speech becomes slurred. Vision and hearing are distorted. Alcohol also reduces people's inhibitions—the ability or desire to control their behavior—sometimes leading them to engage in dangerous activities. Too much alcohol can lead to unconsciousness and death.
- **Ⓑ** *Long-term effects*: Over time, alcohol use can cause brain cell destruction, nervous disorders, and memory loss.

Blood Vessels
- **Ⓐ** *Short-term effects*: Blood vessels are enlarged, bringing blood closer to the skin surface and creating a false sense of warmth. This effect can lead a person whose judgment has been impaired by drinking to go outside without adequate clothing in cold weather.
- **Ⓑ** *Long-term effects*: Drinking alcohol can lead to high blood pressure and stroke.

Liver
- **Ⓐ** *Short-term effects*: Too much alcohol overloads the liver, which filters alcohol from the bloodstream and removes it from the body. Liver poisoning can result.
- **Ⓑ** *Long-term effects*: Heavy drinking can lead to **cirrhosis** (suh·ROH·sis), or *scarring and destruction of liver tissue*, which can cause death. Heavy alcohol consumption is also linked to liver cancer.

Heart
- **Ⓐ** *Short-term effects*: Heart and pulse rate are increased.
- **Ⓑ** *Long-term effects*: Extended use of alcohol can cause irregular heartbeat and damage to the heart muscle.

Stomach
- **Ⓐ** *Short-term effects*: Too much alcohol in the stomach can cause vomiting. If a person vomits while unconscious, choking may result.
- **Ⓑ** *Long-term effects*: Drinking alcohol can lead to ulcers, or open sores, in the stomach lining. Heavy use of alcohol is also linked to stomach cancer.

Lesson 1: Use and Abuse of Alcohol

Chapter 11 Lesson 1

Teach

L1 Identifying Examples
Ask the class to name different alcoholic drinks. Write their suggestions on the board. Stress that beer, wine, and wine coolers are alcoholic drinks; they are not safe or legal for teens.

L1 Critical Thinking
Let students share their responses to these questions: Why do you think so many teens experiment with alcohol, even though underage drinking is illegal and unsafe? What can adults do to discourage drinking among teens?

Teacher Talk

Dealing with Sensitive Issues

Young teens whose parents drink in moderation may be confused and concerned by discussions of the dangers of alcohol. It may be appropriate to emphasize the differences between adolescents and adults in relationship to alcohol; teens whose bodies have not finished growing and developing are at greater risk from the harmful effects of alcohol.

L1 Discussing
After students have described the scene in the photograph near the bottom of page 296, let a volunteer read the caption aloud. Then ask: What are some fun activities you enjoy with friends—without drinking alcohol?

Visual Learning
Let students study Figure 11.1, and call on volunteers to describe the function of each organ highlighted there. Then let students read aloud, and restate in their own words, the labeled effects of alcohol on the body.

More About...

Alcohol in the Home Students should be aware of the alcohol content in common household products. Certain baking ingredients, particularly vanilla extract, contain alcohol, as do many cough medicines and mouthwash products. Encourage students to read the ingredients labels on these products and discuss their findings.

Teens should also be aware of another kind of alcohol: the rubbing alcohol, or isopropyl alcohol, that many families keep in the medicine chest. Be sure students know that this is not the kind of alcohol found in alcoholic drinks. Rubbing alcohol is poisonous if ingested; anyone who drinks it is likely to become seriously ill.

Personal Inventory
ATTITUDES ABOUT ALCOHOL

Your attitudes toward alcohol will affect your decisions and behavior. Use this self-assessment form to examine your attitudes. On a separate sheet of paper, answer yes or no to each of the questions that follow.

1. Do you think that drinking alcohol makes a person seem more mature?
2. Would you go to a party where you knew alcohol would be served?
3. Do you think that drinking alcohol would make you feel more self-confident?
4. Do you think that drinking alcohol would make you more popular?
5. Do you have friends who drink regularly?
6. Do you think that it's less dangerous to drink alcohol than to use other drugs?
7. Would you ride in a car if the driver had been drinking alcohol?
8. Do you think that alcohol helps a person manage stress?

Give yourself 1 point for every time you answered no. A score of 7–8 shows that you have a good sense of the effects of alcohol; a score of 5–6 is fair. If you scored below 5 points, you need to rethink your attitudes about alcohol and its dangers.

Differing Effects

Figure 11.2 shows the amount of alcohol in three typical alcoholic drinks. The level of alcohol in the blood is the primary factor in determining how a person is affected. The impact varies, however, for different people and in different situations. Factors to consider include the following:

Figure 11.2
Alcohol Content of Three Types of Drinks
A typical drink of each of the three most common types of alcoholic beverages contains about the same amount of alcohol.

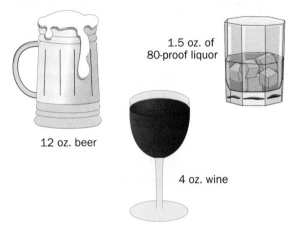

12 oz. beer

1.5 oz. of 80-proof liquor

4 oz. wine

- **Size and gender.** In general, females can tolerate a smaller amount of alcohol than males. Females usually weigh less than males, and their bodies may carry more fat, in which alcohol will not dissolve.
- **Food in the stomach.** Food in the stomach slows down the body's absorption of alcohol.
- **How fast a person drinks.** Gulping down a drink raises the alcohol level in the blood because the body has less time to process it.
- **Other substances in the body.** Drinking alcohol while taking prescription drugs or illegal drugs may have dangerous effects and may even be fatal.

Alcoholism

Alcoholism is *an illness caused by a physical and mental need for alcohol.* Alcoholics differ from moderate or social drinkers in that they are not able to limit the amount of alcohol they drink. In addition, alcoholics need to drink more alcohol in order to experience the same effects from drinking. A person may be an alcoholic if he or she

- drinks increasing amounts of alcohol and becomes drunk often.
- places drinking alcohol ahead of other activities.
- drinks alone.
- experiences blackouts and cannot remember what he or she said or did while drinking.
- shows personality changes when drinking alcohol.
- makes excuses for drinking, promises to quit but does not, or refuses to admit how much he or she is drinking.

If you suspect that a friend or relative is an alcoholic, you should seek help from a trusted adult.

LIFE SKILLS
Helping Someone with a Drinking Problem

You may suspect that someone close to you has a problem with alcohol. How can you be sure? Check for some of the signs of alcoholism listed in this lesson. A person who has even a few of these symptoms may have a drinking problem. You may or may not be able to do anything for this person. However, you can be supportive by

- learning about both alcoholism and sources of treatment.
- guiding the person to treatment, if possible.
- acting to encourage the person during and after treatment.

Remember that alcoholism is an illness. You are not to blame for the person's alcoholism. You are also not responsible for the alcoholic's actions. Although you may want to help, it is important not to put yourself in danger. Here are some strategies for helping someone with a drinking problem.

- Do not confront the person when he or she is drunk.
- Communicate with the person, when he or she is sober, about the damage that is being done.
- Encourage the person to seek help from a health care professional or a support group.
- Show concern and sympathy.
- Join a support group for people who have friends or relatives who are alcoholics.
- Discuss your problem with a counselor or health care worker.
- Do not cover up or make excuses for the person.
- If the person won't get help, consider seeking help for the person. The person may be angry with you at first, but your action could save his or her life.

Follow-up Activity
With a partner, role-play a conversation between two teens about problems with alcohol. One teen should be talking with the other teen (both are sober), expressing concern about the other's use of alcohol.

Lesson 1: Use and Abuse of Alcohol

Chapter 11 Lesson 1

L1 Discussing
Guide students in discussing the effects alcohol can have on daily life: How does drinking alcohol affect students? How does it affect employees and employers? How does it affect family life and friendships?

L1 Critical Thinking
Encourage students to share their responses to these questions: Why do you think an alcoholic cannot begin recovery before he or she admits to having a problem with alcohol? What does this fact indicate about the ability of friends and relatives to help individuals suffering from alcoholism?

L2 Guest Speaker
Invite a counselor who conducts interventions to speak to the class about this process: What are interventions? When and how are they usually used? With what results?

L2 Researching
Assign students to work together to learn more about 12-step programs: What are the 12 steps? How are they intended to be used? What is the main purpose of Alcoholics Anonymous and other 12-step programs?

L2 Language Arts
Encourage students to share the titles of novels they have enjoyed reading in which a teen character deals with the problems of alcohol abuse by parents or friends. Let students offer their analysis of what the teens were able and unable to do.

Making Health Connections
Alcoholism and other forms of drug abuse can have a devastating effect not only on the addict but also on those who love or care about the addict. Helping someone cope with a drinking problem can relieve a difficult situation in the teen's own life and thus improve the teen's mental/emotional health. Taking action, even if only in seeking aid for themselves, can also help teens gain a sense of control over their lives.

Meeting Student Diversity
Students who are still acquiring English skills may have difficulty discussing the problems of alcohol abuse and participating in a role-playing conversation. If possible, arrange for these students to work with others who share the same native language.

Assessing the Work Students Do
Ask students to write a one- or two-sentence summary of what they learned from this activity. Assign credit based on students' participation and attitude.

Chapter 11 Lesson 1

Assess

EVALUATING THE LESSON
Assign Reviewing Terms and Facts and Thinking Critically on this page to review the lesson; then assign the Lesson 1 Quiz in the TCR.

LESSON 1 REVIEW

Answers to Reviewing Terms and Facts
1. Alcohol is a depressant drug created by a chemical reaction in some foods, especially fruits and grains. Examples will vary.
2. Cirrhosis, which is scarring and destruction of liver tissue, is a long-term effect of alcohol on the liver; liver cancer is another long-term effect.
3. Signs of alcoholism include (any three): drinking increasing amounts of alcohol, giving alcohol consumption priority over other activities, drinking alone, experiencing blackouts, experiencing personality changes when drinking, making excuses for drinking, promising but failing to quit drinking, refusing to admit how much one drinks.

Answers to Thinking Critically
4. Responses will vary.
5. Responses will vary.

RETEACHING
▶ Assign Concept Map 48 in the TCR.
▶ Have students complete Reteaching Activity 48 in the TCR.

ENRICHMENT
Assign Enrichment Activity 48 in the TCR.

Close

Have students meet in small groups to review and discuss the most important reasons for avoiding alcohol.

300

 **in your journal**

After reading this lesson, how do you feel about alcohol? In your journal, discuss your feelings. What problems could alcohol cause in your life? How might you avoid these problems? Make a list of health reasons not to drink alcohol.

Peer support can help teens deal with the challenges of living with someone who has a problem with alcohol.

Help for Alcoholism
Although there is no medical cure for alcoholism, the illness can be treated. An alcoholic must go through a treatment process involving several steps to begin the process of recovery from alcoholism. An alcoholic must

- admit that he or she has a problem and decide to seek help.
- go through detoxification, a process of removing all alcohol from the body.
- get counseling and support to learn to avoid alcohol.

Lesson 6 of this chapter discusses in more detail the process of recovery from both alcohol and drug addiction.

Some alcoholics may need to be hospitalized to stop drinking and get the alcohol out of their bodies. After alcoholics stop drinking, they must never touch alcohol again. They may need support for a long time—even for life—to continue to avoid alcohol. Several groups help recovering alcoholics and their friends and family members.

- Alcoholics Anonymous is an organization of people who help each other avoid alcohol. The group promotes a 12-step program.
- Alateen is an organization for teens who have friends or relatives who are alcoholics. At Alateen meetings, teens get peer support to help them cope.
- Al-Anon is a peer support organization of adult friends and relatives of alcoholics.

Lesson 1 Review

Using complete sentences, answer the following questions on a separate sheet of paper.

Reviewing Terms and Facts
1. **Vocabulary** Define the term *alcohol*. Give two examples of beverages that contain alcohol.
2. **Recall** Name and describe a long-term effect of alcohol on the liver.
3. **List** Identify three of the signs of alcoholism.

Thinking Critically
4. **Infer** Why is it easier to avoid alcohol now than to stop drinking later?

5. **Analyze** Why are support groups helpful to recovering alcoholics?

Applying Health Concepts
6. **Consumer Health** Do research to find support organizations for alcoholics and their friends and relatives. Make a pamphlet listing this information, and post it in the classroom.
7. **Growth and Development** With a partner, make a poster listing reasons not to drink alcohol. Include some of the short-term and long-term effects of alcohol on the body. Display your poster in school.

300 Chapter 11: Alcohol and Drugs

Personal Health

Teens who may be facing problems of alcohol abuse by family members can be encouraged to contact Alateen, Al-Anon, or Alcoholics Anonymous. Students who want information—but who do not have a crisis that requires immediate attention—might also contact one of the following organizations:

▶ National Association for Children of Alcoholics
▶ National Council on Alcoholism and Drug Dependence
▶ American Council on Alcoholism
▶ Children of Alcoholics Foundation
▶ Alcohol and Drug Problems Association

Use and Abuse of Drugs

Lesson 2

This lesson will help you find answers to questions that teens often ask about drugs. For example:

▶ How should medicines be used?
▶ In what ways do some people misuse and abuse drugs?
▶ How do various kinds of drugs affect the body?

Using Drugs Properly

You sometimes hear people refer to newly discovered drugs as "miracle drugs." On the other hand, drugs are blamed for causing serious problems in our society. How can these substances be both helpful and harmful? Their effects depend on the types of drugs and how they are used. A **drug** is a *nonfood substance taken into the body that can change the structure or function of the body or mind.* When people talk about drugs that are beneficial, they often call them medications or medicines. A **medicine** is a *drug that is used to treat an illness or relieve pain.*

When you are sick, you might take either prescription or over-the-counter medicine. Prescription medicine is medicine that you can get only with a doctor's written order. You can buy over-the-counter (OTC) medicine without a prescription at pharmacies and other types of stores. Both kinds of medicine can have side effects and should be used with caution.

Words to Know

drug
medicine
stimulant
amphetamine
depressant
inhalant
narcotic
hallucinogen

in your journal

Before reading further in this lesson, think about your feelings toward drugs. In your journal, write whatever comes to mind at the mention of drugs. Do you think that drugs are a serious problem for teens? Why or why not?

When you buy an OTC medicine, read the label carefully to be aware of possible side effects.

Lesson 2: Use and Abuse of Drugs **301**

Lesson 2 Resources

Teacher's Classroom Resources
- Concept Map 49
- Cross-Curriculum Activity 21
- Decision-Making Activity 22
- Enrichment Activity 49
- Lesson Plan 2
- Lesson 2 Quiz
- Reteaching Activity 49
- Transparencies 43, 44

Student Activities Workbook
- Study Guide 11
- Applying Health Skills 49

Lesson 2
Use and Abuse of Drugs

Focus

LESSON OBJECTIVES

After studying this lesson, students should be able to

▶ discuss how people can use drugs properly.
▶ identify ways in which drugs are misused and abused.
▶ explain what marijuana is and discuss some of the dangers associated with its use.
▶ name common stimulants, tranquilizers, and other illegal drugs and discuss some of the dangers associated with their use.
▶ explain why drug abuse is dangerous.

MOTIVATOR

Ask students the following questions: Do students in this school need more information about drugs? Why or why not? Give students a few minutes to jot down their responses.

INTRODUCING THE LESSON

Let volunteers share their ideas in response to the Motivator questions. Then ask: How does having information about drugs help teens avoid drug abuse? Do you think not knowing about drugs is dangerous? Why? Tell students that Lesson 2 presents relevant facts about the use and abuse of drugs.

INTRODUCING WORDS TO KNOW

Help students read aloud the list of Words to Know, and encourage students to share what they already understand about the meaning of each term. Then have students make flash cards with the Word to Know on one side and the definition on the other. Encourage students to use these flash cards as study aids.

Chapter 11 Lesson 2

Teacher Talk
Dealing with Sensitive Issues

It is important to remember that any discussion of the misuse or abuse of drugs may be disquieting for students who are themselves experimenting with drugs or who know or fear that a parent, other family member, or friend abuses drugs. As you discuss drugs with the students, avoid asking questions, even of the class in general, that may seem personal. At the same time, be alert for students who may want or need a private consultation about drugs. Create opportunities to speak with each of those students alone, or ask a school health provider or counselor to talk with them.

L1 Critical Thinking
Let students share their ideas in response to these questions: Do you think there really is such a thing as a miracle drug? Why or why not? Do you think the use of the term *miracle drug* affects our society's attitudes toward drugs? If so, in what way?

L1 Comprehending
Guide students in distinguishing between the terms *drug* and *medicine*. Ask: Are all drugs medicines? Are all medicines drugs? Have volunteers explain their answers.

L1 Role-Playing
After students have discussed the photograph near the bottom of page 301, let several pairs act out the roles of the father and son in that picture.

L2 Demonstrating
Bring to class several empty containers of common OTC medicines. Divide the class into small groups, and give each group at least one container to examine. Instruct group members to find and read on the medicine container each item of information listed near the top of this page.

Did You Know?
Simpler Labels

In February 1997 the Food and Drug Administration (FDA) proposed the simplification of labels on over-the-counter drugs. The FDA's recommendations are designed to make medicine labels easier to read and understand. Crucial information, for example, will be listed at the top of the label, not buried in small print near the end. In addition, language will be simplified. The word "pulmonary," for example, will be replaced with the word "lung," and the phrase "consult a physician" will be replaced by "ask a doctor." The FDA hopes that the changes will take effect by mid-1999.

Reading package labels carefully is important, no matter what kind of medicine you plan to use. The labels on over-the-counter medicines include the following information:

- Directions that explain the medicine's purpose and how to use it
- Warnings about any side effects and when to ask a doctor before using the medicine
- A list of active ingredients
- The expiration date, which is the date after which the medicine should not be used

Figure 11.3 at the bottom of this page identifies the information that appears on a typical prescription medicine label.

Misusing and Abusing Drugs

Medicines can cause harm if misused or abused. Drug misuse means using a legal drug in an improper way. A person is misusing drugs if he or she

- takes more than the recommended or prescribed dosage of a medicine or mixes medicines without asking a doctor.
- continues to take medicine after it is no longer needed.
- stops taking a prescribed medicine sooner than the doctor's instructions indicate.
- uses medicine that was prescribed for someone else (even if he or she has the same symptoms).

Drug abuse is a different and more serious problem in our society. It means using substances that are illegal (such as marijuana, cocaine, or LSD) or that are not intended to be taken into the body. Drug abuse may also involve using a legal drug, such as a painkiller, in an illegal or harmful way.

Figure 11.3
Prescription Medicine Label
If you don't understand the information on a prescription label, ask your doctor or pharmacist to explain it to you.

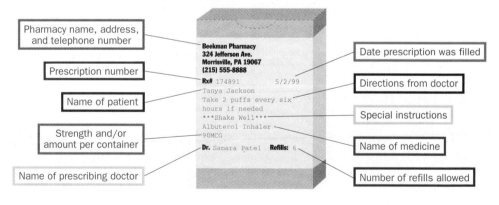

302 Chapter 11: Alcohol and Drugs

Cooperative Learning

Survey How frequently do teens and adults take medicines? What proportion of their medicines are OTC, and what proportion are prescription? How carefully do teens and adults read the directions on medicine labels? How thoroughly do they follow those directions? Have students design a survey that will gather information on these topics. Ask students to revise and proofread their survey carefully, and then make copies of a final written version. Let students devise a method of distributing and collecting the survey, which is to be filled out anonymously by other students, friends, and family members. Remind students to avoid having any person fill out more than one survey. Finally, have students tabulate and publish the results of their survey.

Marijuana

Marijuana (mar·uh·WAH·nuh) is a mood-altering drug made from the leaves, stems, and flowering tops of the hemp plant. It is usually smoked in hand-rolled cigarettes or in pipes. Occasionally it is mixed with food. The effects of marijuana vary and often depend on the user's surroundings and feelings at the time. Some users feel relaxed and unusually sensitive to sights and sounds, while others may feel sad, fearful, and suspicious. **Figure 11.4** shows some short-term and long-term effects of marijuana.

Figure 11.4
Effects of Marijuana Use
Some effects of marijuana are obvious right away, while others occur after repeated use.

The immediate effects of marijuana use may include	Long-term regular marijuana users may experience
• inability to think or speak clearly.	• problems with normal body development (if use begins in adolescence or earlier).
• inability to concentrate.	• damage to lung tissue and the immune cells that fight cancer.
• loss of short-term memory.	• feelings of anxiety and panic.
• lack of coordination and slowed reaction time.	• possible psychological dependence.
• increased heart rate and appetite.	• infertility.

Teen Issues

Drug Abuse Rising

Between 1991 and 1996, the percentage of eighth graders using drugs more than doubled. What is causing this increase? One reason may be that today's students are not aware of the risks involved in drug use.

Healthy teens stay away from drugs and other harmful substances.

Stimulants

Stimulants (STIM·yuh·luhnts) are a type of *drugs that speed up the body's functions.* Amphetamines (am·FE·tuh·meenz), cocaine (koh·KAYN), and crack cocaine are stimulants. Stimulants cause increases in heart rate, breathing rate, and blood pressure. They give users a false sense of energy and power. Abuse of stimulants eventually leaves the user feeling exhausted.

Although stimulants are prescribed for some conditions, the use of most stimulants is illegal. These drugs can affect the body in unpredictable ways, even causing death. Users can also become addicted to them.

Chapter 11 Lesson 2

L1 Discussing
Ask volunteers to name common prescription medicines they have used. Pose these questions: Did you read the prescription medicine label, or did you rely on a parent or other family member to check the information? How does becoming more mature affect the responsibilities you are expected to assume regarding your own medication?

L1 Comparing
Let students share their responses to these questions: Which is more important, checking the label on an OTC medicine or checking the label on a prescription medicine? Why? (Both are very important. A reaction to an OTC medicine can be just as devastating as a reaction to a prescription drug.)

L2 Life Skills
Present the following situation to students: You have just picked up a prescription medicine from a pharmacy. You are confused by the label; the instructions for using the medicine are not what you remember hearing from the doctor. Have students write paragraphs describing how they would handle this situation.

L1 Comprehending
Let volunteers restate in their own words the differences between misusing drugs and abusing drugs. Ask: Which is more dangerous? Why? Is either safe? Why not?

L1 Identifying Examples
Students may be familiar with street names for marijuana. Have volunteers share these names, and record them on the board. Ask: Do you think it is important to know the different names for marijuana? Why or why not?

L1 Critical Thinking
Ask students to imagine that a doctor or a pilot is a marijuana user: What could the effects be? Then have students name other jobs in which marijuana use could lead to severe consequences for society.

Lesson 2: Use and Abuse of Drugs

Personal Health

Most students have probably already experienced the effects of a common, relatively benign stimulant: caffeine. Caffeine has an indirect effect on the brain (unlike other stimulants, which work directly on the brain). It tends to make users alert and wakeful. Too much caffeine, however, can make people jittery and anxious; it can also trigger panic attacks in some individuals.

Caffeine is most widely found in coffee, which is often considered an "adult beverage." However, tea, colas, and some other soft drinks also contain high doses of caffeine. Cocoa, chocolate, and many medicines are also sources of caffeine.

Chapter 11 Lesson 2

L1 Comparing
Ask students how and why the damage to lung tissue caused by marijuana may be similar to that caused by tobacco. Encourage interested students to read more about this comparison.

L2 Guest Speaker
Invite a counselor or doctor who specializes in treating addictions to speak to the class about his or her work, especially with teen patients: To what drugs or other substances are teens commonly addicted? How can those addictions be treated? With what success? What are the best ways to prevent addictions?

L2 Journal Writing
After students have discussed the photograph near the bottom of page 303, suggest that they write journal entries in response to this question: What makes you feel energetic and self-confident?

L2 Listing Examples
Divide the class into four groups, and assign each group one of these drugs or categories of drugs: amphetamines, cocaine, crack cocaine, depressants. Have the members of each group work together to list as many street names as possible for their assigned drug or drug category. Post all the groups' lists on a bulletin board so that all the students can read them.

L2 Current Events
Have students find and bring to class recent news stories (from either newspapers or magazines) about crimes or accidents that involved the use of cocaine, crack cocaine, or other drugs. Let a volunteer use these news stories to create a bulletin board display or a class booklet.

L3 Social Studies
Federal laws make a clear distinction between the possession of cocaine and the possession of crack cocaine. Let a volunteer research this legal distinction and discuss his or her findings with the rest of the class.

Amphetamines

Amphetamines are *strong stimulant drugs that speed up the nervous system.* They come in many forms and can be swallowed, inhaled, smoked, or injected. People may abuse amphetamines to stay awake and to get a temporary feeling of energy. Abuse of amphetamines has serious side effects, however. Some of these are shown in **Figure 11.5**.

Figure 11.5
Effects of Amphetamine Abuse
Amphetamine abuse can cause many serious problems, including death.

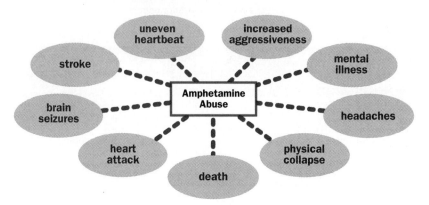

Did You Know?

Speed Kills

Methamphetamine (me·tham·FE·tuh·meen), commonly called "speed," "meth," or "crank," is a stimulant drug that has very strong effects on the nervous system. It can be swallowed in pill form, inhaled or injected in its powdered form, or smoked in crystallized form. This powerful drug can cause paranoia or violent behavior and may also damage the blood vessels in the brain, leading to a stroke.

Cocaine and Crack Cocaine

Cocaine is an illegal stimulant drug. It is a white powder made from the leaves of the coca plant. Cocaine can be inhaled, smoked, or injected. Entering the bloodstream quickly, it gives the user a brief, powerful feeling of well-being. That feeling, however, is quickly replaced by anxiety and depression. In a matter of days, a cocaine user can become addicted to the drug.

The use of cocaine can cause loss of appetite, nausea, sleeplessness, seizures, and stroke. Because of its effect on the heart rate, even a first-time user of cocaine runs the risk of a fatal heart attack. Cocaine can also damage the nasal membranes and lungs. A user may have trouble concentrating, become aggressive, or deny responsibility for his or her actions. In addition, a person who injects cocaine with a shared needle risks becoming infected with the human immunodeficiency virus (HIV) or with hepatitis B.

Crack cocaine, or crack, is a concentrated form of cocaine. When smoked, it reaches the brain within 10 seconds. Crack produces a feeling of energy and excitement, but the feeling lasts only about 15 minutes. Then the user feels depressed and craves more of the drug. The side effects of crack are similar to those of cocaine, including possible heart attack and death. In addition, crack users become dependent on the drug even faster than cocaine users.

304 Chapter 11: Alcohol and Drugs

More About...

Cocaine We often think of cocaine as a modern drug. However, native peoples of South America were probably chewing coca leaves as early as 3000 B.C. Coca use was certainly an important part of the culture of the Incan Empire. The Incas considered the coca plant a gift from their gods; they incorporated coca use into religious ceremonies, including burials.

During the prime of the Incan Empire, coca use was restricted to members of the ruling class and to those participating in special rituals or endeavors. By the time the Spanish arrived in South America, however, many Inca people were apparently using coca daily. Historians believe that this use of coca contributed to the conquest of the Inca by the Spanish invaders.

Depressants

Depressants (di·PRE·suhnts) are *drugs that slow down the body's functions and reactions, including heart and breathing rates.* They come in tablet or capsule form and are swallowed. Doctors sometimes prescribe depressants to treat patients who suffer from anxiety and sleeplessness. Types of depressants include tranquilizers (TRANG·kwuh·ly·zuhrz), hypnotics (hip·NAH·tiks), and barbiturates (bar·BI·chuh·ruhts). Alcohol is also a depressant.

Depressants may make users feel relaxed and less anxious. However, depressants can also cause poor coordination and impaired judgment. People who abuse depressants may experience mood swings or depression. In addition, people can become addicted to depressants. If combined with alcohol, depressants can cause coma and death.

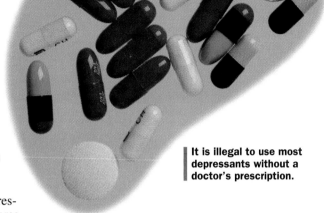

It is illegal to use most depressants without a doctor's prescription.

Other Illegal Drugs

Marijuana, stimulants, and depressants are some of the best-known drugs that people abuse. However, many other illegal and dangerous drugs are also abused. **Inhalants** (in·HAY·luhnts) are *substances whose fumes are breathed in to produce mind-altering sensations.* Most inhalants are substances that were not intended to be taken into the body, such as glue, gasoline, and spray paint. Inhaling these substances can cause nausea or vomiting, dizziness, mental confusion, and loss of motor skills. Inhalant use—even a single use—can lead to permanent brain damage, coma, or death.

Narcotics (nar·KAH·tiks) are *drugs that relieve pain and dull the senses.* Many narcotics, including heroin (HER·uh·wuhn), are very powerful and also illegal. Other narcotics, such as morphine (MAWR·feen) and codeine (KOH·deen), are legal for a person who is under a doctor's care and has a prescription. Narcotics are very carefully controlled because their effects are so strong, and because users can easily become addicted.

Hallucinogens (huh·LOO·suhn·uh·jenz) are *drugs that distort moods, thoughts, and senses.* Two hallucinogens are lysergic (luh·SER·jik) acid diethylamide (dy·e·thuh·LA·myd), or LSD, and phencyclidine (fen·SY·kluh·deen), or PCP. These drugs sometimes cause the user to experience imaginary images and sounds or distortions of real objects. Users may lose control of their actions and behave strangely or violently, possibly endangering their own lives or the lives of others.

Q & A

"Roofies"

Q: I overheard a couple of kids talking about "roofies." What are they?

A: "Roofies" is the street name for Rohypnol, a very powerful sleeping pill. It is not legal in the United States. Roofies are often used with alcohol and other drugs. Recently, roofies have become associated with date rape. A victim is given roofies without her knowledge. She then passes out and may not remember that sexual contact has occurred.

Consumer Health

Researchers are constantly learning more about the functioning of the human body and developing new drugs to help restore or maintain physical health. One rapidly growing group of drugs is antidepressants. Scientists have identified at least some of the neurotransmitters that play a role in depressive disorders; the new prescription antidepressants work on these neurotransmitters. However, depression is not a simple—or even a single—illness. Different antidepressants can be effective for different individuals, and finding the right drug and dose can be challenging. Not all depressive disorders can be treated successfully with drugs currently available, so research continues.

Chapter 11 Lesson 2

Assess

EVALUATING THE LESSON
Assign Reviewing Terms and Facts and Thinking Critically on this page to review the lesson; then assign the Lesson 2 Quiz in the TCR.

LESSON 2 REVIEW

Answers to Reviewing Terms and Facts
1. A drug is a nonfood substance taken into the body that changes the structure or function of the body or mind. A medicine is a drug that is used to treat an illness or relieve pain.
2. Drug misuse involves using a legal drug in an improper way. Drug abuse involves using substances that are illegal, using substances that are not meant to be taken into the body, or using legal substances in an illegal or harmful way.
3. Narcotics relieve pain and dull the senses. Examples of narcotics include heroin, morphine, and codeine (any two).

Answers to Thinking Critically
4. Responses will vary.
5. Responses will vary.

RETEACHING
▶ Assign Concept Map 49 in the TCR.
▶ Have students complete Reteaching Activity 49 in the TCR.

ENRICHMENT
Assign Enrichment Activity 49 in the TCR.

Close

Have each student write and complete the following sentence: *The most important thing for me to remember about drug abuse is…*

---

in your journal

Read your journal entry from the beginning of the lesson. Have your feelings about drugs changed now that you have read the text? In your private journal, describe your thoughts about the need for discussions on drugs among friends, family members, and the school population.

The Dangers of Drug Abuse

All mood-altering drugs have powerful effects on the nervous system. This is why some people turn to drugs when they are anxious, bored, or feeling hopeless. They see drugs as a way to change their feelings and even to escape from them.

Drugs of these kinds are very dangerous, however. Some drugs, such as heroin, LSD, and cocaine, may actually alter the way the nervous system works. Because the nervous system controls the way people receive, process, and transmit information, drug abuse causes individuals to lose control over their own lives. Drugs that are injected can also expose people to deadly infections from viruses such as HIV and hepatitis. Drug addiction destroys many lives. In addition, many of the drugs discussed in this lesson can cause permanent brain damage, coma, and death.

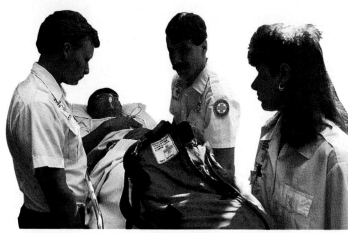

Those who use alcohol and drugs risk their own health and safety and the health and safety of those around them.

Lesson 2 Review

Using complete sentences, answer the following questions on a separate sheet of paper.

Reviewing Terms and Facts
1. **Define** Explain the terms *drug* and *medicine*.
2. **Explain** What is the difference between misuse and abuse of drugs?
3. **Give Examples** Describe the effects of narcotics, and give two examples.

Thinking Critically
4. **Analyze** In what ways are inhalants different from many of the other drugs discussed in this lesson?
5. **Suggest** What advice would you give to a friend who was thinking about taking amphetamines so that she could stay awake to study for a test?

Applying Health Concepts
6. **Health of Others** Make a list of rules telling how to use prescription and over-the-counter medicines correctly. For each rule, explain why it is important. Post your list at home or at school.

306 Chapter 11: Alcohol and Drugs

Health of Others

PCP is an unusually powerful yet inexpensive drug. Often, PCP users feel that their bodies and their minds are separate entities. They are likely to experience both illusions and hallucinations. The body does not rid itself of PCP efficiently, so these disturbing effects can last for days.

PCP use is often associated with acts of violence or other criminal behavior. Suggest that students find and bring to class news stories about the destructive effects of PCP use. Several volunteers might organize a bulletin board on which these news accounts can be displayed.

The Nervous System

Lesson 3

This lesson will help you find answers to questions that teens often ask about the brain and the nervous system. For example:

▶ What are neurons, and what is their function?
▶ What are the parts of the nervous system?
▶ How can I keep my nervous system healthy?

The Body's Control System

As you have learned, the use of drugs and alcohol can produce harmful effects throughout the body. The nervous system, however, is most directly affected by these substances. Abuse of drugs and alcohol can permanently damage the nervous system.

The nervous system is the control center of your body. It responds to changes from inside or outside the body and sends messages to the brain. It also relays instructions from the brain to other parts of the body. The nervous system controls thought processes, senses, movement, and such functions as heartbeat, breathing, and digestion. Therefore, it is essential to keep your nervous system healthy by protecting it from disease, injury, and the effects of alcohol and drugs.

Neurons

The cells that make up the nervous system are called **neurons** (NOO·rahnz). They are also known as nerve cells. Neurons send and receive information in the form of tiny electrical charges. These charges carry messages throughout the body. **Figure 11.6** shows the path that the messages follow to cause the body to react.

Words to Know

neuron
central nervous system (CNS)
peripheral nervous system (PNS)
brain
spinal cord

Figure 11.6
How Neurons Work
The three types of neurons work like a relay team to carry messages throughout the body.

① A receptor—the ear—responds to sound waves. **Sensory neurons** in the ear receive information and send impulses to the brain via the auditory nerve.

② **Connecting neurons,** found in both the brain and the spinal cord, relay information about the sound to motor neurons.

③ The **motor neurons** deliver the message to the muscles. Lauren turns her head toward the sound.

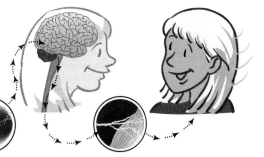

Lauren!

Lesson 3: The Nervous System **307**

Lesson 3 Resources

Teacher's Classroom Resources
- Concept Map 50
- Enrichment Activity 50
- Health Lab 11
- Lesson Plan 3
- Lesson 3 Quiz
- Reteaching Activity 50
- Transparency 45

Student Activities Workbook
- Study Guide 11
- Applying Health Skills 50

Lesson 3 The Nervous System

Focus

LESSON OBJECTIVES

After studying this lesson, students should be able to

▶ explain how the nervous system functions as the central control system of the body.
▶ identify and discuss the two main parts of the nervous system.
▶ identify the factors that may lead to disorders of the nervous system.

MOTIVATOR

Ask students to think about the words *nerve* and *nervous*. What different meanings do they associate with the words? Instruct them to jot down as many phrases or short sentences as they can, using different meanings.

INTRODUCING THE LESSON

Let several volunteers read aloud their lists of phrases and sentences from the Motivator activity. Help students identify the various meanings and shades of meaning they have identified in using the words. Point out any example in which students have used the term *nervous system*, or introduce the term. Encourage volunteers to share some of the facts they already know about that system. Then tell students that they will learn more about the nervous system and its functions as they study Lesson 3.

INTRODUCING WORDS TO KNOW

Read over the list of Words to Know with the class. Instruct students to find the definition of each term in the lesson, and have them record these definitions in their notebooks.

Chapter 11 Lesson 3

Teach

L1 Comprehending
As students begin reading about the nervous system, focus their attention on the fact that abuse of alcohol and drugs can cause permanent damage: How could your life be affected if you suffered permanent damage to the system that controls your thought processes, your movement, your heartbeat, and your digestion?

Visual Learning
Use Figure 11.6 to help students see and discuss how neurons carry messages throughout the body. Let volunteers describe the picture in each frame and discuss what seems to be happening. Then have other volunteers read the labels.

L3 Researching
Ask interested class members to investigate the development of neuron connections within a baby's brain: When and how do these connectors grow? What stimulates their growth? Which connectors later disintegrate? Why? Have the volunteers create a poster presenting their findings.

! Would You Believe?
The neurons in our nervous system produce the same kind of enzyme that the neurons in an insect's nervous system produce. This can be bad news for us, because insecticides kill insects by destroying that enzyme.

L1 Discussing
Help students consider the importance of the spinal cord within the nervous system: Why do spinal cord injuries so often cause permanent paralysis in major portions of the body? How do people injure their spinal cord?

Cultural Connections
The Egyptian View

We know that the brain is the center of the body's control network. The ancient Egyptians, however, did not consider the brain very important. When an Egyptian died, embalmers preserved the liver, intestines, lungs, and stomach, and especially the heart—this organ was thought to be the center of all the emotions. The brain, however, was discarded.

The Nervous System

The nervous system has two main parts, as shown in **Figure 11.7**. The **central nervous system (CNS)** consists of *the brain and the spinal cord*. It coordinates the body's activities. The **peripheral (puh·RIF·uh·ruhl) nervous system (PNS)** is made up of *all the nerves outside the central nervous system*. It connects the brain and the spinal cord to the rest of the body.

The **brain** is the *information center of the nervous system*. It receives and screens information and sends messages to other parts of the body. **Figure 11.7** shows the three main parts of the brain. The **spinal cord** is *a long bundle of neurons that relays messages to and from the brain and all parts of the body*.

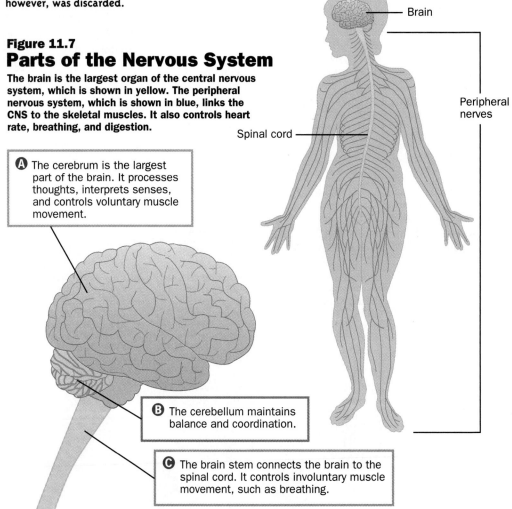

Figure 11.7
Parts of the Nervous System
The brain is the largest organ of the central nervous system, which is shown in yellow. The peripheral nervous system, which is shown in blue, links the CNS to the skeletal muscles. It also controls heart rate, breathing, and digestion.

A The cerebrum is the largest part of the brain. It processes thoughts, interprets senses, and controls voluntary muscle movement.

B The cerebellum maintains balance and coordination.

C The brain stem connects the brain to the spinal cord. It controls involuntary muscle movement, such as breathing.

Chapter 11: Alcohol and Drugs

Growth and Development

The brain of a newborn has much more power than any human can use. That isn't as wasteful as it sounds, however; this great power enables individual human beings to adapt successfully to the environment into which they are born. As humans grow, their brains adapt by losing some of their potential functions.

The human brain requires stimulation to develop properly and to avoid losing those functions it requires. Repeated studies have shown the importance of stimulation during the first year of life. Babies who receive attention, affection, and plenty of interaction with other humans grow up to have higher IQs than babies who lack such stimulation.

Problems of the Nervous System

Several diseases and disorders may affect the nervous system. Healthy lifestyle decisions can prevent many of these problems. The following factors may lead to nervous system disorders:

- **Infections.** Illnesses such as polio, rabies, and meningitis (me·nuhn·JY·tuhs) are caused by viruses that affect the nervous system. Fortunately, vaccines are available for some of these diseases. Others can be treated successfully with medicines.
- **Degenerative disorders.** Some disorders are caused by damage to the nervous system, but they cannot be prevented or cured. These disorders include multiple sclerosis (skluh·ROH·suhs) and cerebral palsy (suh·REE·bruhl PAWL·zee).
- **Injuries.** The most common cause of nervous system damage is physical injury. The results of head, neck, and back injuries can be very severe. A spinal cord injury, for example, can cause paralysis—the loss of feeling and movement in part of the body. Many of these injuries to the nervous system can be prevented by taking safety precautions.
- **Drug abuse.** The misuse and abuse of drugs can cause damage to the nervous system. Drugs act directly on the brain stem, which helps control heart rate, breathing, appetite, and sleeping. Some drugs also create problems with the way messages are sent, received, and responded to by the nervous system.
- **Alcohol abuse.** Drinking alcohol has an immediate effect on the brain. It can impair memory, thought processes, perception, judgment, and attention. Over time, abuse of alcohol can destroy millions of brain cells. Once destroyed, these cells can never be repaired or replaced.

History Connection

Determined to Succeed

Franklin D. Roosevelt, the 31st president of the United States, was unable to walk without assistance for more than 20 years of his life. He had suffered an attack of polio that had left his back, arms, and legs paralyzed. However, Roosevelt was determined not to let the illness destroy his political career. By exercising regularly, he was able to regain the ability to move his arms and back. He went on to be president for 12 years.

Women who are pregnant should avoid consuming alcohol. Drinking during pregnancy may damage the baby's central nervous system, leading to a condition known as fetal alcohol syndrome (FAS).

Health of Others

Every year about half a million Americans suffer a stroke, a loss of blood supply to the brain caused by a clot or a hemorrhage. The effects of a stroke can be devastating; stroke is the third leading cause of death in this country and the most common cause of disabilities that strike adults. Stroke victims who survive may be paralyzed or blinded. They may also be left with an inability to think clearly or an altered personality. Recovery and readjustment following a stroke can be difficult; often physical therapy helps a stroke victim regain at least some abilities. Fortunately, there are effective measures for preventing many strokes. These include detecting and treating high blood pressure and quitting—or never starting—cigarette smoking.

Chapter 11 Lesson 3

L2 Social Studies

Ask students to find out about your state's laws regulating the use of helmets and seat belts: In what situations and for which individuals are these safety devices required? What opposition, if any, has there been to the passage and enforcement of these laws? What additional laws, if any, have been proposed? With what result?

L1 Listing Examples

Have students work with partners to list as many sports safety rules as they can. Then guide the class in discussing those rules: How does each help to protect the nervous system?

Assess

EVALUATING THE LESSON

Assign Reviewing Terms and Facts and Thinking Critically on page 311 to review the lesson; then assign the Lesson 3 Quiz in the TCR.

RETEACHING

▶ Assign Concept Map 50 in the TCR.
▶ Have students complete Reteaching Activity 50 in the TCR.

ENRICHMENT

Assign Enrichment Activity 50 in the TCR.

 in your journal

Do you take good care of your nervous system? In your private journal, list the reasons why you need to protect it. Then make a list of actions you take already, such as avoiding alcohol and drugs, that help in this effort. Do you have any unhealthful habits that you would like to change in order to take better care of your nervous system? Think about the steps you will take to make these changes.

Caring for Your Nervous System

You can take care of your nervous system by practicing good health habits, such as eating properly and getting enough sleep. In addition, the following safety measures will help prevent nervous system damage.

- **Protect yourself from disease.** Most likely, you have already been vaccinated against some diseases, such as polio. To protect yourself from rabies, avoid contact with strange or wild animals. Good hygiene will help protect you from infections such as meningitis.

- **Wear a helmet.** When you are riding a bike, skateboarding, in-line skating, or playing a contact sport, wear a helmet. It will help to protect your head from injury.

- **Wear a seat belt.** Always fasten your seat belt when you are riding in a car.

- **Play it safe.** Be careful when playing sports. For example, never dive into shallow water. If you use gymnastics equipment, such as a trampoline, have spotters watch you.

- **Lift properly.** When lifting heavy objects, use the proper techniques to prevent back injuries. If the object is too heavy, ask someone for help.

- **Observe safety rules.** When walking or riding a bicycle, follow all traffic safety rules.

- **Avoid alcohol and drugs.** By not using alcohol and drugs, you will prevent many disorders of the nervous system.

HEALTH LAB
Nervous System Tricks

Introduction: Your perception of the world is based partly on your senses and partly on past experience. Your perception can be faulty, however, if your nervous system is being fooled—for example, by an optical illusion. Your eyes can play tricks on you because of the way the nerves in your eyes normally react to light and color.

Your retina, the network of nerves that absorbs the light rays that enter your eye, has two kinds of receptors. Cone cells are sensitive to bright light and colors. Rod cells help you see in dim light. The way these cells work together can make the contrast between dark and light confusing.

Objective: View optical illusions, and examine how they occur.

Materials and Method: Look at the picture of the two circles within squares. Which circle looks larger—the white or the black?

Next, look at the picture of black squares. Focus your eyes on one of the squares. What do you perceive in the intersections of the white bars? Now focus closely on one intersection. What happens?

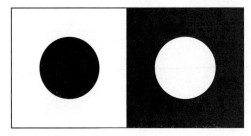

310   Chapter 11: Alcohol and Drugs

HEALTH LAB
Vision Tricks

Time Needed
20 minutes

Supplies
ruler

Focus on the Health Lab

If possible, introduce the Health Lab activity by displaying a color poster that presents an optical illusion—a picture that can be seen in two or more different ways. Let students describe and compare what they see. Can some see "the other" picture or pictures after they have been described? If a poster is not available, encourage students to describe their own previous experiences with optical illusions and to share their current understanding of how these illusions produce their effect.

Understanding Objectives

Students will experience for themselves two optical illusions and will use measurements to help compare the reality of each picture with its effect.

Review

Lesson 3

Using complete sentences, answer the following questions on a separate sheet of paper.

Reviewing Terms and Facts

1. **Vocabulary** Define the term *neuron*. What is the function of neurons?
2. **Compare** What is the difference between the *central nervous system* and the *peripheral nervous system*?
3. **Identify** Name the three main parts of the brain and explain the function of each part.
4. **List** What are four causes of problems in the nervous system?
5. **Recall** Identify five safety measures you can take to protect your nervous system.

Thinking Critically

6. **Describe** Think of any movement that a person might make during a volleyball game. Explain how neurons would enable this action to occur.
7. **Hypothesize** What attitudes might lead to the types of accidents that damage the nervous system?

Applying Health Concepts

8. **Growth and Development** With a small group of classmates, create a skit to show how neurons carry messages and cause a reaction. Decide on an action to portray, choose roles, and act it out for the class. One person can act as the narrator to explain the steps. Your group may need to do some research to create the skit.
9. **Health of Others** With a partner, make a poster showing safety rules to follow to prevent damage to the nervous system. Include drawings that illustrate the dangers, and write brief explanations of how to prevent each problem. Display your poster in the classroom.

Chapter 11 Lesson 3

LESSON 3 REVIEW

Answers to Reviewing Terms and Facts

1. A neuron is one of the cells that make up the nervous system. Neurons send and receive information in the form of tiny electrical charges.
2. The central nervous system consists of the brain and the spinal cord; it coordinates the body's activities. The peripheral nervous system consists of all the nerves outside the central nervous system; it connects the brain and spinal cord to the rest of the body.
3. See Figure 11.7 on page 308.
4. Nervous system problems may be caused by (any four): infections, degenerative disorders, injuries, drug abuse, and alcohol abuse.
5. Safety measures to protect the nervous system include (any five): protecting yourself from disease, wearing a helmet, wearing a seat belt, playing sports safely, lifting heavy objects properly, observing safety rules, and avoiding alcohol and drugs.

Answers to Thinking Critically

6. Responses will vary.
7. Responses will vary.

Ask students: In your opinion, what are the most important facts for teens to know about alcohol, drugs, and the nervous system?

Observations and Analysis:
Measure the diameters of the two circles in the first picture. Did your perception fool you? The way cone cells and rod cells work can make bright objects seem larger than dark ones.

In the second picture, did you perceive dark spots in the intersections when you were looking at a black square? Why do you think the spots seemed to disappear when you looked directly at an intersection? The explanation is that the color white appears whiter when it is next to something black. The white bars appear whiter than the intersections because the bars are right next to black squares, while the intersections are seen as the color white meeting the color white. When, however, you look at the intersection area by itself, you perceive the white as white. Your eye is not comparing it to any other white area.

What other optical illusions have you experienced? Why might they occur? Find another example of an optical illusion in a book or magazine. Write a paragraph about this illusion and how it tricks the eye. Share your observations with your classmates.

Observation and Analysis

Allow students to work with partners as they observe the illusions in the Health Lab feature. Encourage the partners to describe what they see and to compare their perceptions.

Ask the partners to discuss their responses to the questions in the feature. Then have the two work together to write a short description of one of the optical illusions: What do people see when they look at the picture? How is that different from the reality of the picture? Why are most people's perceptions different from that reality?

Further Investigation

Encourage interested students to find other optical illusions; the school or public library may have good resources. Then suggest that students try to draw their own versions of one of the illusions they have found, or perhaps create their own original illusions, using what they have learned about how the eye sees.

Assessing the Work Students Do

In brief conferences, have students summarize what they have learned and assess their own participation in the Health Lab. Assign credit based on students' understanding and participation.

Focus

The Teen Health Digest articles can be used in two ways: as an individual activity for reflection and enrichment or as a cooperative learning activity as described below.

Motivator

What are some of the false ideas teens have about alcohol and other drugs? Ask students for suggestions, and list their ideas on the board. Then have students read and discuss the Teen Health Digest articles: Which of the misconceptions listed on the board are addressed by these articles?

Cooperative Learning Project

Television Ads

Setting Goals

Explain that students will work together to help spread accurate information about alcohol and other drugs. They will plan and put on—and if possible, videotape— advertisements designed to inform teens about the problems of alcohol and other drugs.

Delegating Tasks

Let students discuss their ideas and suggest appropriate working groups. For example, they may decide to form separate groups for researching, writing, performing, planning and playing music, directing, filming, and staging. Have students sign up for the groups they find most interesting. Alternately, let students divide themselves into three or four groups, and have the members of each group work cooperatively on their own television ad.

TEEN HEALTH DIGEST

People at Work
Country Counselor

From her office window, Maria Rodriguez can see farmland and just a few houses. Inside, however, the scene is less peaceful. Maria is a counselor at an alcohol and drug treatment clinic. Her clients are people who have abuse and addiction problems.

Maria sometimes helps on a practical level. For example, she may assist a recovering addict in finding a job or a place to live. She also holds counseling sessions for her clients.

"Many people think that alcohol and drugs are found only in the cities," Maria says. "But people in the country face these kinds of problems, too."

Maria grew up in a nearby small town. She decided to become a counselor after helping a friend who had been abusing drugs. After finishing high school, Maria got an associate degree in counseling and began working at the clinic.

312 Chapter 11: Alcohol and Drugs

CON$UMER FOCU$
On-Line Targets?

It sounds great—a Web site that includes the latest music, interviews with rock stars, and reviews of CDs. The problem is that the site is sponsored by a well-known beer company. Mixed in with the music are numerous promotions for the beer.

Web sites sponsored by other beer and liquor companies include interactive games. Some feature chat rooms and interactive stories. Are these sites designed to appeal to the underage consumer?

Industry representatives claim that the Web sites are geared to adults. Some sites run messages warning that they are for people aged 21 or older. Many people are concerned, however, and feel that these sites are designed to appeal to minors as they promote alcohol products.

Try This:

Do you think that a warning message saying that a Web site is for adults only would be effective at keeping teens out? Write a paragraph explaining your opinion.

TECHNOLOGY OPTIONS

Video Camera If students have access to a video camera, they should produce a taped version of each television advertisement. Remind students to rehearse and polish their ads before actually taping, and encourage students to share their completed tape with the rest of the school, as well as with family members.

On-Line Computer Students may be able to use Internet capabilities to locate research resources in local libraries. In addition, they might find appropriate, reliable resources on the Web, to be used as part of their background research. Review with students the precautions they should exercise in using the Internet, and encourage them to work in pairs or groups.

Myths and Realities

Tough Guy?

Q. I heard some people saying that a guy they know can really "hold his liquor." What does that mean?

A. They probably think that the person can drink a lot of alcohol without getting drunk. They may assume that because they do not see any change in his behavior, the alcohol is not affecting him. In fact, it is the amount of *alcohol consumed* that causes health risks, not how much a person's *behavior* appears to be affected by alcohol. Over time, heavy drinkers get used to alcohol and need greater amounts to produce the same feelings. This heavy drinking causes damage to the liver, heart, and brain. Unfortunately, these people may not realize that they are abusing alcohol and risking their health until it is too late.

Teens Making a Difference

Tricks and Talks

The Shooting Stars are a group of 12- to 15-year-olds who perform basketball tricks. They put on exhibitions during half-time at high school and college games. Their routines include dribbling in unison, spinning balls on the ends of their fingers, and dribbling three balls at once.

Once they have the audience's attention, the Stars talk about the dangers of drugs. They encourage kids to get involved in activities they enjoy instead of using harmful substances. With their talent and such an important message, the Shooting Stars are invited to perform at many schools.

HEALTH UPDATE

Danger to the Unborn

When a pregnant woman is addicted to alcohol or drugs, her baby can be severely damaged. During the early stages of pregnancy, each system in a baby's body is at an important stage of development. Harmful substances used by the mother can cause birth defects, such as low birth weight and heart problems. During the later stages of pregnancy, drug use by the mother can result in slowed growth and brain defects.

A baby born to a drug-addicted mother is sometimes born addicted and can suffer severe and long-lasting withdrawal symptoms. Even if the baby survives the withdrawal process, in later life the child may experience behavioral and psychological disturbances.

Try This:
Do research about alcohol, drugs, and fetal development. Report on your findings to the class.

Teen Health Digest 313

Meeting Student Diversity

Visually Impaired Students with visual impairments can be encouraged to work with a partner in their groups; the partner should be responsible for describing information sources and elements of the ad that cannot be seen clearly. In addition, students should keep in mind the possibility that some teens in their ads' target audience may be visually impaired. Encourage students to use spoken words, music, and/or rhythm—as well as visual images—to communicate their messages.

Learning Styles If necessary, guide the formation of groups to ensure that students with various learning styles work together, each making appropriate contributions.

Chapter 11

The Work Students Do

Following are suggestions for implementing the project:

1. Guide students in keeping a clear focus for each ad they plan and produce. Help them recall effective ads they have seen on television: What is the message of each ad? How is that message conveyed? How do the spoken words, the music, and the visual presentation contribute to that message?
2. Help students understand the impact that a few important statistics or an appropriate factual story can have. A specific message—"Of every ten people who try alcohol, one will become addicted," for example—is more convincing than a general message—such as "Alcohol and other drugs are bad for you."
3. Encourage students to develop several different kinds of ads, imparting different facts and ideas about alcohol and other drugs.
4. Suggest that students "test" their ads with a limited audience, perhaps a group of students from another class. Ask audience members to respond honestly but constructively, and help students revise their ads in response to audience comments.
5. If students are able to videotape their ads, circulate the completed tapes among the other classes at school. If taping equipment is not available, suggest that students perform their ads as part of an all-school assembly.

Assessing the Work Students Do

Ask students to write short reviews of their ads. Then meet briefly with each student to discuss the review and to help the student assess his or her own contribution to the project.

313

Lesson 4

Risks Involved with Alcohol and Drug Use

Focus

LESSON OBJECTIVES

After studying this lesson, students should be able to

► explain why drinking alcohol and taking drugs are risk behaviors.

► discuss the physical, mental/emotional, and social consequences of using alcohol and drugs.

MOTIVATOR

Write the slogan "Just Say NO" on the board. Ask students to write short paragraphs explaining whether or not they feel this is an effective means of encouraging teens to avoid alcohol and drugs.

INTRODUCING THE LESSON

Ask students to vote yes or no on the effectiveness of the "Just Say NO" slogan. Then let several students explain why they do consider it effective, and allow several others to explain why they don't. Remind students that the more they know about the dangers of alcohol and other drugs, the easier it may be to refuse them. These dangers are covered in Lesson 4.

INTRODUCING WORDS TO KNOW

Have several volunteers share their understanding of the Word to Know, *intoxicated*. Then let another volunteer read aloud the formal definition given in the Glossary. Have students write the word and its definition in their notebooks.

Lesson 4 Risks Involved with Alcohol and Drug Use

This lesson will help you find answers to questions that teens often ask about the risks and consequences of using alcohol and drugs. For example:

► What are the physical risks of alcohol and drug use?
► How can alcohol and drug use affect my thoughts and emotions?
► How can alcohol and drug use harm my relationships with others?

Word to Know

intoxicated

in your journal

How do you feel about taking risks? In your journal, describe your attitude on this subject. You may have ideas that contradict one another. For example, you may not want to put yourself in danger. However, you may also feel that taking risks is exciting. What are some ways in which you can experience fun and excitement but not harm yourself or others?

Knowing the Risks

Drinking alcohol and taking drugs are risk behaviors. As you learned in Chapter 1, a risk behavior is an action or choice that may cause injury or harm to you or to others. Alcohol and drug abuse involve many risks. Most are quite serious, and some are even deadly.

These risks affect every part of a person's life and health. Alcohol and drugs can be especially dangerous to teens, who are still growing and developing. These substances cause physical, mental, and social harm.

■ Alcohol and drug use may slow down the time it takes a teen's body to mature physically. Height, weight, and sexual development may be affected.

■ Alcohol and drug use may shorten a person's attention span. Users may have less interest in pursuing their goals.

■ Alcohol and drug use may cause people to lose control and act in ways they later regret. Relationships with family and friends may become strained.

Being aware of the consequences of alcohol and drug use can help you stay away from risky situations. If you do find yourself in such a situation, knowing the dangers will help you make the right decision. You'll feel confident about choosing not to use alcohol or drugs.

Staying away from alcohol and drugs will help you maintain a good relationship with your family.

314 Chapter 11: Alcohol and Drugs

Lesson 4 Resources

Teacher's Classroom Resources
- Concept Map 51
- Enrichment Activity 51
- Lesson Plan 4
- Lesson 4 Quiz
- Reteaching Activity 51
- Transparency 46

Student Activities Workbook
- Study Guide 11
- Applying Health Skills 51

Physical Consequences

You have already read about ways in which using alcohol and drugs can affect a person's short-term and long-term physical health. The intensity of effects produced by these substances varies and may be unpredictable. Short-term effects may include headaches, fever, dizziness, and vomiting. Long-term effects may include serious damage to organs and body systems. Alcohol and drug use can also affect physical fitness by causing

- loss of physical coordination, including difficulty in walking, running, dancing, and playing sports.
- muscle twitches and cramps.
- decreased endurance.
- reduced strength.
- lowered energy level.
- slowed reflexes.

In addition to these health problems, a person may face other physical consequences from alcohol and drug use. Under the influence of alcohol or drugs, people sometimes cause physical harm to themselves or others. An alcohol or drug user may take risks that he or she would normally avoid, such as starting a fight.

Drinking and Driving

Why is it so dangerous to drink and drive? It's because even one alcoholic drink can slow a person's reaction time, impair judgment, and interfere with decision making. The more alcohol a person drinks, the more dangerous he or she is behind the wheel. When a driver has been drinking, the chances that he or she will be involved in an accident are very high. About half of all fatal automobile accidents involve alcohol. In 1995 alcohol-related automobile accidents were the leading cause of death for 15- to 24-year-olds.

Drivers who have been drinking endanger not only themselves but many other people as well. Their passengers face great risk of injury or death. The lives of pedestrians and people in other cars are also threatened. For these reasons, it is illegal for a person to drive while **intoxicated** (in·TAHK·suh·kay·tuhd), also called *drunk*.

Teen Issues

Stopping Drunk Drivers ACTIVITY!

The number of alcohol-related accidents resulting in death has declined since the 1980s. One reason is the rise of organizations that attempt to prevent drunk driving. Students Against Driving Drunk (SADD) was founded in 1981 by a hockey coach who lost two of his players in an alcohol-related car accident. SADD's goal is to end the injury and death caused by driving under the influence of alcohol or drugs. In September of 1997, SADD changed its name to Students Against Destructive Decisions. The change signals a broadening of the group's mission to counter the effects of other negative influences as well.

Find out about the SADD chapter in your area. What activities do they sponsor?

Never accept a ride with a driver who has been drinking. Be firm in refusing to get into the car.

Lesson 4: Risks Involved with Alcohol and Drug Use 315

Health of Others

Even a first experiment with alcohol or other drugs can be dangerous—and indeed lethal. Students should be aware that a teen who has used any form of drug is subject to collapse, an indication that he or she is in urgent need of medical care. If a friend or companion collapses or becomes unconscious, teens should know how to help. Place the ill person in the recovery position, lying on the stomach, with one leg and both arms bent and the head turned to one side. This position will provide protection from choking if the victim vomits. Teens should summon medical assistance immediately, dialing 911 or contacting a qualified adult.

Chapter 11 Lesson 4

Teach

L1 Discussing

Help students identify and discuss other risk behaviors: What are some other risks that people take, knowing that danger may be involved? What risks do people sometimes take, not knowing about the dangers involved? Why do you think teens are especially likely to take risks? What are some good approaches to evaluating risks and deciding whether a behavior is too risky?

L1 Critical Thinking

Ask students: Why are alcohol and drugs especially dangerous for teens? Why do you think it is legal for adults to drink alcohol but not for teens? Why do you think both teens and adults are prohibited from using illegal drugs?

L1 Comprehending

Ask students: If teens are using alcohol or drugs, what kinds of strains can develop between those teens and their parents? between those teens and other family members? between those teens and their friends? How can these relationships be repaired?

L1 Journal Writing

In their private journals, have students write their responses to these questions: What do your parents or other adult family members expect of you in terms of drug and alcohol use? How do you feel about their expectations?

L2 Guest Speaker

Invite a coach from a local high school or college to speak to the class about his or her team, with special attention to alcohol and drug use: What rules restrict team members' use of alcohol and drugs? How are those rules enforced? How does the coach feel about athletes using alcohol and drugs? How does the coach communicate that attitude to the team?

315

Chapter 11 Lesson 4

L1 Discussing
Ask students: At what age will you be allowed to drive legally? What responsibilities go along with the privilege of driving? At what age will you be allowed to drink alcohol? How are the responsibilities of driving and the responsibilities of choosing whether to drink related?

L2 Social Studies
Ask a group of volunteers to find out what the legal definition of *intoxication* is in your state. How is intoxication measured? Under what circumstances are law enforcement officers permitted to check a driver for intoxication? What are the legal consequences of driving while intoxicated? Does your state have a zero-tolerance law for teens? If so, what does it mean? Let the volunteers discuss the results of their research with the rest of the class. Emphasize that the definition of *intoxication* applies only to adults; for teens, drinking any amount of alcohol is illegal.

L1 Life Skills
Ask students to imagine that they are out with friends at night. The person responsible for driving you home has been drinking but does not appear to be drunk. What should you do? Let volunteers role-play various solutions to the situation.

L3 Reporting
Ask a small group of students to read more about how alcohol and drug use by pregnant women affects their unborn babies: What specific kinds of problems can result? How can these problems be treated? What has research shown to be the safe level of alcohol or drug use by pregnant women? Have these students share their findings in an oral presentation.

Teen Issues

False Advertising?

You have probably seen many advertisements for beer and other types of alcoholic beverages. Each year advertisers spend billions of dollars promoting these products. The ads depict happy, healthy people having fun while drinking alcoholic beverages. The advertisers want you to think that drinking their products will make you more popular. They seldom tell you about the risks associated with alcohol use.

The Effects on Others
The physical consequences of alcohol and drug abuse extend beyond the user. If a pregnant woman uses these substances, for example, she can cause disease and birth defects in her unborn child. Children whose mothers abused drugs during pregnancy often have delayed development and learning disabilities. Women who drink alcohol while pregnant may have babies with fetal alcohol syndrome (FAS). This condition causes delays in development and psychological and behavioral problems throughout life.

Mental/Emotional Consequences

As a teen, you are developing your self-concept and trying to find your place in the world. This is an exciting time, but it is also stressful. To try to relieve this stress, some teens turn to alcohol or drugs. These substances actually have the opposite effect, however, making life even more confusing and difficult.

The psychological consequences of alcohol and drug abuse are very serious. A person's ability to think and learn are negatively affected. Alcohol and drug abuse also has many serious emotional consequences. **Figure 11.8** shows some of the mental and emotional consequences of alcohol and drug abuse.

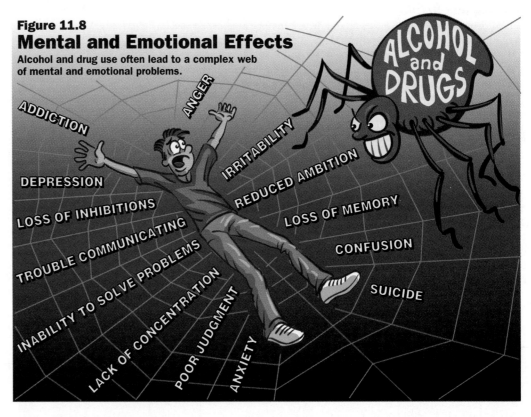

Figure 11.8
Mental and Emotional Effects
Alcohol and drug use often lead to a complex web of mental and emotional problems.

316   Chapter 11: Alcohol and Drugs

Promoting Comprehensive School Health ✓

Drug Resistance What role does the school play in helping students resist drug abuse? in recovering from addiction or other drug-related problems? in dealing with drug-related problems at home? What role can the school play in supporting teachers and other school staff with drug problems? Is there an appropriate role for the school health system in educating and helping other members of the community? These are questions to be considered and resolved in the development of a comprehensive school health program. For more information about developing such a program, consult *Promoting a Comprehensive School Health Program* in the TCR.

Social Consequences

Involvement with alcohol and drugs also leads to problems in social relationships. The use of these substances may cause personality changes. The user may lose control of his or her behavior and do things that he or she later regrets. These actions often cause relationships with others to become strained or even to break apart.

Abusers may become obsessed with alcohol and drugs. This obsession usually causes them to lose interest in family, friends, and even personal hygiene. To avoid being caught, users may lie to loved ones. Some stop seeing their friends and become involved with other people who are also obsessed with alcohol or drugs. Lacking true friendships and honest relationships, drug abusers may feel very lonely.

Alcohol and drug use is illegal and can lead to suspension from a sports team and other lost opportunities.

Risks in School

Teens who use alcohol and drugs also cause problems for themselves in school. They may

- be late or miss school often.
- get bad grades.
- lose the chance to play sports because of missed practices.
- let down classmates or teammates because of poor performance.
- lose opportunities to learn new skills and develop their abilities.

Risks to the Family

When any member of a family abuses alcohol or drugs, the rest of the family is affected. Alcohol and drug abuse cause great disruption in the home.

- A person who uses alcohol or drugs may become violent and hurt other family members.
- Someone who abuses alcohol or drugs may spend more time away from home, lie to other family members, and be moody and unpredictable.
- If parents abuse drugs, they cannot fulfill their obligations to their children or provide emotional support. They may lose their jobs, creating serious financial problems for the family.
- Teens who abuse drugs may disappoint their parents. They usually fail to meet their family responsibilities.
- Alcohol or drug abuse may even cause a family to break up.

Lesson 4: Risks Involved with Alcohol and Drug Use

Chapter 11 Lesson 4

L1 Discussing

Let students share their responses to these questions: Why do you think some teens turn to alcohol or other drugs to relieve stress? In what specific ways do alcohol and other drugs actually increase the stress in a teen's life? What healthier options are available to a teen who feels stressed?

L3 Writing Stories

Ask students to write short stories in which teens have to deal with the mental and emotional effects of alcohol and drugs. Encourage students to share their completed stories with the rest of the class.

L2 Interviewing

Have students talk with older teens or young adults about the use of alcohol and drugs: What alcohol- and drug-related problems have they seen among their peers? Encourage students to share and discuss what they learn from these interviews.

L2 Guest Speaker

Ask a local law enforcement officer to speak with students about the legal consequences of using alcohol and drugs. Encourage students to ask questions after the officer's presentation.

L1 Discussing

After students describe the photograph on this page, let a volunteer read the caption aloud. Ask: What opportunities would you hate to lose?

Teacher Talk

Dealing with Sensitive Issues

Some students may have firsthand experience with the effects of alcohol and drug abuse on the family. As the class considers these risks, keep in mind the fact that some students may be living with—or may be separated from—parents who abuse alcohol and/or drugs.

Meeting Student Diversity

Language Diversity A clear understanding of the risks of using alcohol and other drugs is essential for all students. Those who are still acquiring English skills should be encouraged to consult the Glosario if appropriate and/or to work in groups with others who share a common first language, discussing drug risks and practicing ways to avoid and refuse drugs.

Learning Styles Students who may have difficulties decoding and/or comprehending drug information as presented in the text should be encouraged to work with reading buddies. One student in each pair can be responsible for reading aloud or for helping the other student read aloud. It is important to emphasize that both partners can contribute to learning as they consider and discuss the information in the text.

Chapter 11 Lesson 4

Assess

EVALUATING THE LESSON

Assign Reviewing Terms and Facts and Thinking Critically to review the lesson; then assign the Lesson 4 Quiz in the TCR.

LESSON 4 REVIEW

Answers to Reviewing Terms and Facts

1. *Intoxicated* means drunk.
2. The consequences include (any five): confusion, poor judgment, memory loss, inability to solve problems, lack of concentration, trouble communicating, anxiety, irritability, anger, addiction, reduced ambition, depression, suicide.
3. School problems include (any three): frequent tardiness or absence, bad grades, loss of a position on an athletic team, disappointments to classmates or teammates, loss of opportunities to learn new skills and develop abilities.

Answers to Thinking Critically

4. A teen may act violently, be moody and unpredictable, and fail to meet responsibilities. Dealing with these problems could divide the family.
5. Responses will vary.

RETEACHING

▶ Assign Concept Map 51 in the TCR.
▶ Have students complete Reteaching Activity 51 in the TCR.

ENRICHMENT

Assign Enrichment Activity 51 in the TCR.

Close

Have students complete the following sentence in writing: *In my opinion, the most serious risk of using alcohol or drugs is...*

Did You Know?

Legal Responsibilities ACTIVITY!

The host of a party can be held responsible for guests' drinking. If someone leaves a party and drives while intoxicated, causing an accident, the host may face legal action. Write a paragraph explaining the reasoning behind this law.

Risks with the Law

Teens who use alcohol and drugs often get into serious trouble with the law. It is a crime to buy, sell, or possess marijuana, cocaine, and other illegal substances. The penalties vary from state to state. Breaking the law can lead to arrest, fines, or time spent in a detention center.

It is also illegal for a person under 21 to buy or possess alcohol. An underage person who is caught driving while intoxicated will lose his or her license. Adults who sell liquor to underage individuals are also breaking the law. Tragically, involvement with alcohol and drugs often leads teens to commit other crimes. A person under the influence of alcohol or drugs has an increased chance of becoming violent. For these reasons, even a single use of alcohol or drugs may lead to years in prison.

In addition, people who abuse alcohol or drugs are at risk of becoming crime victims themselves. They may encounter dangerous situations when buying or seeking drugs. Under the influence of alcohol or drugs, people often make unwise decisions. These choices may cause them physical injury and may even result in death.

A drug-related arrest has many serious long-term consequences.

Lesson 4 Review

Using complete sentences, answer the following questions on a separate sheet of paper.

Reviewing Terms and Facts

1. **Vocabulary** Write a definition of the term *intoxicated*.
2. **List** Name five mental and emotional consequences of alcohol and drug use.
3. **Identify** List three problems teen alcohol or drug users may have at school.

Thinking Critically

4. **Hypothesize** How might a teen's use of alcohol or drugs create tension within his or her family?

5. **Analyze** Penalties for selling drugs are usually more severe than those for possessing drugs. Why do you think this is so?

Applying Health Concepts

6. **Personal Health** With a partner, role-play a scene in which a teen refuses to ride in a car with someone who has been drinking. Then role-play a scene in which the teen tries to convince the driver not to drive while under the influence of alcohol.

318 Chapter 11: Alcohol and Drugs

More About...

Drug and Alcohol Risks Alcohol and other drugs are involved, either directly or indirectly, in many teen suicides. Suicide is now the third leading cause of death among high school students, and surveys have shown that as many as one teen in every ten contemplates committing suicide. Some teens attempt suicide because they are—or fear they are—addicted to alcohol or drugs. Many others take drugs in deliberate suicide attempts. Still others die from mixing drugs or from overdosing; friends and family may never be certain whether the death was a suicide or an accident.

Not all drugs involved in suicides are illegal. A small number of common OTC medicinal drugs, such as acetaminophen, can prove fatal if taken improperly.

Avoiding Substance Abuse

Lesson 5

This lesson will help you find answers to questions that teens often ask about ways to avoid alcohol and drugs. For example:

- What are some reasons to avoid substance abuse?
- How can I say no to alcohol and drugs?
- What alternatives are there to alcohol and drug use?

Making Responsible Decisions

When you care about yourself, you take responsibility for your actions. You also feel confident in making your own decisions based on your values. If you value your health, you will protect it by avoiding the risks associated with alcohol and drugs. Making responsible decisions involves an effort on your part. One way to make that effort is to think ahead of time about how you will handle difficult situations.

Using alcohol or drugs will not help you feel good about yourself. In fact, people who use these substances often have a poor self-concept. People with a negative self-concept may think that using alcohol or drugs will make them feel more confident.

Alcohol and drugs, however, are not necessary to give a person self-confidence. In fact, your sense of your own value as an individual is a powerful tool that can help you to choose an alcohol- and drug-free life. When you are in control, you can take pride in the choices you have made.

Words to Know

invulnerable
assertive
alternatives

Q & A

Helping Others

Q: I don't use drugs, and I don't want my friends to, either. How can I help other people avoid drugs?

A: By not using drugs, you show other teens that it's okay to avoid substance abuse. You can encourage your friends to stay away from situations where drugs are used, support one another, and participate in healthful activities. If you feel that a friend needs help, ask a school counselor or a trusted adult about organizations that offer support to troubled teens.

Teens who feel good about themselves and have a sense of purpose can more easily avoid alcohol and drugs.

Lesson 5: Avoiding Substance Abuse   319

Lesson 5 Resources

Teacher's Classroom Resources
- Concept Map 52
- Cross-Curriculum Activity 22
- Enrichment Activity 52
- Lesson Plan 5
- Lesson 5 Quiz
- Reteaching Activity 52

Student Activities Workbook
- Study Guide 11
- Applying Health Skills 52

Lesson 5
Avoiding Substance Abuse

Focus

LESSON OBJECTIVES

After studying this lesson, students should be able to

- discuss the importance of feeling good about themselves.
- list reasons for avoiding substance abuse.
- discuss how teens can say no to substance abuse.
- explain how avoiding alcohol and drugs can help them reach their full potential.

MOTIVATOR

Have students write and complete the following sentence: *Five terrific things about me are...*

INTRODUCING THE LESSON

Ask students to explain their reactions to the Motivator activity: How do you feel about acknowledging some of your special traits? Do you think teens who have a good opinion of themselves are more likely or less likely to use alcohol and other drugs? Explain to students that they will explore these and other, related questions more fully as they study Lesson 5.

INTRODUCING WORDS TO KNOW

Ask: What is the meaning of the base word in each of these Words to Know? What meaning does that prefix or suffix carry? How does this analysis help you understand the meaning of each Word to Know?

Then have students record the Words to Know and their definitions, in their notebooks.

Chapter 11 Lesson 5

Teach

L1 Comprehending
Ask: What actions and attitudes indicate that an individual has high self-esteem? Does a person with a low self-image have a built-in excuse for experimenting with alcohol and other drugs? What do you think teens can do to raise their own self-esteem?

L1 Writing Lists
Have students write two lists: one of the skills they have, and one of the new things they are willing to try. Encourage them to include at least ten items in each list.

Visual Learning
Let students describe the teens in the photograph on page 319: What indicates that these teens feel good about themselves? How do you think these teens' attitudes, activities, and friends will help them resist pressure to try alcohol and other drugs?

L2 Listing Examples
Ask the class to work in groups to list reasons teens give for trying or using alcohol and other drugs. Then guide a class discussion in which students refute each of the listed reasons and suggest other, more healthful activities to meet the needs expressed in each reason.

VIDEODISC/VHS

Teen Health Course 2

You may wish to use video segment 11, "Alcohol and Decision Making," in which a teen uses decision-making skills to avoid being pressured to use alcohol.

Videodisc Side 2, Chapter 11
Alcohol and Decision Making

Search Chapter 11, Play To 12

Reasons to Avoid Substance Abuse

Although most teens are aware that alcohol and drugs are dangerous, some are still tempted to try them. Some teens may want to do something risky; others may just want to satisfy their curiosity. Some teens think that using alcohol or drugs will make them seem more mature. Others begin using these substances because of peer pressure. Still others believe that they are **invulnerable,** or *not able to be hurt.*

In reality, however, there are no good reasons to try alcohol or drugs. As you have learned, substance abuse can have negative consequences in all areas of a person's life. There are many good reasons—even beyond protecting your health—to avoid abusing alcohol or drugs. **Figure 11.9** identifies some of them.

Figure 11.9
Reasons to Avoid Substance Abuse
You should have a ready supply of reasons to avoid abusing alcohol and drugs.

320   Chapter 11: Alcohol and Drugs

Personal Health

Many researchers believe that alcoholism is a familial disease. That is, although it is not directly inherited, it tends to reappear across generations. Studies show that children of alcoholics face a four or five times greater risk of becoming alcoholics than do children whose parents are not alcoholics. Even having a grandparent or great-grandparent with alcoholism increases an individual's risk.

Teens who believe they may be at increased risk for developing alcoholism can take special precautions. Many choose to avoid even experimenting with alcohol and so avoid the risk altogether.

Saying No to Substance Abuse

For many teens, peer pressure is the main reason given for using alcohol or drugs. One of the keys to avoiding substance abuse is to avoid social situations where such abuse might take place. These situations often involve peer pressure. If you think that there will be alcohol or drugs at a party, for example, it's smarter not to go.

It is not always possible, however, to avoid situations involving peer pressure. For this reason, you should prepare yourself ahead of time to say no to alcohol and drugs. You can do this by recognizing possible situations in which you might encounter peer pressure and deciding how you will handle them. **Figure 11.10** describes a few situations and suggests ways to deal with them.

The first step is to decide that you are not going to drink alcohol or take drugs. Sticking to this decision takes determination. You need to be **assertive,** which means *having the determination to stand up for yourself in a firm but positive way.* You also need effective refusal skills. If someone tries to pressure you into trying alcohol or drugs, follow these tips for saying no:

- Use humor to make the situation less tense. Make a joke or give a funny reason for not wanting to use the substance.
- Use "I" messages. Don't accuse or blame the person.
- Take your time. Collect your thoughts if necessary.
- Stand up straight and look the person in the eye.
- Speak in a firm but polite voice.
- Don't apologize for saying no.
- Walk away from the situation.

Brainstorm a list of creative "exit lines"—responses to people who pressure you to take alcohol or drugs. Remember that you don't owe anyone an explanation for refusing to use these substances. You just want to get your message across—that you're not interested.

Figure 11.10
Facing Peer Pressure
Being prepared for peer pressure situations will help you make the right choices.

Situations	Solutions
Some of your friends have started smoking marijuana after school. They are pressuring you to join them.	Tell them that you aren't interested, and find a substitute activity. Join an after-school club, or find volunteer work that you can do during that time.
At a park near your home, some older teens have offered you drugs.	Tell them that you need to leave. Avoid the park in the future when these teens are present.
You go to a friend's house, and no adults are home. Your friend offers you a drink containing alcohol.	You don't have to explain why you don't want to drink. Leave the house, but don't get into a car with anyone who has been drinking. Next time, check first to make sure that adults will be present.

Chapter 11 Lesson 5

L1 Analyzing
Ask: How does your group of friends affect your actions and activities? What risks do you face if some of your friends begin using alcohol or other drugs? What do you think you should do? Why?

L1 Role-Playing
After students have read and discussed the examples in Figure 11.10, let pairs of students role-play similar situations. Encourage some pairs to act out the specific situations shown in the text, offering different solutions. Ask other pairs to make up and act out other situations in which teens resist peer pressure to use alcohol or other drugs.

L2 Language Arts
Ask volunteers to explore the difference in meaning between *assertive* and *aggressive*. Let the volunteers explain the difference to the class and then present a short skit demonstrating that difference.

L1 Demonstrating
Give interested students an opportunity to act out various ways of speaking: timidly, politely but not firmly, firmly and politely, firmly and rudely. Ask: Which is most effective in refusing to try alcohol and other drugs? Why? What messages do the other ways of speaking communicate?

L1 Comprehending
Let students explain what "I" messages are, and encourage volunteers to give examples. How do "I" messages avoid accusation and blame? What "I" messages would students suggest using in place of these statements that express blame?

▶ You're nuts to be smoking marijuana.
▶ You'll be in a lot of trouble if your parents smell beer on your breath.
▶ You don't know what you're doing with those pills.

Cultural Awareness It is important for students to recognize that as many as a third of all adults choose not to drink alcohol at all. Some make this decision only on the basis of health; others follow this practice for religious reasons. For example, all Muslims are expected to avoid alcohol, as are members of The Church of Jesus Christ of Latter-Day Saints. Encourage students who are comfortable with the topic to share what they know about religious groups and the use of alcohol or other drugs. In addition, ask students to read about religions with which they are not familiar: What teachings relate to the use of alcohol and other drugs?

Chapter 11 Lesson 5

L2 Creating Cartoons

Have students draw cartoons that show teen characters offering funny reasons for refusing to try alcohol or other drugs. Display the completed cartoons where other students in the school can see them.

L2 Writing Essays

Ask students to write short essays about their own dreams for the future. Instruct them to begin their essays with one or more paragraphs describing what they hope to do in the years ahead. Have them conclude with one or more paragraphs explaining how their plans and dreams would be affected if they began abusing alcohol or other drugs.

L1 Listing Examples

Have the students work in groups to extend the list of positive alternatives to drug and alcohol use. Then let the groups share their lists.

L2 Creating a Mural

Have students work together to make a mural or other large display urging teens to avoid the use of alcohol and other drugs. One possibility is to use a long sheet of butcher paper. Let each student draw a picture and write a fact or suggestion, signing his or her name to the contribution. Arrange to hang the completed display in a public area of the school.

Assess

EVALUATING THE LESSON

Assign Reviewing Terms and Facts and Thinking Critically on page 323 to review the lesson; then assign the Lesson 5 Quiz in the TCR.

Teen Issues

Reason to Use? ACTIVITY

Some teens say that family or personal problems lead them to drink or take drugs. How could you respond to a friend who gave those reasons as an excuse for using drugs? You might tell your friend that substance abuse only causes more trouble and doesn't solve the original problem.

Make a poster that gives reasons not to use alcohol and drugs. Use illustrations and words to encourage teens to find more constructive solutions to their problems. Be positive!

Reaching Your Potential

What are your plans for the future? You may want to travel to other countries, go to college, become an auto mechanic, or work with computers. Making your dreams come true means not letting yourself be sidetracked by alcohol or drugs.

There are many alternatives to alcohol and drug use. **Alternatives** are *other ways of thinking or acting*. The following list suggests some positive alternatives. What others can you think of?

- Improve your skills in a sport you enjoy, or try a new sport.
- Join a school club or other extracurricular activity.
- Take art classes, or learn to play a musical instrument.
- Study a new language.
- Volunteer in a hospital, homeless shelter, or soup kitchen.

Using your skills and talents, such as musical ability, is a healthy alternative to substance abuse.

MAKING HEALTHY DECISIONS
Offering Alternatives

Emma and Samantha have been friends since second grade. For a long time, they did everything together. Since starting middle school, however, Emma has noticed that Samantha is changing. When Emma speaks with her mother about this, her mother says that maybe it's just time for both of them to develop other friendships. Emma decides to join the school newspaper and begins making new friends. Samantha seems to be finding new friends, too.

One day, Emma and Samantha meet up with each other while walking home from school. Emma is happy to be with her old friend. She asks Samantha to come over to her parents' apartment to listen to a new CD. Samantha has a different idea, though. She suggests that they stop off at the park, where some of Samantha's new friends are smoking marijuana.

Emma knows that she doesn't want to use marijuana or be around people who are using it. She is also concerned about Samantha. She still has some time to think before she and Samantha will reach the park entrance. She uses the time to go through the six-step decision-making process.

322   Chapter 11: Alcohol and Drugs

MAKING HEALTHY DECISIONS
Offering Alternatives

Focus on Healthy Decisions

Let students share their responses to these questions: Once you are committed to avoiding drugs, does it matter whether your friends use drugs? Why or why not? What do you think teens can do to help their friends avoid drugs? Let volunteers read the Making Healthy Decisions feature aloud, and guide students in discussing the situation.

Activity

Use these ideas to discuss Follow-up Activity 1:
Step 1 Emma must decide whether to meet Samantha's new friends who are smoking marijuana.
Step 2 Emma could go along with Samantha but refuse to smoke marijuana with the others. (She has already decided that she doesn't want to smoke marijuana.) Emma could refuse to go along and let Samantha join her new friends. Or, Emma could urge

Review

Using complete sentences, answer the following questions on a separate sheet of paper.

Reviewing Terms and Facts

1. **Vocabulary** What is the meaning of *invulnerable*?
2. **Restate** Why is it a good idea to avoid situations where substance abuse might take place?
3. **Vocabulary** Define *assertive*. Then use the term in an original sentence.
4. **List** Suggest five tips for saying no to alcohol and drugs.
5. **Identify** List four positive alternatives to alcohol and drug use.

Thinking Critically

6. **Analyze** How does a positive self-concept help teens avoid using illegal substances?
7. **Suggest** What advice would you give to a friend who was being pressured to try alcohol or drugs?

Applying Health Concepts

8. **Personal Health** Make a list of all the reasons why you do not want to try alcohol or drugs. Keep the list with you, and look at it when you need encouragement. Add to the list when you think of other good reasons.
9. **Health of Others** With a small group, make up a skit in which teens face peer pressure to use alcohol or drugs. Create a realistic setting for the skit, such as a party or school event, and think of ways for the teens to say no. Present your skit to the class.

❶ State the situation
❷ List the options
❸ Weigh the possible outcomes
❹ Consider your values
❺ Make a decision and act
❻ Evaluate the decision

Follow-up Activities

1. Apply the steps in the decision-making process to Emma's situation.
2. With a partner, role-play a scene in which Emma says no to Samantha. Include several alternative activities that Emma might suggest.
3. As a class, discuss what Emma could do to maintain her friendship with Samantha without using drugs.

Lesson 5: Avoiding Substance Abuse 323

Samantha to stay away from those friends and perhaps do something fun with her instead.

Step 3 If Emma goes with Samantha, she risks facing strong peer pressure to try marijuana. If Emma refuses to go along, she will not have to resist peer pressure, but she will know that her friend is smoking marijuana and putting her health in danger. If Emma urges Samantha to do something else, she risks being rejected, but she might be able to encourage Samantha to engage in more healthy activities.

Step 4 Emma values her own health. She also values her friendship with Samantha, and she values Samantha's health.

Step 5 Emma might choose any of her three options.

Step 6 Encourage students to evaluate all the possible consequences, both short- and long-term, of their decisions.

Chapter 11 Lesson 5

LESSON 5 REVIEW

Answers to Reviewing Terms and Facts

1. *Invulnerable* means not able to be hurt.
2. It is a good idea to avoid situations where substance abuse might take place because you might be pressured to join in the abuse.
3. Being assertive means standing up for yourself in a firm and positive way. Sentences will vary.
4. Tips for saying no to alcohol and drugs include (any five): standing up straight and looking the other person in the eye, speaking in a firm but polite voice, not apologizing, using "I" messages, taking your time, using humor, walking away.
5. Responses will vary.

Answers to Thinking Critically

6. Responses will vary.
7. Responses will vary.

RETEACHING

▶ Assign Concept Map 52 in the TCR.
▶ Have students complete Reteaching Activity 52 in the TCR.

ENRICHMENT

Assign Enrichment Activity 52 in the TCR.

Go around the class, making an individual appeal to each student to try a specific alcoholic drink or other drug. Encourage students to refuse your offers with sincerity and originality.

Assessing the Work Students Do

Ask students to evaluate their own contributions to this activity and to assess what they have learned. Assign credit based on their understanding of the decision-making process and on their own assessment of their efforts.

Lesson 6
Addiction and Recovery

Focus

LESSON OBJECTIVES

After studying this lesson, students should be able to

- explain what addiction is and identify some of the drugs that are addictive.
- list the stages in addiction and recovery.
- discuss the pain and difficulty involved in withdrawal.
- identify some of the signs that a person needs help dealing with drug addiction.
- explain how people can get help for themselves, for friends, or for family members.

MOTIVATOR

Let students respond to the following question by drawing illustrations: What do you think the world looks like to a teen addicted to drugs?

INTRODUCING THE LESSON

Encourage students to share their Motivator pictures. What outlook do the drawings portray? Ask students: How do you think people recover from addiction? Tell students that they will learn more about addiction and recovery in Lesson 6.

INTRODUCING WORDS TO KNOW

Let students find the definitions for the Words to Know. Then divide the class into two teams for a Word-Use Bee. You say a sentence using one of the Words to Know either correctly or incorrectly. A team identifies the word use as correct or incorrect and earns a point for the right response. Correcting an incorrect sentence earns another point. Play several rounds.

Lesson 6 Addiction and Recovery

This lesson will help you find answers to questions that teens often ask about addiction and recovery. For example:

▶ How do people become addicted to alcohol and drugs?
▶ Why is withdrawal from a drug so difficult?
▶ Where can an addicted person get help?

Words to Know

tolerance
recovery

Addiction

As you have learned, addiction is a physical or psychological need for a drug or other substance. With a physical addiction, the body feels a direct need for the drug. With a psychological, or mental, addiction, the mind tells the body that it needs more of the substance.

Addiction starts with alcohol or drug use and then continues with abuse. It then progresses through increased tolerance into total dependency. **Tolerance** (TAHL·er·ens) occurs when *the body becomes used to a drug and needs greater amounts to get the desired effect.* Addiction cannot be cured, but it can be controlled. The only way for a person to control a drug addiction, however, is never to use the substance again.

When people are addicted to a drug, they may put it ahead of everything else in their lives. Many drugs are addictive, including the following.

- Nicotine is considered one of the most addictive substances.
- Marijuana can cause psychological dependence.
- Amphetamines are addictive. Crack and cocaine can be addictive after just one use.
- Alcohol can be addictive. Other types of depressants, such as barbiturates, are also addictive.
- Heroin and other narcotics are highly addictive.

The best way to avoid becoming addicted to alcohol or drugs is never to start using them.

324 Chapter 11: Alcohol and Drugs

Lesson 6 Resources

Teacher's Classroom Resources
- Concept Map 53
- Enrichment Activity 53
- Lesson Plan 6
- Lesson 6 Quiz
- Reteaching Activity 53

Student Activities Workbook
- Study Guide 11
- Applying Health Skills 53
- Health Inventory 11

Substance addiction is like a slide downward, and **recovery,** or *the process of becoming well again,* is a steep climb back up. A recovering addict must continually resist the desire to use the substance again. **Figure 11.11** shows the process.

Figure 11.11
Stages of Addiction and Recovery
A person who is addicted to alcohol or drugs must make an extraordinary effort to become free of the substance.

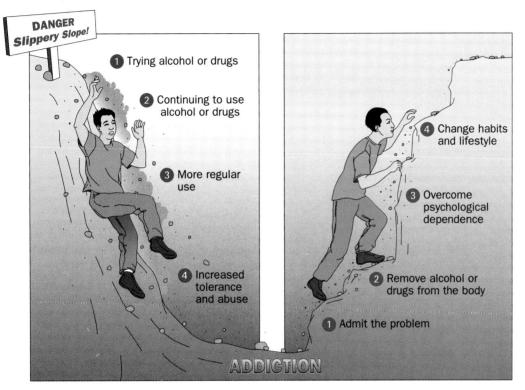

The Pain of Withdrawal

Ending an alcohol or drug habit is much harder than resisting the use of the substance in the first place. The addict must remove all traces of the substance from the body. This process causes withdrawal, a series of painful physical and mental symptoms that occur when a person stops using an addictive substance.

A person going through withdrawal often needs professional help. The withdrawal process strongly affects the nervous system. It can cause very unpleasant symptoms, including

- severe anxiety.
- confusion, memory loss, or hallucinations.
- nausea, vomiting, and diarrhea.
- headaches and chills.

Teen Issues

Danger Zone

Young teens who try alcohol or drugs are at greater risk of addiction than older teens or adults. They are often more vulnerable to the influences of alcohol and drugs because of their physical size and stage of development.

Lesson 6: Addiction and Recovery   **325**

Consumer Health

 Relatively new on the market are various "nonalcoholic" beers. Teens may be interested in these products—perhaps they represent a way to "look cool" without running any risks. However, it should be noted that many "nonalcoholic" drinks do contain very low levels of alcohol. They can represent a threat to the health of a recovering alcoholic, and some experts worry that these beverages may not be completely safe for teens who intend to avoid alcohol altogether. Encourage interested students to read the labels on these products carefully and to look for newspaper or magazine articles evaluating these drinks. Also, suggest that students seek out genuinely nonalcoholic beverages, such as sparkling cider, instead.

Chapter 11 Lesson 6

L1 Comparing
Ask: How are physical addiction and psychological addiction alike? What is the most important difference between the two? Is either kind of addiction easier to overcome than the other? Encourage students to explain their responses.

L2 Language Arts
Ask students to identify at least three other words related to *tolerance,* using a dictionary if necessary. How are the meanings of those words similar to the meaning of *tolerance*? How do the words differ in meaning and in what they imply?

L1 Recalling
Remind students that, in addition to being addictive after just one use, some of the listed drugs, including cocaine, crack cocaine, and heroin, can be deadly when used for the first time.

Visual Learning

Use Figure 11.11 to help students discuss the stages of addiction and recovery: What happens on the way to addiction? Why is that shown as a downward slope? How is addiction shown in the drawing? Why? Why is recovery shown as a steep, uphill slope? What do you think sometimes happens to addicts as they struggle toward recovery?

L1 Comprehending
Help students consider and discuss the final stage in recovery, changing habits and lifestyle: What habits does a recovering addict need to change? What aspects of his or her lifestyle need to change? Why are these changes important? How can family members and friends best help a recovering addict?

325

Chapter 11 Lesson 6

L3 Researching

Ask volunteers to find magazine articles on withdrawal, including first-person accounts if possible. Then have the volunteers compile quotes from these articles into a collage; suggest that they might include photos or drawings in the display as well. Provide time for the volunteers to share their collage with the class.

L1 Critical Thinking

Avoid putting any students on the spot when you ask: What are the differences between seeing the signs of drug addiction in a friend and seeing them in a family member? How do you think the drug addiction of a family member might affect a teen? When a family member has problems with drugs, who should try to help? What kind of help can they offer?

L1 Demonstrating

Provide copies of the local Yellow Pages, and let students work in groups to find the headings "Drug Abuse" and "Alcoholism Treatment." Can group members locate other useful headings as well?

L2 Guest Speaker

Ask a school nurse, school psychologist, counselor, or other staff member to tell the class about school services that support students who face problems with drugs. Encourage the speaker to be as specific as possible, and let students ask relevant questions.

L1 Journal Writing

In their private journals, have students list four adults they would feel comfortable consulting about an alcohol- or drug-related problem.

L2 Life Skills

Direct students to work together to gather information about local support groups for teens whose friends or relatives are addicted to alcohol and/or other drugs. Encourage students to publish a brochure with this information and to make copies available to the rest of the student body.

Q & A

Is My Friend in Trouble?

Q: A friend of mine has started spending a lot of time with some older kids who have a bad reputation. She doesn't include me anymore. She has also been secretive about where she goes and what she is doing. Could she be involved with alcohol or drugs?

A: It is a possibility, but you should not assume that it is true. Check to see if your friend has any of the warning signs of addiction described in this chapter. If you think that there is cause for concern, talk honestly to your friend about your feelings.

Language Arts Connection

Stories of Addiction ACTIVITY

Many novels and short stories for young adults deal with the subject of addiction. Ask your teacher or librarian to recommend an appropriate book for your age group. Read it and write a short review. Describe how the book addresses the problems of alcohol addiction or drug addiction.

Parents and other trusted adults can help you get treatment for an addiction. They can also give you good advice about where to get help for a friend.

Knowing When to Get Help

How can you tell if a friend or family member is addicted to drugs? It can be hard to determine because most people try to hide the problem or deny that one exists. The signs of addiction to drugs are similar to the signs of alcoholism mentioned earlier, such as

- using drugs on a regular basis.
- lying about the amount or frequency of drug use.
- thinking that drugs are a necessary part of having fun.
- avoiding other people in order to use substances.
- giving up activities that the person used to enjoy.
- pressuring others to use substances.
- frequent moods of depression and hopelessness.
- regularly missing school or work.

Getting Help

You can get help for yourself, a friend, or a family member who has an alcohol or drug addiction. First, try going to people who are close at hand, such as parents; teachers, clergy members, and school counselors; trusted friends; or peer counselors. If you do not know anyone who can help you, you can turn to organizations and agencies that offer counseling and treatment. The phone numbers and addresses for these groups are in the Yellow Pages of your telephone directory. They are listed under headings such as "Drug Abuse" or "Alcoholism Treatment." Places to go for help include

- toll-free drug hot line counselors.
- support groups.
- alcohol or drug treatment centers.

326 Chapter 11: Alcohol and Drugs

More About...

Addictive Drugs The most common addictive drug is usually smoked, and it is displayed and available in stores in every community. This drug is nicotine. In addition to cigarettes, nicotine is also found in cigars, pipes, snuff, and chewing tobacco.

Nicotine acts much like stimulants by speeding the functioning of the brain. That's why many smokers say they need a cigarette to focus their thoughts, to feel alert, or to increase their energy level. Interestingly, other smokers say they need a cigarette to help them relax.

Nicotine is a highly addictive drug. Teens who experiment with smoking are likely to become occasional users of the drug nicotine, then regular users, and finally addicted users.

The Challenge of Recovery

As you learned in Lesson 2, recovery from addiction begins with admitting that there is a problem. If you have a friend or family member who has an addiction, he or she may be reluctant to seek help. In that case, join a support group for friends or relatives of addicts. Although you want to help the person, you must remember that you are not responsible for the addiction.

Once an addict has admitted the problem, he or she must receive treatment to stop alcohol or drug use. The person might check into a substance abuse clinic, go to a clinic as an outpatient, or join a long-term support program. The goal is for the person to learn to live without alcohol or drugs.

When teens become addicts, their recovery may include catching up on skills they have missed. They may have dropped out of school, for example, and need to receive tutoring to finish their education. They also need to learn new ways to handle their problems so that they don't return to alcohol or drug use.

In your journal

What would you say to a friend who appeared to have an alcohol or drug problem? Make a list of ways to convince the person to get help. In your journal, write a letter asking the friend to seek counseling.

With help, teens can recover from addiction and achieve their long-term goals.

Review — Lesson 6

Using complete sentences, answer the following questions on a separate sheet of paper.

Reviewing Terms and Facts

1. **Vocabulary** Define *tolerance*. Then use it in an original sentence.
2. **Recall** List four of the symptoms of withdrawal.
3. **List** Name four people or places you might contact to get help for someone with a substance abuse problem.

Thinking Critically

4. **Infer** Why might an addict not seek treatment on his or her own?
5. **Describe** Why is recovery so much more difficult than avoiding drugs in the first place?

Applying Health Concepts

6. **Health of Others** Make a pamphlet about sources of help for people who abuse drugs. Find agencies and organizations in your area where teens can go for help with substance abuse problems. Identify which organizations help teens with their own problems and which ones give support to friends and family members of addicts. If possible, make copies of your pamphlet, and distribute them in the school library, the public library, and other public places in your community.

Lesson 6: Addiction and Recovery   327

Consumer Health

How much *should* an effective recovery program cost? How much *does* it in fact cost? These are important considerations. Some community organizations provide free recovery treatment. Some specialized hospitals offer very extensive and expensive residential treatment. These hospitals often advertise their services on TV and in magazines. Many insurance programs provide recovery treatment or reimburse patients for certain recovery expenses. Encourage teens to explore all of these options. Ask: What choices do teens in your community have? How can they learn more about these choices?

Chapter 11 Lesson 6

Assess

EVALUATING THE LESSON

Assign Reviewing Terms and Facts and Thinking Critically on this page to review the lesson; then assign the Lesson 6 Quiz in the TCR to evaluate students' understanding.

LESSON 6 REVIEW

Answers to Reviewing Terms and Facts

1. Tolerance occurs when the body becomes used to a drug and needs greater amounts to get the desired effect. Sentences will vary.
2. The symptoms of withdrawal include (any four) severe anxiety, confusion, memory loss, hallucinations, nausea, vomiting, diarrhea, headache, chills.
3. Responses will vary.

Answers to Thinking Critically

4. Responses will vary. Possible response: Addicts typically deny that they have a problem with alcohol or drugs.
5. Responses will vary.

RETEACHING

▶ Assign Concept Map 53 in the TCR.
▶ Have students complete Reteaching Activity 53 in the TCR.

ENRICHMENT

Assign Enrichment Activity 53 in the TCR.

Close

Conduct a brief closing discussion, encouraging students to share their opinions in response to these questions: What is the worst part of addiction? What is the most difficult part of recovery? What is the most useful thing you could do for a friend or family member if he or she were addicted to alcohol or other drugs?

327

Chapter 11 Review

CHECKING COMPREHENSION

Use the Chapter Summary and the Chapter 11 Review to help students go over the most important ideas presented in Chapter 11. Encourage students to ask questions and add details as appropriate.

CHAPTER 11 REVIEW ANSWERS

Reviewing Key Terms and Concepts

1. Cirrhosis is the scarring or destruction of liver tissue, usually caused by heavy drinking.
2. Groups that help recovering alcoholics and their family and friends include Alcoholics Anonymous, Alateen, and Al-Anon.
3. Stimulants speed up the body's functions. Depressants slow down the body's functions and reactions.
4. Hallucinogens are drugs that distort moods, thoughts, and senses.
5. The spinal cord carries messages from all parts of the body to the brain, and from the brain to the rest of the body.
6. Drugs act directly on the brain stem, which helps control heart rate, breathing, eating, and sleeping.
7. Alcohol and drug use can cause reduced physical coordination, muscle twitches and cramps, reduced endurance, reduced strength, and slowed reflexes.
8. Alcohol slows a driver's reaction time and impairs judgment and decision making; this greatly increases the likelihood of an accident.
9. You can prepare yourself by having a ready supply of reasons for avoiding alcohol and drugs, and by knowing how to resist peer pressure.

Chapter 11 Review

Chapter Summary

▶ **Lesson 1** Alcohol has many unhealthful effects on the body. Alcoholism is a disease in which a person cannot control his or her drinking.

▶ **Lesson 2** When used properly, medicines are valuable drugs. However, illegal drugs should be avoided.

▶ **Lesson 3** The nervous system sends messages to the brain and from the brain to the rest of the body. Alcohol and drug use can damage this system.

▶ **Lesson 4** The use of alcohol and drugs can damage a person's physical, mental/emotional, and social health. Teens who understand the risks of using these substances will be more likely to avoid them.

▶ **Lesson 5** Having a positive self-concept, staying away from risky situations, and resisting peer pressure are ways to avoid substance abuse.

▶ **Lesson 6** Recovery from alcohol or drug addiction is much more difficult than resisting these substances in the first place. Help is needed.

Reviewing Key Terms and Concepts

Using complete sentences, answer the following questions on a separate sheet of paper.

Lesson 1
1. What is *cirrhosis*?
2. Name three groups that help recovering alcoholics and their families and friends.

Lesson 2
3. Contrast *stimulants* and *depressants*.
4. What are *hallucinogens*?

Lesson 3
5. What is the function of the spinal cord?
6. How can drug abuse cause problems with the nervous system?

Lesson 4
7. How can alcohol and drug use affect physical fitness?
8. Explain why it is so dangerous to drink and drive.

Lesson 5
9. How can teens prepare themselves to say no to alcohol and drugs?
10. What are *alternatives*?

Lesson 6
11. What are the four steps involved in recovery from addiction?
12. What is withdrawal?

Thinking Critically

Using complete sentences, answer the following questions on a separate sheet of paper.

13. **Suggest** How would you help someone who began abusing painkillers?
14. **Analyze** Why is alcohol and drug abuse considered a preventable cause of nervous system disorders?
15. **Hypothesize** Why might someone who knows the risks associated with alcohol and drugs continue to use them?
16. **Invent** Imagine and describe a social situation in which you say no to alcohol.
17. **Apply** What could you do if you felt that you had an alcohol or drug problem?

Meeting Student Diversity

Language Diversity Use the following suggestions to help students who have difficulty with English:

▶ Pair those learners with native speakers of English who can restate the Chapter Summary in language that helps students comprehend important concepts.

▶ Direct auditory learners or those students with language diversity to the Teen Health Audiocassette Program. Available in English and Spanish, this component provides an audio and written summary of the chapter.

Chapter 11 Review

Your Action Plan

In this chapter you have learned how drinking alcohol and using drugs can harm your health. You can set a long-term goal of avoiding alcohol and drugs. Short-term goals can help you keep on track. For example, a short-term goal could be to get involved in a healthful activity with people you like.

Step 1 Review your private journal entries for this chapter. Highlight the journal entries that seem most useful in achieving your long- or short-term goals.

Step 2 Divide a sheet of paper into two columns. List aids to short-term goals in one column and aids to long-term goals in the other. One aid to a short-term goal might be to find out about clubs or sports teams at your school.

Step 3 In your journal, write a letter to your family and friends explaining your goals and how you plan to accomplish them.

Your reward for working toward your goals will be freedom from dependence on alcohol or drugs. You will know that you are managing your own life, not allowing yourself to be controlled by a dangerous substance.

In Your Home and Community

1. **Personal Health** Arrange a time to sit down and talk with members of your family. Tell them that you want to write and sign a contract in which you agree not to use alcohol or drugs. Ask the adults in your family to sign an agreement to support you in keeping to your contract.

2. **Growth and Development** With your classmates, organize a panel discussion on alcohol and drugs. Invite community professionals to take part. Choose an individual to guide the discussion and request questions from audience members.

Building Your Portfolio

1. **Diagram of the Nervous System** Create an illustration that explains how the nervous system works. You will need to include both drawings and text. If necessary, do more research about the nervous system in health and science books. Add the illustration to your portfolio.

2. **Health Care Interview** Interview a professional who helps people overcome addictions. Before conducting the interview, prepare a list of questions to ask. Tape-record or videotape the interview, and add the tape to your portfolio.

3. **Personal Assessment** Look through all the activities and projects you did for this chapter. Choose one or two that you would like to include in your portfolio.

Chapter 11: Chapter Review

Chapter 11 Review

10. Alternatives are other ways of thinking or acting.
11. Admit the problem; remove the alcohol or drug from the body; overcome psychological dependence; change habits and lifestyle.
12. Withdrawal is a series of painful physical and mental symptoms that occur when a person stops using an addictive substance.

Thinking Critically
Responses will vary. Possible responses are given.
13. Suggest getting help.
14. Alcohol and drug abuse are preventable because everyone can choose not to start using these substances.
15. Responses will vary.
16. Responses will vary.
17. You could ask for help.

RETEACHING
Assign Study Guide 11 in the Student Activities Workbook.

EVALUATE
▶ Use the reproducible Chapter 11 Test in the TCR, or construct your own test using the Testmaker Software.
▶ Use Performance Assessment 11 in the TCR.

EXTENSION
Have students work together to compile an annotated bibliography of books written for teens on the subject of alcohol and drug abuse. Encourage students to include all relevant books in the school library, as well as those in local public libraries. Their completed list should present a conventional bibliographic entry for each book, along with a brief summary of the contents and a short comment on the interest level and usefulness of the book. Suggest that students make copies of their annotated bibliography to be shared with other students and/or other members of the community.

Performance Assessment

▶ **Self-evaluation** Direct students to review the activities that are provided throughout the chapter. Encourage each student to select one finished product or activity that demonstrates his or her best work for the chapter. Have students explain what they learned and how the examples they selected show their progress.

▶ **Teacher's Classroom Resources** Assign Performance Assessment 11, "Research Report on Alcohol and Drug Abuse," in the TCR.

Planning Guide

Chapter 12
Understanding Communicable Diseases

	Features	Classroom Resources

Lesson 1: Causes of Communicable Diseases
pages 332–335

Health Lab: Observing Bacteria — page 335

- Concept Map 54
- Cross-Curriculum Activity 23
- Enrichment Activity 54
- Lesson Plan 1
- Lesson 1 Quiz
- Reteaching Activity 54
- Transparency 47

Lesson 2: The Immune System
pages 336–339

- Concept Map 55
- Enrichment Activity 55
- Lesson Plan 2
- Lesson 2 Quiz
- Reteaching Activity 55
- Transparency 48

Lesson 3: Common Communicable Diseases
pages 340–343

- Concept Map 56
- Decision-Making Activity 23
- Enrichment Activity 56
- Lesson Plan 3
- Lesson 3 Quiz
- Reteaching Activity 56

Lesson 4: Avoiding Common Communicable Diseases
pages 344–347

Life Skills: Being a Reliable Information Source — page 346

Personal Inventory: Good Health Habits — page 347

- Concept Map 57
- Cross-Curriculum Activity 24
- Decision-Making Activity 24
- Enrichment Activity 57
- Health Lab 12
- Lesson Plan 4
- Lesson 4 Quiz
- Reteaching Activity 57

TEEN HEALTH DIGEST
pages 348–349

Lesson 5: Sexually Transmitted Diseases and HIV/AIDS
pages 350–357

Making Healthy Decisions: Deciding Whether to Speak Up — page 357

- Concept Map 58
- Enrichment Activity 58
- Lesson Plan 5
- Lesson 5 Quiz
- Reteaching Activity 58
- Transparencies 49, 50

329A

Review

Lesson 1

Using complete sentences, answer the following questions on a separate sheet of paper.

Reviewing Terms and Facts

1. **Vocabulary** What is the difference between a *communicable disease* and a *noncommunicable disease?*
2. **List** Name four kinds of germs.
3. **Identify** List three ways in which germs can be spread.

Thinking Critically

4. **Analyze** How can washing your hands carefully help prevent the spread of germs?

5. **Explain** What is the relationship between germs and infection?

Applying Health Concepts

6. **Health of Others** Find out how to protect yourself against Lyme disease. Create a poster to educate others about this disease. Ask permission to display your poster at a doctor's office or nature center.

HEALTH LAB
Observing Bacteria

Introduction: Bacteria are microscopic in size. However, given the right nutrients, bacteria will increase in number until they form a group (known as a colony) that can be seen with the unaided eye. In this experiment you will take samples of bacteria from various places and watch them grow.

Objective: Determine where the greatest concentrations of bacteria can be found around your school.

Materials and Method: You will need three petri dishes, or other small glass containers with lids, filled with agar or another culture medium—a substance in which bacteria can grow and multiply. You will also need disinfectant soap and paper towels.

Choose two places to collect samples of bacteria. You might use your unwashed hands for one sample. Other areas to collect bacteria could include water fountains, doorknobs, floors, and furniture. Touch the object to the agar in the dish, or wipe the object with a clean paper towel and then wipe the towel on the agar. Immediately after depositing the

sample, cover the dish and label it with the specific location where the sample was collected. Wash and dry your hands between collections.

For the third sample, wash your hands carefully with disinfectant soap, and dry them on a paper towel. Then press your fingers against the agar in the third dish. Cover the dish and label it "clean hands." Put the dishes in a warm, dark place for five days.

Observations and Analysis:
Study the contents of the three dishes. What do you observe? Which jar shows the greatest growth of bacteria? Which shows the least? What conclusions can you draw from your observations?

Lesson 1: Causes of Communicable Diseases 335

Chapter 12 Lesson 1

LESSON 1 REVIEW

Answers to Reviewing Terms and Facts

1. A communicable disease can be passed from person to person; a noncommunicable disease cannot.
2. Viruses, bacteria, fungi, and protozoa are four kinds of germs.
3. Germs can be spread by (any three, or specifics of these): direct contact, indirect contact, contact with animals and insects, consumption of contaminated food or water.

Answers to Thinking Critically

4. Washing can reduce the number of germs on your hands and thus prevent other people's germs from entering your body.
5. An infection occurs when germs invade the body, multiply, and damage body cells.

RETEACHING

▶ Assign Concept Map 54 in the TCR.
▶ Have students complete Reteaching Activity 54 in the TCR.

ENRICHMENT

Assign Enrichment Activity 54 in the TCR.

Ask students: What habit do you need to change or develop to protect yourself from communicable diseases?

---

recognize the importance of avoiding germs and taking measures to halt the spread of germs.

Observation and Analysis

Suggest that students work with partners or in very small cooperative groups to collect the bacteria and to label their containers. Remind students to add their names as well as the identifying information to the labels. After five days, have the same pairs or groups work together to observe and discuss the contents of their containers.

Further Investigation

Provide an opportunity for interested students to gather and observe bacteria samples from areas outside of school. Let them share the results of this investigation with the rest of the class.

Assessing the Work Students Do

Ask students to write one-paragraph summaries of what they learned from this Health Lab activity. Assign credit based on participation and completion.

Lesson 2
The Immune System

Focus

LESSON OBJECTIVES

After studying this lesson, students should be able to
- explain how the body protects itself against germs.
- identify and discuss the general reactions of the immune system.
- identify and discuss the specific reactions of the immune system.

MOTIVATOR

Give students two minutes in which to list familiar symptoms of communicable diseases.

INTRODUCING THE LESSON

Give students an opportunity to share their lists from the Motivator activity. Which symptoms are most frequently mentioned? What do these symptoms indicate about the body's reaction to germs? How does the body recover from these illnesses? After students have shared their ideas, explain that Lesson 2 presents more information about how their body protects itself and fights off communicable diseases.

INTRODUCING WORDS TO KNOW

Divide students into small groups to skim the lesson and find the definition for each Word to Know. Have group members create flash cards for the vocabulary terms. Let them use these cards to check one another's understanding of the Words to Know.

Lesson 2 The Immune System

This lesson will help you find answers to questions that teens often ask about how the body fights germs. For example:
- How does the immune system work?
- What are antibodies, and how do they protect me?
- How do I become immune to a disease?

Words to Know

immune system
lymphatic system
lymphocytes
antigens
antibodies
immunity
vaccine

 in Your journal

Do you usually have some warning when you are getting sick? What signs tell you that you are coming down with something? Write these in your private journal, and describe what you could do to protect your own health and the health of others when you know you are getting sick.

Defending Against Invaders

Each day your body is exposed to countless germs. To protect itself against these germs, your body takes action to repel, trap, or destroy them. The body has natural barriers that keep germs out or destroy them before they can do any damage. The five major barriers, or "first lines of defense," are listed below:

- **Skin.** The skin covering your body acts as a protective barrier.
- **Mucous** (MYOO-kuhs) **membranes.** These tissues line your mouth, nose, throat, eyes, and other body parts. They trap germs. Actions such as coughing and sneezing get rid of the germs trapped by these membranes.
- **Saliva.** Saliva in your mouth destroys many harmful organisms.
- **Tears.** Tears wash away germs. They also contain chemicals that kill some harmful organisms.
- **Stomach acid.** The acid in your stomach kills many germs.

As effective as these defenses are, viruses, bacteria, and other germs do sometimes get through. That's when your immune system springs into action. Your **immune** (i-MYOON) **system** is a *combination of body defenses made up of cells, tissues, and organs that fight off germs and disease.*

Millions of helpful bacteria live on your skin and are part of the skin's defense against germs.

336 Chapter 12: Understanding Communicable Diseases

Lesson 2 Resources

Teacher's Classroom Resources
- Concept Map 55
- Enrichment Activity 55
- Lesson Plan 2
- Lesson 2 Quiz
- Reteaching Activity 55
- Transparency 48

Student Activities Workbook
- Study Guide 12
- Applying Health Skills 55

The Immune System's General Reactions

When germs get inside your body, your immune system launches an attack. Three general reactions may occur, no matter what kind of microorganism has invaded.

- Special white blood cells called phagocytes (FA·guh·syts) attack the invading germs. These cells actually surround the germs and destroy them.

- The cells may release a chemical substance called interferon (in·ter·FIR·ahn) that stops viruses from reproducing.

- Rising body temperature, commonly called a fever, makes it difficult for some microorganisms to reproduce.

The Immune System's Specific Reactions

If invading germs survive the immune system's general reactions, the body responds with more specific defenses. These are responses to certain microorganisms and the toxins, or poisons, that they produce. Often these specific reactions both defend the body when the microorganisms enter and allow the immune system to remember the particular germ. That way it can be destroyed if it enters the body again.

To fight against specific germs, the body calls upon the lymphatic system. The **lymphatic** (lim·FA·tik) **system** is a *secondary circulatory system that helps the body fight germs and maintain its fluid balance.* The lymphatic system carries a watery fluid known as lymph (LIMF). *Special white blood cells in the lymph* are called **lymphocytes** (LIM·fuh·syts). There are two types of lymphocytes: B-cells and T-cells. Both are important in fighting off germs and disease. The lymph also contains phagocytes known as macrophages (MA·kruh·fay·juhz), which help the lymphocytes identify invading germs.

Antigens and Antibodies

Antigens (AN·ti·jenz) are *substances that send your immune system into action when your body is invaded by germs.* The body recognizes antigens as invaders. For example, the toxins produced by bacteria are antigens. Blood cells from a blood type different from yours, as you learned in Chapter 5, are also sensed by the body as antigens. Your body reacts to antigens by producing antibodies. **Antibodies** are *proteins that attach to antigens, keeping them from harming the body.* Your immune system produces specific antibodies to fight each antigen. **Figure 12.3,** on the next page, shows the immune system in action.

Your Total Health

Help Your Immune System ACTIVITY!

To boost your body's immune system, exercise regularly and maintain a healthy diet. Eat plenty of fruits and vegetables, and limit fats. Also, do what you can to minimize stress, which can weaken the immune system.

Look for magazine and newspaper articles on how your diet can benefit your immune system. What nutrients are especially important to immune system function? How can you make sure that you get enough of these nutrients in your diet?

Unfortunately, some people's bodies react to certain pollens as antigens. The body's immune system reacts to the pollen as if it were an invader, producing symptoms such as sneezing or swelling of mucous membranes.

Lesson 2: The Immune System 337

Health of Others

Teens are unlikely to pay much attention to mild communicable diseases that do not cause them unusual discomfort. However, they should recognize the need to avoid spreading even mild diseases to individuals who may be especially susceptible to infection. These people include

▶ infants, especially newborns.

▶ elderly people, especially those in poor health.

▶ people still recovering from major communicable diseases.

▶ people receiving treatment for serious noncommunicable diseases, such as cancer.

▶ people infected with HIV and/or AIDS.

Chapter 12   Lesson 2

Teach

L1 Comprehending

Let volunteers explain why skin, mucus, saliva, tears, and stomach acid are called the "first line of defense." What kind of protection does each provide? What other functions does each serve?

L3 Researching

Have interested volunteers use the library and Internet to research autoimmune diseases: What are they? What are their symptoms and effects? How can they be treated? Have these volunteers write and illustrate short reports on their findings.

L2 Creating Illustrations

Ask students to find and examine enlarged color photographs (or electron micrographs) of skin bacteria, such as *Staphylococcus aureus*. Then have them draw their own illustrations of common skin bacteria.

L3 Science

What is phagocytosis? What are the stages of phagocytosis? What is its purpose? Ask two or three class members to find the answers to these questions and to present an illustrated oral report of their findings.

L2 Life Skills

Ask a group of students to learn about the most effective ways to check and evaluate the temperature of infants, children, and teens. What temperature represents a fever? How and at what point should a fever be treated? In which cases should a health professional be contacted? Have these students share their findings with the rest of the class.

Chapter 12 Lesson 2

L1 Discussing
Encourage students to share their own experiences with allergies and to discuss what they know about the causes and treatments of allergies.

L3 Social Studies
Have interested students research the life and work of Paul Ehrlich, noting particularly his contributions to immunology and other branches of medicine. Let these students summarize their findings in written reports.

Visual Learning
Guide students in describing and discussing the T-cells and B-cells in Figure 12.3. Ask: How do the two kinds of cells work together? How will the responses of these cells be different when a specific virus invades the body a second time?

L2 Reporting
There are actually three different kinds of T-cells, each with a specialized function. Let a volunteer read about these three kinds of T-cells and explain the differences to the rest of the class.

L3 Researching
Those interested in herbal remedies often recommend echinacea for treating infections and for stimulating the immune system. Ask students to learn more about the use of echinacea, seeking both positive and negative points of view. Then have students discuss their findings and share their own opinions about echinacea.

L3 Reporting
Have a small group of students research severe combined immunodeficiency (SCID): What is this disease? What causes it? How does it affect a patient's B-cells and T-cells? What treatments are available? Let group members plan and present a short oral report of their findings.

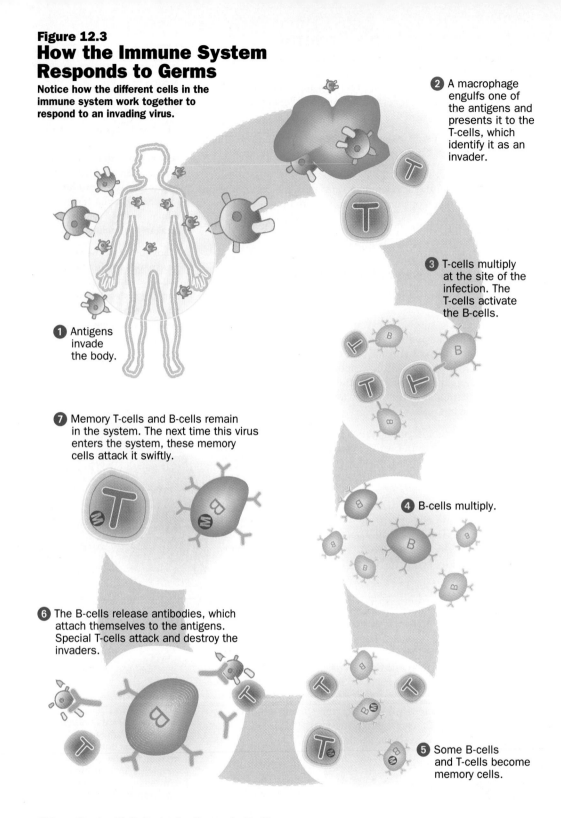

Figure 12.3
How the Immune System Responds to Germs
Notice how the different cells in the immune system work together to respond to an invading virus.

1. Antigens invade the body.
2. A macrophage engulfs one of the antigens and presents it to the T-cells, which identify it as an invader.
3. T-cells multiply at the site of the infection. The T-cells activate the B-cells.
4. B-cells multiply.
5. Some B-cells and T-cells become memory cells.
6. The B-cells release antibodies, which attach themselves to the antigens. Special T-cells attack and destroy the invaders.
7. Memory T-cells and B-cells remain in the system. The next time this virus enters the system, these memory cells attack it swiftly.

338 Chapter 12: Understanding Communicable Diseases

Promoting Comprehensive School Health

Helping students and their families reduce the spread of communicable diseases can be an important goal for a comprehensive school health program. Your school may want to sponsor

▶ vaccination opportunities.
▶ diagnosis of common communicable diseases.
▶ treatment suggestions and/or referrals to doctors or clinics.
▶ informative brochures and/or discussion groups on STDs.
▶ community health awareness programs.

For further information and suggestions, consult *Promoting a Comprehensive School Health Program* in the TCR.

Immunity

Your *body's ability to resist the germs that cause a particular disease* is called **immunity**. You develop immunity in two ways:

- **Natural exposure to germs.** Throughout your life you are exposed to many different germs. As **Figure 12.3** showed, your body produces memory B-cells and T-cells. Often these memory cells stay in your blood. If the same germs return, memory cells allow your body to fight the germs off more quickly.

- **Immunization.** Health care workers administer vaccines to make people immune to certain diseases. A **vaccine** (vak·SEEN) is a *preparation of dead or weakened germs that is injected into the body to cause the immune system to produce antibodies.* Vaccines exist for many diseases, including polio, measles, and mumps. Being vaccinated won't make you ill because the germs used are dead or weak. However, the antibodies your body produces in response to the vaccine will build your immunity.

Many vaccines are given by injection, but the polio vaccine is often given orally.

Review Lesson 2

Using complete sentences, answer the following questions on a separate sheet of paper.

Reviewing Terms and Facts

1. **Explain** What are the body's five barriers against disease?
2. **Identify** List three general reactions of the immune system.
3. **Vocabulary** Define the terms *antigen* and *antibody*.
4. **Recall** List two ways in which people can develop immunity to a disease.

Thinking Critically

5. **Explain** Why are diseases that damage the immune system so dangerous to the body?

6. **Analyze** Vaccines have all but eliminated certain communicable diseases. How is this possible?

Applying Health Concepts

7. **Health of Others** Like people, animals can be vaccinated against certain diseases. Find out what vaccines cats and dogs generally receive. Which of these vaccines indirectly protect the health of people? Share your findings with the class.

Lesson 2: The Immune System   **339**

More About...

Vaccinations By the beginning of the nineteenth century, the smallpox vaccine was being hailed as a safe, effective means of preventing the deadly disease. President Thomas Jefferson supported efforts to have vaccination introduced in major American cities. He also encouraged Native Americans to accept vaccinations. The native peoples, however, were suspicious of the vaccine. Sadly, their decision not to vaccinate had disastrous consequences. In 1837, when a steamboat stopped to trade with the Mandan people in what is now North Dakota, two of the traders were ill with smallpox. Within days, Mandans began to fall ill and die. Of some 2,000 people, only 30 or 40 survived that smallpox epidemic.

Chapter 12 Lesson 2

Assess

EVALUATING THE LESSON

Assign Reviewing Terms and Facts and Thinking Critically to review the lesson; then assign the Lesson 2 Quiz in the TCR to evaluate students' understanding.

LESSON 2 REVIEW

Answers to Reviewing Terms and Facts

1. The five barriers are skin, mucous membranes, saliva, tears, and stomach acid.
2. Three general reactions are phagocytes that surround and destroy germs, the release of interferon, and fever.
3. An antigen is a substance that activates the immune system. The system reacts to an antigen by producing antibodies, proteins that attach to antigens, preventing them from harming the body.
4. You can develop immunity through natural exposure to germs and through vaccines.

Answers to Thinking Critically

5. Responses will vary. Possible response: Anything that damages the immune system opens the body to attack from other germs that people encounter daily.
6. Responses will vary.

RETEACHING

▶ Assign Concept Map 55 in the TCR.
▶ Have students complete Reteaching Activity 55 in the TCR.

ENRICHMENT

Assign Enrichment Activity 55 in the TCR.

Close

Let students work with partners to chart or outline the most important information in this lesson.

339

Lesson 3

Common Communicable Diseases

Focus

LESSON OBJECTIVES

After studying this lesson, students should be able to
- explain why colds are so common.
- identify the most important differences between a cold and the flu.
- identify other common communicable diseases.
- discuss the effects of vaccination on the spread of certain communicable diseases.

MOTIVATOR

Ask students to list the symptoms of their last cold.

INTRODUCING THE LESSON

Let students share their responses to the Motivator activity and discuss answers to these questions: What symptoms tell you that you're getting a cold? That you have a cold? How long does it usually take you to recover from a cold? When and how are you most likely to get a cold? What can you do to protect yourself from getting colds and to protect other people from catching your cold? Conclude by telling students that they will learn more about colds and other common communicable diseases as they study Lesson 3.

INTRODUCING WORDS TO KNOW

Have students work with partners to find the definitions of the Words to Know in the Glossary. Suggest that they record these definitions in their notebooks.

340

Lesson 3 Common Communicable Diseases

This lesson will help you find answers to questions that teens often ask about familiar communicable diseases. For example:
- What is the most common communicable disease?
- How is a cold different from the flu?
- What are some other common communicable diseases?

Words to Know

influenza
mononucleosis
hepatitis
tuberculosis

in your journal

How many colds have you had in the last 12 months? In your private journal, describe what factors you think make you more likely to get a cold (lack of sleep, stress, being indoors, and so on). Which of these factors are ones you can control?

People often catch colds by handling objects that someone with a cold has touched. Viruses can live on surfaces for hours, or even days. Washing your hands often can help you avoid picking up someone else's germs.

Facts About the Common Cold

Of all communicable diseases, the cold is the most common. Consider the following facts.

- Children miss school more often because of colds than for any other medical reason. Colds are also the most common reason for workers to miss days on the job.

- On the average, adults get at least two or three colds a year. Children and young adults get about two to three times as many colds as adults.

Why are colds so common? One reason is that they are spread in several ways. Indirect contact is a common means of spreading cold germs—touching objects or surfaces that someone with a cold has previously handled, and then touching your mouth, nose, or eyes. You can also get cold germs through direct contact—shaking hands, for example. When a cold sufferer coughs or sneezes, germs are expelled into the air. Other people inhale these germs.

Another reason why colds are so common is that colds may be caused by hundreds of different viruses. This is why no cold vaccine has been developed. A vaccine that would give you immunity to one cold virus would not protect you against all the other viruses. You would need a different vaccine for each one!

Chapter 12: Understanding Communicable Diseases

Lesson 3 Resources

Teacher's Classroom Resources
- Concept Map 56
- Decision-Making Activity 23
- Enrichment Activity 56
- Lesson Plan 3
- Lesson 3 Quiz
- Reteaching Activity 56

Student Activities Workbook
- Study Guide 12
- Applying Health Skills 56

Facts About the Flu

Although some of their symptoms are the same, influenza—"flu"—and colds are caused by different viruses. **Influenza** (in·floo·EN·zuh) is *a communicable disease characterized by fever, chills, fatigue, headache, muscle aches, and respiratory symptoms.*

The flu usually begins more abruptly than a cold and lasts longer. The flu is spread by airborne germs as well as through direct and indirect contact. Each year different strains of flu arise. (December through early March are the peak flu months.) Vaccines are generally available in October and November for the strain of flu that doctors expect to be present for the following months. Anyone can get a flu vaccine, but vaccination is particularly recommended for certain groups of people, including

- people who are 65 years of age or older and people who live in nursing homes or other long-term care facilities.
- adults and children with lung or cardiovascular disorders.

Language Arts Connection

Sickness from the Stars?

The word "influenza" comes from the word "influence," probably because astrologers once linked the disease to an evil influence from the stars.

Residents of nursing homes should have flu shots yearly. Flu is easily passed from one person to another in any residential facility.

Other Communicable Diseases

Several communicable diseases are discussed in **Figure 12.4** on the next page. The three diseases listed below are caused by viruses and can be spread by direct or indirect contact. People who have these diseases should stay at home during the time when they are able to infect others. Vaccines are available for all three:

- **Chicken pox.** This illness involves an itchy rash, fever, headache, and body aches. This disease can be passed to others from two days before the rash appears until about six days after.
- **Measles.** Measles is characterized by a rash with fever, runny nose, and coughing. People with measles can infect others from several days before the rash appears until about five days after.
- **Mumps.** This disease causes a fever, headache, and swollen salivary glands. It is most easily passed to others around the time when the symptoms appear, but may be passed on from as much as seven days before the symptoms appear until nine days after.

Q & A

No Sure Shortcuts

Q: If I start taking cold medicines when I feel as if I'm getting a cold, will that keep me from getting sick?

A: There's an old saying that a cold will last about a week if you treat it and seven days if you don't. That's because cold medicines can't cure the illness or kill the germs that cause it. All they can do is treat the symptoms. To treat a cold, get plenty of rest and let your immune system do its job.

Lesson 3: Common Communicable Diseases   341

Chapter 12 Lesson 3

Teach

L2 Math

Suggest that students work in small groups to devise surveys about the frequency and/or severity of colds among a certain population (kindergartners or seventh graders, for example). Have group members distribute the survey and summarize the collected information in an appropriate graph or chart.

Would You Believe?

On the day before the launch of *Apollo 9* in 1969, all three astronauts scheduled for the mission developed cold symptoms. The launch was delayed a week to allow the astronauts to recover from their colds—but the delay cost the U.S. government half a million dollars.

L2 Reporting

Ask interested volunteers to read about OTCs that are designed to treat or prevent colds. Have the volunteers present their findings to the rest of the class.

L2 Life Skills

Headaches can be a symptom of colds, flu, or other communicable diseases. However, headaches can also result from other causes, including too much or too little sleep, certain food triggers, and stress. Suggest that students read more about the causes of headaches and then keep a log of their own headaches: Can they determine what causes most of their own headaches? Can they use their findings to reduce the number of headaches they suffer?

L2 Current Events

Have students find, read, and clip recent newspaper articles about flu epidemics and/or flu vaccines. Let students summarize their articles for the rest of the class and then post them on a bulletin board.

Growth and Development

Aspirin in some form has been used for centuries to treat pain, including the pains associated with cold and flu. By inhibiting the production of body chemicals that cause inflammation, aspirin does reduce aches and pains. However, researchers have established a link between the use of aspirin during viral infections and the onset of Reye's syndrome. Reye's syndrome affects the liver and the brain, and it can prove fatal in as many as a quarter of all its victims. Reye's strikes only children and adolescents; doctors now recommend that no one under 18 take aspirin. Acetaminophen, a common aspirin substitute, is recommended for children and adolescents.

Chapter 12 Lesson 3

L2 Family and Consumer Sciences

Ask a group of interested volunteers to read about chicken pox, measles, and mumps in several child care books: How are these diseases prevented? For whom are vaccines advised? Under what circumstances are children likely to be infected with these diseases? What treatments are recommended?

L2 Guest Speaker

Mononucleosis is most likely to strike older teens, usually between the ages of 15 and 17, although younger teens certainly can contract this viral disease. If you know a teen or young adult who has had mononucleosis, invite him or her to speak with the class, describing the symptoms, the diagnosis, the course of the disease, and the probable extended recovery period.

L3 Researching

To learn more about the form of inoculation that preceded smallpox vaccines, suggest that students read about the work of Lady Mary Wortley Montagu. Ask them to share their findings in brief written or oral reports.

L3 Social Studies

Why do so many people die from tuberculosis each year? Where are people most likely to contract tuberculosis? What prevention techniques and treatments are available in those areas? Let students use library resources to find answers to these questions and then summarize their findings in short reports.

L2 Reporting

Louis Pasteur proved that tuberculosis is caused by bacteria. Ask a volunteer to read at least one encyclopedia article about Pasteur's life and work; have the volunteer summarize the information for the rest of the class.

Figure 12.4
Some Communicable Diseases

Many communicable diseases require a doctor's diagnosis and treatment. For some, a doctor may prescribe antibiotics or other medication.

Disease	Symptoms	Transmission Method	Treatment
Mononucleosis	Known as "mono" or "the kissing disease," **mononucleosis** (MAH·noh·noo·klee·OH·sis) is *a virus-caused disease characterized by swelling of the lymph nodes in the neck and throat*. It is most common in teens and young adults. Other symptoms may include fatigue, appetite loss, fever, headache, and sore throat.	Mononucleosis is usually spread through kissing or by sharing drinking glasses, eating utensils, or toothbrushes.	Mononucleosis is treated with rest and pain relievers. Recovery may take 3 to 12 weeks. Fatigue may linger much longer.
Hepatitis	**Hepatitis** (he·puh·TY·tis) is *a virus-caused liver inflammation characterized by yellowing of the skin and the whites of the eyes*. There are several types. The two main ones are hepatitis A and hepatitis B, caused by different viruses. Hepatitis B may permanently damage the liver. Hepatitis symptoms may include weakness, nausea, fever, headache, sore throat, and appetite loss. A vaccine is available for both types.	Hepatitis A is transmitted through contaminated food or water. Hepatitis B is usually spread through direct contact with the body fluids of an infected person. Both types can also be spread through sexual contact.	Hepatitis is treated with rest and a healthful diet. Although most people recover completely, those infected with hepatitis B can remain contagious for the rest of their lives.
Tuberculosis	**Tuberculosis** (too·ber·kyuh·LOH·sis), or TB, is *a bacteria-caused disease that usually affects the lungs*. Symptoms may include cough, fatigue, night sweats, fever, and weight loss. Although a TB vaccine is available, as many as 3 million people around the world die from the disease each year.	Tuberculosis is spread by water droplets in the air from coughs and sneezes.	Tuberculosis is treated with antibiotics for a period of several months.

342   Chapter 12: Understanding Communicable Diseases

More About...

The Flu A severe form of flu occasionally affects people around the world. The worst such outbreak occurred in 1918 and 1919. It was known in most parts of the world as the Spanish flu, although it was called the American flu in Japan and the Japanese flu in China. About a fifth of the population was afflicted, and public activities, including schools and religious services, were halted in an effort to reduce contagion. In New York City, it was illegal to cough or sneeze without covering your face with a handkerchief. In spite of such efforts, about 20 million people died as a result of this flu.

Vaccination Schedules

Some communicable diseases, such as measles and mumps, used to be much more common than they are now. The vaccination of infants and children is making such diseases increasingly rare. **Figure 12.5** shows a typical vaccination schedule.

Figure 12.5
Vaccination Schedule
These are the vaccinations recommended by the American Academy of Pediatrics.

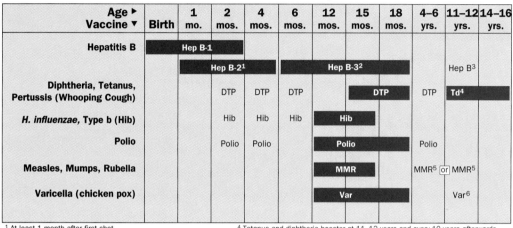

[1] At least 1 month after first shot
[2] At least 4 months after first shot and 2 months after second shot
[3] Children who were not vaccinated in infancy should begin series of shots at 11–12 years
[4] Tetanus and diphtheria booster at 11–12 years and every 10 years afterwards
[5] Second dose of MMR at either 4–6 years or 11–12 years
[6] For children who were not vaccinated in infancy and who have not had the disease

Review — Lesson 3

Using complete sentences, answer the following questions on a separate sheet of paper.

Reviewing Terms and Facts

1. **Recall** How can you tell the difference between a cold and influenza?
2. **Vocabulary** What is *mononucleosis*? How is it spread?
3. **Vocabulary** What is *tuberculosis*? How is it treated?

Thinking Critically

4. **Explain** Why do you think children get more colds than adults?
5. **Infer** Why would it be important to know when a disease can be passed on to others?

Applying Health Concepts

6. **Consumer Health** Go to a pharmacy, and compare over-the-counter medicines that are meant to relieve cold and flu symptoms. Read the labels carefully. How do the various medicines differ? How are they alike? Which would you consider buying? Make a chart that compares at least three different brands in terms of their ingredients and claims.

Lesson 3: Common Communicable Diseases   343

Health of Others

Not long ago, polio was a communicable disease widely feared in this country. Since the introduction of effective polio vaccines, however, new cases of polio are unknown in the United States and in other countries that offer high levels of medical care.

Typical symptoms of a relatively mild case of polio include fever, headache, muscle pain and spasms, nausea and vomiting, and stiffness of the neck and back. More severe cases can result in permanent paralysis and even death.

Vaccination against polio is still recommended for all children and adults in the United States. Polio continues to be a serious health problem in other parts of the world, including Africa, Asia, and the Indian subcontinent.

Lesson 4
Avoiding Common Communicable Diseases

Focus

LESSON OBJECTIVES
After studying this lesson, students should be able to
- discuss how they can protect themselves and others from the spread of communicable diseases.
- identify health habits that lead to a healthy lifestyle, and explain the importance of practicing those habits.

MOTIVATOR
Display some or all of the following items: a container of liquid hand soap, a box of facial tissues, a pillow (representing sleep), a jump rope, a container of prescription medicine. Ask students to explain briefly, in writing, how each item can help them avoid or treat communicable diseases.

INTRODUCING THE LESSON
Let volunteers share their responses to the Motivator activity. Ask: Why is it important to know how to use these and other items to protect yourself and others from communicable diseases? How can helping to prevent the spread of diseases (physical health problems) improve your mental/emotional health? your social health? Conclude by telling students that Lesson 4 presents more information on avoiding communicable diseases.

INTRODUCING WORDS TO KNOW
Let a volunteer find and read aloud the definition of the Word to Know. Then have each student write an original sentence using that term.

Lesson 4
Avoiding Common Communicable Diseases

This lesson will help you find answers to questions that teens often ask about staying healthy. For example:
- How can I avoid communicable diseases?
- How can I protect other people against diseases?

Words to Know
contagious period

Cultural Connections
Blocking Germs

In Japan, people who have a cold or the flu may wear a mask to keep germs from spreading. While this measure is not always effective against colds—which are more often spread by hands—it may help to prevent the spread of airborne flu viruses.

Preventing the Spread of Disease

Anyone who's suffered through a cold, flu, or other disease would certainly agree that it's far better to avoid an illness than to treat one. In this chapter you have read about many communicable diseases. This lesson will help you make decisions that will protect you against disease and protect those around you as well.

Protecting Yourself

The first rule in staying healthy is to use your common sense. Be smart. Don't take risks with your health. Here are some tips.

- Avoid close contact with people who are ill, especially during the contagious period of a disease. The **contagious period** is *the length of time when a particular disease can spread from person to person.* Avoid even shaking hands with someone who you know has a contagious disease.
- Do not share eating utensils, dishes, or glasses with others.
- Wash your hands often, especially when you have been around people who are ill. **Figure 12.6**, on the next page, presents some handwashing tips.
- Keep your hands away from your mouth, nose, and eyes.
  - Encourage cold and flu sufferers to cover their mouths and noses when they cough or sneeze.
    - Take proper precautions against known dangers, such as exposure to ticks that may carry disease.
      - To avoid food poisoning and other diseases, follow safe practices in handling, preparing, and storing food.

When you walk or hike in the woods, be sure to protect yourself against tick bites.

344 Chapter 12: Understanding Communicable Diseases

Lesson 4 Resources

Teacher's Classroom Resources
- Concept Map 57
- Cross-Curriculum Activity 24
- Decision-Making Activity 24
- Enrichment Activity 57
- Health Lab 12
- Lesson Plan 4
- Lesson 4 Quiz
- Reteaching Activity 57

Student Activities Workbook
- Study Guide 12
- Applying Health Skills 57

Figure 12.6
Washing Away Germs
Washing your hands is an important way to protect yourself against germs.

Always wash with soap. Consider buying liquid soap for home use. Bar soaps may accumulate germs. Special germicidal or antibacterial soaps are also available.

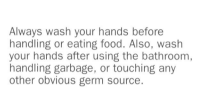

To do a thorough job, spend at least 20 seconds washing your hands. One to three minutes would be even better.

Use a brush to scrub under nails.

Always wash your hands before handling or eating food. Also, wash your hands after using the bathroom, handling garbage, or touching any other obvious germ source.

Protecting Others

Part of behaving responsibly involves doing whatever you can to avoid spreading germs to other people. The following are some examples of ways to protect others.

- Seek medical treatment if you need it. Delaying a visit to the doctor can allow your condition to worsen while increasing the chances that you'll spread germs to others.

- Cover your mouth and nose when you cough or sneeze. Use disposable tissues—and dispose of them properly.

- If you are ill, stay home from school during the contagious period of the disease, and avoid close contact with other people. Wash your hands frequently.

- When taking a prescribed medication, follow the directions carefully. Take all of the medication—don't stop taking it just because you feel better.

- Encourage other people to follow wise health practices.

Lesson 4: Avoiding Common Communicable Diseases 345

Chapter 12 Lesson 4

Teach

L1 Discussing
Help students discuss the contagious period for common diseases: Why is it important to recognize the very first symptoms of a cold, the flu, and other diseases?

L1 Critical Thinking
Ask: How do people usually cover their mouth and nose when they cough or sneeze? What are some of the benefits of using your hand for that purpose? What risks are involved with this practice? How can you avoid those risks?

L2 Family and Consumer Sciences
Ask a group of interested students to learn more about how to handle, prepare, and store foods safely. Let these students give an oral report to the class, with demonstrations and/or visual aids.

L2 Presenting Skits
Have students work with partners to plan and present short skits in which one teen encourages another to follow a safe health practice.

L1 Discussing
Ask students to reveal, by a show of hands, whether they usually wash their hands just before eating lunch at school. Ask students who don't to explain their reasons. Pose questions: What could the school do to make it more practical for students to wash their hands before eating?

Consumer Health

Students should recognize that washing their hands with soap is an essential part of protecting themselves against germs. Now, though, they may wonder which kind of soap is best. They have an increasing number of options, including various brands of bar soaps, liquid soaps, and germicidal soaps. Encourage students to select several soaps to study. (As the text points out, bar soaps may actually accumulate germs. Let students decide whether to include bar soaps in their investigation.) Have students compare the ingredients, the size, the price, and the approximate cost per use of various soaps. Can they reach a conclusion about the best soap? Which do they think is the best buy?

Chapter 12 Lesson 4

L2 Creating Displays
Allow students to form eight groups, and assign one of the healthy lifestyle habits to each group. Have group members discuss and perhaps even learn more about that habit, noting the particular health benefits of establishing and practicing the habit. Instruct group members to work together to plan and create a collage, a poster, a bulletin board, or another kind of display promoting that habit.

L1 Discussing
Let students read aloud and discuss the statements in the Personal Inventory: Good Health Habits. Be sure students understand just what each statement involves: Why is it a good health habit? What can teens do to help themselves establish or maintain that habit? Provide time for students to respond to the Personal Inventory statements in writing, or ask them to do so at home.

L1 Journal Writing
Suggest that students record their reactions to taking the Personal Inventory in their journals: How did you feel while discussing the inventory? How did you feel while taking it? How do you feel about your score? What have your responses taught you?

Assess

EVALUATING THE LESSON
Assign Reviewing Terms and Facts and Thinking Critically to review the lesson; then assign the Lesson 4 Quiz in the TCR.

Make a list in your journal of positive habits that you have that contribute to a healthy lifestyle. Then make a mental list of habits that you have that work against a healthy lifestyle. How can you change your negatives to positives? You might want to discuss your ideas for change with an encouraging adult whom you respect and admire.

Developing a Healthy Lifestyle

To feel well and keep your body in top condition, practice good health habits. Having a healthy body increases your ability to resist germs and disease. The following health habits are part of a healthy lifestyle.

- Eat a well-balanced, nutritious diet.
- Exercise regularly.
- Get an adequate amount of sleep.
- Take a shower or bath daily.
- Don't use alcohol or other drugs.
- Don't smoke.
- Be sure that you are vaccinated against diseases for which a vaccine exists. If you're not sure which vaccinations you need, ask your doctor.
- Learn to manage stress.

LIFE SKILLS
Being a Reliable Information Source

People share information constantly. They ask one another questions and try to give helpful answers. One topic about which people seek a great deal of information is health. You can make a valuable contribution to your community if you are a source of accurate health-related information. Here are some guidelines.

▶ You've learned about a wide range of health topics and concerns, from nutrition to communicable diseases, from sports and conditioning to drug abuse. Share your knowledge with others when the opportunity arises.

▶ Give sound advice based on your knowledge. For example, explain the benefits of remaining tobacco-free to a friend who is thinking of taking up smoking.

▶ Substitute information for misinformation. For example, if you hear someone repeat myths about how to treat a cold, provide facts.

▶ Know where to go for additional information. You can find up-to-date health information at the library, in magazines and newspapers, at doctors' offices, and from other reliable sources. Identify resources in your community.

▶ If you don't know, don't guess. Instead, find out the facts. Too often people mistake myths and half-truths for reliable information. If you need the answer to a medical question, look it up, or ask a health care worker.

Follow-up Activity
Recall a time when someone you know needed health-related information. Were you able to help? What other useful suggestions could you have made?

346 Chapter 12: Understanding Communicable Diseases

LIFE SKILLS
Being a Reliable Information Source

Focus on Life Skills
Let students share their responses to these questions: When you have a question about health issues, whom do you ask? Why? If you wanted to read information on your own, rather than asking someone, where would you look? When people ask you questions about health-related issues, how do you feel? What do you do if you don't know or aren't sure of the answer? After this introductory discussion, explain that the Life Skills feature focuses on how teens can be reliable sources of health information. Have students read the feature with partners or as a class.

Making Health Connections
Becoming knowledgeable about health issues can support all aspects of students' health. It enables teens to protect their own physical health, to enhance their mental/emotional health by encouraging confi-

Personal Inventory
GOOD HEALTH HABITS

You can protect yourself and others from germs and disease by developing good health habits. Keeping your body healthy helps you feel your best and enables you to recover quickly if you do get sick. Have you developed good health habits? To find out, write yes or no for each statement below. Use a separate sheet of paper.

1. I always wash my hands after using the bathroom and before handling food.
2. I take a shower or bath daily.
3. I cover my mouth and nose when I cough or sneeze.
4. I avoid contact with anyone who has a contagious disease.
5. When I am sick, I take precautions to avoid spreading germs.
6. I generally eat a well-balanced, nutritious diet.
7. I get an adequate amount of rest.
8. I exercise regularly.
9. I don't use alcohol or other drugs.
10. I don't smoke.

Give yourself 1 point for each yes. A score of 9–10 is very good. A score of 7–8 is good. If you score below 7, you need to improve your health habits.

Review — Lesson 4

Using complete sentences, answer the following questions on a separate sheet of paper.

Reviewing Terms and Facts

1. **Vocabulary** Use your own words to define the term *contagious period*.
2. **State** List five ways to protect yourself against germs and disease.
3. **Recall** Identify three ways to protect others from disease.

Thinking Critically

4. **Explain** Many people do not recognize the importance of frequent hand-washing. Why do you think this is true?
5. **Evaluate** A friend tells you that he's planning to go to school, even though he's running a fever and thinks he may have the flu. What would you say?

Applying Health Concepts

6. **Personal Health** With a partner, choose one of the guidelines for healthy living mentioned in this chapter. Write a short story depicting two characters: one who regularly follows this guideline and one who does not. Show how their different habits affect their overall health and the health of others.

Lesson 4: Avoiding Common Communicable Diseases 347

Chapter 12 Lesson 4

LESSON 4 REVIEW

Answers to Reviewing Terms and Facts

1. Responses will vary.
2. You can protect yourself by (any five): avoiding close contact with people who are ill; not sharing eating utensils, dishes, or glasses with others; washing hands often; keeping hands away from mouth, nose, and eyes; encouraging others to cover their mouth and nose when coughing or sneezing; taking precautions against dangers like ticks; using safe food-handling practices.
3. You can protect others by (any three): seeking medical treatment; covering your mouth and nose when you cough or sneeze; staying home during the contagious period of a disease; following directions for taking medication; encouraging others in safe health practices.

Answers to Thinking Critically

4. Responses will vary.
5. Responses will vary.

RETEACHING

▶ Assign Concept Map 57 in the TCR.
▶ Have students complete Reteaching Activity 57 in the TCR.

ENRICHMENT

Assign Enrichment Activity 57 in the TCR.

Ask: What fact from this lesson do you want other people to know? Why?

---

dence in their own abilities, and to improve their social health by improving their relationships with peers.

Meeting Student Diversity

Sharing information with others can be a special challenge for students who are shy or withdrawn. Encourage these students to begin slowly, simply by agreeing or disagreeing with information presented by others in small group settings. Remind students to focus on their own growth and their progress toward increased willingness and ability to share information.

Assessing the Work Students Do

Let students role-play various situations in which they share relevant health information. Then ask students to write brief assessments of their own participation and learning.

Focus

The Teen Health Digest articles can be used in two ways: as an individual activity for reflection and enrichment or as a cooperative learning activity as described below.

Motivator

Write the words *prevention* and *treatment* on the board, and help students review basic approaches to preventing and treating communicable diseases. Then ask students to read the title of each Teen Health Digest article and predict whether the information focuses on prevention or treatment. Write the title under the heading students identify. Guide students in reading and discussing the articles.

Cooperative Learning Project

Beware: A Board Game

Setting Goals

Tell students that they will use what they have learned about preventing and treating communicable diseases to create their own board game, "Beware."

Delegating Tasks

Have students work together to discuss and decide on the basic format of the game: its object, the form of the board, and the method of moving playing pieces on the board. Then let students sign up to participate in one of these groups: Game-Rule Writers, Researchers/Fact-Checkers, Game-Question Writers, Artists, or Designers.

TEEN HEALTH DIGEST

CON$UMER FOCU$

Health Hype

Q: Ads for cold remedies toss around phrases like "extra strength," "new and improved," and "special formula." Do such phrases mean anything?

A: Catchy phrases aren't much help to comparison shoppers because there are no standards regarding their meaning. For example, "extra strength" or "maximum strength" may just mean that a product contains more medication than similar products made by the same company. Suppose that the product is a multisymptom cold remedy. How can the buyer know which of the many ingredients has been increased to provide the "extra strength"? The smart way to shop for cold remedies is to compare ingredients and amounts. Then buy the simplest medications for your particular symptoms.

Try This: Examine the various over-the-counter cold remedies sold at the pharmacy. Which phrases do you see repeated most often on the product packaging?

348 Chapter 12: Understanding Communicable Diseases

Teens Making a Difference

Hoping to Save a Life

Although she's only 12, Hydeia Broadbent is one of the nation's best-known AIDS activists. She's spread her message about the dangers of AIDS to thousands of people. Hydeia has appeared in a National Institutes of Health video about AIDS, spoken with the president, and even addressed the Republican National Convention.

Hydeia knows about AIDS from personal experience. Her mother, who abandoned her shortly after Hydeia was born, was infected with HIV. Hydeia, too, is infected with the virus.

Hydeia looks forward to speaking to many more groups of people. Her reason is simple: "It might save someone's life."

Try This: Find out about some other ways in which individuals are fighting the spread of AIDS. Report this information to your class.

TECHNOLOGY OPTIONS

Computer If students have access to appropriate computer programs, they may want to use clip art or computer-generated pictures on their game board and game cards. In addition, they can use word processing programs to create and print the questions and answers for the game.

Camera or Video Camera Students may want to create a visual record of their work on the board game, photographing various steps in the development or videotaping parts of group meetings. If students take still photos, they can create an album to accompany their board game. Encourage students to borrow the album or videotape and to share this record with family members.

HEALTH UPDATE

Fleeing the Flu

Flu vaccines are given every year because flu viruses change from one year to the next. Some people resist getting annual vaccinations because they hate shots. However, studies suggest that flu vaccines—which work about 90 percent of the time—may, for many people, be worth a twinge of discomfort.

About one in five people gets the flu each year. Studies have shown that workers who get flu shots have 25 percent fewer cases of upper respiratory illness than those who don't. In addition, they lose 43 percent fewer days from work for illness.

Already used by millions of people in the United States each year, flu vaccines may soon become even more common. Within three years, the vaccines may be available in nasal sprays—a great relief for those who hate needles!

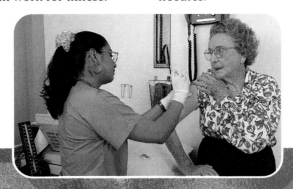

Myths and Realities

Feed a Cold? Not Too Much

Q. How much truth is there to the old saying, "Feed a cold, starve a fever"?

A. This idea shouldn't be taken too seriously. However, it's probably best to eat easily digestible foods when you're sick. Liquids are especially good choices, because they will prevent dehydration. Liquids also loosen nasal secretions, so that you'll feel better and may even get better sooner.

People at Work

Disease Detective

Alex Wong is a disease detective. His job is to study diseases to figure out what causes them and how they can be prevented. Alex is an epidemiologist (e·puh·dee·mee·AH·luh·jist).

Alex first became interested in studying diseases when he was in high school. He studied biology in college and then went on to get a master's degree in public health. Now he works for his state's health department, gathering information about how often different diseases occur in the state.

Asked what he likes about his job, Alex says, "You get to see the whole picture. When a doctor treats an individual for tuberculosis, he or she is interested in just that one person—what the patient's symptoms are, how the patient is responding to the treatment, and so on. An epidemiologist can look at *all* the cases of tuberculosis and see that more and more cases are not responding to the drugs used to treat it. That's important information that can help doctors with their specific cases."

Teen Health Digest 349

Chapter 12

The Work Students Do

Following are suggestions for implementing the project:

1. The different groups will want to coordinate their efforts; provide some class time for group-to-group conferences and for general meetings.
2. Remind students to keep game questions focused on the prevention and treatment of communicable diseases. Researchers/fact-checkers should make sure that each game question has a clear and correct answer.
3. Game-Rule Writers may want to work closely with Game-Question Writers: Should correct answers to certain questions be more "valuable" than others? How will these high-value questions be designated? Can players choose whether to draw a high-value question, or should the determination be left to chance? Encourage students to resolve these kinds of questions early in the project.
4. Suggest that students try playing their game at several points during the project. These run-throughs will allow them to identify problems and help them make the final version of their game fun and easy to play.
5. Be sure students have an opportunity to share their board game with other students and with family members.

Assessing the Work Students Do

Observe students as they work in their project groups, and note each student's contributions and interactions with others. At the end of the project, let the members of each group work together to evaluate the game itself and the group's contribution to the game.

Meeting Student Diversity

Language Diversity If several students share the same first language (other than English), suggest that the class may want to create a bilingual version of their board game. If non-native speakers represent a variety of language backgrounds, let students who may need assistance work as partners with capable students whose native language is English.

Gifted If the class includes several students gifted in understanding and using computers, suggest that they may want to develop a second version of the class board game to be played as a computer game.

349

Lesson 5

Sexually Transmitted Diseases and HIV/AIDS

Focus

LESSON OBJECTIVES

After studying this lesson, students should be able to
- identify and refute some common misconceptions about STDs.
- name and discuss common STDs.
- explain the relationship between HIV and AIDS.
- discuss the effects of AIDS.
- explain how HIV is—and is not—spread.

MOTIVATOR

Write *AIDS* on the board. Ask students to free-associate, listing all the words and phrases that come to mind.

INTRODUCING THE LESSON

Have volunteers share their responses to the word *AIDS*. Ask: Which responses are emotional? Which are facts about the disease? Why is it important to acknowledge your emotional responses to AIDS? to gather facts about the disease? Why do you think some teens don't want to know about this disease? What are some of the dangers associated with not having all the facts about AIDS? Tell students that they will learn more about AIDS and other communicable diseases as they study Lesson 5.

INTRODUCING WORDS TO KNOW

Have students skim the lesson to find the definition provided for each Word to Know. Then have them record the terms and their definitions in their notebooks.

Lesson 5: Sexually Transmitted Diseases and HIV/AIDS

This lesson will help you find answers to questions that teens often ask about sexually transmitted diseases and AIDS. For example:

- What is a sexually transmitted disease?
- What treatments are there for sexually transmitted diseases?
- What is the difference between HIV infection and AIDS?
- How is AIDS transmitted?
- What are the benefits of practicing abstinence before marriage?

Words to Know

STD (sexually transmitted disease)
chlamydia
gonorrhea
genital herpes
syphilis
AIDS (acquired immunodeficiency syndrome)
HIV (human immunodeficiency virus)

What You Should Know About STDs

STD stands for **sexually transmitted disease**—that is, *a disease spread from person to person through sexual contact*. Talking about sexually transmitted diseases makes many people feel uncomfortable or embarrassed. However, knowing about STDs gives you the power to avoid them. **Figure 12.7** shows some common misconceptions about STDs as well as the facts.

Figure 12.7
Get the Facts!

A lot of people have an "It can't happen to me" attitude about sexually transmitted diseases. In fact, STDs can happen to anyone who doesn't practice abstinence.

Myth: Young teens don't get STDs.

Fact: Some 12 million new cases of STDs are reported annually in the United States. About 3 million occur in teens.

Myth: Getting an STD is no big deal.

Fact: STDs can have serious health consequences, including sterility and even death.

Myth: You can always just take medicine to get rid of an STD.

Fact: Some STDs are curable, while others are not. All STDs, however, are preventable.

Myth: If you're careful, you won't get STDs.

Fact: The only sure protection against STDs is sexual abstinence—avoidance of sexual contact.

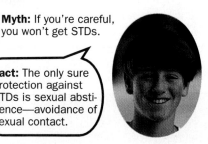

350 Chapter 12: Understanding Communicable Diseases

Lesson 5 Resources

Teacher's Classroom Resources
- Concept Map 58
- Enrichment Activity 58
- Lesson Plan 5
- Lesson 5 Quiz
- Reteaching Activity 58
- Transparencies 49, 50

Student Activities Workbook
- Study Guide 12
- Applying Health Skills 58
- Health Inventory 12

Common STDs

Sexually transmitted diseases differ from most other communicable diseases in a number of ways.

- Most STDs can be spread only through sexual contact, especially sexual intercourse.
- Someone who has an STD may not have visible symptoms, or may have symptoms that come and go.
- Having an STD once does not make you immune to the disease in the future.
- There are no vaccines available against STDs.
- STDs can make a person sterile or infertile.
- STDs can be prevented by avoiding sexual activity.

Many people who think that they may have an STD delay seeking medical attention. Some are embarrassed. Others hope that the symptoms will somehow just go away. Postponing medical treatment, however, is dangerous. Prompt diagnosis and treatment is the first step toward curing or controlling an STD *and* preventing its spread to others.

Chlamydia

Chlamydia (kluh·MI·dee·uh) is *a bacteria-caused STD that may affect the reproductive organs, urethra, and anus.* In 1995 chlamydia was the most frequently reported infectious disease in the United States. Nearly 4 million cases of the disease are reported each year, the majority of them in women.

About half of the people who have chlamydia do not even know that they have it. Symptoms, when they are present, may include a genital discharge and burning during urination. If it is not treated, chlamydia can cause pelvic pain, infertility, and various additional infections. Antibiotics can cure chlamydia.

Language Arts Connection

Infectious = Contagious? ACTIVITY!

Many people think that the terms "infectious" and "contagious" have the same meaning. Actually, an infectious disease is one that can be spread by any means, such as by infected animals or through contaminated food. A contagious disease, however, is one that can be directly transmitted from one person to another.

Find out which infectious diseases pose a danger to people. Share your findings with the class.

Abstaining from sexual activity before marriage is the only healthy choice for teens.

Lesson 5: Sexually Transmitted Diseases and HIV/AIDS   **351**

Personal Health

Teens should be concerned about the possibilities of contracting sexually transmitted diseases, but some teen girls may need to be informed that a vaginal discharge is not necessarily an indication of an STD, or even of an infection. A discharge of clear or white odorless fluid from the vagina is a normal part of adolescent development. The discharge is most noticeable during the year or two before a teen has her first menstrual period, but it shouldn't stop altogether, even after a teen is menstruating regularly. It may be helpful to remind students that, just as no one who has sexual contact is safe from STDs, no one who abstains from sexual contact is at risk from STDs.

Chapter 12 Lesson 5

L1 Applying Life Skills
Have the students identify appropriate sources of help for teens who wonder whether they might have a sexually transmitted disease.

L1 Critical Thinking
Ask: Why are teens—and adults—so often reluctant to acknowledge the symptoms of STDs? Why do many postpone seeking treatment? What are the dangers of this kind of behavior?

L1 Discussing
To help students consider the risks of chlamydia, ask questions such as these: Why do you think so many people contract chlamydia each year? What problems are associated with the lack of symptoms? In your opinion, who is likely to worry about having chlamydia? What would you recommend for those people?

L1 Discussing
Pose these questions to the class. What are possible long-term consequences of gonorrhea? Why is it important for anyone who might have gonorrhea to seek medical treatment? Why is it also important for people with gonorrhea (and any other STD) to discuss the disease with their partners?

L2 Guest Speaker
Invite a local health care provider who works with adolescents to visit the class. The provider might be a general practitioner, a family nurse-practitioner, or a physician's assistant. Ask the health professional to discuss with students the problems he or she sees with STDs among adolescents: How does he or she try to screen patients for STDs? How does he or she encourage patients to talk about STDs and recognize the risks? What advice does he or she give teen patients? Encourage students to listen attentively and to ask appropriate questions.

352

Gonorrhea
Gonorrhea (gah·nuh·REE·uh) is *a bacteria-caused STD that affects the genital mucous membrane and sometimes other body parts, such as the heart or joints.* Symptoms include a thick, yellow genital discharge and burning during urination. People often catch chlamydia and gonorrhea at the same time. If left untreated, gonorrhea can cause serious damage, including sterility in men and infertility in women. Doctors can treat gonorrhea with antibiotics.

Genital Herpes
Genital herpes (jen·i·tuhl HER·peez) is *an STD caused by a virus that produces painful blisters in the genital area.* An estimated 30 million Americans have this disease, although many may not realize that they do. While the signs of herpes may temporarily go away, the disease has no cure, and symptoms often recur. Herpes can be passed on to others when symptoms are present. It may also be contagious for some time before symptoms appear and after they are gone. Medications can relieve herpes symptoms.

Syphilis
Syphilis (SI·fuh·lis) is *a bacteria-caused STD that can affect many parts of the body.* Symptoms vary as the disease progresses. Early symptoms may include only a painless sore at the site where the disease entered the body and swollen lymph glands in the genital area. However, syphilis is a very dangerous disease. Over time, untreated syphilis may spread to the central nervous system and the heart, causing heart disease, insanity, and eventually death. If treated early, syphilis can be cured by antibiotics.

Your parents, doctor, and school nurse are good sources of health care information.

352 Chapter 12: Understanding Communicable Diseases

More About...

STDs Another common STD that students may need information about is pubic lice. These are tiny insects that thrive in warm, hairy environments such as the pubic area, and that can be spread through sexual contact. Pubic lice are sometimes called crabs because the insects have six legs. A case of pubic lice can be treated quite effectively with topical medicines, but the medicines do not prevent repeat infections. Like any other STD, the only sure protection against getting pubic lice is to avoid sexual contact.

Students should recognize that pubic lice are different from head lice. You may want to remind students that head lice are not an STD.

HIV and AIDS

AIDS, or **acquired immunodeficiency syndrome,** is *a deadly disease that interferes with the body's ability to fight infection. The virus that causes AIDS* is called **HIV,** which stands for **human immunodeficiency virus.** In 1996 more than 1.5 million people around the world died of AIDS.

A person who is infected by HIV may not show any signs of illness for a long time. In fact, an average of 10 years may pass before HIV infection leads to AIDS. Nevertheless, during this time the virus is doing serious damage to the infected person's immune system. When the system's defenses are critically weakened, AIDS develops. The body becomes unable to fight off other infections and diseases, which eventually prove fatal. **Figure 12.8** shows how HIV attacks the body.

Did You Know?
Frightening Numbers

AIDS has taken an estimated 4.5 million lives worldwide. Among Americans between the ages of 25 and 44, it is a leading cause of death.

Figure 12.8
How HIV Attacks the Body

AIDS is a progressive disease. Early symptoms of HIV infection may include fatigue, rash, fever, swollen lymph nodes, and diarrhea. Over time, the immune system of the HIV–infected person weakens. The person becomes vulnerable to other infections.

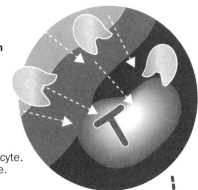

1 After the virus enters the body through the mucous membranes or through a break in the skin, HIV invades a "host" cell—a T-cell lymphocyte. The virus uses the cell's resources to reproduce. Eventually the host cell is destroyed.

KEY
- HIV
- Lymphocytes
- Germs

2 The virus multiplies within the body. More and more lymphocytes are destroyed. These cells are a key part of the immune system.

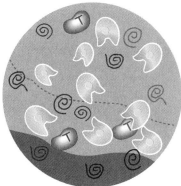

3 As the number of T-cells drops, the immune system gets weaker and weaker. The body loses its ability to resist infections and diseases that a healthy immune system could fight off.

Lesson 5: Sexually Transmitted Diseases and HIV/AIDS **353**

Chapter 12 Lesson 5

L2 Reporting
Let a pair of volunteers use library and Internet resources to learn about recent developments in the search for drugs to treat and/or cure genital herpes. Ask them to share their findings with the rest of the class. Then ask: If a cure for genital herpes is found, can people stop worrying about contracting this STD? Why or why not?

L2 Researching
Have a small group of volunteers read about the effects of untreated syphilis. Let them prepare and display a chart that presents their findings.

L1 Discussing
As students begin reading about AIDS, let volunteers share their own perceptions of the disease, either from knowing people who have it or from reading or hearing about the disease. Correct any misconceptions students express.

L1 Recalling
Ask students to recall what T-cells are and what they do. (If necessary, refer students to Lesson 2, Figure 12.3, on page 338.) Why are T-cell counts so important to AIDS patients?

Visual Learning
Use Figure 12.8 to help students study and discuss the attack HIV makes on the body.

Home and Community Connection

Does your community have a clinic or health center that deals with STDs and/or AIDS? If so, invite a representative from that organization to speak with students about the dangers of STDs and the value of abstinence from sexual activity before marriage.

If such a center is not available, students might be encouraged to contact the city, county, or state health department, or a local chapter of the Red Cross. These organizations usually provide information, including brochures and pamphlets, and answer questions for individuals.

Chapter 12 Lesson 5

L2 Current Events
Tell students to find, read, and clip newspaper articles about AIDS research, AIDS clinics, and AIDS patients. Let students display these articles on a classroom bulletin board.

L3 Current Events
Assign interested volunteers to compile the most recent statistics on HIV infection and AIDS deaths in the United States. What trends are noticeable? Ask these volunteers to prepare a graph or other display and to discuss their findings and conclusions with the rest of the class.

L3 Social Studies
Have students research the rates of AIDS infection and death around the world. In which areas is AIDS becoming an increasingly serious threat? Why? In which areas does the AIDS epidemic seem to be in its early stages? How are governments, doctors, and citizens trying to deal with the epidemic? Let students meet in groups to compare and discuss their findings.

L1 Comprehending
Help students discuss and understand the dangers of contracting HIV through sexual relations: Are heterosexual people safe from HIV? Why not? Why are people with multiple partners at highest risk?

L1 Critical Thinking
Ask the class why a person who knows he or she is infected with HIV might not tell a potential sexual partner. Encourage students to explore a variety of reasons why the person might not disclose that information.

L1 Analyzing
Ask students to explain why drug abusers are especially likely to share used needles.

Did You Know?

Stopping the Virus

Some of the most promising new drugs used to treat AIDS are drugs called protease inhibitors (PRO•tee•ays in•HI•buh•terz). These are drugs that prevent the virus from reproducing itself. Protease inhibitors, taken along with other AIDS-fighting drugs, have made it impossible to detect the virus in the blood of a large number of the patients treated. Unfortunately, these drugs are much too expensive for most AIDS patients to afford.

How HIV Is Spread

Infection with HIV nearly always occurs in one of the following four ways:

- **Having sexual relations with an infected person.** This is the most common way of getting AIDS. HIV circulates in the bloodstream and in body fluids. Therefore, an infected person can pass on the disease even if he or she has no symptoms. People who have sex with multiple partners are at highest risk. *Avoiding sexual contact is the only sure way to protect yourself against this kind of transmission.*

- **Using a contaminated needle.** A tiny amount of blood left in the needle can pass the virus from person to person. To protect yourself, avoid illegal drugs, and be sure that any procedure involving needles or blood is performed by a health professional.

- **Getting blood from an infected person.** Most people who were infected through blood transfusions were infected before 1986. In 1985 the Red Cross began routinely testing all donated blood for HIV.

- **Passing HIV from infected mother to fetus.** About one third of HIV-infected mothers transmit the infection to their children. The child usually dies within a few years.

How HIV Is *Not* Spread

People are afraid of getting AIDS—and they should be. However, don't let fear cause you to mistake myth for truth. HIV is *not* spread in the following ways:

- Through the air, by coughing or sneezing, for example
- By casual contact with an HIV-infected person—for example, by shaking hands or hugging
- By using the same sports equipment, clothing, towel, comb, or furniture as an infected person
- By using the same telephone, shower, bathtub, or toilet as an infected person
- By sharing eating utensils or dishes with an infected person (although other kinds of germs are capable of being spread this way)
- Through the bites of mosquitos, ticks, or other insects
- By swimming in the same pool as an infected person
- By donating blood

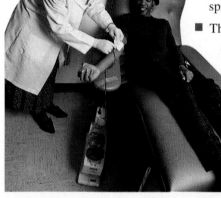

People do not risk getting HIV when they donate blood. A fresh, sterile needle is used for each donor.

354 Chapter 12: Understanding Communicable Diseases

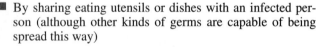

More About...

Sensitive Issues on the Internet
Although a wide range of educational resources are available on the Internet, some sites offer inappropriate materials for teens. Content-filtering software blocks access to these sites. Generally, these products come with a basic set of filters that screen out inappropriate sites and searches. The software vendors then offer updates of barred sites and services to keep the filters current. You can also customize the software to block additional sites and word searches. Some products even block or filter access to chat groups, which prevents teens from divulging personal information. See reviews in leading computer magazines for the best products and features for handling this sensitive issue.

The Continuing Battle Against AIDS

Scientists, health workers, educators, and many other people around the world are fighting against the AIDS epidemic. AIDS is a fierce enemy, but much progress has been made:

- **Drug therapy.** Combinations of powerful new drugs can dramatically reduce HIV levels in the blood and prolong the life expectancy of many HIV-infected people. Medical researchers have also reported some success in improving the immune systems of some HIV-infected people. However, many of the drugs now in use have serious side effects, and they do not work for all patients. The drugs are also much too expensive for everyone to afford. Research continues, with well over a hundred drugs currently under development.

- **Vaccine research.** Scientists are working intensively to develop an AIDS vaccine. However, the virus that causes the disease occurs in many forms. A vaccine that works against one form of HIV may not work against another form. Most researchers feel that the development of an effective vaccine for AIDS is still many years away.

- **AIDS education.** With neither a cure for AIDS nor a vaccine, educating people about AIDS is crucial. Health professionals, educators, and workers in federal, state, and local governments are making an all-out effort to teach the public how AIDS is spread and how the disease can be prevented.

Q & A

Testing for AIDS

Q: My cousin was tested for AIDS. The test was negative. Does that mean that he's okay?

A: That depends on the timing. A blood test searches for antibodies to HIV to determine whether you are infected. However, it may take three to six months after infection before antibodies can show up on the test.

Government and community programs are available to help people with HIV and AIDS.

Lesson 5: Sexually Transmitted Diseases and HIV/AIDS

Chapter 12 Lesson 5

L2 Reporting
Allow a small group of volunteers to prepare and present an oral report on injected illegal drugs. Ask: In addition to the risk of AIDS, what health risks do these drugs pose?

L3 Researching
Suggest that interested students investigate hemophilia: What is this disease? How is it treated? Why have so many hemophiliacs become infected with HIV? What has been done to prevent the spread of HIV to hemophiliacs? Ask these students to work together in preparing a written report, and make the report available to other class members.

L2 Language Arts
Have students read about the life and death of Ryan White, either in the book *Ryan White: My Own Story* or in articles or selections from books about AIDS. Provide class time in which students can discuss their reactions to Ryan White's experiences.

L2 Investigating
Suggest that a volunteer find out about the policy of your local school district toward students infected with HIV and/or AIDS: What restrictions apply? What special support services are available? How many infected students now attend—or have attended—schools in the district? What policy does the district have regarding teachers and/or other faculty members with HIV or AIDS? What reasons does the district give to support that policy? Let the student discuss his or her findings with the rest of the class.

L3 Current Events
Ask students to investigate laws that require pregnant women to be tested for AIDS: What purposes are such laws intended to serve? What objections are raised to these legal requirements? Have such laws been passed and enforced in your area? Have volunteers share their findings with the rest of the class.

TECHNOLOGY UPDATE

The Internet is widely used as a means of gathering and distributing information on many health-related subjects. Many organizations, both private and public, maintain health-related Web sites. Students can find these sites, as well as articles and personal stories, by using a search engine and inputting a key health term. However, students should be warned to evaluate the source of any information they find on the Internet. Suggest that students should "surf" the Internet with partners; they can help each other assess the reliability of the information they locate. Otherwise, students can be encouraged to discuss what they read with a parent or another trusted adult.

Chapter 12 Lesson 5

L2 Presenting Skits
Have students work in small groups to plan and present skits involving a teen who is infected with HIV. Students may want to show the infected teen discussing his or her health status with friends; they may want to show teens interacting with their infected friend; or they may have other ideas they want to explore. Provide time for students to watch and discuss all the skits.

L3 Researching
Ask students to read about the most recent International AIDS Conference: Where was it held? What issues were discussed, and what new research was presented? What controversies arose? With what results? Let students meet in groups to discuss what they have learned from this research.

L3 Current Events
Have class members use library and/or Internet resources to find reports on the most recent research into drug therapies and vaccine development for AIDS. Then have students meet in groups to summarize and discuss their findings. What do group members predict for advancements in AIDS research in the year ahead? in the coming decade?

L1 Critical Thinking
Let students share and discuss their responses to these questions: Why do teens so often feel pressured to become sexually active? What influence do TV shows, movies, and advertisements exert? What are the best reasons for resisting pressure to become sexually active? What are the most effective ways to resist pressure?

Teen Issues

Abstinence on the Rise

Across the country, more and more teens are practicing abstinence. The 1995 Survey of Family Growth—a government survey conducted every five years—found that the number of teen girls who had had sexual intercourse had dropped nearly 10 percent since the 1990 survey. Researchers believe that education about AIDS may be responsible for the decline.

Abstinence Before Marriage

STDs are extremely dangerous. They can even be deadly. The only sure way to protect yourself from STDs is to practice sexual abstinence. Because most STDs can be transmitted only through sexual contact, by avoiding sexual activity you will be able to avoid the dangers of STDs.

Although the pressure to engage in sexual activity can be very strong, abstinence is the only healthy decision. Every day you are exposed to messages in the media that give the impression that everyone is having sex. Sex is often portrayed as an activity that is fun, exciting, safe, and very important. Sex, however, is the wrong choice for teens. It can lead to unintended pregnancy as well as STDs. It can also cause emotional problems for people who decide to have sex before marriage.

It is important to recognize the many benefits of choosing sexual abstinence before marriage. Some of these include:

- You will avoid the risk of unplanned pregnancy.
- You will avoid the risk of getting a sexually transmitted disease.
- You will be respecting the wishes of your parents.
- You will be focused and committed to long-term goals such as education, career, and family.

Physical attraction to another person is a normal part of life. However, it is important not to let these feelings control your decisions. You can acknowledge your feelings without acting irresponsibly on them. You can also express affection for another person in ways that don't risk your health or compromise your values. By practicing abstinence, you can be sure that your relationships with others are built on shared interests and mutual trust, rather than on sex. When you are an adult, you may choose to marry. By choosing abstinence now, you will know that you can safely enjoy sexual activity within a mutually faithful relationship with your husband or wife.

Teens can express their feelings in many healthy ways.

356 Chapter 12: Understanding Communicable Diseases

MAKING HEALTHY DECISIONS

Deciding Whether to Speak Up

Focus on Healthy Decisions
Let students share their responses to questions such as these: What are you likely to do if, while you and a friend are discussing a homework assignment, you realize that your friend has written down the wrong page number? Suppose you and some friends are planning to go to a movie, but one friend thinks the movie starts at 4:30 and you know that it starts at 5:15. What will you say? When is it easy to correct a friend's misinformation or misunderstanding? Why is it harder to try to correct friends when they are misinformed about AIDS? After this introductory discussion, let students read the Making Healthy Decisions feature independently.

Activity
Use the following suggestions to discuss Follow-up Activity 1.

Step 1 Michael must decide whether to speak up and explain some of the facts about AIDS to his friends, who misunderstand how the disease is spread.

Review — Lesson 5

Using complete sentences, answer the following questions on a separate sheet of paper.

Reviewing Terms and Facts

1. **Vocabulary** What is *chlamydia*? How is it treated?
2. **Compare** What is the difference between *HIV* and *AIDS*?
3. **Recall** Identify four ways in which HIV is spread.

Thinking Critically

4. **Analyze** What factors do you think make STDs especially dangerous?
5. **Explain** Why should teens abstain from sexual activity?

Applying Health Concepts

6. **Growth and Development** Chlamydia, AIDS, herpes, and many other STDs can spread from a pregnant woman to her baby. Do some research to find out what kinds of effects these STDs have on a child during its development. Write a report summarizing your findings.

MAKING HEALTHY DECISIONS
Deciding Whether to Speak Up

At school, Michael eats lunch at a table with four other students. His friends have recently been talking about AIDS and how the disease is spread. Michael's neighbor has AIDS, so Michael knows that much of what his friends are saying is wrong. They think that AIDS can be spread in all sorts of ways that are actually impossible. One student even said that anyone who is infected with HIV should not be allowed in school because "then everyone would catch it." No one at the table disagreed—or at least, no one said anything.

Michael wants to explain to his friends that they are mistaken. However, he's afraid that they will all tell him that *he's* wrong. They might even reject him as a know-it-all. Still, Michael feels that speaking up would be the right thing to do. To figure out what to do, Michael uses the six-step decision-making process:

❶ **State the situation**
❷ **List the options**
❸ **Weigh the possible outcomes**
❹ **Consider your values**
❺ **Make a decision and act**
❻ **Evaluate the decision**

Follow-up Activities

1. Apply the six steps of the decision-making process to Michael's dilemma.
2. As a class, discuss the best ways to go about correcting myths about AIDS when you hear people spreading them.

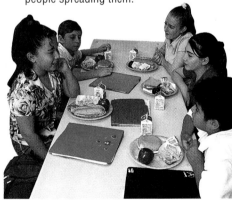

Lesson 5: Sexually Transmitted Diseases and HIV/AIDS

Chapter 12 Review

CHECKING COMPREHENSION

Use the Chapter Summary and the Chapter 12 Review to help students go over the most important ideas presented in Chapter 12. Encourage students to ask questions and add details as appropriate.

CHAPTER 12 REVIEW ANSWERS

Reviewing Key Terms and Concepts

1. A disease is a condition that interferes with the proper functioning of the body or mind.
2. See Figure 12.1 on page 333.
3. Lymphocytes are special white blood cells in the lymph. There are two types: B-cells and T-cells.
4. The lymphatic system carries B-cells and T-cells. B-cells produce antibodies that attach themselves to an invading germ. The antibodies attract T-cells, which surround and destroy the germ.
5. Influenza is a communicable disease typically characterized by fever, chills, fatigue, headache, muscle aches, and respiratory problems.
6. Hepatitis is a virus-caused liver inflammation characterized by yellowing of the skin and of the whites of the eyes.
7. Responses will vary. Possible response: To wash your hands correctly, you should use liquid soap, wash for at least 20 seconds, and use a nail brush.
8. Responses will vary. Possible response: Healthy lifestyle choices include exercising regularly, getting an adequate amount of sleep, not using alcohol or other drugs, and not smoking.
9. Four STDs are chlamydia, gonorrhea, genital herpes, and syphilis.
10. See page 354.

Chapter 12 Review

Chapter Summary

- **Lesson 1** Communicable diseases are caused by germs. These germs can be spread in several ways, including direct and indirect contact with other people and contact with animals or insects.
- **Lesson 2** The body's immune system protects against disease. People develop immunity through exposure to germs and through vaccination.
- **Lesson 3** The cold is the most common communicable disease. Other common communicable diseases include the flu, mononucleosis, hepatitis, and tuberculosis.
- **Lesson 4** Good health habits can help protect you and other people against infection and disease.
- **Lesson 5** STDs are sexually transmitted diseases, which include chlamydia, gonorrhea, and AIDS. HIV, the virus that causes AIDS, weakens and eventually destroys the immune system.

Reviewing Key Terms and Concepts

Using complete sentences, answer the following questions on a separate sheet of paper.

Lesson 1

1. Define the word *disease*. Use it in an original sentence.
2. List three diseases that are caused by viruses.

Lesson 2

3. What are *lymphocytes*?
4. How does the lymphatic system protect the body against disease?

Lesson 3

5. Describe the disease referred to as *influenza*.

Thinking Critically

Using complete sentences, answer the following questions on a separate sheet of paper.

11. **Analyze** How can understanding the ways in which germs are spread help people stay healthy?

6. What is *hepatitis*?

Lesson 4

7. List three of the procedures you should follow to wash your hands thoroughly.
8. Give four examples of healthy lifestyle choices you can make to keep your body resistant to disease.

Lesson 5

9. List four sexually transmitted diseases besides AIDS.
10. Identify four ways in which AIDS is *not* spread.

12. **Explain** How does exposure to germs enable the body to develop immunity?
13. **Describe** Explain how HIV damages the immune system and causes this system to stop functioning.

358 Chapter 12: Understanding Communicable Diseases

Meeting Student Diversity

Language Diversity Use the following suggestions to help students who have difficulty with English:

- Pair those learners with native speakers of English who can restate the Chapter Summary in language that helps students comprehend important concepts.
- Direct auditory learners or those students with language diversity to the Teen Health Audiocassette Program. Available in English and Spanish, this component provides an audio and written summary of the chapter.

Chapter 12 Review

Your Action Plan

Protecting yourself against communicable diseases is an important part of your total health. Are you doing all you can to prevent disease?

Step 1 Review your journal entries for this chapter. Make a list of ways in which you can help yourself remain healthy.

Step 2 Prioritize the elements of the list. Give each one a rating of 1 for "very important," 2 for "fairly important," and 3 for "less important."

Step 3 Set a long-term goal that will help you address all the points that you have rated with a 1. Set several short-term goals that will help you gauge your progress toward that goal.

As you achieve your goals, you will be rewarded by getting sick less often and feeling healthy and strong. Next, try setting a new long-term goal—one that will cover the other points on your list.

In Your Home and Community

1. **Personal Health** Are you and your family members doing all you can to protect yourselves and others against germs and disease? Which behaviors need to be changed? What additional actions might be taken? Work with family members to develop a checklist of health habits to follow in your home.

2. **Health of Others** Create an illustrated poster entitled "AIDS: Fact and Fiction." Include basic information about the disease, explaining how it is—and is not—spread. List at least two reliable sources of additional information. Discuss with your teacher where you might display your poster.

Building Your Portfolio

1. **Vaccine Research** Do research to learn more about vaccines that scientists are working to develop. For example, what progress have researchers made in developing a vaccine against Lyme disease? Against the different strains of flu? Summarize your findings in a report. Include the report in your portfolio.

2. **True or False Test** Create a true-false test asking ten basic questions about STDs. Give the test to at least six peers (not in your class) or adults. Evaluate the results. Make a chart summarizing your findings. At the bottom, write a paragraph suggesting ways to communicate information about STDs to teens. Add your chart to your portfolio.

3. **Personal Assessment** Look through all the activities and projects you did for this chapter. Choose one or two that you would like to include in your portfolio.

Chapter 12: Chapter Review **359**

Performance Assessment

▶ **Self-evaluation** Direct students to review the activities that are provided throughout the chapter. Encourage each student to select one finished product or activity that demonstrates his or her best work for the chapter. Have students explain what they learned and how the examples they selected show their progress.

▶ **Teacher's Classroom Resources** Assign Performance Assessment 12, "Skit on Avoiding Communicable Diseases," in the TCR.

Chapter 12 Review

Thinking Critically
Responses will vary. Possible responses are given.

11. You can use your knowledge to take actions necessary to avoid germs and disease. If you know, for example, that direct contact such as shaking hands and then touching your nose can transmit disease, you will keep your hands away from your nose, especially after shaking hands.

12. Your body produces antibodies against germs. In many cases, the antibodies stay in your blood. White blood cells "remember" the germs encountered. If the same germs return, your body can fight them off more quickly.

13. HIV destroys T-cells in order to reproduce. As HIV reduces the number of T-cells, fewer remain to fight off the other germs that invade the body. Eventually the body cannot fight off any infections.

RETEACHING
Assign Study Guide 12 in the Student Activities Workbook.

EVALUATE
▶ Use the reproducible Chapter 12 Test in the TCR, or construct your own test using the Testmaker Software.

▶ Use Performance Assessment 12 in the TCR.

EXTENSION
What programs does your community offer to help prevent and treat communicable diseases? These may include free or low-cost vaccinations in malls, information about STDs at health fairs, and special clinics designed to meet the needs of certain populations, including teens. Depending on your community, emergency rooms, medical centers, and doctors' offices can also provide help. Let students work together to compile a complete list of local resources and then to publish and distribute that list.

Planning Guide

Chapter 13
Understanding Noncommunicable Diseases

	Features	Classroom Resources

Lesson 1: Understanding Heart Disease
pages 362–366

- Concept Map 59
- Cross-Curriculum Activity 25
- Enrichment Activity 59
- Lesson Plan 1
- Lesson 1 Quiz
- Reteaching Activity 59
- Transparency 51

Lesson 2: Understanding Cancer
pages 367–372

Life Skills: Dietary Guidelines to Lower Cancer Risk
pages 370–371

- Concept Map 60
- Cross-Curriculum Activity 26
- Decision-Making Activity 25
- Enrichment Activity 60
- Lesson Plan 2
- Lesson 2 Quiz
- Reteaching Activity 60
- Transparencies 52, 53, 54

Lesson 3: Allergies and Asthma
pages 373–377

Health Lab: Determining Lung Capacity
pages 376–377

pages 378–379

- Concept Map 61
- Decision-Making Activity 26
- Enrichment Activity 61
- Lesson Plan 3
- Lesson 3 Quiz
- Reteaching Activity 61
- Transparency 55

Lesson 4: Other Noncommunicable Diseases
pages 380–385

Making Healthy Decisions: Coping with a Relative Who Has Alzheimer's Disease
pages 384–385

- Concept Map 62
- Enrichment Activity 62
- Health Lab 13
- Lesson Plan 4
- Lesson 4 Quiz
- Reteaching Activity 62

National Health Education Standards

One of the main goals of the National Health Education Standards is to move students toward "health literacy." Health literacy is the capacity of individuals to obtain, interpret, and understand basic health information and services and the competence to use such information and services in ways which promote health. The health standards were developed by applying the characteristics of a well-educated, literate person within the context of health. The health literate person is:
- a critical thinker and problem solver
- a responsible, productive citizen
- a self-directed learner
- an effective communicator

Listed below are the Health Education Standards Performance Indicators addressed in each lesson of this chapter.

Lesson	Health Standards Performance Indicators
1	(1.1, 1.3, 1.6, 1.7, 1.8, 3.1)
2	(1.1, 1.6, 1.7, 1.8, 3.1)
3	(1.7, 1.8, 3.1)
4	(1.7, 1.8, 3.1, 6.1)

ABCNEWS InterActive Videodisc Series

You may wish to use the videodisc *Tobacco* with this chapter. See side one, video segments 4, 5, 6, 7, 8, 9. Use the *ABCNews InterActive™ Correlation Bar Code Guide* for title reference. Also available in VHS format.

Chapter Resources

Teacher's Classroom Resources
- Chapter 13 Test
- Parent Letter and Activities 13
- Performance Assessment 13
- Testmaker Software

Student Activities Workbook
- Study Guide 13
- Applying Health Skills 59–62
- Health Inventory 13

Student Diversity Strategies
- Audiocassette Program (English)
- Audiocassette Program (Spanish)
- Spanish Parent Letters
- Spanish Summaries, Quizzes, and Activities

Multimedia Components
- English Audiocassette Program
- Spanish Audiocassette Program
- *Teen Health* Videodisc/VHS Series
- *Teen Health* Video Kit: *Teens with Cancer*

Other Resources

Readings for the Teacher
The American Lung Association Asthma Advisory Group. *Family Guide to Asthma and Allergies.* Boston: Little, Brown & Co., 1997.

Baldwin, J. *To Heal the Heart of a Child: Helen Taussig, M.D.* New York: Walker and Co., 1992.

LeVert, Suzanne. *Teens Face to Face with Chronic Illness.* New York: J. Messner, 1993.

Readings for the Student
Gold, John Coopersmith. *Heart Disease.* Parsippany, NJ: Crestwood House, 1996.

Semple, Carol McCormick. *Diabetes.* Parsippany, NJ: Crestwood House, 1996.

Silverstein, Alvin, et al. *Asthma.* Springfield, NJ: Enslow Publishers, 1997.

Out of Time?

If time does not permit teaching this chapter, you may use these features: Life Skills on pages 370–371; Health Lab on pages 376–377; Teen Health Digest on pages 378–379; Making Healthy Decisions on pages 384–385; and the Chapter Summary on page 386.

Chapter 13
Understanding Noncommunicable Diseases

CHAPTER OVERVIEW

Chapter 13 introduces the most common noncommunicable diseases and helps students understand their causes, prevention, and treatment.

Lesson 1 discusses the various forms of heart disease and explains what teens can do to protect the health of their hearts.

Lesson 2 helps students understand the most common types of cancer and discusses what they can do to reduce their cancer risks.

Lesson 3 presents information on the causes, effects, and treatment of allergies and asthma.

Lesson 4 discusses the diagnosis and treatment of rheumatoid arthritis, osteoarthritis, both types of diabetes, and Alzheimer's disease.

PATHWAYS THROUGH THE PROGRAM

For most students, the major noncommunicable diseases introduced in this chapter are not a current concern. However, knowing about the causes and treatments of such diseases can encourage students to take positive steps that will help protect their physical health as they mature and even as they age. In addition, understanding these diseases will make students more compassionate for others—peers, family members, and acquaintances—who are dealing with such diseases. Thus, knowledge and understanding can enhance teens' mental/emotional health and social health.

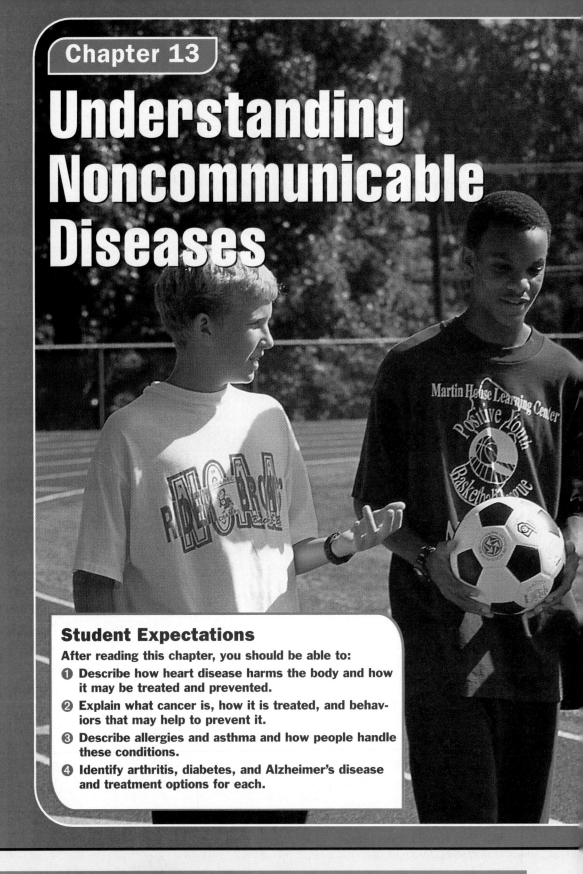

Chapter 13
Understanding Noncommunicable Diseases

Student Expectations
After reading this chapter, you should be able to:
1. Describe how heart disease harms the body and how it may be treated and prevented.
2. Explain what cancer is, how it is treated, and behaviors that may help to prevent it.
3. Describe allergies and asthma and how people handle these conditions.
4. Identify arthritis, diabetes, and Alzheimer's disease and treatment options for each.

Key to Ability Levels

Teaching strategies that appear throughout the chapter have been identified by one of three codes to give you an idea of their suitability for students of varying learning styles and abilities.

L1 **Level 1** strategies should be within the ability range of all students. Full class participation is often required. Teacher direction is usually needed.

L2 **Level 2** strategies are for average to above-average students or for small groups. Some teacher direction is necessary.

L3 **Level 3** strategies are designed for students able and willing to work independently. Minimal teacher direction is necessary.

Teen Chat Group

Tia: What's the matter, Zach?

Zach: I'm worried about my sister. We just found out that she has arthritis.

Troy: But she's only eight. I thought only older people got arthritis.

Zach: The doctor told my parents that people of any age can get arthritis—even kids.

Tia: That's not something you can catch, right?

Zach: No, you can't catch it from someone else.

Troy: Will your sister be okay?

Zach: I think so. My parents and I have to make sure that she takes her medicine, does her exercises, and gets plenty of rest.

Tia: Will she be able to play sports?

Zach: I don't know. Her doctor said we'll have to see how her disease responds to the medication.

Read the dialogue on this page. How much do you know about conditions such as arthritis, heart disease, and cancer? Start your private journal entries on noncommunicable diseases by responding to these questions:

▶ Have you ever known someone young who had arthritis?

▶ Do you know anyone who has any other noncommunicable disease? How does he or she manage the condition?

▶ Which noncommunicable diseases do you think that people can prevent with a healthy lifestyle?

When you reach the end of the chapter, you will use your journal entries to make an action plan.

Chapter 13: Understanding Noncommunicable Diseases

Chapter 13

INTRODUCING THE CHAPTER

▶ Let students recall TV shows or movies in which characters have to deal with a heart attack, cancer, or another major noncommunicable disease. How does the patient respond? How do the patient's family and friends respond? What medical treatment is made available? With what effect? How realistic do you consider the depiction of the disease and of the characters' responses to it? After students have shared their ideas, explain that they will learn more about such diseases as they read the lessons in Chapter 13.

▶ Ask students to recall what they have already learned about communicable and noncommunicable diseases: Why is it important to understand the differences between these two categories of disease? What specific noncommunicable diseases can you name? What do you know about the causes and treatment of those diseases? How can knowing more about noncommunicable diseases help you protect your own health? How can it help you encourage friends and family members to protect their health? Tell students this is the kind of information they will acquire as they study Chapter 13.

Cooperative Learning Project

Now Hear This: Public Service Announcements

The Teen Health Digest on pages 378 and 379 provides students with high-interest articles related to the content of this chapter.

The material in the Teacher's Wraparound Edition presents suggestions for a class project in which students plan and present a series of public service announcements on prevention and treatment of noncommunicable diseases.

Let three volunteers read the Teen Chat Group conversation aloud for the rest of the class, and ask students to respond to Zach's situation: Why is he upset? What do you think of his reaction to his sister's illness? Of his friends' concerns for her? Then encourage students to name other noncommunicable diseases—those you can't catch. What would a diagnosis of one of those other diseases mean for the patient? for family members? Finally, let volunteers share their ideas in response to the In Your Journal questions, and provide time in which students can write their private journal entries.

Lesson 1
Understanding Heart Disease

Focus

LESSON OBJECTIVES

After studying this lesson, students should be able to

▶ explain how noncommunicable diseases differ from communicable diseases and explain the three categories of noncommunicable disease.

▶ discuss the dangers of heart disease and identify the four major problems of the cardiovascular system.

▶ explain methods of combating heart disease.

MOTIVATOR

Ask students: What does your heart do for the rest of your body? Give students time to list in writing as many responses as possible.

INTRODUCING THE LESSON

Let volunteers share their responses to the Motivator activity. Encourage students to share what they may already know about heart disorders, and then explain that they will learn more about heart disease—and how they can help avoid it—as they study Lesson 1.

INTRODUCING WORDS TO KNOW

Have volunteers read the Words to Know aloud. Then let pairs of students work together to make two cards for each Word to Know, writing the term on one card and the definition on the other. Provide time for students to use these cards to play "Concentration," matching each term with its definition.

Lesson 1 Understanding Heart Disease

This lesson will help you find answers to questions that teens often ask about heart disease. For example:

▶ How does heart disease harm the body?
▶ How is heart disease treated?
▶ What can I do to avoid heart disease?

Words to Know

chronic
stroke
heart attack
hypertension
atherosclerosis
arteriosclerosis
angioplasty

Cultural Connections

Inherited Conditions

Some diseases are found almost only among people of a particular origin. Tay-Sachs disease, a disorder that destroys brain cells, generally affects only Jewish people whose families came from a particular area of Eastern Europe. Sickle-cell anemia is a blood disorder that affects mainly people whose families come from Africa, India, and Middle Eastern and Mediterranean countries.

Noncommunicable Disease

A noncommunicable disease is a disease that cannot be spread from person to person. Common noncommunicable diseases include heart disease, cancer, asthma, and Alzheimer's disease. These diseases are *not* caused by germs that can be spread by contact. Instead, they are caused by changes within the body.

Most noncommunicable diseases are **chronic** (KRAH·nik), or *present continuously or on and off over a long time*. These diseases generally fall into one of the following three categories:

■ Diseases present from birth
■ Diseases resulting from lifestyle behaviors
■ Diseases caused by the person's environment

Many noncommunicable diseases involve one or more of these factors. Some conditions, however, do not fit into any of these categories because their causes have not been determined. **Figure 13.1** on the following page provides more information about noncommunicable diseases.

Some noncommunicable diseases are caused by environmental factors, such as leaks of hazardous chemicals.

362 Chapter 13: Understanding Noncommunicable Diseases

Lesson 1 Resources

Teacher's Classroom Resources

📁 Concept Map 59
📁 Cross-Curriculum Activity 25
📁 Enrichment Activity 59
📁 Lesson Plan 1
📁 Lesson 1 Quiz

📁 Reteaching Activity 59
🖨 Transparency 51

Student Activities Workbook

📁 Study Guide 13
📁 Applying Health Skills 59

Heart Disease

Heart disease is any condition that reduces the strength or functioning of the heart and blood vessels. Heart disease kills more American adults than any other cause of death. In addition to those who die from it, more than 60 million Americans are living with some form of heart disease.

How Heart Disease Kills

Heart disease is responsible for more than 42 percent of all deaths each year. In the United States, someone dies from heart disease every 33 seconds. **Figure 13.2** on the following page shows the types of heart disease and their causes.

interNET CONNECTION
Good health habits are key to preventing many diseases. Unlock a treasure chest of health tips on the Internet.
http://www.glencoe.com/sec/health

Figure 13.1
Types of Noncommunicable Disease
Noncommunicable diseases can be organized into several different categories, based on their causes.

Type	Description	Examples
Present at birth	Diseases that are caused by hereditary factors or that occur because of problems during a baby's development or birth.	• Cerebral palsy • Cystic fibrosis • Muscular dystrophy • Sickle-cell anemia • Tay-Sachs disease
Lifestyle behaviors	Diseases that occur more often in people with unhealthful habits or lifestyles, including poor food choices, obesity, lack of exercise, smoking, drinking alcohol, and poor stress management.	• Heart disease • Many types of cancer • High blood pressure
Environmental causes	Diseases that are caused by exposure to the hazards around us, including air pollution, toxic wastes, asbestos, and secondhand smoke.	• Allergies • Asthma • Many types of cancer
Unknown causes	Diseases whose causes remain unknown.	• Alzheimer's disease • Arthritis • Chronic fatigue syndrome • Multiple sclerosis

Lesson 1: Understanding Heart Disease **363**

Chapter 13 Lesson 1

Teach

L1 Comprehending
Help students discuss the differences between noncommunicable and communicable disease: Is there ever any need to worry about "catching" a noncommunicable disease? Do you think you should be careful about exposing a person with a noncommunicable disease to your own cold or other communicable disease? Why?

L1 Life Skills
Let students describe the scene in the photograph near the bottom of page 362: Are there any chemical leaks or other hazards in our community? If so, who is responsible for them? What can you and other community members do to help get them removed or cleaned up?

Visual Learning
Guide students in reading and discussing Figure 13.1. Allow volunteers to share any experiences they have had with the specific diseases listed: Do you know anyone with cystic fibrosis (or another specific disease)? What health problems does that person have? How do those health problems affect the person's everyday life?

L3 Researching
Let interested volunteers research diseases listed in Figure 13.1. Have them write and illustrate short reports to be displayed in the classroom.

L1 Critical Thinking
Have students consider the unhealthful habits noted in the *Lifestyle behaviors* section of Figure 13.1. Ask: Can you be certain of preventing these diseases if you avoid all the unhealthful habits? What *can* you accomplish by avoiding the unhealthful habits?

Personal Health

Students should recall (or be reminded) that the heart is a muscle. Like other muscles, the heart grows stronger and healthier with appropriate exercise. An active lifestyle—less time in front of the TV and more time engaged in sports and other activities—is the key for most teens. They should also work on establishing a habit of regular aerobic exercise, such as brisk walking, running, cycling, swimming, or aerobic dancing, several times each week. Teens should be encouraged to find forms of exercise they enjoy and can share with friends or teammates; these are the exercises that are most likely to become lifelong habits.

Chapter 12 Lesson 1

L2 Math
Assign a small group of volunteers to research the ages at which men (or women) die of heart disease. Ask them to plan and draw a circle graph that shows the percent of deaths caused by heart disease among patients of various age ranges (such as 25 to 40, 40 to 50, 50 to 60, and over 60).

Visual Learning
Help students read and discuss the information in Figure 13.2. Guide them in locating the correct organ or system as volunteers read aloud about each problem of the cardiovascular system. Encourage other volunteers to restate the explanation in their own words.

L2 Reporting
A transient ischemic attack, or TIA, is a very serious warning that a patient may suffer a stroke. In fact, a TIA is actually a tiny stroke caused by a brief, temporary loss of blood supply to part of the brain. Let a group of students read about the symptoms of TIAs and the importance of seeking treatment. Have them report their findings to the rest of the class.

L2 Reporting
Ask students to research the relationship between salt in the diet and high blood pressure. How much salt do we need? How much do most people consume? Let students share and discuss their findings.

Health Minute
The primary indicator of a heart attack is usually described as a crushing pain in the middle of the chest; the pain may spread into the neck and jaw and along the arms. There are other causes of chest pain that teens are much more likely to experience: strained muscles, inflamed muscles, and anxiety, which can cause periods of breathlessness as well as pain.

364

Figure 13.2
Problems of the Cardiovascular System
Heart disease affects more than just the heart itself.

A Stroke
A **stroke** occurs when *part of the brain is damaged because the blood supply to the brain is cut off.* Damage occurs because blood normally carries oxygen to the brain, and when the blood supply is cut off, the brain does not receive enough oxygen. Some strokes are caused by a blood clot that blocks an artery leading to the brain. Other strokes are caused when an artery in the brain breaks. Depending on what part of the brain is affected, a person who has a stroke may have trouble moving or speaking. A stroke can also cause death.

B Heart Attack
In healthy arteries, blood flows to the heart muscle and supplies it with oxygen. Sometimes, however, an artery becomes blocked. This blockage is often caused by a buildup of fatty substances on the artery walls. A **heart attack** occurs when *the blood supply to the heart slows or stops and the heart muscle is damaged.* A heart attack is a sign of coronary (KOR·uh·nehr·ee) heart disease.

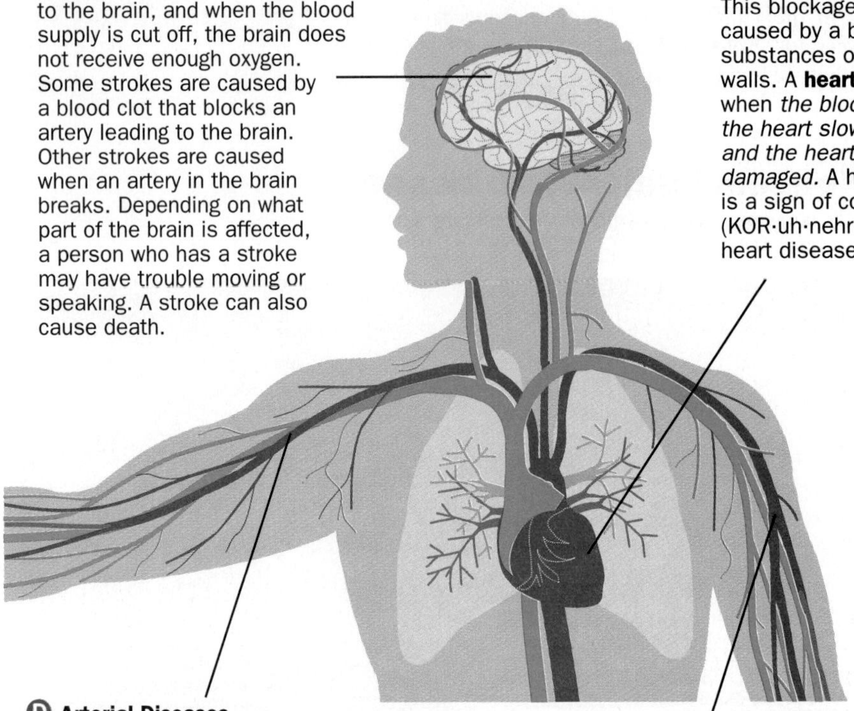

D Arterial Diseases
For the heart to function normally, blood must be able to flow through the arteries freely. A slowing or stopping of the blood flow can cause serious problems. *A condition in which fatty substances build up on the inner lining of arteries* is called **atherosclerosis** (a·thuh·roh·skluh·ROH·sis). High levels of cholesterol are associated with atherosclerosis.

Another arterial disease is **arteriosclerosis** (ar·tir·ee·oh·skluh·ROH·sis), or *hardening of the arteries.* Some hardening of the arteries is a natural part of the aging process. Arteriosclerosis causes the blood flow through the arteries to slow down. If atherosclerosis is also present, it worsens the problems caused by arteriosclerosis. These arterial diseases can lead to a heart attack or stroke.

C Hypertension
As the heart beats, it is pumping blood throughout the body. The force of the blood on the inside walls of the blood vessels is called blood pressure. Your blood pressure varies with certain factors. For example, it may go down when you are sleeping, and will increase when you are exercising. If, however, *a person's blood pressure stays at a level that is higher than normal,* he or she has **hypertension** (hy·per·TEN·shuhn). Another term for hypertension is *high blood pressure.* This condition can lead to serious health problems, including heart attack and stroke.

364 Chapter 13: Understanding Noncommunicable Diseases

Cooperative Learning

Approximately 1 baby in every 125 is born with some form of heart defect. These are some of the most common problems:
- patent ductus arteriosus
- septal defects
- coarctation of the aorta
- heart valve abnormalities
- tetralogy of Fallot
- transposition of the great arteries

Have students form cooperative groups, and ask each group to select one kind of congenital heart defect. Let group members work together to learn more about that defect and to prepare a poster or model explaining the defect and how it can be treated.

Combating Heart Disease

Medical science has made great strides in the treatment of heart disease. Today, medications are used to treat some heart problems and to help prevent others. In addition, surgical procedures have been developed to correct some of the more severe types of heart disease.

What Doctors Can Do

When treating heart disease, doctors have several options. The choice depends on many factors, such as the person's age, the type of heart disease, and the extent of the problem. Possible treatments include the following:

- **Angioplasty.** Sometimes used to treat severe atherosclerosis, **angioplasty** (AN·jee·uh·plas·tee) is *a surgical procedure in which an instrument with a tiny balloon attached is inserted into the artery to clear a blockage.* When the balloon is inflated, it crushes the deposit that was blocking the artery. At the end of the procedure, doctors deflate and remove the balloon.

- **Medication.** In some people, blood vessels become blocked, and blood clots form. Doctors may use medication to dissolve these clots and to stop new clots from forming. Medication is also prescribed for people with hypertension. Regular exercise and a reduction in stress are usually suggested as well.

- **Pacemaker.** Some people have an irregular or weak heartbeat. A pacemaker (PAYS·may·ker) is an electrical device that can be surgically inserted in the body. It sends pulses to the heart to make it beat regularly.

- **Bypass surgery.** When an artery is blocked, doctors can perform bypass surgery to create new paths for the blood to flow around the blockage. During this procedure, surgeons take a vein from the person's leg. They reattach it in the area around the heart so that it takes the blood on a detour around the blocked area.

- **Heart transplant.** If a person's heart has been severely damaged by heart disease, he or she may need a new heart. Surgeons can replace a diseased heart with a healthy one that has been donated.

Did You Know?

Grape Aid

A recent study showed that drinking a glass of grape juice every day is beneficial to your heart. Purple grape juice has a positive effect on platelets, a type of blood cell. The juice makes the platelets less likely to form blood clots. This, in turn, decreases the risk of a heart attack.

Science Connection

Another Risk Reducer? ACTIVITY

Scientists recently discovered that folic acid, a B vitamin, helps protect against heart disease. Folic acid lowers blood levels of homocysteine (hoh·moh·SIS·tuh·een), an amino acid that has been linked to arteriosclerosis, heart attacks, and strokes. Examine a book on nutrition to find out which foods are good sources of folic acid.

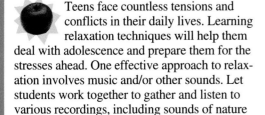

High blood pressure (hypertension) can usually be kept under control with medication.

Lesson 1: Understanding Heart Disease 365

Personal Health

Teens face countless tensions and conflicts in their daily lives. Learning relaxation techniques will help them deal with adolescence and prepare them for the stresses ahead. One effective approach to relaxation involves music and/or other sounds. Let students work together to gather and listen to various recordings, including sounds of nature and various forms of soothing music. Encourage students to discuss their own responses to each selection. Remind students that they should be listening for sounds that help them feel relaxed; not everyone will have the same responses. Then suggest that students choose the selections they find most relaxing and record them to make their own personalized relaxation tapes.

Chapter 12 Lesson 1

L2 Creating Illustrations

Suggest that a group of students do further reading on atherosclerosis. Then have them work together to draw a cross section of an artery in which blockages have begun to build up.

L2 Guest Speaker

Invite a local cardiologist to speak with the class about his or her practice: What heart problems do patients usually have? How does the cardiologist treat those problems? What advice can he or she give to help teens protect the health of their hearts?

L2 Researching

Suggest that students use library or Internet resources to find articles about heart transplantation. Articles from a doctor's point of view and those about the patient's experience can be appropriate. Let students meet in groups to summarize and discuss the articles.

L1 Family and Consumer Sciences

Ask a group of interested students to compile two lists of snack foods and lunch foods: one list of foods dangerously high in fat and cholesterol and the other of foods low in fat and cholesterol.

L1 Discussing

Guide students in discussing the importance of regular exercise: How can you choose a good form of exercise? How will exercise help you manage your weight? How will it help manage stress?

L1 Listing Examples

Have the class work in groups to make a list of responses to this question: What can you and other teens do to reduce stress?

Would You Believe?

Of the people under age 50 who die from heart attacks, about 90 percent are smokers.

Chapter 13 Lesson 1

Assess

EVALUATING THE LESSON

Assign Reviewing Terms and Facts and Thinking Critically to review the lesson; then assign the Lesson 1 Quiz in the TCR.

LESSON 1 REVIEW

Answers to Reviewing Terms and Facts

1. A chronic disease is one that is present continuously or on and off over a long period of time.
2. Hypertension is high blood pressure.
3. Both are arterial diseases; atherosclerosis is a condition in which fatty substances build up on artery walls, and arteriosclerosis is hardening of the arteries.
4. Doctors can combat heart disease with angioplasty, medication, pacemakers, bypass surgery, and heart transplants.

Answers to Thinking Critically

5. Responses will vary.
6. Responses will vary.

RETEACHING

▶ Assign Concept Map 59 in the TCR.
▶ Have students complete Reteaching Activity 59 in the TCR.

ENRICHMENT

▶ Assign Enrichment Activity 59 in the TCR.
▶ Let students work in groups to develop a program of specific behaviors that will help them maintain good heart health. Suggest that each group outline its program and then discuss it with the rest of the class.

Close

Have students respond to the three objective questions on page 362.

366

---

in your journal

Are you "heart smart"? Do you practice healthful behaviors—such as exercise and stress management—that can help prevent heart disease? In your journal, make a list of your "heart smart" habits. Then list any risk behaviors that you would like to change. Describe how you might make those changes.

Healthy relationships with others help teens reduce stress.

What You Can Do

You may think that you are too young to be concerned about heart disease. Children and teens with certain risk behaviors, however, increase their chances of developing heart disease as adults. Following these tips will help you maintain a healthy heart.

- **Eat healthful foods.** Choose foods that are low in fat and cholesterol and high in fiber. Be sure to include plenty of fruits, vegetables, and grains in your diet.
- **Get plenty of exercise.** Regular exercise strengthens your heart. It also helps build muscle tone and keep you trim.
- **Maintain a healthy weight.** By keeping your weight within a healthy range, you put less strain on your heart. If you are overweight, talk to your doctor about a weight-loss program.
- **Manage stress.** Learn to cope with stressful situations. Good stress management will help you keep your blood pressure and cholesterol level down, which is good for your heart.
- **Don't smoke.** Smokers greatly increase their risk of heart disease and stroke. Almost 20 percent of deaths from heart disease are directly related to smoking.

Lesson 1 Review

Using complete sentences, answer the following questions on a separate sheet of paper.

Reviewing Terms and Facts

1. **Vocabulary** What is a *chronic* disease?
2. **Relate** In what way is blood pressure related to *hypertension*?
3. **Compare** What is the difference between the following conditions: *atherosclerosis* and *arteriosclerosis*?
4. **Identify** List five treatment options that doctors can use to combat heart disease.

Thinking Critically

5. **Hypothesize** Why might some people continue to practice unhealthy lifestyle behaviors after these behaviors have been linked to diseases?
6. **Analyze** What lifestyle behaviors help to prevent both communicable and noncommunicable diseases?

Applying Health Concepts

7. **Growth and Development** Create a poster to teach children about healthful behaviors that protect the heart. Use colorful drawings or magazine clippings to illustrate each of the positive behaviors mentioned in this lesson. Ask permission to display your poster in a public area in your community where children gather.

Chapter 13: Understanding Noncommunicable Diseases

More About...

Tobacco and the Heart Everyone is familiar with the connection between lung disease, especially cancer, and tobacco. However, teens may need to be reminded of the great risks that tobacco poses to their heart health.

▶ Smokers usually have faster resting heart rates and are more likely than nonsmokers to have irregular heartbeats.

▶ Smoking typically raises blood pressure, which increases the likelihood of stroke.

▶ Arterial walls become roughened, leading to increased risks of clogged arteries among smokers.

▶ Smoking contributes to the hardening of arteries.

▶ Smoking apparently makes blood platelets more likely to clump, thus restricting blood flow.

Understanding Cancer

Lesson 2

This lesson will help you find answers to questions that teens often ask about cancer. For example:

- What is cancer?
- What are some of the causes of cancer?
- How can cancer be treated?
- What can I do to avoid cancer?

Words to Know

cancer
tumor
benign
malignant
carcinogen
radiation therapy
chemotherapy

Cancer

Cancer is the second leading cause of death in the United States, just behind heart disease. What exactly is cancer? **Cancer** occurs when *abnormal cells grow out of control*. It is not just one disease, however. Cancer is actually a group of over 200 different diseases that affect many different parts of the body.

The human body is made up of many trillions of cells. These cells are continually growing and reproducing themselves. Although the majority of cells are normal, some are abnormal. The body's natural defenses usually destroy these abnormal cells. Sometimes, however, an abnormal cell survives and starts to reproduce itself.

A **tumor** (TOO·mer) is *a group of abnormal cells that forms a mass*. Tumors can be either benign (bi·NYN) or malignant (muh·LIG·nuhnt). A **benign** tumor is *not cancerous* and does not spread. A **malignant** tumor is *cancerous* and may spread to other parts of the body.

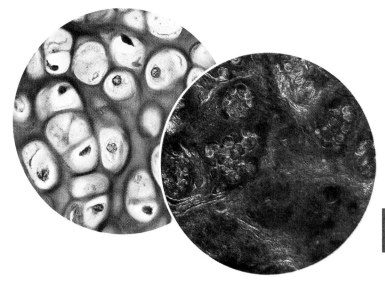

The photo on the left shows normal cells, while the photo on the right shows cancer cells.

Lesson 2: Understanding Cancer 367

Chapter 13 Lesson 2

Teach

L1 Language Arts
Let students use dictionaries to find the nonmedical definitions of *benign* and *malignant*. How do these meanings relate to the medical meanings presented in the lesson?

Visual Learning
Guide students in describing and comparing the normal cells and the cancer cells shown in the photographs near the bottom of page 367.

L1 Comprehending
Remind the class that cancer cells grow "out of control." Ask: How do you think the growth of a tumor affects the organ on which the tumor is growing?

L1 Critical Thinking
Ask: How does recognizing and treating cancer early help improve a patient's chances of recovering from cancer? Why is it important to know the signs and symptoms of various kinds of cancer?

! Would You Believe?
The only parts of the body where cancer can *not* strike are the hair and the nails.

Visual Learning
Help students study and discuss the information presented in Figure 13.3. Let them share their reactions to the relative numbers of new cases of various types of cancer. Ask volunteers to read aloud and restate the description of each type of cancer.

L2 Reporting
Divide the class into seven groups, one for each of the types of cancer named in Figure 13.3. Let group members prepare an illustrated report of their research on the cancer.

368

Types of Cancer

Cancer can affect many different parts of the body. Sometimes it begins in one area and then spreads to other areas. **Figure 13.3** shows the number of new cases in 1997 of some common types of cancer. It also provides information about these forms of cancer, including their risk factors.

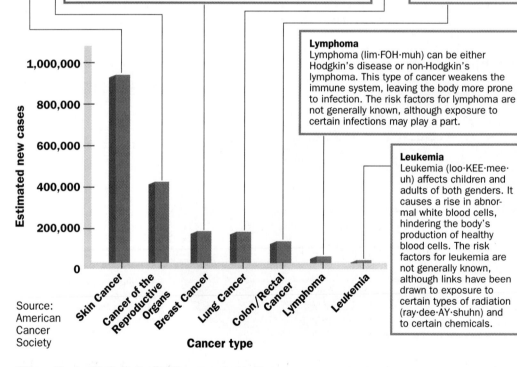

Figure 13.3
Estimated Number of New Cancer Cases, 1997
By understanding risk factors, you can take steps to lower your chances of developing some types of cancer.

Skin Cancer
This is the most common form of cancer in the United States. It occurs most often in people with light skin, hair, and eyes. The most important risk factor is exposure to the sun. There are several different types of skin cancer, some of which are easier to treat than others.

Cancer of the Reproductive Organs
In females, this type of cancer can affect the ovaries, cervix, and uterus. In males, it can affect the testicles and prostate. The risk factors are varied but include age, family history of cancer, obesity, and cigarette smoking.

Breast Cancer
This is the second major cause of cancer death in women. Age is a risk factor for breast cancer. It is more commonly found in women over 50 years old, but it also occurs in younger women and men. A family history of breast cancer is another risk factor.

Lung Cancer
Cigarette smoking is clearly the biggest risk factor in the development of lung cancer, which is the largest cause of cancer death in men and women. Nearly 90 percent of lung cancer cases in men and 79 percent in women are related to smoking.

Colon/Rectal Cancer
This type of cancer has declined somewhat in recent years, probably because of increased cancer screening. Risk factors include a high-fat, low-fiber diet and lack of exercise.

Lymphoma
Lymphoma (lim·FOH·muh) can be either Hodgkin's disease or non-Hodgkin's lymphoma. This type of cancer weakens the immune system, leaving the body more prone to infection. The risk factors for lymphoma are not generally known, although exposure to certain infections may play a part.

Leukemia
Leukemia (loo·KEE·mee·uh) affects children and adults of both genders. It causes a rise in abnormal white blood cells, hindering the body's production of healthy blood cells. The risk factors for leukemia are not generally known, although links have been drawn to exposure to certain types of radiation (ray·dee·AY·shuhn) and to certain chemicals.

Source: American Cancer Society

368 Chapter 13: Understanding Noncommunicable Diseases

Personal Health

The best way to protect yourself against skin cancer is to avoid prolonged exposure to the sun. Sunscreens, hats, and other clothing, when used in combination, can be relatively effective.

Skin cancer is most likely to develop on areas frequently exposed to the sun, including the face, the tops of the ears, and the backs of the hands. Bald scalps are also particularly vulnerable.

Students should be reminded that the appearance of one's skin during exposure to the sun is not a good indicator of danger. The full extent of a sunburn will usually not be evident for several hours.

Causes of Cancer

Like other noncommunicable diseases, cancer can be linked to inherited traits, lifestyle behaviors, and environmental factors. **Carcinogens** (kar·SI·nuh·juhnz) are *substances in the environment that cause cancer.* Usually a person must be exposed to a carcinogen over a long period in order for cancer to develop. The following is a list of substances and conditions that have been linked to the development of cancers:

- Tobacco in any form
- Ultraviolet rays from the sun
- Certain types of radiation
- Certain minerals and chemicals used in construction and manufacturing, including asbestos and benzene
- Air and water pollution
- A diet high in fat and low in fiber

Some carcinogens may be difficult to avoid, such as those found in air and water pollution. Other carcinogens, however, can be avoided by choosing a healthful lifestyle.

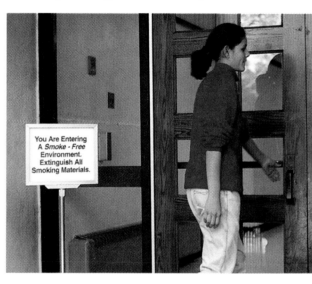

Teen Issues

Check Your ABCD's

Early detection of skin cancer is essential, even for teens. Check your skin for new growths or changes in growths. The American Cancer Society has devised the following ABCD rule that outlines the warning signs of skin cancer in moles.

A Asymmetry: One side of a mole looks different from the other side.

B Border irregularity: The edges are jagged or blurred.

C Color: The color is not uniform or the same throughout.

D Diameter: The diameter is greater than six millimeters (about the size of a pencil eraser). A growth that has grown to this size over time should be checked.

Smoke from other people's cigarettes is one environmental factor that you should avoid as much as possible.

Combating Cancer

The best weapon in the fight against an existing cancer is early detection. To find cancer early, you need to have regular physical checkups. Doctors look for warning signs of cancer during a routine examination.

You also play a role in early detection of cancer. Train yourself to become aware of any unusual changes in your body. One way to do this is to perform regular self-examinations. If you notice something abnormal, be sure to tell your doctor.

Lesson 2: Understanding Cancer 369

Health of Others

A person who smokes cigarettes is ten times more likely to get lung cancer than a nonsmoker. Diagnosed lung cancer is curable in only 10 percent of the cases, which means it is fatal in 90 percent of the cases. In addition to these frightening statistics, teens should know that smoking is directly related to cancer of the

- larynx
- mouth
- lip
- tongue
- pharynx
- esophagus
- pancreas
- bladder

Chapter 13 Lesson 2

L2 Researching

Have students find articles on individuals who have been diagnosed with cancer and have recovered. What form of cancer did the patient have? How did the patient react to the diagnosis? What course of treatment did the patient undergo? Let students meet in groups to discuss the articles they read.

L2 Listing Examples

Have students work together either as a class or in several groups, to list the carcinogens in cigarettes.

L2 Reporting

Have volunteers research these questions: What dangers does asbestos pose? What restrictions are now placed on the use of asbestos? Where can asbestos still be found? Ask the volunteers to share their findings with the class.

L1 Comprehending

Ask students: What are some examples of foods that are high in fat and/or low in fiber? What health risks other than heart disease does a diet containing those foods pose?

L1 Discussing

After class members have described the scene in the photograph on page 369 and read the caption, ask: What buildings and areas in our community have been designated smoke-free environments? What are the benefits of such areas? How can more areas be designated?

Teacher Talk

Dealing with Sensitive Issues

Avoiding other people's cigarette smoke is nearly impossible for teens if their parents or other family members smoke. Keep in mind that students in such a situation may be uncomfortable discussing secondhand smoke as an avoidable risk factor.

Chapter 13 Lesson 2

L1 Life Skills
Suggest that students discuss cancer and its warning signs with their health care provider during their next regular physical checkup.

L1 Analyzing
Let students share their responses to these questions: What excuses do teens—and adults—give for avoiding self-examinations? What would you say in response to each excuse?

L2 Creating Collages
In groups, have students learn more about a specific cancer treatment: surgery, radiation therapy, or chemotherapy. Group members should work together to create a collage, using photographs, words, drawings, and/or other images, to share information about that kind of treatment.

L2 Language Arts
Encourage students to identify and discuss novels or short stories they have read in which characters have to deal with cancer: In each work, what decisions does the cancer patient face? What treatment does he or she receive? With what results?

L3 Researching
Bone marrow transplants are now being used to treat several different forms of cancer: What do these transplants involve? How are they intended to help the cancer patient? What dangers are associated with the transplant? What is the current rate of success? Ask interested volunteers to research the procedure and write reports summarizing their findings.

Science Connection
Miracle Mineral?

A recent scientific study found that a mineral called selenium helps fight cancer. This mineral seems to help stop some types of tumors from growing and cause cancerous cells to die. Most Americans do not need to take selenium supplements, however, because they get enough of the mineral in their diet. Good sources of selenium include fish, meats, and poultry.

What Doctors Can Do
Doctors treat cancer in several ways. The most common cancer treatments are surgery, radiation therapy, and chemotherapy (kee·moh·THEHR·uh·pee). Often a combination of these methods is used.

- **Surgery** is the primary treatment for many types of cancer. These types include breast, skin, lung, and colon cancer. During surgery, doctors remove tumors and other cancerous cells. Surgery is most effective when the cancer is limited to just one part of the body.

- **Radiation therapy** is *a treatment for some types of cancer that uses X-rays or other forms of radioactivity.* These rays can be used to destroy cancer cells. Radiation is often used to kill any cancer cells that may remain after surgery.

- **Chemotherapy** is *the use of chemicals to destroy cancer cells.* This treatment is used to stop cancers that have spread throughout the body.

All three of these treatments have side effects, including the destruction of some healthy cells along with cancerous ones. Both radiation and chemotherapy often cause temporary nausea, fatigue, and hair loss.

What You Can Do
Although there are no guaranteed ways to avoid cancer, there are many precautions you can take. You can, for example, make lifestyle decisions that will lower your risk of developing the disease. You can also take action to find cancer early, thereby improving your chances of recovering from it. By following the guidelines on the next pages, you will be helping to protect yourself from cancer.

LIFE SKILLS
Dietary Guidelines to Lower Cancer Risk

For several decades, scientists have known about the relationship between diet and heart disease. In recent years they have also found a link between diet and cancer. Experts estimate that 35 percent of all deaths from cancer can be linked with food choices. Improving your eating habits can improve your chances of preventing cancer and other diseases.

The American Institute for Cancer Research has created four dietary guidelines. By following them, you can help reduce your risk of getting cancer.

1. **Limit the amount of fat in your diet.** Your total intake of fat should be no more than 30

370 Chapter 13: Understanding Noncommunicable Diseases

LIFE SKILLS
Dietary Guidelines to Lower Cancer Risk

Focus on Life Skills
Ask students to recall what they have eaten so far today. List the foods they name on the board, without associating any student's name with any of the foods. Then ask: Which of these foods do you think is part of a healthy diet that will help you reduce your risk of cancer? Let students share their ideas, and then have them read the Life Skills article.

Making Health Connections
Understanding how to make healthy food choices will enable students to reduce their cancer risks, especially as they mature. The ability to plan healthful meals and snacks has immediate benefits, too: The same healthy diet that reduces their long-term cancer risks also contributes to increased energy and to the maintenance of a healthy weight.

- **Avoid tobacco.** In the United States, tobacco use is the single most preventable cause of disease and death. Avoiding tobacco is the easiest way to lower your risk of cancer. If you never start using tobacco, you will never have to break the habit.
- **Eat healthful foods.** A diet low in fat and rich in fruits, vegetables, and whole grains can help prevent cancer. The Life Skills feature in this lesson provides more information about dietary guidelines that will lower cancer risk.
- **Limit sun exposure.** Skin cancer is the most common form of cancer in the United States. To reduce your risk, it is best to avoid being outdoors when the sun's rays are at their strongest—between 10:00 a.m. and 3:00 p.m. If you must be in the sun, wear a hat and use a sunscreen with a sun protection factor (SPF) of at least 15.
- **Perform self-examinations.** Once a month, females should perform a breast self-examination. Males should perform a testicular self-examination. Ask your doctor how to check for lumps that may be cancerous. In addition, both males and females should check their skin for changes in moles or other growths.
- **Know the seven warning signs.** The American Cancer Society has identified seven warning signs that may signal cancer. The first letters of these warning signs spell the word *caution,* as shown in **Figure 13.4** on the following page. If you notice one of these signs, check with your doctor.

If you plan an outdoor activity during the day, take precautions to protect your skin from the sun's rays. Use a sunscreen with an SPF of at least 15.

percent of total calories. Specifically, limit the amount of saturated fat to less than 10 percent of total calories. A large amount of fat in the diet has been shown to increase a person's chances of developing cancer.

2. **Eat fruits, vegetables, and whole grains.** Including more fruits, vegetables, and whole grains in your diet can lower your risk of developing cancer. The nutrients in these foods can assist your body in getting rid of carcinogens before they cause cancer. A healthy diet should include at least five servings of fruits and vegetables per day.

3. **Be moderate in your consumption of salt-cured, salt-pickled, smoked, and grilled foods.** Curing, pickling, and smoking are methods of preserving foods. Food preserved in these ways, however, often contain substances that turn into carcinogens in the body. The charred surface of some grilled foods also contains a large number of carcinogens.

4. **Avoid alcoholic beverages.** For teens, of course, drinking alcoholic beverages is illegal. Adults, however, should keep in mind that consumption of alcohol seems to increase the likelihood of getting certain types of cancer. For this reason and many others, it is best not to start drinking alcohol at all.

Follow-up Activity

Think about the four dietary guidelines to lower cancer risk. Which ones do you follow? Which ones do you need to work on? Make an action plan to improve your eating habits and improve your health.

Chapter 13 Lesson 2

Assess

EVALUATING THE LESSON

Assign Reviewing Terms and Facts and Thinking Critically to review the lesson; then assign the Lesson 2 Quiz in the TCR to evaluate students' understanding.

LESSON 2 REVIEW

Answers to Reviewing Terms and Facts

1. Cancer is a group of diseases in which abnormal cells grow out of control. Sentences will vary.
2. A benign tumor is not cancerous; a malignant tumor is cancerous.
3. The main risk factor for lung cancer is cigarette smoking. The risk factors for colon/rectal cancer are a high-fat, low-fiber diet and lack of exercise.
4. Examples of carcinogens will vary. Possible response: tobacco, ultraviolet rays from the sun, certain types of radiation, and air pollution.
5. The three main ways to treat cancer are surgery, radiation therapy, and chemotherapy.

Answers to Thinking Critically

6. Responses will vary.
7. Responses will vary.

RETEACHING

- Assign Concept Map 60 in the TCR.
- Have students complete Reteaching Activity 60 in the TCR.

ENRICHMENT

Assign Enrichment Activity 60 in the TCR.

Close

Let each student identify at least one lifestyle habit he or she intends to change to help reduce the risk of developing cancer.

---

in Your Journal

Think about the lifestyle behaviors you have learned about that can help you prevent cancer. In your journal, describe the actions you already take to reduce your risk of cancer. Then write down any behaviors you would like to improve, and describe ways in which you can make those changes.

Figure 13.4
The Seven Warning Signs of Cancer

If you notice any of the following signs, speak with a physician. Although these signs could be connected to a condition that is not cancer, it is safest to check with a doctor.

Change in bowel or bladder habits
A sore that does not heal
Unusual bleeding or discharge
Thickening or lump in breast or elsewhere
Indigestion or difficulty swallowing
Obvious change in a wart or mole
Nagging cough or hoarseness

Lesson 2 Review

Using complete sentences, answer the following questions on a separate sheet of paper.

Reviewing Terms and Facts

1. **Vocabulary** Define the term *cancer*. Use it in an original sentence.
2. **Compare** Explain the difference between a *benign* tumor and a *malignant* tumor.
3. **Recall** What is the main risk factor for lung cancer? What are the risk factors for colon/rectal cancer?
4. **Give Examples** Identify four examples of carcinogens.
5. **Identify** What are the three main ways to treat cancer?

Thinking Critically

6. **Hypothesize** Why do you think skin cancer is the most common form of cancer in the United States?
7. **Analyze** How might knowing the seven warning signs of cancer help people protect themselves from the disease?

Applying Health Concepts

8. **Health of Others** Write a script for a public service announcement for radio or television. Your goal is to convince people of the importance of practicing healthy lifestyle behaviors, such as avoiding sun exposure, to reduce the risk of cancer.

372 Chapter 13: Understanding Noncommunicable Diseases

Growth and Development

It is not unusual for teens to find new moles on their skin. Almost everyone has at least a few moles, and most appear during childhood and adolescence. Teens who already have moles are likely to notice that they become darker and perhaps larger during adolescence. As people age, it is not unusual for their moles to fade away or even to be rubbed off; however, no one should attempt to rub a mole away or otherwise remove it.

Teens should be reminded that most moles are—and will remain—harmless. Still, they should be encouraged to examine their moles regularly and to discuss any changes with their health care providers.

Allergies and Asthma

Lesson 3

This lesson will help you find answers to questions that teens often ask about allergies and asthma. For example:

- What are allergies?
- How are allergies treated?
- What is asthma?
- How do people who have asthma manage their condition?

Words to Know

allergy
allergen
pollen
histamines
antihistamine
asthma
bronchodilator

Allergies

In the United States, between 40 and 50 million people suffer from allergies. An **allergy** is *an extreme sensitivity to a substance.* As you have learned, the human body's immune system fights and destroys germs. In a person with allergies, however, the immune system is overly sensitive to certain substances that would normally be harmless. *The substances that cause an allergic reaction* are called **allergens** (AL·er·juhnz). Some common allergens are shown in **Figure 13.5**.

Figure 13.5
Common Allergens
Allergens can be found both indoors and outdoors and in any season.

A Pollen
Allergy to **pollen**, *a powdery substance released by certain plants*, is very common. Pollen is released by flowers, grasses, and weeds.

B Dust
Household dust contains several types of allergens.

C Pets
A number of animals, from various birds to rabbits and cats, can cause an allergic reaction.

D Plants
Many people are allergic to plants such as poison ivy and poison oak. Recognizing these plants can help avoid an allergic reaction.

E Insects
The venom of some insects, such as yellow jackets and wasps, can cause an allergic reaction.

F Food
Almost any food can be an allergen. Common food allergies include milk, eggs, wheat, seafood, and nuts.

Chapter 13 Lesson 3

Teach

L2 Math

Let some volunteers use an almanac or other resource to find the total population of the United States. Using that figure along with statistics on page 373, have students calculate the percentage of people in this country who suffer from allergies. According to that percentage, what number of students in the class should have allergies? What number actually do?

Visual Learning

Help students describe the illustrations and discuss the information in Figure 13.5. Then ask: What specific plants in our community release pollen that causes allergies? Why do you think allergies to household dust can be so difficult to deal with? Why do pets, rather than wild animals, cause allergies? Do poison ivy or poison oak grow in our community? If so, where are you likely to find them? What is the difference between an unpleasant reaction to an insect sting or bite and an allergic reaction to that same sting or bite? How do you think people deal with food allergies?

L3 Reporting

What is anaphylactic shock? Which allergens are most likely to trigger this reaction? What are the dangers of anaphylactic shock? How should it be treated? Let a volunteer research this topic and report to the class.

L1 Discussing

Discuss reactions to allergens: Which symptoms may also be symptoms of diseases or other problems? How do you think parents recognize allergic reactions in young children? How might you recognize a new allergic reaction in your own body?

Allergic Reactions

When an allergen enters the body, the body releases histamines. **Histamines** (HIS·tuh·meenz) are *chemicals in the body that cause the symptoms of an allergic reaction.* Figure 13.6 illustrates how the body reacts to allergens. Allergic reactions may involve the following parts of the body:

- **Eyes.** Allergies can make the eyes red, watery, and itchy.
- **Nose.** Common allergy symptoms include a runny nose and sneezing.
- **Throat.** Some food allergies cause difficulty in swallowing. In extreme reactions, the throat can close up.
- **Skin.** An allergic reaction can cause the skin to break out in a rash. It can also cause hives, which are raised, itchy welts.
- **Respiratory system.** Allergies can cause coughing and difficulty in breathing.
- **Digestive system.** An allergic reaction to food may cause stomach pain, cramps, and diarrhea.

Figure 13.6 The Body's Response to Allergens

Allergens can enter the body in three ways: through breathing, touching, or swallowing.

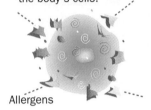

1. When a person who has an allergy comes in contact with an allergen, the allergen locks onto the body's cells.
2. These cells then release histamines, the chemicals that cause the symptoms of an allergic reaction.

in Your Journal

Do you have any allergies? In your journal, make a list of any known allergies and describe your body's reactions. Maybe you think that you have allergies but aren't sure what substances you are allergic to. If so, keep a record of any suspected allergic reactions in your journal. Note when and where the reactions occurred and what you were doing at the time.

If you can pinpoint your allergens, you will be better able to avoid future reactions. Discuss any severe symptoms with your doctor.

Treating Allergies

How do you know what you are allergic to? Sometimes the answer is obvious. Perhaps you develop a rash if you touch poison ivy. Sometimes, however, the source of the allergy is not obvious. In these cases, doctors can perform tests. One test involves injecting small amounts of possible allergens into the skin. In another test, patches are soaked with possible allergens and taped to the skin. With either test, the skin becomes red and itchy if the person is allergic to the substance.

Once the specific allergy is known, there are several possible ways to cope with it.

- **Avoid the allergen.** If you are allergic to peanuts, for example, you can easily prevent a reaction by not eating them (or any foods that contain them). Some allergens, such as dust and pollen, are more difficult to avoid.
- **Take medication.** If you can't avoid the allergen, antihistamines may provide relief of symptoms. **Antihistamines** are *medications that relieve the symptoms of allergic reactions.*
- **Get injections.** For people with severe allergies, doctors may recommend a stronger treatment plan. This involves getting shots of tiny amounts of the allergen to help the individual overcome his or her sensitivity to it.

374 Chapter 13: Understanding Noncommunicable Diseases

Consumer Health

People who have serious allergies—or other potentially life-threatening diseases—sometimes wear medical identification tags or bracelets. These can take the form of "dog tags" or of attractive pieces of jewelry. They can be essential in treating patients with allergies, especially in the case of an accident or if the patient is unconscious.

Suggest that students find out more about this special form of identification:

- Who should have a medical identification tag?
- What information should be included?
- Where and how can the tags be obtained?

Asthma

The number of people with asthma (AZ·muh), especially children, has risen in the past two decades. About 5 million children and 10 million adults in the United States are affected by this disorder. **Asthma** is *a chronic respiratory disease that causes air passages to become narrow or blocked,* making breathing difficult.

What causes asthma? Many of the same substances that cause allergies also cause asthma. These substances, along with several specific conditions, are called asthma triggers. Common asthma triggers include

- certain allergens, such as pollen, dust, pets, and mold.
- strenuous exercise, especially in cold weather.
- infections of the respiratory system, such as colds and flu.
- irritants, such as cigarette smoke, and fumes, such as those from paint and gasoline.
- situations in which the breathing rate increases, such as stressful events and hard laughing or crying.
- changes in the weather and cold air.

A Typical Asthma Attack

The signs of an asthma attack may include wheezing, shortness of breath, a high-pitched whistling noise in breathing, a feeling of gagging or choking, and chest tightness. **Figure 13.7** shows how an asthma attack affects the airways to the lungs.

Q & A

Exercise-Induced Asthma

Q: My friend has exercise-induced asthma. Can he still play sports?

A: This condition can usually be treated and controlled with medication. Your friend's doctor might also suggest some coping strategies, such as the following.

- Extend warm-up time before exercising and cool-down time afterwards.
- Develop physical fitness with a regular aerobic workout.
- Drink plenty of fluids before, during, and after exercising.
- When the weather is very cold, don't exercise outdoors.

Figure 13.7
Effects of an Asthma Attack
The changes caused by an asthma attack make the airways more narrow. This narrowing makes breathing difficult.

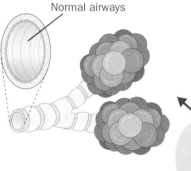
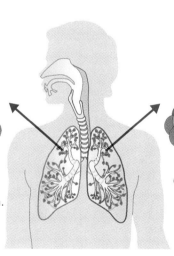
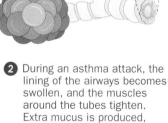

Normal airways

Swollen airway and contracted muscle

Narrowed airways

Mucus

1 During normal breathing, the lungs' airways are wide open. Air passes freely in and out through these tubes.

2 During an asthma attack, the lining of the airways becomes swollen, and the muscles around the tubes tighten. Extra mucus is produced, further clogging the airways.

Lesson 3: Allergies and Asthma

More About...

Asthma Students who have asthma or whose friends or family members have asthma may be interested in gathering more information about this breathing disorder. Suggest that they can contact one or more of the following organizations, using reference works or directories to find addresses and/or telephone numbers:

- American Academy of Allergy and Immunology
- American Lung Association
- Asthma and Allergy Foundation of America
- National Allergy and Asthma Network/Mothers of Asthmatics
- National Asthma Education Program

Chapter 13   Lesson 3

L1 Discussing
Let students share their responses to these questions: How can avoiding a food to which you are allergic be difficult? How can friends and family members help a teen who has food allergies?

L2 Reporting
Ask volunteers to read about some of the common side effects of antihistamines: When are these side effects most likely to cause problems? How can patients deal with or avoid these side effects? Let the volunteers discuss their findings with the rest of the class.

L1 Discussing
Encourage students to share their reactions to allergy shots: How do most people feel about having shots? What do you think motivates some teens to have regular injections to treat their allergies?

L2 Researching
Have students research the relationship between allergies and asthma: Do all asthmatics suffer from allergies?

Teacher Talk

Dealing with Sensitive Issues

Young teens can be especially sensitive about any physical conditions that they perceive as "different." Although you are likely to have asthma sufferers in every class, and although you may know about those students' conditions, don't assume that those students are willing to discuss their asthma publicly.

Would You Believe?

Health care for asthma costs $6 billion per year. Each year 100,000 new cases are diagnosed!

Chapter 13 Lesson 3

L1 Discussing
Help students consider and discuss the listed asthma triggers. Ask: Which triggers can be somewhat easily avoided? Which triggers are nearly impossible to avoid? Why?

L1 Language Arts
Have a volunteer use a dictionary to find the Greek word from which *asthma* derives: What is the meaning of that Greek word? Students may be interested to know that the word was apparently first used by Hippocrates, who lived around 400 B.C. and who is often called "the father of medicine."

L2 Guest Speaker
Ask the school nurse or another health professional to speak with students about asthma attacks. How does the care provider recognize an asthma attack? Ask the speaker to explain when and how an inhaler is used.

L3 Researching
Suggest that students read about famous people who have (or had) asthma. Have them write short reports about the person's accomplishments. Possible subjects include Theodore Roosevelt, Robert Joffrey, and Jackie Joyner-Kersee.

Assess

EVALUATING THE LESSON
Assign Reviewing Terms and Facts and Thinking Critically on page 377 to review the lesson; then assign the Lesson 3 Quiz in the TCR.

Did You Know?
The Real Culprits ACTIVITY

Contrary to popular belief, dust itself is not the main problem for allergy sufferers. The real culprits are dust mites, microscopic insects that live in beds, linens, and carpets. Find out some ways to cope with dust mites and other allergens. Share your findings with the class.

Asthma did not keep swimmer Tom Dolan from winning a gold medal in the 1996 Olympics.

Coping with Asthma

Most people who have asthma can take part in normal activities, including sports. Under the direction of a doctor, individuals with asthma can develop a treatment plan that will keep the condition under control. The following coping strategies are common:

- **Monitoring the condition.** Because asthma is a chronic disease, people who have asthma should always remain aware of their condition. Many doctors suggest using an instrument called a peak flow meter to monitor lung capacity. By breathing into this device, a person can detect when airways are becoming narrower—even before there are any symptoms. In addition, people who have asthma learn to recognize the warning signs of a severe attack.

- **Managing the environment.** This involves avoiding or getting rid of asthma triggers. In the home, floors and carpeting should be vacuumed regularly. If necessary, carpeting can be removed to reduce the amount of dust. Bed linens should be washed frequently in hot water. Feather pillows may be replaced with ones stuffed with cotton or synthetic fibers.

- **Managing stress.** Stress can sometimes trigger an asthma attack or make the condition worse. It is therefore important for people with asthma to control their stress levels. Relaxation techniques and special breathing exercises may also be helpful.

- **Taking medication.** Two main types of medication are used to treat asthma. One type helps prevent asthma attacks by making the airways less sensitive to triggers. The other type is a **bronchodilator** (brahng·koh·dy·LAY·ter), or *medication that relaxes the muscles around the bronchial air passages.* Bronchodilators help relieve asthma symptoms.

HEALTH LAB
Determining Lung Capacity

Introduction: During an asthma attack, a person's airways become blocked. This blockage reduces the air capacity of the lungs and makes breathing difficult.

Objective: With a partner, measure the air capacity of your lungs.

Materials and Method: You will need a clean 1-gallon plastic milk jug with a cap, masking tape, a two-foot length of plastic tubing, a plastic dishpan, two plastic drinking straws, and two pens in different colors. Tape a long strip of masking tape up the side of the plastic milk jug. Completely fill the jug with water, and put the cap on. Then fill a dishpan about halfway with water. Turn the jug upside down, put it in the dishpan, and remove the cap (keeping the top underwater).

Have your partner hold the jug, being careful not to let in air bubbles. Take the two-foot length of tubing,

376 Chapter 13: Understanding Noncommunicable Diseases

HEALTH LAB
Lung Capacity

Time Needed
1 class period

Supplies
- clean plastic milk jug, with cap
- masking tape
- 2-foot length of plastic tubing
- plastic dishpan
- 2 drinking straws
- pen

Focus on the Health Lab
Ask students to concentrate on their breathing for a minute or two: How much air do you think goes in and out of your lungs with a normal breath? with a deep breath? What does that air do for your body? How would you feel if you had trouble taking a full breath? Why? After this brief introductory discussion, let students read the Health Lab feature.

Review

Chapter 13 Lesson 3

Using complete sentences, answer the following questions on a separate sheet of paper.

Reviewing Terms and Facts
1. **Vocabulary** Define the term *allergy*.
2. **Give Examples** Name four common *allergens*.
3. **Compare** Explain the difference between *histamines* and *antihistamines*.
4. **List** What three methods are used for treating allergies?
5. **Identify** List four ways to cope with asthma.

Thinking Critically
6. **Hypothesize** Why do you think allergies often become worse over time?

7. **Synthesize** How would you help a friend who was undergoing an asthma attack?

Applying Health Concepts
8. **Personal Health** Many people think that they have a cold when they actually have allergies. Find out how to recognize the difference between a cold and an allergy. Then create a chart that compares the two.
9. **Consumer Health** In magazines and newspapers and on television, find advertisements of products for people who have allergies or asthma. Discuss the ads with a classmate. What techniques did the advertisers use? What claims did they make? Do you think that the ads are effective?

LESSON 3 REVIEW
Answers to Reviewing Terms and Facts
1. An allergy is the body's sensitivity to certain substances.
2. Common allergens include (any four) pollen, dust, pets, plants, insects, and foods.
3. A histamine is a chemical in the body that causes the symptoms of an allergic reaction. An antihistamine is a medication that works against the effects of histamines.
4. Methods for treating allergies are avoiding the allergen, taking medication, and getting injections.
5. Ways to cope with asthma include monitoring the condition, managing the environment, managing stress, and taking medication.

Answers to Thinking Critically
6. Responses will vary.
7. Responses will vary.

RETEACHING
► Assign Concept Map 61 in the TCR.
► Have students complete Reteaching Activity 61 in the TCR.

ENRICHMENT
Assign Enrichment Activity 61 in the TCR.

Have students write sentences explaining at least one new fact they now know about allergies or asthma.

and put one end into the jug opening. In the other end of the tubing, insert a drinking straw. (Put your hand around the top of the tubing to seal it and to keep the straw from sliding down the tubing.) Take a normal breath of air, and then exhale it into the straw. The air from your lungs will displace some of the water in the jug. Take time to note the water level, and mark it on the tape.

Refill the jug with water, and put it back into the dishpan. Then take a deep breath, and try to exhale all of the air from your lungs into the straw. Mark this water level on the tape.

Using a clean drinking straw, let your partner follow the same procedures. In a different color of ink, mark his or her water levels on the tape.

Observations and Analysis:
After you and your partner have completed this experiment, discuss your findings. Did the water level drop more when you exhaled a deep breath than when you exhaled a normal breath? Which partner's lung capacity was greater? Imagine that you were having an asthma attack and could not get a deep breath. What does this experiment suggest to you about lung capacity during an asthma attack?

Lesson 3: Allergies and Asthma

Understanding Objectives
Students will gain an understanding of the capacity of their own lungs to hold air. This understanding will help them appreciate the risks and discomforts associated with asthma and other disorders of the respiratory system.

Observation and Analysis
You may want to perform the experiment yourself, as a model for students. Then, if enough supplies are available, students can perform the experiment simultaneously. Once both partners have measured their lung capacity, conduct a brief class discussion in which students share and compare their results.

Further Investigation
Interested students should be encouraged to read more about lung capacity: How is it increased by regular exercise? How is it reduced by cigarette smoking? by asthma? by other respiratory diseases?

Assessing the Work Students Do
Ask partners to plan and draw a graph showing the four readings of their experiment. Meet with partners to discuss their work together, their understanding of the results, and their graph. Assign credit based on participation and comprehension.

Teen Health Digest

Focus
The Teen Health Digest articles can be used in two ways: as an individual activity for reflection and enrichment or as a cooperative activity as described below.

Motivator
Ask students to name some of the major noncommunicable diseases, and list those names on the board. Then divide the class into five groups. Assign one of the Teen Health Digest articles to each group, and let the group members read and discuss their assigned article. Ask: Which noncommunicable disease does it discuss? Have group members summarize their article for the rest of the class.

Cooperative Learning Project

Now Hear This: Public Service Announcements

Setting Goals
Ask students to recall public service announcements they have heard on radio or TV: What kinds of information do they present? Then tell students that they are going to write and "broadcast" their own public service announcements for teens about preventing and treating noncommunicable diseases.

Delegating Tasks
Refer students to the noncommunicable diseases listed on the board, and have them form working groups for each disease. The members of each group will work cooperatively to plan, research, write, edit, and present a public service announcement about that disease.

378

TEEN HEALTH DIGEST

Sports and Recreation
Super Mario

His fans call him Super Mario. Mario Lemieux is an ice hockey player who lives up to his nickname. Lemieux led his team to two championships in a row.

Winning these championships wasn't Lemieux's biggest challenge, however. In 1993 he faced a more personal one. After Lemieux noticed a lump in his neck, doctors found that he had Hodgkin's disease, a type of cancer.

Lemieux had radiation treatments every day for five weeks. These treatments make people feel weak and tired. Even so, on the day of his final treatment, Lemieux rejoined his teammates on the ice. He even scored a goal!

Now Lemieux wants to help other people who have cancer. He started the Mario Lemieux Foundation to support cancer research. Each year he hosts a golf tournament that raises thousands of dollars.

Try This: Find out about another sports star who overcame a noncommunicable disease. Write a brief report, and share it with the class.

378 Chapter 13: Understanding Noncommunicable Diseases

Teens Making a Difference
Cancer Crusader

When Robert Abramson was only ten years old, doctors discovered that he had a cancerous brain tumor. After more than 25 hospital stays and five operations, doctors removed the tumor and got rid of the cancer. Today, Abramson is a healthy young man in his twenties.

Grateful to be alive, Abramson is eager to help in the fight against cancer. He is one of the top youth fund-raisers for the American Cancer Society. Each year he takes part in the walkathon called Making Strides.

Abramson also volunteers as a peer counselor. He visits a children's hospital to talk to children who have cancer. He encourages them to have a positive attitude.

TECHNOLOGY OPTIONS

Public Address System Students may be able to obtain permission to use the school's public address system for broadcasting their public service announcements live to all the classrooms (or to selected classrooms). In this case, be sure students have rehearsed their announcements thoroughly and have had a chance to become familiar with the system itself.

Tape Recorder As an alternate to using the school's public address system, students may want to make polished audio recordings of their announcements. These taped announcements can be shared with other classes and with family members.

People at Work

Physical Therapist

Michelle Huarez is a physical therapist at Oakcrest Rehabilitation Center.

Q. What is your main responsibility?

A. I help people improve their physical mobility, or ability to move around, after an illness or injury.

Q. What types of problems do your patients typically have?

A. Some people have had an illness, such as a stroke. Other people have had surgery, such as a knee replacement. Still others have had an injury, such as a broken leg or a dislocated shoulder.

Q. How do you help these people?

A. I teach them exercises and show them how to apply heat and cold to relieve pain. I also teach them how to use aids, such as crutches, and exercise equipment, such as a stationary bicycle.

Q. What type of education does a person need for this job?

A. You need a bachelor's degree. It usually takes about 18 months after that to complete your certification. But different states have different rules, so you have to check with the state you expect to practice in.

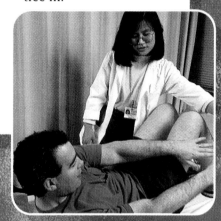

Myths and Realities

A Mysterious Illness

Q. My neighbor has chronic fatigue syndrome. What is it? I've heard people say that it's not really a disease.

A. Chronic fatigue syndrome, or CFS, is a real illness. People who have CFS feel extremely tired all the time. They also suffer from muscle pain and depression. CFS is often misunderstood because many people feel tired and depressed at times. CFS, however, is a chronic, or long-lasting, illness. With CFS, people have trouble working or leading normal lives.

HEALTH UPDATE

No More Shots?

Diabetes is a disease in which the body has problems with the production and function of insulin, a hormone. Many people who have diabetes must get a shot of insulin every day.

Now, however, some people with diabetes have another option. The Federal Drug Administration (FDA) recently approved a drug called troglitazone. It is taken in pill form. For some diabetics, the drug reduces or eliminates the need for insulin shots.

Try This: Learn more about the FDA. What is its function? How does it protect consumers? Discuss your findings with the class.

Teen Health Digest 379

Chapter 13

The Work Students Do

Following are suggestions for implementing the project:

1. Students in each group may want to divide up the duties—a few to do research, several to write, and a few others to broadcast, for example. Or, they may prefer to work cooperatively on all aspects.
2. Members of each group should review the most important facts about the noncommunicable disease they are covering: Which of these facts are most important for teens to know? Why? How can we communicate these facts most effectively?
3. Remind students that their public service announcements should resemble advertisements in certain ways: They should attract and hold the listeners' interest; they should appeal to listeners' minds as well as to their emotions; they should convey a single, clear message that listeners will remember.
4. Help students consider various methods of conveying their messages. Members of different groups may want to write their own skits, narratives, and songs. They may also wish to use existing songs or background music.
5. Help students select and use the most appropriate means of sharing their completed announcements with other teens at school or with the rest of the community.

Assessing the Work Students Do

Observe group members at work, and evaluate each student's contributions and participation. After the announcements have been made, meet briefly with the members of each group: What were the best aspects of your announcement? What would you do differently if you were starting this project over?

Meeting Student Diversity

Learning Styles This kind of cooperative project provides a good opportunity for auditory, kinesthetic, and visual learners to work together productively. If necessary, guide students in using and valuing contributions from all group members.

Hearing Impaired Students with hearing impairments should be encouraged to participate fully in the cooperative work of the project. Remind group members to arrange themselves so that everyone can see whichever student is speaking. In addition, encourage students to consider any teens with hearing impairments who may be part of the target audience for these announcements; it may be appropriate to prepare written versions of the announcement for these teens.

Lesson 4

Other Noncommunicable Diseases

Focus

LESSON OBJECTIVES

After studying this lesson, students should be able to

- distinguish between the two forms of arthritis and explain how arthritis is treated.
- explain the difference between the two types of diabetes and discuss how people with diabetes can manage the disease and lead normal lives.
- describe the effects of Alzheimer's disease and explain how family members can deal with this disease.

MOTIVATOR

Have the students imagine themselves as "senior citizens." What kinds of health problems do you think you may have? Why? Let students jot down words or phrases in response to the questions.

INTRODUCING THE LESSON

Allow several volunteers to share their responses to the Motivator questions. If some students mention that they expect to feel stiff and achy when they are older, suggest that they may be thinking of arthritis, which afflicts many older people but which can be handled quite successfully. Tell students that they will learn more about arthritis and other noncommunicable diseases as they read Lesson 4.

INTRODUCING WORDS TO KNOW

Let students read aloud and discuss the Words to Know. Then have each student find the definitions and write six original sentences, using a different Word to Know in each.

Lesson 4 Other Noncommunicable Diseases

This lesson will help you find answers to questions that teens often ask about other noncommunicable diseases. For example:

- What is arthritis, and how is it treated?
- What is diabetes, and how is it treated?
- What is Alzheimer's disease, and how do people cope with it?

Words to Know

arthritis
rheumatoid arthritis
osteoarthritis
diabetes
insulin
Alzheimer's disease

in Your journal

Do you know anyone who has arthritis? What common, everyday tasks might be difficult for someone with this condition? In your journal, describe these potential problems. Then think of at least one possible solution for each.

Arthritis

Arthritis (ar-THRY-tuhs) affects about 40 million people in the United States. The term *arthritis* actually refers not just to one disease but to a group of more than 100 conditions. **Arthritis** is *a disease of the joints marked by painful swelling and stiffness.* Although arthritis is more common among older people, anyone can be affected by it, including children.

Types of Arthritis

Although the term *arthritis* refers to many different diseases, there are two main types. **Rheumatoid** (ROO-muh-toyd) **arthritis** is *a chronic disease characterized by pain, inflammation, swelling, and stiffness of the joints.* It is the most serious and disabling form of arthritis. **Osteoarthritis** (ahs-tee-oh-ar-THRY-tuhs) is *a chronic disease, common in elderly people, that results from the breakdown of cartilage in the joints.* It is the more common type of arthritis. **Figure 13.8** shows the causes, characteristics, symptoms, and diagnoses for both types.

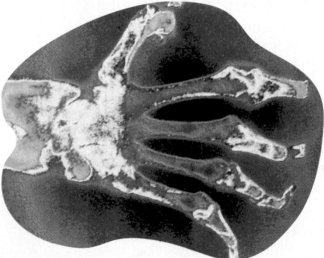

Arthritis can affect any joints, making them stiff and painful.

380 Chapter 13: Understanding Noncommunicable Diseases

Lesson 4 Resources

Teacher's Classroom Resources
- Concept Map 62
- Enrichment Activity 62
- Health Lab 13
- Lesson Plan 4
- Lesson 4 Quiz
- Reteaching Activity 62

Student Activities Workbook
- Study Guide 13
- Applying Health Skills 62
- Health Inventory 13

Figure 13.8
A Comparison of Rheumatoid Arthritis and Osteoarthritis
Although they have similarities, rheumatoid arthritis and osteoarthritis also have many differences.

Rheumatoid Arthritis	Osteoarthritis
Cause: The cause of rheumatoid arthritis is not known. This disease causes the immune system to attack healthy joint tissue, causing swelling and damage to the joints. Children with arthritis usually have this type.	*Cause:* The cause of osteoarthritis is not known. Because it is more common among older people, age is considered a major risk factor. Obesity and joint injuries are also risk factors.
Characteristics: Rheumatoid arthritis can affect any joints, including those of the hands, elbows, knees, hips, and feet. The disease usually progresses symmetrically—when one knee shows symptoms, the other knee will soon also show symptoms. Joints feel stiff and painful and may become deformed.	*Characteristics:* Osteoarthritis most often affects weight-bearing joints, such as the knees and hips. With this disease, the cartilage surrounding the joints breaks down. Because cartilage normally cushions the end of the bones, a breakdown causes bone to rub against bone. This causes pain and may limit movement.
Symptoms: The symptoms of rheumatoid arthritis include soreness, stiffness, aching, and general tiredness.	*Symptoms:* The symptoms of osteoarthritis include joint pain or swelling and pain and stiffness in the morning.
Diagnosis: Doctors diagnose rheumatoid arthritis based on symptoms, a physical examination, and blood tests.	*Diagnosis:* Doctors diagnose osteoarthritis based on symptoms and a physical examination. They may also take X-rays.

Coping with Arthritis

Although there is no cure for arthritis, there are ways to cope with the disease. A doctor can develop an effective treatment plan. Most plans involve a combination of some or all of the following:

- **Medication.** For some types of arthritis, doctors can prescribe medication to slow down the spread of the disease.
- **Painkillers.** Other medications can be taken to ease pain and reduce swelling.
- **Diet.** A balanced diet is necessary to maintain overall health and control weight.
- **Exercise.** Daily exercise helps keep joints flexible and improve muscle strength.
- **Rest.** Getting enough rest is an important way to relieve stress on the affected joints.
- **Heat/cold treatments.** Hot baths or heating pads can help to relieve pain. Cold treatments are sometimes recommended to reduce swelling.
- **Joint replacement.** In severe cases, surgeons can replace a diseased joint, such as a knee or hip, with an artificial one.

Lesson 4: Other Noncommunicable Diseases

Health of Others

Students may want to know some of the serious complications associated with diabetes:

▶ For young adults, diabetes is the leading cause of blindness.

▶ People with diabetes are two to four times as likely as non-diabetics to have heart disease.

▶ The risk for stroke is two and a half times as great for diabetics as for non-diabetics.

▶ More than half of all diabetics have high blood pressure.

▶ Many diabetics suffer from nerve damage associated with the disease.

▶ Diabetics are likely to suffer from kidney/urinary tract disease and to require dialysis or kidney transplantation.

Chapter 13 Lesson 4

L1 Discussing
Swimming is sometimes called the perfect exercise for arthritis patients. Ask students to explain this. Then let them suggest other forms of exercise that might be appropriate for those with arthritis. What forms are likely to be difficult?

L2 Guest Speaker
Arthritis patients who have had successful knee or hip replacement surgery often have very interesting—and positive—stories to tell. If possible, invite such a patient to visit the class and discuss how the surgery improved his or her life.

L1 Discussing
Explain to to the class that the complications associated with diabetes, including heart disease and stroke, can be especially dangerous. Ask: Why do these complications make early detection and treatment of diabetes so important?

! Would You Believe?
Researchers estimate that there are some 8 million cases of undiagnosed diabetes in the United States.

L2 Creating Posters
Let students work in groups to plan and make posters publicizing the importance of knowing the warning symptoms of diabetes.

VIDEODISC/VHS
Teen Health Course 2
You may wish to use video segment 13, "Understanding Noncommunicable Diseases," in which a group of teens learn more about diabetes and help to raise funds to find a cure.

Videodisc Side 2, Chapter 13 Understanding Noncommunicable Diseases

Search Chapter 13, Play to 14

382

Teen Issues
Teens and Chronic Illness
During the teen years, you are developing your self-concept and striving for greater independence. For healthy teens, these can be trying times. For a teen with a chronic illness, however, these years can be even more difficult. Teens in such a situation may feel depressed about the illness or develop a negative self-concept. They may feel that they are different from other teens. The support and encouragement of family members and friends can make these trying times a little easier.

Diabetes

Diabetes mellitus, usually called just *diabetes* (dy·uh·BEE·teez), affects about 16 million Americans. About half of these people don't know that they have the condition and are not being treated for it. **Diabetes** is *a disease that prevents the body from converting food into energy*. In people who have diabetes, the body has trouble with the production and use of insulin (IN·suh·lin). **Insulin** is *a hormone produced in the pancreas that regulates the level of sugar in the blood*.

There are two main types of diabetes: insulin-dependent diabetes and non-insulin-dependent diabetes. Insulin-dependent diabetes, also called Type I, generally develops during childhood or adolescence. With this condition, the body produces little or no insulin. Non-insulin-dependent diabetes, also called Type II, affects about 90 to 95 percent of the people who have diabetes. This type of diabetes usually develops in people who are over 40 years old and overweight. With this condition, the body produces insulin but does not use it effectively.

What causes diabetes? People with a family history of the disease have an increased risk of developing it. For non-insulin-dependent diabetes, obesity is a major risk factor. Anyone who has the following symptoms of diabetes should check with a doctor:

- Excessive production of urine
- Excessive thirst and hunger
- Unexplained weight loss
- Shortness of breath
- Blurred vision
- Dry, itchy skin
- Lack of energy

Coping with Diabetes
There is no cure for diabetes. However, people who have the disease can usually keep it under control and lead normal lives. **Figure 13.9** on the following page shows methods for managing diabetes.

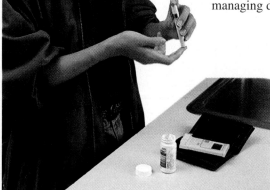

People with diabetes learn how to manage their condition.

382   Chapter 13: Understanding Noncommunicable Diseases

More About...

Diabetes Students who want to obtain further information about diabetes can be encouraged to write to one of the following organizations:

▶ American Diabetes Association
1660 Duke Street
Alexandria, VA 22314

▶ Juvenile Diabetes Research Foundation International
4332 Park Avenue South
New York, NY 10016

▶ National Institute of Diabetes and Digestive and Kidney Diseases
National Institutes of Health
Building 31, Room 9A0
Bethesda, MD 10892

Figure 13.9
Managing Diabetes
People who have diabetes must take responsibility for the management of their disease.

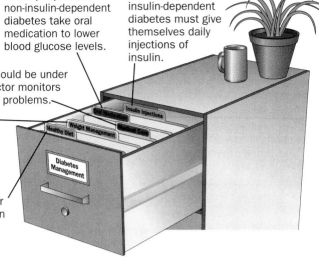

Medical Care
Anyone who has diabetes should be under the care of a doctor. The doctor monitors the condition and checks for problems.

Weight Management
It is important for people who have diabetes to exercise and maintain a healthy weight.

Healthy Diet
People who have diabetes must monitor their diet to keep the amount of sugar in their blood within safe limits.

Oral Medication
Some people with non-insulin-dependent diabetes take oral medication to lower blood glucose levels.

Insulin Injections
Most people with insulin-dependent diabetes must give themselves daily injections of insulin.

Alzheimer's Disease

Alzheimer's disease affects approximately 4 million people in the United States. **Alzheimer's** (AHLTS·hy·merz) **disease** is *an illness that attacks the brain and affects thinking, memory, and behavior.* The effects of the disease gradually worsen over time. The cause of Alzheimer's disease is unknown. Although this condition usually arises in older people, no one has proven that the condition is caused by the aging process.

Symptoms

Alzheimer's disease causes a loss of live nerve cells in the brain. The affected parts of the brain are those associated with memory and thinking. As a result, a person with Alzheimer's disease may have the following symptoms:

- **Memory loss and mental confusion.** Everyone forgets things, such as a name or telephone number, now and then. A person who has Alzheimer's disease, however, forgets things more often and may not ever remember them again. A person with the disease may also get lost—even in a familiar area—and not know how to get home.

- **Personality changes and mood swings.** It is not unusual for a person's personality to change slightly as he or she grows older. With Alzheimer's disease, however, the change is extreme. The person may become very confused and afraid. Alzheimer's disease may also cause a person to have drastic mood swings. He or she may be calm, then angry, and then tearful—all within a matter of minutes.

- **Altered speech.** Alzheimer's disease may cause a person to forget common words. In place of these words, she or he may use other words that are inappropriate. This can make it very difficult for others to understand what the person is trying to say.

Your Total Health

The Diet Connection ACTIVITY

A study by Harvard researchers, published in the journal of the American Medical Association in February 1997, found that people who eat a high-starch, low-fiber diet were more likely to develop diabetes. High starch content is found in white rice, white bread, and potatoes. High-fiber alternatives, such as brown rice and whole-grain pasta, breads, and cereals, are much healthier choices. Keep track of how much of these high-fiber foods you eat each day for the next week. If you are eating none or very little, make a plan to change your eating habits.

Lesson 4: Other Noncommunicable Diseases 383

Home and Community Connection

Community Resources What resources are available in your community to family members and others who have assumed the care of Alzheimer's patients? What facilities are available to provide care for Alzheimer's patients? Encourage class members to explore local organizations. Perhaps a group of students could arrange for a guest speaker from a local residence for Alzheimer's patients. Likewise, interested students might arrange a class field trip to a local day care center where Alzheimer's patients can spend part of each day. As students gather information, ask: What can you do to help? Remind them that they can make an important contribution to others with simple efforts, such as regular visits or weekend volunteering.

Chapter 13 Lesson 4

L2 Current Events

Have students read and summarize articles about research into the causes and/or treatment of Alzheimer's disease.

L1 Discussing

Ask: What special problems are family members likely to face in taking care of a relative who has Alzheimer's? Why are many people reluctant to let non-relatives take care of a family member?

L3 Family and Consumer Sciences

What are the costs of providing part-time in-home care for an Alzheimer's patient? of full-time in-home care? of placing an Alzheimer's patient in a residential facility? What proportion of each cost is likely to be covered by Medicaid? Ask volunteers to research these questions and report to the class.

L1 Critical Thinking

Ask: Why are family members likely to feel depressed if a loved one has Alzheimer's? guilty? feel angry? Why is it important to acknowledge these and other emotional responses?

Assess

EVALUATING THE LESSON

Assign Reviewing Terms and Facts and Thinking Critically on page 385 to review the lesson; then assign the Lesson 4 Quiz in the TCR to evaluate students' understanding.

ENRICHMENT

Assign Enrichment Activity 62 in the TCR.

Did You Know?

Helpful Hounds

For many years, dog guides have been helping people who are visually impaired. These dogs are specially trained to act as their masters' "eyes." Now an organization in the Milwaukee area is training dogs to help people who have Alzheimer's disease. The dogs are taught to let the person's caregiver know if the person is in danger. They also provide comfort and companionship to both the patient and the caregiver.

Coping with Alzheimer's Disease

Researchers have not yet found a cure for Alzheimer's disease. The average life expectancy of someone who has the disease is five to ten years. It is difficult for family members to deal with this type of illness in a loved one. Support groups have been formed to help families deal with the following issues:

- **Medical concerns.** Family members should find out all they can about any available medication or treatment. They will probably have to help the person with Alzheimer's disease follow the treatment plan.

- **Practical concerns.** Family members need to know how to care for the person who has Alzheimer's disease, including making the home safe. Eventually, however, the person will need constant care. The family usually must get live-in help or place the person in a health care facility.

- **Emotional concerns.** Family members must learn how to deal with their emotions regarding the loved one and the disease. They may feel depressed, guilty, or even angry. Learning how to handle these emotions is an important part of dealing with the disease.

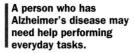

A person who has Alzheimer's disease may need help performing everyday tasks.

MAKING HEALTHY DECISIONS

Coping with a Relative Who Has Alzheimer's Disease

Next month Selena's parents will be celebrating their 20th wedding anniversary. Selena and her sister are planning a big surprise party for the occasion. They are going to invite a lot of people, including relatives, friends, and neighbors. The girls have been saving their money for the past six months. They want to make the party really special.

Selena and her sister have a problem, however. They are having trouble deciding whether or not to invite their grandfather to the party. He has Alzheimer's disease, and he needs constant care. He often can't remember who people are—even people he knows well. Sometimes he forgets where he is, and then he becomes very confused and upset. Large numbers of people, especially if he doesn't know some of them very well, sometimes make him irritable.

Selena and her sister would like to invite their grandfather to their parents' party. They love him, and he is their mother's father. However, they don't want to frustrate or upset him. They also realize that someone would have to be with him at all times.

384 Chapter 13: Understanding Noncommunicable Diseases

MAKING HEALTHY DECISIONS

Coping with a Relative Who Has Alzheimer's Disease

Focus on Healthy Decisions

Let students share their responses to these questions: How do you think your life would change if an older relative suffered from Alzheimer's disease? What would be the advantages of including that relative in daily family life? In special celebrations? What problems might arise? After this introductory discussion, let students read and discuss the Making Healthy Decisions feature.

Activity

Use the following suggestions to discuss Follow-up Activity 1:

Step 1 Selena and her sister must decide whether to include their grandfather in the party to celebrate their parents' twentieth anniversary.

Step 2 They can choose not to invite him to the party, or they can choose to invite him and arrange for someone to stay with him during the party.

Review

Lesson 4

Using complete sentences, answer the following questions on a separate sheet of paper.

Reviewing Terms and Facts

1. **Vocabulary** Define the term *arthritis*. Then use it in an original sentence.
2. **Compare** What is the main difference between *rheumatoid arthritis* and *osteoarthritis*?
3. **Recall** Explain the role of insulin in diabetes.
4. **Identify** List five of the symptoms of diabetes.
5. **List** Name three symptoms of Alzheimer's disease.

Thinking Critically

6. **Suggest** What would you say to a friend who had diabetes, but who didn't want to take her insulin injection?

7. **Infer** Choose one of the diseases discussed in this lesson. Describe the possible effects of the disease on the person who has it and on the person's family or close friends.

Applying Health Concepts

8. **Consumer Health** Look in the Yellow Pages for organizations that offer support groups for people who have arthritis, diabetes, or Alzheimer's disease. Make a list of these organizations, and combine your list with those of your classmates. Post the complete list on the bulletin board.

How should the girls handle this situation? If they invite him, who will take care of him during the party? Selena and her sister decide to use the six-step decision-making process to solve their problem:

1. State the situation
2. List the options
3. Weigh the possible outcomes
4. Consider your values
5. Make a decision and act
6. Evaluate the decision

Follow-up Activities

1. Apply the six steps of the decision-making process to Selena's story.
2. With another student, role-play a scene in which Selena and her sister discuss their problem.
3. Because they want the party to be a surprise, the girls can't ask their parents for advice. Whom else might they ask to help them with their decision?

Lesson 4: Other Noncommunicable Diseases 385

Step 3 If he is not included in the party, the girls' grandfather may not mind; however, their parents may be disappointed. If he is included in the party, the grandfather may well be irritable and confused; however, his being there will complete the family.

Step 4 Selena and her sister value their family as a group, including their parents and their grandfather.

Step 5 Selena and her sister may decide not to invite their grandfather and to explain their decision to their parents after the party. Otherwise, they may decide to invite him and to take turns staying with him.

Step 6 Guide students in considering both the short- and long-term consequences of their decisions.

Chapter 13 Review

CHECKING COMPREHENSION

Use the Chapter Summary and the Chapter 13 Review to help students go over the most important ideas presented in Chapter 13. Encourage students to ask questions and add details as appropriate.

CHAPTER 13 REVIEW ANSWERS

Reviewing Key Terms and Concepts

1. Both a stroke and a heart attack occur when the blood supply slows or stops and damage results.
2. Positive behaviors for lowering the risk of heart disease include (any four): eating healthful foods, getting plenty of exercise, controlling weight, managing stress, and not smoking.
3. In radiation therapy, X-rays or other radioactivity are directed at a tumor. In chemotherapy, chemicals are used to destroy cancer cells.
4. Ways in which you can help protect yourself from developing cancer include (any four): avoiding tobacco, eating healthful foods, limiting sun exposure, performing self-examinations, and knowing the seven warning signs of cancer.
5. Asthma is a chronic respiratory disease that causes tiny air passages to become narrow or blocked.
6. A bronchodilator relieves asthma symptoms by delivering medication that relaxes the muscles around the bronchial air passages.
7. Methods of coping with arthritis include (any five): taking medication, taking painkillers, eating a balanced diet, exercising daily, using heat/cold treatments, and having joint replacement surgery.

Chapter 13 Review

Chapter Summary

► **Lesson 1** Noncommunicable diseases cannot be spread from person to person. They can be present from birth, result from lifestyle behaviors, or be caused by the person's environment. Heart disease, which includes diseases of the heart and blood vessels, kills more Americans than any other cause.

► **Lesson 2** Cancer is a group of diseases that occurs when abnormal cells grow out of control. You can lower your risk of developing cancer by practicing healthy lifestyle behaviors.

► **Lesson 3** Many substances in the environment may cause people to have allergic reactions. Some of these same substances cause asthma, a chronic respiratory disease.

► **Lesson 4** Arthritis is a disease of the joints. Diabetes is a disease in which the body cannot properly convert food into energy. Alzheimer's disease attacks thought processes, memory, and behavior. Although no cures exist for these illnesses, methods of coping with them are available.

Reviewing Key Terms and Concepts

Using complete sentences, answer the following questions on a separate sheet of paper.

Lesson 1

1. In what way is a *stroke* similar to a *heart attack*?
2. Identify four lifestyle behaviors that can help you lower your risk of heart disease.

Lesson 2

3. Explain the difference between *radiation therapy* and *chemotherapy*.

Thinking Critically

Using complete sentences, answer the following questions on a separate sheet of paper.

9. **Deduce** Why do you think heart disease is the number one killer in the United States?
10. **Analyze** Which healthy behavior do you think is the most important for reducing the risk of cancer? Explain your answer.

4. What are four ways in which you can reduce your risk of developing cancer?

Lesson 3

5. What is *asthma*?
6. Explain the purpose of a *bronchodilator*.

Lesson 4

7. List five methods of coping with *arthritis*.
8. Identify the two main types of *diabetes*, and explain the difference between them.

11. **Synthesize** Why is it a good idea to know what substances you are allergic to?
12. **Infer** In what way is a chronic illness, such as arthritis, different from other illnesses, such as the flu?

Chapter 13: Understanding Noncommunicable Diseases

Meeting Student Diversity

Language Diversity Use the following suggestions to help students who have difficulty with English:

► Pair those learners with native speakers of English who can restate the Chapter Summary in language that helps students comprehend important concepts.

► Direct auditory learners or those students with language diversity to the Teen Health Audiocassette Program. Available in English and Spanish, this component provides an audio and written summary of the chapter.

Chapter 13 Review

Your Action Plan

A healthy lifestyle can help you reduce your risk of developing some types of noncommunicable diseases.

Step 1 Review your private journal entries for this chapter. Do you have any lifestyle behaviors that you would like to change? Decide on a long-term goal that you would like to achieve. For example, you might want to start eating a low-fat diet.

Step 2 Now write down short-term goals that will help you achieve your long-term goal. A short-term goal for reducing fat in your diet might be to replace three high-fat foods you would otherwise eat this week with low-fat choices.

Plan a schedule and a method for reaching each short-term goal. For example, you might decide to try one new low-fat food each week. When you reach your long-term goal, reward yourself for adopting a healthier lifestyle.

In Your Home and Community

1. **Community Resources** Contact an organization in your community that helps people cope with a noncommunicable disease (for example, the American Heart Association or the American Cancer Society). Ask for brochures on ways to reduce the risk of developing the disease or ways to cope with it. Display the brochures.

2. **Health of Others** Talk to your family about the importance of a low-fat, high-fiber diet to reduce the risk of heart disease and cancer. Together, make a list of foods to buy the next time you go to the grocery store. Then develop an action plan to achieve the goal of healthier food choices.

Building Your Portfolio

1. **Interview** Talk to someone you know who has allergies or asthma. Ask the person for tips on how he or she copes with the illness. Audiotape or videotape the interview, and add the tape to your portfolio.

2. **Research and Report** Find out about the latest medical research on arthritis, diabetes, or Alzheimer's disease. Read current articles in newspapers or magazines. Do researchers have ideas about what causes the illness? Are they close to finding a cure? Summarize your findings in a one-page report. Add the report to your portfolio.

3. **Personal Assessment** Look through all the activities and projects you did for this chapter. Choose one or two that you would like to include in your portfolio.

Performance Assessment

▶ **Self-evaluation** Direct students to review the activities that are provided throughout the chapter. Encourage each student to select one finished product or activity that demonstrates his or her best work for the chapter. Have students explain what they learned and how the examples they selected show their progress.

▶ **Teacher's Classroom Resources** Assign Performance Assessment 13, "Noncommunicable Diseases Chart," in the TCR.

Chapter 13 Review

8. The two main types are insulin-dependent (or Type I) diabetes, which generally develops in childhood and in which the body produces little or no insulin, and non-insulin-dependent (or Type II) diabetes, in which the body produces insulin but does not use it effectively.

Thinking Critically
Responses will vary. Possible responses are given.

9. Heart disease is the number one killer because many Americans practice unhealthy lifestyle behaviors that have a long-term negative impact on the health of their hearts.
10. Avoiding tobacco is an important behavior for reducing the risk of cancer; tobacco use is the single most preventable cause of disease and death.
11. Knowing what you are allergic to helps you avoid the allergens and thus avoid allergic reactions.
12. A chronic illness is present continuously or on and off over a long period of time; you cannot look forward to recovering from a chronic illness as you can the flu.

RETEACHING
Assign Study Guide 13 in the Student Activities Workbook.

EVALUATE
▶ Use the reproducible Chapter 13 Test in the TCR, or construct your own test using the Testmaker Software.
▶ Use Performance Assessment 13 in the TCR.

EXTENSION
Have students select one or more healthy life-style behaviors that teens can adopt to reduce the risks of noncommunicable diseases. Then have them plan and practice a short play or other production about the benefits of those behaviors. Schedule an opportunity for students to perform their work for other classes.

Unit 5
Personal Safety

UNIT OBJECTIVES
Students will learn to recognize the indications of violence and abuse, and will learn how they can prevent, avoid, or deal with those problems. Students will also become familiar with behaviors and attitudes that can help them protect their own safety, the safety of other people, and the safety of the environment.

UNIT OVERVIEW

Chapter 14 Preventing Violence and Abuse
In Chapter 14, students explore the causes of violence and consider what they can do to prevent or reduce violence in their own lives. The chapter also presents information about the four major types of abuse and explains when and how teens should seek help for abuse victims.

Chapter 15 Safety and Recreation
Chapter 15 focuses on issues of safety from accidents. Students learn how to recognize and break the chain of events that can lead to accidents. They become familiar with safety practices they should follow at home, during outdoor activities, and in the case of weather emergencies or natural disasters. Finally, students examine steps they can take to help protect the safety of their environment.

Bulletin Board Suggestion

Playing It Safe Post several photographs or drawings of teens who are making an effort to "play it safe." You might show two teens using words to resolve a disagreement, a teen talking with a mental health professional, a teen wiping up spilled water from a kitchen floor, a group of teens wearing safety equipment while in-line skating, and a teen and a parent putting together a disaster supplies kit. Let students describe and discuss the pictures: What safety measures are these teens taking? As students study the lessons in Chapters 14 and 15, encourage them to add other appropriate photographs or drawings to the bulletin board.

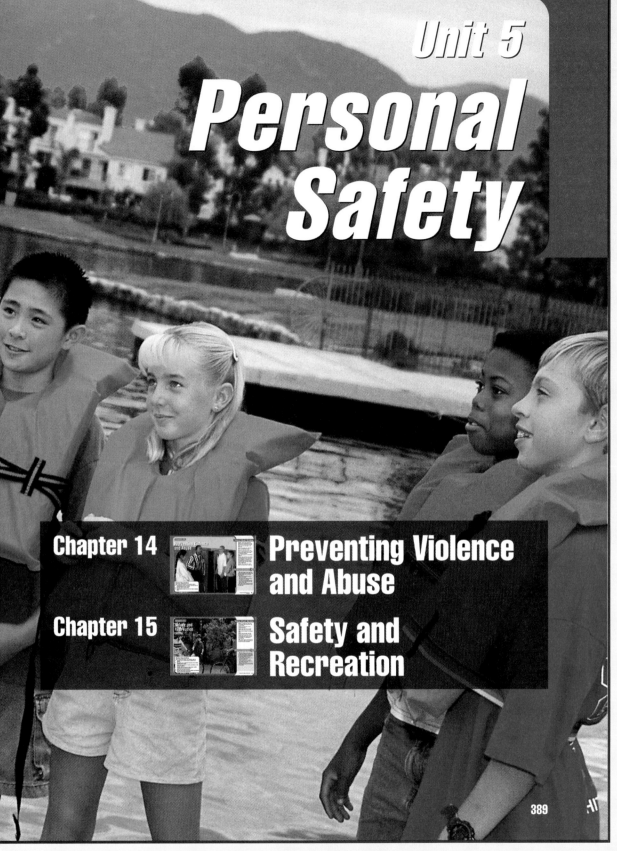

Unit 5
Personal Safety

Chapter 14 Preventing Violence and Abuse

Chapter 15 Safety and Recreation

Unit 5
INTRODUCING THE UNIT

Write the following sentence on the board, or say the sentence with or without inflection: *It's an emergency.* Ask students to jot down notes telling what comes to mind and how they would react. Then ask volunteers to share their ideas with the rest of the class. Guide students in recognizing the wide variety of emergencies they envisioned, and help students see how important it is to know—in advance—how to handle an emergency situation. Explain that students will consider many different kinds of emergencies and problems, as well as ways to avoid and deal with them, as they study the chapters in Unit 5.

VIDEODISC/VHS
Teen Health Course 2

You may wish to use video segment 14, "Preventing Violence," to show teens how listening attentively and responding calmly can help defuse conflicts.

Videodisc Side 2, Chapter 14 Preventing Violence

Search Chapter 14, Play to 15

DEALING WITH SENSITIVE ISSUES

Enhancing Self-esteem You can help students build self-esteem by convincing them they have potential. Urge them to maximize their potential by working to increase their skills. Convey to them that achievements, such as good grades and athletic awards, can make them feel better about themselves. Everyone has value; it is how people apply their energies and talents that enhances self-esteem. Encourage students to realize that their problems are separate from who they are. Make them aware that although they may have difficult problems or may have made poor choices in the past, they themselves are good. Reassure students that most problems can be resolved and that whatever pain they may feel now will not last forever.

Planning Guide

Chapter 14
Preventing Violence and Abuse

	Features	Classroom Resources

Lesson 1

Dealing with Violence
pages 392–397

Life Skills:
Protecting Yourself
pages 394–395

- Concept Map 63
- Cross-Curriculum Activity 27
- Decision-Making Activity 27
- Enrichment Activity 63
- Health Lab 14
- Lesson Plan 1
- Lesson 1 Quiz
- Reteaching Activity 63
- Transparencies 56, 57

TEEN HEALTH DIGEST
pages 398–399

Lesson 2

Understanding Abuse
pages 400–403

Making Healthy
Decisions: Deciding
to Report Abuse
page 402

- Concept Map 64
- Enrichment Activity 64
- Lesson Plan 2
- Lesson 2 Quiz
- Reteaching Activity 64
- Transparencies 58, 59

Lesson 3

Finding Help
pages 404–407

Health Lab:
Abuse and the Media
pages 406–407

- Concept Map 65
- Cross-Curriculum Activity 28
- Decision-Making Activity 28
- Enrichment Activity 65
- Lesson Plan 3
- Lesson 3 Quiz
- Reteaching Activity 65

National Health Education Standards

One of the main goals of the National Health Education Standards is to move students toward "health literacy." Health literacy is the capacity of individuals to obtain, interpret, and understand basic health information and services and the competence to use such information and services in ways which promote health. The health standards were developed by applying the characteristics of a well-educated, literate person within the context of health. The health literate person is:

- a critical thinker and problem solver
- a self-directed learner
- a responsible, productive citizen
- an effective communicator

Listed below are the Health Education Standards Performance Indicators addressed in each lesson of this chapter.

Lesson	Health Standards Performance Indicators
1	(1.6, 3.1, 3.6, 5.7)
2	(5.7, 6.1)
3	(2.2, 2.4, 2.6)

ABCNEWS InterActive Videodisc Series

You may wish to use the videodisc *Violence Prevention* with this chapter. Use the *ABCNews InterActive™ Correlation Bar Code Guide* for title reference. Also available in VHS format.

Chapter Resources

Teacher's Classroom Resources
- Chapter 14 Test
- Parent Letter and Activities 14
- Performance Assessment 14
- Testmaker Software

Student Activities Workbook
- Study Guide 14
- Applying Health Skills 63–65
- Health Inventory 14

Student Diversity Strategies
- Audiocassette Program (English)
- Audiocassette Program (Spanish)
- Spanish Parent Letters
- Spanish Summaries, Quizzes, and Activities

Multimedia Components
- English Audiocassette Program
- Spanish Audiocassette Program
- *Teen Health* Videodisc/VHS Series

Other Resources

Readings for the Teacher
Arbetter, Sandra. "Family Violence: When We Hurt the Ones We Love," Current Health 2, November 1995, pp. 6–12.

Guernsey, JoAnn Bren. *Youth Violence: An American Epidemic?* Minneapolis, MN: Lerner Publications, 1996.

Rogers, Anita M., and Diane Guerin. *Conflict Resolution: A Facilitator's Guide.* W. Greenwich, RI: The Consortium, 1994.

Readings for the Student
Hyde, Margaret O. *Know About Abuse.* New York: Walker and Company, 1992.

Lang, Susan. *Teen Violence.* New York: Franklin Watts, 1994 (revised edition).

Newton, David E. *Gun Control: An Issue for the Nineties.* Springfield, NJ: Enslow Publishers, 1992.

Out of Time?

If time does not permit teaching this chapter, you may use these features: Life Skills on pages 394–395; Teen Health Digest on pages 398–399; Making Healthy Decisions on page 402; Health Lab on pages 406–407; and the Chapter Summary on page 408.

Chapter 14
Preventing Violence and Abuse

CHAPTER OVERVIEW
Chapter 14 examines the issues of violence and abuse in our society. It helps students recognize, prevent, and—when necessary—deal with these problems.

Lesson 1 focuses on the causes of violence in society and shows teens what they can do to help prevent violence in their own lives.

Lesson 2 identifies the four major types of abuse and discusses their effects.

Lesson 3 explains how teens can find help in dealing with abuse.

PATHWAYS THROUGH THE PROGRAM
Of the numerous threats to teens' total health, few are as disturbing as violence and abuse. Students who have suffered directly from violent crimes or abuse are affected in more ways than they might suspect. Consequently, as they begin to understand the interrelationship between physical, mental/emotional, and social health, their reactions to this chapter's content may range from discovery to trauma. Throughout the lessons, stress the positive steps of recognition, seeking help, and recovery. Provide students with risk-free outlets such as private journal writing to explore their feelings. Remind them that setbacks in maintaining any aspect of good health—physical, mental/emotional, or social—can be dealt with in a healthy way. The journey to recovery might be rough, but recovery is possible.

Chapter 14
Preventing Violence and Abuse

Student Expectations
After reading this chapter, you should be able to:
1. Recognize the causes of violence and know how to prevent violence in your school.
2. Describe abuse and its effects.
3. Identify places people can go to for help in stopping abuse.

Key to Ability Levels

Teaching strategies that appear throughout the chapter have been identified by one of three codes to give you an idea of their suitability for students of varying learning styles and abilities.

L1 **Level 1** strategies should be within the ability range of all students. Full class participation is often required. Teacher direction is usually needed.

L2 **Level 2** strategies are for average to above-average students or for small groups. Some teacher direction is necessary.

L3 **Level 3** strategies are designed for students able and willing to work independently. Minimal teacher direction is necessary.

Teen Chat Group

Charles: Did you hear about that fight at school yesterday? One of the guys had to be taken to the hospital.

Greg: That's the third fight so far this week!

Luis: It was pretty scary. We talked about it in class this morning.

Susan: So what did everyone say?

Luis: We talked about ways we could help stop the fighting, like starting an antiviolence program.

Greg: What exactly is an antiviolence program?

Susan: I just read about one of them yesterday. Students are trained to be peer mediators so that people who have a problem can sit down and talk about it with someone who isn't involved. It helps solve the problem before a fight starts.

Charles: That sounds like a good idea. Let's find out how to start one.

in Your Journal

Read the dialogue on this page. Start your private journal entries on preventing violence and abuse by responding to these questions:

- What violence prevention programs exist in your school?
- Why do you think violence and abuse are problems in our society?
- If someone you knew experienced violence or abuse, where would you suggest they go for help?

When you reach the end of the chapter, you will use your journal entries to make an action plan.

Chapter 14: Preventing Violence and Abuse

Chapter 14

INTRODUCING THE CHAPTER

▶ Prepare a bulletin board with headlines and/or photographs from news articles about violent crimes. Encourage students to examine and discuss the display. Ask: What role do you think violence plays in your lives? How does violence or the fear of violence affect your attitudes? How does it affect your actions? How do you think we can reduce the threat of violence in the lives of children, teens, and adults? Explain that students will learn more about these issues as they study the lessons in Chapter 14.

▶ Write the following sentence on the board: *Violence results in unlived lives.* Ask students to read the sentence and consider its meaning: What do you think it says about the physical effects of violence? about the mental/emotional effects? about the social effects? After students have shared their reactions, tell them that they will learn more about violence—and how it can be prevented—as they study Chapter 14.

Cooperative Learning Project

Keeping the Peace: A Video

The Teen Health Digest on pages 398 and 399 provides students with high-interest articles related to the content of this chapter.

The material in the Teacher's Wraparound Edition presents suggestions for a class project in which students plan and produce their own video showing how teens in their community are resisting violence.

in Your Journal

Let students meet in groups to read and discuss the Teen Chat Group dialogue: How does this situation seem similar to situations at our school? What are the most important differences? Who should take responsibility for reducing fighting and other forms of violence at school? Why? Do you think violence can be completely prevented at school? If so, how? If not, why not? Suggest that group members role-play the dialogue, adding their own ideas and responses.

Then have group members discuss their responses to the In Your Journal questions, in preparation for writing their own private journal entries.

Lesson 1
Dealing with Violence

Focus

LESSON OBJECTIVES

After studying this lesson, students should be able to
- discuss the role of violence in our society.
- explain how teens are affected by violence.
- identify and discuss factors that contribute to violence.
- list ways to reduce the chances of becoming the victim of violent crime.

MOTIVATOR

Write the word *violence* on the board. Ask: What comes to mind when you read this word? Let students write or draw their responses.

INTRODUCING THE LESSON

Ask volunteers to share and discuss their responses to the Motivator question. Then ask students to name as many violent crimes as they can; record their ideas on the board. Explain that violent crime is a far greater problem in the United States than it is in many other countries. Tell students that in this lesson, they will learn about the causes of violence in our society and about the steps people can take to prevent violence in their own lives.

INTRODUCING WORDS TO KNOW

Have students write original short compositions using all three Words to Know. Ask for volunteers to read their compositions aloud. Help correct any misconceptions about the terms as they have used them.

Lesson 1
Dealing with Violence

This lesson will help you find answers to questions that teens often ask about violence and violence prevention. For example:

▶ What are the causes of violence?
▶ What can I do to protect myself so I do not become a victim?
▶ How can I help to prevent violence in my school?

Words to Know

homicide
hate crime
gang

Violence in Society

The images of violence are all around us—from music lyrics to the daily news reports. Grim stories of beatings, stabbings, gang wars, and family abuse are reported on the news and highlighted in the movies. Violence is a major public health problem in the United States. **Figure 14.1** shows how frequently violent crimes are committed in the nation.

Some people blame the increase in violence on television. By the age of 18, the average teen has viewed 200,000 acts of violence on television. Because some television programs glamorize violence, they may lead people to believe that violence is an acceptable way to settle disagreements.

Other people blame the increase of violent acts on the breakdown of the family, the decline in moral values, and the availability of weapons. Whatever the cause, there are many factors that contribute to the violence in society. In the end, we are all victims—either directly or indirectly.

Figure 14.1
Violent Crime Watch

Violent crime is a problem that affects everyone in our society. A violent crime occurs every 18 seconds. The statistics shown in this figure are for the four main types of violent crime.

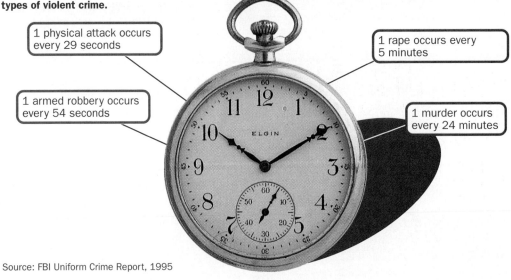

- 1 physical attack occurs every 29 seconds
- 1 armed robbery occurs every 54 seconds
- 1 rape occurs every 5 minutes
- 1 murder occurs every 24 minutes

Source: FBI Uniform Crime Report, 1995

Lesson 1 Resources

Teacher's Classroom Resources
- Concept Map 63
- Cross-Curriculum Activity 27
- Decision-Making Activity 27
- Enrichment Activity 63
- Health Lab 14
- Lesson Plan 1
- Lesson 1 Quiz
- Reteaching Activity 63
- Transparencies 56, 57

Student Activities Workbook
- Study Guide 14
- Applying Health Skills 63

The Victims of Violence

Who are the victims of violence? Each and every one of us is a victim of the violence in society. To solve this problem, communities, schools, and individuals need to work together.

In the past, an argument might lead to a shouting match or a fistfight. Today, the results may be more tragic. Simple arguments or disagreements might end in gunfire, stabbings, and possibly death. The fact is that violence today is more serious and more random than in the past. Random violence is committed for no particular reason and against anyone who happens to be around at the time. As a result, innocent people may be the victims of violence. Long after the violence has occurred, the victims and their families experience prolonged emotional trauma. In addition, victims may undergo the stress of testifying at long, costly legal trials.

Perhaps you have not been a direct victim of violence. However, you still have felt the impact of violence. Every year, crime in the United States costs an estimated $500 billion. This includes police protection, prisons, lost lives and wages, and medical costs. Even within some schools, security measures have been initiated to prevent violent acts. Locker searches, metal detectors, and security in schools increase costs to schools and communities.

Teens and Violence

The majority of teens are not violent and they do not commit crimes. However, teens are twice as likely as other age groups to be the victims of violence. In fact, the second leading cause of death of all people between the ages of 15 and 24 is homicide. A **homicide** (HAH·muh·syd) is *a violent crime that results in the death of another individual.*

Teens are also more likely than any other age group to commit violent crimes. In fact, more than a third of all violent crime in the United States is committed by young people under the age of 21. What drives these teens to commit such crimes?

interNET CONNECTION
Break the cycle of violence and the silence of abuse. Connect with strategies on the Internet.
http://www.glencoe.com/sec/health

in Your Journal

Do you know someone who has been the victim of a violent crime? In your private journal, describe how you felt when you heard about the crime. Could the victim have prevented the crime? Explain your response.

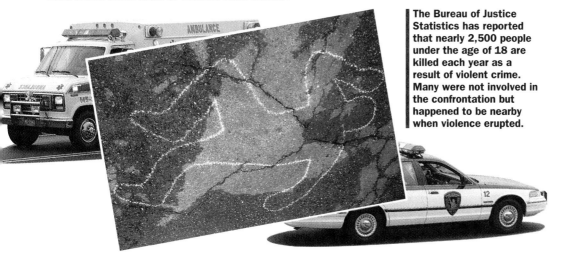

The Bureau of Justice Statistics has reported that nearly 2,500 people under the age of 18 are killed each year as a result of violent crime. Many were not involved in the confrontation but happened to be nearby when violence erupted.

Lesson 1: Dealing with Violence

Chapter 14 Lesson 1

Teach

L1 Critical Thinking

Ask students to share their ideas about the effects of violence in TV shows and movies on real-life acts of violence. Are you able to distinguish the glamour of media violence from the agony of real-life violence? Do you think other teens can make that distinction? How do you think media violent affects younger children? Do you think it is possible now to reduce the violence in TV shows and movies? Why or why not?

L2 Interviewing

Ask each student to speak with at least three adults—if possible, older adults—about their opinions on the cause of increased violence. How have society and daily life changed since their childhood? Which of these changes might have contributed to the rapid increase in violence? Have students write brief essays summarizing what they have learned from these adults.

L1 Discussing

Remind students that when a family member is the victim of a violent crime, the whole family is affected. What effects do you think violence against a parent might have on other family members? How do you think family members are affected when a teen or child is the victim of violence? How would the relationships among family members probably change? Why?

L1 Critical Thinking

Ask students to imagine themselves as victims of a violent crime. What emotions would you be likely to experience during the attack? after the attack? How do you think your self-confidence and self-esteem would be affected?

Meeting Student Diversity

Gifted Suggest to gifted students that they present a debate for the rest of the class. Each student should adopt one of the following points of view in the debate:

▶ Violence should be permitted on television at any time because censoring violence would be denying freedom of speech.

▶ Networks should reduce violence during prime-time hours, but it is up to parents to control their children's exposure to violence on television at other times.

▶ Networks should take an active role in protecting children from violence by reducing or eliminating violence in all television programming.

Chapter 14 Lesson 1

L2 Language Arts

Have students find and read at least two magazine articles dealing with television violence and its effects on violent behavior in children and teens. Let them write brief summaries of each article.

L1 Identifying Examples

Ask students to name specific things that they do or that teens can do to control anger. Write their responses on the board.

L2 Role-Playing

Let volunteers role-play situations in which a disagreement between two people leads to anger. Have the volunteers present two outcomes: one in which the anger erupts into violence and the other in which the anger is dealt with in a healthier way.

L1 Comprehending

Tell the class that prejudice often involves a negative feeling toward a person or group of people that is based not on experience but on stereotypes. Define stereotypes as exaggerated and oversimplified beliefs about entire groups. Help students understand that stereotypes deny members of the group the opportunity to be judged as individuals. Encourage students to suggest ways to recognize and overcome prejudiced attitudes.

L1 Guest Speaker

Invite a physically challenged individual to speak to the class about prejudice based on disability. Ask the speaker to describe the challenges he or she has experienced as a result of this kind of prejudice.

Teen Issues

Be Gun-Shy

Guns cause more than 1,300 accidental deaths annually. Most of these deaths occur at home, and many occur because people think that the guns are unloaded. To protect yourself and the people around you, never aim a gun at anyone, even as a joke. In addition, avoid spending time with anyone who carries a gun illegally. If you find a gun, report it immediately to the police.

Causes of Violence

People who commit violent acts usually have not learned how to deal with their feelings. Other factors that contribute to violence are discussed below.

Anger

Anger is a normal emotion. Learning how to control it is the most important step in preventing violence. Here are some ways to control your anger.

- Count to ten before you say or do anything.
- Talk to someone you trust and respect about your feelings.
- Exercise to get rid of some of your pent-up feelings.
- Channel your energy into a worthwhile activity.
- Find a nonviolent way to deal with the situation.

Prejudice

Prejudice is an opinion that has been formed without careful consideration. Prejudice is often based on a person's gender, race, religion, or country of origin. Prejudice sometimes leads to **hate crime,** which is an *illegal act against someone just because he or she is a member of a particular group.*

Possession of Weapons

An important relationship exists between access to weapons and the rise in violent crime. As anger increases during a dispute, a weapon may be used as an easy solution. See **Figure 14.2** on the next page.

LIFE SKILLS
Protecting Yourself

People who commit violent crimes seek out people who look vulnerable. You can reduce the chances of becoming a victim of violent crime by learning to protect yourself. The following tips will help you to stay safe.

In General:

- Do not look like an easy target. Stand up straight and walk with a confident stride.
- If someone bothers you, use direct eye contact and a forceful voice and say "Leave me alone," or shout "Fire!"
- If you are attacked by someone, get away in any way that you can.

LIFE SKILLS
Protecting Yourself

Focus on Life Skills

Introduce the Life Skills feature by telling students that they are going to learn some tips for protecting themselves against violence. Ask: Do you think information about protecting yourself from violence would have been included in a health textbook published 20 years ago? 10 years ago? 5 years ago? Why or why not? Why is it included now?

Then guide students in reading and discussing the feature. Encourage volunteers to give their reactions to each of the tips for personal safety. Why is it important to be safety conscious regardless of the size of your community? Why does the textbook recommend that a person should yell "Fire!" rather than "Help me!" if he or she is being attacked?

Making Health Connections

Taking steps to reduce violence benefits both the individual and the community. Emphasize to students that one of the first steps to take in staying safe is

Figure 14.2
Causes of Gun-Related Death Among Young People

When young people use guns, the result is often death.

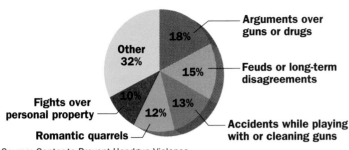

- Arguments over guns or drugs — 18%
- Feuds or long-term disagreements — 15%
- Accidents while playing with or cleaning guns — 13%
- Romantic quarrels — 12%
- Fights over personal property — 10%
- Other — 32%

Source: Center to Prevent Handgun Violence

in your journal

Think about a time when you felt really afraid. How did your body react? Did your hands turn cold? Did you tremble? Did the hair on the back of your neck stand on end? Use your private journal to write about your reactions to the fear you felt.

Peer Pressure

Many teens want to be accepted by a group and will take part in any conflict to show loyalty to the group. Sometimes, pressure from the group causes a teen to do something that goes against his or her own values.

Alcohol and Other Drugs

Alcohol and other drugs can contribute to violence. Substance abuse makes people act in unpredictable and dangerous ways. It can also prevent a person from making good decisions and judgments. Almost half of all violent crimes are committed by people under the influence of alcohol or other drugs.

Outside:

▶ Do not walk alone at night or near alleyways. Walk in lighted areas.

▶ If you think someone is following you, go into a store or other public place.

▶ When entering your house, make sure your keys are ready so that you do not have to fumble for them at the door.

▶ Do not hitchhike or ride in a car with strangers.

▶ If someone wants your money or jewelry and you are in danger, throw your purse, wallet, or jewelry away from you. Then run in the opposite direction.

Inside:

▶ Avoid entering an elevator alone with a stranger.

▶ At home, keep the doors and windows locked. Do not open the door for someone you do not know. Do not tell strangers on the phone that you are home alone.

Follow-up Activity

Using these tips as a checklist, examine the way you protect yourself in general, on the street, and in your home. In what areas are you safety conscious? In what areas do you need to improve? Make an action plan to keep yourself safe.

Lesson 1: Dealing with Violence

Chapter 14 Lesson 1

L1 Applying Knowledge

Ask students: If you know that another student has a weapon at school, what are you expected to do? Why? What should you do if you've heard a rumor about a weapon? What should you do if you suspect someone might have a weapon?

Visual Learning

Help students read and discuss the information presented in Figure 14.2. Ask: Why do you think arguments about guns and drugs are more often fatal than other kinds of arguments? When is it safe for teens to play with guns? *(never)*

L1 Discussing

Ask students: How do teens use words to pressure others into conflict and violence? How do they use actions? How can you resist those kinds of pressure?

L2 Current Events

Have students find, read, and bring to class news articles about violent crimes. In how many of the crimes did alcohol or other drugs play a role?

VIDEODISC/VHS

Teen Health Course 2

You may wish to use video segment 14, "Preventing Violence," to show teens how listening attentively and responding calmly can help defuse conflicts.

Videodisc Side 2, Chapter 14
Preventing Violence

Search Chapter 14, Play to 15

---

learning to protect themselves and practicing nonviolent conflict resolution.

Meeting Student Diversity

Learning to avoid looking like a target may be difficult for students who are shy or insecure or who have a negative self-concept. Encourage these students to practice good posture and forceful walking until the effort seems more natural.

Assessing the Work Students Do

Let students meet in groups to share and discuss their action plans. In addition, have group members plan short lessons on personal safety to be presented to young children. If possible, arrange for the groups to practice their lessons and then present them to young children at a local elementary school. Assign credit based on students' performance and participation.

Chapter 14 Lesson 1

L2 Creating Posters

Help students discuss alternatives to gangs. What school groups are available to teens who want to spend more time with their peers? What community groups and organized clubs are available? What volunteer opportunities are available? What else can teens do to spend more time with peers? After this discussion, let students work with partners or in small groups to make posters promoting alternatives to gang involvement.

L1 Critical Thinking

Encourage students to share their own thoughts and feelings about violence—or the potential for violence—at school. Do you feel safe here? What could you and other students do to make our school a safer place? What do you think teachers and other staff members could do? What do you think of programs such as dress codes and locker searches? Are you willing to give up some of your rights to free choice and to privacy? Do you think such programs actually make a school safer? Why or why not?

L2 Applying Knowledge

Direct students to apply what they are learning about violence prevention to problems within their own community. Let the class work together to brainstorm a list of activities and programs that could make the community safer. Then have students work in small groups to develop plans for making the suggested changes.

Assess

EVALUATING THE LESSON

Assign Reviewing Terms and Facts and Thinking Critically on page 397 to review the lesson; then assign the Lesson 1 Quiz in the TCR to evaluate students' understanding.

Your Total Health

Feeling Better ACTIVITY!

Taking action can help to boost your self-esteem and make you feel that the violence in our society is not a hopeless situation. Write to lawmakers about the violence in your school or community, or describe a personal experience if you have been a victim. Urge your representatives to introduce and pass laws that prevent violence and help victims of violent crime. You can write to your U.S. senator at the U.S. Senate, Washington, DC 20510 or your U.S. representative at the U.S. House of Representatives, Washington, DC 20525.

As a way to remember loved ones and friends who have been victims of violence, people in a community can work together. They can help to make their neighborhoods safer.

Gangs and Gang-Related Violence

A **gang** is *a group of people who associate with one another because they have something in common.* Although teen gangs are not all alike, many are involved in criminal activities. Most of the crimes involve some type of violence, intimidation, drive-by shootings, robbery, gang warfare, or rape.

Young people join gangs for many reasons, including racism, poverty, boredom, anger, lack of family support, peer pressure, or the need for companionship. However, there are many safer, non-criminal alternatives to joining a gang.

- If you are lonely or bored, look for a youth group, sports team, or church group in your neighborhood.
- If you are being harassed by gang members and are scared, get help. A family member, community group, school counselor, or police officer can give you support and help protect you.
- If peers are pressuring you, band together with other teens to start a community group that works for positive change.

Stopping and Preventing Violence

Although violence in America is on the rise, communities and individuals are working together to address the risk factors and reduce the incidence of violent crimes. Here is a list of what some communities are doing to make their neighborhoods safer.

- More police on the streets
- Stricter gun laws
- Nighttime sports programs
- Improved lighting in parks and playgrounds
- Neighborhood Watch programs
- Teen curfews
- Tougher punishments for violent crimes

Violence and You

Figure 14.3 on page 397 shows how many students and teachers are victims of violence. Do you, like other students, feel that you would learn more in school if you felt safer? If that is true, then there are many ways that you can work with others to help make your school a safer place.

396 Chapter 14: Preventing Violence and Abuse

Personal Health

How people express anger and other strong emotions can affect not only their relationships with peers but also their self-esteem. For example, people who don't express their anger but keep it bottled up inside tend to feel frustrated and resentful. Those who take their anger out on others tend to feel guilty and negative toward themselves. In both cases, self-esteem suffers. Emphasize to students that they need to learn positive ways of expressing their anger for the sake of their self-esteem and overall mental health.

Figure 14.3

Violence in School

Violence in schools does not always involve physical injury. Violence also includes shoving another person, sending threatening notes, making physical threats, and offering insults.

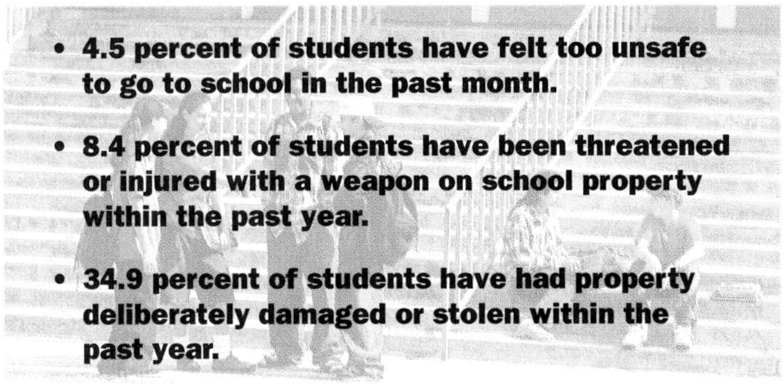

- **4.5 percent of students have felt too unsafe to go to school in the past month.**
- **8.4 percent of students have been threatened or injured with a weapon on school property within the past year.**
- **34.9 percent of students have had property deliberately damaged or stolen within the past year.**

Source: Youth Risk Behavior Surveillance, United States, 1995 (Centers for Disease Control)

Toward Safer Schools

Principals, school board members, teachers, parents, and students are working together to stop violence in the schools. Here is a list of what some schools are trying.

- Peer mediation programs
- Violence prevention programs
- Stricter dress codes
- Security guards
- Metal detectors
- Locker searches
- Video surveillance cameras
- Drug- and gun-sniffing dogs

Schools are also trying to eliminate some of the causes of violence by teaching respect for others and providing counseling.

Teen Issues

Fighting Crime ACTIVITY!

Crime Stoppers International, Inc., is an organization that encourages students with information about school crimes to report what they know to the proper authorities. The source of the tip is kept confidential. For more information about starting one in your school, call 1-800-245-0009. SAVE (Students Against Violence Everywhere) is an organization in many schools. Ask your guidance counselor or principal about forming a SAVE chapter in your school.

Review Lesson 1

Using complete sentences, answer the following questions on a separate sheet of paper.

Reviewing Terms and Facts

1. **Vocabulary** Define the terms *homicide* and *hate crime*.
2. **Give Examples** Identify ways in which peer pressure might lead a teen to do something she or he would not ordinarily do.

Thinking Critically

3. **Hypothesize** Why do you think some young people join gangs?
4. **Analyze** Do you feel safe at school? What actions do you take to protect yourself? In what ways do you help to protect the safety of others?

Applying Health Concepts

5. **Health of Others** Organize a group to brainstorm ways to prevent school violence. Attend a student council meeting. Share your group's ideas with the students.
6. **Consumer Health** Find out about movie ratings and warnings on tapes and compact discs. Evaluate the need for them, and share your findings.

LESSON 1 REVIEW

Answers to Reviewing Terms and Facts

1. Homicide is a violent crime that results in the death of another person. A hate crime is an illegal act against someone just because he or she is a member of a particular group.
2. Responses will vary. Possible response: A teen might do something he or she would not ordinarily do in order to be accepted by a group or gang and to show loyalty.

Answers to Thinking Critically

3. Students may offer reasons given in the text, as well as ideas of their own, such as that a teen might have an older family member who is a gang member or feel the need for protection.
4. Responses will vary. Suggested actions may be similar to those in the Life Skills feature on pages 394 and 395.

RETEACHING

- Assign Concept Map 63 in the TCR.
- Have students complete Reteaching Activity 63 in the TCR.

ENRICHMENT

- Assign Enrichment Activity 63 in the TCR.
- Have students make posters depicting specific things teens can do to protect themselves from violence on the street, such as avoiding dark alleys and staying with a group.

Close

Ask volunteers to identify how they can change their own behavior to reduce the risk of violence in their lives.

Cooperative Learning

Images of Violence Let several groups of students each select one of the following topics:

- Violence in comic books
- Violence in cartoons
- Violence in computer games
- Violence in live-action television
- Violence in advertising

How important a role does violence play in that medium? What special dangers does violence in that medium pose for children? for teens? What recommendations would you make regarding limiting violence in that medium? Have group members share their findings and conclusions with the rest of the class.

Focus

The Teen Health Digest articles can be used in two ways: as an individual activity for reflection and enrichment or as a cooperative learning activity as described below.

Motivator

Ask: Where do teens face the risk of violence or abuse? List the students' responses (home, school, and neighborhood, for example) on the board. Then ask students to read the titles of the Teen Health Digest articles and speculate about the contents of each article. Which places listed on the board do you think are locations for the violence or abuse discussed in this article? Have students read each article independently and summarize its contents.

Cooperative Learning Project

Keeping the Peace: A Video

Setting Goals

Explain that seeing the examples of others can help teens resist violent behavior and keep their homes, schools, and communities safe. Tell students that they are going to identify some positive examples and make their own video urging other teens to make nonviolent choices.

Delegating Tasks

Let students brainstorm a list of the responsibilities involved in planning and producing their own video. For example, they might list these tasks: researching, interviewing, writing, directing, filming, and editing. Then let students sign up for a task or tasks.

TEEN HEALTH DIGEST

CON$UMER FOCU$

Rating TV Programs

In January 1997, six categories of ratings began to appear on network and cable channels. TVY and TVY7 apply only to children's programming. A rating of TVY means that the program is considered suitable for all ages, and a program rated TVY7 is intended for children seven years and older. If an FV accompanies a Y7 symbol, there are scenes of fantasy violence. Other ratings that may accompany the basic ratings are V (violence), S (sexual content), L (coarse language), and D (suggestive dialogue). The rest of the ratings are defined below:

- **TVG:** Material suitable for all audiences.
- **TVPG:** Parental guidance suggested. May contain some coarse language, limited violence, or suggestive sexual dialogue and situations.
- **TV14:** Material may be inappropriate for children under 14. May contain sophisticated themes, strong language, or sexual content.
- **TVMA:** Program is designed to be viewed by mature audiences. May contain profane language and graphic violence or sexual content.

398 Chapter 14: Preventing Violence and Abuse

Myths and Realities

The Truth About Abuse

There is plenty of talk about abuse these days. Can you separate the facts from the false statements?

Myth: If a man abuses a young girl sexually, that is rape, but a woman's sexual abuse of a young boy is not rape.

Reality: Any adult who exploits a child sexually is an abuser. No matter what the gender of the abuser or the victim, such an act *is* a form of rape and causes lasting harm.

Myth: People who are sexually abused always go on to abuse others.

Reality: Not true. While it is true that most abusers have histories of being abused, most victims do *not* go on to abuse others. In fact, telling about abuse and getting help often prevents victims from abusing others.

Try This:

For one week, take note of the ratings of various shows you normally watch. Write a short report about your views of the rating system.

TECHNOLOGY OPTIONS

Video Camera and VCR Students will need one or more video cameras (for filming) and a VCR/TV (for viewing) to complete this project. If this equipment is not available for students' use, they can devise their own version of the project: a brochure, a newsletter, or a series of oral presentations, for example. Be sure students have an opportunity to share their project, whatever its final form, with others.

Computer Students may want to use word processing programs for composing their interview questions, recording interview responses, and writing and revising the final script for their video.

People at Work

Helping People

Job: Social worker
Responsibilities: To help people with problems related to poverty, sickness and family matters; to find foster homes for children; to begin legal action in child abuse cases
Education: Four to six years of college
Workplace: A wide range of settings, including schools, hospitals, offices, agencies, jails, and courts
Positive: Social workers can experience the satisfaction that comes from helping others.
Negative: This type of job may seem frustrating when important work cannot be accomplished because agencies lack the money for supplies and services.

Sports and Recreation

A Very Un-Martial Art

Aikido (eye·ki·DOH) is a Japanese martial art that is becoming popular in the United States. Developed originally as a means of self-defense, aikido has as one of its primary purposes the peaceful resolution of conflict.

Even when used as a form of self-defense, aikido retains a nonviolent focus. Rather than kicking or pushing opponents, people who practice aikido use their opponents' energy against them to ward off attack. In addition, aikido is primarily cooperative rather than competitive. People who learn this technique seek to match their partners in ability rather than overpower them.

Teens Making a Difference

A Dramatic 47 Seconds

About five years ago Anna Akbari heard a frightening statistic: that a child was abused every 47 seconds. The problem of abuse suddenly became real for her, and she decided to devote her time and talent to educating people about child abuse.

Akbari started a dramatic troupe called 47 Seconds. The troupe performs a mix of music, dance, and drama at schools, churches, and foster parent groups in the Midwest, Akbari's home region. Some of the performers are at-risk teens themselves.

"Our main focus is to let people know there is a way to get help and there are lots of people who care," Anna says.

Try This: Find out what groups in your area give presentations on the topic of abuse and violence. Arrange for one to visit your school.

Teen Health Digest 399

Chapter 14

The Work Students Do

Following are suggestions for implementing the project:

1. Suggest that researchers develop a list of teens and organizations that might be included in the video. Ask them to discuss these possibilities with the rest of the class, and let all students work together to select those to be video taped.
2. Help students determine how long they want their finished video to be. Then help them keep time constraints in mind as they develop their ideas.
3. Remind students of the importance of writing effective interview questions in advance of the interview. These questions should be designed to encourage thoughtful and detailed responses; any question that calls for a simple *yes* or *no* answer should be revised. Also discuss with students the importance of asking follow-up questions, inspired by the interviewee's responses, in addition to the planned interview questions.
4. Students may want to film scenes and activities, as well as interviews, to be included in their final work. Remind them that they will be able to "edit out" parts of the video, but in most cases they will not have an opportunity to return and reshoot a scene or an interview segment; for that reason, they should be as thorough as possible when filming.

Assessing the Work Students Do

Monitor the progress of each group, and be sure all members are participating actively and cooperatively. At the end of the project, have each student write a brief review of the class's completed video. What is the video's message? How effectively does it communicate that message? What might they do differently if they were given another opportunity?

Meeting Student Diversity

Language Diversity If languages other than English are spoken in your community, students may want to consider conducting interviews in the language that is most comfortable for the interviewee. They may then include an audio translation of the interview as part of the completed video; or, if a significant portion of the school or local community speaks a single language other than English, students may choose to create a bilingual video or to make two versions of their video, one in English and the other in the language native to many in the audience.

Lesson 2
Understanding Abuse

Focus

LESSON OBJECTIVES

After studying this lesson, students should be able to
- define *abuse* and identify four major types of abuse.
- explain some of the reasons that people become abusive.
- describe how abuse affects people.

MOTIVATOR

Ask: What do parents and other adults do to discipline their children? Have students list as many techniques as they can in four or five minutes.

INTRODUCING THE LESSON

Encourage students to compare and discuss their responses to the Motivator question. Then ask students to define the word *discipline*. Note that there are several accurate definitions and that the word is often used to mean "punishment." However, our word comes from the Latin word *disciplina*, which actually means "teaching" or "learning." Ask students: Which of the discipline approaches that you identified seem designed to teach a child correct behavior? Which seem designed to punish a child? Which of those punishments involve mistreatment? Tell students that they will learn more about various forms of mistreatment (called abuse) as they read this lesson.

INTRODUCING WORDS TO KNOW

Let volunteers share what they already know about the three Words to Know. Have students record each term, along with its definition as presented in the Glossary, in their notebooks.

400

Lesson 2 Understanding Abuse

This lesson will help you find answers to questions that teens often ask about abuse and violence. For example:

- What is abuse and why does it happen?
- How does abuse affect a person?
- What can be done to prevent abuse?

Words to Know

abuse
victim
battery

What Is Abuse?

You may find it difficult to understand how anyone can abuse a child, a family member, another relative, or a stranger. However, abuse in the United States is a significant problem. See **Figure 14.4** for some alarming statistics on abuse.

Abuse (uh·BYOOS) is *the physical, emotional, or mental mistreatment of another person,* and it can take several forms. Abuse may be physical, emotional, sexual, or a result of neglect. It can cause obvious physical wounds, such as bruises, scratches, or broken bones. It can also cause emotional wounds, such as anger, sadness, and fear. Abuse can happen in all kinds of families—from the richest to the poorest, from the most well educated to those with no schooling at all. Abuse can happen anywhere—in the largest cities and in the smallest towns.

Wherever abuse occurs, and whatever form it takes, abuse always does long-lasting damage. It is a crime to abuse another person. This is so important that physicians and teachers are required by law to report suspected cases of abuse. Abuse is never the fault of the **victim**, *the person against whom a crime is committed.*

Figure 14.4
Abuse Statistics
The incidence of abuse in families is evidence of the number of troubled families in our society. These are figures for a single year.

Reported cases of child abuse and neglect in the United States—3.1 million
Estimated number of women physically abused by their husbands or boyfriends—almost 4 million
Estimated number of elderly abused or neglected—820,000

Chapter 14: Preventing Violence and Abuse

Lesson 2 Resources

Teacher's Classroom Resources
- Concept Map 64
- Enrichment Activity 64
- Lesson Plan 2
- Lesson 2 Quiz
- Reteaching Activity 64
- Transparencies 58, 59

Student Activities Workbook
- Study Guide 14
- Applying Health Skills 64

Types of Abuse

There are four major types of abuse. They are physical abuse, emotional abuse, sexual abuse, and neglect. Anyone in a family may be the victim of abuse: a child or adolescent, a husband or wife, a brother or sister, or a grandparent.

Physical Abuse

Physical abuse is mistreatment that results in injury to the body. In some cases, physical abuse results in burns or broken bones. In severe cases, the victims of physical abuse die. The most common form of physical abuse is **battery,** *the beating, hitting, or kicking of another person.*

Emotional Abuse

Emotional abuse is the use of words and gestures to mistreat another person. Angry insults, repeated threats, constant teasing, and harsh criticism are examples of emotional abuse. When people are treated this way over a long period of time, they end up feeling worthless and helpless.

Sexual Abuse

Sexual abuse is any sexual contact that is forced upon a victim. It is a very serious crime. The abuser is often someone the victim knows and trusts, such as a parent, stepparent, older brother or sister, or family friend. When sexual abuse occurs between family members, it is called incest.

The victim of sexual abuse may feel guilty that somehow he or she is responsible for what has happened. However, no matter who the abuser is, sexual abuse is *never the victim's fault*—it is always the abuser's. The person who commits sexual abuse needs professional help. One way to get that help is for the victim to talk to a trusted adult—someone who can assist in arranging for the counseling that is needed.

Neglect

Neglect is the failure to meet the basic physical and emotional needs of a person. Young people need proper food, clothing, and housing to help them grow and develop. They need love and encouragement to help them feel safe and secure. Children who do not have these needs satisfied are neglected.

Did You Know?

Statistics Tell So Much

- In 1995 an estimated 1,215 children died as a result of abuse or neglect. Almost 82 percent of these children were under the age of five.
- One common form of child abuse is Shaken-Baby Syndrome, in which a baby is shaken to make it stop crying. Brain damage and even death may result from this form of abuse.
- Since 1985 the rate of child abuse has increased by 39 percent.

Teaching young children about child abuse is one way to help prevent it.

Lesson 2: Understanding Abuse   **401**

More About...

Discipline Properly approached, discipline in the home and at school can help children and teens learn to work productively within the limits of society's laws. When parents or teachers discipline a child, they should follow these guidelines to avoid damaging a child's self-esteem:

- Be consistent so that the child knows what to expect.
- Distinguish between the inappropriate behavior and the child. Help the child realize that although the behavior may be inappropriate, he or she is still a good person.
- Praise appropriate behavior.
- Remain friendly, fair, and firm, even when disciplining.
- Give children choices whenever possible.

Chapter 14 Lesson 2

L1 Discussing

Ask students about the role of communication in preventing abuse: How does communication help people work out their problems and disagreements? What are some examples of situations in which people have used communication to solve a problem? How does communication help people express strong emotions, such as anger? How might better communication decrease the incidence of abuse?

L2 Speculating

Ask class members to study the list of causes that may contribute to abusive behavior. Let students speculate about the relationship between each factor and abuse. For example, ask: Why do you think a person who was abused as a child is likely to be an abusive adult? After students have considered each factor in this way, ask: Is any of these factors—or any combination of them—an excuse for abuse? Why or why not?

L2 Analyzing

Help students analyze the effects of abuse by asking: How does abuse damage the victim? How does it damage the abuser? Does the damage to the abuser in any way reduce the abuser's responsibility for his or her actions?

Assess

EVALUATING THE LESSON

Assign Reviewing Terms and Facts and Thinking Critically on page 403 to review the lesson; then assign the Lesson 2 Quiz in the TCR.

 Your Total Health

Stemming the Tide of Abuse

Many cases of abuse are related to problems with alcohol or drugs. See Chapter 11 for more information on where to get help if someone in your family has these problems. Organizations such as Alcoholics Anonymous help people to stop drinking and reduce the chances that they will abuse someone else. Find out about organizations in your community that offer help to alcohol and drug abusers. Share your findings in class.

Causes of Abuse

All families have problems from time to time. An important key to dealing with these problems is communication. In a healthy family, members learn how to express emotions in a nonviolent way. Often the cause of abuse is that the person does not know how to control his or her own frustrations and emotions, or does not know how to handle problems in a positive way. People who abuse others often do not intend to do harm. Factors that seem to increase the risk that a person will become abusive are listed below:

- History of having been abused as a child
- Alcohol or other drug abuse
- Unemployment and poverty
- Illness
- Divorce
- Feelings of worthlessness
- Emotional immaturity
- Lack of parenting skills
- Inability to deal with anger
- Lack of communication and coping skills

MAKING HEALTHY DECISIONS
Deciding to Report Abuse

Jenny is Tasha's best friend. She knows that Tasha's father drinks too much, and when he does, he gets really mean. Several times Tasha has told Jenny about her father's "fits"—how he hollers, throws things, and once put his fist through a door.

This afternoon, in the locker room, Jenny noticed some awful bruises on Tasha's back and arms. At first Tasha explained the cause as a "skating accident." Later, though, she confided to Jenny that her father had beaten her. This was not the first time, either. Tasha made Jenny promise not to tell anyone.

Jenny has to make a decision. She knows that it is important to report abuse. She is afraid that the next beating may be worse. Still, she does not want to break a promise to her best friend. To help her choose the best solution, Jenny will use the six-step decision-making process.

❶ **State the situation**
❷ **List the options**
❸ **Weigh the possible outcomes**
❹ **Consider your values**
❺ **Make a decision and act**
❻ **Evaluate the decision**

Follow-up Activities

1. Apply the six steps of the decision-making process to Jenny's dilemma.
2. Imagine that you are Jenny. Write a diary entry for the day she decides what to do.

402   Chapter 14: Preventing Violence and Abuse

MAKING HEALTHY DECISIONS

Deciding to Report Abuse

Focus on Healthy Decisions

Ask students: Is there any way to recognize an abuser? What are the signs that indicate a person might be an abuse victim? Is there any way to know for sure whether a person is being—or has been—abused? Be sure students understand that abuse is not always physically obvious, and emphasize that any person, of any age, may be a victim. Then let students read the feature, and ask: How does Jenny's promise to Tasha complicate her decision?

Activity

Use the following suggestions to discuss Follow-up Activity 1:

Step 1 Jenny must decide whether to keep her promise to Tasha.

Step 2 Jenny might decide to keep her promise not to tell anyone but encourage Tasha to tell a trusted adult about the abuse, or Jenny might decide to break her promise and report the abuse herself.

People at Work

Helping People

Job: Social worker
Responsibilities: To help people with problems related to poverty, sickness and family matters; to find foster homes for children; to begin legal action in child abuse cases
Education: Four to six years of college
Workplace: A wide range of settings, including schools, hospitals, offices, agencies, jails, and courts
Positive: Social workers can experience the satisfaction that comes from helping others.
Negative: This type of job may seem frustrating when important work cannot be accomplished because agencies lack the money for supplies and services.

Sports and Recreation

A Very Un-Martial Art

Aikido (eye·ki·DOH) is a Japanese martial art that is becoming popular in the United States. Developed originally as a means of self-defense, aikido has as one of its primary purposes the peaceful resolution of conflict.

Even when used as a form of self-defense, aikido retains a nonviolent focus. Rather than kicking or pushing opponents, people who practice aikido use their opponents' energy against them to ward off attack. In addition, aikido is primarily cooperative rather than competitive. People who learn this technique seek to match their partners in ability rather than overpower them.

Teens Making a Difference

A Dramatic 47 Seconds

About five years ago Anna Akbari heard a frightening statistic: that a child was abused every 47 seconds. The problem of abuse suddenly became real for her, and she decided to devote her time and talent to educating people about child abuse.

Akbari started a dramatic troupe called 47 Seconds. The troupe performs a mix of music, dance, and drama at schools, churches, and foster parent groups in the Midwest, Akbari's home region. Some of the performers are at-risk teens themselves.

"Our main focus is to let people know there is a way to get help and there are lots of people who care," Anna says.

Try This: Find out what groups in your area give presentations on the topic of abuse and violence. Arrange for one to visit your school.

Teen Health Digest **399**

Chapter 14

The Work Students Do

Following are suggestions for implementing the project:

1. Suggest that researchers develop a list of teens and organizations that might be included in the video. Ask them to discuss these possibilities with the rest of the class, and let all students work together to select those to be video taped.
2. Help students determine how long they want their finished video to be. Then help them keep time constraints in mind as they develop their ideas.
3. Remind students of the importance of writing effective interview questions in advance of the interview. These questions should be designed to encourage thoughtful and detailed responses; any question that calls for a simple *yes* or *no* answer should be revised. Also discuss with students the importance of asking follow-up questions, inspired by the interviewee's responses, in addition to the planned interview questions.
4. Students may want to film scenes and activities, as well as interviews, to be included in their final work. Remind them that they will be able to "edit out" parts of the video, but in most cases they will not have an opportunity to return and reshoot a scene or an interview segment; for that reason, they should be as thorough as possible when filming.

Assessing the Work Students Do

Monitor the progress of each group, and be sure all members are participating actively and cooperatively. At the end of the project, have each student write a brief review of the class's completed video. What is the video's message? How effectively does it communicate that message? What might they do differently if they were given another opportunity?

Language Diversity If languages other than English are spoken in your community, students may want to consider conducting interviews in the language that is most comfortable for the interviewee. They may then include an audio translation of the interview as part of the completed video; or, if a significant portion of the school or local community speaks a single language other than English, students may choose to create a bilingual video or to make two versions of their video, one in English and the other in the language native to many in the audience.

Lesson 2 Understanding Abuse

Focus

LESSON OBJECTIVES

After studying this lesson, students should be able to

- define *abuse* and identify four major types of abuse.
- explain some of the reasons that people become abusive.
- describe how abuse affects people.

MOTIVATOR

Ask: What do parents and other adults do to discipline their children? Have students list as many techniques as they can in four or five minutes.

INTRODUCING THE LESSON

Encourage students to compare and discuss their responses to the Motivator question. Then ask students to define the word *discipline*. Note that there are several accurate definitions and that the word is often used to mean "punishment." However, our word comes from the Latin word *disciplina*, which actually means "teaching" or "learning." Ask students: Which of the discipline approaches that you identified seem designed to teach a child correct behavior? Which seem designed to punish a child? Which of those punishments involve mistreatment? Tell students that they will learn more about various forms of mistreatment (called abuse) as they read this lesson.

INTRODUCING WORDS TO KNOW

Let volunteers share what they already know about the three Words to Know. Have students record each term, along with its definition as presented in the Glossary, in their notebooks.

400

Lesson 2 Understanding Abuse

This lesson will help you find answers to questions that teens often ask about abuse and violence. For example:

▶ What is abuse and why does it happen?
▶ How does abuse affect a person?
▶ What can be done to prevent abuse?

Words to Know

abuse
victim
battery

What Is Abuse?

You may find it difficult to understand how anyone can abuse a child, a family member, another relative, or a stranger. However, abuse in the United States is a significant problem. See **Figure 14.4** for some alarming statistics on abuse.

Abuse (uh·BYOOS) is *the physical, emotional, or mental mistreatment of another person,* and it can take several forms. Abuse may be physical, emotional, sexual, or a result of neglect. It can cause obvious physical wounds, such as bruises, scratches, or broken bones. It can also cause emotional wounds, such as anger, sadness, and fear. Abuse can happen in all kinds of families—from the richest to the poorest, from the most well educated to those with no schooling at all. Abuse can happen anywhere—in the largest cities and in the smallest towns.

Wherever abuse occurs, and whatever form it takes, abuse always does long-lasting damage. It is a crime to abuse another person. This is so important that physicians and teachers are required by law to report suspected cases of abuse. Abuse is never the fault of the **victim,** *the person against whom a crime is committed.*

Figure 14.4
Abuse Statistics
The incidence of abuse in families is evidence of the number of troubled families in our society. These are figures for a single year.

Reported cases of child abuse and neglect in the United States —3.1 million
Estimated number of women physically abused by their husbands or boyfriends—almost 4 million
Estimated number of elderly abused or neglected—820,000

400   Chapter 14: Preventing Violence and Abuse

Lesson 2 Resources

Teacher's Classroom Resources
- 📁 Concept Map 64
- 📁 Enrichment Activity 64
- 📁 Lesson Plan 2
- 📁 Lesson 2 Quiz
- 📁 Reteaching Activity 64
- Transparencies 58, 59

Student Activities Workbook
- 📁 Study Guide 14
- 📁 Applying Health Skills 64

Types of Abuse

There are four major types of abuse. They are physical abuse, emotional abuse, sexual abuse, and neglect. Anyone in a family may be the victim of abuse: a child or adolescent, a husband or wife, a brother or sister, or a grandparent.

Physical Abuse

Physical abuse is mistreatment that results in injury to the body. In some cases, physical abuse results in burns or broken bones. In severe cases, the victims of physical abuse die. The most common form of physical abuse is **battery,** *the beating, hitting, or kicking of another person.*

Emotional Abuse

Emotional abuse is the use of words and gestures to mistreat another person. Angry insults, repeated threats, constant teasing, and harsh criticism are examples of emotional abuse. When people are treated this way over a long period of time, they end up feeling worthless and helpless.

Sexual Abuse

Sexual abuse is any sexual contact that is forced upon a victim. It is a very serious crime. The abuser is often someone the victim knows and trusts, such as a parent, stepparent, older brother or sister, or family friend. When sexual abuse occurs between family members, it is called incest.

The victim of sexual abuse may feel guilty that somehow he or she is responsible for what has happened. However, no matter who the abuser is, sexual abuse is *never the victim's fault*—it is always the abuser's. The person who commits sexual abuse needs professional help. One way to get that help is for the victim to talk to a trusted adult—someone who can assist in arranging for the counseling that is needed.

Neglect

Neglect is the failure to meet the basic physical and emotional needs of a person. Young people need proper food, clothing, and housing to help them grow and develop. They need love and encouragement to help them feel safe and secure. Children who do not have these needs satisfied are neglected.

Did You Know?

Statistics Tell So Much

- In 1995 an estimated 1,215 children died as a result of abuse or neglect. Almost 82 percent of these children were under the age of five.
- One common form of child abuse is Shaken-Baby Syndrome, in which a baby is shaken to make it stop crying. Brain damage and even death may result from this form of abuse.
- Since 1985 the rate of child abuse has increased by 39 percent.

Teaching young children about child abuse is one way to help prevent it.

Lesson 2: Understanding Abuse

More About...

Discipline Properly approached, discipline in the home and at school can help children and teens learn to work productively within the limits of society's laws. When parents or teachers discipline a child, they should follow these guidelines to avoid damaging a child's self-esteem:

- Be consistent so that the child knows what to expect.
- Distinguish between the inappropriate behavior and the child. Help the child realize that although the behavior may be inappropriate, he or she is still a good person.
- Praise appropriate behavior.
- Remain friendly, fair, and firm, even when disciplining.
- Give children choices whenever possible.

Chapter 14 Lesson 2

Teach

L1 Critical Thinking

After the class has read the explanations of abuse in general and of physical abuse in particular, ask: How can you—or a parent—distinguish among discipline, punishment, and abuse? Do you think parents should ever spank or hit their children? If so, when and why? If not, why not?

L1 Discussing

Ask students to identify situations in which a teen may face physical or emotional abuse. How might a teen avoid those situations?

L2 Language Arts

Have students write short dialogues in which one character abuses another emotionally, using only words. Indicate that insulting, belittling, or sarcastic remarks can be as abusive as vulgar language, which is not to be used in this exercise. Encourage volunteers to read their dialogues aloud; have other students explain what makes the language abusive.

L2 Guest Speaker

Invite the director of a shelter for battered women or a police officer who deals with spousal abuse to speak to the class. Encourage the speaker to discuss the impact of spousal abuse on children in the family.

Would You Believe?

Sibling abuse may be the most common form of family violence, yet many parents underreact to it because they consider it normal. It is now known that people who were subjected to serious sibling abuse as children are at increased risk for low self-esteem, anxiety, and depression as young adults.

L1 Critical Thinking

Ask: Do you think teen parents are more likely to abuse their children than are older parents? Why or why not?

Chapter 14 Lesson 2

L1 Discussing
Ask students about the role of communication in preventing abuse: How does communication help people work out their problems and disagreements? What are some examples of situations in which people have used communication to solve a problem? How does communication help people express strong emotions, such as anger? How might better communication decrease the incidence of abuse?

L2 Speculating
Ask class members to study the list of causes that may contribute to abusive behavior. Let students speculate about the relationship between each factor and abuse. For example, ask: Why do you think a person who was abused as a child is likely to be an abusive adult? After students have considered each factor in this way, ask: Is any of these factors—or any combination of them—an excuse for abuse? Why or why not?

L2 Analyzing
Help students analyze the effects of abuse by asking: How does abuse damage the victim? How does it damage the abuser? Does the damage to the abuser in any way reduce the abuser's responsibility for his or her actions?

Assess

EVALUATING THE LESSON
Assign Reviewing Terms and Facts and Thinking Critically on page 403 to review the lesson; then assign the Lesson 2 Quiz in the TCR.

Your Total Health

Stemming the Tide of Abuse

Many cases of abuse are related to problems with alcohol or drugs. See Chapter 11 for more information on where to get help if someone in your family has these problems. Organizations such as Alcoholics Anonymous help people to stop drinking and reduce the chances that they will abuse someone else. Find out about organizations in your community that offer help to alcohol and drug abusers. Share your findings in class.

Causes of Abuse

All families have problems from time to time. An important key to dealing with these problems is communication. In a healthy family, members learn how to express emotions in a nonviolent way. Often the cause of abuse is that the person does not know how to control his or her own frustrations and emotions, or does not know how to handle problems in a positive way. People who abuse others often do not intend to do harm. Factors that seem to increase the risk that a person will become abusive are listed below:

- History of having been abused as a child
- Alcohol or other drug abuse
- Unemployment and poverty
- Illness
- Divorce
- Feelings of worthlessness
- Emotional immaturity
- Lack of parenting skills
- Inability to deal with anger
- Lack of communication and coping skills

MAKING HEALTHY DECISIONS
Deciding to Report Abuse

Jenny is Tasha's best friend. She knows that Tasha's father drinks too much, and when he does, he gets really mean. Several times Tasha has told Jenny about her father's "fits"—how he hollers, throws things, and once put his fist through a door.

This afternoon, in the locker room, Jenny noticed some awful bruises on Tasha's back and arms. At first Tasha explained the cause as a "skating accident." Later, though, she confided to Jenny that her father had beaten her. This was not the first time, either. Tasha made Jenny promise not to tell anyone.

Jenny has to make a decision. She knows that it is important to report abuse. She is afraid that the next beating may be worse. Still, she does not want to break a promise to her best friend. To help her choose the best solution, Jenny will use the six-step decision-making process.

❶ State the situation
❷ List the options
❸ Weigh the possible outcomes
❹ Consider your values
❺ Make a decision and act
❻ Evaluate the decision

Follow-up Activities
1. Apply the six steps of the decision-making process to Jenny's dilemma.
2. Imagine that you are Jenny. Write a diary entry for the day she decides what to do.

402 Chapter 14: Preventing Violence and Abuse

MAKING HEALTHY DECISIONS
Deciding to Report Abuse

Focus on Healthy Decisions
Ask students: Is there any way to recognize an abuser? What are the signs that indicate a person might be an abuse victim? Is there any way to know for sure whether a person is being—or has been—abused? Be sure students understand that abuse is not always physically obvious, and emphasize that any person, of any age, may be a victim. Then let students read the feature, and ask: How does Jenny's promise to Tasha complicate her decision?

Activity
Use the following suggestions to discuss Follow-up Activity 1:
Step 1 Jenny must decide whether to keep her promise to Tasha.
Step 2 Jenny might decide to keep her promise not to tell anyone but encourage Tasha to tell a trusted adult about the abuse, or Jenny might decide to break her promise and report the abuse herself.

402

Signs of Abuse

A child who has been beaten will usually show signs of physical abuse. The signs of emotional abuse, sexual abuse, and neglect are more difficult to recognize. Some of them are listed here:

- Frequent absences from school
- Poor grades and lack of interest in school
- Dirty or neglected appearance
- Extreme shyness, sadness, or fear
- Aggressive behavior toward others
- Inability to communicate

Effects of Abuse

Abuse is always harmful. It causes damage to the abuser and the victim. Some teens try to escape from abuse by leaving home. Running away, however, often leads to other problems. Runaways usually have no way to support themselves. They have no place to live and no money for food. Life on the street is rough. Some runaways turn to crime. Many runaways become the victims of crime.

People who were abused as children often have low self-esteem, a high level of stress, and other problems. They may find themselves in abusive relationships as adults, and they sometimes become abusers themselves. However, with help, people can break the cycle of abuse.

Teen Issues

Sexual Harassment at School

Sexual harassment (huh·RAS·muhnt) is any unwelcome sexual comment, contact, or behavior. This may include jokes, looks, notes, touching, or gestures. Sexual harassment is a type of sexual abuse. It can happen to boys or to girls. Find out if your school has a policy for dealing with sexual harassment.

Review Lesson 2

Using complete sentences, answer the following questions on a separate sheet of paper.

Reviewing Terms and Facts

1. **Vocabulary** What is a *victim?*
2. **Vocabulary** Define the term *battery.* Use it in an original sentence.
3. **Give Examples** Identify four signs of abuse.

Thinking Critically

4. **Compare and Contrast** In what ways are neglect and emotional abuse similar? How are they different?
5. **Hypothesize** Why do you think the effects of abuse are often so deeply felt and so long-lasting?

Applying Health Concepts

6. **Personal Health** Contact the National Clearinghouse on Child Abuse and Neglect Information. Use some of their most current statistics to create a chart of the various forms of abuse. Share your findings with your class.

Lesson 2: Understanding Abuse 403

Lesson 3
Finding Help

Focus

LESSON OBJECTIVES

After studying this lesson, students should be able to
- discuss the importance of breaking the cycle of abuse.
- identify sources of help for dealing with and preventing abuse.
- explain the concerns that keep some victims from reporting abuse, and discuss the need to overcome these concerns.

MOTIVATOR

Display photographs showing adults of various ages and backgrounds. Include representatives of different races and ethnic groups as well as people who appear wealthy, middle class, and poor. Ask: Which of these adults might be abusers? What makes you think that? Let students write their responses using brief notes or complete sentences.

INTRODUCING THE LESSON

Encourage volunteers to compare and discuss their responses to the Motivator questions. Guide students as necessary to recognize that abuse takes place in any and every kind of family; education, money, and drug-free living do not guarantee against abuse. Conclude by assuring students that help is available for abuse victims; Lesson 3 tells where.

INTRODUCING WORDS TO KNOW

Let students discuss the meaning of each of the words that make up the Words to Know term. Then have them find the definition. How does each part of the term contribute to its complete meaning?

Lesson 3 Finding Help

This lesson will help you find answers to questions that teens often ask about finding help for abuse victims. For example:
- What help is available for abused and troubled people?
- Why do some people fail to report abuse?

Words to Know

family violence shelter

Breaking the Cycle of Abuse

The longer abuse continues, the greater the damage will be. The key to breaking the cycle of abuse is reporting it and talking about it. If someone has been abused or is in danger of being abused, it is important for that person to tell someone, such as a family member, a teacher, a school nurse, a doctor, a counselor, or another adult he or she trusts. The victim may be afraid to tell, however, for fear that the information will break up the family, that the abuser will go to jail, or that no one will believe him or her.

There are many different types of community programs available to help teens break the cycle of abuse.

404   Chapter 14: Preventing Violence and Abuse

Lesson 3 Resources

Teacher's Classroom Resources
- Concept Map 65
- Cross-Curriculum Activity 28
- Decision-Making Activity 28
- Enrichment Activity 65
- Lesson Plan 3
- Lesson 3 Quiz
- Reteaching Activity 65
- Transparency 58

Student Activities Workbook
- Study Guide 14
- Applying Health Skills 65
- Health Inventory 14

Reporting abuse to law enforcement authorities is one way to get help for both victims and abusers.

Where to Get Help

What kind of help is available for abused children and troubled families? Some of the community resources for dealing with and preventing abuse are described below:

- **Police department.** This is the place to call for help if someone is in immediate danger. In many communities the emergency number for the local police department is 911. Dial 0 to call the operator if you are not sure how to reach the police department in your community.

- **Local hospital.** Hospitals provide emergency medical treatment for people who are injured, hurt, or seriously ill.

- **Family violence shelters.** These are *places where family members in danger of being abused can stay while they get their lives in order.* Counselors at the shelters help family members find solutions to their problems.

- **Family counseling programs.** These programs help family members identify their problems and work together to solve them. School guidance counselors, youth counselors, hospital social workers, and clergy members also provide support to family members on an individual basis or as a group.

- **Support or self-help groups.** In these groups, people have a chance to talk with and listen to others with similar problems. Some support groups are for victims of abuse; others are for abusers. For example, Parents Anonymous is for parents who have abused their children or are afraid that they might begin to do so. Members help one another to understand and change their behavior.

- **Home health visitors.** Some communities arrange for nurses to visit families to help parents improve their parenting skills.

- **Crisis hot lines.** These are telephone services that parents and abused children can call to get help. Some people may be reluctant to talk about their problems to strangers. However, hot line workers have received special training to help people in trouble. All conversations are kept confidential, and the caller does not have to give his or her name.

Cultural Connections

"Space Invaders"

Personal space refers to the space you need between yourself and another person to feel comfortable when you are talking. People from some cultures feel comfortable with less distance between them. People from other cultures may feel comfortable only with greater distances between them. They may feel that people who stand too close are invading their space and showing disrespect. Think of ways to maintain your personal space comfort zone. If you feel that someone is trying to get too close or is touching, patting, or grabbing you, you have a right to ask them to stop. If they do not, walk away or yell for help.

Lesson 3: Finding Help   405

Home and Community Connection

Reporting Abuse Some children and teens try to escape abuse at home by running away. Unfortunately, they often end up being exploited for pornography and prostitution. Abused children should be helped before they become runaways. By law, a child or a teen can be removed from an abusive home environment until the situation is corrected. First, however, the abuse must be reported. All states require doctors and health professionals to report suspected child abuse, and many states require anyone who suspects child abuse to report it. Remind students that the law in your state is incorporated into your school's policy. Discuss how the law affects other community organizations, such as sports groups.

Chapter 14   Lesson 3

Teach

L1 Discussing
Point out the term *cycle of abuse*. Ask: Why do you think this term is used so often? In what sense is abuse a cycle? What do you think a victim can do to break the cycle? What do you think would motivate an abuser to break the cycle?

Visual Learning
Let students describe the group of teens and the counselor in the photograph on page 404: What do you imagine they are discussing? How do the teens seem to feel? Explain that meeting in a support group can be very important to the self-esteem of an abused teen. Encourage students to discuss the benefits of this kind of group work.

L1 Identifying Examples
Let students identify specific situations in which individuals might be in immediate danger. For each situation, ask: Who should call the police? What information should the caller give?

L1 Discussing
Help students discuss the benefits of family counseling programs: Who suffers when an abuser is part of the family? How can working together as a family help the abuser? the victim of abuse? the other family members?

L1 Guest Speaker
Does the school have support groups for students who are victims of abuse? If so, ask the counselor or psychologist who facilitates the groups to speak to the class, explaining the benefits and assuring students of the protection of their privacy. Be sure the speaker tells students what steps they can take if they want more information or if they want to be included in a support group.

405

Chapter 14   Lesson 3

L1 Discussing

Help students consider the importance of reporting abuse: Why might it be difficult for a parent or other family member to believe a son's or daughter's report of abuse? What should a teen do if a parent or other relative refuses to believe the report? What adults at school could a teen turn to?

L1 Analyzing

Let students briefly share their experiences in hearing stories of abuse on talk shows. Then ask: Why do you think some people are willing to talk about abuse in this very public setting? Do you think shows like this help some people? Do you think they hurt some people? Encourage students to explain their responses.

L2 Applying Knowledge

Ask students to recall TV shows or movies in which a family member was abused. What consequences of the abuse did the program portray? What kind of help did the victim get? How realistic do you think the show or movie was?

Assess

EVALUATING THE LESSON

Assign Reviewing Terms and Facts and Thinking Critically on page 407 to review the lesson; then assign the Lesson 3 Quiz in the TCR.

RETEACHING

▶ Assign Concept Map 65 in the TCR.

▶ Have students complete Reteaching Activity 65 in the TCR.

---

in your journal

When you are angry, do you ever use words or actions that hurt someone else? If you do, write in your private journal about what makes you angry enough to do this. Suggest how you might deal with your anger in a healthier way. If you have never expressed your anger in a hurtful way, write some advice that you would give to a friend who does.

Difficult Issues

Though there are many places to go for help, abuse victims often remain silent. The number of reported abuse cases is much lower than the number of actual cases. Why is this true?

- **Victims may feel that nobody will believe them.** Witnesses and victims of abuse often don't report abuse because they fear that people will think they're not telling the truth. Children who have been victims of abuse are especially likely to think this way. However, the only way to stop abuse is to report it. If one person doesn't believe you, tell someone else.

- **People may feel that abuse is a private or personal issue.** Many victims feel uncomfortable sharing something that affects them so personally. For many years, abuse was a subject that no one wanted to talk about. Today, however, the subject of abuse is more openly discussed in the media as well as in the rest of society. Many people have realized how widespread abuse is and that silence only allows it to do more terrible damage.

- **Males may feel that they cannot or should not be victims.** Some males may feel that reporting abuse is an admission of personal weakness. They may have been told that they should be able to protect themselves. Abusers, however, always have an advantage over their victims. Even when they are not physically stronger, they are usually older or in a position of authority.

- **Victims may be afraid of their abusers.** Many abuse victims are afraid to come forward because they think their abusers might get even with them for telling, by hurting them or abusing them again. Abusers often tell children that they will be punished if they tell anyone else what has occurred, or that no one will believe them. However, confiding in a trusted adult can often allow the victim to get help—and get away from the abuser.

HEALTH LAB
Abuse and the Media

Introduction: At one time abuse was off-limits as a topic for discussion. Today abuse is a common subject for television shows and for movies, books, and many other media. Some people feel that more widespread coverage has helped a greater number of people to come forward and report incidents of abuse. Other people are concerned that the wrong kinds of media attention will keep people from taking abuse seriously.

Objective: Work with a group of three or four to study the coverage of abuse in the media and hypothesize about how increased coverage has affected the problem of abuse.

Materials and Method: You will need access to newspapers and news magazines, television, radio, and a community library. Within your group, decide which members will cover each of the media to be studied. For one week, examine media reporting on events of that week that concern abuse. Also, search past reports in your assigned area for coverage on incidents of abuse and sources of help for abuse victims.

Observations and Analysis: As you examine media reports on the subject of abuse, analyze their treatment of the subject. Does the reporting seem responsible and thorough? Is

406 Chapter 14: Preventing Violence and Abuse

HEALTH LAB
Abuse and the Media

Time Needed
1 week

Supplies
▶ access to newspapers, news magazines, television, radio, and a community library

Focus on the Health Lab

Let students share their ideas in response to questions such as these: What is the difference between abuse as it is depicted in fiction (books and movies, for example) and as it is depicted in the news? How realistic are both those depictions? Do you think we need to hear and see more or less about abuse? Why? After this introductory discussion, tell students to form cooperative working groups; have group members read the Health Lab feature together.

Understanding Objectives

This activity focuses students' attention on specific media coverage of abuse. It allows them to analyze that coverage and helps them to begin drawing their own conclusions about possible relationships

Review — Lesson 3

Using complete sentences, answer the following questions on a separate sheet of paper.

Reviewing Terms and Facts

1. **Vocabulary** What is a *family violence shelter*?
2. **Give Examples** List three different places where an abuse victim could go for help.
3. **Recall** List three reasons why people might not report abuse.

Thinking Critically

4. **Analyze** Why might young children be afraid to tell someone about an abusive parent?
5. **Hypothesize** Why do you think the number of reported cases of abuse has been increasing in recent years?

Applying Health Concepts

6. **Growth and Development** Abuse is very harmful to the normal development of a child. Research this subject to find out how people who have been abused deal with and recover from their experiences. Prepare a short report about your findings.

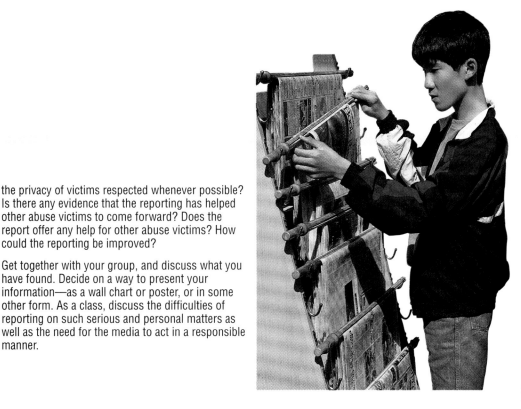

the privacy of victims respected whenever possible? Is there any evidence that the reporting has helped other abuse victims to come forward? Does the report offer any help for other abuse victims? How could the reporting be improved?

Get together with your group, and discuss what you have found. Decide on a way to present your information—as a wall chart or poster, or in some other form. As a class, discuss the difficulties of reporting on such serious and personal matters as well as the need for the media to act in a responsible manner.

Lesson 3: Finding Help

Chapter 14 Lesson 3

LESSON 3 REVIEW

Answers to Reviewing Terms and Facts

1. A family violence shelter is a place where family members in danger of abuse can stay while they figure out what to do.
2. An abuse victim could go to (any three): the police, a hospital, a family violence shelter, a family counseling program, a support group, a crisis hot line.
3. People may avoid reporting abuse because (any three): they feel no one will believe them; they feel that abuse is a private or personal issue; they feel that males cannot, or should not, be victims; they are afraid of the abuser.

Answers to Thinking Critically

4. Responses will vary.
5. Responses will vary.

ENRICHMENT

- Assign Enrichment Activity 65 in the TCR.
- Suggest students contact an abuse-prevention hot line. Ask them to find out the number and types of calls received, the kind of training given to hot-line workers, and the effects the hot line has on preventing abuse.

Close

Ask students to write private journal entries identifying at least three people they or a friend could turn to if faced with abuse.

between media coverage and the number of victims reporting abuse.

Observation and Analysis

Suggest that group members meet several times during the week to discuss their observations: Is everyone finding appropriate media reports? Does anyone need help from other group members? Encourage students to make daily notes of their observations. At the end of the week, provide time during which group members can discuss their responses to the feature questions and plan their presentations.

Further Investigation

Suggest that students discuss their findings with parents and other adults: What have the adults observed about media coverage of abuse? What are their opinions about the effects of that coverage?

Assessing the Work Students Do

Let the members of each group present their display to the class, and encourage students to compare and discuss their findings. Then have each student write a short evaluation of his or her part in the group's investigation and presentation. Assign credit based on students' efforts and participation.

407

Chapter 14 Review

CHECKING COMPREHENSION

Use the Chapter Summary and the Chapter 14 Review to help students go over the most important ideas presented in Chapter 14. Encourage students to ask questions and add details as appropriate.

CHAPTER 14 REVIEW ANSWERS

Reviewing Key Terms and Concepts

1. A gang is a group of people who associate with each other because they believe they have something in common.
2. Efforts to make neighborhoods safer include (any four): more police on the streets, stricter gun laws, nighttime sports programs, improved lighting in parks and playgrounds, Neighborhood Watch programs, teen curfews, tougher punishments for violent crimes.
3. The four main types of abuse are physical abuse, emotional abuse, sexual abuse, and neglect.
4. People who are abused as children often have low self-esteem and experience high levels of stress; they are likely to become involved in abusive relationships as adults.
5. An abused person can turn to (any three): a family member, a teacher, a school nurse, a doctor, a counselor, any trusted adult.
6. Crisis hot lines are telephone services that parents and abused children can call to get help.

Chapter 14 Review

Chapter Summary

▶ **Lesson 1** Violence is a major public health problem in the United States, particularly youth violence. Some factors that contribute to violence are anger, prejudice, possession of weapons, poor coping and communication skills, and abuse of alcohol and other drugs.

▶ **Lesson 2** Abuse is the physical, emotional, or mental mistreatment of another person. The four main types of abuse are physical abuse, emotional abuse, sexual abuse, and neglect.

▶ **Lesson 3** To break the cycle of abuse, people must seek help. There are many community programs available that offer help to abused and troubled families.

Reviewing Key Terms and Concepts

Using complete sentences, answer the following questions on a separate sheet of paper.

Lesson 1
1. What is a *gang*?
2. List four ways in which communities are working together to make their neighborhoods safer.

Lesson 2
3. What are the four main types of abuse?
4. What effects can abuse have on a person?

Lesson 3
5. List three people an abused person might go to in order to break the cycle of abuse.
6. What are crisis hot lines?

Thinking Critically

Using complete sentences, answer the following questions on a separate sheet of paper.

7. **Evaluate** How strong an effect do you think violence on television and in movies has on people in our society? Explain your answer.
8. **Explain** How can reporting abuse help the abuser as well as the victim?
9. **Compare and Contrast** Some of the places listed on page 405 offer short-term help, while others offer long-term help. When might short-term help be needed? When might a person seek long-term help?

408   Chapter 14: Preventing Violence and Abuse

Meeting Student Diversity

Language Diversity Use the following suggestions to help students who have difficulty with English:

▶ Pair those learners with native speakers of English who can restate the Chapter Summary in language that helps students comprehend important concepts.

▶ Direct auditory learners or those students with language diversity to the Teen Health Audiocassette Program. Available in English and Spanish, this component provides an audio and written summary of the chapter.

Chapter 14 Review

Your Action Plan

Although violence is present throughout our society, there are many ways to protect yourself from becoming a victim of violence. To do this you will need to make an action plan.

Step 1 Look through your private journal entries for this chapter. What do they tell you about the choices you can make to keep yourself from becoming a victim of violence?

Step 2 Set a long-term goal, such as not getting involved in fights or working to reduce violence in your school or community.

Step 3 Write down at least two short-term goals that will help you reach the long-term goal you have chosen. If your long-term goal is to work to reduce violence in your community, one of your short-term goals might be to find out what violence prevention programs are in place at your school.

Plan a timetable for taking the steps that will help you reach your short-term and long-term goals. When you have reached your long-term goal, you will have the personal satisfaction of knowing that you have protected yourself and others from violence in your community.

In Your Home and Community

1. **Community Resources** Find out about crime prevention programs in your community. The National Crime Prevention Council, 1700 K Street NW, 2nd Floor, Washington, DC 20006-3817, can provide advice on how to start one.

2. **Health of Others** Collect free pamphlets that contain information about places where victims of abuse and violence can go for help. Ask permission to put them in a place where other teens go, such as the school library.

Building Your Portfolio

1. **Local Profile** Collect articles from a newspaper about violence in your area. Put together a scrapbook of the articles, along with suggestions for avoiding the kind of violent act described. Place the scrapbook in your portfolio.
2. **Short Story** Write a story in which a teen helps a younger child who has been abused. Be sure to describe how the teen finds out about the abuse and what he or she does to help. Place a copy of your story in your portfolio.
3. **Personal Assessment** Look through all the activities and projects you did for this chapter. Choose one or two that you would like to include in your portfolio.

Chapter 14: Chapter Review

Chapter 14 Review

Thinking Critically

Responses will vary. Possible responses are given.

7. Seeing acts of violence, especially when realistic consequences are not shown, makes violence seem less shocking and more acceptable as a means of resolving grievances.
8. Reporting the abuse may allow the abuser to get help with the problems that led to the abuse.
9. Short-term help is needed when abuse is happening or threatened. Long-term help is needed after an incident is reported or when a person becomes aware of his or her own abusive behavior and wants to seek help.

RETEACHING

Assign Study Guide 14 in the Student Activities Workbook.

EVALUATE

▶ Use the reproducible Chapter 14 Test in the TCR, or construct your own test using the Testmaker Software.

▶ Use Performance Assessment 14 in the TCR.

EXTENSION

Have students work together to compile a list of local agencies and individuals who can help teens who have been or are being abused. Let students decide on the best way to share this information with other teens in the school and in the rest of their community.

Performance Assessment

▶ **Self-evaluation** Direct students to review the activities that are provided throughout the chapter. Encourage each student to select one finished product or activity that demonstrates his or her best work for the chapter. Have students explain what they learned and how the examples they selected show their progress.

▶ **Teacher's Classroom Resources** Assign Performance Assessment 14, "A Victim's Story," in the TCR.

Planning Guide

Chapter 15
Safety and Recreation

	Features	**Classroom Resources**

Lesson 1: Building Safe Habits
pages 412–415

Making Healthy Decisions: Preventing Accidents While Baby-sitting
page 414

- Concept Map 66
- Enrichment Activity 66
- Lesson Plan 1
- Lesson 1 Quiz
- Reteaching Activity 66
- Transparency 60

Lesson 2: Safety at Home and in School
pages 416–421

Life Skills: How to Extinguish Fires Correctly
pages 418–419

- Concept Map 67
- Cross-Curriculum Activity 29
- Decision-Making Activity 29
- Enrichment Activity 67
- Lesson Plan 2
- Lesson 2 Quiz
- Reteaching Activity 67
- Transparency 61

Lesson 3: Safety on the Road and Outdoors
pages 422–427

TEEN HEALTH DIGEST
pages 428–429

- Concept Map 68
- Enrichment Activity 68
- Health Lab 15
- Lesson Plan 3
- Lesson 3 Quiz
- Reteaching Activity 68
- Transparency 62

Lesson 4: Safety in Weather Emergencies
pages 430–434

- Concept Map 69
- Enrichment Activity 69
- Lesson Plan 4
- Lesson 4 Quiz
- Reteaching Activity 69

Lesson 5: First Aid
pages 435–439

- Concept Map 70
- Enrichment Activity 70
- Lesson Plan 5
- Lesson 5 Quiz
- Reteaching Activity 70
- Transparency 63

	Features	Classroom Resources

Lesson 6 — Protecting Our Planet
pages 440–443

Features
- Health Lab: The Dangers of Water Pollution, *page 443*

Classroom Resources
- Concept Map 71
- Cross-Curriculum Activity 30
- Decision-Making Activity 30
- Enrichment Activity 71
- Lesson Plan 6
- Lesson 6 Quiz
- Reteaching Activity 71
- Transparency 64

National Health Education Standards

Listed below are the Health Education Standards Performance Indicators addressed in each lesson of this chapter.

Lesson	Health Standards Performance Indicators
1	(1.1, 1.5, 1.6, 1.8, 3.1, 6.1)
2, 6	(1.1, 1.5, 1.6, 1.8, 3.1, 7.5)
3	(1.1, 1.5, 1.6, 1.8, 3.1, 3.6)
4	(1.1, 1.5, 1.6, 1.8, 3.1)
5	(1.1, 1.7, 3.1)

Chapter Resources

Teacher's Classroom Resources
- Chapter 15 Test
- Parent Letter and Activities 15
- Performance Assessment 15
- Testmaker Software

Student Activities Workbook
- Study Guide 15
- Applying Health Skills 66–71
- Health Inventory 15

Student Diversity Strategies
- Audiocassette Program (English)
- Audiocassette Program (Spanish)
- Spanish Parent Letters
- Spanish Summaries, Quizzes, and Activities

Multimedia Components
- English Audiocassette Program
- Spanish Audiocassette Program
- *Teen Health* Videodisc/VHS Series

Other Resources

Readings for the Teacher

American Red Cross and Kathleen A. Handal. *The American Red Cross First Aid and Safety Handbook.* Boston: Little, Brown, 1992.

Micheli, Lyle J., and Mark D. Jenkins. *Sportswise: An Essential Guide for Young Athletes, Parents, and Coaches.* Boston: Houghton Mifflin, 1990.

O'Connor, Raymond. *The Healthy Home Environment Guide.* New York: Berkley Books, 1996.

Readings for the Student

Aaseng, Nathan, and Jay Aaseng. *Head Injuries.* New York: Franklin Watts, 1996.

First Aid for Children Fast. London and New York: Dorling Kindersley, 1995.

Pediatric Emergency Department, Children's Hospital at Yale-New Haven. *Now I Know Better: Kids Tell Kids about Safety.* Brookfield, CT: Millbrook Press, 1996.

Out of Time?

If time does not permit teaching this chapter, you may use these features: Making Healthy Decisions on page 414; Life Skills on pages 418–419; Teen Health Digest on pages 428–429; Health Lab on page 443; and the Chapter Summary on page 444.

Chapter 15
Safety and Recreation

CHAPTER OVERVIEW

Chapter 15 teaches students what they can do to protect their own safety, to protect and aid others, and to contribute to the safety of the environment.

Lesson 1 explains the sequence of events that form an accident chain and helps students recognize how they can avoid common accidents and injuries.

Lesson 2 focuses students' attention on safety hazards in the home and helps them protect their safety at home and at school.

Lesson 3 reminds students of the importance of knowing and following safety guidelines while walking, cycling, or skating, and while enjoying other outdoor activities.

Lesson 4 discusses various kinds of weather emergencies, as well as earthquakes, and explains how students can protect themselves in these situations.

Lesson 5 presents first-aid information that will help students recognize emergency situations and provide immediate care for victims of accidents and serious illness.

Lesson 6 guides students in understanding what they can do to help reduce pollution of the air, water, and land.

PATHWAYS THROUGH THE PROGRAM

The lessons in this chapter help students develop their ability both to *be* safe and to *feel* safe. The chapter guides students in assuming responsibility for themselves and for others, and thus enhances their mental/emotional health and their social health.

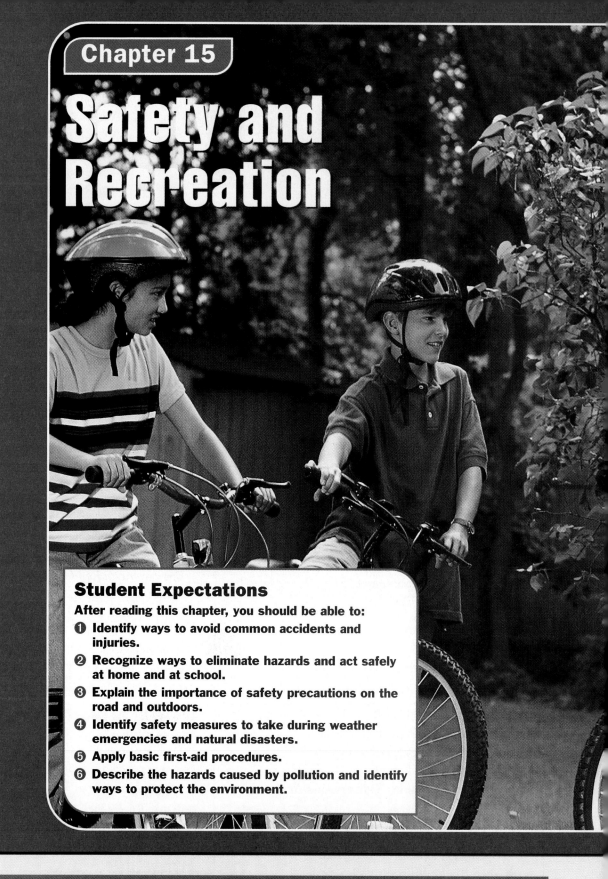

Chapter 15
Safety and Recreation

Student Expectations
After reading this chapter, you should be able to:
1. Identify ways to avoid common accidents and injuries.
2. Recognize ways to eliminate hazards and act safely at home and at school.
3. Explain the importance of safety precautions on the road and outdoors.
4. Identify safety measures to take during weather emergencies and natural disasters.
5. Apply basic first-aid procedures.
6. Describe the hazards caused by pollution and identify ways to protect the environment.

Key to Ability Levels

Teaching strategies that appear throughout the chapter have been identified by one of three codes to give you an idea of their suitability for students of varying learning styles and abilities.

L1 Level 1 strategies should be within the ability range of all students. Full class participation is often required. Teacher direction is usually needed.

L2 Level 2 strategies are for average to above-average students or for small groups. Some teacher direction is necessary.

L3 Level 3 strategies are designed for students able and willing to work independently. Minimal teacher direction is necessary.

Teen Chat Group

Bryan: I'm really bored. Why don't we go explore that construction site on River Road?

Jamal: Don't they have a fence around that place?

Bryan: Sure, but I'll bet we could climb it without too much trouble.

Mika: Are you kidding? We could get killed.

Bryan: Come on, Mika. How dangerous can it be? We'll be careful.

Jamal: Mika's right, Bryan. One of us could get hurt, and we could all get in trouble. It's definitely not worth the risk.

Mika: There must be something else we could do. Why don't we go over to the park and play basketball?

in your journal

Read the dialogue on this page. Have you ever been tempted to do something unsafe? Start your private journal entries on ways to act safely and avoid injuries by responding to these questions:

- How would you convince Bryan not to explore an unsafe area?
- Why is it important to act safely?
- What safety rules do you follow when you are active outdoors?

When you reach the end of the chapter, you will use your journal entries to make an action plan.

Chapter 15: Safety and Recreation **411**

Chapter 15

INTRODUCING THE CHAPTER

▶ Show students slides or photographs of different situations that require safety-related action: accidents "waiting to happen," serious storms, and a teen with a serious gash after a bicycle fall, for example. Let students describe the scene in each picture. Ask: What has happened here? What needs to be done? If you were on the spot, would you be able to protect your own safety? Would you like to be able to protect other people? help people in need? Encourage students to recognize that they need specific information to deal effectively with specific kinds of emergencies. Explain that they will find that kind of information in the lessons of Chapter 15.

▶ Display several different safety posters and let students discuss them: What are the safety practices suggested by each poster? Who needs to be reminded of those practices? Why? Then encourage students to recall other safety reminders they have learned. Ask: Why is it important to know these safety guidelines? How can they help you when you have to decide whether to take a risk or what to do in an emergency? Tell students they will learn more about safety in Chapter 15.

Cooperative Learning Project

Living Safe: An Original Play

The Teen Health Digest on pages 428 and 429 provides students with high-interest articles related to the content of this chapter.

The material in the Teacher's Wraparound Edition presents suggestions for a class project in which students plan, write, and perform an original play that focuses on important safety considerations.

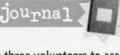

Ask three volunteers to assume the roles of the three teens in the photograph and read the Teen Chat Group aloud. Guide students in discussing the conversation: What is wrong with Bryan's suggestion? What are other unsafe activities that teens sometimes suggest? What would you say to a friend who suggested one of those activities? Why? What are fun outdoor activities you and your friends can enjoy without taking unnecessary risks? Why is it easier for Mika and Jamal to say no to Bryan together? Why is it important to avoid risky situations, even if you have to do so on your own? After this discussion, let students read the In Your Journal questions and write their own private journal entries.

Lesson 1
Building Safe Habits

Focus

LESSON OBJECTIVES

After studying this lesson, students should be able to
- discuss the importance of being safety conscious.
- explain what an accident chain is and how to break it.

MOTIVATOR

Provide one picture of an accident for each small group of students. Let students write a sentence explaining what happened, followed by a list of actions that might have prevented the accident.

INTRODUCING THE LESSON

Have each group describe their accident and how it might have been prevented. Explain that students will learn more about accident prevention in Lesson 1.

INTRODUCING WORDS TO KNOW

Let students read aloud and discuss the Words to Know. Then have them look up the terms and write each one in a sentence.

VIDEODISC/VHS

Teen Health Course 2

You may wish to use video segment 2, "Personal Responsibility," to show students some habits that will benefit their health throughout their lives.

Videodisc Side 1, Chapter 2
Personal Responsibility

Search Chapter 2, Play to 3

Lesson 1
Building Safe Habits

This lesson will help you find answers to questions that teens often ask about avoiding accidents and injuries. For example:

▶ What does it mean to be safety conscious?
▶ Why do accidents and injuries occur?
▶ How can I avoid common accidents and injuries?

Words to Know

safety conscious
accident chain

Safety First

As soon as you were old enough to understand safety rules, a parent or some other family member probably began to teach them to you. For example, you were most likely told not to run into the street, talk to strangers, or play with matches.

Although safety rules are important, you also need to know more. You must become **safety conscious,** which means being *aware that safety is important and careful to act in a safe manner.* Being safety conscious also means that you plan ahead and spot possible hazards. Throughout this book you have learned ways to prevent accidents and injuries by avoiding risk behaviors. Safety consciousness involves using these prevention skills in your everyday habits and actions.

Many young people believe that bad things happen only to *other* people, but this just isn't true. Each year, about 2 million people are hospitalized, and more than 140,000 die, because of accidents and injuries. Many of these people are teens. **Figure 15.1** shows the types of accidents that cause the largest number of deaths in the United States.

Figure 15.1
Leading Causes of Death from Accidents, 1993

Almost half of the deaths of people aged 15 to 24 resulted from accidents. Many of these accidents could be prevented.

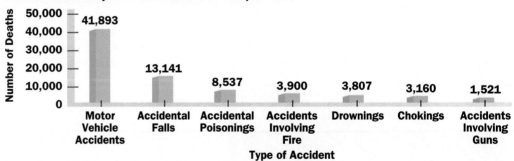

Source: Statistical Abstracts of the United States, 1996

412 Chapter 15: Safety and Recreation

Lesson 1 Resources

Teacher's Classroom Resources
- Concept Map 66
- Enrichment Activity 66
- Lesson Plan 1
- Lesson 1 Quiz
- Reteaching Activity 66
- Transparency 60

Student Activities Workbook
- Study Guide 15
- Applying Health Skills 66

The Accident Chain

Avoiding risks and preventing injuries doesn't mean that you have to give up riding a bike or playing sports. You do need to be careful, plan ahead, and take precautions to protect yourself. These precautions include wearing protective gear and practicing common sense. Most accidents and injuries occur because of carelessness. They are often the result of an **accident chain,** which is *a series of events that include a situation, an unsafe habit, and an unsafe action.* **Figure 15.2** illustrates an accident chain.

> **interNET CONNECTION**
> Learn on-line how to recognize and avoid safety hazards and how to treat injuries when they occur.
> http://www.glencoe.com/sec/health

Figure 15.2
Example of an Accident Chain
The unsafe actions of both Nicole and her brother Dylan contributed to this accident chain.

❶ **The Situation**
It is Nicole's turn to wash the dinner dishes. She wants to finish quickly because she has lots of homework to do.

❷ **The Unsafe Habit**
Nicole rinses the dishes with the water running at full volume. She is not careful about splashes, drips, and spills on the floor.

❸ **The Unsafe Action**
Finished with her chore, Nicole rushes upstairs to do her homework. She does not wipe up the spilled water on the floor.

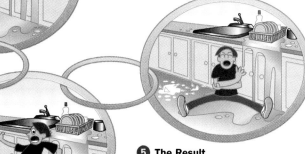

❹ **The Accident**
During a commercial in the middle of his favorite TV show, Nicole's brother Dylan comes into the kitchen to get a glass of milk. Rushing back out, he doesn't see the water on the floor, and he slips.

❺ **The Result**
When Dylan falls, he bruises his elbow and turns his ankle. He drops and breaks the glass of milk.

Lesson 1: Building Safe Habits **413**

Chapter 15 Lesson 1

Teach

L1 Identifying Examples
Ask the students to identify specific safety rules they were taught as preschoolers and as young schoolchildren. Then help students discuss these safety rules: What was the purpose of each rule? How are the rules relevant to your lives today?

L1 Discussing
Guide students in discussing what being safety conscious might mean in these situations:

▶ An older teen who has been drinking offers to give you a ride home.

▶ Two other students urge you to smoke marijuana with them.

▶ Your friends want you to go along in exploring a construction site.

▶ Older teens make fun of the safety equipment you wear while in-line skating.

Then let students identify other common situations and explain how being safety conscious would affect their behavior in those situations.

L2 Math
Suggest that interested students research the causes of death, including accidents, among teens in the United States. Have them plan, draw, and present a circle graph that shows their findings.

Visual Learning
Guide students in discussing the information presented in the graph, Figure 15.1. Pose questions such as these: Which kind of accident caused the most deaths? How many deaths resulted from accidental poisoning? from drowning? How many more deaths were caused by accidental falls than were caused by accidents involving fire?

Personal Health

Part of being safe is knowing how to approach police officers and other figures of authority with respect and confidence. What if an accident does result in injury? What if a friend does become unconscious after experimenting with a drug? Students will know that they need to summon help, but they may not think ahead: They will have to be prepared to speak clearly to a representative of the police department or a medical facility. Let students discuss appropriate responses and then role-play various encounters of this kind. If possible, invite a local officer to speak with the class, describing the kinds of questions that might be asked after an accident and discussing the helpful answers that teens can give.

413

Chapter 15 Lesson 1

Visual Learning

Let students read and discuss each step in the accident chain shown in Figure 15.2. Have pairs of students take turns acting out the sequence of events and then identifying how they might change each step to avoid the accident.

L2 Creating Models

Divide the whole class into five groups, and provide each group with five long strips of construction paper. Ask the members of each group to think of a situation that might be the beginning link in an accident chain. Have them write a sentence about that situation on one strip of paper and then tape or glue the two ends of that strip together to form a circle. Instruct each group to pass its situation link on to another group. Have the members read the situation on the link they have been given and think of an unsafe habit that might be the next link in that accident chain. Let group members write and add a second link to the chain and then pass the chain on. Continue in this way until the class has produced five complete accident chains. Finally, let students read aloud the links in each accident chain and make specific suggestions for ways to break that chain.

L2 Writing Stories

Ask students to write short stories in which teens avoid accidents by breaking one of the links in a potential accident chain. Encourage students to share their stories with the rest of the class.

Did You Know?

Needless Drownings

Each year many young toddlers drown after climbing into cleaning pails that are partially filled with water or other liquids. If there are small children in your home, or if you baby-sit, keep cleaning pails out of reach. For similar reasons, keep all toilet lids closed, and don't let babies or toddlers play in bathrooms.

Breaking the Accident Chain

Most accidents and injuries can be prevented by breaking the accident chain. For example, either Nicole or Dylan could have acted differently during at least one step in the series of events. If so, Dylan's accident and injury might not have happened (see **Figure 15.3**).

Figure 15.3
Breaking the Accident Chain
By removing any link from the accident chain, you can prevent most accidents and injuries.

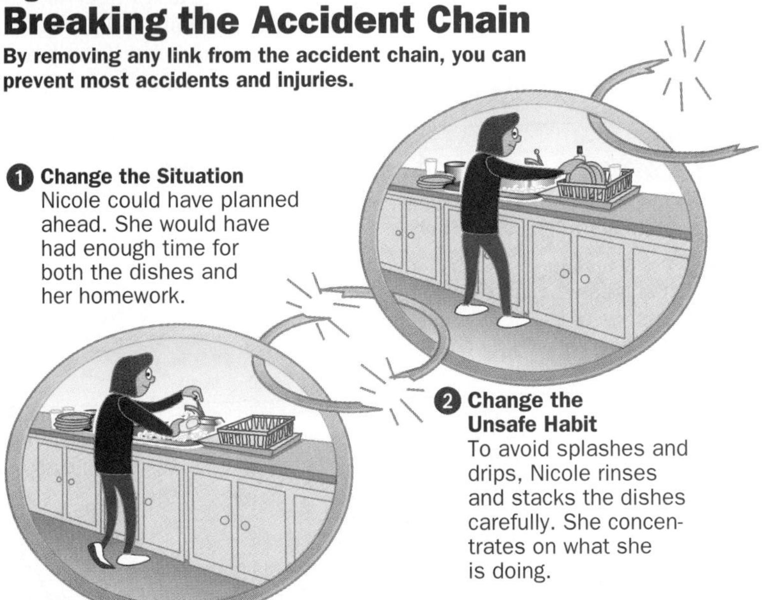

① Change the Situation
Nicole could have planned ahead. She would have had enough time for both the dishes and her homework.

② Change the Unsafe Habit
To avoid splashes and drips, Nicole rinses and stacks the dishes carefully. She concentrates on what she is doing.

MAKING HEALTHY DECISIONS
Preventing Accidents While Baby-sitting

Mrs. Salenas has hired Brandon to baby-sit her four-year-old son, Tommy. She asks Brandon to give Tommy a bath before putting him to bed. Mrs. Salenas gives Brandon the telephone number where she can be reached.

An hour before bedtime, Brandon takes Tommy to the bathroom and starts to turn on the water in the tub. "I do it myself!" Tommy shouts. "I'm big now! You go away."

Brandon has learned in health class that small children can get burned from hot tap water. He also knows that they can easily drown in even a small amount of water. "No, Tommy," he says. "I have to stay with you and make sure that you're safe."

Tommy begins to cry loudly. Brandon doesn't know what to do, so he decides to use the six-step decision-making process.

414 Chapter 15: Safety and Recreation

① **State the situation**
② **List the options**
③ **Weigh the possible outcomes**
④ **Consider your values**
⑤ **Make a decision and act**
⑥ **Evaluate the decision**

Follow-up Activities

1. Apply the six steps of the decision-making process to Brandon's situation.
2. Role-play a scene in which Brandon acts on his decision.
3. In small groups, discuss other options that Brandon might have chosen.

MAKING HEALTHY DECISIONS

Preventing Accidents While Baby-sitting

Focus on Healthy Decisions

Let students share some of their own ideas about baby-sitting: What special responsibilities do you have when you baby-sit? How do those responsibilities change with the ages of the children you are caring for? Why is it especially important to be safety conscious when you baby-sit?

Activity

Use these ideas to discuss Follow-Up Activity 1:

Step 1 Brandon has to decide whether and how to give four-year-old Tommy a bath.

Step 2 Brandon could let Tommy take a bath by himself. He could insist on giving Tommy a bath, running the water and supervising Tommy throughout. He could call Mrs. Salenas and ask what to do. Or, he could put Tommy to bed without bathing him.

Step 3 If Brandon lets Tommy take a bath by

3 Change the Unsafe Action Finished with her chore, Nicole checks for spills on the floor and wipes them up.

4 No Accident Dylan does not rush from the kitchen. He uses a plastic cup instead of a glass for his milk.

5 No Injury Dylan drinks his milk and enjoys the rest of the show.

in your journal

Have you, or has a member of your family, ever been injured? How might the injury have been prevented? In your private journal, describe a possible injury, and draw a diagram to break it down into the links of an accident chain. Then describe ways in which the chain might have been broken.

Review — Lesson 1

Using complete sentences, answer the following questions on a separate sheet of paper.

Reviewing Terms and Facts

1. **Vocabulary** Define *safety conscious*. Explain how being safety conscious can help you to prevent common accidents and injuries.
2. **List** Name the links in an accident chain.
3. **Restate** How can an accident chain be broken?

Thinking Critically

4. **Explain** What do you think is meant by "safety first"?
5. **Hypothesize** People aged 15 to 24 are involved in more accidents than any other age group. Why do you think this might be true?

Applying Health Concepts

6. **Health of Others** Make a poster to show an accident chain that might occur during a sporting event or bike trip. Include a second illustration that shows how the accident chain could have been broken. Display your poster on the classroom bulletin board.

Lesson 1: Building Safe Habits **415**

Lesson 2

Safety at Home and in School

Focus

LESSON OBJECTIVES

After studying this lesson, students should be able to

- identify fire-safety measures to be taken at home.
- explain how to prevent falls, poisonings, electrical shocks, and gun accidents at home.
- discuss guidelines for being safe at school.

MOTIVATOR

Pose the following question: What specific accidents can people have in their own homes? Give students two or three minutes to list as many examples as they can.

INTRODUCING THE LESSON

Let students compare and discuss the lists they wrote in response to the Motivator question. Encourage students to discuss some of the accidents included in their lists: What might cause that accident? What consequences could it have? How could it be prevented? Tell students that Lesson 2 presents information that will help them prevent accidents and injuries at home and at school.

INTRODUCING WORDS TO KNOW

Encourage students to discuss what they already understand about the four terms listed as Words to Know. Then have volunteers read aloud the definitions provided in the Glossary. Finally, have each student draw a simple web illustrating possible relationships among these Words to Know.

Lesson 2 Safety at Home and in School

This lesson will help you find answers to questions that teens often ask about staying safe at home and at school. For example:

▶ How can I protect my home from fire?
▶ How can I eliminate other common safety hazards at home?
▶ How can I avoid accidents and injuries in school?

Words to Know

flammable
smoke detector
electrocution
electrical overload

in your journal

Make a list of all the fire safety rules you can think of in three minutes. As you read this lesson, add to your list.

Safety at Home

Most people feel safest at home. However, even this safest of places has its hazards, or possible sources of harm. In fact, each year millions of people are seriously injured in their own homes. Like other accidents and injuries, most of these that occur at home could be prevented.

Home Fire Safety

According to the American Red Cross, fires cause more deaths than any other type of disaster. The leading causes of home fires include

- careless cooking habits.
- cigarette smoking.
- improperly used or poorly maintained heating devices.

- improper storage of materials that are **flammable**, ones that are *able to catch fire easily*, such as paint, old newspapers, and rags.
- overloaded or damaged electrical circuits and wiring.

Figure 15.4 on the next page lists several precautions that will prevent fires in the home.

| Careful cooking habits can help prevent a kitchen fire.

416 Chapter 15: Safety and Recreation

Lesson 2 Resources

Teacher's Classroom Resources
- 📁 Concept Map 67
- 📁 Cross-Curriculum Activity 29
- 📁 Decision-Making Activity 29
- 📁 Enrichment Activity 67
- 📁 Lesson Plan 2
- 📁 Lesson 2 Quiz

- 📁 Reteaching Activity 67
- 🖨 Transparency 61

Student Activities Workbook
- 📁 Study Guide 15
- 📁 Applying Health Skills 67

Figure 15.4
Effective Fire Safety Measures

Taking these fire safety measures can help prevent fires, or save lives if a fire should occur.

Action

1. Install **smoke detectors**, or *devices that sound an alarm when they sense smoke*. Place a smoke detector outside each sleeping area and on each additional level of the home. Test smoke detectors once a month. Replace batteries at least once a year.

2. Keep stoves and ovens clean. Wipe up spills promptly, and clean the stove thoroughly at least once a week.

3. Install a fire extinguisher in the kitchen, away from heat sources and near an exit.

4. If anyone smokes, make sure that he or she does not smoke in bed and that ashtrays are completely cold before their contents are thrown away.

5. Store matches and cigarette lighters in safe places, out of the reach of children.

6. Follow directions when using portable heating devices. Keep them away from flammable materials. Unplug the heater before leaving the room.

7. Check electrical appliances for loose or damaged cords. Repair or replace damaged appliances and broken outlets. Never run electrical cords under carpets.

Reason

1. Many home fires occur in the middle of the night, when everyone is asleep. Smoke detectors save lives by waking people up. The alarm will not sound, however, if the detector's batteries are dead, or if connections inside are dirty or worn.

2. All food ingredients burn. The heat of the oven, the burners, or an open flame can cause spills to catch on fire. Spattered grease or oil can be especially dangerous.

3. Oil, grease, and electrical fires are among the most common and most serious kitchen fires. Water cannot be used to put out these types of fires. Instead, a fire extinguisher labeled *ABC* must be used. (See the Life Skills feature in this lesson.)

4. Many home fires have been caused by people who fell asleep while smoking or put smoldering cigarettes into wastebaskets.

5. Children are naturally curious about fire. Many injuries and deaths have occurred because children were playing with matches or cigarette lighters.

6. Portable heaters are a common cause of home fires. If they come into contact with flammable materials, a fire can start.

7. Loose or damaged cords release heat that can cause a fire. The movement of people and the weight of furniture can damage electrical cords that are under rugs, causing wires to be exposed.

Planning an Escape Route

Taking action to prevent fires is one of the most important ways to protect your family and your home. A fire could still occur, however, so you and your family must be prepared.

Work together to plan an escape route for each family member. Since most fatal home fires occur during the night, every escape route should begin in the bedroom. A window with a fire escape or a fire ladder may save a life if the bedroom door is blocked by flames, heat, or smoke.

After all family members understand their escape routes, agree on a place to meet outside your home. That way you can make sure that everyone is safe. All family members should promise that during a fire they will concentrate on saving *themselves*—not their belongings or even pets. Finally, practice your escape plan by holding a family fire drill twice a year.

Did You Know?

Smoke Detectors

According to surveys cited in the "University of California at Berkeley Wellness Letter," as many as half of the smoke detectors installed in homes are not in working order. Batteries should be tested regularly and replaced as needed. All smoke detectors should be replaced after they have been used for ten years.

Home and Community Connection

What information and public relations programs does the local fire department offer? In many communities, fire stations present weekend fairs, during which younger children can examine the firefighting equipment and teens and adults can learn more about fire safety. Ask several volunteers to look into such programs in your area and to share their findings with the rest of the class. If such programs are available, students might volunteer to help advertise and promote them. In addition, students may be interested in gathering and distributing relevant fire safety information, making their own brochures and posters to be handed out to families or displayed in public areas.

Chapter 15 Lesson 2

L1 Discussing
Guide the students in discussing each of the steps in escaping a fire, as presented in Figure 15.5: Why is it wise not to call 911 before leaving the building? If the door to your room is closed and it feels hot, what other options do you have? How can you keep yourself calm if you have to crawl through smoke or stay in your room? What does it mean to "stop, drop, and roll"? (You may want to let a few volunteers demonstrate.) Why is it essential to go to the prearranged meeting place?

L2 Guest Speaker
Invite a representative of the local firefighting department to speak with students about fire safety at home. Request that the speaker address issues both of prevention and of escape. Remind students to listen attentively and to ask appropriate follow-up questions.

L3 Researching
Suggest that interested students research arson as a cause of home fires: What is arson? What motive underlies many acts of arson? What percentage of fires in your area or state are caused by arson? What is the punishment for arson in your state? Have students present their findings in written reports.

L1 Discussing
Encourage students to share their ideas about the dangers of falls: What kinds of injuries can result from a fall? Which family members are most likely to fall?

In Case of Fire

You have learned how to prevent fires. Figure 15.5 outlines the steps to take if a fire does break out. Knowing what to do ahead of time will help you stay calm and could save your life and the lives of other family members.

Figure 15.5
Effective Steps for Escaping a Fire
If a fire breaks out, follow the escape route that you planned with your family.

1 Leave quickly. Get out of the building *before* calling 911 or the fire department.

2 Before opening a closed door, feel to see if it is hot. If it is, *do not* open it. There may be flames just outside the door.

3 If you must exit through smoke, crawl along the floor. Smoke and hot air rise. Stay as low as possible, breathing in the cleaner air.

4 If you can't get out, stay in the room with the door closed. Roll up a blanket and put it across the bottom of the door to keep out smoke. If there is a telephone in the room, call 911 or the fire department. If possible, open the window and yell for help.

5 If your clothing catches on fire, stop, drop, and roll. Rolling on the ground will help smother the flames. Never run—the rush of air will fan the flames.

6 Once outside, go to the prearranged meeting place. Let everyone know that you are safe. Then someone should go to a neighbor's home and call 911 or the fire department. *Never* re-enter a burning building.

LIFE SKILLS
How to Extinguish Fires Correctly

For a fire to occur, three elements must be present: fuel, oxygen, and heat. To prevent or extinguish a fire, one of these three factors must be eliminated. Most fires must be extinguished by fire fighters. If the fire is small, however, you may be able to put it out yourself. Follow these rules.

1. **Use water wisely.** Water will extinguish fires in which paper, wood, or cloth is burning. However, *never* use water on the following types of fires:

▶ **Oil or grease fires.** Putting water on burning grease will spread the fire. If grease is burning in a frying pan, turn off the heat, and smother the flames with a lid.

▶ **Electrical fires.** Putting water on an electrical fire can result in **electrocution,** or *death by electrical shock.* A fire extinguisher must be used to put out electrical fires.

2. **Install a fire extinguisher.** Fire extinguishers are labeled A, B, or C. *Class A* will put out

418 Chapter 15: Safety and Recreation

LIFE SKILLS
How to Extinguish Fires Correctly

Focus on Life Skills
Introduce this feature by letting students respond to questions such as these: What should you do if a smoke detector awakens you at night? if you see smoke coming from a building you are passing? if you see that a field in your neighborhood is on fire? Are there ever any fires that you can—or should—try to put out on your own? What would you need before you could try to put out a small fire? Tell students that they will get that specific information from this Life Skills feature; then let students read the feature independently.

Making Health Connections
Knowing whether and how to extinguish a small fire independently is an essential safety skill. A teen who makes the wrong choice can put himself or herself at risk, and may endanger many other lives as well. This Life Skills activity teaches students how

Preventing Other Injuries at Home

Fires are not the only events that can cause serious injuries or deaths in the home. You and your family must also take care to prevent falls, poisonings, electrical shocks, and accidents involving guns. The following sections will provide this safety information.

Preventing Falls

Falls account for the largest number of nonfatal injuries in homes. Most falls are caused by tripping or slipping. The kitchen, bathroom, and stairs are the most common sites for falls. You can help prevent such accidents in your home by following these rules.

By using a stepladder instead of a chair to reach items on high shelves, you can prevent a fall.

- **Safety in the kitchen.** Wipe up spills promptly. Use a stepladder, not a chair, to reach items on upper shelves. After washing the floor, alert family members, or display a "Caution: Wet Floor" sign.

- **Safety in the bathroom.** Put nonslip strips on the bottom of the tub or shower. Install a secure handgrip on the side of the tub. Anchor rugs with nonslip tape, or use rugs with latex backings.

- **Safety on the stairs.** Do not leave objects on the steps. Keep all staircases well lighted. Check the handrails to be sure that they are stable. If young children live in the house, put gates at the top and bottom of the staircase.

- **Safety in other rooms.** Secure electrical cords to baseboards with tape. Repair or replace all loose or worn carpeting. Arrange furniture so that it is out of heavy-traffic areas.

---

ordinary fires, such as those involving paper, wood, or cloth. *Class B* should be used for fires involving fast-burning liquids, such as oil, grease, and gasoline. *Class C* will extinguish electrical fires. All-purpose fire extinguishers are labeled *ABC*, meaning they are safe and effective to use on all three types of fires. Follow these tips for the proper use and storage of fire extinguishers.

- Make sure that your home has at least one ABC fire extinguisher.
- Install the fire extinguisher in a handy place, away from heat, and near an exit.
- Read and understand the instructions after purchasing the extinguisher so that you are prepared in case of fire.
- Check the pressure gauge periodically to make sure that the fire extinguisher is ready to use. If not, it should be recharged or replaced.
- When using the extinguisher, always aim at the *base* of the flames, not at the top.

Follow-up Activity

Discuss with family members the rules for extinguishing fires safely. If your home does not have a fire extinguisher, urge your parents or other adult members to buy one.

Chapter 15 Lesson 2

L1 Critical Thinking
Ask the students to discuss how they would respond in this situation: You go to another student's home to work on a class project. You notice a small gun sitting on a high bookshelf but in clear view. What—if anything—will you say and do? Why?

L3 Social Studies
Ask the class to research the issue of curfews for teens: Does your community have a curfew? If so, what is that curfew? When, how, and by whom is it enforced? What purpose is it intended to serve? If the community does not have a curfew, has one ever been considered? Why? Why was it rejected? Let students share the results of their research and then discuss their reactions to curfews for teens: Do curfews help reduce violence? Are the restrictions on individual liberties justified by the increased safety?

L1 Comprehending
Guide students in restating and discussing the importance of each guideline for being safe at school. How does each guideline affect you and other students at school?

L2 Creating Posters
Students at a school can be truly safe only when all students follow the school's safety guidelines. Let the class work in cooperative groups, each to plan and make a poster reminding others to follow a specific safety guideline. Display the completed posters in the school hallways or other public areas.

Assess

EVALUATING THE LESSON
Assign Reviewing Terms and Facts and Thinking Critically on page 421 to review the lesson; then assign the Lesson 2 Quiz in the TCR.

Q & A

Poisonous Paint?

Q: I heard that young children can be poisoned by lead-based paint. How do people prevent this?

A: Most lead-based paint in good condition is not a hazard. However, children can become seriously ill if they breathe dust or eat chips from peeling or crumbling paint that contains lead. Inexpensive kits are available to test the lead content of paint. If lead-based paint is in poor condition, only qualified professionals should handle repairing it.

Unsafe

Safe

To avoid an electrical overload, plug only two electrical appliances into each wall outlet.

420 Chapter 15: Safety and Recreation

Preventing Poisonings

Children are naturally curious, and young children frequently put objects into their mouths. Each year nearly 1 million children under the age of six are poisoned because they have found and swallowed harmful substances. Follow these rules to prevent such tragedies in your home or when you are baby-sitting.

- Never refer to a child's medicine or vitamins as "candy."
- Make sure that all medicines have child-resistant caps.
- Put all medicines or poisonous substances away immediately after use.
- Keep all cleaning products in their original, labeled containers.
- Store all potentially poisonous substances in high cabinets, out of children's reach. If possible, keep the cabinets locked.

Preventing Electrical Shocks

Electricity provides energy for heat, lights, cooking, and entertainment. Electricity can be extremely dangerous, however. Improper use or maintenance of electrical appliances, wiring, or outlets can cause severe electric shock and even death. To prevent such injuries at home, follow these rules.

- Never plug more than two electrical appliances into a wall outlet. More than two may cause an **electrical overload,** *a dangerous situation in which too much electric current flows along a single circuit.* An electrical overload can cause a fire.
- Never use an electrical appliance near water or if you are wet.
- Unplug small appliances, such as hair dryers and toasters, when they are not in use. Have broken appliances repaired or replaced.
- Pull out electrical plugs by the plug itself, *not* by the cord.
- Have loose or damaged cords repaired or replaced.

Preventing Gun Accidents

In 1994 more than 1,300 people were killed by firearms that were accidentally discharged. Many of those people were children. More than one-third of all American households have guns. If your family or anyone you know keeps guns at home, follow these guidelines.

- All guns should be treated as if they are loaded.
- Guns should always be stored unloaded and in locked cabinets. The bullets should be kept in a separate locked cabinet.
- A gun should never be pointed at anyone.
- The barrel of a gun should point downward whenever it is being carried.

Health of Others

Many students may already be baby-sitting, either taking care of younger children occasionally or having daily responsibility for their younger siblings. Point out that, even though they are experienced in caring for children, they can increase their skills, especially in dealing with emergency situations, by taking a baby-sitting class. These are often offered by community schools, local hospitals, or organizations such as the American Red Cross. Suggest that a volunteer gather information about such classes and publish that information in a flier or on a bulletin board. Encourage interested students to enroll in a convenient course, and ask them to discuss what they have learned with other students.

Safety at School

In some schools, teens may feel the need to protect themselves against violence, some of which may involve guns and other weapons. To be safe at school:

- Cooperate with your school's efforts to keep weapons out of the building. These efforts may include metal detectors and video cameras.
- If you suspect that another student is carrying a gun or any other dangerous weapon, report it to a teacher or school administrator. You can request that your name be held in confidence.
- If someone harasses you, follows you, threatens you, or frightens you in any way, report the incident.
- Don't be a target for theft. Carry only enough money for lunch and other small items. Avoid wearing clothing or jewelry that is very expensive.
- At evening events, such as games, meets, or dances, stay with a group of friends in well-lighted areas. Avoid arguments, and don't take sides in other people's conflicts.

in your journal

Have you ever felt unsafe at school? In your private journal, describe a time when you were concerned about possible violence or injury at school. How might you avoid such situations in the future? What could school administrators do to help protect you and other students?

Staying with a group will help keep you and your friends safe.

Review — Lesson 2

Using complete sentences, answer the following questions on a separate sheet of paper.

Reviewing Terms and Facts

1. **Vocabulary** Define the word *flammable*. Give three examples of flammable materials.
2. **Identify** List five ways to prevent fires in the home.
3. **Outline** What steps should you take if a fire breaks out in your home?
4. **List** Identify two ways to prevent falls in the kitchen and two ways to prevent falls in the bathroom.
5. **Identify** Name five ways to prevent electrical shocks.

Thinking Critically

6. **Explain** Why is it important to establish and practice a fire escape plan at home? Describe the main elements of an effective plan.
7. **Infer** Why do you think people should always treat guns as if they are loaded?

Applying Health Concepts

8. **Personal Health** Evaluate your home for safety. Make a list of any potential safety hazards. With your family, discuss possible changes and improvements that might create a safer home environment.

Lesson 2: Safety at Home and in School

Chapter 15 Lesson 2

LESSON 2 REVIEW

Answers to Reviewing Terms and Facts

1. Flammable means able to catch fire easily. Three types of flammable materials are paint, newspapers, and rags.
2. See Figure 15.4 on page 417.
3. See Figure 15.5 on page 418.
4. See page 419.
5. You can prevent electrical shocks by never plugging more than two appliances into a wall outlet; never using an electrical appliance near water or if you are wet; unplugging small appliances when they're not in use; pulling out electrical plugs by the plug itself, not the cord; and having damaged cords repaired or replaced.

Answers to Thinking Critically

6. When you have a plan, you are better able to remain calm and act carefully and safely. Elements of an effective plan include identifying two possible escape routes for each family member; agreeing on an outside location where family members will meet; having all members agree to save themselves, not their belongings.
7. You can't tell by looking at a gun whether it is loaded, and firing a loaded gun could be fatal.

RETEACHING

- Assign Concept Map 67 in the TCR.
- Have students complete Reteaching Activity 67 in the TCR.

ENRICHMENT

Assign Enrichment Activity 67 in the TCR.

More About...

Gun Safety Students may be interested in gathering more information about gun safety. They can be encouraged to contact one of the following organizations:

- Center to Prevent Handgun Violence
  1225 I Street NW, Suite 1100
  Washington, DC 20005
- National Council for a Responsible Firearms Policy
  7216 Stafford Road
  Alexandria, VA 22307

Close

Have students respond in writing to this question: What is the most important thing you learned from this lesson?

Lesson 3

Safety on the Road and Outdoors

Lesson 3

Safety on the Road and Outdoors

This lesson will help you find answers to questions that teens often ask about safe participation in outdoor activities. For example:

▶ What should I know about pedestrian safety?
▶ What are the basic rules for safety on wheels?
▶ How can I be safe in my community?
▶ How can I avoid injuries outdoors and in the water?

Words to Know

pedestrian
hypothermia

Traffic Safety

Can you imagine what would happen on the roads if there were no traffic signals, speed limits, or stop signs? To maintain order and ensure safety, every area enforces traffic laws. The laws must be followed by the drivers of cars, trucks, buses, and motorcycles, as well as by bicycle riders.

Pedestrians, or *people who travel on foot,* must also be aware of traffic laws. They must take responsibility for their own safety on the road. Part of being responsible involves following traffic signs. **Figure 15.6** explains some common signs.

Figure 15.6
Traffic Signs
Understanding the meaning of traffic signs will help you stay safe.

Cars and bicycles must slow down and check for traffic. They must give the right-of-way to pedestrians and approaching cross traffic before proceeding.

Cars and bicycles must wait for the green traffic signal before proceeding. This sign is used at intersections where oncoming traffic receives an early green light.

Cars and bicycles are *not* permitted to turn right at this red light. (In most states, you may turn right at a red light if this sign is *not* posted. Before turning, however, you must first stop and yield to pedestrians and cross traffic.)

422 Chapter 15: Safety and Recreation

Focus

LESSON OBJECTIVES

After studying this lesson, students should be able to

▶ discuss how they can be safe as pedestrians, cyclists, skaters, and skateboarders.
▶ explain how they can keep themselves safe from crime.
▶ identify behaviors and attitudes that will protect their safety outdoors.

MOTIVATOR

Ask students: What safety signs do you need to look for when you're out walking or riding your bike? Let students draw their answers or write brief descriptions of the signs and their meanings.

INTRODUCING THE LESSON

Ask volunteers to share their drawings or descriptions from the Motivator activity. Let students explain the signs, telling what they mean and how the information on those signs can help protect them. Encourage students to discuss why remaining safety conscious is so important when they are "on the road." Then explain that they will learn more about safety in traffic, safety from crime, and safety during outdoor activities as they read Lesson 3.

INTRODUCING WORDS TO KNOW

Guide students in pronouncing the Words to Know, and let volunteers find and read aloud the definitions provided in the lesson. Have students record each Word to Know, along with its definition and a mnemonic sketch, in their notebooks.

Lesson 3 Resources

Teacher's Classroom Resources
- 📁 Concept Map 68
- 📁 Enrichment Activity 68
- 📁 Health Lab 15
- 📁 Lesson Plan 3
- 📁 Lesson 3 Quiz

- 📁 Reteaching Activity 68
- 🖨 Transparency 62

Student Activities Workbook
- 📁 Study Guide 15
- 📁 Applying Health Skills 68

Pedestrian Safety

Motor vehicle drivers are often encouraged to "drive defensively." For pedestrians, "walk defensively" is excellent advice as well. When walking, be alert, self-protective, and cautious. *Don't* assume that drivers will always obey the laws or follow the traffic signals and signs. Obey traffic signals when crossing the street, and cross only at intersections or crosswalks. Do not walk or run into the street from between parked cars. Drivers may not see you in time to stop. Finally, *never* assume that motorists or cyclists

- can see you.
- know what you plan to do—for example, that you are about to cross the street.
- are paying attention.
- will act in a safe and capable manner.
- will signal before they turn.
- will act according to their signals.

Safety on Wheels

Many people enjoy bicycling, in-line skating, and skateboarding. Not only are these activities fun, but they also provide good exercise. They do involve risks, however. According to the U.S. Department of Transportation, 830 cyclists were killed and about 61,000 were seriously injured in this country in 1995. During the same year, nearly 100,000 in-line skaters were seriously injured. Most were between the ages of 10 and 15. You can have fun and still be safe, however, if you follow these important guidelines.

- **Check your equipment.** Check your bike regularly for cracks in the frame. Make sure that the wheels spin freely. Check to see that the handlebars, seat, chain, and spokes are tight and straight. Be sure that the tires have the correct air pressure and show no leaks or excessive wear.

- **Wear a helmet and other protective gear.** Head injuries are involved in 70 to 80 percent of all bicycle accident deaths. Wearing a helmet reduces the risk of head injury by 85 percent. When in-line skating or skateboarding, wear a helmet, wrist guards, and elbow and knee pads. Doing so will reduce your risk of serious injury by 90 percent.

- **Follow the rules.** Obey traffic laws and follow common sense. Do not use bicycles, skateboards, or in-line skates where they are not permitted, such as on sidewalks. **Figure 15.7** on pages 424 and 425 summarizes other important safety rules.

Teen Issues
Pedestrians Beware

Crossing the street is a greater risk for pedestrians these days, due to the growing number of drivers with cellular telephones. According to a study reported in the February 13, 1997, issue of the "New England Journal of Medicine," a driver who is talking on a cell phone is just about as dangerous as a driver who has been drinking alcohol. A driver's risk of having an accident becomes four times greater when he or she picks up a cellular phone. Keep this in mind, and use caution when crossing the street.

Checking your equipment will help you prevent accidents and injuries.

Cooperative Learning

Present the following statistics:
- ▶ Every year more than 6,000 pedestrians are killed in traffic accidents.
- ▶ An additional 90,000 pedestrians are injured each year.
- ▶ Pedestrians between the ages of five and nine are at the greatest risk.

Ask: What is the impact of these statistics? What can you do to help improve the safety of children in your community? Guide students in brainstorming various ideas. Then let the class work as a group or form several smaller groups in which to implement one or more projects for improving pedestrian safety.

Chapter 15 Lesson 3

Teach

L1 Language Arts
Ask a volunteer to look up *pedestrian* in a dictionary: From what Latin root does the word derive? What other words with related meaning derive from the same root?

L1 Discussing
Ask: How does an understanding of traffic laws protect you as a pedestrian? as a cyclist? How does it help prepare you to be a safe driver? Whom do you endanger—besides yourself—if you fail to follow traffic laws?

Visual Learning
Guide students in describing and discussing each traffic sign in Figure 15.6: Where have you seen this kind of sign? Why do you think those signs are posted there? What would happen if drivers, riders, or pedestrians ignored those signs?

L2 Presenting Skits
Let students work with partners or in small groups to plan short skits about pedestrian safety. Provide time for students to present their skits and for audience members to comment on them.

L2 Demonstrating
Have a pair of students bring a bicycle to class and demonstrate the procedure for checking its safety. Let them identify the kinds of problems to look for and the best ways to remedy those problems.

L1 Discussing
If some students ride bicycles to school, encourage a class discussion of safe riding routes in the area and the best procedures for entering and leaving the campus. Also ask: Where do you leave your bike during the school day? What safety precautions do you take there?

Chapter 15 Lesson 3

L1 Applying Knowledge
Let students discuss the kinds of loads they can carry safely when they ride their bikes. Encourage them to share their experiences in trying to carry books, sports equipment, groceries, and/or other loads on their bikes.

L1 Discussing
Let volunteers read aloud the safety rules on skates and bikes in Figure 15.7. Ask students to explain the importance of each rule. In addition, when appropriate, let students demonstrate and/or give specific examples of situations in which each rule might apply.

L1 Applying Knowledge
Ask: Are there special areas in this neighborhood set aside for in-line skating and/or skateboarding? If so, what are the advantages of using those areas? If not, how can you find safe areas for these activities? What community groups might work to establish special park sections or other public areas for skating and skateboarding?

L1 Discussing
Let students share their own feelings about crime in the neighborhood: How safe do you feel in the areas around school and home? Why? What do you think you can do to make these areas safer for yourselves and other teens? for young children?

L2 Current Events
Have students start a scrapbook with newspaper and magazine articles about community organizations that are responding to crime in their neighborhoods.

L1 Life Skills
Pose these questions to the class: What forms of personal identification can you carry? How can you get official identification cards? Encourage students to take the necessary steps so that they will have—and be able to carry—formal identification.

Cultural Connections
Roller Cops

In Amsterdam, the capital of the Netherlands, the streets are very narrow. This has always made it difficult for the police to use cars and vans to capture criminals. Recently several young police officers left their cars at the station and strapped on in-line skates. Skating through the pedestrian malls of the city, they were able to do their jobs better.

Figure 15.7
Be Safe on Skates
Why is it important to follow these safety rules when skating?

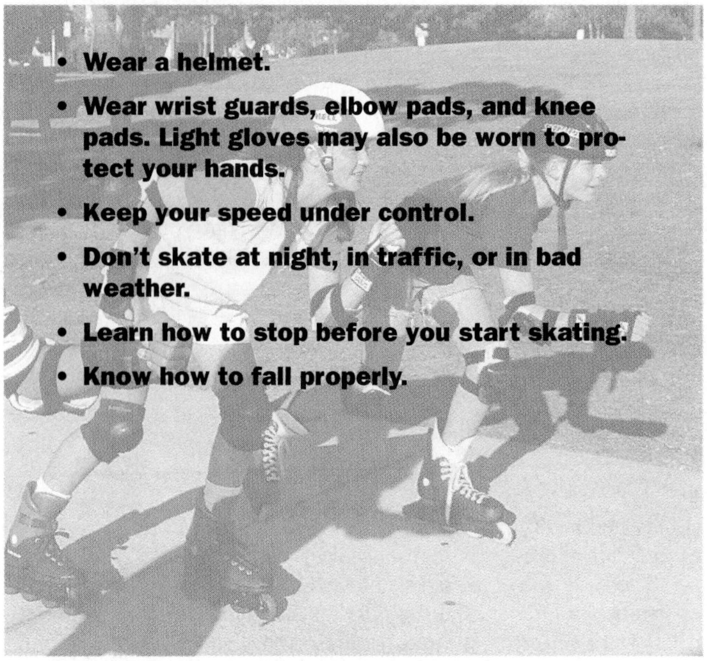

- Wear a helmet.
- Wear wrist guards, elbow pads, and knee pads. Light gloves may also be worn to protect your hands.
- Keep your speed under control.
- Don't skate at night, in traffic, or in bad weather.
- Learn how to stop before you start skating.
- Know how to fall properly.

Safety Against Crime

Pedestrians and cyclists must protect themselves not only from injuries but also from violence and crime. Stay away from areas that you think are dangerous or where you feel unsafe. To avoid looking like an easy target, walk with a confident and purposeful step. In addition, follow these tips for safety on the street.

- **Avoid potential trouble.** At night, do not go out alone. Stay in well-lighted public places and on safe, well-lighted sidewalks. Tell a parent where you are going, what route you are taking, and when you will be home. Do not carry valuables, but always carry some sort of identification. Always carry change or some other means of making an emergency phone call. Do not talk to strangers or approach a stranger who calls to you from a car.

- **Be smart and aware, and protect yourself.** When standing in line or in other crowded places, be aware of the people around you. If someone stands too close and makes you feel uncomfortable, move away from that person.

- **Get help when you need it.** If a stranger tries to touch you or says anything that frightens you, scream and run away. Go to a nearby public place and ask for help. Call 911 or the police. Try to remember details about the stranger, such as clothing, physical appearance, and type and color of car.

424   Chapter 15: Safety and Recreation

Personal Health

In-line skating enthusiasts should, of course, learn to wear safety gear whenever they skate. They should also learn as much as they can about their sport, taking classes when possible and practicing in safe, in-line-friendly areas. Students can learn more by contacting the International In-Line Skating Association, an association with a mission "to develop in-line safety and education programs" and "to protect the right-to-skate and expand public skate ways."

International In-Line Skating Association
3720 Farragut Avenue, Suite 400
Kensington, MD 20895

Figure 15.7 (continued)
Be Safe on Bikes
Why is it important to follow these safety rules when bicycling?

- Wear a helmet.
- Obey traffic laws.
- Ride in single file in the same direction as traffic.
- Have reflectors on spokes, rear fender, and pedals.
- Before turning left, look back for traffic behind you.
- Learn and use hand signals.
- At night, use lights and wear reflective clothing.
- Wear clothing that will not become caught in the bicycle chain.
- Don't ride in bad weather.

in Your Journal

Have the people in your neighborhood formed a group to make the streets safer? If so, describe the actions they have taken. If not, suggest a way to encourage your neighbors to pitch in. What might teens do to help?

Outdoor Safety

Most people enjoy a chance to get outdoors to explore, exercise, or simply enjoy the fresh air. No matter which outdoor activities you choose, ensure your safety with these commonsense rules.

- **Take a buddy.** Outdoor activities are safer and more fun when they're shared. Agree with a friend that you will stay together. In an emergency, you and your friend can help each other.

- **Stay aware.** Learn the signs of oncoming weather emergencies. When necessary, move to safe shelter quickly.

- **Know your limits.** Most outdoor injuries occur when people are tired or try to stretch themselves beyond their abilities. Don't set unreasonable goals for yourself. If you're a beginning swimmer, for example, don't try to swim farther than you can handle.

- **Use your judgment.** Of course you wouldn't dive into a swimming pool before checking to make sure that the pool contained water. Not all safety rules are so obvious, however. Always ask yourself, "Do I have the equipment I need, and am I acting safely?" If you're unsure, ask a parent or other trusted adult.

- **Be sure to warm up and cool down.** Simple stretching exercises will prepare your body for outdoor activities. When you're finished with the activity, cool down slowly with additional stretching.

Lesson 3: Safety on the Road and Outdoors   425

More About...

Crime Prevention The Missouri Crime Prevention Information Center says:
1. Crime prevention is everyone's business.
2. Crime prevention is more than security.
3. Crime prevention is a responsibility of all levels of government.
4. Crime prevention is linked with solving social problems.
5. Crime prevention is cost effective.
6. Crime prevention requires a central position in law enforcement.
7. Crime prevention requires cooperation by all elements of the community.
8. Crime prevention requires education.
9. Crime prevention requires tailoring to local needs and conditions.
10. Crime prevention improves the quality of life for every community and its residents.

Chapter 15 Lesson 3

Would You Believe?

Today's 12-year-olds face an 83 percent chance of being the victim of at least one violent crime during their lifetimes.

L1 Critical Thinking

Present the following statistics to students:
- Teens are more likely than members of any other age group to be the victims of crime.
- Teens are less likely than members of any other age group to report crimes to the police.

Let students discuss their responses to these statistics: What do you think they indicate? Why do you think teens are at such high risk of being crime victims? What accounts for their low likelihood of reporting crimes? What can you and other teens do to change that?

L2 Role-Playing

Ask students to role-play situations in which teens turn down invitations or suggest changes in plans so they can avoid being out alone at night.

L1 Comprehending

Ask students: Who is "a stranger"? What should you do if a stranger frightens you? What might you do when you are acquainted with the person who frightens you?

L1 Discussing

Guide a discussion of the advantages of enjoying outdoor activities with one or more friends. Then ask: What excuses do you think teens give for swimming, boating, hiking, or camping alone? What would you say to a teen who used those excuses?

L2 Demonstrating

Let a group of volunteers demonstrate effective stretching movements or exercises that will help in warming up and cooling down for specific outdoor activities, such as swimming, hiking, and biking.

Chapter 15 Lesson 3

L2 Reporting
Ask a student or a small group of students to read about safety in making, burning, and putting out campfires. Have the student or group present an oral report, with illustrations or models, to the rest of the class.

L1 Comprehending
Let class members share their ideas about winter sports: What winter sports and activities are popular in your area? Do they involve snow and/or cold weather? If not, do you need to know guidelines for winter sports safety? Why?

L2 Reporting
Have a volunteer do further reading on hypothermia: What are the symptoms? What can the effects be? What are the most effective treatments? Let this student discuss his or her findings with the rest of the class.

L2 Researching
You may want to have students research another danger associated with cold weather activities: frostbite. What is it? How can it be prevented? What are the symptoms? How should it be treated? Have students meet in groups to compare and discuss their findings.

L2 Home Economics
Ask students to find and bring to class ads (from circulars, magazines, and newspapers) for cold-weather apparel. Have students meet in groups to discuss the ads: What special advantages—if any—does each item offer for protection during winter activities? What alternatives can you suggest?

Assess

EVALUATING THE LESSON
Assign Reviewing Terms and Facts and Thinking Critically on page 427 to review the lesson; then assign the Lesson 3 Quiz in the TCR to evaluate students' understanding.

Science Connection

Dangers in the Wild

Many people are fascinated when they catch sight of an animal in the wild. Wild animals can be dangerous, however, and one reason is the risk of rabies (RAY•beez). This is a severe, often fatal disease, spread by contact with the saliva of an infected animal. To avoid rabies infection and the danger of other injury, never approach an animal in the wild. If an animal seems either dazed or very aggressive, it may be carrying the rabies virus. Leave the area immediately, and then report the incident.

Safety in the Water
Swimming, boating, and diving are pleasant outdoor activities. The most important way to be safe in the water is to know how to swim well. **Figure 15.8** provides additional guidelines for water safety.

Hiking and Camping Safety
Hiking and camping are great ways to exercise and enjoy nature. When you hike or camp, being prepared and acting safely are the keys to preventing injuries. Follow these guidelines.

- Dress appropriately for both the activity and the weather conditions. Wear thick socks and comfortable, sturdy shoes. To prevent blisters, break in hiking shoes ahead of time. Tuck pants into socks to avoid insect bites or stings.

- Take necessary equipment, including a first-aid kit, a flashlight with extra batteries, a supply of fresh water, and a compass.

- Hike on marked trails and camp in specified areas.

- Never camp or hike by yourself. Make sure that family members know your schedule and your route.

- Learn to identify poisonous plants, such as poison ivy and poison oak, so that you can avoid contact with them.

- Make sure that all campfires are extinguished completely. Drench them with water or smother them with sand or dirt.

Figure 15.8
Water Safety
Water activities can be safe, healthful fun if you follow these rules.

Activity	Safety Tips
Swimming	• Swim with a friend and only when a lifeguard or adult is present. • Don't swim when you're tired. • If you ever feel that you are in trouble, do not panic. Wave or call for help. Then take a deep breath and let your body sink vertically beneath the water. Relax. Take several "steps" forward with one leg in front of the other. At the same time, push down with cupped hands and lift your head out of the water. Take a breath and then let your body sink again. Repeat until help arrives.
Diving	• Don't try diving unless you have been taught the proper technique. • Check the depth of the water before diving. Walk into the water first. • Never dive into an above-ground pool or into the shallow end of any pool.
Boating	• Always wear a life jacket. • Know how to handle the boat. • Make sure that the boat is in good working condition.

426 Chapter 15: Safety and Recreation

TECHNOLOGY UPDATE

Students can find a variety of facts, tips, and activities on the Internet that relate to crime prevention. One approach is to use an available search engine and input the phrase *crime prevention*. Information about specific kinds of crimes or forms of crime prevention can be located by inputting more specialized requests.

One useful site is maintained by the National Crime Prevention Council, the creator of McGruff the Crime Dog and the Take a Bite Out of Crime program. The site includes, among many other items, "Activities, Games, and Tips for Kids Ages 5–12" and "Materials for Youth Ages 13–21."

Winter Sports Safety

Outdoor activities in winter require special safety precautions. Always dress in layered, warm clothing, and stay dry. Protect your hands, head, and feet with snug, warm coverings. If you begin to shiver, go inside and warm up quickly. Such precautions will prevent **hypothermia** (hy·poh·THER·mee·uh), *a sudden and dangerous drop in body temperature.* In addition, remember the tips shown in **Figure 15.9**.

Figure 15.9
Tips for Winter Sports Safety
Here are some tips for enjoying winter sports safely.

A When skating, make sure that the ice is frozen solid. Do not skate on ice that has not been tested for thickness.

B Take lessons before you ski or snowboard. Make sure that your equipment is in good condition and fits you correctly.

C Sled or snowboard only on hills that are away from roads and have no obstacles (such as trees).

D Ski or snowboard only on slopes that you can handle safely. Check the condition of slopes before heading down, and stay alert to condition updates and avalanche warnings.

Review — Lesson 3

Using complete sentences, answer the following questions on a separate sheet of paper.

Reviewing Terms and Facts

1. **List** Recall six safety rules to follow when bicycling.
2. **Outline** Describe the steps to take if you feel that you are in trouble while swimming.
3. **Vocabulary** Define the term *hypothermia*. Then explain how to avoid it.

Thinking Critically

4. **Explain** What does it mean to "walk defensively"? Why is this an important rule for pedestrians?
5. **Compare** Give reasons why the buddy system is important in each of these activities: swimming, hiking, going out at night.

Applying Health Concepts

6. **Growth and Development** Select a sport or outdoor activity mentioned in this lesson that children could participate in. Create a poster that explains the safety rules to young children. Use drawings or magazine clippings to illustrate each rule.

Lesson 3: Safety on the Road and Outdoors **427**

Personal Health

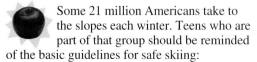

Some 21 million Americans take to the slopes each winter. Teens who are part of that group should be reminded of the basic guidelines for safe skiing:

▶ Ski under control at all times. You must be able to stop or turn out of the way if you encounter an unexpected obstacle or another skier.

▶ Stay off the ski trails unless you're skiing. Don't block the path of other skiers.

▶ Observe and follow all posted warning signs.

▶ Avoid skiing too close to other skiers.

Focus

The Teen Health Digest articles can be used in two ways: as an individual activity for reflection and enrichment or as a cooperative learning activity as described below.

Motivator

Encourage students to speculate about the contents of each Teen Health Digest article, using the title and photograph or image as guides. Then guide students in reading and discussing the articles: What does each article tell you about safety for yourself, for other people, and for the environment? Why is this information important? How might you and other teens use this information?

Cooperative Learning Project

Living Safe: An Original Play

Setting Goals

Ask students: What do you wish other people knew about safety? Explain that they will have an opportunity to share those facts and practices with others in an original play that they will write and perform.

Delegating Tasks

Encourage students to work together to list the tasks involved in this project. The following are suggestions: writers, actors, set designers and decorators, prop handlers, musicians, directors. Let students sign up to work on the task or tasks of their choice.

TEEN HEALTH DIGEST

People at Work
A Lifesaving Job

Job: Emergency medical technician (EMT)
Responsibilities: To give on-the-spot emergency medical care to victims of accidents, injuries, and illnesses; to transport victims to the hospital and care for them on the way

Education and Training: EMTs must complete a training course and field training, pass an examination, and be certified in accordance with state regulations.
Workplace: EMTs spend much of their time driving or riding in ambulances.
Positive: An EMT has the personal satisfaction of helping others and even saving lives.
Negative: The work is physically and emotionally demanding; hours are irregular.

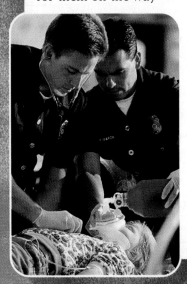

428   Chapter 15: Safety and Recreation

Myths and Realities

The Facts About Safety Belts

Car safety belts save lives. Even so, some people still don't buckle up. Do you know the facts from the myths?

Myth: If a car has air bags, you don't need to wear a safety belt.
Reality: Air bags are truly effective only when used *with* safety belts. Children aged 12 and under should always sit in the back seat, away from air bags. Infants less than a year old should be in the back seat and secured in a rear-facing child safety seat.
Myth: You need to wear a safety belt only on long trips or during highway driving.
Reality: The majority of serious car accidents occur when cars are going less than 40 miles per hour and when occupants are less than 25 miles from home.

Try This:

Use one of these myths, or another that you've heard about safety belts, to make a poster. Point out the truth about the need to buckle up.

TECHNOLOGY OPTIONS

Computer Students working on writing and revising the play's script can benefit from the use of a word processing program. Saving the drafts of the script on a computer allows all group members to edit and revise. Final versions of the script can be printed with each actor's part set off in boldface, underscoring, or different fonts.

Audio Recorders Students who are preparing to act in the play may want to use tape recorders as an aid to memorizing their lines. They might also record their performances and listen to the tapes, noting how they can improve their own enunciation, projection, and expression.

Sports and Recreation
Cycling Tips from a Coach

As coach of the U.S. Cycling Team, Chris Carmichael helps young cyclists improve their skills. Here are his top four safety tips for bicycle riders.

- **Wear a helmet.** Helmets help prevent head injuries, the number one cause of cyclists' deaths and serious injuries.
- **Follow the rules of the road.** Cyclists need to follow the same traffic rules as drivers. These rules include riding with traffic and obeying traffic lights and signs.
- **Be visible.** It's the cyclist's responsibility to be seen by motorists. Wear bright clothing, avoid riding at night, and equip your bike with reflectors.
- **Share the road.** Be considerate. "Using hand signals, making eye contact, smiling and waving—just being courteous—all help keep the roads safe for everyone," Chris advises.

CON$UMER FOCU$
Safe Toys for Tots

Q. How can I buy a safe toy for my two-year-old brother?

A. Check the age recommendations on the product label. Because young children put things into their mouths, they can choke on small toys or toys that have removable parts. In addition, stay away from toys made of brittle plastic that may break into sharp pieces.

Teens Making a Difference
Tree Musketeers

When Tara Church and Sabrina Alimahomed were eight years old, they planted a young sycamore tree. They named the tree Marcie the Marvelous. Planting Marcie was so inspiring to them that they went on to form a group they called the Tree Musketeers. Their goal was to plant trees in their hometown of El Segundo, California.

Ten years later, Tree Musketeers has become a nonprofit corporation run *by* teens and *for* teens. The group's projects include Partners for the Planet, a series of national and regional meetings where teens can share ideas on how to help the environment, and Tree House, in which teens are converting a vacant lot into an environmental learning area.

About her work Tara has stated, "My wish is to make other teens see that every single one of them is extremely important and extremely powerful in their ability to effect change. It all starts with just one individual's determination to make a difference."

Try This: Marbles are an example of unsafe toys for small children. What others can you think of? Make a warning list to share with your family and neighbors.

Teen Health Digest 429

Chapter 15
The Work Students Do

Following are suggestions for implementing the project:

1. The students can begin by working as a class to decide on a general plan and format for their play. They may want to tell one safety-related story in a single act, or they might choose to write and perform a series of short scenes, each with a separate safety-related message.
2. Remind the writers to use action as well as dialogue to make the story (or stories) interesting and entertaining. Also, suggest that writers read their dialogue aloud as they compose and revise it; they will want the script to sound as realistic and natural as possible.
3. Students handling sets and props will want to begin working on their parts of the production as soon as possible. To make their work more effective, they should stay in close touch with those who are preparing the script.
4. Students who are acting in the play should practice not only their lines but also their movements, gestures, and expressions. Provide several opportunities for the performers to rehearse in front of other group members; encourage students to respond with positive comments and constructive suggestions.
5. Be sure students have an opportunity to perform their play for an interested audience, either other students at the school or invited family members and friends.

Assessing the Work Students Do

Observe students working in their groups, and encourage active participation as necessary. After the performance, let each student write an assessment of his or her own contributions to the project.

Meeting Student Diversity

Language Diversity If many of the students—and many in their homes and/or the rest of the school—speak languages other than English, suggest that a bilingual or even a multilingual script be composed and performed. This kind of script can be an accurate presentation of life and languages as students experience them.

Hearing Impaired Students with hearing impairments can participate successfully in any or all aspects of the project. You may want a student with serious impairment to participate in the performance; other "actors" can communicate with this student as they do in daily life, through American Sign Language, for example.

Lesson 4

Safety in Weather Emergencies

Focus

LESSON OBJECTIVES

After studying this lesson, students should be able to

- discuss general guidelines for protecting themselves before, during, and after a weather emergency or natural disaster.
- explain how they can remain safe in a flood.
- discuss safety during and after an earthquake.
- explain how to be safe before and during a tornado.
- explain how to be safe before and during a hurricane.
- discuss safety measures to take during blizzards and thunderstorms.

MOTIVATOR

Ask: What is the worst weather emergency or natural disaster you have ever experienced? Let students write short paragraphs or draw pictures depicting their responses.

INTRODUCING THE LESSON

Allow students to share their Motivator responses: How dangerous was the situation? What did you and your family do to stay safe? Close by telling students that, as they read Lesson 4, they will learn how to protect themselves in various kinds of weather emergencies and natural disasters.

INTRODUCING WORDS TO KNOW

Let partners work together to read the Words to Know and their Glossary definitions. Then ask them to create a set of flash cards and use these cards to quiz each other on the meanings of the Words to Know.

430

Lesson 4 · Safety in Weather Emergencies

This lesson will help you find answers to questions that teens often ask about how to stay safe during weather emergencies. For example:

▶ What should I do in case of a flood?
▶ What are the basic safety rules for surviving an earthquake, tornado, or hurricane?
▶ What should I do if I am caught in a blizzard or thunderstorm?

Words to Know

earthquake
aftershocks
tornado
tornado watch
tornado warning
hurricane
blizzard

in your journal

What weather emergencies and natural disasters are most common in your area? In your journal, write about what you might do to help others in your community who were affected by such a natural disaster.

Weather Emergencies and Natural Disasters

Floods, earthquakes, and violent storms can strike without warning, causing tremendous damage. These weather emergencies and natural disasters cannot be prevented. Therefore, people should learn the most effective ways to protect themselves before, during, and after a natural disaster. Many of the same protective measures apply in all weather emergencies and natural disasters. These measures include the following.

- **Be prepared.** Assemble a disaster supplies kit with your family. It should contain a first-aid kit, canned food and a can opener, and bottled water. Also include a battery-operated portable radio, one or more working flashlights (check them regularly), and a supply of extra batteries.

- **Stay alert to information.** The National Weather Service monitors the progress of storms and weather emergencies. Listen to the radio or watch television to receive news bulletins and advisories about oncoming storms.

- **Stay calm and be smart.** During a weather emergency, the worst thing you can do is to panic. Pay attention to news bulletins. Do not hesitate to evacuate your home if you are instructed to do so.

Floods are the most common natural disaster. They often occur after heavy rain or snow, a hurricane, or a break in a dam or levee.

Lesson 4 Resources

Teacher's Classroom Resources
- Concept Map 69
- Enrichment Activity 69
- Lesson Plan 4
- Lesson 4 Quiz
- Reteaching Activity 69

Student Activities Workbook
- Study Guide 15
- Applying Health Skills 69

Floods

Floods can happen almost anywhere and at any time. People who live in low-lying coastal regions or river valleys are at greatest risk. Floods may be caused by severe rainfalls or damage to dams.

During a flood, never try to walk, swim, ride a bike, or ride in a car through the water. Stay tuned to National Weather Service news bulletins and follow their advice. Although many floods take hours or days to develop, a flash flood can occur in minutes. **Figure 15.10** explains the three types of flood advisories broadcast by the National Weather Service. Follow these guidelines.

- Do not walk in flooded areas. Drowning poses a severe risk, as does electrocution caused by downed power lines.
- Because flood waters often pollute the water supply, do not drink tap water. Drink bottled water instead.
- If you have been evacuated, do not try to return to your home until you are advised that it is safe to do so.
- Once you have returned home, discard contaminated food. Disinfect anything that has come into contact with the flood waters.

Figure 15.10
Flood Advisories
The National Weather Service issues three types of flood advisories: flood or flash flood watch, flood warning, and flash flood warning.

Type of Advisory	Meaning	What to Do
Flood or Flash Flood Watch	A flood is possible in your area. Stay tuned.	Move valuables to higher levels of your home. Listen to the radio. Watch for rising water and other signs of flash flooding. Keep your disaster supplies kit with you, and prepare to evacuate.
Flood Warning	Flooding is already occurring or will occur soon in your area.	Follow the advice of the National Weather Service and local officials. If told to evacuate your home, do so as quickly as possible. Move to higher ground, away from rivers, streams, and creeks.
Flash Flood Warning	Sudden, violent flooding is already occurring or will occur soon in your area.	Evacuate immediately. Follow the guidelines listed above for flood warnings, but act quickly.

Earthquakes

An **earthquake** is *a violent shaking of the earth's surface*. In the United States, earthquakes are most common along the Pacific Coast, but they can occur elsewhere. An earthquake is usually not a single event. Following the initial shaking, several **aftershocks,** or *secondary earthquakes,* often occur. **Figure 15.11** on the next page provides guidelines for personal safety during an earthquake.

Lesson 4: Safety in Weather Emergencies 431

Q & A

The Richter Scale

Q: What is the Richter Scale?

A: The Richter (RIK•ter) Scale is a method of recording the amount of ground motion caused by earthquakes. It was developed in 1935 by an American scientist named Charles F. Richter. It rates the ground motion on a scale of 1 to 10, with 10 representing the maximum force. An earthquake with a magnitude of 4.5 usually causes minor damage, while one with a rating of 6 can be very destructive. Major earthquakes are those with a magnitude of 7 or more.

More About...

Floods Particularly devastating floods can be caused by huge, destructive waves called tsunami. A tsunami is typically triggered by an earthquake or by severe weather conditions, usually storms at sea. Tsunami are perhaps most familiar in Japan, where earthquakes are quite common. A massive tsunami struck the United States in 1964, following a major earthquake in Alaska. It created waves as high as 35 feet along the southern coast of Alaska and on the Kodiak Islands. That tsunami killed 122 people—including 4 in Oregon and 11 in California.

Chapter 15 Lesson 4

Teach

L2 Demonstrating
Let a small group of volunteers prepare at least one disaster supply kit. The kit should include the items named on page 430 and any additional items recommended for the kinds of disasters likely to strike in your area. Have the volunteers display and discuss the kit with the rest of the class.

L1 Applying Knowledge
Ask students to work with their own family members to assemble—or to check the contents of—a family disaster supplies kit. Have students list the contents of their family's kit and identify its location.

L1 Critical Thinking
Encourage students to recall news stories they may have read or heard about people who refused to evacuate their homes when instructed to do so. Ask: What do you think motivates some people to take that kind of risk? What would you say to someone who didn't want to evacuate in the face of a weather emergency?

L1 Discussing
Ask students: How common are floods in your area? Why? If flooding is unlikely in your community, ask students to identify the closest areas in which flooding is likely to occur. Then ask: Is any community ever so safe from flooding that the people there don't need to know how to respond to a flood?

L2 Language Arts
Suggest that students collect traditional sayings about the weather (such as "Red sky at morning, sailors take warning") and natural signs about the weather (cows lying down in a field means that rain is on the way, for example). Let students illustrate these sayings and signs. Post these on a classroom bulletin board.

Chapter 15 Lesson 4

L2 Researching

Asssign students to research information on this topic: How and where do storms at sea cause flooding? Let them write one- or two-paragraph summaries of their findings.

L1 Discussing

Ask the class: Why is it especially important to be prepared in advance for an earthquake? How do you think an understanding of aftershocks helps earthquake victims?

L1 Critical Thinking

Let students share their ideas about "earthquake weather": Is there such a thing? Why not? Why do stories about an association between earthquakes and a particular kind of weather persist?

L2 Demonstrating

The American Red Cross recommends the "Drop, cover, and hold on" technique during an earthquake. Let a volunteer read about this technique and then explain and demonstrate it to the rest of the class.

L3 Researching

Ask the students to read accounts of major earthquakes that have struck around the world in recent years: Which areas seem to have suffered the most deaths, injuries, and destruction? What accounts for that high level of devastation?

L2 Reporting

Have a volunteer investigate and report on waterspouts: How are they related to tornadoes? How does the wind speed of most waterspouts compare to that of most tornadoes?

L2 Language Arts

Instruct students to read at least one nonfiction account of a damaging tornado. Then ask them to incorporate some of the details they gathered from their reading into an original short story or brief play about teens in a tornado.

432

Figure 15.11
Protecting Yourself During an Earthquake

Most injuries and deaths that occur during an earthquake are caused by falling objects or crumbling buildings.

If you are indoors . . .

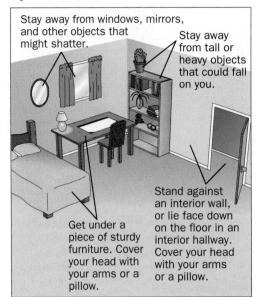

Stay away from windows, mirrors, and other objects that might shatter.

Stay away from tall or heavy objects that could fall on you.

Get under a piece of sturdy furniture. Cover your head with your arms or a pillow.

Stand against an interior wall, or lie face down on the floor in an interior hallway. Cover your head with your arms or a pillow.

If you are outdoors . . .

Stay away from trees, buildings, or power lines. They may fall.

Find a clear, open area. Drop to the ground and protect your head with your arms.

Did You Know?

Mobile Homes

Mobile homes, because of their relatively light weight, are particularly vulnerable to severe damage from tornadoes. If you live in a mobile home, evacuate immediately at the first mention of a tornado watch. Go as quickly as possible to a nearby building.

Tornadoes

A **tornado** is *a whirling, funnel-shaped windstorm that drops from the sky to the ground.* Of all types of storms, tornadoes can cause the most severe destruction. States in the Midwest and states bordering the Gulf of Mexico have more tornadoes than other regions of the United States. Tornadoes most often occur during the spring and summer. **Figure 15.12** on the next page explains how tornadoes develop.

The National Weather Service monitors the possibility of tornadoes and issues two types of news bulletins. If *weather conditions indicate that a tornado may develop,* a **tornado watch** is issued. If your area receives a tornado watch, turn on a battery-powered radio for further updates, and prepare to take shelter. The Weather Service may also issue a **tornado warning.** This means that *a tornado has been sighted and people in the area are in immediate danger.* If you receive a tornado warning, take shelter at once.

- **Where to go.** The safest place to be is underground—in a cellar or basement. If you can't get underground, go to an interior room or hallway. If you are outdoors, lie in a ditch or flat on the ground. Stay away from trees, cars, and buildings.

- **What to do.** Cover yourself with whatever protection you can find, such as a mattress or heavy blankets and clothing. Then stay where you are. Tornadoes move along a narrow path at about 25 to 40 miles per hour. The storm will pass quickly.

432   Chapter 15: Safety and Recreation

More About...

Earthquakes This chart outlines some of the most destructive earthquakes in history.

Year	Location	Approx. Deaths	Year	Location	Approx. Deaths
365	Crete	50,000	1868	Ecuador	70,000
856	Iran	200,000	1920	Central China	200,000
1201	Northern Egypt	1,100,000	1976	Northeast China	240,000
1556	Central China	830,000	1988	Armenia	25,000
1693	Sicily	100,000	1990	Iran	40,000
			1993	India	10,000
			1995	Japan	5,500

Figure 15.12
How a Tornado Develops
Winds within the spiral of a tornado often swirl at more than 200 miles per hour.

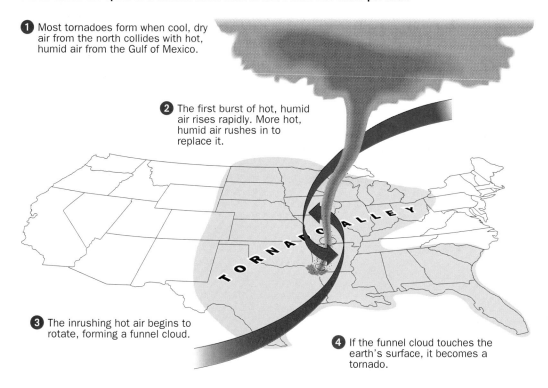

① Most tornadoes form when cool, dry air from the north collides with hot, humid air from the Gulf of Mexico.

② The first burst of hot, humid air rises rapidly. More hot, humid air rushes in to replace it.

③ The inrushing hot air begins to rotate, forming a funnel cloud.

④ If the funnel cloud touches the earth's surface, it becomes a tornado.

Hurricanes

A **hurricane** is *a strong windstorm with driving rain that originates at sea.* Hurricanes begin in tropical areas and often move toward land and along the coast. A whirling, circular cloud mass, a hurricane has its fiercest strength in the area surrounding the center. The center of the hurricane itself, called its eye, remains calm.

Hurricanes most often occur during the late summer and early fall, and coastal areas receive the greatest damage. However, strong winds and driving rain often extend to interior areas as well. The National Weather Service can often determine a storm's path well in advance. Listen to news bulletins, and follow these guidelines.

- Safeguard your home. Board up windows and doors. Bring inside any items such as outdoor furniture and bicycles.
- If local officials advise you to evacuate, do so immediately.
- If no evacuation is advised, remain indoors. Be prepared for power loss.
- Don't be fooled by the eye of the hurricane. As the eye passes overhead, the storm may appear to be over. It is not. Stay indoors until news bulletins notify you that the storm has passed.

Cultural Connections
Storm Names

In the United States and other regions of the Northern Hemisphere, hurricanes swirl in a counterclockwise direction. In the Southern Hemisphere, they swirl in a clockwise direction. These southern storms are often called cyclones (SY·klohnz). Similarly, hurricanes that originate in the western Pacific, equal in strength and characteristics to Atlantic storms, are often called typhoons (ty·FOONZ).

Lesson 4: Safety in Weather Emergencies

Promoting Comprehensive School Health

As part of your comprehensive health program, you will want to be sure that trained staff members and adequate supplies are available to provide care in case of an accident or injury at school. You will also want to know that staff and students are trained and prepared to deal with a weather emergency or other kind of natural disaster. It may be appropriate to have large stores of emergency supplies that can be used if, for example, a major earthquake should strike during school hours. Are there also ways in which the school can help families and other community members train and/or prepare for emergencies? For more information on these topics, consult *Promoting a Comprehensive School Health Program* in the TCR.

Chapter 15 Lesson 4

L3 Social Studies
Have students use reference works to find the specific locations of tornadoes and hurricanes during a recent period (such as the past decade). Let them work together to mark those locations on a map. Based on their map, what generalizations can students make?

L2 Career Education
What are the career opportunities for people interested in studying and/or predicting the weather? What education and other preparation are required? Let interested students investigate this topic and summarize their findings in short written reports.

L2 Science
Have students read about lightning: What causes it? What are the different kinds of lightning? What makes lightning especially dangerous? Ask students to plan, draw, and label posters or other illustrations presenting their answers to these questions.

L2 Language Arts
Let students meet in groups to discuss novels or short stories in which characters have been caught in blizzards or thunderstorms. Which books and stories do they recommend to others? Why?

Assess

EVALUATING THE LESSON
Assign Reviewing Terms and Facts and Thinking Critically on page 434 to review the lesson; then assign the Lesson 4 Quiz in the TCR.

RETEACHING
- Assign Concept Map 69 in the TCR.
- Have students complete Reteaching Activity 69 in the TCR.

ENRICHMENT
Assign Enrichment Activity 69 in the TCR.

Chapter 15 Lesson 4

LESSON 4 REVIEW

Answers to Reviewing Terms and Facts

1. After a flood, do not try to return to your home until you are advised that it is safe; do not walk in flooded areas; do not drink tap water; and discard contaminated food.
2. See Figure 15.11 on page 432.
3. A tornado watch means that weather conditions are such that a tornado may develop. A tornado warning means that a tornado has been sighted and people in the area are in immediate danger.
4. A hurricane is a strong windstorm with driving rain. Hurricanes occur most often during the late summer and early fall.
5. During a blizzard, stay inside if possible and be prepared for power loss. If you are outside, keep moving and reach shelter quickly. Keep your nose, mouth, and ears covered. Pay attention to landmarks along the way.

Answers to Thinking Critically

6. Hurricanes and tornadoes are both dangerous storms with severe winds. Tornadoes are the most violent type of storm; they are funnel shaped and drop to the ground with little warning. The paths of hurricanes can often be determined well in advance.
7. Being prepared will help you remain calm during a weather emergency. Advance preparation can also help ensure your safety because you will have necessary supplies on hand.

Close

Let each student identify, either orally or in writing, one important new safety fact he or she learned from this lesson.

Blizzards and Thunderstorms

If you are swimming or boating when a thunderstorm approaches, get to land immediately, and seek shelter.

Blizzards and thunderstorms are more common and usually less violent than hurricanes and tornadoes. They can still pose serious risks, however. A **blizzard** is *a very heavy snowstorm with winds of up to 45 miles per hour.* During a blizzard, follow these safety rules.

- If possible, stay inside. Be prepared for power loss.
- If you are outside, keep moving, and reach shelter as quickly as possible. Keep your nose, mouth, and ears covered. Because it is easy to get lost during a blizzard, pay attention to landmarks along the way.

Thunderstorms usually pass quickly with only minor damage. However, lightning has the potential for extreme danger. Protect yourself by following these safety guidelines.

- Stay inside, or seek shelter. Be prepared for power loss. To avoid electric shock as well as damage to electric appliances, unplug them. Do not use the telephone or running water until the storm has passed.
- If you are caught outdoors, squat low to the ground in an open area. Keep away from electric poles and wires, tall trees, water, and metal objects.

Lesson 4 Review

Using complete sentences, answer the following questions on a separate sheet of paper.

Reviewing Terms and Facts

1. **Describe** What actions should you take *after* a flood?
2. **Identify** List two ways to protect yourself if you are inside during an earthquake and two ways to protect yourself if you are outside.
3. **Compare** Explain the difference between a *tornado watch* and a *tornado warning*.
4. **Vocabulary** Define the term *hurricane*. During what time of year do hurricanes most often occur?
5. **Recall** What safety rules should you follow during a blizzard?

Thinking Critically

6. **Compare and Contrast** How are hurricanes and tornadoes similar? How are they different?
7. **Analyze** Why is it so important to be prepared *before* a weather emergency strikes?

Applying Health Concepts

8. **Personal Health** Write a plan of action for your family in case of a natural disaster most common to your area. Include details about where in your home you might be safest. Share the plan with your family.

434 Chapter 15: Safety and Recreation

Cooperative Learning

Almost everyone has heard about Benjamin Franklin's famous experiment with lightning—but what did Franklin really do? What risks did he take? Could or should he have known about those risks? Could or should he have performed a different kind of experiment to avoid those risks? Would it be reasonable for anyone to reenact Franklin's experiment now?

Let students work in cooperative groups to explore answers to these questions. Then have group members plan and rehearse a skit about one or more aspects of Franklin's experiment. Provide time for all groups to perform their skits for the rest of the class.

First Aid

Lesson 5

This lesson will help you find answers to questions that teens often ask about how to administer first aid. For example:

- What should I do first in an emergency?
- What is the proper technique for rescue breathing?
- What techniques help someone who is choking?
- How can I help someone who has been poisoned or burned?

First Steps

Illness and injury can occur at any time. If a person becomes seriously ill or injured, someone else must step forward to provide immediate emergency care. Time is often critical and may mean the difference between the victim's recovery and his or her death. It is important for everyone to know basic first-aid techniques. **First aid** is *the immediate care given to someone who becomes injured or ill until regular medical care can be provided.* If someone needs first aid, take these steps.

- Recognize that an emergency exists.
- Call for help or dial 911.
- Provide immediate ABC care (see **Figure 15.13**) until medical help arrives.

Words to Know

first aid
cardiopulmonary resuscitation (CPR)
rescue breathing
choking
abdominal thrusts
chest thrusts
first-degree burn
second-degree burn
third-degree burn
fracture

Figure 15.13
The ABCs of First Aid

In an emergency situation, move the victim *only* if he or she is not safe. Check the ABCs until help arrives.

A = Airway
Check the person's airway to make sure that it is not blocked. If it is blocked, it must be cleared. Gently roll the person onto his or her back. Tilt the head back as you lift up on the chin. This will open the airway.

B = Breathing
Look, listen, and feel to make sure that the person can breathe. Look for the rise and fall of the chest. Listen for the sound of air moving in and out of the mouth and nose. Feel for exhaled breath on your hand or cheek. If the person is not breathing, perform rescue breathing (see **Figures 15.14** and **15.15**).

C = Circulation
Press two fingers on either side of the person's neck to check for a pulse. If you feel a pulse, continue with rescue breathing. If you do not feel a pulse, the person needs **cardiopulmonary resuscitation (CPR)**. This is *a first-aid procedure to restore breathing and circulation.* To perform CPR, a person must have special training. Call to passersby to find someone who has CPR training. In the meantime, continue with rescue breathing.

Lesson 5: First Aid   **435**

Lesson 5 Resources

Teacher's Classroom Resources
- Concept Map 70
- Enrichment Activity 70
- Lesson Plan 5
- Lesson 5 Quiz
- Reteaching Activity 70
- Transparency 63

Student Activities Workbook
- Study Guide 15
- Applying Health Skills 70

Lesson 5
First Aid

FOCUS

LESSON OBJECTIVES

After studying this lesson, students should be able to

- identify the first steps they should take to help someone who is injured or seriously ill.
- explain how to respond effectively to breathing emergencies.
- explain how to help a victim who is bleeding.
- discuss the dangers of choking and explain how to provide help.
- discuss how to help victims of poisoning, burns, broken bones, and sprains or bruises.

MOTIVATOR

Show students a photograph of a teen who has collapsed (or describe a scene in which a person has collapsed unexpectedly). Ask: What will you do? Have students respond by writing lists or several sentences.

INTRODUCING THE LESSON

Provide an opportunity for all interested students to share their responses to the Motivator question. Then ask: What do you wish you could do in a situation like that? What special skills do you need in order to help? Explain that Lesson 5 presents information that will help students develop and use the kinds of skills they need in a medical emergency.

INTRODUCING WORDS TO KNOW

Help students with the pronunciation of the Words to Know if necessary. Then have partners work together to skim the lesson and find the definition of each term. Ask partners to write a one- or two-paragraph narrative using at least five of the Words to Know in context.

Chapter 15 Lesson 5

Teach

L1 Role-Playing
Guide students in discussing how to call 911 or another source of emergency help: What kinds of information should you be prepared to give? How should you speak? Let students work with partners to role-play such calls.

L1 Identifying Examples
Instruct the class to think of various situations in which first aid might be necessary: Where could the victim be? Which situations might be unsafe and require that the victim be moved? Which situations would be safe enough that you should avoid moving the victim?

L1 Demonstrating
Have students choose partners and demonstrate the ABCs of first aid. Provide supervision and help to be sure students can check successfully for a clear airway, breathing, and circulation. Let volunteers explain, in their own words, what they would do if they found problems with the victim's airway, breathing, or circulation.

L2 Guest Speaker
Invite a representative from the American Red Cross (or from another local organization that offers CPR classes) to speak to the students: What special techniques are taught in CPR classes? Who can take these classes? Who should? When are they offered? How have teens been able to use CPR techniques?

L2 Applying Knowledge
Let students meet in small groups to read and discuss the information in Figures 15.14 and 15.15. Then let each group member practice using the techniques on an imaginary adult and on a baby-sized doll (if available). Have other group members offer directions as necessary and constructive criticism.

Teen Issues
Learning About CPR

To perform cardiopulmonary resuscitation (CPR), a person must be trained. Done incorrectly, this procedure can cause severe injuries. The American Red Cross and other community agencies give courses in CPR.

Breathing Emergencies

If a person is not breathing, he or she will survive for only a few minutes unless help is provided. If the person is not breathing but has a pulse, you need to perform rescue breathing. **Rescue breathing** is *a substitute for normal breathing, in which someone forces air into the victim's lungs.* **Figure 15.14** shows how to perform rescue breathing on adults and children. **Figure 15.15** on the next page shows how to perform rescue breathing on infants.

Figure 15.14
Rescue Breathing for Adults and Children

Knowing how to perform rescue breathing can help you save a life. Although not shown here, special masks and latex gloves are often used by emergency workers for this procedure.

1 Tilt the person's head back. To do this, place one hand under the chin and gently lift up. At the same time, place your other hand on the person's forehead and gently press down.

2 With your fingers, pinch the person's nostrils shut. Take a deep breath and place your mouth over his or her mouth, forming a seal. Give two slow breaths, each about 1½ seconds long. Breathe only until the chest gently rises.

3 For an adult or child 8 years or older, continue this procedure by giving one slow breath every 5 seconds for 1 minute. For a child 1–8 years old, give one slow breath every 3 seconds for 1 minute. After each breath, remove your mouth to allow the person to exhale.

4 Using your hands to keep the person's head tilted back, check for signs that the person is breathing on his or her own. Look for the rise and fall of the chest. Listen for the sound of exhaling. Then feel the person's neck for a pulse. If the person has not started breathing, or if you do not feel a pulse, continue rescue breathing. Check the breathing and pulse after each minute.

436 Chapter 15: Safety and Recreation

Consumer Health

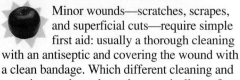

Minor wounds—scratches, scrapes, and superficial cuts—require simple first aid: usually a thorough cleaning with an antiseptic and covering the wound with a clean bandage. Which different cleaning and covering products do students typically use? Why? Suggest that students work in groups to compare several different brands of antiseptics and several different brands of bandages, including store brands when available. How do the products compare in terms of ingredients or materials? in terms of cost? in terms of effectiveness or reliability as reported in journals or magazine articles? Let the members of each group share their findings and conclusions with the rest of the class.

Figure 15.15
Rescue Breathing for Infants

When performing rescue breathing on infants, it is important to be very gentle. Although not shown here, special masks and latex gloves are often used by emergency workers for this procedure.

1 Gently tilt the child's head back slightly—not as far back as you would tilt an adult's head. Support the head with both hands. Do *not* pinch the nostrils shut.

2 Take a breath and place your mouth over the infant's nose and mouth, forming a seal. Give two slow, long, very gentle breaths. Each breath should last about 1½ seconds. Remove your mouth and wait 3 seconds. Repeat, giving 1 slow breath every 3 seconds. Remove your mouth after each breath to allow the child to exhale.

3 Check the child's pulse. Look, listen, and feel with your cheek for signs that he or she is breathing. If not, repeat rescue breathing. After each minute, recheck the pulse and breathing signs.

Severe Bleeding

The victim is now stable—he or she has a pulse and is breathing. Medical help has been called and is on the way. You must now check for any injuries that may involve severe bleeding. Remember that whenever blood is present, there is a risk of HIV transmission. The use of latex gloves can minimize this risk. To stop or slow the victim's loss of blood, use one of the following three methods.

- Cover the wound with a clean cloth and press firmly against the wound with your hand. If the cloth becomes soaked with blood, add a second cloth. Do not remove the first one, however.

- Elevate the wound above the level of the heart. Elevation will force the blood to move uphill, thus slowing its flow. If you think that the injury might also involve a broken bone, however, do not move or elevate it.

- If bleeding does not stop, use your other hand to apply pressure to a main artery leading to the wound. Squeeze the artery against a bone. **Figure 15.16** shows the location of these arteries and the points at which you should apply pressure.

After the bleeding has stopped, cover the wound with a clean cloth to prevent infection. Remain with the victim until professional medical help arrives.

Figure 15.16
Pressure Points

The dots on this drawing indicate the pressure points for the main arteries.

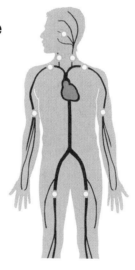

Lesson 5: First Aid **437**

Chapter 15 Lesson 5

L1 Discussing
Ask: What kinds of injuries are likely to cause severe bleeding in a victim? What should you do before you try to stop the bleeding? What do you think you should do if the sight of blood frightens you or makes you feel ill?

L1 Comprehending
Suggest that students find, on their own bodies, the pressure points indicated in Figure 15.16.

L1 Critical Thinking
Present the following situation: You are alone with a friend who has been injured and is bleeding severely. You are able to stop the bleeding, and now you are waiting for medical help to arrive. Your friend is conscious but weak and frightened. What will you say to your friend? Then present an alternative situation in which the injured teen is a stranger: What will you say to him or her?

L1 Comprehending
Help students discuss situations in which children, teens, and adults might choke. How should you respond? Emphasize that a person who is coughing forcefully should not be given first aid; the victim's body will probably be able to dislodge the food or other item on its own. However, a person who has trouble making sounds and may be coughing weakly needs immediate help.

L1 Reporting
What is the universal sign for choking? What is the best response to that sign? If students do not know the answer, ask a volunteer to find out and demonstrate the sign to the rest of the class.

Would You Believe?

A four-year-old was able to use first-aid abdominal thrusts to save the life of his two-year-old sister, who was choking on a piece of food.

Growth and Development

Students should recognize that although knowing first-aid techniques is essential, the basis of remaining safe is preventing accidents and injuries. They should also recognize that prevention is usually guided by a set of safety rules, often those set down by families. Family rules typically cover where, when, and with whom teens go; they also relate to risk behaviors, such as experimenting with tobacco and drugs. Remind students that as they mature, reasonable revisions to these rules—always proposed with safety and responsibility in mind—can reassure parents that their adolescent sons and daughters are taking an active role in regulating their own behavior.

Chapter 15 Lesson 5

L2 Home Economics
Babies and toddlers are at special risk for choking. Let a group of interested volunteers compile two lists: one of foods that should not be given to very young children, the other of toys and household objects that should be kept out of young children's reach. Have the volunteers prepare a bulletin board display of their findings.

L1 Life Skills
Have students contact a local poison control center for posters or pamphlets on emergency responses to poisoning. Suggest that students post these in their kitchen or near the main home telephone.

L2 Reporting
Ask a volunteer to find and share answers to these questions: What is syrup of ipecac? Where can you buy it? How should it be used? In what kinds of poisonings should it be avoided? Which households should always have syrup of ipecac on hand?

L1 Discussing
Guide students in discussing the symptoms of broken bones and sprains and bruises: How can you tell the difference? What should you do if you are not sure whether an injury is a fracture or a sprain?

Assess

EVALUATING THE LESSON
Assign Reviewing Terms and Facts and Thinking Critically on page 439 to review the lesson; then assign the Lesson 5 Quiz in the TCR.

RETEACHING
▶ Assign Concept Map 70 in the TCR.
▶ Have students complete Reteaching Activity 70 in the TCR.

ENRICHMENT
Assign Enrichment Activity 70 in the TCR.

438

If a Person Is Choking

in your journal

Every family should post a list of emergency telephone numbers in an easy-to-find location in the home. These numbers include the poison control center, fire department, other local health and rescue agencies, and close neighbors and relatives. Make sure that such a list is posted by every phone in your house.

Choking is *a condition that occurs when a person's airway becomes blocked.* Every year nearly 4,000 people in the United States die from accidental choking. When a person is choking, an obstruction prevents air from entering the lungs. If the obstruction is not removed, the person can die in just a few minutes.

Be alert to signs that a person is choking. Clutching at the throat is the universal sign for "I am choking." The person may also wheeze, have difficulty speaking, or turn red or blue in the face. Immediate first aid is needed.

For an adult or older child who is choking, use **abdominal thrusts.** This procedure consists of *quick, upward pulls into the diaphragm to force the obstruction out of the airway.* For an infant or young child who is choking, use **chest thrusts.** These are *quick presses into the middle of the breastbone to force the obstruction out of the airway.* Both of these procedures are shown in **Figure 15.17** below.

Figure 15.17
First Aid for Choking Victims
Why do you think first aid for a choking adult is different from first aid for a choking infant or young child?

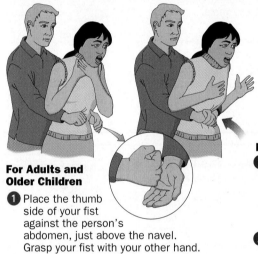

For Adults and Older Children
1. Place the thumb side of your fist against the person's abdomen, just above the navel. Grasp your fist with your other hand.
2. Give quick, upward thrusts. Continue until the person coughs up the object. If necessary, have the person lie down, face up, and begin rescue breathing.

For Infants and Young Children
1. Hold the child in one arm. While supporting the child's head with your hand, point the head down. Give the child five blows with the heel of your hand, between the shoulder blades. If the object is not dislodged, proceed to chest thrusts (step 2).
2. Turn the child over onto his or her back, supporting the head with one hand. With two or three fingers, press into the middle of the child's breastbone—directly between and just below the nipples—four times. Once the object is dislodged, begin rescue breathing if necessary.

Poisoning, Burns, and Other Emergencies

Be prepared to help out in all kinds of emergencies. Always start by checking the ABCs of first aid. Then follow the steps described in **Figure 15.18** on the next page.

438   Chapter 15: Safety and Recreation

More About...

Safety Information The National Safety Council is a large, private organization with the mission of educating and influencing people to adopt practices and policies that will protect personal safety and health, as well as the environment. The council publishes several different newsletters and magazines; it conducts education programs for various audiences; and it publishes booklets, brochures, and posters. Students interested in learning more about the National Safety Council and its offerings can write or browse its site:

National Safety Council
444 N. Michigan Avenue
Chicago, IL 60611
http://www.nsc.org

Figure 15.18
First Aid for Other Emergencies
Each type of emergency requires a different type of first aid.

Type of Emergency	First-Aid Treatment
Poisoning	Call 911 or a poison control center and follow directions carefully. Save the container of poison. Check the victim's breathing and pulse once a minute. Perform rescue breathing if needed. Keep the victim warm and still.
Burns	A **first-degree burn** is *a burn in which only the outer layer of the skin is burned and turns red.* A **second-degree burn** is *a serious burn in which the burned area blisters.* For these types of burns, cover the burned area with cool water (not ice) for 10 to 15 minutes. Then wrap loosely in a clean, dry dressing. Do not pop blisters or peel loose skin. Elevating the burned area may reduce pain. A **third-degree burn** is *a very serious burn in which all layers of the skin are damaged.* For this type of burn, call 911 or an ambulance. Victims of these burns, which are usually caused by fire, electricity, or chemicals, need immediate medical help. Do *not* apply cold water to the burn or attempt to remove burned clothing. While waiting for medical help, keep the victim still, and have him or her sip fluids.
Broken bones	*A break in a bone* is a **fracture.** Have the victim remain still while you wait for medical help. Apply a cold pack, but do not attempt to move the injured part. Moving broken bones can cause further injury.
Sprains and bruises	Tell the victim not to use the injured part of the body. Elevate it, and apply cold packs for 24 hours.

Review — Lesson 5

Using complete sentences, answer the following questions on a separate sheet of paper.

Reviewing Terms and Facts

1. **List** What three steps should be taken if someone needs first aid?
2. **Restate** Explain clearly the ABCs of first aid.
3. **Recall** When would you use abdominal thrusts on a victim, and when would you use chest thrusts?
4. **Identify** List and describe the three levels of burns.
5. **Vocabulary** Define the term *fracture.* How should a fracture be treated?

Thinking Critically

6. **Compare and Contrast** How are the rescue breathing techniques for adults similar to and different from those for infants?
7. **Hypothesize** If someone has been poisoned, why is it important to save the container of poison to show to the medical team?

Applying Health Concepts

8. **Health of Others** With one or two classmates, create a skit about one of the emergencies discussed in this lesson. Use dialogue, narrative, and movement to show correct and effective first aid.

Lesson 5: First Aid **439**

Personal Health

Insect bites and stings are, at best, irritating. In other cases, they can present real health risks. Students should be familiar with the basic steps in treating the bite or sting of any insect:

▶ In the case of a sting, remove the stinger by scraping it away with a fingernail or a plastic credit card.
▶ Wash the area of the insect bite or sting carefully, using soap and water.
▶ If the bite or sting is on a hand or finger, remove rings and other jewelry in case swelling develops.
▶ Apply cold compresses to the area.
▶ If any breathing problems develop, call 911 immediately.

Lesson 6
Protecting Our Planet

Focus

LESSON OBJECTIVES

After studying this lesson, students should be able to

▶ discuss their relationship to the environment and their responsibility for reducing pollution.

▶ identify steps they can take to help in the fight against air and water pollution.

▶ explain how they can reduce, reuse, and recycle to fight land pollution.

MOTIVATOR

Display several photographs showing various aspects of our environment, including people from around the world; lakes, rivers, oceans, or other bodies of water; animals that inhabit land, water, and air; and various forms of plant life. Ask students to look at the pictures and list everything they see that is part of their environment.

INTRODUCING THE LESSON

Have students compare and discuss their responses to the Motivator activity. Then ask: Why does the good health of our environment matter? Who is responsible for protecting our environment? Close this discussion by telling students that Lesson 6 presents information on our environment and on what everyone can do to help protect it.

INTRODUCING WORDS TO KNOW

Let students work in small groups to read and discuss the definitions of the Words to Know. Then have group members work together to draw a web or other organizational device, showing relationships among the Words to Know.

440

Lesson 6 Protecting Our Planet

This lesson will help you find answers to questions that teens often ask about protecting the environment. For example:

▶ What causes air pollution, and what can I do about it?
▶ What causes water pollution, and what can I do about it?
▶ How can I cut down on trash and other waste materials?

Words to Know

pollution
Environmental Protection Agency (EPA)
Occupational Safety and Health Administration (OSHA)
fossil fuels
acid rain
sewage
conservation
biodegradable

Science Connection

Endangered Animals ACTIVITY!

Many wild animals in the United States are endangered species. They have become endangered for several reasons, including pollution and destruction of the wilderness. Find out which species of animals and plants are endangered in your area. Write a brief report on your findings.

Pollution and the Environment

Your environment consists of all the living and nonliving elements that surround you. In a local sense, your environment includes your neighborhood, family, friends, and pets. It also includes trees, flowers, the air you breathe, the water you drink, and animals such as squirrels and birds. In a global sense, your environment includes the sun, world populations, natural resources, and wild animals. Because we all share this global environment, we all have a responsibility to protect it.

People haven't always protected the global environment. Many human actions and lifestyles have caused environmental problems. One major problem is **pollution,** or *dirty or harmful substances in the environment.* Fortunately, the governments of many nations are working to fight and prevent pollution. The **Environmental Protection Agency (EPA),** for example, is *an agency of the United States government that is committed to protecting the environment.*

Another government agency working in this area is the **Occupational Safety and Health Administration (OSHA).** OSHA is *a branch of the U.S. Department of Labor whose job is to ensure the protection of American workers.* OSHA's responsibilities include creating and enforcing regulations to lessen hazards, such as carcinogens, in workplaces.

EPA regulations have helped to protect such endangered species as the American bald eagle.

440   Chapter 15: Safety and Recreation

Lesson 6 Resources

Teacher's Classroom Resources
- 📁 Concept Map 71
- 📁 Cross-Curriculum Activity 30
- 📁 Decision-Making Activity 30
- 📁 Enrichment Activity 71
- 📁 Lesson Plan 6
- 📁 Lesson 6 Quiz

- 📁 Reteaching Activity 71
- 🗒 Transparency 64

Student Activities Workbook
- 📁 Study Guide 15
- 📁 Applying Health Skills 71
- 📁 Health Inventory 15

Air and Water Pollution

Most air pollution results from the burning of **fossil** (FAH·suhl) **fuels,** which are *the oil, coal, and natural gas used to provide energy.* People use fossil fuels to operate motor vehicles and factories and to provide heat and electricity to homes and other buildings. When fossil fuels are burned, poisonous gases are released into the atmosphere. These gases mix with moisture in the atmosphere to form **acid rain.** This is *rain that is more acidic than normal.* Over time, acid rain can harm forests by changing the chemistry of the soil. It can also contaminate fresh water supplies.

Most water pollution results from chemical wastes from factories, oil spills, and sewage. **Sewage** (soo·ij) is *garbage, detergents, and other household wastes that are washed down drains.* These waste materials can pollute drinking water and kill plants and animals in rivers, oceans, and lakes. In addition, polluted drinking water can lead to very serious human diseases, including cholera and typhoid fever.

What You Can Do

No single person can solve all the problems that threaten the global environment. However, each person can help. We can all work to prevent air and water pollution. We can also do our part in **conservation,** or *the saving of natural resources.* These resources include oil, natural gas, and water. You can start by following the guidelines presented in **Figure 15.19.**

Did You Know?
Free Bikes

To encourage people to use bicycles instead of cars for short trips, some cities have begun "free bike" programs. Officials in Boulder, Colorado, and Austin, Texas, have placed specially marked bicycles throughout these cities. Anyone can use the bikes. When riders get to their destinations, they must leave the bikes unlocked and out in the open. In this way, people can take turns using the bicycles.

Figure 15.19
Preventing Pollution
There are many ways to help in the fight against air and water pollution.

To Help Prevent Air Pollution
- Carpool with others, or use public transportation.
- Ride your bike or walk to nearby activities.
- Never start smoking. Cigarette smoke pollutes not only your body but also the environment of those around you.
- Conserve energy. Turn off lights when you leave a room. Turn off televisions, radios, and computers when you are not using them. If you are cold, put on a sweater rather than turn up the heat.

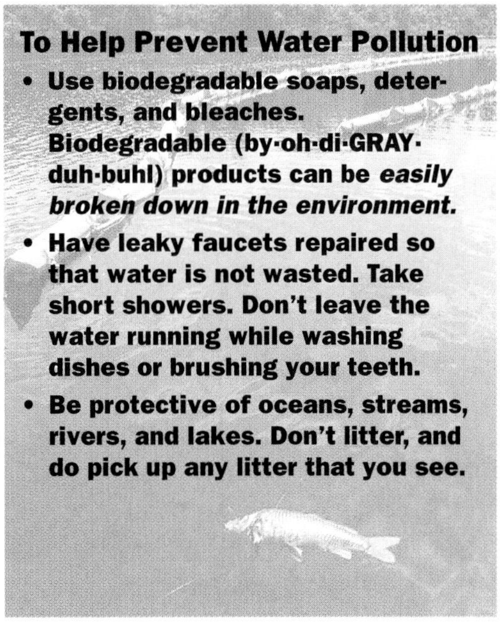

To Help Prevent Water Pollution
- Use biodegradable soaps, detergents, and bleaches. Biodegradable (by·oh·di·GRAY·duh·buhl) products can be *easily broken down in the environment.*
- Have leaky faucets repaired so that water is not wasted. Take short showers. Don't leave the water running while washing dishes or brushing your teeth.
- Be protective of oceans, streams, rivers, and lakes. Don't litter, and do pick up any litter that you see.

Lesson 6: Protecting Our Planet **441**

Chapter 15 Lesson 6

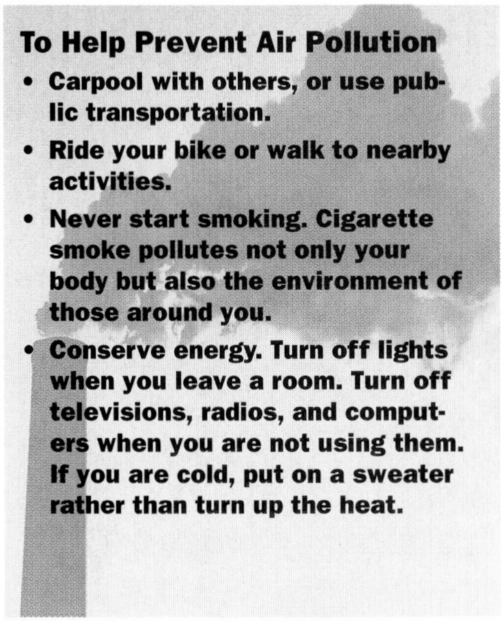

Teach

L1 Discussing

Let students share their responses to questions such as these: What difference does clean water in other parts of the world make to you? Why does it matter if a few species of plants and animals become extinct? How are you involved in the health and life expectancy of people you will never know?

L2 Reporting

Have each student do further reading on either the EPA or OSHA. Then ask students to write short reports or prepare posters about those agencies.

L3 Social Studies

The organization of groups to protect and preserve the environment, sometimes called the Green Movement, is relatively new. Ask volunteers to research the development and activities of this movement. Have them summarize their findings in a written report.

L3 Art

Suggest that interested volunteers research the effects of air pollution on paintings, sculpture, and other works of art. What damage has been done? What restorations have been attempted? What protective measures are being taken? Let these volunteers discuss their findings with the class.

VIDEODISC/VHS

Teen Health Course 2

You may wish to use video segment 15, "Protecting Our Planet," to show how teens can have an impact on pollution by choosing to consume less and reuse more, in addition to recycling materials.

Videodisc Side 2, Chapter 15
Protecting Our Planet

Search Chapter 15, Play to 16

Cooperative Learning

What chemicals are in the household cleaners your family uses? How do those chemicals affect the environment? Let students work in cooperative groups to find the answers to those questions. Have group members publish their findings and their recommendations in graphs, posters, or articles for the school paper.

Then suggest that group members find "recipes" for basic household cleaners that are environmentally friendly—usually a mix of hot water with ammonia, vinegar, and/or baking soda. Suggest that students make up batches of these cleaners and "test" them at home. (Warn students that they should not try to mix commercially prepared products; such mixing could release toxic fumes.)

Chapter 15 Lesson 6

L2 Applying Knowledge
Ask the class: Where do you see litter in this school? in our community? What can you do to help clean this litter up? To prevent further littering? Encourage students to work together to develop and carry out an appropriate program.

L2 Science
Have students read about "natural recycling," such as the decomposition of animal and plant waste and of dead animals and plants and the renutrition of the soil they provide. Provide time for students to discuss their findings and compare this "natural recycling" with human recycling efforts.

L1 Discussing
Let students describe their efforts—and their families'—to reduce, reuse, and recycle.

L2 Life Skills
Ask: What products can students recycle at school? How can our school recycling programs be improved or expanded? Encourage students to work together on devising and implementing a workable plan.

Assess

EVALUATING THE LESSON
Assign Reviewing Terms and Facts and Thinking Critically on page 443 to review the lesson; then assign the Lesson 6 Quiz in the TCR.

RETEACHING
▶ Assign Concept Map 71 in the TCR.
▶ Have students complete Reteaching Activity 71 in the TCR.

in your journal

Make a list in your journal of actions you already take to protect our natural resources of air, water, and land. Make a second list of additional ways in which you could help protect the environment. Create an action plan to follow through on your ideas.

Protecting the Land

Along with air and water, the land is a natural resource that is threatened by pollution. In the United States alone, people throw away hundreds of millions of tons of trash and garbage every year. Much of it goes to landfills—huge pits where waste materials are dumped and buried. Eventually most of the trash will break down into tiny particles in the soil. However, a number of commonly used materials take many years to break down.

Land pollution is dangerous to human beings in many ways. Homes are sometimes built on top of old landfill sites that have been contaminated by hazardous wastes. Forests and agricultural lands are often contaminated as well. When forests and wilderness areas are destroyed, wild animals and plants are threatened. When agricultural land is destroyed, our food supply is threatened.

What You Can Do
People can fight the problem of land pollution by changing their waste disposal habits. Individuals need to cut down on the amount of trash and garbage they create. **Figure 15.20** lists some strategies to follow.

Figure 15.20
Ways to Fight Land Pollution
By following the "Three Rs"—reduce, reuse, recycle—people can significantly cut back the amount of trash being created.

Strategy	Definition	Tips
Reduce	Cut down on the amount of trash and garbage you throw away.	Use your own baskets or cloth bags to carry groceries home. Avoid using disposable plates, cups, tableware, or napkins. Use cloth towels instead of paper towels. Buy products in bulk—they use less packaging per individual item. Do not buy items with unnecessary packaging, such as tomatoes that come in a plastic tray *and* are covered by plastic wrap.
Reuse	Find a practical use for an item that you might otherwise throw away.	Wash glass and plastic food containers, and use them for storage. Reuse plastic or paper bags when you go grocery shopping, or use them as trash bags. Donate good clothing that you no longer use to charity. Save worn-out clothing to use as cleaning rags. When possible, have broken items repaired rather than throw them away.
Recycle	Change an item in some way so that it can be used again.	Most communities have recycling programs. Find out which items are recycled in your area, how they should be separated, and how they are collected. Commonly recycled items include aluminum and tin cans; green, brown, and clear glass bottles; plastic containers; newspapers, magazines, and telephone books; paper and cardboard; and batteries.

442   Chapter 15: Safety and Recreation

HEALTH LAB

The Dangers of Water Pollution

Time Needed
half an hour

Supplies
▶ fresh water
▶ 1-quart clear glass bowl
▶ 1 teaspoon cooking oil
▶ 2 teaspoons powdered dishwashing detergent
▶ long-handled wooden spoon

Focus on the Health Lab
Let students share their responses to these questions: Why should you be concerned about water pollution? How does polluted water affect you? How does it affect people, animals, and plants? What do you think you and other teens can do to help prevent water pollution? Why should you bother?

Review

Using complete sentences, answer the following questions on a separate sheet of paper.

Reviewing Terms and Facts

1. **Restate** What are the roles of the EPA and OSHA in the protection of the environment and public welfare?
2. **Recall** Name three *fossil fuels*. How are fossil fuels used?
3. **Explain** How is acid rain formed?
4. **Identify** List three common causes of water pollution.
5. **Vocabulary** Define the term *conservation*. Then use this term in an original sentence.

Thinking Critically

6. **Analyze** Why is it so important to protect our environment from pollution?

7. **Predict** What might happen if people don't reduce, reuse, and recycle?

Applying Health Concepts

8. **Consumer Health** Next time you go grocery shopping, examine the packaging of all the products you buy. Make a list of all the types of packaging you encounter and the environmental advantages and disadvantages of each one. At the bottom of the page, suggest ways in which poor packaging could be improved.

HEALTH LAB

The Dangers of Water Pollution

Introduction: Water pollution is hazardous to many types of wildlife, including water birds. These birds have oil on their feathers. Because the oil does not mix with water, it acts as a natural raincoat, preventing water from soaking into the feathers. This protective coating makes birds lighter and helps them stay afloat. The oil also insulates the birds, keeping them warm.

Waste materials such as detergents are sometimes dumped into rivers and lakes. Fertilizers can also end up in lakes when they wash off lawns in a heavy rain. Some of the chemicals destroy the birds' natural foods. Others break down the oil on their feathers. When this happens, water soaks into the feathers and makes the birds heavier. Often they can no longer swim or fly.

Objective: With a partner, observe how water pollution can endanger wildlife.

Materials and Method: You will need 1 cup of fresh water; a 1-quart clear glass bowl; 1 teaspoon cooking oil (to represent the oil on a duck's feathers); 2 teaspoons powdered dishwashing detergent (to represent a water-polluting chemical); and a long-handled wooden spoon.

Pour the water into the bowl, and add the oil. Look at the oil. Does it appear to be mixing with the water? Using the long-handled spoon, gently stir in the detergent. Be careful not to let the detergent come into contact with your skin or eyes.

Observations and Analysis: What changes occurred after the detergent was added? Explain why. Use your observations to write a paragraph describing how detergents that get into waterways could harm wildlife.

Lesson 6: Protecting Our Planet

Chapter 15 Review

CHECKING COMPREHENSION

Use the Chapter Summary and the Chapter 15 Review to help students go over the most important ideas presented in Chapter 15. Encourage students to ask questions and add details as appropriate.

CHAPTER 15 REVIEW ANSWERS

Reviewing Key Terms and Concepts

1. Responses will vary. They should include the five links in the accident chain: the situation, the unsafe habit, the unsafe action, the accident, and the result.
2. Responses will vary.
3. Putting water on an electrical fire can cause electrocution.
4. An electrical overload is a dangerous situation in which too much electric current flows along a single circuit. This hazard can be prevented by plugging no more than two electric appliances into a wall outlet.
5. A pedestrian is a person who travels on foot.
6. Rules to follow when using skateboards and in-line skates include (any five): wear a helmet; wear wrist guards, elbow pads, and knee pads; keep your speed under control; don't skate at night, in traffic, or in bad weather; learn how to stop before you start skating; know how to fall properly.
7. An earthquake is a violent shaking of the earth's surfaces. An earthquake is usually a chain of events; aftershocks, or secondary earthquakes, often follow the initial quake.
8. Tornadoes are whirling, funnel-shaped windstorms that drop from the sky to the ground. Tornadoes most often occur in states in the Midwest and those bordering the Gulf of Mexico.

Chapter 15 Review

Chapter Summary

▶ **Lesson 1** Most accidents and injuries are caused by a chain of circumstances. Changing unsafe habits and actions can break the chain.

▶ **Lesson 2** Following safety procedures and eliminating hazards can prevent injuries in the home. At school, be alert and aware.

▶ **Lesson 3** Pedestrians and bicycle riders should follow the rules of the road. Following safety guidelines will help you avoid injuries outdoors.

▶ **Lesson 4** During weather emergencies and natural disasters, act quickly, seek shelter, and take safety precautions to protect yourself.

▶ **Lesson 5** Effective first aid involves recognizing an emergency, calling for medical help, and providing immediate care for the victim.

▶ **Lesson 6** Everyone can help to protect the environment by reducing pollution of the air, water, and land.

Reviewing Key Terms and Concepts

Using complete sentences, answer the following questions on a separate sheet of paper.

Lesson 1
1. Give an example of an accident chain.
2. Give an example of a situation in which changes in habits could prevent an injury.

Lesson 2
3. Why is it dangerous to use water on an electrical fire?
4. Define the term *electrical overload*. How can this hazard be prevented?

Lesson 3
5. What is a *pedestrian*?
6. List five rules to follow when using skateboards and in-line skates.

Lesson 4
7. How are *earthquakes* and *aftershocks* related?
8. Describe a *tornado*. Where do tornadoes most often occur?

Lesson 5
9. Explain the difference between CPR and rescue breathing.
10. What are the signs that a person is choking?

Lesson 6
11. How are *pollution* and the environment related?
12. What is a *biodegradable* product?

Thinking Critically

Using complete sentences, answer the following questions on a separate sheet of paper.

13. **Analyze** How can safety consciousness help you prevent accidents and injuries?
14. **Synthesize** Suppose that you are hiking in the woods and see a thunderstorm approaching. What should you do?
15. **Infer** If you came upon a serious car accident, what action would you take first? Why would you do this?
16. **Explain** Why should everyone take some degree of responsibility for protecting the environment?

444 Chapter 15: Safety and Recreation

Meeting Student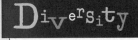

Language Diversity Use the following suggestions to help students who have difficulty with English:

▶ Pair those learners with native speakers of English who can restate the Chapter Summary in language that helps students comprehend important concepts.

▶ Direct auditory learners or those students with language diversity to the Teen Health Audiocassette Program. Available in English and Spanish, this component provides an audio and written summary of the chapter.

Chapter 15 Review

Your Action Plan

You can prevent many injuries by being safety conscious at home, at school, and outdoors. This chapter has provided many tips for staying safe.

Step 1 Review your private journal entries for this chapter. Highlight points that will enable you to remember and practice safety precautions at home, at school, and outdoors. What habits or actions should you change to reduce your risk of injury?

Step 2 Set a long-term goal that will improve your personal safety. For example, you might decide to make your home as safe as possible from fire.

Step 3 Next, write down short-term goals that will help you achieve your long-term goal. A short-term goal could be to create a checklist for possible hazards around the house or to find out where a fire extinguisher might be purchased.

Create a schedule for reaching each short-term goal. Review your schedule regularly to make sure that you are on track. When you have reached your goal, consider becoming safety conscious in a different area of your life.

In Your Home and Community

1. **Health of Others** Draw a floor plan of your home, including bedrooms, doors, windows, hallways, and stairs. Indicate two escape routes for each family member. The escape routes should begin at each bedroom. Share the plan with the members of your family. Then practice your escape plan by conducting a family fire drill.

2. **Community Resources** Invite an emergency medical technician or a police officer to speak to your class. Ask the person to demonstrate common first-aid techniques. Also ask the speaker about safety tips for preventing accidents and injuries.

Building Your Portfolio

1. **Persuasive Essay** Do you feel that bicycle riders, like drivers, should have to pass a test before using public roads? Take a stand on this issue, and write a one-page essay that supports your views. Put the essay in your portfolio.

2. **Report of an Event** Find newspaper articles about weather emergencies and natural disasters. Select one event, and pretend that you are there. Describe what actions you would take. Add the articles and your description to your portfolio.

3. **Personal Assessment** Look through all the activities and projects you did for this chapter. Choose one or two that you would like to include in your portfolio.

Chapter 15: Chapter Review **445**

Performance Assessment

▶ **Self-evaluation** Direct students to review the activities that are provided throughout the chapter. Encourage each student to select one finished product or activity that demonstrates his or her best work for the chapter. Have students explain what they learned and how the examples they selected show their progress.

▶ **Teacher's Classroom Resources** Assign Performance Assessment 15, "Safety Handbook," in the TCR.

Chapter 15 Review

9. CPR is required if a victim is not breathing and has no pulse. A person must have special training to perform CPR. Rescue breathing is needed when a victim is not breathing but does have a pulse. A person without special training can perform rescue breathing successfully.
10. Signs that a person is choking include grabbing the throat, coughing or wheezing, difficulty speaking, face turning red or blue.
11. Pollution contaminates the environment.
12. A biodegradable product is one that can be easily broken down in the environment.

Thinking Critically

Responses will vary. Possible responses are given.
13. Acting safely prevents accidents and injuries.
14. Seek shelter. If you cannot find shelter before the storm starts, squat low to the ground in an open area. Keep away from tall trees, water, and metal objects.
15. Dial 911 or call for medical help immediately. Then you can provide appropriate first aid.
16. Because we all live in the environment and use its natural resources, including air and water, we should all take responsibility for protecting it.

RETEACHING

Assign Study Guide 15 in the Student Activities Workbook.

EVALUATE

▶ Use the reproducible Chapter 15 Test in the TCR, or construct your own test using the Testmaker Software.
▶ Use Performance Assessment 15 in the TCR.

EXTENSION

Have the students collect brochures and fliers on emergency preparedness, first aid, and other aspects of safety. Suggest that the students make these available in the school.

445

Glossary

The Glossary contains all the important terms used throughout the text. It includes the **boldfaced** terms listed in the "Words to Know" lists at the beginning of each lesson and that appear in text, captions, and features.

The Glossary lists the term, the pronunciation (in the case of difficult terms), the definition, and the page on which the term is defined. The pronunciations here and in the text follow the system outlined below. The column headed "Symbol" shows the spelling used in this book to represent the appropriate method.

Pronunciation Key

Sound	As in	Symbol	Example
ă	hat, map	a	abscess (AB·sess)
ā	age, face	ay	atrium (AY·tree·uhm)
a	care, their	ehr	capillaries (KAP·uh·lehr·eez)
ä, ŏ	father, hot	ah	biopsy (BY·ahp·see)
ar	far	ar	cardiac (KAR·dee·ak)
ch	child, much	ch	barbiturate (bar·BI·chuh·ruht)
ĕ	let, best	e	vessel (VE·suhl)
ē	beat, see, city	ee	acne (AK·nee)
er	term, stir, purr	er	nuclear (NOO·klee·er)
g	grow	g	malignant (muh·LIG·nuhnt)
ĭ	it, hymn	i	bacteria (bak·TIR·ee·uh)
ī	ice, five	y	benign (bi·NYN)
		eye	iris (EYE·ris)
j	page, fungi	j	cartilage (KAR·tuhl·ij)
k	coat, look, chorus	k	defect (DEE·fekt)
ō	open, coat, grow	oh	aerobic (e·ROH·bik)
ô	order	or	organ (OR·guhn)
ȯ	flaw, all	aw	palsy (PAWL·zee)
oi	voice	oy	goiter (GOY·ter)
ou	out	ow	fountain (FOWN·tuhn)
s	say, rice	s	dermis (DER·mis)
sh	she, attention	sh	conservation (kahn·ser·VAY·shuhn)
ŭ	cup, flood	uh	bunion (BUHN·yuhn)
u	put, wood, could	u	pulmonary (PUL·muh·nehr·ee)
ü	rule, move, you	oo	attitudes (AT·i·toodz)
w	win	w	warranty (WAWR·uhn·tee)
y	your	yu	urethritis (yur·i·THRY·tuhs)
z	says	z	hormones (HOR·mohnz)
zh	pleasure	zh	transfusion (trans·FYOO·zhuhn)
ə	about, collide	uh	asthma (AZ·muh)

Abdominal thrusts Quick, upward pulls into the diaphragm to force an obstruction out of a person's airway. (page 438)

Abstinence Refusing to participate in unsafe behaviors or activities. (page 255)

Abuse (uh·BYOOS) The physical, emotional, or mental mistreatment of another person. (page 400)

Accident chain A series of events that includes a situation, an unsafe habit, and an unsafe action. (page 413)

Acid rain Rain that is more acidic than normal. (page 441)

Action plan A series of steps for reaching a goal. (page 54)

Addiction (uh·DIK·shuhn) A physical or psychological need for a drug or other substance. (page 283)

Adolescence (a·duhl·E·suhns) Time of life between childhood and adulthood. (page 176)

Adrenaline (uh·DRE·nuhl·in) A hormone that increases the level of sugar in the blood, which gives the body extra energy. (page 220)

Advertisement A message designed to get consumers to buy a product or service. (page 61)

Aerobic (e·ROH·bik) **exercise** Nonstop, rhythmic, vigorous activity that increases breathing and heartbeat rates. (page 121)

Aftershocks Secondary earthquakes that may occur after an initial earthquake. (page 431)

AIDS (acquired immunodeficiency syndrome) A deadly disease that interferes with the body's ability to fight infection. (page 353)

Alcohol A drug created by a chemical reaction in some foods, especially fruits and grains. (page 296)

Alcoholism An illness caused by a physical and mental need for alcohol. (page 299)

Allergen (AL·er·juhn) A substance that causes an allergic reaction. (page 373)

Allergy An extreme sensitivity to a substance. (page 373)

Alternatives Other ways of thinking or acting. (page 322)

Alveoli (al·VEE·uh·ly) Microscopic air sacs in the lungs where gases are exchanged. (page 277)

Alzheimer's (AHLTS·hy·merz) **disease** An illness that attacks the brain and affects thinking, memory, and behavior. (page 383)

Amphetamine (am·FE·tuh·meen) A strong stimulant drug that speeds up the nervous system. (page 304)

Anabolic steroids (a·nuh·BAH·lik STIR·oydz) Synthetic compounds that cause muscle tissue to develop at an abnormally high rate. (page 158)

Anaerobic (an·e·ROH·bik) **exercise** Intense physical activity that requires short bursts of energy. (page 121)

Angioplasty (AN·jee·uh·plas·tee) A surgical procedure in which an instrument with a tiny balloon attached is inserted into an artery to clear a blockage. (page 365)

Anorexia nervosa (an·uh·REK·see·uh ner·VOH·suh) An eating disorder in which a person has an intense fear of weight gain and starves himself or herself. (page 145)

Antibodies Proteins that attach to antigens, keeping them from harming the body. (page 337)

Antigens (AN·ti·jenz) Substances that send your immune system into action when your body is invaded by germs. (page 337)

Antihistamine A medication that relieves the symptoms of an allergic reaction. (page 374)

Anxiety disorder A disorder in which real or imagined fears keep a person from functioning normally. (page 229)

Arteriosclerosis (ar·tir·ee·oh·skluh·ROH·sis) Hardening of the arteries. (page 364)

Artery A type of blood vessel that carries blood away from the heart to all parts of the body. (page 123)

Arthritis (ar·THRY·tuhs) A disease of the joints marked by painful swelling and stiffness. (page 380)

Assertive Having the determination to stand up for yourself in a firm but positive way. (page 321)

Glossary 447

Asthma (AZ·muh) A chronic respiratory disease that causes tiny air passages in the lungs to become narrow or blocked. (page 375)

Astigmatism (uh·STIG·muh·tiz·uhm) An eye condition in which images are distorted, causing objects to appear wavy or blurry. (page 77)

Atherosclerosis (a·thuh·roh·skluh·ROH·sis) A condition in which fatty substances build up on the inner lining of arteries. (page 364)

Bacteria (bak·TIR·ee·uh) Tiny one-celled organisms, some causing disease, that live nearly everywhere. (page 333)

Battery The beating, hitting, or kicking of another person. (page 401)

Benign (bi·NYN) Not cancerous. (page 367)

Biodegradable (by·oh·di·GRAY·duh·buhl) Easily broken down in the environment. (page 441)

Blizzard A very heavy snowstorm with winds of up to 45 miles per hour. (page 434)

Blood pressure The force of blood pushing against the walls of the blood vessels. (page 126)

Body composition The proportions of fat, bones, muscle, and fluid that make up body weight. (page 120)

Body language Nonverbal communication that includes posture, gestures, and facial expressions. (page 241)

Body Mass Index (BMI) A measure of weight based on comparing body weight to height. (page 141)

Body system A group of organs working together to carry out related tasks. (page 196)

Brain The information center of the nervous system, which receives and screens information and sends messages to other parts of the body. (page 308)

Bronchi (BRAHN·ky) Two tubes that branch from the trachea, one to each lung. (page 276)

Bronchodilator (brahn·koh·dy·LAY·ter) Medication that relaxes the muscles around the bronchial air passages. (page 376)

Bulimia nervosa (boo·LEE·mee·uh ner·VOH·suh) An eating disorder in which a person repeatedly eats large amounts of food and then purges. (page 145)

Burnout A sense of exhaustion caused by exerting too much energy for too long a time. (page 167)

Calories (KA·luh·reez) Units of heat that measure the energy available in foods. (page 95)

Cancer A disease that occurs when abnormal cells grow out of control. (page 367)

Capillary The smallest blood vessel, which provides body cells with blood and connects an artery with a vein. (page 123)

Carbohydrates (kar·boh·HY·drayts) The starches and sugars that provide the body with most of its energy. (page 96)

Carbon monoxide (KAR·buhn muh·NAHK·syd) A colorless, odorless, poisonous gas produced when tobacco burns. (page 272)

Carcinogen (kar·SI·nuh·juhn) A substance in the environment that causes cancer. (page 369)

Cardiopulmonary resuscitation (CPR) A first-aid procedure to restore breathing and circulation. (page 435)

Cardiovascular (KAR·dee·oh·VAS·kyoo·ler) **system** Another name for the circulatory system. (page 123)

Cartilage (KAR·tuhl·ij) Strong, flexible tissue that covers the ends of bones and also supports some structures. (page 130)

Cavity A hole that begins in a tooth's enamel. (page 70)

Cell The basic unit, or building block, of life. (page 196)

Central nervous system (CNS) The brain and the spinal cord. (page 308)

Chemotherapy (kee·moh·THEHR·uh·pee) The use of chemicals to destroy cancer cells. (page 370)

Chest thrusts Quick presses into the middle of the breastbone to force an obstruction out of a person's airway. (page 438)

Chlamydia (kluh·MI·dee·uh) A bacteria-caused STD that may affect the reproductive organs, urethra, and anus. (page 351)

Choking A condition that occurs when a person's airway becomes blocked. (page 438)

Cholesterol (kuh·LES·tuh·rawl) A waxlike substance used by the body to build cells and make other substances. (page 99)

Chromosomes (KROH·muh·sohmz) Threadlike structures within a cell that carry the codes for inherited traits. (page 199)

Chronic (KRAH·nik) Present continuously or on and off over a long time. (page 362)

Circulatory system The group of organs and tissues that transport essential materials to body cells and remove their waste products. (page 123)

Cirrhosis (suh·ROH·sis) Scarring and destruction of liver tissue. (page 297)

Colon (KOH·luhn) The large intestine, which is a storage tube for solid wastes. (page 113)

Commitment A pledge or promise. (page 150)

Communicable (kuh·MYOO·ni·kuh·buhl) **disease** A disease that can be passed from one person to another. (page 332)

Communication The exchange of thoughts, feelings, and beliefs among people. (page 240)

Comparison (kuhm·PEHR·i·suhn) **shopping** Comparing products, evaluating their benefits, and choosing the products that offer the best value. (page 68)

Competition Rivalry between two or more individuals or groups trying to reach the same goal. (page 152)

Compromise A method in which each person gives up something in order to reach a solution that satisfies everyone. (page 244)

Conditioning Training to get into shape. (page 159)

Conflict A disagreement between people with opposing viewpoints. (page 260)

Consequences The effects or results of actions. (page 32)

Conservation The saving of natural resources. (page 441)

Consumer (kuhn·SOO·mer) Anyone who buys goods and services. (page 60)

Contagious period The length of time when a particular disease can spread from person to person. (page 344)

Contract (Kuhn·TRAKT) To shorten, often referring to muscle function. (page 130)

Cool-down Gentle exercise that lets your body adjust to ending a workout. (page 138)

Cornea (KOR·nee·uh) The clear section that lets in light at the front of the eye. (page 76)

Criteria (kry·TIR·ee·uh) Standards on which to base a decision. (page 48)

Cross-training Any fitness program that includes a variety of physical activities to promote balanced fitness. (page 161)

Cumulative (KYOO·myuh·luh·tiv) **risks** Related risks that increase in effect with each added risk. (page 41)

Dandruff Flaking of the outer layer of dead skin cells on the scalp. (page 75)

Decibel The unit for measuring the loudness of sound waves. (page 79)

Decision making The process of making a choice or finding a solution. (page 46)

Defense mechanism (di·FENS MEK·uh·nizm) A short-term way of dealing with stress. (page 222)

Dehydration (dee·hy·DRAY·shuhn) Excessive loss of water from the body. (page 155)

Depressant (di·PRE·suhnt) A drug that slows down the body's functions and reactions, including heart and breathing rates. (page 305)

Depression A mood disorder involving feelings of hopelessness, helplessness, worthlessness, guilt, and extreme sadness that continue for periods of weeks. (page 229)

Dermis (DER·mis) The inner layer of skin that contains blood vessels, nerve endings, and hair follicles. (page 73)

Diabetes (dy·uh·BEE·teez) A disease that prevents the body from converting food into energy. (page 382)

Diaphragm (DY·uh·fram) Large dome-shaped muscle below the lungs that draws air in and pushes air out. (page 276)

Diet All the things you regularly eat and drink. (page 92)

Digestion (dy·JES·chuhn) The process by which the body breaks food down into smaller components that can be absorbed by the bloodstream and sent to each cell in the body. (page 108)

Digestive (dy·JES·tiv) **system** A series of organs that work together to break down foods into substances that your cells can use. (page 108)

Discount store A store that offers special reduced prices. (page 68)

Disease A condition that interferes with the proper functioning of the body or mind. (page 332)

Distress Stress that keeps you from doing the things you need to do or that causes you discomfort. (page 219)

Drug A nonfood substance taken into the body that can change the structure or function of the body or mind. (page 301)

Earthquake A violent shaking of the earth's surface. (page 431)

Eating disorder An extreme eating behavior that can lead to serious illness or even death. (page 145)

Electrical overload A dangerous situation in which too much electric current flows along a single circuit. (page 420)

Electrocution Death by electrical shock. (page 418)

Embryo (EM·bree·oh) The name for an organism from fertilization to about the eighth week. (page 197)

Emotions Feelings such as love, joy, or fear. (page 4)

Empathy The ability to identify and share another person's feelings. (page 243)

Endocrine (EN·duh·krin) **system** Glands throughout your body that regulate body functions. (page 182)

Endorsement (en·DOR·smuhnt) A statement of approval. (page 63)

Endurance (in·DUR·uhnts) The ability to perform vigorous physical activity without getting overly tired. (page 121)

Environment (en·VY·ruhn·ment) All the living and nonliving things that surround you. (page 9)

Environmental Protection Agency (EPA) An agency of the United States government that is committed to protecting the environment. (page 440)

Epidermis (e·puh·DER·mis) The outermost layer of skin. (page 73)

Excretion (ek·SKREE·shuhn) The process by which the body gets rid of liquid waste materials. (page 112)

Excretory (EK·skruh·tor·ee) **system** The system that removes wastes from the body and controls water balance. (page 112)

Exercise Physical activity that develops fitness. (page 118)

Exercise frequency How often a person works out each week. (page 139)

Exercise intensity How much energy a person uses when working out. (page 139)

Extend To lengthen, often referring to muscle function. (page 131)

Family The basic unit of society. (page 245)

Family violence shelter A place where family members in danger of abuse can stay while they get their lives in order. (page 405)

Fatigue Extreme tiredness. (page 221)

Fats Sources of energy that also perform other functions, such as vitamin storage and body insulation. (page 97)

Fertilization (fer·til·i·ZAY·shuhn) The joining of a male sperm cell and a female egg cell to form a new human life. (page 190)

Fetus (FEE·tuhs) The name for the developing child from about the ninth week until the time of birth. (page 197)

Fiber The part of grains, fruits, and vegetables that the body cannot break down. (page 98)

First aid The immediate care given to someone who becomes injured or ill until regular medical care can be provided. (page 435)

First-degree burn A burn in which only the outer layer of the skin is burned and turns red. (page 439)

Fitness The ability to handle the physical work and play of everyday life without becoming overly tired. (page 118)

Flammable Able to catch fire easily. (page 416)

Flexibility The ability to move joints fully and easily. (page 123)

Fluoride (FLAWR·eyed) A substance that helps teeth resist decay. (page 70)

Food Guide Pyramid A guide for making healthful daily food choices, developed by the U.S. Department of Agriculture. (page 93)

Fossil (FAH·suhl) **fuels** Oil, coal, and natural gas used to provide energy. (page 441)

Fracture A break in a bone. (page 439)

Friendship The relationship between people who like each other and who have similar interests and values. (page 249)

Fungi (FUHN·jy) Primitive life forms that cannot make their own food. (page 333)

Gang A group of people who associate with one another because they have something in common. (page 396)

Gene The basic unit of heredity that determines the traits you inherit from your parents. (page 199)

Generic (juh·NER·ik) **products** Goods sold in plain packages at lower prices than brand name goods. (page 68)

Genital herpes (JEN·i·tuhl HER·peez) An STD caused by a virus that produces painful blisters in the genital area. (page 352)

Germ A microorganism that causes disease. (page 332)

Gland A group of cells, or an organ, that produces a chemical substance. (page 182)

Glucose (GLOO·kohs) A sugar that is the body's main source of energy. (page 102)

Gonorrhea (gah·nuh·REE·uh) A bacteria-caused STD that affects the genital mucous membrane and sometimes other body parts, such as the heart or joints. (page 352)

Grief Sorrow caused by the loss of something precious. (page 205)

Gynecologist (gy·nuh·KAH·luh·jist) A doctor who specializes in the female reproductive system. (page 192)

Habit A pattern of behavior repeated frequently enough to be performed almost without thinking. (page 34)

Hallucinogen (huh·LOO·suhn·uh·jen) A drug that distorts moods, thoughts, and senses. (page 305)

Hate crime An illegal act committed against someone just because he or she is a member of a particular group. (page 394)

Hazard A potential source of danger. (page 41)

Head lice Parasitic insects that live in the hair and cause itching. (page 75)

Health A combination of physical, mental/emotional, and social well-being. (page 4)

Health insurance (in·SHUR·uhns) A plan in which people pay a set fee to an insurance company in return for the company's agreement to pay some or most medical costs. (page 84)

Health maintenance (MAYN·tuh·nuhns) **organization (HMO)** A form of managed health care that offers its members the services of many different types of health care providers. (page 84)

Heart attack Condition that occurs when the blood supply to the heart slows or stops and the heart muscle is damaged. (page 364)

Hepatitis (he·puh·TY·tis) A liver inflammation, caused by a virus, characterized by yellowing of the skin and the whites of the eyes. (page 342)

Heredity (huh·RED·i·tee) The passing on of traits from biological parents to children. (page 9)

Histamines (HIS·tuh·meenz) Chemicals in the body that cause the symptoms of an allergic reaction. (page 374)

HIV (human immunodeficiency virus) The virus that causes AIDS. (page 353)

Homicide (HAH·muh·syd) A violent crime that results in the death of an individual. (page 393)

Hormones (HOR·mohnz) Chemical substances, produced by glands, which help to regulate the way the body functions. (page 176)

Hurricane A strong windstorm with driving rain that originates at sea. (page 433)

Hygiene (HY·jeen) Cleanliness. (page 19)

Hypertension (hy·per·TEN·shuhn) Condition in which a person's blood pressure stays at a level that is higher than normal; also called *high blood pressure*. (page 364)

Hypothermia (hy·poh·THER·mee·uh) A sudden and dangerous drop in body temperature. (page 427)

Immune (i·MYOON) **system** A combination of body defenses made up of cells, tissues, and organs that fight off germs and disease. (page 336)

Immunity The body's ability to resist the germs that cause a particular disease. (page 339)

Individual sports Physical activities you can take part in by yourself or with a friend, without being part of a team. (page 151)

Infancy (IN·fuhn·see) The first year of a baby's life. (page 201)

Infection Condition that occurs when germs get inside the body, multiply, and damage body cells. (page 332)

Influenza (in·floo·EN·zuh) A communicable disease typically characterized by fever, chills, fatigue, headache, muscle aches, and respiratory symptoms. (page 341)

Infomercial (IN·foh·mer·shuhl) A longer television commercial whose main purpose appears to be to present information rather than to sell a product. (page 63)

Inhalant (in·HAY·luhnt) A substance whose fumes are breathed in to produce mind-altering sensations. (page 305)

Insulin (IN·suh·lin) A hormone produced in the pancreas that regulates the level of sugar in the blood. (page 382)

Intoxicated (in·TAHK·suh·kay·tuhd) Drunk. (page 315)

Invulnerable Not able to be hurt. (page 320)

Iris (EYE·ris) The colored part of the eye. (page 76)

Joint A place where two or more bones meet. (page 128)

Kidney One of a pair of bean-shaped organs that filter water and waste materials from the blood. (page 112)

Lens (LENZ) The part of the eye that focuses light on the retina. (page 76)

Life-altering Capable of changing a person's day-to-day existence. (page 38)

Life-threatening Possibly causing death. (page 36)

Ligament (LI·guh·ment) A strong band of tissue that holds a bone in place at a joint. (page 130)

Liver The body's largest gland, which secretes a liquid called bile that helps to digest fats. (page 111)

Long-term goal A goal that you plan to reach over an extended length of time. (page 53)

Lymphatic (lim·FA·tik) **system** A secondary circulatory system that helps the body fight germs and maintain its fluid balance. (page 337)

Lymphocytes (LIM·fuh·syts) Special white blood cells in the lymph that are important in fighting off germs and disease. (page 337)

Mainstream smoke Smoke exhaled by a smoker. (page 290)

Malignant (muh·LIG·nuhnt) Cancerous. (page 367)

Managed care A health care plan that emphasizes preventive medicine and works to manage the cost and quality of health care. (page 84)

Mass media Media that can reach large groups of people, such as newspapers, books, television, movies, and the Internet. (page 61)

Media The various methods for communicating information. (page 61)

Mediation (mee·dee·AY·shuhn) Resolving conflicts by using a neutral person to help reach a solution that is acceptable to both sides. (page 265)

Medicaid (MED·i·kayd) A public health insurance program for low-income families and individuals. (page 85)

Medicare (MED·i·kehr) A federal government health insurance program that provides health insurance to people who are 65 years old and over. (page 85)

Medicine A drug that is used to treat an illness or relieve pain. (page 301)

Melanin (ME·luh·nuhn) A substance made by cells in the epidermis that gives skin its color. (page 73)

Menstruation (men·stroo·WAY·shuhn) The flow of the uterine lining material from the body. (page 191)

Mental health Your ability to deal in a reasonable way with the stresses and changes of everyday life. (page 210)

Minerals Substances that strengthen the muscles, bones, and teeth; enrich the blood; and keep the heart and other organs operating properly. (page 97)

Mononucleosis (MAH·noh·noo·klee·OH·sis) A virus-caused disease characterized by swelling of the lymph nodes in the neck and throat. (page 342)

Mood disorder A disorder in which a person undergoes changes in mood that seem inappropriate or extreme. (page 229)

Mortality (mor·TA·luh·tee) Death. (page 39)

Muscle endurance The measure of how long a group of muscles can exert a force without tiring. (page 4)

Muscular system The body system that consists of tissues that move parts of the body and operate internal organs. (page 128)

Narcotic (nar·KAH·tik) A drug that relieves pain and dulls the senses. (page 305)

Negotiation (ni·goh·shee·AY·shuhn) The process of discussing problems face-to-face in order to reach a solution. (page 265)

Neuron (NOO·rahn) One of the cells that make up the nervous system. (page 307)

Neutrality (noo·TRA·luh·tee) Not taking sides when others are arguing. (page 265)

Nicotine (NIK·uh·teen) An addictive drug found in tobacco. (page 272)

Noncommunicable disease A disease, such as asthma and cancer, that cannot be spread from one person to another. (page 332)

Nonverbal communication All the ways in which you can get a message across without using words. (page 241)

Nonviolent confrontation Resolving a conflict by peaceful methods. (page 264)

Nurture To provide for the physical, emotional, mental, and social needs of a person. (page 246)

Nutrient density The amount of nutrients in a food relative to the number of calories. (page 104)

Nutrients (NOO·tree·ents) The substances in foods that your body needs in order to grow, have energy, and stay healthy. (page 95)

Nutrition (noo·TRI·shuhn) The body's process of taking in food and using it for growth and good health. (page 92)

Obesity Condition in which a person weighs 20 percent more than his or her ideal weight. (page 142)

Objective Based on facts. (page 41)

Occupational Safety and Health Administration (OSHA) A branch of the U.S. Department of Labor whose job is to ensure the protection of American workers. (page 440)

Ophthalmologist (ahf·thahl·MAH·luh·jist) A physician who specializes in the structure, functions, and diseases of the eye. (page 77)

Optic (AHP·tik) **nerve** A bundle of nerve fibers that send messages from the eye to the brain. (page 76)

Optometrist (ahp·TAH·muh·trist) An eye care professional trained to examine the eyes for vision problems and to prescribe corrective lenses. (page 77)

Organ A body part made up of different tissues joined together to perform a function. (page 196)

Osteoarthritis (ahs·tee·oh·ar·THRY·tuhs) A chronic disease, common in elderly people, that results from the breakdown of cartilage in the joints. (page 380)

Ovaries (OH·vuh·reez) The female reproductive glands. (page 190)

Overcommitment Obligating yourself to more people or projects than you have time or energy to follow through on. (page 167)

Overtraining Exercising too hard or too often, without enough rest in between. (page 161)

Pancreas (PAN·kree·uhs) A gland that helps the small intestine by producing pancreatic juice, a blend of enzymes that breaks down proteins, carbohydrates, and fats. (page 111)

Passive smoker A nonsmoker who breathes secondhand smoke. (page 290)

Pedestrian A person who travels on foot. (page 422)

Peer pressure The influence that your friends have on you to believe and act like them. (page 251)

Peers People close to your age who are similar to you in many ways. (page 251)

Peripheral (puh·RIF·uh·ruhl) **nervous system (PNS)** All the nerves outside the central nervous system, which connect the brain and the spinal cord to the rest of the body. (page 308)

Personality An individual's special mix of traits, feelings, attitudes, and habits. (page 212)

Physical fatigue Extreme tiredness of the whole body. (page 221)

Physiological (fi·zee·uh·LAH·ji·kuhl) **dependence** A type of addiction in which the body itself feels a direct need for a drug. (page 284)

Pituitary (pi·TOO·i·tehr·ee) **gland** Gland, located at the base of the brain, that produces several hormones that control other glands. (page 183)

Plaque (PLAK) A thin, sticky film that contributes to tooth decay. (page 70)

Plasma (PLAZ·muh) A yellowish fluid; the watery part of the blood. (page 125)

Pollen A powdery substance released by certain plants. (page 373)

Pollution Dirty or harmful substances in the environment. (page 440)

Pores Tiny openings in the skin. (page 73)

Precaution An action taken to avoid danger. (page 41)

Prejudice (PRE·juh·dis) A negative and unjustly formed opinion, usually against people of a different racial, religious, or cultural group. (page 261)

Preschooler A child between the ages of three and five. (page 202)

Prevention Taking steps to make sure that something unhealthy does not happen. (page 14)

Primary care provider A doctor or other health professional who provides checkups and general care. (page 83)

Proteins (PROH·teenz) Essential nutrients used to build and repair body cells and tissues. (page 96)

Protozoa (proh·tuh·ZOH·uh) One-celled, animal-like organisms. (page 333)

Psychological (sy·kuh·LAH·ji·kuhl) **dependence** An addiction in which the mind sends the body the message that it needs more of a substance. (page 285)

Psychological fatigue Extreme tiredness caused by a person's mental state. (page 221)

Puberty (PYOO·ber·tee) The time when you start to develop certain physical characteristics of adults of your own gender. (page 177)

Pulmonary (PUL·muh·nehr·ee) **circulation** Circulation that carries blood from the heart, through the lungs, and back to the heart. (page 124)

Pupil (PYOO·puhl) The dark opening in the center of the iris that allows the right amount of light to enter the eye. (page 76)

Quackery (KWAK·uh·ree) The sale of worthless products and treatments through false claims. (page 66)

Radiation (ray·dee·AY·shuhn) **therapy** A treatment for some types of cancer that uses X-rays or other forms of radioactivity. (page 370)

Recovery The process of becoming well again. (page 325)

Refusal skills Ways to say no effectively. (page 23)

Reproduction (ree·pruh·DUHK·shuhn) The process by which living organisms produce new individuals of their kind. (page 186)

Reproductive (ree·pruh·DUHK·tiv) **system** Body organs and structures that make possible the production of offspring. (page 186)

Rescue breathing A substitute for normal breathing in which someone forces air into an accident victim's lungs. (page 436)

Respiratory system The set of organs that supply the body with oxygen and rid the body of carbon dioxide. (page 276)

Responsibility The ability to make choices and to accept the results of those choices. (page 28)

Retina (RE·tin·uh) The light-sensing part of the inner eye. (page 76)

Rheumatoid (ROO·muh·toyd) **arthritis** A chronic disease characterized by pain, inflammation, swelling, and stiffness of the joints. (page 380)

Risk behavior An action or choice that may cause injury or harm to you or others. (page 14)

Safety conscious Aware that safety is important, and careful to act in a safe manner. (page 412)

Saliva (suh·LY·vuh) A digestive juice produced by the salivary glands in the mouth. (page 109)

Saturated fats Fats found mostly in meats and dairy products. (page 99)

Schizophrenia (skit·zoh·FREE·nee·uh) A serious mental disorder in which a person's perceptions lose their connection to reality. (page 230)

Second-degree burn A serious burn in which the burned area blisters. (page 439)

Secondhand smoke Air that has been contaminated by tobacco smoke. (page 290)

Self-assessment Careful examination and evaluation of your own patterns of behavior. (page 14)

Self-concept The view that you have of yourself. (page 214)

Self-esteem Confidence in yourself. (page 53)

Self-management The ability to take care of your overall health and to take control of your behavior and actions. (page 22)

Semen (SEE·muhn) A mixture of sperm and fluids produced by the male reproductive system. (page 187)

Sewage (SOO·ij) Garbage, detergents, and other household wastes that are washed down drains. (page 441)

Short-term goal A goal that you can reach right away. (page 53)

Sidestream smoke Smoke that comes from the burning tip of a cigarette. (page 290)

Skeletal system The framework of bones and other tissues that support the body. (page 128)

Small intestine A coiled tube, about 20 feet long, where most of the digestive process takes place. (page 111)

Smoke detector A device that sounds an alarm when it senses smoke. (page 417)

Specialist (SPE·shuh·list) A doctor trained to treat particular types of patients or health matters. (page 83)

Sperm The male reproductive cells. (page 187)

Spinal cord A long bundle of neurons that relays messages to and from the brain and all parts of the body. (page 308)

STD (sexually transmitted disease) A disease spread from person to person through sexual contact. (page 350)

Stimulant (STIM·yuh·luhnt) A drug that speeds up the body's functions. (page 303)

Stomach A muscular organ in which some digestion occurs. (page 110)

Strength The ability of your muscles to exert a force. (page 120)

Stress Your body's response to changes around you. (page 220)

Stressor Something that triggers stress. (page 220)

Stroke Condition that occurs when part of the brain is damaged because the blood supply to the brain is cut off. (page 364)

Subjective Coming from a person's own views and beliefs, not necessarily from facts. (page 41)

Suicide Intentionally killing oneself. (page 230)

Support system A network of people available to help when needed. (page 233)

Syphilis (SI·fuh·lis) A bacteria-caused STD that can affect many parts of the body. (page 352)

Systemic (sis·TE·mik) **circulation** Circulation that sends oxygen-rich blood to all the body tissues except the lungs. (page 124)

Tact The quality of knowing what to say to avoid offending others. (page 242)

Tar A thick, dark liquid that forms when tobacco burns. (page 272)

Target heart rate The number of heartbeats per minute a person should aim for during vigorous exercise for cardiovascular benefit. (page 139)

Tartar (TAR·ter) Plaque that has hardened on teeth, threatening gum health. (page 70)

Team sports Organized physical activities with specific rules, played by opposing groups of people. (page 151)

Teen hot line A special telephone service that teens can call when feeling stressed. (page 233)

Tendon (TEN·duhn) Strong, flexible, fibrous tissue that joins muscle to muscle or muscle to bone. (page 130)

Third-degree burn A very serious burn in which all layers of the skin are damaged. (page 439)

Tissue A group of similar cells that do a particular job. (page 196)

Toddler A child between the ages of one and three. (page 202)

Tolerance (TAHL·er·ens) Situation in which the body becomes used to a drug and needs greater amounts to get the desired effect. (page 324)

Tornado A whirling, funnel-shaped windstorm that drops from the sky to the ground. (page 432)

Tornado warning A news bulletin warning that a tornado has been sighted and that people in the area are in immediate danger. (page 432)

Tornado watch A news bulletin warning that weather conditions indicate that a tornado may develop. (page 432)

Trachea (TRAY·kee·uh) The tube in the throat that takes air to and from the lungs. (page 276)

Tuberculosis (too·ber·kyuh·LOH·sis) A bacteria-caused disease that usually affects the lungs. (page 342)

Tumor (TOO·mer) A group of abnormal cells that forms a mass. (page 367)

Unsaturated fats Liquid fats that come mainly from plants. (page 99)

Uterus (YOO·tuh·ruhs) A pear-shaped female organ in which a developing child is nourished. (page 190)

Vaccine (vak·SEEN) A preparation of dead or weakened germs that is injected into the body to cause the immune system to produce antibodies. (page 339)

Values The beliefs that guide the way a person lives, such as beliefs about what is right and wrong and what is most important. (page 22)

Vein (VAYN) A type of blood vessel that carries blood from all parts of the body back to the heart. (page 123)

Verbal communication Using words to express yourself, either in speaking or in writing. (page 241)

Victim The person against whom a crime is committed. (page 400)

Virus (VY·ruhs) The smallest and simplest life form, causing a wide range of diseases. (page 333)

Vitamins Substances that help to regulate the body's functions. (page 97)

Warm-up Gentle exercise you do to prepare your muscles for vigorous activity. (page 138)

Warranty A company's or a store's written agreement to repair a product or refund your money if the product does not work properly. (page 68)

Wellness An overall state of well-being involving regular behaviors that have a positive result over time. (page 6)

Withdrawal Unpleasant symptoms that occur when someone stops using an addictive substance. (page 283)

Glosario

Abdominal thrusts/presiones abdominales Presiones rápidas hacia arriba para forzar la salida de un objeto que esté bloqueando las vías respiratorias.

Abstinence/abstinencia Negarse a comportarse o tomar parte en actividades peligrosas.

Abuse/abuso El maltrato físico, emocional o mental de otra persona.

Accident chain/cadena de accidentes Una serie de sucesos que incluyen una situación, una costumbre peligrosa y una acción peligrosa.

Acid rain/lluvia ácida Lluvia que es más ácida de lo normal.

Action plan/plan de acción Una serie de pasos para alcanzar una meta.

Addiction/adicción La necesidad física o mental de una droga u otra substancia.

Adolescence/adolescencia El período de vida entre la niñez y la adultez.

Adrenaline/adrenalina Una hormona que aumenta el nivel de azúcar en la sangre, lo cual da energía adicional al cuerpo.

Advertisement/anuncio Un mensaje diseñado para hacer que los consumidores compren un producto o servicio.

Aerobic exercise/ejercicio aeróbico Actividad rítmica y vigorosa que aumenta la velocidad de la respiración y de los latidos del corazón.

Aftershocks/ondas posteriores Terremotos secundarios que ocurren después del terremoto principal.

AIDS (acquired immunodeficiency syndrome)/SIDA (síndrome de inmunodeficiencia adquirida) Una enfermedad mortal que interfiere con la habilidad del cuerpo de defenderse contra infecciones.

Alcohol/alcohol Una droga creada por una reacción química en ciertos alimentos, especialmente frutas y granos.

Alcoholism/alcoholismo Una enfermedad causada por la necesidad física y mental de consumir alcohol.

Allergen/alergeno Una substancia que causa una reacción alérgica.

Allergy/alergia Sensibilidad extrema a una substancia.

Alternatives/alternativas Diferentes maneras de pensar o actuar.

Alveoli/alvéolos Sacos microscópicos en los pulmones, donde se intercambian gases.

Alzheimer's disease/enfermedad de Alzheimer Una enfermedad que ataca el cerebro y afecta el razonamiento, memoria y conducta.

Amphetamine/anfetamina Una droga estimulante fuerte que acelera el sistema nervioso.

Anabolic steroids/esteroides anabólicos Compuestos sintéticos que hacen que el tejido muscular se desarrolle con rapidez anormal.

Anaerobic exercise/ejercicio anaeróbico Intensa actividad física que requiere arranques de energía.

Angioplasty/angioplastia Una operación quirúrgica en la que un instrumento con un globo diminuto se introduce en una arteria para desatascar algo que la bloquea.

Anorexia nervosa/anorexia nerviosa Un trastorno alimenticio en que la persona tiene miedo intenso de ganar de peso y se priva de alimentos.

Antibodies/anticuerpos Proteínas que se pegan a los antígenos, impidiendo que éstos le hagan daño al cuerpo.

Antigens/antígenos Substancias que hacen que el sistema de inmunidad reaccione cuando el cuerpo es invadido por gérmenes.

Antihistamine/antihistamínico Una medicina que alivia los síntomas de una reacción alérgica.

Anxiety disorder/trastorno de ansiedad Un trastorno en el que el temor de problemas reales o imaginarios no deja que la persona funcione normalmente.

Arteriosclerosis/arteriosclerosis Endurecimiento de las arterias.

Artery/arteria Un tipo de vaso sanguíneo que lleva sangre del corazón a todas partes del cuerpo.

Arthritis/artritis Una enfermedad de las articulaciones caracterizada por inflamación y rigidez.

Assertive/resuelto Que tiene la determinación para actuar con confianza y firmeza, de manera positiva.

Asthma/asma Una enfermedad crónica que hace que se estrechen o se atasquen las diminutas vías respiratorias en los pulmones.

Astigmatism/astigmatismo Una condición del ojo en que las imágenes se deforman, haciendo que los objetos aparezcan ondulados o borrosos.

Atherosclerosis/aterosclerosis Una condición en que substancias grasosas se acumulan en las paredes internas de las arterias.

Bacteria/bacterias Organismos diminutos de una sóla célula que viven en casi todas partes, algunos de los cuales causan enfermedades.

Battery/golpeadura Golpear o dar patadas repetidas a otra persona.

Benign/benigno Que no tiene cáncer.

Biodegradable/biodegradable Que se descompone fácilmente en el medio ambiente.

Blizzard/nevasca Tormenta de nieve fuerte, con vientos que llegan a 45 millas por hora.

Blood pressure/presión arterial La fuerza de la sangre presionando contra las paredes de los vasos sanguíneos.

Body composition/composición del cuerpo La proporción de grasa, hueso, músculo y fluido que contiene el cuerpo.

Body language/lenguaje corporal Comunicación no verbal que incluye la postura, los gestos y las expresiones faciales.

Body Mass Index (BMI)/Índice de masa corporal (IMC) Una medida del peso del cuerpo basada en una comparación del peso con la estatura.

Body system/sistema corporal Un grupo de órganos que trabajan juntos para ejecutar tareas relacionadas.

Brain/cerebro El centro de información del sistema nervioso que recibe y revisa información y envía mensajes a otras partes del cuerpo.

Bronchi/bronquios Dos tubos que salen de la traquea y conducen a los pulmones.

Bronchodilator/broncodilatador Medicina que relaja los músculos alrededor de los bronquios.

Bulimia nervosa/bulimia nerviosa Un trastorno alimenticio en que la persona consume grandes cantidades de comida seguida de purgantes.

Burnout/agotamiento Sentirse extenuado por haber gastado demasiada energía durante demasiado tiempo.

Calories/calorías Unidades de calor que miden la energía que contienen los alimentos.

Cancer/cáncer Una enfermedad causada por el crecimiento incontrolado de células anormales.

Capillary/capilar El tipo de vaso sanguíneo más pequeño, que proporciona sangre a las células y conecta una arteria con una vena.

Carbohydrates/carbohidratos El azúcar y las féculas que le proporcionan al cuerpo la mayor parte de su energía.

Carbon monoxide/monóxido de carbono Un gas sin color, sin olor y venenoso que se produce al quemarse el tabaco.

Carcinogen/carcinógeno Una substancia en el medio ambiente que produce cáncer.

Cardiopulmonary resuscitation (CPR)/resucitación cardiopulmonar Un procedimiento de primeros auxilios para restaurar la respiración y circulación.

Cardiovascular system/sistema cardiovascular El sistema circulatorio.

Cartilage/cartílago Un tejido fuerte y flexible que cubre las terminaciones de los huesos y soporta ciertas estructuras.

Cavity/cavidad Un hueco que se empieza a formar en el esmalte del diente.

Cell/célula La unidad básica o bloque formativo de la vida.

Central nervous system (CNS)/sistema nervioso central (SNC) El cerebro y la espina dorsal.

Chemotherapy/quimioterapia El uso de substancias químicas para destruir células cancerosas.

Chest thrusts/presiones torácicas Presiones rápidas en el centro del esternón para forzar a que salga algo que esté bloqueando las vías respiratorias.

Chlamydia/clamidia Una enfermedad causada por una bacteria, que se transmite sexualmente y afecta los órganos de reproducción, la uretra y el ano.

Choking/asfixia La condición que ocurre cuando las vías respiratorias de una persona están bloqueadas.

Cholesterol/colesterol Una substancia parecida a la cera que el cuerpo usa para crear células y otras substancias.

Chromosomes/cromosomas Estructuras en forma de hilos dentro de la célula que contienen los códigos de las características hereditarias.

Chronic/crónico Que está presente o reaparece repetidamente durante un largo período de tiempo.

Circulatory system/sistema circulatorio El grupo de órganos y tejidos que transportan materiales esenciales a las células y se llevan los desechos de éstas.

Cirrhosis/cirrosis Condición en la que los tejidos del hígado presentan cicatrices y destrucción.

Colon/colon El intestino largo, donde se almacenan los desechos sólidos.

Commitment/compromiso Una promesa u obligación.

Communicable disease/enfermedad contagiosa Una enfermedad que se puede transmitir de una persona a otra.

Communication/comunicación El intercambio de pensamientos, sentimientos y creencias entre personas.

Comparison shopping/compras comparadas Comparar productos, evaluar sus beneficios y escoger los que ofrecen el mejor valor por el precio.

Competition/competencia Rivalidad entre dos o más personas o grupos que están tratando de alcanzar la misma meta.

Compromise/transigir Un modo de resolver diferencias en que cada persona renuncia a algo que desea para llegar a un acuerdo que satisfaga a todos.

Conditioning/preparación física Entrenamiento para ponerse en forma.

Conflict/conflicto Un desacuerdo entre personas con puntos de vista opuestos.

Consequences/consecuencias Los efectos o resultados de los actos.

Conservation/conservación El ahorro de recursos naturales.

Consumer/consumidor El que compra bienes y servicios.

Contagious period/período de contagio El período de tiempo en que se puede transmitir una enfermedad de una persona a otra.

Contract/contraer Hacerse más corto, a menudo se refiere a los músculos en función.

Cool-down/enfriamiento Ejercicios suaves que permiten que el cuerpo se acostumbre a dejar de ejercitarse.

Cornea/córnea La parte transparente anterior del ojo que permite que pase la luz.

Criteria/criterios Principios en que se basa una decisión.

Cross-training/entrenamiento variado Cualquier programa de ejercicio que incluya distintos tipos de actividades físicas para estar en buena condición física balanceada.

Cumulative risks/riesgos acumulativos
Riesgos relacionados cuyos efectos aumentan con cada riesgo que se añade.

Dandruff/caspa Escamas que forman las células muertas de la capa externa del cuero cabelludo.

Decibel/decibel La unidad que se usa para medir el volumen de las ondas del sonido.

Decision making/tomar decisiones El proceso de hacer una selección o hallar una solución.

Defense mechanism/mecanismo de defensa Una forma temporal de manejar el estrés.

Dehydration/deshidratación Pérdida excesiva de agua por el cuerpo.

Depressant/depresor Una droga que disminuye las funciones y reacciones del cuerpo, incluyendo el ritmo del corazón y la respiración.

Depression/depresión Un trastorno del ánimo en el que una persona se siente indefensa, despreciable, culpable, sin esperanza y extremadamente triste durante períodos que continuan varias semanas.

Dermis/dermis La capa interna de la piel que contiene vasos sanguíneos, terminaciones nerviosas y folículos pilosos.

Diabetes/diabetes Enfermedad que impide que el cuerpo convierta alimentos en energía.

Diaphragm/diafragma Músculo grande en forma de domo, situado debajo de los pulmones, que aspira y despide el aire.

Diet/dieta Todo lo que comes y bebes con regularidad.

Digestion/digestión El proceso corporal de descomponer los alimentos en componentes más pequeños para que la sangre los pueda absorber y llevar a cada célula del cuerpo.

Digestive system/sistema digestivo Una serie de órganos que trabajan juntos para descomponer los alimentos en substancias que las células puedan usar.

Discount store/tienda de descuento Una tienda que vende mercancía a precios rebajados.

Disease/enfermedad Una condición que interfiere con el funcionamiento normal del cuerpo o la mente.

Distress/angustia Estrés que impide que hagas lo que tienes que hacer o que te causa molestia.

Drug/droga Una substancia no alimenticia que al tomarse puede causar cambios en la estructura o funcionamiento del cuerpo o la mente.

Earthquake/terremoto El sacudimiento violento de la superficie de la Tierra.

Eating disorder/trastorno alimenticio Costumbres de alimentación exageradas que pueden causar enfermedades graves o muerte.

Electrical overload/sobrecarga eléctrica Una situación peligrosa en que demasiada corriente eléctrica fluye a través de un solo circuito.

Electrocution/electrocución Muerte causada por una descarga eléctrica.

Embryo/embrión El nombre de un organismo desde la fertilización hasta la octava semana.

Emotions/emociones Sentimientos como el amor, la alegría o el miedo.

Empathy/empatía La compenetración o habilidad de identificarse y compartir los sentimientos de otra persona.

Endocrine system/sistema endocrino Las glándulas del cuerpo que regulan sus funciones.

Endorsement/aprobación Una declaración de aceptación o consentimiento.

Endurance/resistencia La habilidad de hacer actividades físicas vigorosas sin cansarse demasiado.

Environment/medio ambiente Todas las cosas vivas y no vivas que te rodean.

Environmental Protection Agency (EPA)/Agencia de Protección Ambiental Una agencia del gobierno de los Estados Unidos que está encargada de proteger el medio ambiente.

Epidermis/epidermis La capa más externa de la piel.

Excretion/excreción El proceso mediante el cual el cuerpo se deshace de los desechos líquidos.

Excretory system/sistema excretorio El sistema que despide los desechos del cuerpo y controla el balance del agua.

Exercise/ejercicio Actividad física que aumenta la buena forma física.

Exercise frequency/frecuencia del ejercicio El número de veces en que una persona hace ejercicios semanalmente.

Exercise intensity/intensidad del ejercicio La cantidad de energía que una persona consume al ejercitarse.

Extend/extender Estirar, a menudo se refiere a la función muscular.

Family/familia La unidad básica de la sociedad.

Family violence shelter/refugio para familias maltratadas Un sitio donde personas en peligro de ser abusadas por otros miembros de su familia pueden quedarse mientras que ponen sus vidas en orden.

Fatigue/fatiga Cansancio extremo.

Fats/grasas Fuentes de energía que también cumplen otras funciones, como el almacenamiento de vitaminas y el aislamiento térmico del cuerpo.

Fertilization/fertilización La unión de un espermatozoide masculino y un óvulo femenino para formar una vida nueva.

Fetus/feto El organismo desde la novena semana hasta el momento del nacimiento.

Fiber/fibra La parte de los granos, frutas y vegetales que el cuerpo no puede descomponer.

First aid/primeros auxilios El cuidado inmediato que se da a una persona herida o enferma hasta que se le pueda proporcionar ayuda médica regular.

First-degree burn/quemadura de primer grado Una quemadura en que sólo la capa exterior de la piel se quema y enrojese.

Fitness/buena condición física La habilidad de tomar parte en el trabajo y las diversiones de la vida diaria sin cansarse demasiado.

Flammable/inflamable Que se quema con facilidad.

Flexibility/flexibilidad La habilidad de mover las articulaciones completa y fácilmente.

Fluoride/fluoruro Substancia que ayuda a que los dientes resistan las caries.

Food Guide Pyramid/Pirámide de los alimentos Una guía para seleccionar alimentos sanos diariamente, diseñada por el Departamento de Agricultura de EE.UU.

Fossil fuels/combustibles fósiles Petróleo, carbón y gas natural usados para proporcionar energía.

Fracture/fractura Una rotura en el hueso.

Friendship/amistad La relación entre personas que simpatizan y tienen intereses y valores similares.

Fungi/hongos Seres vivos primitivos que no pueden hacer sus propios alimentos.

Gang/pandilla Un grupo de personas que se relacionan porque tienen algo en común.

Gene/gene La unidad básica de la herencia que determina las características que heredas de tus padres.

Generic products/productos genéricos Productos que se venden en envolturas sencillas y a menor precio que los de marca.

Genital herpes/herpes genital Una enfermedad transmitida sexualmente, causada por un virus, que produce ampollas dolorosas en el área genital.

Germ/germen Un microorganismo que causa una enfermedad.

Gland/glándula Un grupo de células o un órgano que produce una substancia química.

Glucose/glucosa El tipo de azúcar que es la fuente principal de energía del cuerpo.

Gonorrhea/gonorrea Una enfermedad transmitida sexualmente que es causada por una bacteria que afecta la membrana mucosa genital y a veces otras partes del cuerpo, como el corazón o las articulaciones.

Grief/pena Tristeza profunda causada por la pérdida de algo muy querido.

Gynecologist/ginecólogo Médico que se especializa en el sistema reproductor femenino.

Habit/hábito Un patrón de conducta que se repite con gran frecuencia, casi sin pensar.

Hallucinogen/alucinógeno Una droga que altera el estado de ánimo, los pensamientos y los sentidos.

Hate crime/crimen por odio Un acto ilegal cometido en contra de alguien, sólo porque pertenece a un grupo en particular.

Hazard/peligro Una amenaza posible.

Head lice/piojos Insectos parasíticos que viven en el pelo y causan picazón.

Health/salud Una combinación de bienestar físico, mental/emocional y social.

Health insurance/seguro de salud Un plan en el que la gente paga una cantidad fija a una compañía de seguros la cual, a cambio, paga algunos o la mayoría de sus gastos médicos.

Health maintenance organization (HMO)/organización para el mantenimiento de la salud Un tipo de servicio de salud controlado que le ofrece a sus miembros distintas clases de servicios de salud.

Heart attack/ataque al corazón Una condición que se presenta cuando el flujo de sangre al corazón disminuye o se para, dañando el músculo cardíaco.

Hepatitis/hepatitis Una inflamación del hígado, causada por un virus, en que la piel y esclerótica del ojo se ponen amarillas.

Heredity/herencia La transferencia de rasgos de los padres biológicos a sus hijos.

Histamines/histaminas Substancias químicas en el cuerpo que causan los síntomas de una reacción alérgica.

HIV (human immunodeficiency virus)/VIH (virus de inmunodeficiencia humana) El virus que causa el SIDA.

Homicide/homicidio Un crimen violento que resulta en la muerte de una persona.

Hormones/hormonas Substancias químicas, producidas por glándulas, que ayudan a regular las funciones del cuerpo.

Hurricane/huracán Tormenta de fuertes vientos y lluvia que se origina en alta mar.

Hygiene/higiene Limpieza.

Hypertension/hipertensión Una condición en que la presión arterial de una persona se mantiene a niveles más altos de lo normal; también se llama presión alta.

Hypothermia/hipotermia Una disminución repentina y peligrosa de la temperatura del cuerpo.

Immune system/sistema de inmunidad Las defensas del cuerpo combinadas, compuestas de células, tejidos y órganos que combaten gérmenes que causan enfermedades.

Immunity/inmunidad La habilidad del cuerpo de resistir los gérmenes que causan una enfermedad en particular.

Individual sports/deportes individuales Actividades físicas en que puedes participar solo o con un amigo, sin formar parte de un equipo.

Infancy/infancia El primer año de vida de un ser humano.

Infection/infección La condición que se produce cuando gérmenes invaden el cuerpo, se multiplican y dañan las células.

Influenza/influenza Una enfermedad contagiosa, caracterizada por fiebre, escalofríos, fatiga, dolor de cabeza, dolores musculares y síntomas respiratorios.

Infomercial/comercial informativo Un anuncio de televisión largo cuyo propósito aparenta ser informar en vez de vender un producto.

Inhalant/inhalante Una substancia cuyos gases se respiran para obtener sensaciones alucinantes.

Insulin/insulina Una hormona producida en el páncreas que regula el nivel de azúcar en la sangre.

Intoxicated/embriagado Borracho.

Invulnerable/invulnerable Que no puede ser lastimado.

Iris/iris La parte coloreada del ojo.

Joint/articulación Un lugar en donde se unen dos o más huesos.

Kidney/riñón Uno de la pareja de órganos en forma de frijol que filtran el agua y los desechos de la sangre.

Lens/cristalino La parte del ojo que enfoca la luz en la retina.

Life-altering/que cambia la vida Capaz de modificar la vida diaria de una persona.

Life-threatening/amenaza a la vida Que puede causar la muerte.

Ligament/ligamento Una banda de tejido fuerte que sostiene el hueso en su sitio en una articulación.

Liver/hígado La glándula más grande del cuerpo que secreta un líquido llamado bilis que ayuda a digerir las grasas.

Long-term goal/meta a largo plazo Una meta que piensas alcanzar durante un largo período de tiempo.

Lymphatic system/sistema linfático Un sistema circulatorio secundario que ayuda al cuerpo a defenderse de gérmenes y a mantener el equilibrio de fluido.

Lymphocytes/linfocitos Glóbulos blancos especiales en la linfa que son importantes en la defensa contra gérmenes y enfermedades.

Mainstream smoke/humo directo El humo inhalado por el fumador.

Malignant/maligno Canceroso.

Managed care/asistencia médica manejada Un plan para el cuidado de la salud que hace hincapié en la medicina preventiva y trata de controlar el costo y la calidad de la asistencia médica.

Mass media/medios de comunicación de masas Medios de comunicación que llegan a gran número de personas, como periódicos, libros, televisión, películas y la Internet.

Media/medios de comunicación Los distintos métodos de comunicar información.

Mediation/mediación La resolución de conflictos por medio de una persona imparcial que ayuda a llegar a una solución aceptable para ambos bandos.

Medicaid/Medicaid Un programa público de seguro médico para familias pobres.

Medicare/Medicare Un programa de seguro médico del gobierno que proporciona seguro a personas mayores de 65 años.

Medicine/medicina Una droga que se usa para tratar enfermedades o aliviar el dolor.

Melanin/melanina Una substancia producida por las células de la epidermis que dan el color a la piel.

Menstruation/menstruación La eliminación de material de las paredes del útero.

Mental health/salud mental La habilidad de ocuparse, de manera razonable, del estrés y los cambios de la vida diaria.

Minerals/minerales Substancias que fortalecen los músculos, huesos y dientes; enriquecen la sangre y ayudan a que el corazón y otros órganos funcionen debidamente.

Mononucleosis/mononucleosis Una enfermedad causada por un virus, caracterizada por inflamación de los nódulos linfáticos en el cuello y la garganta.

Mood disorder/trastorno del estado de ánimo Un trastorno en que la persona cambia de humor de manera inapropiada o extrema.

Mortality/mortalidad La muerte.

Muscle endurance/resistencia muscular Una medida de la cantidad de tiempo durante el cual un grupo de músculos puede trabajar sin cansarse.

Muscular system/sistema muscular El sistema del cuerpo que consiste en tejidos que mueven las distintas partes del cuerpo y hacen funcionar los órganos internos.

Narcotic/narcótico Una droga que alivia el dolor y entorpece los sentidos.

Negotiation/negociación El proceso de discutir un problema cara a cara para llegar a una solución.

Neuron/neurona Una de las células que componen el sistema nervioso.

Neutrality/neutralidad No ponerse de parte de nadie cuando otros están en desacuerdo.

Nicotine/nicotina Una droga que se encuentra en el tabaco que causa adicción.

Noncommunicable disease/enfermedad no contagiosa Una enfermedad,

Glosario

como el asma o cáncer, que no se puede transmitir de una persona a otra.

Nonverbal communication/comunicación no verbal Todos los modos de comunicar un mensaje sin palabras.

Nonviolent confrontation/confrontación no violenta Resolver un conflicto por medios pacíficos.

Nurture/criar y cuidar Ocuparse de las necesidades físicas, emocionales, mentales y sociales de una persona.

Nutrient density/densidad de nutrientes La cantidad de nutrientes que contiene un alimento en proporción a la cantidad de calorías.

Nutrients/nutrientes Las substancias en los alimentos que el cuerpo necesita para crecer, tener energía y mantenerse sano.

Nutrition/nutrición El proceso corporal de tomar alimentos y usarlos para crecer y mantener buena salud.

Obesity/obesidad La condición de estar el 20 porciento o más sobre el peso ideal.

Objective/objetivo Basado en los hechos.

Occupational Safety and Health Administration (OSHA)/Administración de Salud y Seguridad Ocupacional Una rama del Departamento de Trabajo que protege la seguridad de los trabajadores estadounidenses.

Ophthalmologist/oftalmólogo Un médico que se especializa en la estructura, función y enfermedades del ojo.

Optic nerve/nervio óptico Un grupo de fibras nerviosas que envían mensajes del ojo al cerebro.

Optometrist/optometrista Un profesional de la salud que está preparado para examinar la vista y recetar lentes correctivos.

Organ/órgano Una parte del cuerpo que comprende distintos tipos de tejidos que hacen una función particular.

Osteoarthritis/osteoartritis Una enfermedad crónica, común en los ancianos, que es el resultado de la degeneración del cartílago de las articulaciones.

Ovaries/ovarios Las glándulas reproductoras femeninas.

Overcommitment/demasiados compromisos Comprometerse a cumplir con más personas o proyectos de los que uno puede por falta de tiempo o energía.

Overtraining/entrenamiento excesivo Ejercitarse con demasiada fuerza o frecuencia, sin descanso suficiente.

Pancreas/páncreas Una glándula que ayuda al intestino delgado, a través de la producción de jugo pancreático, el cual está formado por una mezcla de varias enzimas que descomponen las proteínas, carbohidratos y grasas.

Passive smoker/fumador pasivo Una persona que no fuma pero inhala humo secundario.

Pedestrian/peatón Una persona que viaja a pie.

Peer pressure/presión de contemporáneos La presión que tus amigos ejercen sobre ti para que creas y actúes igual que ellos.

Peers/contemporáneos Personas de la misma edad que se parecen a ti en muchas maneras.

Peripheral nervous system (PNS)/sistema nervioso periférico Todos los nervios fuera del sistema nervioso central que conectan el cerebro y la espina dorsal al resto del cuerpo.

Personality/personalidad La mezcla de rasgos, sentimientos, actitudes y hábitos de un individuo.

Physical fatigue/fatiga física Cansancio extremo de todo el cuerpo.

Physiological dependence/dependencia fisiológica Un tipo de adicción en que el cuerpo tiene una necesidad directa de una droga.

Pituitary gland/glándula pituitaria Glándula situada en la base del cerebro que produce distintas hormonas que controlan otras glándulas.

Plaque/placa Una película pegajosa que contribuye a las caries dentales.

Plasma/plasma Un líquido amarillento; la parte líquida de la sangre.

Pollen/polen Una substancia en forma de polvo, despedida por ciertas plantas.

Pollution/contaminación Substancias sucias o dañinas en el medio ambiente.

Pores/poros Pequeñas aberturas en la piel.

Precaution/precaución Una medida que se toma para evitar peligro.

Prejudice/prejuicio Una opinión negativa e injusta, generalmente en contra de personas de otro grupo racial, religioso o cultural.

Preschooler/niño preescolar Un niño entre las edades de tres y cinco años.

Prevention/prevención Tomar medidas para asegurarse de, que no suceda algo que no es saludable.

Primary care provider/profesional primario de la salud El médico u otro profesional de la salud que proporciona chequeos médicos y cuidado general.

Proteins/proteínas Nutrientes esenciales que se usan para crear y reparar las células y tejidos del cuerpo.

Protozoa/protozoos Organismos de una célula, semejantes a animales.

Psychological dependence/dependencia psicológica Una adicción en que la mente envía mensajes al cuerpo comunicándole que necesita más de una substancia.

Psychological fatigue/fatiga psicológica Cansancio extremo causado por el estado mental de una persona.

Puberty/pubertad El período en que comienzas a desarrollar las características físicas de los adultos de tu sexo.

Pulmonary circulation/circulación pulmonar La circulación que lleva la sangre del corazón, a través de los pulmones y de regreso al corazón.

Pupil/pupila La abertura oscura en el centro del ojo que permite que la cantidad de luz apropiada entre en el ojo.

Quackery/curanderismo La venta, por medio de engaños, de productos y tratamientos que no valen nada.

Radiation therapy/radioterapia Tratamiento para ciertos tipos de cáncer que utiliza rayos X u otro tipo de radiación.

Recovery/recuperación El proceso de ponerse bien de nuevo.

Refusal skills/habilidad de negarse Modos efectivos de decir que no.

Reproduction/reproducción El proceso por el cual los organismos vivos producen otros de su especie.

Reproductive system/sistema reproductor Los órganos y estructuras del cuerpo que hacen posible producir hijos.

Rescue breathing/respiración de rescate Forzar aire en los pulmones de la víctima de un accidente que no puede respirar normalmente.

Respiratory system/sistema respiratorio El conjunto de órganos que proporcionan oxígeno al cuerpo y eliminan el bióxido de carbono.

Responsibility/responsabilidad La habilidad de tomar una determinación y de aceptar las consecuencias.

Retina/retina La parte del ojo sensible a la luz.

Rheumatoid arthritis/artritis reumatoidea Una enfermedad crónica, caracterizada por dolor, inflamación y rigidez de las articulaciones.

Risk behavior/conducta arriesgada Un acto o selección que puede causar lesiones o daño a uno mismo o a otros.

Safety conscious/consciente de la seguridad Que se da cuenta de la importancia de la seguridad y actúa con cuidado.

Saliva/saliva El líquido con enzimas digestivas producido por las glándulas salivales de la boca.

Saturated fats/grasas saturadas Grasas que se encuentran principalmente en carnes y productos lácteos.

Schizophrenia/esquizofrenia Un trastorno mental serio en el que las percepciones de la persona pierden su conexión con la realidad.

Second-degree burn/quemadura de segundo grado Quemadura grave que forma ampollas en el área quemada.

Secondhand smoke/humo secundario Aire que está contaminado por el humo del tabaco.

Self-assessment/autoevaluación Examinación y valoración minuciosas de los patrones de conducta de uno mismo.

Self-concept/autoimagen La manera en que te ves a ti mismo.

Self-esteem/autoestima Confianza en ti mismo.

Self-management/manejo personal La habilidad de ocuparte de tu salud total y controlar tu conducta y actos.

Semen/semen Una mezcla de espermatozoides con los fluidos producidos por el sistema reproductor masculino.

Sewage/aguas residuales Basura, detergentes y otros desechos caseros que se llevan las tuberías de desagüe.

Short-term goal/meta a corto plazo Una meta que puedes alcanzar rápidamente.

Sidestream smoke/humo indirecto Humo que procede de un cigarrillo prendido.

Skeletal system/sistema esquelético El armazón de huesos y otros tejidos que sostienen el cuerpo.

Small intestine/intestino delgado Un tubo enrollado, de unos 20 pies de largo, donde se produce la mayor parte de la digestión.

Smoke detector/detector de humo Un aparato que hace sonar una alarma cuando descubre humo.

Specialist/especialista Un médico que ha estudiado para tratar determinados tipos de pacientes o problemas de la salud.

Sperm/espermatozoides Las células reproductoras masculinas.

Spinal cord/espina dorsal Un largo conjunto de neuronas que transmite mensajes entre el cerebro y todas las otras partes del cuerpo.

STD (sexually transmitted disease)/enfermedad transmitida sexualmente Una enfermedad que se pasa de una persona a otra a través del contacto sexual.

Stimulant/estimulante Una droga que acelera las funciones del cuerpo.

Stomach/estómago Un órgano musculoso en que ocurre parte de la digestión.

Strength/fuerza La capacidad que tienen tus músculos para producir un efecto.

Stress/estrés La reacción del cuerpo a los cambios a su alrededor.

Stressor/estresante Algo que provoca el estrés.

Stroke/embolia Condición que se produce cuando parte del cerebro queda dañado porque disminuye el flujo de sangre a éste.

Subjective/subjetivo Que proviene de las opiniones y creencias de una persona y que no necesariamente está basado en los hechos.

Suicide/suicidio Matarse intencionalmente.

Support system/sistema de apoyo Una red de personas dispuestas a ayudar cuando sea necesario.

Syphilis/sífilis Una enfermedad transmitida sexualmente, causada por una bacteria, que afecta muchas partes del cuerpo.

Systemic circulation/circulación sistémica Circulación que lleva sangre rica en oxígeno a todos los tejidos del cuerpo, menos a los pulmones.

Tact/tacto La habilidad de saber lo que se debe decir para no ofender a los demás.

Tar/alquitrán Un líquido espeso y oscuro que se forma al quemarse el tabaco.

Target heart rate/ritmo deseado del corazón El número de latidos del corazón, por minuto, que una persona debe tratar de alcanzar durante el ejercicio vigoroso, para obtener beneficio cardiovascular.

Tartar/sarro Placa dental que se ha endurecido en los dientes, amenazando la salud de las encías.

Team sports/deportes en equipo Actividades físicas organizadas, con reglas particulares, practicadas por grupos opuestos de personas.

Teen hot line/línea de emergencia para adolescentes Un servicio telefónico especial a donde los adolescentes pueden llamar cuando sienten estrés.

Tendon/tendón Tejido fibroso, fuerte y flexible que une un músculo a otro o a una articulación.

Third-degree burn/quemadura de tercer grado Una quemadura muy grave en que todas las capas de la piel quedan dañadas.

Tissue/tejido Un grupo de células similares que tienen una función en particular.

Toddler/niño pequeño Un niño entre uno y tres años.

Tolerance/tolerancia Situación en que el cuerpo se acostumbra a una droga y necesita mayores cantidades para obtener el efecto deseado.

Tornado/tornado Tormenta de viento, en forma de torbellino, que cae del cielo a la tierra.

Tornado warning/aviso de tornado Un boletín de noticias comunicando que un tornado se acerca y que la gente del área corre peligro.

Tornado watch/alerta de tornado Un boletín del estado del tiempo anunciando que un tornado se puede estar formando.

Trachea/tráquea El tubo en la garganta por donde el aire entra y sale de los pulmones.

Tuberculosis/tuberculosis Una enfermedad causada por bacterias que generalmente afecta los pulmones.

Tumor/tumor Una masa de células anormales.

Unsaturated fats/grasas no saturadas Grasas líquidas que provienen principalmente de plantas.

Uterus/útero El órgano femenino en forma de pera en que se desarrolla y nutre un niño antes de nacer.

Vaccine/vacuna Una preparación de gérmenes muertos o debilitados que se inyecta en el cuerpo para hacer que el sistema de inmunidad produzca anticuerpos.

Values/valores Las creencias que guían la manera en que una persona vive, como creencias sobre el bien y el mal y lo que es importante.

Vein/vena El tipo de vaso sanguíneo que lleva la sangre de todas partes del cuerpo de regreso al corazón.

Verbal communication/comunicación verbal El uso de palabras, escritas o habladas, para expresarse.

Victim/víctima La persona contra quien se comete un crimen.

Virus/virus Los seres vivos más pequeños y simples, que causan una gran variedad de enfermedades.

Vitamins/vitaminas Substancias que ayudan al cuerpo a regular sus funciones.

Warm-up/calentamiento Ejercicios suaves que se hacen para preparar el cuerpo para hacer ejercicios vigorosos.

Warranty/garantía La promesa escrita de un fabricante o una tienda de reparar un producto o devolver el dinero al comprador, si el producto no funciona debidamente.

Wellness/bienestar El estado general de buena salud, que incluye patrones de conducta que a la larga tienen buenos resultados.

Withdrawal/retirada Síntomas desagradables que se producen cuando alguien deja de usar una substancia a la que está adicto.

Index

Note: Page numbers in *italics* refer to art and marginal features.

ABCD rule, *369*
Abilities, using your, 30
Abstinence, 254–257, 356
Abuse, 400–7
   breaking cycle of, 404
   causes of, 402
   defined, 400
   effects of, 403
   getting help for, 405
   and media, 406–7
   myths about, 398
   reporting, 402, 406
   signs of, 403
   statistics on, *400, 401*
   types of, 401
Acceptance, silence as, 46
Accident chains, *413*, 413–14, *414*
Accidents
   death from, *412*
   falls, 419
   as family challenge, *247*
   gun accidents, 420
   poisonings, 420, *439*
Acne, 74
Acrophobia, 229
Action plans, 54, *55*
Acupuncture, *83*
Addiction, 36, 283, 324–27
   getting help for, 326
   recovery from, *325*, 327
   stages of, *325*
   and withdrawal, 325
Adolescence, 176–81, 202–203, *203*
   and advancement toward adulthood, 181
   defined, 176
   emotional changes during, 178–80
   and hormones, 176
   mental growth during, 178
   physical changes during, 177, *177*
   social development during, 180
Adrenal glands, *183*
Adrenaline, 220
Adulthood, *203*
Advertising/advertisements
   for alcohol, 312, *316*
   analyzing claims in, *64*
   for cigarettes, 281, 285
   defined, *61*
   endorsements, 63
   and food choices, 92
   health hype in, 348
   health-related, 62–63
   image, 63
   infomercials, 63
   informational, 63
   in magazine articles, *63*
   for phony diets, 134
   and purchasing decisions, *61*
   and self–image, 227
   for tobacco products, *41*
   unrealistic images in, *66*
   value of, 62
Aerobic exercise, 121
AIDS (acquired immunodeficiency syndrome), 353–55, *353–55*
Aikido, 399
Air bags, 45
Air pollution, 441, *441*
Al–Anon, 300
Alateen, 300
Alcohol, 296–99
   advertising for, 312, *316*
   attitudes toward, 298
   avoiding, *320*, 320–21
   body, effects on, 297, *297–98*, 315
   and cancer, 370–71
   defined, 296
   and driving, 315
   and family, 317
   and the law, 318
   and liver, *111*
   mental/emotional consequences of using, 316, *316*
   and nervous system, 309
   others, effects on, 316
   and peer pressure, *321*
   and personal responsibility, 319
   and pregnancy, *200*
   risks of using, 314
   and school, 317
   social consequences of, 317
   and violence, 395
   and your potential, 322
Alcohol/drug counselors, 312
Alcoholics Anonymous, 300
Alcoholism, *299*, 299–300
Allergens, 373, *373*, *374*
Allergies, 373–74, *374*
   food, 6
Allergists, *83*
Alternatives to substance abuse, 322–23
Alzheimer's disease, 383–85
American Cancer Society, 13
Amphetamines, 304, *304*
Anabolic steroids, 158
Anaerobic exercise, 121
Anger, *204*
   as cause of violence, 394
   controlling, 263
Angioplasty, 365
Anorexia nervosa, 145, *230*
Antibacterial soaps, 12
Antibodies, 337
Antigens, *126*, 337, *338*
Antihistamines, 374
Antioxidants, 195
Antiperspirants, 73
Antismoking laws, *283*
Anus, 113
Anvil (of ear), 78, *78*
Anxiety disorders, 229
Appetite, and exercise, 13
Arguments, 261
Arterial disease, *364*
Arteries, *123*
Arteriosclerosis, *364*
Arthritis, 380–81
Assertiveness, of leaders, *30*
Asthma, *278*, *375*, 375–76
Astigmatism, 77
Astronauts, exercise by, *138*
Atherosclerosis, *364*
Atrium (of heart), right/left, *124*
Auditory nerve, 78, *78*

Baby-sitting, preventing accidents while, 414
Bacteria, 73, *332*, 332–35, *333*
Ball-and-socket joints, *129*
Barbiturates, 305
Basketball, *159*
Bathing, 73

B-cells, 337, *338*
Benign tumors, 367
Bicycle helmets, *40*
Bicycling, 429
　as risk behavior, 14
Bipolar disorder, 229
Birth, 198, *201*
Birthing centers, 83
Blackheads, 74
Bladder cancer, *274*
Bleeding, severe, 437, *437*
Blended family, *245*
Blizzards, 434
Blodgett, Cindy, 135
Blood, 111, 125–26. *See also* Circulatory system
　clotting of, 125
　parts of, 125, *125*
Blood pressure, 126, *364*
Blood types, 126, *126*
Blood vessels, alcohol's effect on, *297*
BMI. *See* Body mass index
Body composition, 120
Body image, 143, *145*
Body mass index (BMI), 141, *142*
Body odor, 73
Body piercing, 81
Body systems, *196*
Bone marrow, *126*
Bones, 128, *129, 132*
　of children, 133
　fractures of, *439*
Bowling, in ancient Egypt, *166*
Brain, 308, *308*
　alcohol's effect on, *297*
　of baby, 195
　of embryo, *197*
Bread, Cereal, Rice, and Pasta Group, recommended daily servings for, *93, 94*
Breakfast, 102–3, 104, 155
　European, *102*
　quick, *17*
Breast cancer, 368
Breast self-exam (BSE), *188*
Breathing, 276–77, *276, 277*
　and smoking, 275
Breathing emergencies, 436, *436–37*
Breathing exercises, 226
Bronchi, *276*
Bronchodilators, 376
Brushing
　of hair, 75
　of teeth, 70, *71*
BSE. *See* Breast self-exam
Bulimia nervosa, *109*, 145
Burnout, 167
Burns, first aid for, *439*
Bush, George, *30*
Bypass surgery, 365

Calcium, in diet, *98,* 107
Calf stretch, *138*
Calories, 95, 143, *143, 154*
Camping safety, 426
Cancer, 367–72
　causes of, 369
　combating, 369–71
　defined, 367
　and female reproductive system, 193
　testicular, 189
　types of, 368, *368*
　warning signs of, *372*
Capillaries, *123, 124*
Carbon monoxide, in tobacco smoke, 272, *273*
Carcinogens, 290, 369
Cardiac muscle, *130*
Cardiologists, *83*
Cardiopulmonary resuscitation (CPR), *436*
Cardiovascular system. *See also* Circulatory system
　problems of the, *364*
Career, choosing a, *47*
Career goals, *52*
Cars. *See also* Seat belts
　accidents involving, *412*
　air bags in, 45
Carter, Jimmy, *30*
Cartilage, 130, *130*
Cavities, *70*
CDC. *See* Centers for Disease Control
Cells, 196, *196*
Cellular phones, and pedestrian safety, *423*
Centers for Disease Control (CDC), 40, *121*
Central nervous system (CNS), 308
Cereals, *94,* 102
Cerebral palsy, 309
Cervix, *190*
CFS. *See* Chronic fatigue syndrome
Checkups, *29*
　dental, 71
　eye, 77
　during pregnancy, *200*
Chemotherapy, 370
Chewing tobacco, 273
Chicken pox, 341
Chicken soup, as cure for colds, *14*
Childhood, 202, *202*
Chlamydia, 351
Choices. *See* Decision making
Choking, 438, *438*
Cholesterol, 99
Chromosomes, 199
Chronic disease, 362, *382*
Chronic fatigue syndrome (CFS), 379
Chyme, 110

Cigarettes, 273, *273*
　advertising for, 281, 285
Cigars, 274
Circulatory system, *123,* 123–24, *124. See also* Blood
　caring for, 127
　defined, 123
Cirrhosis, *111,* 297
Clergy, members of, *233,* 405
Climate, and health, 9
Clinton, Bill, *30*
Clotting, of blood, 125, *125*
CNS. *See* Central nervous system
Cocaine, 304
Cochlea, 78, *78*
Codeine, 305
Colds, 14, *278,* 340, 348
"Cold turkey," quitting smoking, 289
Collagen, *110,* 129
Colon, 113
Colon cancer, *368*
Commercials. *See* Advertising/advertisements
Commitment, to sports, 150
Communicable diseases, 332, 340–43. *See also* Sexually transmitted diseases (STDs)
　avoiding, 344–46
　chicken pox, 341
　colds, 340
　flu, 341
　hepatitis, *342*
　measles, 341
　mononucleosis, *342*
　mumps, 341
　tuberculosis, *342*
　vaccination against, 343
Communication/communication skills, 22, 240–44
　clarity in, 242
　and compromise, 244
　criticism, taking, 242–43
　effective, 241
　with family, 248
　of leaders, 30
　listening, 243
　and tact, 242
　and teamwork, *153*
　verbal vs. nonverbal, 241, *241*
Comparison shopping, 68
Competition, and sports, 152, *152*
Compromise, 244
Conditioning, 159–63, *161, 162*
　chart, *162*
　cross-training, 160–161, *160–161*
　and goals, 159, 162
　and keeping records, 162
　overtraining, avoiding, 161, *161*
Cone cells, 310
Confidence. *See* Self-esteem
Conflicting feelings, 179
Conflicting goals, *53*

Index **469**

Conflict(s), 260–65
  causes of, 261
  defined, 260
  disagreeing without, *22*
  helping others avoid, 264–65
  ignoring, 263
  mediation of, 265
  and nonviolent confrontation, 264
  preventing, 262
  recognizing, 262–63
  resolution of, 262–65
Connecting neurons, *307*
Connecting tissues, 130, *130*
Consequences
  considering, *42*
  defined, 32
  and responsibility, 32
  and risks, 39–40
Consumer Reports, *60*
Consumers, 60–64. *See also* Personal products, buying
  and advertising, 62–64
  awareness of, 64
  defined, 60
  of health information, 62–63
  of information, 60
  and media messages, 61
  and purchasing decisions, 60, *61*
  unit pricing, using, *66*
Contact lenses, 78, *78*
Contagious diseases, infectious vs., *351*
Contagious period, 344
Contraction, muscle, 130
Control numbers, product, *67*
Cool-down, 138
Coping with stress, 224
Cornea, *76*
Coronary heart disease, *364*
Cost(s)
  of health care, 84
  and purchasing decisions, *61*, 68
  of smoking, 274
Couple family, *245*
CPR. *See* Cardiopulmonary resuscitation
Crack cocaine, 304
Cranberry juice, *112*
Crank, *304*
Crime
  safety against, 424
  violent, 393
Crime Stoppers International, Inc., *397*
Crisis hot lines, 405
Criteria, for decision making, 48
Criticism, giving/taking, 242–43
Cross-training, 161, *161*
Crown (of tooth), *69*
Cultural factors/issues
  customs, *36*
  and decision making, 47, *47*

and food choices, 92
and values, 47, *47*
Cumulative risks, 41
Curl-ups, 120
Customs, habits vs., *36*
Cyanide, in tobacco smoke, *273*
Cyclones, *433*

Daily schedule, *18*
Dandruff, 75
Death
  from accidents, *412*
  and anger, *204*
  as family challenge, *247*
Decibel, 79
Decision making, 46–50
  and chains of decisions, 50
  criteria for, 48
  difficult choices, making, 50–51
  minor vs. major decisions, 46
  about personal products, 65, *65*
  positive choices, 29, *29*
  as process, 48, *49*, 50
  purchasing decisions, 60, *61*
  and values, 47–48
Defense mechanisms, 222, *222*
Degenerative disorders, of nervous system, 309
Demographers, 195
Denial, 222
Dental care. *See* Teeth
Dental floss, 71
Dental sealants, 80
Dentin, *69*
Dentists, 80
Deodorants, 73
Dependence
  physiological, 284
  psychological, 285
Depressants, 305
Depression, 229
Dermatologists, 74, *83*
Dermis
  and hair, *74*
  of skin, 73, *73*
Development, human. *See* Human development
Diabetes, 44, *142*, 185, 379, 382, *383*
Diaphragm, *276*
Diet. *See also* Nutrition
  balanced, 20
  and cancer prevention, 370–71
  and cancer risk, 13
  and circulatory system, 127
  defined, 92
  and diabetes, *383*
  factors affecting, 92
  and heart disease risk, 45
  and immune system, *337*
  and skin, 73
  for sports, 155–56, *156*
  and teeth, *70*, 71

variety in, 95, 98
and weight management, 144
Diet plans, phony, 134
Difficult choices, making, 50–51
Digestion/digestive system, 108–13
  absorption, 111, *111*
  caring for, 113
  definitions of, 108
  esophagus, 109, *109*
  and excretory system, 112–13
  mouth/throat, 109, *109*
  process of, *108*, 108–11
  of proteins, 110–11
  stomach/small intestine, *110*, 110–11
Disagreements, 22
Disease(s), 332–34. *See also* Communicable diseases
  causes of, *333*, 333–34
  chronic, 362
  noncommunicable, 362, *362*
  risk factors for, 41
Displacement, 222
Distress, 219
Divorce, as family challenge, *247*
Dolan, Tom, *376*
Down syndrome, *199*
Drinking. *See* Alcohol
Driving, drinking and, 315, *315*
Drowning, *414*
Drug/alcohol counselors, 312
Drug treatment centers, 83
Drug use/abuse, 301–6. *See also* Addiction; Alcohol
  amphetamines, 304, *304*
  avoiding, *320*, 320–21
  body, effects on, 315
  cocaine/crack, 304
  dangers of, 306
  definition of drugs, 301
  depressants, 305
  and family, 317
  hallucinogens, 305
  inhalants, 305
  and the law, 318
  marijuana, 303, *303*
  mental/emotional consequences of, 316, *316*
  misuse of drugs, 302
  narcotics, 305
  and nervous system, 309
  others, effects on, 316
  and peer pressure, *321*
  and personal responsibility, 319
  and pregnancy, *200*
  prescription medicines, *302*
  proper use of drugs, 301–2
  rise in, *303*
  risks of, 314
  and school, 317
  social consequences of, 317
  stimulants, 303–4

470 Index

and violence, 395
and your potential, 322
Dust, *373, 376*

Eardrum, *78*
Ears, 78–79
   caring for, 79
   and direction of sound, *79*
   structure of, *78*
Earthquakes, 431, *432*
Eating disorders, *109*, 230
   and sports, *161*
Eating well. *See* Nutrition
Egg cells, *183,* 190–91, *190, 191,* 199
Ejaculation, 187, *187*
Electrical fires, extinguishing, *418–19*
Electrical overloads, 420
Electrical shocks, preventing, 420
Embryo, 197, *197*
Emergency medical technicians (EMTs), 428
Emergency rooms, 83, *83*
Emotional abuse, 401
Emotional growth, during adolescence, 178–80
Emotional health. *See* Mental/emotional health
Emotional needs, providing for, by family, *246*
Emotions
   control of, by leaders, *31*
   defined, 4
Empathy, 243
Emphysema, *278*
EMTs. *See* Emergency medical technicians
Enamel (of tooth), *69*
Endangered animals, *440*
Endocrine system, 182–85
   defined, 182
   feedback in, 184, *184*
   glands of, 183, *183*
   problems in, 185
Endometrium, *190*
Endorsements, 63
Endurance, 121
Energy, sports and, *154*
Environment
   defined, 9
   and noncommunicable diseases, 363
   physical, 9
   protecting the, 440–42
Enzymes, 108
Epidemiologists, 349
Epidermis
   and hair, *74*
   of skin, 73, *73*
Epididymis, *187*
Epiglottis, 109, *109, 276*

Esophagus, 109, *109*
Estrogen, *190*
Ethanol, *296*
European breakfast, *102*
Eustachian tubes, *78*
Excretion, defined, 112
Excretory system, 112–13
Exercise(s). *See also* Fitness programs
   aerobic, 121
   anaerobic, 121
   and appetite, 13
   asthma induced by, 375
   by astronauts, *138*
   before bed, *5*
   breathing, 226
   circulatory system, effect on, 127
   curl-ups, 120
   defined, 118
   and diabetes, 44
   and diet, 135
   frequency/intensity of, 139
   heartbeat during, 127, *127*
   and immune system, *337*
   music for, *122*
   and pain, 137
   push-ups, 120
   routine of, 120–21
   and skeletal/muscular systems, 133
   step-ups, 120
   water workout, 134
   weight-bearing, 107
   and weight management, 98, 143
Exercise videos, 13
Exhaling, *226*
"Exit lines," 321
Extended family, *245*
Extended-wear contact lenses, *78*
Extension, muscle, 131
External auditory canal, 78, *78*
Eyes, 76–78, *310*
   caring for, 77
   parts of, *76*
   problems with, 77–78
   rubbing, 77

Fact-finding skills, developing, 21
Fallopian tubes, *190, 191*
Falls, preventing, 419
Family counseling programs, 405
Family(-ies)
   challenges faced by, 247, *247*
   defined, 245
   nurturing by, *246*
   and purchasing decisions, *61*
   relationships within, 245–48
   and society, 245
   substance abuse and effects on, 317
   types of, *245*
   and values, 47, *246*
Farsightedness, 77

FAS. *See* Fetal alcohol syndrome
Fast-food restaurants, making healthful choices at, 100
Fat-free snacks, *103*
Fatigue, and stress, 221
Fat(s)
   calories vs., *143*
   in diet, 99
   and heart disease risk, 45
   limiting intake of, 370
Fat tissue
   and hair, *74*
   in skin, *73*
FDA. *See* Food and Drug Administration
Feces, 113
Feelings, conflicting, 179
Female reproductive system, *190,* 190–93
   caring for, 192
   fertilization, 191, *192*
   menstruation, 191, *191*
   problems of, 192–93
Females, physical changes of adolescence in, *177*
Femur, *129*
Fertilization, 190
   development after, 197
Fetal alcohol syndrome (FAS), *309,* 316
Fetal environment, 199, *200*
Fetus, 197
   and smoking by mother, 291
   and substance abuse by mother, 313
Fever, 337, 349
Fiber, 98
Fight-or-flight response, 220, *220*
Fights, *40*
Fingerprints, *198*
Fires, extinguishing, 418–19
Fire safety, 416–18, *416–18*
First aid, *118,* 435–39
   ABCs of, *435*
   bleeding, severe, 437, *437*
   breathing emergencies, 436, *436–37*
   burns, *439*
   choking, 438, *438*
   defined, 435
   poisoning, *439*
First-degree burns, *439*
Fitness, 118–22
   benefits of, 119, *119*
   and circulatory system, 127
   defined, 118
   elements of, 120–22
   and endurance, 121
   and exercise, 118
   and flexibility, 122
   and helping others, *118*
   and lifestyle, 120–21
   and strength, 120

Index   **471**

Fitness programs, 136–40
  aerobics vs. strength training, 138
  choice of exercises for, 136, *136*
  cool-down, 138
  frequency/intensity of exercise, 139
  goals for, 135, 136, *139*
  progress, checking, 140
  and safety, 137, *137*
  stretching, *138*
  and target heart rate, 139
  warm-up, 138
Fitness testing, 162–63
Flexibility, 122
Floods, 431, *431*
Flossing, 71, *71*
Flu (influenza), *278*, 341, *341,* 349
Fluoride, 70
Folic acid, *365*
Food allergies/allergens, *6,* 373
Food and Drug Administration (FDA), *41, 302*
Food Guide Pyramid, *93,* 93–95, 103
Formaldehyde, in tobacco smoke, *273*
Fractures, *439*
Friends/friendships, 249–53
  choosing, 256
  defined, 249
  influence of, on health, *10*
  and peer pressure, 251–53, *252*
  qualities of, 250
  self-assessment form for evaluating, 250
Fruit(s), 107
  antioxidants in, 195
  and cancer prevention, 371
  preserving vitamins in, *99*
  rinsing, 107
Fruit Group, recommended daily servings for, *93, 94*
Fungi, *333*

Gallbladder, 111
Gangs, 396
Genes, 199
Genital herpes, 352
Germs, 332–34, *333*
  defined, 332
  spreading of, 334, *334*
Giantism, *185*
Glands, 183, *183*
Gliding joints, *129*
Glucose, 102
Goals, 52–55
  and action plans, 54, *55*
  career, *52*
  conflicting, *53*
  long-term, 53, *54*
  reasons for having, 52
  and self-esteem, 53
  short-term, 37, 53

Goiter, 185
Gonorrhea, 352
Government, and health care, 85
Grandparents, 248
Grape juice, *365*
Grease fires, extinguishing, *418*
Greeting others, 34, *34*
Gretzky, Wayne, 258
Grief, dealing with, 205
Gun accidents, preventing, 420
Guns, and violence, *394, 395*
Gynecologists, *83,* 192

Habits, 34–38
  addictions vs., 36
  analyzing, 36–37
  and avoiding risks, 42
  breaking, 38
  changing harmful, 36–38
  customs vs., *36*
  defined, 34
  forming healthful, 35, *35,* 347
  identifying, 36, *286*
  life-altering, 38
  life-threatening, 36
  and repetition, 36
  safe, 412–14
Hair, 74–75
  caring for, 75
  problems with, 75
  structure of, *74*
Hair follicles, *73, 74*
Hallucinogens, 305
Hammer (of ear), 78, *78*
Handwashing, 12, *19, 345*
Hard contact lenses, *78*
Hate crimes, 394
Hazards, 41, *41*
Head lice, 75
Health
  assessing your, 14–15
  defined, 4
  and environment, 9
  and heredity, 9
  improving your, 16–18
  influences on, *8,* 8–11, *10, 11*
  mental/emotional. *See* Mental/emotional health
  physical, 4, 19–20, *119*
  responsibility for others', 31, *31*
  responsibility for your own, *29,* 29–30
  and self-esteem, 217
  social, 5, 22–23, *119*
  taking charge of your, 19, 44
  total, 6, *6*
  and weight, 142, *142*
  wellness vs., 6
Health care, 82–85
  cost of, 84
  facilities for, 83
  government programs, 85

  insurance, 84
  managed care, 84
  medical specialists, *83*
  and prevention, 82
  primary care providers, 83
  by specialists, 83, *83*
  and treatment, 82
Healthful habits, 35, *35*
Health information, being a consumer of, 62–63
Health insurance, 84
Health risks. *See* Risks
Health triangle, *119*
Heart, *124*
  alcohol's effect on, *297*
  and exercise, 127, *127*
  muscle in, 130
Heart attack, *364*
Heart disease, 363, *364,* 365–66
  risk for, 45
Heart rate, target, 139
Heart transplants, 365
Help, getting, 232–35, *233*
HELP criteria, 48
Helping others, 23, *23*
Hepatitis, *342*
Heredity, 9, 199
  and noncommunicable diseases, 363
Hernia, 188
Heroin, 305
Herpes, genital, 352
Hiking safety, 426
Hinge joints, *129*
Histamines, 374
HIV (human immunodeficiency virus), 353–55
Home, safety at, 416–20
Homicide, 393
Homocysteine, *365*
Honest, being, 47
Hormones, 176, 182, *183, 184*
Hospice, 83
Hospice workers, 194
Hospital emergency rooms, *83*
Hot lines
  crisis, 405
  teen, *233*
Human development, 196–200. *See also* Life stages
  after fertilization, 197, *197*
  birth, 198
  and cell-to-system structure, 196, *196*
  and fetal environment, 199, *200*
  and heredity, 199
Hurricanes, 433
Hurt pride, 261
Hygiene, 19
Hyperactivity, sugar and, 106
Hypertension, *364*
Hypnotics, 305

Illegal risks, 40
Illness, as family challenge, *247*
Image ads, 63
"I" messages, 242
Immune system, 336–39, *337*
Immunity, 339
Immunization, 339
Individual sports, 151
Infancy, 201, *201*
Infants
   choking first aid for, *438*
   rescue breathing for, *437*
   speaking to, by parents, *198*
Infections, 193, 309, 332–34
Infectious diseases, contagious vs., *351*
Infertility, 193
Influenza. *See* Flu
Infomercials, 63
Information, health, 62–63, *346*
Informational ads, 63
Information resources, 21, *21*
Ingredients, product, *67*
Inhalants, 305
Inhaling, *277*
Injuries
   skeletal/muscular, 133
   sports, 157
Inner ear, 78, *78*
Insects, allergens from, *373*
Instructions, product, *67*
Insulin, 382
Insurance, health, 84
Interferon, 337
Internet, *10*
Intoxication, *315*
Iris, *76*
Isopropyl alcohol, *296*

Jaeger, Andrea, 44
Japan, *344*
Job loss, as family challenge, *247*
Jogging, 121
Joints, 128
Jumping rope, *120,* 121

Kids' Stuff Foundation, 44
Kwan, Michelle, 12

Labels
   Nutrition Facts, 100, *101*
   reading product, 66, *67*
Land pollution, 442, *442*
Law, substance abuse and the, 318
Leadership skills, 30–31
Lead paint, *420*
Left atrium (of heart), *124*
Left ventricle, *124*
Lemieux, Mario, 378
Lens (of eye), *76*
Leukemia, *368*
Lice, head, 75

Life-altering habits, 38
Life changes, and stress, 223
Life stages, *201–3,* 201–4
   adolescence, 203
   adulthood, 203, *203*
   childhood, 202
   death, 204
   infancy, 201
Lifestyle, 120–21, *346*
   and noncommunicable diseases, 363
Life-threatening habits, 36
Ligaments, 130, *130*
Listening, 243
Liver, 111
   and alcohol, *111*
   alcohol's effect on, *297*
Long-term goals, 53, *54*
Loud sounds, hearing damage from, 79
Low-fat snacks, *143*
LSD, 305
Lung cancer, *278, 368*
Lungs, *124, 276, 277,* 376–77
Lymphatic system, 337
Lymphocytes, 337
Lymphoma, *368*
Lynch, Nnenna, 164

Macrophages, *337, 338*
Magazine articles, advertising in, *63*
Mainstream smoke, 290
Male reproductive system, 186–89
   caring for, 188
   parts of, *187*
   problems of, 188–89
Males, physical changes of adolescence in, *177*
Malignant tumors, 367
Managed care, 84
Marijuana, 303, *303*
Mass media, 61
Meal planning, 102–5
Measles, 341
Meat, Poultry, Fish, Dry Beans, Eggs, and Nuts Group, recommended daily servings for, *93, 94*
Media. *See also* Advertising/advertisements; Television
   abuse and, 406–7
   defined, 61
   influence of, on health, *10*
   mass, 61
Media literacy, 20–21
Mediation, 265
Medicaid, 85
Medicare, 85
Medications, and pregnancy, *200*
Medicine, defined, 301
Melanin, 74
Menstruation/menstrual cycle, 191, *191, 192*

Mental disorders, 228–31
   anxiety disorders, 229
   causes of, 229
   getting help for, 232–35, *233*
   and handling problems, 228, *228*
   mood disorders, 229
   schizophrenia, 230
   teen suicide, 230–31, *234*
   treatment of, 230
   warning signs of, *232*
Mental/emotional health, 5, 20–22, 210–13. *See also* Mental disorders; Self-esteem
   assessing your, 15
   defined, 210
   fitness and, *119*
   and personality, *212,* 212–13
   and positive outlook, *210–11*
   and substance abuse, 316, *316*
Mental growth, during adolescence, 178
Mental health professionals, *233*
Mental needs, providing for, by family, 246
Messier, Mark, 258
Metamphetamine, *304*
Methanol, *296*
   in tobacco smoke, *273*
Microorganisms, 332
Middle ear, 78, *78*
Milk, Yogurt, and Cheese Group, recommended daily servings for, *93, 94*
Mobile homes, and tornadoes, *432*
Mononucleosis, *342*
Mood disorders, 229
Morphine, 305
Mortality, defined, 39
Motor neurons, *307*
Motor vehicle accidents, *412*
Mouth, digestion in, 109, *109*
Moving, as family challenge, *247*
Mucous membranes, 336
Multiple sclerosis, 309
Mumps, *189,* 341
Muscles/muscular system, *131*
   and bones, *132*
   caring for, 133
   contraction, 130
   defined, 128
   extension, 131
   pairs, muscle, 131, *132*
   soreness in, after exercise, *162*
   types of, *130*
Music, for exercising, *122*
Music therapists, 226

Narcotics, 305
Natural disasters
   earthquakes, 431, *432*
   floods, 431, *431*
   weather emergencies, 432–34

Index   **473**

Nearsightedness, 77
Neck (of tooth), *69*
Neglect, 401
Negotiation, 265
Nervous system, 307–10
   caring for, 310
   central vs. peripheral, 308
   parts of, *308*
   problems of, 309
Neurons, 307, *307*
Neutrality, 265
Nicotine, 272, *273*, 280, *282*, 283–84, *284*
Noncommunicable diseases, 332, 362, *362*, 363
Nonverbal communication, 241, *241*
Nonviolent confrontation, 264
Nuclear family, 245
Nursing homes, 83
Nurturing, 246, *246*
Nutrients, 95, 96, *96–97*
Nutrition, 20, 44–45, 92–105
   defined, 92
   and exercise, 98
   and Food Guide Pyramid, *93*, 93–95, 103
   and healthful choices, 98–100
   major nutrients, 96, *96–97*
   Nutrition Facts labels, 100, *101*
   recommended daily servings, *94*
   and skeletal/muscular systems, 133
   and snacks, *104*, 104–5, *105*
   and sports, 155–56, *156*
   variety in diet, 95, 98
   and weight management, 143
Nutrition Facts labels, 100, *101*
Nuts, *93*

Obesity, 142
Objective information, 41
Obsessive-compulsive disorder, 229
Occupational Safety and Health Administration (OSHA), 440
Oil fires, extinguishing, 418
Oldest old, 195
Olestra, *103*
Ophthalmologists, 77, 83
Optic nerve, *76*
Options, listing, in decision-making process, *49*
Optometrists, 77
Organs, *196*
Orthodontists, 83
Orthopedists, *83*
OSHA. *See* Occupational Safety and Health Administration
Osteoarthritis, 380, *381*
Osteoporosis, 98
Outdoor safety, 425–27
Outer ear, 78, *78*
Oval window (in ear), 78

Ovaries, *183, 190, 191*
Overcommitment, 167, *168*
Over-the-counter medicine, 301, *302*
Overtraining, avoiding, 161, *161*
Ovulation, 191

Pacemakers, 365
Pain, and exercise, *137*
Palamarchuk, Howard, 165
Pancreas, 111, *183*
Papain, 110
Pap test, *192*
Parathyroid glands, *183*
Parents, separation/divorce of, as family challenge, 247
Parents Anonymous, 405
Passive smokers, 290
Paying attention, *42*
PCP (phencyclidine), 305
Peace murals, *261*
Pedestrian safety, 423
Pediatricians, *83*
Peer pressure, 251–53, *252*
   and conflict, 261
   handling, 252–53
   positive vs. negative, 251, *251*
   recognizing, 251
   and refusal skills, 256, *257*
   and substance abuse, *321*
   to use tobacco, 287
   and violence, 395
Penis, 187, *187*
Peripheral nervous system (PNS), 308
Personal experiences, and values, 47
Personal hygiene, 19
Personality, *212*, 212–13
Personal products, buying
   comparison shopping, 68
   decisions, reaching, 67
   product labels, 66, *67*
   and quackery, 66
   questions to ask, 65
Personal space, 405
Personal wellness contract, *17*, 17–18
Pets, allergens from, *373*
Phagocytes, 337
Phencyclidine (PCP), 305
Philabundance, 106
Phobias, 229
Physical abuse, 401
Physical changes, during adolescence, 177, *177*
Physical environment, 9, *9*
Physical fatigue, 221
Physical fitness. *See* Fitness
Physical health, 4, 19–20
   assessing your, 15
   fitness and, *119*
Physical needs, providing for, by family, *246*

Physical therapists, 379
Physiological dependence, 284
Pimples, 74
Pipes, 274
Pituitary gland, *183*
Pivot joints, *129*
Pizza, 100
Plants, allergens from, *373*
Plasma, 125, *125*
Plastic surgeons, 83
Platelets, 125, *125*
PMS. *See* Premenstrual syndrome
Pneumonia, 278
PNS. *See* Peripheral nervous system
Poisonings
   first aid for, *439*
   preventing, 420
Police departments, 405
Polio, 310
Pollen, *373*
Pollution, 440
   air/water, 441
   land, 442, *442*
   preventing, *441*
   water, *443*
Pores (of skin), 73, *73*
Positive choices, making, 29
Positive stress, 219
Posture, and skeletal/muscular systems, 133
Precautions, 41
Pregnancy
   cigarette smoking during, 291, *291*
   and fetal environment, 199, *200*
   teen, *255*
Prejudice, 261
   as cause of violence, 394
Premenstrual syndrome (PMS), 192
Preschoolers, 202
Prescription medicine, 301–2, *302*
Pressure from others, resisting, *42*
Pressure points, *437*
Pressure statements, *257*
Prevention, 14, 82
Primary care providers, 83
Problems, identifying, in decision-making process, *49*
Problem solving, by leaders, 30
Product labels, 66–67, *67*
Projection, 222
Promises, keeping, 47
Prostate gland, *187*
Protease inhibitors, *354*
Proteins, digestion of, 110–11
Psychiatrists, 83
Psychological dependence, 285
Psychological fatigue, 221
Puberty, 177
Pull-ups, *162*
Pulmonary circulation, 124, *124*
Pulmonary vein, *124*
Pulp (of tooth), *69*

474   Index

Pupil (of eye), 76
Purchasing decisions, 60, *61*
Push-ups, 120

Quackery, 66

Rabies, 310, *426*
Radiation therapy, 370
Rationalization, 222
RDAs. *See* Recommended Dietary Allowances
Reading, light for, 77
Recommended Dietary Allowances (RDAs), 98
Recovery, 325, *325,* 327
Rectal cancer, *368*
Recycling, 442
Red blood cells, 125, *125,* 126, *126*
Refusal skills, 23, 256, *257*
Relationships. *See* Family(-ies); Friends/friendships
Relaxing
 and stress, *225,* 226
Religious beliefs, and values, 47
Repetition, 36–37
Reproduction, 186
Reproductive organs, cancer of, *368*
Reproductive systems, 186. *See also* Female reproductive system; Male reproductive system
Rescue breathing, 436, *436–37*
Respiratory system, 276–79
 and breathing, 276–77
 caring for, 278, *279*
 parts of, 276, *276*
 problems of, 278
Responsibility. *See also* Decision making
 and consequences, 32
 defined, 28
 and factors beyond your control, 33
 for others' health, 31, *31*
 taking, 22, 28
 and teamwork, 153
 for your own health, 29–30
Retina, *76,* 310
Revenge, 261, *261*
Rheumatoid arthritis, 380, *381*
Rh factor, 126
RICE, 140
Richter scale, *431*
Right atrium (of heart), *124*
Right ventricle, *124*
Risk behaviors
 and abstinence, 255, *255*
 avoiding, 254
 changing, 43
 defined, 14
Risks, 39–43
 avoiding, 42, *42*
 and consequences, 39–40

cumulative, 41
defined, 39
illegal, 40
and managing your life, 43
and objective vs. subjective information, 41
taking, 40, *40*
Robinson, David, *168*
Rod cells, *310*
"Roofies," *305*
Roosevelt, Franklin D., *309*
Root (of tooth), *69*
Rubbing alcohol, *296*
Rubella, 200

SAD. *See* Seasonal Affective Disorder
SADD. *See* Students Against Destructive Decisions
Safety
 from crime, 424
 electrical shocks, preventing, 420
 falls, preventing, 419
 fire, 416–18, *416–18*
 gun accidents, preventing, 420
 habits, 412–14
 hiking/camping, 426
 at home, 416–20
 outdoor, 425–27
 pedestrian, 423
 poisonings, preventing, 420
 on the road, 422–24
 at school, 421
 traffic, 422, *422*
 in water, 426, *426*
 in weather emergencies. *See* Weather emergencies
 on wheels, 423, *424*
 winter sports, 427
Safety belts. *See* Seat belts
Safety consciousness, 412
SAIL program, 227
Salespeople, and purchasing decisions, *61*
Saliva, 109, *109,* 336
Salt, 99
Saturated fats, 99
Schedule, daily, *18*
Schizophrenia, 230
School nurses, *233*
Sclera, *76*
Seasonal Affective Disorder (SAD), 229
Seat belts, 16, *40,* 45, 428
Second-degree burns, *439*
Secondhand smoke, 290
Selenium, *370*
Self-assessment
 defined, 14
 example of, 15
 as habit, 20
 value of, 16

Self-concept, 214–16
Self-esteem, 214–18
 behaviors indicating, *216*
 defined, 53
 and goals, 53
 and health, 217
 improving your, 217–18
 low, 216–17
 and self-concept, 214–16
Self-image
 and advertising, 227
 and values, 47
Self-management, 22
Self-respect, 47
Semen, 187
Semicircular canals, *78*
Sensation, through skin, *72*
Sensory neurons, *307*
Sexual abuse, 401
Sexual harassment, at school, *403*
Sexually transmitted diseases (STDs), 350–56
 abstinence as protection against, 356
 chlamydia, 351
 defined, 350
 genital herpes, 352
 gonorrhea, 352
 HIV and AIDS, *353,* 353–55
 myths about, *350*
 syphilis, 352
Shaken-Baby Syndrome, *401*
Shampooing, 75
Shoes, sports, 164
Shooting Stars, 313
Short-term goals, 37, 53
Shoulder stretch, *138, 162*
Showering, 73
Shriver, Eunice Kennedy, 227
Sickle-cell anemia, *362*
Sidestream smoke, 290
SIDS. *See* Sudden Infant Death Syndrome
Signs, traffic, *422*
Silence, as acceptance, *46*
Single-parent family, *245*
Skeletal muscle, *130*
Skeletal system, 128, *129*
 bones, 128, *129*
 caring for, 133
 connecting tissues, 130, *130*
 defined, 128
 joints, 128, *129*
Skin, 72–74. *See also* Hair
 and acne, 74
 caring for, 73–74
 functions of, 72, *72*
 layers of, *73*
 as organ, 72
 as protective barrier, 336
 and sunscreens, 72
 weight of, *73*

Skin cancer, 368
Sleep, getting enough, 4
Small intestine, *110,* 111
Smallpox vaccine, *343*
Smoke detectors, *417*
Smokeless tobacco, 273
Smoking. *See also* Tobacco
　body, effects on, *274,* 280, 281
　and breathing, *275*
　and circulatory system, 127
　cost of, 274
　and diabetes, 44–45
　and pregnancy, *200*
　as risk behavior, 14, *40*
　and rituals, 285, 286
　starting, 283
　and tobacco, 272–74
Smooth muscle, *130*
Snacks, *104,* 104–5, *105*
　fat-free, *103*
　healthful, *155, 290*
　low-fat, *143*
　and teeth, *70*
Snuff, 273
Soccer, *159*
Social development, during adolescence, 180
Social health, 5
　assessing your, 15
　fitness and, *119*
　and substance abuse, 317
Social needs, providing for, by family, *246*
Social phobia, 229
Social workers, 399
Sodium, in diet, 99
Soft contact lenses, *78*
Sounds, determining direction of, *79*
Specialists, 83, *83*
Special Olympics, 227
Speed, *304*
Sperm, 187, *187,* 191, 199
SPF (sun protection factor), 74
Sphygmomanometer, *126*
Spitting up, by babies, *110*
Sports, 150–68
　balancing school and home life with, 166–68, *167, 168*
　choosing, 150
　commitment to, 150
　and competition, 152, *152*
　conditioning for, 159–63, *160–1, 162*
　and diet/nutrition, 155–56, *156*
　and eating disorders, *161*
　and energy, 154, *154*
　individual, 151
　injuries, preventing, 157, *157*
　overtraining, avoiding, 161
　and physical wellness, 154
　protective equipment for, *157*
　shoes for, 164
　and steroids, 158
　team, 151
Sports drinks, *156*
STDs. *See* Sexually Transmitted Diseases
Step-ups, 120, 163
Sterility, 189, *189*
Steroids, 158
Stethoscope, *126*
Stimulants, 303–4
Stirrup (of ear), 78, *78*
Stomach, 110, *110*
　alcohol's effect on, *297*
Stomach acid, 336
Strength, 120
Stress, 219–25
　avoiding, 224
　coping with, 224
　and defense mechanisms, 222, *222*
　defined, 219
　effective management of, 225
　and fatigue, 221
　and fight-or-flight response, 220, *220*
　and life changes, 223
　negative vs. positive, 219
　reducing, *124,* 221, *221*
　and stressors, 220
Stressors, 220
Stretching, 122, *122, 138*
Stroke, *364*
Students Against Destructive Decisions (SADD), *315*
Subjective ideas, 41
Substance abuse. *See* Alcohol; Drug use/abuse
Sudden Infant Death Syndrome (SIDS), 291
Sugar(s), 99, *99,* 106
　and teeth, *70,* 71
Suicide, teen, 230–31, *234*
Sun, protecting skin against, 74, 371
Sunburn, 74
Sunlamps, 81
Sun protection factor. *See* SPF
Sunscreens, 72, *72,* 74
Support systems, *233*
Sweat glands, 73, *73*
Swimming, as exercise, 122, *122*
Syphilis, 352
Systemic circulation, 124, *124*

Tact, 242
Talking, and stress, 225
Tanning beds, 81
Tar (in tobacco smoke), 272, *273*
Tay-Sachs disease, *362*
T-cells, 337, *338,* 353
TDDs. *See* Telecommunication devices for the deaf
Team sports, 151
Teamwork skills, 153
Tears, 336
Teen pregnancy, *255*
Teeth, 69–71
　brushing, 70, *71*
　caring for, 70–71, *71*
　decay, tooth, 70, *70*
　and dental sealants, 80
　flossing, 71, *71*
　parts of tooth, *69*
　and personal hygiene, 69
Telecommunication devices for the deaf (TDDs), 258
Television
　light from watching, 77
　smoking on, *288*
Television ratings, 398
Temperature control, by skin, *72*
Tendons, 130, *130*
Tension, and circulatory system, 127
Testes, *183,* 187, *187*
Testicular cancer, 189
Testicular self-examination (TSE), *188*
Thailand, antismoking laws in, *283*
Therapists, 259
Third-degree burns, *439*
Throat, as part of the digestive system, 109, *109*
Thunderstorms, 434
Thyroid gland, *183*
Time, managing your, *18*
Time management, 221, *221*
Time outs, *179*
Tissues, *196*
Tobacco, 272–74, 282–86. *See also* Smoking
　addiction to, 283–86
　avoiding, 371
　body, effects on, *274,* 284
　choosing to be free of, *287, 287*
　and nonsmokers, 290–91
　quitting use of, 288–89
　regulation of, 274
　smokeless, 273
　starting to use, 283
　substances in, 272, *273*
Tobacco products, advertising for, 41, *41*
Toddlers, 202
Tooth decay, 70, *70*
Tornadoes, *432, 433*
Total health, 6, *6*
Toxic shock syndrome, 193
Toys, unsafe, 429
Trachea, 109, *109,* 276
Traffic safety, *422,* 422–24
Training, physical. *See* Conditioning
Traits, 9
Tranquilizers, 305
Travel safety, *413*
Treatment, 82
Tree Musketeers, 429

TSE. *See* Testicular self-examination
Tuberculosis, *278, 342*
Tumors, 367
Typhoons, *433*

Ultraviolet rays, 74
Unit pricing, *66*
Unsaturated fats, 99
Urethra, 187, *187*
Urologists, *83*
U.S. Department of Agriculture (USDA), 93
Uterus, *190,* 191, *191,* 193
Uvula, 109, *109*

Vaccination, 343, *343*
Vaccines, 339, *343*
Vagina, *190,* 191, 193
Vaginitis, 193
Values
   and decision making, 47–48, *49*
   defined, 22, 213
   and family, 246
   of friends, 250
   and peer pressure, *252*
   and purchasing decisions, *61*
Vegetables, 107
   antioxidants in, 195
   and cancer prevention, *370*
   preserving vitamins in, *99*
   rinsing, 107
Vegetable Group, recommended daily servings for, *93, 94*
Veins, *123*

Ventricles, *124*
Verbal communication, 241, *241*
Vestibule (of ear), *78*
Victims, of abuse, 400, 406
Violence, 392–97
   causes of, 394–95
   gang-related, 396
   preventing, 396
   random, 393
   and schools, 397, *397*
   in society, 392
   and teens, 393
   victims of, 393
Viruses, 333, *333*
Vision problems, 77–78. *See also* Eyes
Vitamin D formation, by skin, *72*
Vitamins, preserving, in fruits and vegetables, *99*
Volunteering, *30,* 45, 81, 106

Walking, as exercise, 121
Warm-up, 138
Warnings, *48*
   product, *67*
Washing hands, 12, *19, 345*
Water, drinking, 155, 156
Water pollution, 441, *441,* 443
Water safety, 426, *426*
Water workout, 134
Weapons, and violence, 394, *395*
Weather, and exercise, 137
Weather emergencies, 432–34
Weight, and circulatory system, 127

Weight-bearing exercise, 106–7
Weight management, 44, 45, 141–45
   and body mass index, 141, *142*
   and calories, 143, *143*
   eating disorders, 145
   and exercise, 143
   and health, 142, *142*
   and nutrition, 143
   obesity, 142
   through exercise, 98
   tips for, *144*
   and wellness, 141
Wellness
   defined, 6
   health vs., 6
   high level of, 6
   range of, *7*
   sports and, *154*
   and weight management, 141
Wheels, safety on, 423, *424*
White blood cells, 125, *125,* 126, 337
Whiteheads, 74
Winter blues, *229*
Winter sports safety, 427, *427*
Withdrawal, 283, 325
Womb, *190*
Wood alcohol, *296*

Yoga, *19, 131*
Y.O.U.T.H. *See* Youth Organization Unites to Help
Youth Organization Unites to Help (Y.O.U.T.H.), 45

Credits

PHOTOGRAPHS

Cover photo: Tim Fuller Photography.

© Bruce Bennett Studio: Scott Levy, page 258.

Camp Nejeda, Stillwater, NJ: page 185.

David Kelly Crow: pages xi (bottom), 88–89, 260, 262, 263, 264, 265, 360–361, 365, 366, 369, 370, 371, 377, 382, 384, 385, 390–391, 393, 394, 401, 404, 407.

Focus on Sports: page 188; John Cordes, page 378; Jerry Wachter, page 227.

© FPG International: Frank Cezus, page 327; Ron Chapple, page 393; Michael Goodman, page 441 (right).

Gamma Liaison: Jean-Marc Giboux, pages xii (top), 396.

Richard Hutchings: pages ix (bottom), xv, xvi–1, 116–117, 119, 120, 121, 122 (both), 127, 133, 137 (all), 138 (both), 139, 140, 155, 163, 166, 176, 177, 222 (both), 230, 238–239, 240, 243, 244, 248, 249, 253, 254, 256, 259, 303, 321, 422 (all).

© Index Stock Photography Inc.: page 151.

Ken Lax: pages ix (top), x (bottom), xii (bottom), 178, 181, 199, 204, 208–209, 232, 268–269, 294–295, 299, 301, 309, 312, 314, 315, 317, 318, 319, 323, 326, 410–411, 416, 419, 423, 443.

© Medical Images Inc.: Leslie O'Shaughnessy, page 305; Dan Reynolds, page 194; Frederick C. Skvara, M.D., page 367 (both); Howard Sochurek, page 380.

Cliff Moore: pages 210 (both), 211 (both), 215, 217, 218, 219, 234, 235.

© New England Stock Photo: Eric R. Berndt, page 153; Roger Bickel, page 441 (left); Grant Klotz, page 440; Brent Parrett, page 434.

© Peter Arnold Inc.: Ed Reschke, page 337.

© Photo Network: Mark Miller, page 275.

© Photo Researchers: Michael Abbey, page 130 (left); Bill Bachmann, page 213; Dr. Tony Brain, page 189; Scott Camazine, page 128; D. W. Fawcett, page 192; A. Glauberman/Science Source, page 272; Eric V. Grave, page 130 (right); Hank Morgan/Science Source, page 355; M. I. Walker, page 130 (middle).

© PhotoEdit: pages 41, 182, 313; Bill Aron, page 350 (top right); Bill Bachmann, page 9; Vic Bider, page 152 (top right); Michelle Bridwell, pages 16, 349; Peter Byron, page 48; Mary Kate Denny, pages vii (both), 18, 63, 145, 159, 224, 300, 336; Laura Dwight, page 186; Amy Etra, page 291; Myrleen Ferguson Cate, pages xi (top), 2–3, 11, 19, 26–27, 60, 62, 106, 283, 322, 350 (bottom left), 351, 352, 356, 388–389; Tony Freeman, pages 43, 66, 68, 111, 113, 160, 287, 335, 339, 346; Jose Galvez, pages 22, 85, 168, 203 (left, right); Spencer Grant, pages 174–175, 286, 324; Jeff Greenberg, page 118; Richard Hutchings, page 152 (bottom right); Bonnie Kamin, page 14; Felicia Martinez, page 105; Tom McCarthy, pages 246, 405; Michael Newman, pages viii (top), x (top), 13, 20, 71 (both), 77, 79, 80, 92, 95, 102, 148–149, 156 (all), 193, 270–271, 279 (bottom), 288, 289, 290, 357; Dwayne Newton, pages 23, 150, 344; Novastock, pages 141, 172–173, 280, 330–331, 341; Jonathan Nourok, pages 84, 282, 425; A. Ramey, page 362; Robin L. Sachs, page 126; James Shaffer, pages 83, 399, 430; Rhoda Sidney, page 306; David Young-Wolff, pages 8, 58–59, 75, 151, 152 (top left, bottom left), 169, 203 (middle), 279 (left, right), 296, 340, 350 (top left, bottom right), 424.

© Phototake: CNRI, page 200; David M. Grossman, page 379; Institut Pasteur/CNRI, page 332; Dr. Dennis Kunkel, page 125; Yoav Levy, page 354.

Skjold Photographs: Steve Skjold, page 205.

© Sports Illustrated: Dick Durrance II, page 44; John Iacono, page 164.

© SportsChrome: Rob Tringali, Jr., pages 12, 376.

Terry Wild Studio: pages ii–iii, 28, 31, 32, 33, 34, 37 (both), 38, 39, 45, 46, 47, 50, 51, 52, 90–91, 98, 100, 103, 104, 107, 212.

© The Stock Market: Jeff Zaruba, page 82.

© Tony Stone Images: page 53; Bruce Ayres, page 428; David Young-Wolff, page 397.

© Unicorn Stock Photos: Jeff Greenberg, pages 281, 421; Martin R. Jones, pages viii (bottom), 158; Aneal F. Vohra, page 81.

University of Maine, Department of Public Affairs, Photographic Services: page 135.

ILLUSTRATIONS

Max Crandall: pages xiv, 4, 5 (both), 6, 7, 10, 31, 42, 49, 55, 64, 65, 70 (insets), 72, 96–97, 108, 154, 180, 179, 184, 201–203, 214, 220, 225, 226, 241, 255, 257, 273, 307, 316, 320, 348, 398, 413, 414–415, 427.

David Kelley Design: pages 69, 70, 73, 74, 76, 78, 109 (both), 110, 111, 112, 123, 124, 129, 131, 132 (top), 183, 187, 190, 191, 196, 277, 297, 307 (insets), 308, 333, 338, 353, 364, 374, 375, 437.

Illustrious Interactive: pages 29, 35, 54, 61, 67, 71, 93, 99, 132 (bottom), 134, 139, 144, 157, 160–161, 165, 195, 198, 250, 325, 334, 345, 373, 383, 417, 418, 432, 433, 435, 436, 437, 438.

Network Graphics: pages 216, 284, 285.

Parrot Graphics: pages 17, 21, 40, 94, 101, 119, 136, 142, 143, 298, 302, 304, 310, 311, 343, 368, 372, 392, 395, 412, 420, 429.